T0289249

GEOTECHNICAL ENGINEERING HANDBOOK

Editor-in-Chief

Braja M. Das

Copyright © 2011 by J. Ross Publishing, Inc.

ISBN 978-1-932159-83-7

Printed and bound in the U.S.A. Printed on acid-free paper
10   9   8   7   6   5   4   3   2   1

Library of Congress Cataloging-in-Publication Data

Geotechnical engineering handbook / edited by Braja M. Das.
      p. cm.
   Includes bibliographical references and index.
   ISBN 978-1-932159-83-7 (hardcover : alk. paper)
  1. Engineering geology—Handbooks, manuals, etc. 2. Soil mechanics—
Handbooks, manuals, etc. I. Das, Braja M., 1941-
   TA705.G4275 2010
   624.1′51—dc22   2010025992

Phone: (954) 727-9333
Fax: (561) 892-0700
Web: www.jrosspub.com

Contents

4   Foundation-Soil Interaction   *Priti Maheshwari*

5   Design of Pile Foundations   *Sanjeev Kumar*

6   Retaining Walls   *Aniruddha Sengupta*

7 Slope Stability   *Khaled Sobhan*

8 Expansive Clays   *Thomas M. Petry*

9 Ground Improvement   *Thomas M. Petry*

Preface

The record of the first use of soil as a construction material is lost in antiquity. For years, the art of soil engineering was based only on past experience. With the growth of science and technology, the need for better and more economical structural design and construction became critical. This led to a detailed study of the nature and properties of soil as it relates to engineering during the early part of the 20th century. The publication of *Erdbaumechanik* by Karl Terzaghi in 1926 gave birth to modern soil mechanics. The term *geotechnical engineering* is defined as the science and practice of that part of civil engineering which involves natural materials found close to the surface of the earth. In a general sense it includes the application of the fundamental principles of soil mechanics and rock mechanics to foundation design problems.

This handbook on geotechnical engineering is designed for use by geotechnical engineers and professionals in other civil engineering disciplines as a ready reference. It consists of 15 chapters which cover a wide range of topics including engineering properties of soil, site investigation, lateral earth pressure, shallow and deep foundations, slope stability, expansive soil and ground improvement, geosynthetics and environmental geotechnology, railroad base foundations, and other special foundations. For complete coverage, a chapter on foundation-soil interactions and a chapter on the vibration of machine foundations also are included. All the chapters were written by various authors well recognized in their areas of specialty.

As is the case in all handbooks, final equations are presented in the text without detailed mathematical derivations in many instances. The reader can, however, refer to the references provided at the end of each chapter for further elaboration.

I sincerely hope that this handbook will be a useful tool for practicing engineers and others interested in the field of geotechnical engineering.

I am truly grateful to all the authors for their contributions. Thanks also are due to Tim Pletscher, Senior Acquisitions Editor and Stephen Buda, Vice President for New Business Development at J. Ross Publishing for their initiative and patience during the development of this book.

Braja M. Das

Editor-in-Chief

Dr. Braja M. Das, Professor and Dean Emeritus, College of Engineering and Computer Science, California State University, Sacramento, is presently a geotechnical consulting engineer in the state of Nevada. He earned his M.S. in civil engineering from the University of Iowa and his Ph.D. in geotechnical engineering from the University of Wisconsin-Madison. He is a Fellow of the American Society of Civil Engineers and is a registered professional engineer.

Dr. Das is the author of several geotechnical engineering texts and reference books, including *Principles of Geotechnical Engineering, Principles of Foundation Engineering, Fundamentals of Geotechnical Engineering, Introduction to Geotechnical Engineering, Principles of Soil Dynamics, Shallow Foundations: Bearing Capacity and Settlement, Advanced Soil Mechanics, Earth Anchors,* and *Theoretical Foundation Engineering.*

He has authored more than 250 technical papers in the area of geotechnical engineering, has served on the editorial board of several international journals, and is currently the editor-in-chief of the *International Journal of Geotechnical Engineering.*

Contributors

Ahmet H. Aydilek, Ph.D.
Associate Professor, Department of Civil and Environmental Engineering, University of Maryland, College Park, Maryland

Michael Burrow, Ph.D.
Lecturer, Civil Engineering, The University of Birmingham, Birmingham, U.K.

Braja M. Das, Ph.D.
Dean Emeritus, College of Engineering and Computer Science, California State University, Sacramento, California

Tuncer B. Edil, Ph.D., P.E.
Professor, Department of Civil and Environmental Engineering, University of Wisconsin-Madison, Madison, Wisconsin

Gurmel S. Ghataora, Ph.D., MIMM, MILT
Lecturer, Civil Engineering, The University of Birmingham, Birmingham, U.K.

R.L. Handy, Ph.D.
Distinguished Professor Emeritus, Department of Civil, Construction and Environmental Engineering, Iowa State University, Ames, Iowa

Sanjeev Kumar, Ph.D., P.E.
Professor and Distinguished Teacher, Department of Civil and Environmental Engineering, Southern Illinois University Carbondale, Carbondale, Illinois

Priti Maheshwari, Ph.D.
Assistant Professor, Department of Civil Engineering, Indian Institute of Technology Roorkee, India

Marcus Pacheco, Ph.D.
Professor, Department of Civil Engineering, Universidade do Estado do Rio de Janeiro, Brazil

Thomas M. Petry, Ph.D., D.GE., F.ASCE
Professor Emeritus, Department of Civil, Architectural and Environmental Engineering, Missouri University of Science and Technology, Rolla, Missouri

Aniruddha Sengupta, Ph.D., P.E.
Associate Professor, Department of Civil Engineering, Indian Institute of Technology, Kharagpur, India

Charles D. Shackelford, Ph.D., P.E.
Professor, Department of Civil and Environmental Engineering, Colorado State University, Fort Collins, Colorado

Sanjay Kumar Shukla, Ph.D., FIGS, FIE(India), MIRC, MISRMTT, MISTE, MCAII
Associate Professor and Program Leader, Discipline of Civil Engineering, School of Engineering, Edith Cowan University, Perth, Australia; Adjunct Associate Professor, School of Engineering and Physical Sciences, James Cook University, Townsville, Australia; and Associate Professor, Department of Civil Engineering, Institute of Technology, Banaras Hindu University, Varanasi, India

Nagaratnam Sivakugan, Ph.D., CPEng, FIEAust., RPEQ
Associate Professor and Head, Civil and Environmental Engineering, School of Engineering and Physical Sciences, James Cook University, Townsville, Australia

Khaled Sobhan, Ph.D.
Associate Professor, Department of Civil, Environmental and Geomatics Engineering, Florida Atlantic University, Boca Raton, Florida

David J. White, Ph.D., P.E.
Associate Professor and holder of Wegner Professorship Director, Earthworks Engineering Research Center, Department of Civil, Construction and Environmental Engineering, Iowa State University, Ames, Iowa

Nazli Yesiller, Ph.D.
Director, Global Waste Research Institute, California Polytechnic State University, San Luis Obispo, California

1

Engineering Properties of Soil

by
Nagaratnam Sivakugan
James Cook University, Townsville, Australia

1.1 Introduction

The earth is about 12,500 km in diameter. All geotechnical activities including underground excavations, tunneling, etc. are limited to the upper part of the crust, which consists primarily of oxygen (49.2%), silicon (25.7%), aluminum (7.5%), and other elements such as iron, calcium, sodium, potassium, and magnesium. These are present mostly in the form of aluminum silicates. All clay minerals are made primarily of two distinct structural units, namely *tetrahedrons* and *octahedrons,* which contain silicon and aluminum ions, respectively, at the center of the units. Several of these units can form *tetrahedral* or *octahedral sheets* that can be stacked on each other, forming different clay minerals. Clay particles are *colloidal,* where surface forces have greater influence than the body forces, less than 2 μm in size, and have net negative charges. They look like flakes or needles under a microscope. Depending on their charge imbalance, mineralogy, and pore fluid characteristics, they can form a *flocculated* (*random*) or *dispersed* (*oriented*) matrix, which can influence their fundamental behavior. *Kaolinite, montmorillonite,* and *illite* are three of the most common clay minerals. Other clay minerals include *chlorite, attapulgite, halloysite,* and *vermiculite.* Montmorillonites have the largest *cation exchange capacity* and *specific surface* (surface area per unit mass) and can swell significantly in the presence of water, thus posing a serious threat to the structural integrity of buildings and roads due to intermittent swelling and shrinking. Montmorillonitic clays are known as *expansive* or *reactive clays* and cause millions of dollars worth of damage every year worldwide.

Soils are primarily of two types: *residual* or *transported. Residual soils* are formed by disintegration of the parent rock. Depending on the geologic process by which the parent rock is formed, it is called igneous, sedimentary, or metamorphic. *Igneous rocks* (e.g., granite) are formed by cooling of lava. *Sedimentary rocks* (e.g., limestone, shale) are formed by gradual deposition of fine particles over long periods. *Metamorphic rocks* (e.g., marble) are formed by altering igneous or sedimentary rocks by pressure or temperature.

Transported soils are soils that are transported by glacier, wind, water, or gravity and deposited away from their geological origin. Depending on whether they are transported by wind, sea, lake, river, ice, or gravity, the soils are called *aeolian, marine, lacustrine, alluvial, glacial,* or *colluvial,* respectively. Some special terms used to describe certain soils are:

- *Boulder clay*—Unstratified mixture of clay and rock fragments of all sizes
- *Calcareous soil*—Soil that contains calcium carbonate
- *Conglomerate*—Cemented sand and gravel
- *Dispersive clay*—A clay that is easily erodible under low-velocity water
- *Fat clay*—Highly plastic clay
- *Hardpan*—Very dense soil layer, often cemented, that is difficult to excavate
- *Loam*—Mixture of sand, silt, and clay used as topsoil
- *Loess*—Uniform silt-sized wind-blown deposits
- *Laterite*—Red-colored residual soil in the tropics
- *Reactive clay*—Expansive clay that swells when in contact with water
- *Varved clay*—Thin alternating layers of silts and fat clays of glacial origin

1.2 Phase Relations

Soil contains soil grains, water, and air, making it a three-phase material. Two extreme cases are dry soils and saturated soils, both of which have only two phases. The relative proportions of these three phases play an important role in the engineering behavior of a soil. In geotechnical problems, including earthworks and laboratory tests, it is sometimes necessary to compute weights and volumes of the three phases.

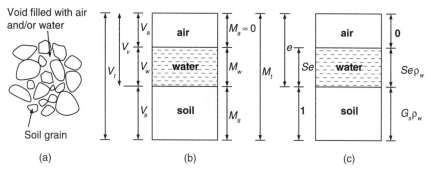

FIGURE 1.1 Phase relations: (a) soil skeleton, (b) phase diagram, and (c) phase diagram for $V_s = 1$.

Let's consider the soil mass shown in Figure 1.1a, where all three phases are present. The *soil grains* (*s*), *water* (*w*), and *air* (*a*) are separated in Figure 1.1b, known as a *phase diagram*, where *volume* (*V*) is shown on the left and *mass* or *weight* (*M*) is shown on the right. *Water content* (*w*) is the ratio of the mass of water (M_w) to the mass of the soil grains (M_s) and often is expressed as a percentage. *Void ratio* (*e*) is the ratio of the void volume (V_v) to the soil grain volume (V_s). *Porosity* (*n*) is the ratio of the void volume (V_v) to the total volume (V_t), expressed as a percentage. *Degree of saturation* (*S*) is defined as the ratio of the water volume (V_w) to the void volume (V_v), expressed as a percentage. *Air content* (*a*), as defined in compaction, is the ratio of air volume (V_a) to total volume (V_t).

Assuming the soil is homogeneous, if all parameters discussed are ratios, they should be the same irrespective of the quantity of soil under consideration. Let's consider a portion of the soil where $V_s = 1$ (Figure 1.1c), which makes $V_v = e$ and $V_w = Se$. The masses of soil grains (M_s) and water (M_w) are $G_s\rho_w$ and $Se\rho_w$, respectively, where ρ_w is the density of water. Here, G_s is the specific gravity of the soil grains, which is generally in the range of 2.6–2.8. It can be slightly lower for organic clays and significantly higher for mine tailings rich in minerals. It is determined using density bottles or a pycnometer (ASTM D854; AS1289.3.5.1). Based on the above definitions and Figure 1.1c, it can be deduced that:

$$n = \frac{e}{1 + e} \qquad (1.1)$$

$$w = \frac{Se}{G_s} \qquad (1.2)$$

$$a = \frac{(1 - S)e}{1 + e} \qquad (1.3)$$

Different forms of densities are used in geotechnical engineering. *Dry density* (ρ_d) is the density assuming the soil is dry and is M_s/V_t. *Bulk density* (ρ_m), also known as *wet, moist,* or *total* density, is M_t/V_t. *Saturated density* (ρ_{sat}) is the bulk density of the soil assuming it is saturated. *Submerged density* (ρ') is the effective buoyant density when the soil is submerged. It is obtained by subtracting ρ_w from ρ_{sat}. From Figure 1.1c, it can be deduced that:

$$\rho_d = \frac{G_s \rho_w}{1 + e} \tag{1.4}$$

$$\rho_m = \frac{(G_s + Se)\rho_w}{1 + e} \tag{1.5}$$

$$\rho_{\text{sat}} = \frac{(G_s + e)\rho_w}{1 + e} \tag{1.6}$$

$$\rho' = \rho_{\text{sat}} - \rho_w = \frac{(G_s - 1)\rho_w}{1 + e} \tag{1.7}$$

When dealing with weight (e.g., kN) instead of mass (e.g., g, kg, t), density (ρ) becomes unit weight (γ). It is helpful to remember that $\rho_w = 1 \text{ g/cm}^3 = 1 \text{ t/m}^3 = 1000 \text{ kg/m}^3$ and $\gamma_w = 9.81$ kN/m^3.

1.3 Soil Classification

Soils can behave quite differently depending on their geotechnical characteristics. In *coarse-grained soils,* where the grains are larger than 75 μm, the engineering behavior is influenced mainly by the relative proportions of the different sizes and the density of the packing. These soils are also known as *granular soils.* In *fine-grained soils,* where the grains are smaller than 75 μm, the mineralogy of the grains and the water content will have greater influence than the grain size on the soil properties. The borderline between coarse- and fine-grained soils is 75 μm, which is the smallest grain size one can distinguish with the naked eye.

1.3.1 Coarse-Grained Soils: Grain Size Distribution

The relative proportion of grain sizes within a coarse-grained soil generally is determined through sieve analysis, using a stack of sieves of different sizes (ASTM C136; AS1289.3.6.1). A hydrometer is used for fine-grained soils (ASTM D422; AS1289.3.6.3). In soils that contain both coarse and fine grains, both sieve and hydrometer analyses are required to generate the complete *grain size distribution curve,* as shown in Figure 1.2. A logarithmic scale is used for grain sizes that vary over a very wide range. In Europe and Australia, the grain size axis is shown in reverse order, increasing from left to right. In samples that contain substantial fines, it may be necessary to carry out *wet sieving* (ASTM C117), where the samples are washed through the sieves. Laser sizing also has become quite popular for determining grain size distribution of fines.

In North America, sieves are also numbered based on the number of openings per inch, instead of the size of the openings in the mesh. This number is known as the *U.S. Standard* or *ASTM Standard.* A No. 40 sieve has 40 openings per inch, or 1600 openings per square inch, and the openings are 0.425 mm in diameter. This is slightly different than the Tyler Standard or British Standard. Some common sieve numbers and the size of their openings are given in Table 1.1.

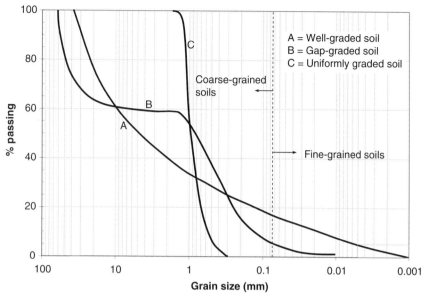

FIGURE 1.2 Grain size distribution curves.

TABLE 1.1 ASTM Sieve Numbers and Size of Openings

Sieve number	4	8	10	20	40	60	100	200
Opening (mm)	4.75	2.36	2.00	0.850	0.425	0.250	0.150	0.075

Coefficient of uniformity (C_u) and *coefficient of curvature* (C_c) are two parameters that reflect the shape of the grain size distribution curve and are used in classifying a coarse-grained soil. They are defined as:

$$C_u = \frac{D_{60}}{D_{10}} \tag{1.8}$$

$$C_c = \frac{D_{30}^2}{D_{10} D_{60}} \tag{1.9}$$

D_{10}, D_{30}, and D_{60} are the grain sizes that correspond to 10, 30, and 60% passing, respectively, and can be read off the grain size distribution plot. A *well-graded soil* contains a wide range of grain sizes that fill up the voids very effectively and form a rather dense assemblage of grains. The grain size distribution curve generally is smooth and concave upward, as shown in Figure 1.2 for soil A. Fuller and Thompson (1907) suggested that a well-graded soil can be represented by

$$p = \left(\frac{D}{D_{\max}} \right)^n \times 100\% \tag{1.10}$$

where p = percentage passing, D = grain size, D_{max} = maximum grain size in the soil, and n = 0.3–0.6. Equation 1.10 is sometimes used in pavement engineering to select the aggregates for roadwork. In *gap-graded soils*, a range of grain sizes is missing, similar to soil B in Figure 1.2. In *uniformly graded soils*, all grains are about the same size, similar to soil C in Figure 1.2. Uniformly graded and gap-graded soils are special cases of *poorly graded soils*.

A sandy soil is classified as well graded if $C_u > 6$ and $C_c = 1$–3. A gravelly soil is classified as well graded if $C_u > 4$ and $C_c = 1$–3. D_{10}, also known as the effective grain size, is an indirect measure of the pore sizes within the soil and is related to the permeability of a coarse-grained soil. Grain size distribution is of little value in a fine-grained soil.

The deformation characteristics such as strength or stiffness of a granular soil, with any specific grain size distribution, depend on how closely the grains are packed. The density of packing is quantified through a simple parameter known as *relative density* (D_r) or the *density index* (I_D), defined as

$$D_r = \frac{e_{max} - e}{e_{max} - e_{min}} \times 100\% \qquad (1.11)$$

where e_{max} and e_{min} are the *maximum* (ASTM D4254; AS1289.5.5.1) and *minimum* (ASTM 4253; AS1289.5.5.1) *possible void ratios* at which the grains can be packed and e is the void ratio at which the relative density is being computed. The maximum and minimum void ratios reflect the *loosest* and *densest possible states*, respectively.

The shape of the grains in a coarse-grained soil can be *angular, subangular, subrounded*, or *rounded*. When the grains are angular, there is more interlocking between them, and therefore the strength and stiffness of the soils will be greater. In roadwork, angular aggregates would provide better interlocking and good resistance to becoming dislodged by traffic.

1.3.2 Fine-Grained Soils: Atterberg Limits

As the water content of a fine-grained soil is increased from 0%, it goes through different consistencies, namely *brittle solid, semisolid, plastic*, and *liquid* states. The borderline water content between two states is known as the *Atterberg limits* (Figure 1.3). Atterberg limits originally were developed by the Swedish scientist A. Atterberg in the early 1900s, working in the ceramics industry. They were modified by K. Terzaghi (in the late 1920s) and A. Casagrande (in the early 1930s) to suit geotechnical work. The three Atterberg limits are *liquid limit* (LL or w_L), *plastic limit* (PL or w_P), and *shrinkage limit* (SL or w_S). LL is the lowest water content at which the soil behaves like a viscous mud, flowing under its own weight with very little

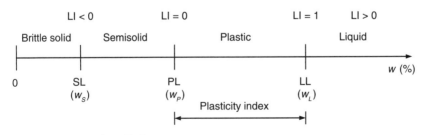

FIGURE 1.3 Atterberg limits.

strength. PL is the lowest water content at which the soil exhibits plastic characteristics. The range of water content over which the soil remains plastic is known as the *plasticity index* (PI), which is the difference between LL and PL (i.e., PI = LL − PL). SL is the water content below which soil will not shrink when dried. LL and PL tests in the laboratory are done on samples passing 425-µm (No. 40) sieves that contain some fine sands as well (ASTM D4318; AS1289.3.1.1, AS1289.3.9.1, AS1289.3.2.1). Burmister (1949) classified cohesive soils based on PI as listed in Table 1.2.

TABLE 1.2 Classification of Clays Based on PI

PI	Classification
0	Nonplastic
1–5	Slightly plastic
5–10	Low plastic
10–20	Medium plastic
20–40	High plastic
>40	Very high plastic

After Burmister (1949).

Similar to relative density in granular soils, *liquidity index* (LI *or* I_L) is a parameter used to define the consistency of a fine-grained soil with respect to LL and PL. It is defined as:

$$\text{LI} = \frac{w - \text{PL}}{\text{LL} - \text{PL}} \tag{1.12}$$

It takes a value of 0 at PL and 1 at LL. Fine-grained soils contain clays and silts, where the clays are plastic and silts are nonplastic. The plasticity of fine-grained soil is derived mainly from the clay fraction. *Activity* (A) is a term used to quantify the plasticity of the clay fraction in a fine-grained soil and is defined as:

$$A = \frac{\text{PI}}{\% \text{ of clay}} \tag{1.13}$$

Activity is a good indicator of potential shrink-swell problems associated with expansive clays. Clays with $A > 1.25$ are generally expansive and those where $A < 0.75$ are inactive. Clays with $A = 0.75$–1.25 are known as normal clays.

1.3.3 Unified Soil Classification System

A soil classification system is a universal language that all geotechnical engineers understand, where soils of similar behavior are grouped together, and systematic and rational ways are in place to classify and describe them, using standardized symbols. The use of such standard and precise terms eliminates the ambiguity in communicating the soil characteristics. Several soil classification systems are currently in use. The Unified Soil Classification System (USCS) is the one that is used the most in geotechnical engineering worldwide. The American Association of State Highway Transportation Officials (AASHTO) system is used mainly with roadwork.

The major soil groups in the USCS are defined on the basis of grain size (see Figure 1.4) as gravel (G), sand (S), silt (M), and clay (C). Two special groups are organic clays (O) and peats (Pt). Organic clays are clays where the LL reduces by more than 25% when oven dried.

USCS recommends a symbol in the form of XY for a soil, where the prefix X is the major soil group and the suffix Y is the descriptor. Coarse-grained soils (G or S) are described on the basis of the grain size distribution as well graded (W) or poorly graded (P), and fine-grained soils (M or C) are classified on the basis of their plasticity as low (L) or high (H).

A fine-grained soil is classified as clay or silt depending on the Atterberg limits and not based on the relative proportions. Casagrande (1948) proposed the PI-LL chart shown in

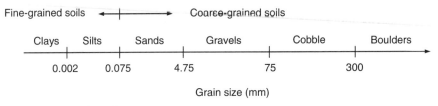

FIGURE 1.4 Major soil groups.

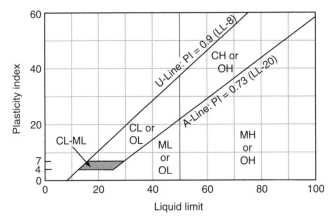

FIGURE 1.5 Casagrande's PI-LL chart.

Figure 1.5, where the A-line separates the clays and silts. If the LL and PI values of a fine-grained soil plot below the A-line, it is a silt, and if above, it is a clay. For a fine-grained soil, the descriptor L or H is used, depending on whether the LL is less or greater than 50. The U-line in Figure 1.5 gives the upper limit, and all fine-grained soils are expected to lie below this line.

There are borderline soils that cannot adequately be described by the XY symbol. A fine-grained soil that plots within the hatched area in Figure 1.5 is classified as CL-ML. A coarse-grained soil which contains fines that fall within this hatched area is classified as GC-GM or SC-SM. When there are 5–12% fines within a coarse-grained soil, it is given a dual symbol in the form of XY-XZ, where X denotes the major coarse-grained soil type, Y indicates whether it is well or poorly graded, and Z indicates whether the fines are clays or silts. The possible USCS symbols and a simple way to remember the USCS are shown in Figure 1.6.

1.3.4 Visual Identification and Description of Soils

Very often in the field, it is necessary to identify soil without any instrument or laboratory facility and then describe it in a systematic manner. This is fairly straightforward in the case of granular soils, where the qualitative field descriptions include the grain size (fine, medium, or coarse), shape (angular, subangular, subrounded, or rounded), color, gradation (well or poorly), state of compaction, and presence of fines. Fine-grained soils are identified based on *dry strength* and *dilatancy*. Dry strength is a measure of how hard it is to squeeze a dry lump between the fingers and crush. The standard terms used are very low, low, medium, high, and very high. A dilatancy test involves placing a moist pat of soil in the palm and shaking it vigorously to see how quickly the water rises to the surface, making it shiny. The standard

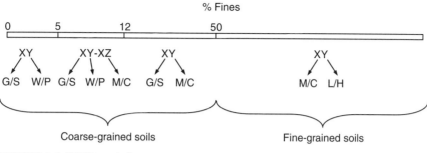

FIGURE 1.6 USCS symbols.

descriptors are very quick, quick, medium, slow, and very slow. Silts have low dry strength and quick dilatancy. Clays have high dry strength and slow dilatancy. Fines also can be identified by feeling a moist pat; clays feel sticky and silts feel gritty.

1.4 Compaction

Very often, the existing ground conditions are not suitable for the proposed engineering work. Poor ground conditions can lead to shear failure within the subsoil and/or excessive deformation. Compaction is one of the oldest, simplest, and most economical means of ground improvement and is still very popular in the modern world. The objective of compaction is to bring the soil grains closer, by applying an external effort, using some compaction equipment such as rollers. Water is added to the soil during compaction to act as a "lubricant," making the process more effective.

1.4.1 Compaction Curve and Zero Air Void Curve

Water content is one of the major variables in compaction. The relative volumes of soil grains, water, and air at five different water contents are shown in Figure 1.7a. At *optimum water content* (w_{opt}), shown by point 3 in Figure 1.7b, the soil attains the densest possible packing (see Figure 1.7c) under the applied *compactive effort*. The corresponding dry density is known as the *maximum dry density* ($\rho_{d,max}$). Increasing the compactive effort leads to a reduction in the optimum water content and an increase in the maximum dry density.

Every point in the ρ_d-w space in Figure 1.7b corresponds to a specific value of the degree of saturation (S) or air content (a). The *zero air void curve* is the locus of the points that correspond to $S = 100\%$ and $a = 0\%$. The equation for this curve is

$$\rho_d = \frac{G_s \rho_w}{1 + wG_s} \tag{1.14}$$

The zero air void curve is sensitive to the value of G_s, which must be determined precisely. Similar contours can be drawn for any value of S or a, using the following equations:

$$\rho_d = \frac{G_s \rho_w}{1 + \dfrac{wG_s}{S}} \tag{1.15}$$

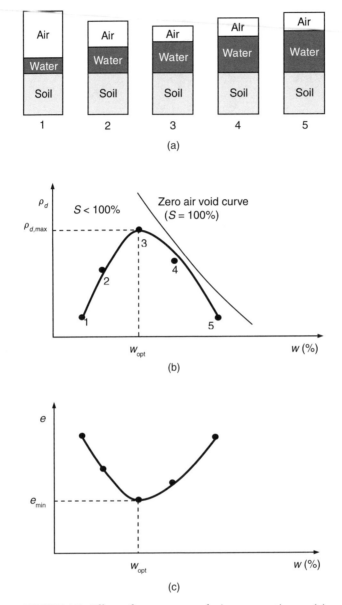

FIGURE 1.7 Effects of water content during compaction on (a) phase diagram, (b) dry density, and (c) void ratio.

$$\rho_d = \frac{G_s(1 - a)\rho_w}{1 + wG_s} \qquad (1.16)$$

The theoretical zero air void curve, drawn using Equation 1.14, provides a good check on the laboratory compaction tests and the field control tests. All test points should lie to the left of the zero air void curve.

1.4.2 Laboratory Compaction Tests

Laboratory compaction tests originally were proposed by Proctor (1933). The objective of these tests is to develop the compaction curve and determine the optimum water content and the maximum dry density of a soil, at a specific compactive effort. The tests require the soil to be placed in a 1000-ml cylindrical mold in layers, with each layer compacted using a standard hammer, simulating the field compaction process, where the soil is compacted in layers. This is repeated at different water contents, and the compaction curve is developed. *Standard Proctor* (ASTM D698; AS1289.5.1.1) and *modified Proctor* (ASTM D1557; AS1289.5.2.1) are the two compactive efforts commonly used. The details of these two tests are summarized in Table 1.3. The mold volume, hammer weight, and drop can vary slightly depending on the country of use.

TABLE 1.3 Standard and Modified Compaction Tests

Variable	Standard Proctor	Modified Proctor
Hammer		
Mass	2.7 kg	4.9 kg
Drop	300 mm	450 mm
No. of layers	3	5
Blows per layer	25	25
Energy/m^3	596 kJ	2703 kJ

1.4.3 Field Compaction

Compaction in the field is carried out by placing the soil in 100- to 300-mm-thick *lifts* at appropriate water contents that would meet the specifications. Water is brought in trucks and sprinkled as necessary (Figure 1.8a). Rollers or equipment that would suit the soil are used, providing a static or dynamic compactive effort. Granular soils are compacted most effectively by vibratory loads, such as vibrating rollers, plates, or rammers. Clays are compacted most effectively by *sheepsfoot* rollers that provide a good kneading action. Smooth-wheeled rollers are used for the finishing touch.

(a)   (b)

FIGURE 1.8 Field compaction: (a) watering the soil layers for compaction and (b) nuclear densometer measuring water content and density.

Relative compaction or dry density ratio is defined as

$$\text{Relative compaction} = \frac{\rho_{d,\text{field}}}{\rho_{d,\text{max-lab}}} \times 100\% \tag{1.17}$$

where $\rho_{d,\text{field}}$ = dry density measured in the field and $\rho_{d,\text{max-lab}}$ = maximum dry density from the laboratory compaction curve at the specific compactive effort. Hausmann (1990) suggested that $\rho_{d,\text{max-lab}}$ for a standard Proctor compaction test is approximately 90 and 95% that of a modified Proctor compaction test for clays and sands, respectively. It is quite common to specify relative compaction of 90–105% with respect to modified Proctor compactive effort, with water content of ±2% within the optimum water content.

The geotechnical characteristics of compacted clays are influenced significantly by the molding water content (Lambe 1958a, 1958b). The clay fabric will become more oriented (dispersed) when the water content or compactive effort is increased. Clays compacted to the *dry of optimum* have flocculated fabric and higher strength and permeability. While the clays compacted to the dry of optimum are prone to more swelling, the ones compacted to the *wet of optimum* are prone to more shrinkage.

The dry density and the water content of the compacted earthwork are checked through a *sand replacement test* (ASTM D1556; AS1289.5.3.1) or *nuclear density test* (ASTM D2922; AS1289.5.8.1). These *control tests* are carried out for every 500–1500 m³; in the case of backfills behind retaining walls, etc., where the volume is relatively small, tests are carried out for every 100–200 m³. In a sand replacement test, also known as a *sand cone test,* a hole is dug into the compacted earthwork and the soil removed is weighed and the water content measured. The volume is computed by filling the hole with uniform sand of known density. *Nuclear densometers* (Figure 1.8b) are quite popular nowadays due to several advantages. The measurements are so rapid that the density and water content measurements are available within minutes, enabling corrective measures to the compacted earthwork to be taken at once. The frequency of tests can be increased at a relatively modest cost.

Dynamic compaction is a relatively recent method to compact loose granular soils, sanitary landfills, waste dumps, sinkhole-weakened terrain, and sometimes clays too, where a 100- to 400-kN weight is raised to a height of 5–30 m and dropped repeatedly in a well-planned grid at appropriate spacing, with few passes (Figure 1.9a). The soil is densified by the stress waves generated by the impact. The large craters formed during the process are backfilled. The effectiveness of compaction is assessed through *in situ* static or dynamic penetration tests (see Figure 1.9b), carried out before and after the dynamic compaction. The dynamic compaction process effectively compacts the soil to a depth given by (Leonards et al. 1980)

$$D \ (\text{m}) \approx 0.5 \sqrt{WH} \tag{1.18}$$

where W is the weight in metric tons and H is the drop in meters. Dynamic compaction and other ground improvement techniques are covered in Chapter 9.

1.5 Flow through Soils

When water flows through soils beneath a concrete dam or a sheet pile, sometimes it is necessary to estimate the flow rate and assess the stability of the structure with respect to any

(a)   (b)

FIGURE 1.9 (a) Dynamic compaction and (b) JCU heavy dynamic cone penetration test rig.

potential problems such as piping or uplift. Here, it becomes necessary to separate the stresses caused by the soil skeleton and the water.

1.5.1 Effective Stresses and Capillary

Total normal stresses (σ) applied to a saturated soil are carried partly by the soil skeleton and the rest by the pore water. The component carried by the soil skeleton is known as *effective stress or intergranular stress* (σ'), and the pressure of the water within the voids is known as the *neutral or pore water pressure* (u). Therefore,

$$\sigma = \sigma' + u \tag{1.19}$$

in all directions, at all times, in all saturated soils. The pore water pressure is the same in all directions at a given time, whereas σ and σ' vary with direction.

In fine-grained soils, the interconnected voids act like capillary tubes and let the water rise above the phreatic surface or water table, saturating the soil within this height; this is known as capillary rise (h_c). Generally, the finer the grains, the finer the pore sizes and the larger the capillary rise. The diameter of the capillary tube (d) is approximately one-fifth of D_{10} and the capillary rise h_c is given by:

$$h_c \ (\text{m}) \ \approx \ \frac{0.03}{d \ (\text{mm})} \ \approx \ \frac{0.15}{D_{10} \ (\text{mm})} \tag{1.20}$$

In clays, several meters of capillary rise can be expected. The capillary pore water pressures are negative (i.e., suction) and can increase the effective stresses significantly.

1.5.2 Permeability

Bernoulli's equation in fluid mechanics states that for steady incompressible flow, the total head at a point P can be expressed as the summation of three independent components—pressure head, elevation head, and velocity head, as given below

Total head = Pressure head + Elevation head + Velocity head

$$h \;=\; \frac{p}{\rho_w g} \;+\; z \;+\; \frac{v^2}{2g} \tag{1.21}$$

where p is the pressure and v is the velocity at point P and z is the height of point P above the *datum*. The elevation head and therefore the total pressure head at a point depend on the selected datum. In the case of flow through soils, the seepage velocity is very low and the velocity head is negligible. The pressure is simply the pore water pressure. Therefore, Equation 1.21 becomes:

Total head = Pressure head + Elevation head

$$h \;=\; \frac{u}{\rho_w g} \;+\; z \tag{1.22}$$

Flow takes place from higher head to lower head. The energy dissipated in overcoming the frictional resistance provided by the soil matrix results in the head loss between two points. The *hydraulic gradient* (i) between two points A and B on the flow path is the ratio of the total head loss between the two points to the distance between the two points, measured along the flow path. It is a dimensionless quantity and is the head loss per unit length and therefore a constant within a homogeneous soil.

Darcy's law states that when the flow through soils is laminar, the discharge velocity is proportional to the hydraulic gradient, and therefore,

$$v \;=\; ki \tag{1.23}$$

where k is known as the *permeability* or *hydraulic conductivity* of the soil, which is expressed in units of velocity. Typical values for permeability of soils are given in Figure 1.10 (Terzaghi et al. 1996). Hazen (1930) showed that for clean filter sands in a loose state,

$$k \text{ (cm/s)} \;=\; C \times D_{10}^2 \text{ (mm)} \tag{1.24}$$

where C is about 1.

When water flows through soils, the flow takes place only through the voids. Therefore, the effective cross-sectional area (A_e) should be used in calculating the flow instead of the total cross-sectional area (A). This leads to the definition of two different velocities: *discharge velocity* (v) and *seepage velocity* (v_s). They are simply $v = Q/A$ and $v_s = Q/A_e$, where Q is the

Permeability (m/s)											
10^0	10^{-1}	10^{-2}	10^{-3}	10^{-4}	10^{-5}	10^{-6}	10^{-7}	10^{-8}	10^{-9}	10^{-10}	10^{-11}

Drainage	Good	Poor	Practically impervious
Soil Types	Clean gravel	Clean sands, clean sand & gravel mixtures	Very fine sands, organic & inorganic silts, mixtures of sand, silt & clay, glacial till, stratified clay	Impervious soils, e.g., homogeneous clays below zone of weathering
			"Impervious" soils modified by effects of vegetation & weathering	

FIGURE 1.10 Typical permeability values (after Terzaghi et al. 1996; reprinted with permission of John Wiley & Sons, Inc.).

flow rate. Seepage velocity is always greater than the average discharge velocity. In geotechnical engineering, especially when dealing with Darcy's law, discharge velocity is used:

$$\frac{v}{v_s} = \frac{A_e}{A} = n \tag{1.25}$$

Physicists define a more general form of permeability known as *intrinsic permeability* (K), which is not influenced by fluid properties such as density or viscosity. Intrinsic permeability depends only on the porosity of the soil and is expressed in units of area (e.g., m^2, Darcy). In rocks and in the oil industry, Darcy is often used for intrinsic permeability, where 1 Darcy = 0.987 μm^2. In sandstones, where the pores are well connected, the intrinsic permeability is large and can be of the order of 1 Darcy. In impermeable rocks such as siltstones, the intrinsic permeability can be of the order of 1 milli-Darcy.

K and k are related by

$$k = \frac{\gamma}{\eta} K \tag{1.26}$$

where η and γ are the dynamic viscosity (N·s/m^2) and unit weight (N/m^3), respectively, of the permeant fluid, which depend on the temperature. It can be deduced from the above equation that the heavier the fluid, the larger the permeability, and the higher the viscosity, the lower the permeability, which makes sense intuitively.

What geotechnical engineers refer to as *permeability* or *hydraulic conductivity* (k) is expressed in units of velocity. It is specifically for flow of *water through soils*. Assuming $\eta_w = 1.002 \times 10^{-3}$ N·s/m^2 and $\gamma_w = 9810$ N/m^3 at 20°C,

$$K \text{ (cm}^2) = k \text{ (cm/s)} \times 1.02 \times 10^{-5}$$

$$K \text{ (Darcy)} = k \text{ (cm/s)} \times 1.035 \times 10^3$$

In laminar flow, fluid flows in parallel layers without mixing. In turbulent flow, random velocity fluctuations result in mixing of fluid and energy dissipation. When water flows through soils, laminar flow becomes turbulent flow when the *Reynolds number* (R) is of the order of 1–12 (Harr 1962). Harr (1962) and Leonards (1962) conservatively suggest using a lower limit of 1.0 as the cutoff between laminar and turbulent flow in soils. The Reynolds number is defined as

$$R = \frac{vD\rho_w}{\eta} \tag{1.27}$$

where D is the characteristic dimension, which is the average diameter of the soil grains.

Permeability of coarse-grained soils and fine-grained soils can be determined in the laboratory through *constant head* (ASTM D2434; AS1289.6.7.1, AS1289.6.7.3) and *falling head* (ASTM D5856; AS1289.6.7.2) *permeability tests,* respectively. In a constant head test, carried out mostly on reconstituted samples of granular soils, flow takes place through the sample under a constant head (h_L), as shown in Figure 1.11a, and the flow rate is measured. Based on Darcy's law, permeability is computed using the following equation:

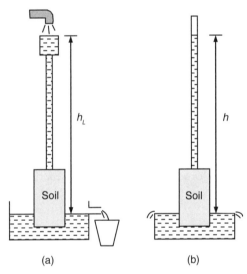

$$k = \frac{QL}{h_L At} \tag{1.28}$$

where Q = water collected in time t, L = sample length, A = sample cross section, and h_L = head loss.

In the laboratory, falling head tests can be carried out on reconstituted silt-sized soils such as mine tailings or undisturbed clay samples. Here, the time (t) taken for the water column in Figure 1.11b to drop from the head of h_1 to h_2 is measured. The permeability of the soil sample is given by

(a)   (b)

FIGURE 1.11 Permeability tests: (a) constant head and (b) falling head.

$$k = \frac{aL}{At} \ln\left(\frac{h_1}{h_2}\right) \tag{1.29}$$

where a = a cross-sectional area of the standpipe. Permeability also can be measured *in situ,* through pump-in or pump-out tests, where water is pumped into or out of a well until steady state is achieved. Permeability is determined from the flow rate, pipe diameter, and other geometric dimensions.

When there is upward flow within a granular soil, the hydraulic gradient reduces the effective vertical stresses. When the hydraulic gradient becomes equal to the *critical hydraulic gradient* (i_{cr}), the effective vertical stress becomes 0, and the soil grains are barely in contact. This situation is known as a *quick condition,* where the granular soil has no strength. The critical hydraulic gradient is given by:

$$i_{cr} = \frac{\rho'}{\rho_w} = \frac{G_s - 1}{1 + e} \tag{1.30}$$

1.5.3 Seepage

When seepage takes place beneath a concrete dam or a sheet pile, a flow net is used for computing the *flow rate*, pore water pressures within the flow domain, and *maximum exit hydraulic gradient*. The flow net for seepage beneath a sheet pile is shown in Figure 1.12. The soil properties are $k = 6.5 \times 10^{-5}$ cm/s, $G_s = 2.65$, and $e = 0.72$. The flow rate per unit length, perpendicular to this plane, can be computed using

$$Q = kh_L \frac{N_f}{N_d} \qquad (1.31)$$

where h_L = head loss within the flow domain, from upstream to downstream; N_f = number of flow channels in the flow net; and N_d = number of equipotential drops. In Figure 1.12, h_L = 9.0 m, N_f = 4, and N_d = 8. Therefore, using Equation 1.30, the flow rate becomes 0.253 m³/day/m.

Taking the downstream water level as the datum, the total heads at *upstream* and *downstream* become 9 m and 0, respectively. This implies that 9 m of head is lost along each *stream line* during the flow from upstream to downstream. The total head difference (Δh) between two adjacent *equipotential lines* is 9/8 = 1.125 m. Therefore, the total head at any point within

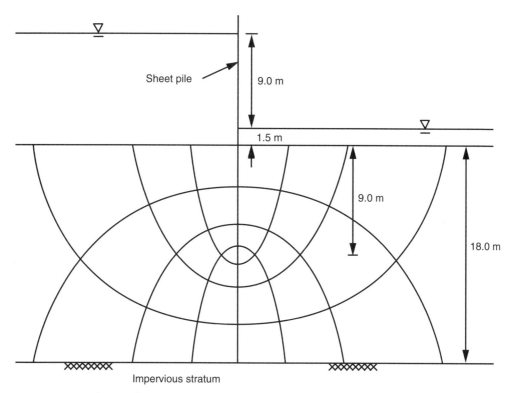

FIGURE 1.12 Flow net for seepage beneath a sheet pile.

the flow domain can be estimated. Knowing the elevation head, pressure head can be determined. Pore water pressure is simply the product of pressure head and unit weight of water.

The maximum exit hydraulic gradient ($i_{exit,max}$) which occurs next to the sheet pile can be estimated as 0.35. The critical hydraulic gradient can be computed as 0.96, using Equation 1.30. The safety factor with respect to piping is generally defined as:

$$F_{piping} = \frac{i_{cr}}{i_{exit,max}} \qquad (1.32)$$

Piping can become catastrophic, putting property and lives downstream at risk; therefore, safety factors as high as 5 often are recommended. A safety factor of 2.74 in the above example is inadequate, unless the structure is temporary.

1.5.4 Design of Granular Filters

When seepage takes place within the soil beneath embankments or behind retaining walls, often drains are installed to collect the water. In the past, the drains were made mostly of granular soils, which act as filters. Lately, geosynthetics have become increasingly popular as drainage materials.

The granular filter material has to satisfy *permeability criteria* and *retention criteria*. Permeability criteria ensure that the filter is porous enough and facilitates quick drainage without buildup of pore water pressure. To ensure that the filter pores are large enough compared to those of the surrounding soils, the following rule is enforced:

$$D_{15,filter} \geq 4D_{15,soil}$$

Retention criteria ensure that the filter pores are small enough to prevent migration of fines from the surrounding soil into the filter and eventually clogging it. This is ensured through the following rule:

$$D_{15,filter} \leq 5D_{85,soil}$$

It should be noted that $D_{15,filter}$ is the average pore size of the filter. These two criteria will establish the upper and lower bounds for the grain size distribution of the filter material. Traditionally, the grains are selected such that the grain size distribution curves of the filter material and surrounding soil are approximately parallel. The U.S. Navy (1971) suggests the following two additional conditions to reinforce retention criteria:

$$D_{15,filter} \leq 20D_{15,soil}$$

$$D_{50,filter} \leq 25D_{50,soil}$$

1.6 Consolidation

When buildings or embankments are constructed on saturated clays, the settlement is not instantaneous. Settlement occurs due to expulsion of water from the voids, and this process, known as consolidation, takes place over a long period of time in clays. During consolidation, pore water pressure decreases and effective stress increases at a point within the clay. In the case of granular soils, the consolidation process is almost instantaneous.

1.6.1 Void Ratio vs. Effective Stress

Let's assume that the applied loading at the ground level is of large lateral extent, as shown in Figure 1.13a, and therefore the deformations and drainage are only vertical (i.e., one-dimensional). The consolidation behavior of a clay can be studied through laboratory testing on an undisturbed sample in an *odometer*, as shown in Figure 1.13b, replicating the one-dimensional *in situ* loading (ASTM D2435; AS1289.6.6.1).

The void ratio versus effective stress (in log scale) plot, shown in Figure 1.13c, known as an $e - \log \sigma'_v$ *plot*, is developed through several incremental loadings in an odometer, allowing full consolidation during each increment. The loading part of the curve consists of two approximate straight lines AB and BC, with slopes of C_r and C_c, known as the *recompression index* and *compression index*, respectively. The unloading part CD has approximately the same slope as AB. The value of σ'_v at B is known as the *preconsolidation pressure* (σ'_p), which is the maximum pressure the soil element has experienced in the past. These three parameters are required for the settlement calculations and can be determined from an $e - \log \sigma'_v$ plot derived from a consolidation test. In the absence of consolidation test data, C_c can be estimated from some of the empirical equations available in the literature, which relate C_c to LL, natural water content, and *in situ* void ratio. Based on the work by Skempton (1944) and others, Terzaghi and Peck (1967) suggested that for undisturbed clays

$$C_c = 0.009(\text{LL} - 10) \tag{1.33}$$

and for remolded clays

$$C_c = 0.007(\text{LL} - 7) \tag{1.34}$$

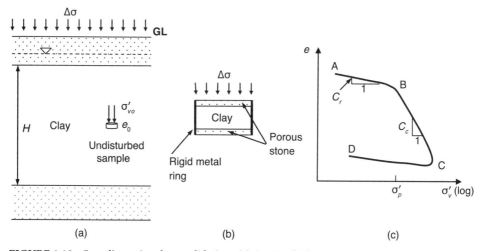

FIGURE 1.13 One-dimensional consolidation: (a) *in situ*, (b) laboratory, and (c) $e - \log \sigma'_v$ plot.

The recompression index, also known as *swelling index,* can be estimated as:

$$C_r \approx (0.1 - 0.2)C_c \qquad (1.35)$$

Typical values of C_r range from 0.01 to 0.04, where the lower end of the range applies to low-plastic clays. C_c values for inorganic clays range from 0.2 to 1.0, but for organic clays and sensitive clays, this can even exceed 5.

The *final consolidation settlement* (s_c) of a clay layer with thickness H is computed from one of the following two equations:

$$s_c = \Delta H = m_v \Delta \sigma H \qquad (1.36)$$

$$s_c = \Delta H = \frac{\Delta e}{1 + e_0} H \qquad (1.37)$$

where m_v is the *coefficient of volume compressibility,* defined as the volumetric strain per unit increase in effective stress. The *initial void ratio* of the clay layer is e_0, and the *vertical normal stress increase at the middle of the layer* is $\Delta \sigma$. Δe and ΔH are the reductions in the void ratio and layer thickness, respectively. The problem with Equation 1.36 is that m_v is not a constant and it varies with σ_v'. Therefore, it is necessary to use the value of m_v appropriate to the stress level to estimate the consolidation settlements more realistically. Settlement computations using Equation 1.37 are discussed in detail in Chapter 3. The ratio of preconsolidation pressure (σ_p') to the *initial effective overburden pressure of the sample* (σ_{vo}') gives the *overconsolidation ratio* (OCR) of the clay.

The *constrained modulus,* also known as the *odometer modulus* (D), is related to m_v and Young's modulus (E) by

$$D = \frac{1}{m_v} = \frac{(1 - v)}{(1 + v)(1 - 2v)} E = K + \frac{4}{3} G \qquad (1.38)$$

where v is Poisson's ratio. K and G are the bulk and shear moduli, respectively. D or m_v can be determined in an odometer, and assuming a value for v, E can be estimated. For saturated clays, theoretically, $v = 0.5$. For partially saturated clays, $v = 0.3$–0.4. Typical values of v for silts and sands vary from 0.2 in a loose state to 0.4 in a dense state. m_v can be less than 0.05 MPa^{-1} for very stiff clays and can exceed 1.5 MPa^{-1} for soft clays and peats. Classification of clays based on m_v is given in Table 1.4.

For linearly elastic material, K and G are related to E and v by:

$$K = \frac{E}{3(1 - 2v)} \qquad (1.39)$$

$$G = \frac{E}{2(1 + v)} \qquad (1.40)$$

The constrained modulus D is approximately related to the preconsolidation pressure by (Canadian Geotechnical Society 1992)

TABLE 1.4 Classification of Clays Based on m_v

Type of Soil	m_v (MPa^{-1})	Compressibility
Heavily overconsolidated clays	<0.05	Very low
Overconsolidated clays	0.05–0.3	Low to medium
Normally consolidated clays	0.3–1.5	High
Organic clays and peats	>1.5	Very high

$$D = (40 \sim 80)\sigma'_p \qquad (1.41)$$

where the lower end of the range is for soft clays and the upper end is for stiff clays.

From the definition of C_c and m_v, it can be shown that in normally consolidated clays they are related by

$$m_v = \frac{0.434 C_c}{(1 + e_0)\sigma_{\text{avg}}} \qquad (1.42)$$

where e_0 is the void ratio at the beginning of consolidation and σ_{avg} is the average vertical stress during consolidation. If the loading is entirely on the recompression line, C_c can be replaced by C_r and the above equation still can be used.

Young's modulus derived from *in situ* tests often is obtained under undrained conditions (E_u), and it is useful to relate this to the drained Young's modulus (E). By equating the shear moduli for undrained and drained conditions,

$$G_u = \frac{E_u}{2(1 + v_u)} = G = \frac{E}{2(1 + v)} \qquad (1.43)$$

Substituting $v_u = 0.5$ in Equation 1.41,

$$E_u = \frac{3}{2(1 + v)} E \qquad (1.44)$$

1.6.2 Rate of Consolidation

The settlements computed using Equations 1.36 and 1.37 are the final consolidation settlements that are expected to take place after a very long time, at the end of the consolidation process. In practice, when an embankment or a footing is placed on clay, it is necessary to know how long it takes the settlement to reach a certain magnitude, or how much settlement will take place after a certain time. Terzaghi (1925) developed the *one-dimensional* consolidation theory, based on the following assumptions:

1. Soil is homogeneous and saturated.
2. Soil grains and water are incompressible.
3. Strains and drainage are both one-dimensional.
4. Strains are small.
5. Darcy's law is valid.

6. Coefficients of permeability and volume compressibility remain constant during consolidation.

For the same clay layer discussed in Figure 1.13, the excess pore water pressure (Δu) distribution with depth z at a specific time t is shown in Figure 1.14. When the surcharge pressure $\Delta\sigma$ is applied at the ground level, it is immediately transferred to the pore water at every depth within the clay layer, in the form of excess pore water pressure that takes the initial value of Δu_0. Assuming the clay layer is sandwiched between two free-draining granular soil layers, the excess pore water pressure dissipates instantaneously at the top and bottom of the clay layer.

Terzaghi (1925) showed that the governing differential equation for the excess pore water pressure can be written as

$$\frac{\partial u}{\partial t} = c_v \frac{\partial^2 u}{\partial z^2} \tag{1.45}$$

where c_v is the *coefficient of consolidation*, defined as

$$c_v = \frac{k}{m_v \gamma_w} \tag{1.46}$$

with preferred units of m²/yr. By solving the above differential equation (Equation 1.45) with the appropriate boundary conditions, it can be shown that the excess pore water pressure at a depth z at time t can be expressed as

$$\Delta u(z,t) = \Delta u_0 \sum_{m=0}^{m=\infty} \frac{2}{M} \sin(MZ) e^{-M^2 T} \tag{1.47}$$

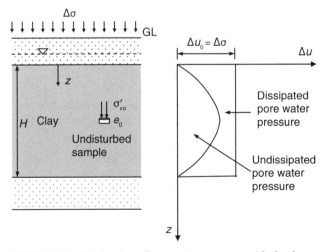

FIGURE 1.14 Dissipation of pore water pressure with depth.

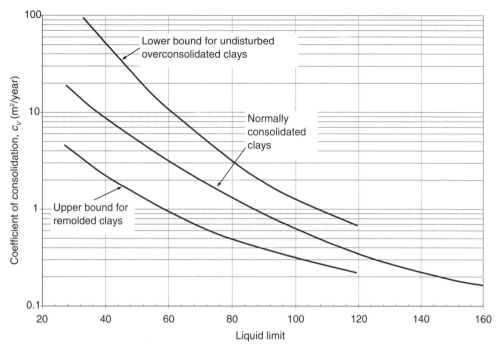

FIGURE 1.15 Approximate values of c_v (after U.S. Navy 1986).

where $M = (\pi/2)(2m + 1)$, and Z and T are the dimensionless *depth* and *time factors*, defined as

$$Z = \frac{z}{H_{dr}} \tag{1.48}$$

$$T = \frac{c_v t}{H_{dr}^2} \tag{1.49}$$

where H_{dr} is the *maximum length of the drainage path* within the clay layer. In Figures 1.13 and 1.14, where the clay is sandwiched between two granular soil layers, the clay is doubly drained and $H_{dr} = \frac{1}{2}H$. When the clay is underlain by an impervious stratum, it is singly drained and $H_{dr} = H$. The value of c_v can vary from less than 1 m²/yr for low-permeability clays to as high as 1000 m²/yr for sandy clays of very high permeability. Figure 1.15, proposed by the U.S. Navy (1986), can be used as a guide for estimating c_v from LL or as a check on measured c_v values. When a clay becomes overconsolidated, c_v increases by an order of magnitude. Therefore, overconsolidated clays consolidate significantly quicker than normally consolidated clays.

1.6.3 Degree of Consolidation

The fraction of excess pore water pressure that has dissipated at a specific depth z at a specific time t is the *degree of consolidation* (U) and is often expressed as a percentage. It is given by:

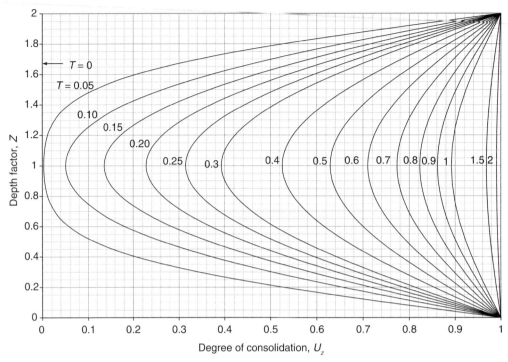

FIGURE 1.16 *U-Z-T* relationship.

$$U(z,t) = 1 - \sum_{m=0}^{m=\infty} \frac{2}{M} \sin(MZ) e^{-M^2 T} \qquad (1.50)$$

The interrelationship among *U*, *Z*, and *T* is shown graphically in Figure 1.16. The *average degree of consolidation* (U_{avg}) for the overall depth, at a specific time, is the area of the dissipated excess pore water pressure distribution diagram in Figure 1.14 divided by the initial excess pore water pressure distribution diagram. It is given by:

$$U_{avg} = 1 - \sum_{m=0}^{m=\infty} \frac{2}{M^2} e^{-M^2 T} \qquad (1.51)$$

The U_{avg}-*T* relationship is shown graphically in Figure 1.17. This can be approximated as:

$$T = \frac{\pi}{4} U_{avg}^2 \qquad \text{for } U_{avg} \leq 60\% \qquad (1.52)$$

$$T = 1.781 - 0.933 \log(100 - U_{avg}) \qquad \text{for } U_{avg} \geq 60\% \qquad (1.53)$$

1.6.4 Secondary Compression

Once the excess pore water pressure dissipates fully, there will be no more consolidation settlement. However, due to realignment of the clay particles and other mechanisms, there will be a continuous reduction of the void ratio, which leads to further settlement. This process,

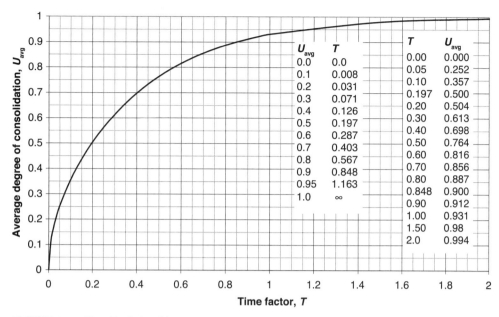

FIGURE 1.17 U_{avg}-T relationship.

which occurs at constant effective stress, is known as *secondary compression* or *creep*. The void ratio decreases linearly with logarithm of time, and the *coefficient of secondary compression* (C_α) is defined as:

$$C_\alpha = \frac{\Delta e}{\Delta(\log t)} \tag{1.54}$$

C_α can be determined from the dial gauge reading vs. log time plot in a consolidation test. Mesri and Godlewski (1977) observed that C_α/C_c varies within a narrow range of 0.025–0.10 for all soils, with an average value of 0.05. Lately, Terzaghi et al. (1996) suggested a slightly lower range of 0.01–0.07, with an average value of 0.04. The upper end of the range applies to organic clays, muskeg, and peat, and the lower end applies to granular soils. For normally consolidated clays, $C_{\alpha\varepsilon}$ lies in the range of 0.005–0.02. For highly plastic clays and organic clays, it can be as high as 0.03 or even more. For overconsolidated clays with OCR > 2, $C_{\alpha\varepsilon}$ is less than 0.001 (Lambe and Whitman 1979). Here, $C_{\alpha\varepsilon}$ is the modified secondary compression, defined as $C_\alpha/(1 + e_p)$, where e_p is the void ratio at the end of consolidation.

1.7 Shear Strength

Soil derives its shear strength from two distinct components: *friction* and *cohesion*. In granular soils, where there is no cohesion, the shear strength is purely frictional. The maximum shear stress that can be applied on a plane before failure is given by

$$\tau_f = c + \sigma \tan \phi \tag{1.55}$$

where τ_f = *shear strength* (or shear stress at failure), c = *cohesion*, σ = normal stress acting on the plane, and ϕ = friction angle. The cohesive component c of the shear strength is independent of the normal stress and remains the same at all stress levels. The frictional component $\sigma \tan \phi$ is proportional to the normal stress σ. At failure, the *major* and *minor principal stresses* are related by

$$\sigma_1 = \left(\frac{1 + \sin \phi}{1 - \sin \phi} \right) \sigma_3 + 2c \left(\frac{1 + \sin \phi}{1 - \sin \phi} \right)^{0.5}$$

$$= \tan^2 (45 + \phi/2)\sigma_3 + 2c \tan(45 + \phi/2)$$

(1.56)

and

$$\sigma_3 = \left(\frac{1 - \sin \phi}{1 + \sin \phi} \right) \sigma_1 - 2c \left(\frac{1 - \sin \phi}{1 + \sin \phi} \right)^{0.5}$$

$$= \tan^2 (45 - \phi/2)\sigma_1 - 2c \tan(45 - \phi/2)$$

(1.57)

1.7.1 Drained and Undrained Loading

When a footing or embankment is built on saturated clays, it is necessary to ensure that it remains safe at all times. Immediately after construction, known as the *short-term* situation, when there is no drainage taking place, the clay is under an undrained condition. Here, the clay is generally analyzed in terms of *total stresses,* using undrained shear strength parameters (c_u and $\phi_u = 0$). At a long time after loading, the clay would have fully consolidated and drained, and there will be no excess pore water pressure. Under this circumstance, known as the *long-term* situation, the clay is generally analyzed in terms of *effective stresses,* using shear strength parameters in terms of effective stresses (c' and ϕ'). Short term and long term are two extreme loading conditions which are easy to analyze. In reality, the loading situation is in between these two and the soil will be partially drained. Granular soils drain quickly and therefore are analyzed using effective stress parameters at all times, assuming drained conditions.

The two common laboratory tests to determine the shear strength parameters are the *triaxial test* and the *direct shear test.* They can be done under drained or undrained conditions, to determine c', ϕ', c_u, and ϕ_u. The tests on clays generally are carried out on undisturbed samples obtained from boreholes or trial pits. It is difficult to obtain undisturbed samples of granular soils, which often are reconstituted in the laboratory at specific relative densities representing the *in situ* conditions. *In situ* tests are a better option in the case of granular soils. For undrained analysis of clays in terms of total stresses, $\phi_u = 0$, with the failure envelope being horizontal in the τ-σ plane. Therefore, the undrained shear strength (τ_f) is simply the undrained cohesion (c_u). Equation 1.44 relates the undrained and drained Young's modulus.

1.7.2 Triaxial Test

A triaxial test is carried out on an undisturbed clay sample or reconstituted sand sample that is cylindrical in shape, with a length-to-diameter ratio of 2.0. The diameter of the sample can be 38–100 mm; the larger samples are used mainly for research purposes. An oversimplified schematic diagram of a triaxial test setup is shown in Figure 1.18. The soil specimen is loaded in two stages. In the first stage, the specimen is subjected to isotropic confinement, where only

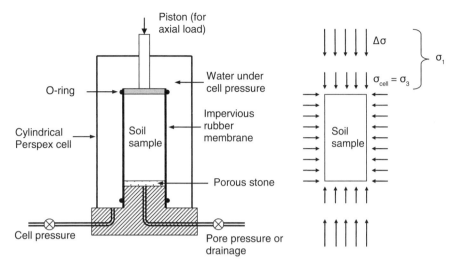

FIGURE 1.18 Schematic diagram of triaxial test setup.

the cell pressure (σ_{cell}) is applied. The drainage valve can be open to consolidate the sample, or it can be closed to prevent any consolidation. During the second stage of loading, the axial load is applied through a piston, while the cell pressure is maintained constant at σ_{cell}. The loading can be drained or undrained, depending on whether the drainage valve is open or closed during this second stage. At any time of the loading during the second stage, the vertical and horizontal stresses are assumed to be major ($\sigma_1 = \sigma_{cell} + \Delta\sigma$) and minor ($\sigma_3 = \sigma_{cell}$) principal stresses. $\Delta\sigma$ is the axial stress applied through the piston, known as the *deviator stress*. At least three samples are tested at different confining pressures, with Mohr circles drawn for each of them at failure and the failure envelope drawn tangent to the Mohr circles.

Depending on whether the drainage valve is open or closed during the isotropic confinement (stage 1) and the application of deviatoric stress (stage 2), three different test setups are commonly used to study different loading scenarios in the field:

1. Consolidated drained (CD) test (ASTM D4767)
2. Consolidated undrained (CU) test (ASTM D4767; AS1289.6.4.2)
3. Unconsolidated undrained (UU) test (ASTM D2850; AS1289.6.4.1)

Sometimes, a back pressure is applied through the drainage line to ensure full saturation of the specimen, and when the drainage valve is open, the drainage takes place against the back pressure. Generally, porous stones are provided at the top and bottom of the specimen, and a separate line is provided from the top of the sample for drainage, back pressure application, or pore pressure measurement. The effective stress parameters c' and ϕ' can be obtained from CD or CU tests, and the total stress parameters c_u and ϕ_u are determined from UU tests.

An *unconfined compressive strength test* or *uniaxial compression test* is a simple form of a triaxial test conducted on clay specimens under undrained conditions, without using the triaxial setup (ASTM D2166). This is a special case of a UU test with $\sigma_{cell} = 0$. Here, the specimen is loaded axially without any confinement. The failure pressure is known as the *unconfined compressive strength* (q_u), which is twice the *undrained shear strength* (c_u). Approximate measurements of unconfined compressive strength can be obtained from a pocket penetrometer or torvane on laboratory samples or exposed excavations in the field.

Skempton (1957) suggested that for *normally consolidated clays*:

$$\frac{c_u}{\sigma'_{vo}} = 0.0037\,\text{PI} + 0.11 \tag{1.58}$$

For *overconsolidated clays*, Ladd et al. (1977) suggested that

$$\left(\frac{c_u}{\sigma'_{vo}}\right)_{oc} = \left(\frac{c_u}{\sigma'_{vo}}\right)_{nc} (\text{OCR})^{0.8} \tag{1.59}$$

and Jamiolokowski et al. (1985) suggested that

$$\left(\frac{c_u}{\sigma'_{vo}}\right)_{oc} = (0.19 \sim 0.27)\,(\text{OCR})^{0.8} \tag{1.60}$$

Mesri (1989) suggested that for all clays

$$\frac{c_u}{\sigma'_p} = 0.22 \tag{1.61}$$

For normally consolidated clays, the effective friction angle (ϕ') decreases with increasing PI (Holtz and Kovacs 1981). A *very rough estimate* of ϕ' can be obtained from the following equation (McCarthy 2007):

$$\sin \phi' = 0.8 - 0.094 \ln \text{PI} \tag{1.62}$$

For normally consolidated clays, $c' \approx 0$.

1.7.3 Direct Shear Test

A direct shear test is carried out mostly on sands but sometimes also on clays. The soil sample is placed in a square box, approximately 60 mm × 60 mm in plan, which is split into lower and upper halves, as shown in Figure 1.19a (ASTM D3080; AS1289.6.2.2). The lower box is fixed, and the upper one can move horizontally, with provisions for measuring the horizontal displacement and load. Under a specific normal load (N), the horizontal shear load (S) is increased until the sample fails along the *horizontal failure plane* separating the two boxes. At any time during loading, the normal (σ) and shear (τ) stresses on the failure plane are given by $\sigma = N/A$ and $\tau = S/A$, where A is the cross-sectional area of the sample, which varies with the applied load.

Variations in shear stress with *vertical displacement* (δ_{ver}) and *horizontal displacement* (δ_{hor}) during the test for loose and dense sand are shown in Figure 1.19b. The plots for overconsolidated clays are similar to those for dense sands, and the ones for normally consolidated clays are similar to those for loose sands. Failure, and thus friction angle, can be defined in terms of "peak," "ultimate," "critical state," or "residual" values. *Peak shear strength* is the maximum shear stress that can be sustained. In loose sands and soft clays, where there is no pronounced peak, it is necessary to define a limiting value of strain, of the order of 10–20%,

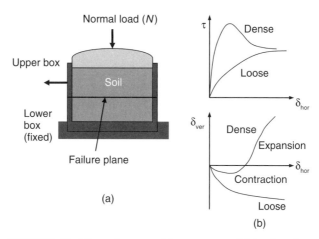

FIGURE 1.19   Direct shear test: (a) apparatus and (b) τ-δ_{hor}-δ_{ver} variations for sands.

and define the corresponding shear stress as the *ultimate shear strength. Critical state strength* is the shear stress when the critical state or constant volume state is achieved and is also known as *constant volume strength. Residual strength* is the shear strength at very large strain. In most geotechnical problems, which involve small strains, peak values generally are used. Only in large-strain problems, such as landslides, are residual or ultimate values more appropriate. Typical values of peak and ultimate friction angles for some granular soils are given in Table 1.5. It can be seen from the table that peak friction angle is always greater than residual friction angle. It also should be noted that effective cohesion in terms of residual stresses is approximately 0. A clay can be assumed to be sufficiently remolded at the residual state and hence will have negligible cohesion. The friction angle is 1–2° lower in wet sands than in dry sands. At the same relative density, peak friction angle is greater for coarse sands than fine sands by 2–5° (U.S. Army 1993). The friction angle is about 5° larger for angular grains than rounded grains (Lambe and Whitman 1979).

The *dilation angle* (ψ) is a useful parameter in numerical modeling of geotechnical problems. It is one of the input parameters in some *constitutive models,* to account for the volume change behavior of soils under stresses. It is defined as

TABLE 1.5   Friction Angles of Granular Soils

	Friction angle, ϕ (degrees)	
Soil Type	Ultimate	Peak
Medium-dense silt	26–30	28–32
Dense silt	26–30	30–34
Medium-dense uniform fine to medium sand	26–30	30–34
Dense uniform fine to medium sand	26–30	32–36
Medium-dense well-graded sand	30–34	34–40
Dense well-graded sand	30–34	38–46
Medium-dense sand and gravel	32–36	36–42
Dense sand and gravel	32–36	40–48

After Lambe and Whitman (1979).

TABLE 1.6 Sensitivity Classification

S_t	1	2–4	4–8	8–16	>16
Term	Insensitive	Low sensitivity	Medium sensitivity	High sensitivity	Quick

$$\tan \psi \; = \; - \frac{d\varepsilon_{\text{vol}}}{d\gamma} \tag{1.63}$$

where ε_{vol} and γ are the *volumetric* and *shear strains*, respectively. ψ is significantly smaller than the friction angle, with typical values varying from 0 for very loose sands to 15° for very dense sands. For plane strain situations, it can be estimated as (Bolton 1986)

$$\psi \; = \; 1.25(\phi_{\text{peak}} - \phi_{cv}) \tag{1.64}$$

where ϕ_{peak} and ϕ_{cv} are the friction angles at the peak and critical state, respectively. Salgado (2008) extended Equation 1.64 to triaxial compression as follows:

$$\psi \; = \; 2.0(\phi_{\text{peak}} - \phi_{cv}) \tag{1.65}$$

Clays can become weak when remolded, and the ratio of undisturbed shear strength to remolded shear strength is known as *sensitivity* (S_t). Some marine clays are very sensitive, showing high peak strength and very low remolded strength. At very large strains, clay becomes remolded, and therefore, the ratio of peak to residual shear strength is a measure of sensitivity. Classification of clays based on sensitivity is given in Table 1.6. Highly sensitive clays generally have flocculated fabric.

1.7.4 Skempton's Pore Pressure Parameters

Skempton (1954) introduced a simple method for estimating the pore water pressure response of saturated soils to *undrained loading*, in terms of *total stress changes*. He proposed that

$$\Delta u \; = \; B[(\Delta\sigma_3 + A(\Delta\sigma_1 - \Delta\sigma_3)] \tag{1.66}$$

where $\Delta\sigma_1$ and $\Delta\sigma_3$ are the largest and smallest of the total stress increments, respectively. (Note that they are not the increments of major and minor principal stresses, respectively.) A and B are known as Skempton's pore pressure parameters. B is a measure of the degree of saturation; it varies in the range of 0 (for dry soils) to 1 (for saturated soils). In soils with very stiff soil skeletons, B can be less than 1 even when fully saturated.

The parameter A depends on several factors, including the stress path, strain level, OCR, etc. At failure, it is denoted by A_f and can vary between –0.5 (for heavily overconsolidated clays) and 1 (for normally consolidated clays).

1.7.5 Stress Paths

A stress path is a simple way of keeping track of the loading on a soil element. It is the locus of the stress point, which is simply the top of the Mohr circle, as shown in Figure 1.20.

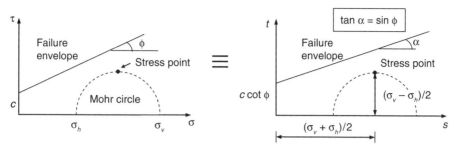

FIGURE 1.20 Mohr circle and stress point.

Compared to plotting a series of Mohr circles, stress paths are much neater and less confusing. Here, the σ-axis and τ-axis used for Mohr circles are replaced by the s-axis and t-axis, where s and t are defined as

$$s = \frac{\sigma_v + \sigma_h}{2} \tag{1.67}$$

$$t = \frac{\sigma_v - \sigma_h}{2} \tag{1.68}$$

where σ_v = vertical normal stress and σ_h = horizontal normal stress. The failure envelope in the s-t plane has an intercept of $c \cos \phi$ and slope of $\sin \phi$. In three-dimensional problems, s and t are replaced by p and q, the average principal stress and deviator stress, respectively, defined as

$$p = \frac{\sigma_1 + \sigma_2 + \sigma_3}{3} \tag{1.69}$$

$$q = \sigma_1 - \sigma_3 \tag{1.70}$$

where σ_1, σ_2, and σ_3 are the major, intermediate, and minor principal stresses.

1.8 Site Investigation

Unlike other civil engineering materials, soils have significant variability associated with them. Their engineering properties can vary dramatically within a few meters. A thorough site investigation is a prerequisite for all large geotechnical projects. A typical site investigation includes *in situ* tests, sampling, and laboratory tests, in addition to the preliminary studies such as a desk study and site reconnaissance. All the findings are presented in the form of a site investigation report, which consists of a site plan, several bore logs which summarize the soil properties at each borehole and trial pit, and the associated laboratory and *in situ* test data. A site investigation project can cost about 0.1–0.5% of the total cost of a project.

A desk study involves collection of as much existing information as possible about the site, through geological maps, aerial photographs, soil survey reports, site investigation reports of nearby sites, etc. Site reconnaissance consists of a walk-over survey to visually assess the local conditions, such as site access, adjacent properties and structures on them, topography, drainage, etc.

Trial pits are excavated using a backhoe and can be as deep as 4–5 m. They enable visual inspection of the top soil and are inexpensive. Boreholes are about 50–100 mm in diameter and can extend to depths greater than 30 m. Every meter a borehole is advanced costs money; therefore, care is required in selecting the right number of boreholes and trial pits and limiting the depth to what is absolutely necessary. Deciding on the number of boreholes for a project is governed by budget limitations, type of project, loadings, and site conditions. Typically, boreholes are spaced at intervals of 15 m (for very heavy column loads) to 50 m (for very light loads). For residential subdivisions, trial pits often are adequate. If boreholes are required, they can be spaced at 250- to 500-m intervals. Along proposed highways, boreholes can be located at 150- to 500-m intervals. The boreholes should be advanced to depths where the average vertical stress increase due to the proposed structure is about 10% of what is applied at the surface or depths where the additional vertical stress increase is about 5% of the current effective overburden stress (the lesser value should be used).

Undisturbed samples of cohesive soils are collected from boreholes in thin-walled samplers (e.g., Shelby tubes) for laboratory tests such as triaxial and consolidation tests. The degree of disturbance is quantified through the area ratio (A_R) of the sampling tube, defined as

$$A_R \ (\%) \ = \ \frac{D_o^2 \ - \ D_i^2}{D_i^2} \tag{1.71}$$

where D_o and D_i are the external and internal diameter, respectively, of the sampling tube. For a good-quality undisturbed sample, A_R must be less than 10%.

The standard penetration test and static cone penetration test are the most popular *in situ* tests used for deriving soil parameters for most geotechnical and foundation engineering designs. Other specialized *in situ* tests include the vane shear test, pressuremeter test, dilatometer test, plate load test, borehole shear test, and K_0 stepped blade test. The standard penetration test, cone penetration test, and vane shear test are discussed below in detail.

1.8.1 Standard Penetration Test

The standard penetration test (SPT) originally was developed in 1927 in the U.S. for granular soils and is one of the oldest and most popular *in situ* tests (ASTM D3441; AS1289.6.3.1). A schematic diagram of an SPT setup, using an old-fashioned rotating cathead, is shown in Figure 1.21a. These days, an automatic tripping mechanism (Figure 1.21b) is used instead of the rope and cathead arrangement to raise and release the hammer. A 35-mm-inside-diameter × 50-mm-outside-diameter *split-spoon sampler* at the bottom of the borehole, connected to the anvil through drill rods, is driven into the ground by repeatedly dropping a 63.5-kg hammer over a distance of 760 mm onto the anvil. The number of blows required to achieve three subsequent 150-mm penetrations is recorded. The number of blows it takes to penetrate the final 300 mm is known as the *blow count, standard penetration number,* or *N-value* at that depth. The test is carried out in a borehole, at 1- to 2-m-depth intervals, and complements the drilling work. The *N*-values are plotted with depth, where the points are connected by *straight lines.*

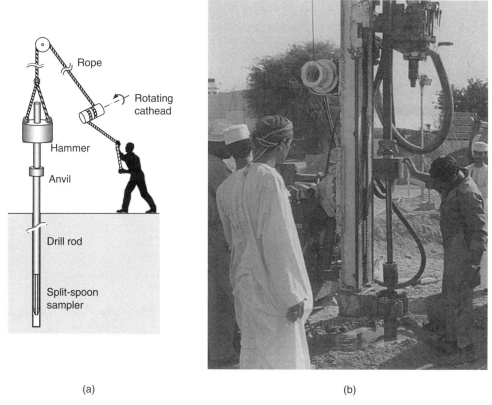

(a)   (b)

FIGURE 1.21 SPT setup: (a) schematic of rotating cathead arrangement and (b) photograph of automatic tripping arrangement.

In very fine or silty sands below the water table, the blow count is overestimated due to the buildup of excess pore water pressures during driving, which in turn reduces the effective stresses. Here, the measured blow count must be reduced using the following equation (Terzaghi and Peck 1948):

$$N = 15 + 0.5(N_{\text{measured}} - 15) \qquad (1.72)$$

Due to the variability worldwide associated with the choice of SPT equipment and the test procedure, various correction factors are applied to the measured blow count (N). The two most important correction factors are the *overburden pressure correction* (C_N) and *hammer efficiency correction* (E_h). The blow count corrected for overburden pressure and hammer efficiency is expressed as:

$$(N_1)_{60} = C_N E_h N \qquad (1.73)$$

C_N is the ratio of the measured blow count to what it would be at an overburden pressure of 1 ton/ft^2 (or 1 kg/cm^2). Several expressions have been proposed for C_N, the most popular of which is (Liao and Whitman 1986):

$$C_N = 9.78 \sqrt{\frac{1}{\sigma'_{vo} \text{ (kPa)}}} \tag{1.74}$$

Without the overburden correction, Equation 1.73 becomes:

$$N_{60} = E_h N \tag{1.75}$$

The actual energy delivered by the hammer to the split-spoon sampler can be significantly less than the theoretical value, which is the product of the hammer weight and the drop. Kovacs and Salomone (1982) reported that the actual efficiency of the system is of the order of 30–80%. Because most SPT correlations are based on a hammer efficiency of 60%, the current practice is to accept an efficiency of 60% as the standard (Terzaghi et al. 1996). Therefore, E_h is defined as:

$$E_h = \frac{\text{Hammer efficiency}}{60} \tag{1.76}$$

Two other correction factors are the *borehole diameter correction* (C_b) and *drill rod length correction* (C_d), given in Tables 1.7 and 1.8. These are discussed in detail by Skempton (1986). When using samplers with liners, the blow count is overestimated, and a multiplication factor of 0.8 is recommended in dense sands and clays and 0.9 in loose sands (Bowles 1988).

While an SPT gives the blow count, laboratory tests on sands are carried out on the basis of relative density. The interrelationships among blow count, relative density, friction angle, and Young's modulus are discussed below.

Using Meyerhof's (1957) approximation,

$$\frac{N_{60}}{D_r^2} = a + b\sigma'_{vo} \tag{1.77}$$

Skempton (1986) suggested that for sands where $D_r > 35\%$:

$$\frac{(N_1)_{60}}{D_r^2} \approx 60 \tag{1.78}$$

Here, $(N_1)_{60}$ should be multiplied by 0.92 for coarse sands and 1.08 for fine sands.

TABLE 1.7 Borehole Diameter Correction Factor (Skempton 1986)

Borehole Diameter (mm)	Correction Factor, C_b
60–115	1.00
150	1.05
200	1.15

TABLE 1.8 Drill Rod Length Correction Factor (Skempton 1986)

Rod Length (m)	Correction Factor, C_d
0–4	0.75
4–6	0.85
6–10	0.95
>10	1.00

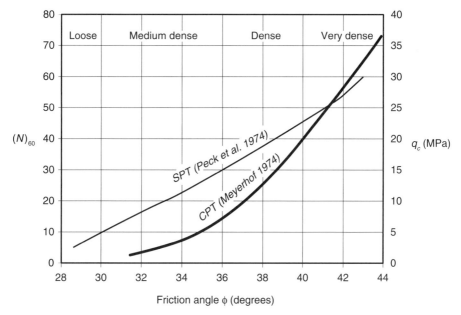

FIGURE 1.22 Penetration resistance vs. friction angle.

Peck et al. (1974) proposed a relationship between N_{60} and ϕ for granular soils, shown in Figure 1.22, which is widely used in granular soils to estimate the friction angle from the blow count. Wolff (1989) expressed this relation as:

$$\phi \text{ (deg)} = 27.1 + 0.3N_{60} - 0.00054N_{60}^2 \tag{1.79}$$

Hatanaka and Uchida (1996) provided a simple correlation between ϕ and $(N_1)_{60}$ for granular soils:

$$\phi \text{ (deg)} = \sqrt{20(N_1)_{60}} + 20 \tag{1.80}$$

Schmertmann (1975) proposed an N_{60}-ϕ-σ'_{vo} relation graphically for granular soils, which can be expressed as (Kulhawy and Mayne 1990):

$$\phi = \tan^{-1} \left[\frac{N_{60}}{12.2 + 20.3 \left(\dfrac{\sigma'_{vo}}{p_a} \right)} \right]^{0.34} \tag{1.81}$$

where p_a is the atmospheric pressure (= 101.3 kPa).

Young's modulus is an essential parameter for computing deformations, including settlement, of foundations. Leonards (1986) suggested that for normally consolidated sands:

$$E \text{ (kg/cm}^2) \approx 2q_c \text{ (kg/cm}^2) \approx 8N_{60} \tag{1.82}$$

Kulhawy and Mayne (1990) suggested that

$$E/p_a = \alpha N_{60} \tag{1.83}$$

where $\alpha = 5$ for sands with fines, 10 for clean normally consolidated sands, and 15 for clean overconsolidated sands.

A 35-mm-inside-diameter $\times$ 457- to 610-mm-long split-spoon sampler enables recovery of disturbed samples that can be used for soil classification and any further testing. The area ratio for the split-spoon sampler is about 110%, implying the samples from an SPT split-spoon are highly disturbed.

In spite of its simplicity and the large historical database, the SPT has numerous sources of uncertainty and error, making it less reproducible. Lately, static cone penetration tests, using piezocones, have become increasingly popular because they offer better rationale, reproducibility, and continuous measurements. An SPT is not very reliable in cohesive soils, due to the pore pressure developments during driving that may affect the effective stresses temporarily, and any correlations in clays should be used with caution. A rough estimate of undrained shear strength can be obtained from (Hara et al. 1971; Kulhawy and Mayne 1990):

$$\left(\frac{c_u}{p_a}\right) = 0.29N_{60}^{0.72} \tag{1.84}$$

1.8.2 Static Cone Penetration Test and Piezocones

The static cone penetration test (CPT), also known as the Dutch cone penetration test, originally was developed in the Netherlands in 1920 and can be used in most soils (ASTM D3441; AS1289.6.5.1). The split-spoon sampler is replaced by a probe that consists of a solid cone with a 60° apex angle and a base area of 10 cm², attached to a drill rod with a friction sleeve that has a surface area of 150 cm². It is advanced into the soil at a rate of 20 mm/s. A mini test rig is shown in Figure 1.23a. A *piezocone*, shown in Figures 1.23b and 1.23c, is simply a static cone with a piezometer built in for pore pressure measurements. Here, the three measurements that are taken continuously as the cone is pushed into the soil are *cone resistance* (q_c), *sleeve friction* (f_s), and *pore water pressure* (u). The friction ratio, defined as

$$f_R = \frac{f_s}{q_c} \times 100\% \tag{1.85}$$

is a useful parameter in identifying the soil. Values for f_R are in the range of 0–10%, with the granular soils at the lower end and cohesive soils at the upper end of the range. Using the pair of values for q_c and f_R, the soil type can be identified from Figure 1.24.

The undrained shear strength (c_u) of clays can be estimated from (Schmertmann 1975)

$$c_u = \frac{q_c - \sigma_{vo}}{N_k} \tag{1.86}$$

where N_k is known as the cone factor, which varies in the range of 14–25 and can be obtained through calibration. The lower end of the range applies to normally consolidated clays and the upper end to overconsolidated clays. It depends on the penetrometer and the type of clay and

(a) (b) (c)

FIGURE 1.23 Static CPT: (a) mini test rig, (b) piezocone, and (c) inside a piezocone.

increases slightly with the PI. Based on the test data from Aas et al. (1984), N_k can be estimated by (Bowles 1988):

$$N_k = 13 + 0.11\text{PI} \pm 2 \tag{1.87}$$

Classification of clays based on undrained shear strength and their corresponding consistency terms are given in Table 1.9. Also given in the table are the approximate borderline values of $(N_1)_{60}$ and q_c/p_a and a field identification guide.

Variation of ϕ with q_c for granular soils, as proposed by Meyerhof (1974), is shown in Figure 1.22. The dependence of q_c on overburden stress is not incorporated here, and therefore this must be used with caution.

Kulhawy and Mayne (1990) showed that the q_c-σ'_{vo}-ϕ relationship in sands, proposed by Robertson and Campanella (1983), can be approximated by:

$$\phi = \tan^{-1}\left[0.1 + 0.38 \log\left(\frac{q_c}{\sigma'_{vo}} \right) \right] \tag{1.88}$$

Schmertmann (1970) proposed that $E = 2q_c$ and later (Schmertmann et al. 1978) suggested that $E = 2.5q_c$ for axisymmetric loading in sands and $E = 3.5q_c$ for plane strain loading.

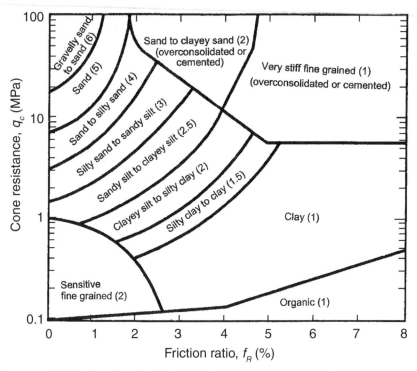

FIGURE 1.24 Soil classification from a piezocone (adapted from Robertson et al. 1986). $(q_c/p_a)/N_{60}$ values shown in parentheses.

TABLE 1.9 Consistency Terms for Clays with $(N)_{60}$ and q_c Values

Consistency[a]	c_u (kPa)[a]	$(N_1)_{60}$[a]	q_c/p_a[b]	Field Identification Guide[c]
Very soft	<12	0–2	<5	Exudes between fingers when squeezed in hand; can easily be penetrated several centimeters by fist
Soft	12–25	2–4		Can be molded by light finger pressure; can easily be penetrated several centimeters by thumb
Firm	25–50	4–8	5–15	Can be molded by strong finger pressure; can be penetrated several centimeters by thumb with moderate effort
Stiff	50–100	8–15	15–30	Cannot be molded by fingers; can be indented by thumb but penetrated only with great effort
Very stiff	100–200	15–30	30–60	Readily indented by thumbnail
Hard	>200	>30	>60	Can be indented by thumbnail with difficulty

[a] Terzaghi and Peck (1948).
[b] McCarthy (2007).
[c] Australian Standards (1993), Canadian Geotechnical Society (1992).

TABLE 1.10 Ratio of q_c/N

Soil	q_c (kg/cm²)/N_{60}
Silts, sandy silts, slightly cohesive silt-sand mix	2 [a] (2–4) [b]
Clean fine to medium sands and slightly silty sands	3–4 [a] (3–5) [b]
Coarse sands and sands with little gravel	5–6 [a] (4–5) [b]
Sandy gravel and gravel	8–10 [a] (6–8) [b]

After Sanglerat (1972) and Schmertmann (1970, 1978).

[a] Values proposed by Sanglerat (1972) and reported in Peck et al. (1974).

[b] Values suggested by Schmertmann (1970, 1978) and reported by Holtz (1991) in parentheses.

Geotechnical engineers do not always have the luxury of availability of both SPT and CPT data. When only one type of data is available, it is useful to have some means of converting it to the other. Ratios of q_c/N for different soils, as given by Sanglerat (1972) and Schmertmann (1970, 1978), are shown in Table 1.10. Robertson et al. (1983) presented the variation of q_c/N with a median grain size of D_{50}, and the upper and lower bounds are shown in Figure 1.25. The soil data were limited to D_{50} less than 1 mm. Also shown in the figure are the upper and lower bounds proposed by Burland and Burbidge (1985) and the average values suggested in the *Canadian Foundation Engineering Manual* (Canadian Geotechnical Society 1985) and by Kulhawy and Mayne (1990) and Anagnostopoulos et al. (2003).

All the curves in Figure 1.25 take the form

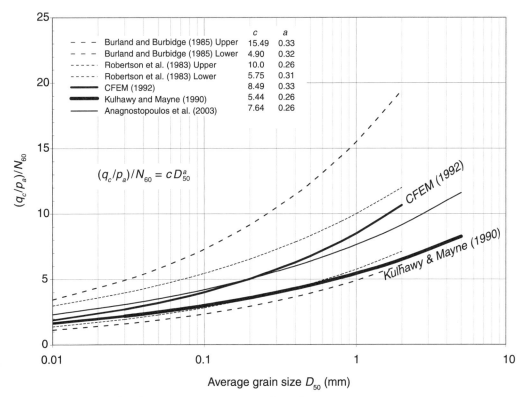

FIGURE 1.25 $(q_c/p_a)/N_{60}$-D_{50} variation in granular soils.

$$\left(\frac{q_c}{p_a} \right) / N_{60} \approx cD_{50}^{a} \tag{1.89}$$

where the values of c and a are given in Figure 1.25. Kulhawy and Mayne (1990) approximated the dependence of the q_c/N_{60} ratio on D_{50} (mm) as:

$$\left(\frac{q_c}{p_a} \right) / N_{60} \approx 5.44 D_{50}^{0.26} \tag{1.90}$$

Based on an extensive database of 337 points, with test data for D_{50} as high as 8 mm, Anagnostopoulos et al. (2003) noted that for Greek soils:

$$\left(\frac{q_c}{p_a} \right) / N_{60} \approx 7.64 D_{50}^{0.26} \tag{1.91}$$

Kulhawy and Mayne (1990) also suggested that q_c/N_{60} can be related to the fine content in a granular soil as:

$$\left(\frac{q_c}{p_a} \right) / N_{60} \approx 4.25 - \frac{\% \text{ fines}}{41.3} \tag{1.92}$$

Based on relative density, granular soils can be classified as shown in Figure 1.26. Also given in the figure are the values of N_{60}, $(N_1)_{60}$, ϕ', and $(N_1)_{60}/D_r^2$.

1.8.3 Vane Shear Test

The vane shear test is used for determining undrained shear strength in clays that are particularly soft. The vane consists of two rectangular blades that are perpendicular to each other, as shown in Figure 1.27.

	*Very loose	Loose	Medium dense	Dense	Very dense	
#D_r (%)	0	15	35	65	85	100
*N_{60}		4	10	30	50	
##$(N_1)_{60}$		3	8	25	42	
**ϕ' (deg)		28	30	36	41	
##$(N_1)_{60}/D_r^2$			65	59	58	

FIGURE 1.26 Classification of granular soils based on relative density: # = Gibbs and Holtz (1957), * = Terzaghi and Peck (1948), ## = Skempton (1986), and ** = Peck et al. (1974).

The vane is pushed into the borehole at the required depth, where the test is carried out. It is rotated at a rate of $0.1°/s$ by applying a torque at the surface through a torque meter that measures the torque. This rotation will initiate shearing of the clay along a cylindrical surface surrounding the vanes. The undrained shear strength of the undisturbed clay can be determined from the applied torque (T) using the following equation:

$$c_u = \frac{2T}{\pi d^2(h + d/3)} \quad (1.93)$$

where h and d are the height and breadth, respectively, of the rectangular blades (i.e., height and diameter of the cylindrical surface sheared), which are typically of a 2:1 ratio, with d in the range of 38–100 mm for field vanes. Miniature vanes are used in laboratories to determine undrained shear strength of clay samples still in sampling tubes. The test can be continued by rotating the vane rapidly after shearing

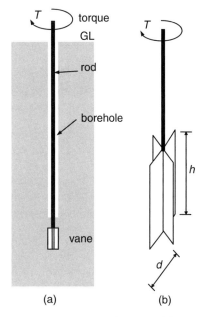

(a)   (b)

FIGURE 1.27 Vane shear test: (a) in a borehole and (b) vane.

the clay, to determine the remolded shear strength. The test can be carried out at depths greater than 50 m.

Back analysis of several failed embankments, foundations, and excavations in clays has shown that the vane shear test overestimates the undrained shear strength. A reduction factor (λ) has been proposed to correct the shear strength measured by the vane shear test; the correct shear strength is given by

$$c_{u(\text{corrected})} = \lambda c_{u(\text{vane})} \quad (1.94)$$

for which Bjerrum (1972) has proposed that:

$$\lambda = 1.7 - 0.54 \log[\text{PI}] \quad (1.95)$$

Morris and Williams (1994) suggested that for PI > 5:

$$\lambda = 1.18 \exp(-0.08\text{PI}) + 0.57 \quad (1.96)$$

1.9 Soil Variability

Unlike other civil engineering materials (e.g., concrete, steel), soils are nonhomogeneous, three-phase particulate materials, and theories often are oversimplified. With limited site investigation data available, it becomes even more difficult to arrive at realistic deterministic solutions with confidence. A probabilistic approach to geotechnical problems is becoming increasingly popular, with risk analysis and reliability studies quite common in traditional geotechnical and mining engineering.

TABLE 1.11 Suggested COV Values

Parameter	COV (%)	Reference
e or n	20–30	Lee et al. (1983), Baecher and Christian (2003)
D_r (sand)	10–40	Baecher and Christian (2003)
G_s	2–3	Harr (1987), Baecher and Christian (2003)
ρ or γ	3–10	Lee et al. (1983), Harr (1987), Baecher and Christian (2003)
LL	10–20	Lee et al. (1983), Baecher and Christian (2003)
PL	10–20	Lee et al. (1983)
PI	30–70[a]	Lee et al. (1983)
$w_{optimum}$	20–40[b]	Lee et al. (1983)
$\rho_{d,maximum}$	5	Lee et al. (1983)
k	200–300	Lee et al. (1983), Baecher and Christian (2003)
c_v	25–50	Baecher and Christian (2003), Lee et al. (1983)
C_c and C_r	20–40	Baecher and Christian (2003)
CBR[c]	25	Lee et al. (1983)
ϕ-sand	10	Harr (1987), Lee et al. (1983)
ϕ-clay	10–50	Lee et al. (1983), Baecher and Christian (2003)
ϕ-mine tailings	5–20	Baecher and Christian (2003)
c_u or q_u	40	Harr (1987), Lee et al. (1983)
N (from SPT)	25–40	Harr (1987), Lee et al. (1983), Baecher and Christian (2003)
q_c (from CPT)	20–50[a]	Baecher and Christian (2003)
$c_{u,VST}$[c]	10–40	Baecher and Christian (2003)

After Lee et al. (1983) and Baecher and Christian (2003).
[a] Lower values for clays and higher ones for sandy/gravelly clays.
[b] Lower values for clays and higher ones for granular soils.
[c] CBR = California bearing ratio, VST = vane shear test.

With very limited geotechnical data coming from the laboratory and *in situ* tests for a project, it is not possible to get realistic estimates of the standard deviation of the soil parameters. Typical values of the *coefficient of variation* (COV) reported in the literature can be used as the basis for estimating the standard deviation of the soil parameters. COV is defined as:

$$\text{COV} = \frac{\text{standard deviation}}{\text{mean}} \times 100\% \tag{1.97}$$

Harr (1987), Lee et al. (1983), and Baecher and Christian (2003) collated test data from various sources and presented the COV values. These are summarized in Table 1.11.

1.10 Geotechnical Instrumentation

Use of instruments to monitor the performances of earth and earth-supported structures is increasingly becoming the way of life in large geotechnical projects. Geotechnical instrumentation is used for one or more of the following reasons:

1. *Design verification*—To verify new or unconventional designs, particularly when simplified theories and assumptions are involved (e.g., pile load test)
2. *Construction control*—To monitor performance during construction so as to be able to alter or modify the design and procedures (e.g., deep excavations near buildings)

3. *Safety*—To warn of any impending failure (e.g., early warning systems for landslides)
4. *Legal protection*—To document strong evidence against any possible lawsuit (e.g., noise and vibrations due to pile driving)
5. *Verification of long-term performance*—To monitor in-service performance (e.g., drainage behind retaining walls)
6. *Advancing state-of-the-art*—To verify new developments in R&D and new design methodologies or construction techniques (e.g., new geosynthetic products)
7. *Quality control*—To verify compliance by the contractor (e.g., settlement of a compacted fill)

Whereas site investigation offers one-off measurements prior to construction, geotechnical instrumentation is used during or after construction, to monitor the ongoing performance of bridge abutments, retaining walls, foundations, and embankments. This includes monitoring deformation, pore water pressure, load, pressure, strain, and temperature. Brief descriptions of some of the common instruments used in geotechnical engineering are given below.

Piezometers are used for measuring water level and pore water pressure. Piezometers range from the simple and inexpensive Casagrande's open standpipe piezometer to more complex pneumatic, vibrating wire, or hydraulic piezometers. *Settlement cells* or *plates* can be placed within embankments or foundations to monitor ongoing settlement. *Vertical inclinometers* are quite useful for monitoring lateral deformation near embankments on soft soils, landslides, and deflection of piles under lateral loading. *Horizontal inclinometers* can be used to determine the settlement profile beneath an embankment cross section. *Load cells* are used to monitor the load on tiebacks, rock anchors, soil nails, and piles. *Extensometers* measure deformation along the axis and can be used for measuring deformation in any direction, such as settlement, heave, or lateral displacement. *Strain gauges* can be mounted onto steel or concrete structures such as piles, bridges, and tunnel linings to monitor strain while in service. Pressure cells are useful for measuring horizontal or vertical earth pressure within the soil beneath foundations and embankments.

In using geotechnical instrumentation, it is necessary to understand properly what the following terms mean: accuracy, precision, resolution, and sensitivity. *Accuracy* is how close the mean value of a measurement is to the *true* value. *Precision* is how close a set of measurements is to the mean value (not necessarily the true value). It is a measure of repeatability or reproducibility. Precision should not be confused with accuracy. Precise measurements need not be accurate and vice versa. *Resolution* is the smallest change that can be detected by a readout device, whether digital or analog (e.g., 0.01 g on a digital balance). *Sensitivity* refers to the response of a device to a unit input (e.g., 100 mV/mm in a linear variable differential transformer). Dunnicliff (1993) discussed geotechnical instrumentation in great detail.

References

Aas, G., Lacasse, S., Lunne, T., and Madshus, C. (1984). In situ testing: new developments. *NGM-84,* Linkoping, 2, 705–716.

Anagnostopoulos, A., Koukis, G., and Sabatakakis, N. (2003). Empirical correlations of soil parameters based on cone penetration tests (CPT) for Greek soils. *Geotech. Geol. Eng.,* 21(4):377–387.

Australian Standards (1993). *Geotechnical Site Investigations,* AS1726-1993, 40 pp.

Baecher, G.B. and Christian, J.T. (2003). *Reliability and Statistics in Geotechnical Engineering,* John Wiley & Sons, 605 pp.

Bjerrum, L. (1972). Embankments on soft ground. *Proceedings of ASCE Specialty Conference on Performance on Earth and Earth Supported Structures,* Purdue University, West Lafayette, IN, 2, 1–54.

Bolton, M.D. (1986). The strength and dilatancy of sands. *Geotechnique,* 36(1):65–78.

Bowles, J.E. (1988). *Foundation Analysis and Design,* 4th edition, McGraw-Hill, 1004 pp.

Burland, J.B. and Burbidge, M.C. (1985). Settlements of foundations on sand and gravel. *Proc. Inst. Civ. Eng.,* 78(1):1325–1381.

Burmister, D.M. (1949). Principles and techniques of soil identification. *Proceedings of Annual Highway Research Board Meeting,* National Research Council, Washington, D.C., 29, 402–433.

Canadian Geotechnical Society (1992). *Canadian Foundation Engineering Manual,* 3rd edition, 511 pp.

Casagrande, A. (1948). Classification and identification of soils. *ASCE Trans.,* 113:901–930.

Dunnicliff, J. (1993). *Geotechnical Instrumentation for Monitoring Field Performance,* John Wiley & Sons, 577 pp.

Fuller, W.B. and Thompson, S.E. (1907). The laws of proportioning concrete. *ASCE Trans.,* 59:67–143.

Gibbs, H.J. and Holtz, W.G. (1957). Research on determining the density of sands by spoon penetration testing. *Proceedings of 4th International Conference on Soil Mechanics and Foundation Engineering,* Butterworths, London, 1, 35–39.

Hara, A., Ohata, T., and Niwa, M. (1971). Shear modulus and shear strength of cohesive soils. *Soils Found.,* 14(3):1–12.

Harr, M.E. (1962). *Groundwater and Seepage,* McGraw-Hill, New York, 315 pp.

Harr, M.E. (1987). *Reliability-Based Design in Civil Engineering,* McGraw-Hill, 290 pp.

Hatanaka, M. and Uchida, A. (1996). Empirical correlation between penetration resistance and internal friction angle of sandy soils. *Soils Found.,* 36(4):1–10.

Hausmann, M.R. (1990). *Engineering Principles of Ground Modification,* McGraw-Hill, 632 pp.

Hazen, A. (1930). Water supply. *American Civil Engineers Handbook,* John Wiley & Sons, New York, 1444–1518.

Holtz, R.D. (1991). Stress distribution and settlement of shallow foundations. *Foundation Engineering Handbook,* 2nd edition, H.-Y. Fang, Ed., Van Nostrand Reinhold, New York, 166–222.

Holtz, R.D. and Kovacs, W.D. (1981). *An Introduction to Geotechnical Engineering,* Prentice Hall, 733.

Jamiolokowski, M., Ladd, C.C., Germaine, J.T., and Lancellotta, R. (1985). New developments in field and laboratory testing of soils. *11th International Conference on Soil Mechanics and Foundation Engineering,* San Francisco, 1, 57–154.

Kovacs, W.D. and Salomone, L.A. (1982). SPT hammer energy measurement. *J. Geotech. Eng. Div. ASCE,* 108(4):599–620.

Kulhawy, F.H. and Mayne, P.W. (1990). *Manual on Estimating Soil Properties for Foundation Design,* Final Report EL-6800, Electric Power Research Institute, Palo Alto, CA.

Ladd, C.C., Foote, R., Ishihara, K., Schlosser, F., and Poulos, H.G. (1977). Stress deformation and strength characteristics. *Proceedings of the 9th International Conference on Soil Mechanics and Foundation Engineering,* Tokyo, 2, 421–494.

Lambe, T.W. (1958a). The structure of compacted clay. *J. Soil Mech. Found. Div. ASCE,* 84(SM2):1654.

Lambe, T.W. (1958b). The engineering behavior of compacted clay. *J. Soil Mech. Found. Div. ASCE,* 84(SM2):1655.

Lambe, T.W. and Whitman, R.V. (1979). *Soil Mechanics,* John Wiley & Sons, 553 pp.

Lee, I.K., White, W., and Ingles, O.G. (1983). Soil variability. *Geotechnical Engineering,* Pitman, chap. 2.

Leonards, G.A. (1962). *Foundation Engineering,* McGraw-Hill, New York, 1136 pp.

Leonards, G.A. (1986). *Advanced Foundation Engineering—CE683,* Lecture Notes, Purdue University, West Lafayette, IN.

Leonards, G.A., Cutter, W.A., and Holtz, R.D. (1980). Dynamic compaction of granular soils. *J. Geotech. Eng. Div. ASCE,* 106(1):35–44.

Liao, S.S. and Whitman, R.V. (1986). Overburden correction factors for SPT in sand. *J. Geotech. Eng. Div. ASCE,* 112(3):373–377.

Lunne, T., Robertson, P.K., and Powell, J.J.M. (1997). *Cone Penetration Testing in Geotechnical Practice,* Blackie Academic & Professional, 312 pp.

McCarthy, D.F. (2007). *Essentials of Soil Mechanics and Foundations,* 7th edition, Pearson Prentice Hall, 850 pp.

Mesri, G. (1989). A re-evaluation of $s_{u(\text{mob})}$ 0.22 σ'_p using laboratory shear tests. *Can. Geotech. J.,* 26(1):162–164.

Mesri, G. and Godlewski, P. (1977). Time- and stress-compressibility interrelationships. *J. Geotech. Eng. Div. ASCE,* 103:417–430.

Meyerhof, G.G. (1957). Discussion on research on determining the density of sands by spoon penetration testing. *Proceedings of 4th International Conference on Soil Mechanics and Foundation Engineering,* London, 3, 110.

Meyerhof, G.G. (1974). Penetration testing outside Europe. General report. *Proceedings of European Symposium on Penetration Testing,* Stockholm, 2.2, 40–48.

Morris, P.M. and Williams, D.T. (1994). Effective stress vane shear strength correction factor correlations. *Can. Geotech. J.,* 31(3):335–342.

Peck, R.B., Hanson, W.E., and Thornburn, T.H. (1974). *Foundation Engineering,* John Wiley & Sons, New York.

Proctor, R.R. (1933). Design and construction of rolled earth dams. *Eng. News-Rec.,* 3:245–248, 286–289, 348–351, 372–376.

Robertson, P.K. and Campanella, R.G. (1983). Interpretation of cone penetration tests. I. Sand. *Can. Geotech. J.,* 20(4):718–733.

Robertson, P.K., Campanella, R.G., and Wightman, A. (1983). SPT-CPT correlations. *J. Geotech. Eng. Div. ASCE,* 109(11):1449–1459.

Robertson, P.K., Campanella, R.G., Gillespie, D., and Greig, J. (1986). Use of piezometer cone data. *Use of In Situ Tests in Geotechnical Engineering,* Geotechnical Special Publication No. 6, ASCE, 1263–1280.

Sanglerat, G. (1972). *The Penetrometer and Soil Exploration,* Elsevier, Amsterdam, 464 pp.

Schmertmann, J.H. (1970). Static cone to compute static settlement over sand. *J. Soil Mech. Found. Div. ASCE,* 96(SM3):1011–1043.

Schmertmann, J.H. (1975). Measurement of in situ shear strength. *Proceedings of Specialty Conference on In Situ Measurement of Soil Properties,* Raleigh, NC, 2, 57–138.

Schmertmann, J.H. (1978). *Guidelines for Cone Penetration Test Performance and Design,* Report FHWA-TS-78-209, Federal Highway Administration, Washington, D.C., 145 pp.

Schmertmann, J.H., Hartman, J.P., and Brown, P.R. (1978). Improved strain influence factors diagram. *J. Geotech. Eng. Div. ASCE,* 104(GT8):1131–1135.

Skempton, A.W. (1944). Notes on compressibility of clays. *Q. J. Geol. Soc. London,* 100:119–135.

Skempton, A.W. (1954). The pore-pressure coefficients A and B, *Geotechnique,* 4:143–147.

Skempton, A.W. (1957). Discussion on "the planning and the design of the new Hong Kong airport." *Proc. Inst. Civ. Eng. London,* 7:305–307.

Skempton, A.W. (1986). Standard penetration test procedures and the effects in sands of overburden pressure, relative density, particle size, ageing and over consolidation. *Geotechnique,* 36(3):425–447.

Terzaghi, K. (1925). *Erdbaumechanik auf Bodenphysikalischer Grundlage*, Franz Deuticke, Leipzig und Wein, 399 pp.

Terzaghi, K. and Peck, R.B. (1948). *Soil Mechanics in Engineering Practice*, John Wiley & Sons, New York.

Terzaghi, K. and Peck, R.B. (1967). *Soil Mechanics in Engineering Practice*, 2nd edition, John Wiley & Sons, New York.

Terzaghi, K., Peck, R.B., and Mesri, G. (1996). S*oil Mechanics in Engineering Practice*, 3rd edition, John Wiley & Sons, New York.

U.S. Army (1993). *Bearing Capacity of Soils*, Technical Engineering and Design Guides, ASCE, 142 pp.

U.S. Navy (1971). Soil mechanics, foundations and earth structures. *NAVFAC Design Manual DM-7*, Washington, D.C.

U.S. Navy (1986). Soil mechanics—7.01. *NAVFAC Design Manual*, U.S. Government Printing Office, Washington, D.C.

Wolff, T.F. (1989). Pile capacity prediction using parameter functions. *Predicted and Observed Axial Behaviour of Piles, Results of a Pile Prediction Symposium*, Evanston, IL, ASCE Geotechnical Special Publication, 96–106.

2

Lateral Earth Pressure

by
Braja M. Das
California State University, Sacramento, California

2.1 Introduction

The design of earth retaining structures such as retaining walls, basement walls, bulkheads, and other structures requires a thorough knowledge of the lateral pressures that act between the retaining structures and the soil masses being retained. This lateral pressure is generally called the *lateral earth pressure*. The magnitude of lateral earth pressure at any depth will depend on the type and amount of wall movement, the shear strength of the soil, the unit weight of the soil, and the drainage conditions. Figure 2.1 shows a retaining wall of height H supporting a soil mass whose shear strength can be defined as

$$s = c' + \sigma' \tan \phi' \tag{2.1}$$

where s = shear strength, c' = cohesion, σ' = effective normal stress, and ϕ' = effective stress angle of friction.

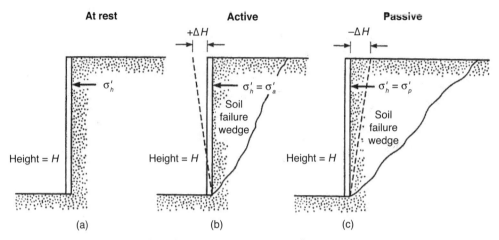

FIGURE 2.1 Nature of lateral earth pressure on retaining wall.

Three conditions may arise related to the degree of wall movement:

1. The wall is restrained from moving, as shown in Figure 2.1a. The effective lateral earth pressure σ'_h for this condition at any depth is referred to as *at-rest earth pressure*.
2. The wall may tilt away from the soil that is retained (Figure 2.1b). With sufficient wall tilt, a triangular soil wedge behind the wall will fail. The effective lateral pressure for this condition is referred to as *active earth pressure*.
3. The wall may be pushed into the soil that is retained (Figure 2.1c). With sufficient wall movement, a soil wedge will fail. The effective lateral pressure for this condition is referred to as *passive earth pressure*.

The relationships for estimation of at-rest, active, and passive earth pressures are elaborated upon in the following sections.

2.2 At-Rest Earth Pressure

Figure 2.2a shows a wall of height H supporting a soil mass that has a unit weight of γ. A uniformly distributed load of q per unit area is applied at the ground surface. If the wall is restrained from moving, the effective lateral pressure σ'_h at a depth z can be expressed as

$$\frac{\sigma'_h}{\sigma'_o} = K_o \qquad (2.2)$$

where σ'_o = vertical effective stress at depth z and K_o = coefficient of at-rest earth pressure.
For normally consolidated soil (Jaky 1944):

$$K_o = 1 - \sin \phi' \qquad (2.3)$$

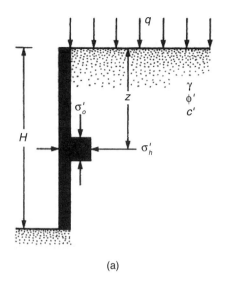 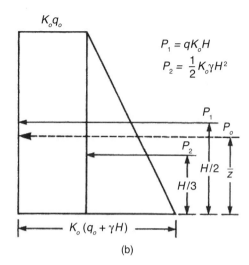

(a) (b)

FIGURE 2.2 At-rest pressure.

For overconsolidated soil (Mayne and Kulhawy 1982):

$$K_o = (1 - \sin \phi') \text{OCR}^{\sin \phi'} \qquad (2.4)$$

where OCR = overconsolidation ratio.

For normally consolidated cohesive soil (Massarsch 1979):

$$K_o = 0.44 + 0.42 \left[\frac{\text{PI (\%)}}{100} \right] \qquad (2.5)$$

where PI = plasticity index of the soil.

For overconsolidated cohesive soil:

$$K_{oc} = K_{nc} \sqrt{\text{OCR}} \qquad (2.6)$$

where K_{oc} and $K_{nc} = K_o$ for overconsolidated and normally consolidated soils, respectively.

If the groundwater table is present, the total lateral pressure at any depth z can be expressed as

$$\sigma_h = \sigma_h' + u = K_o \sigma_o' + u \qquad (2.7)$$

where u = pore water pressure and σ_o' = effective vertical stress.

Figure 2.2b shows the variation of σ_h' with depth. The force per unit length of the retaining wall P_o can be obtained by calculating the area of the pressure diagram, or

$$P_o = qK_oH + \frac{1}{2}K_o\gamma H^2 \qquad (2.8)$$

The location of the line of action of the resultant can be obtained by taking the moment of the areas about the bottom of the wall, or

$$\bar{z} = \frac{(qK_oH)\left(\dfrac{H}{2}\right) + \left(\dfrac{1}{2}K_o\gamma H^2\right)\left(\dfrac{H}{3}\right)}{P_o}$$

$$= \frac{\dfrac{1}{2}qK_oH^2 + \dfrac{1}{6}K_o\gamma H^3}{P_o} \qquad (2.9)$$

2.3 Rankine Active Pressure

Figure 2.3a shows a *frictionless retaining wall*. If the wall is allowed to yield sufficiently to the left (away from the soil mass), a triangular wedge of soil mass (*ABC*) will fail, and *BC* will make an angle $45 + \phi'/2$ with the horizontal. The lateral earth pressure when the failure occurs $\sigma_h' = \sigma_a'$ is the *Rankine active earth pressure* (Rankine 1857), and it can be given by the expression

$$\sigma_a' = \sigma_o'K_a - 2c'\sqrt{K_a} \qquad (2.10)$$

where

$$K_a = \tan^2\left(45 + \frac{\phi'}{2}\right) = \text{coefficient of Rankine active earth pressure} \qquad (2.11)$$

$\sigma_o' = \gamma H$ (for the case shown in Figure 2.3a)

The variation of σ_a' with depth is shown in Figure 2.3b. Note that from $z = 0$ to z_o, the value of σ_a' is negative (that is, tension). In such case, a tensile crack develops with time up to a depth of $z = z_o$, or

$$z_o = \frac{2c'}{\gamma\sqrt{K_a}} \qquad (2.12)$$

The Rankine active force per unit length of the wall can then be given as follows. *Before the occurrence of the tensile crack*:

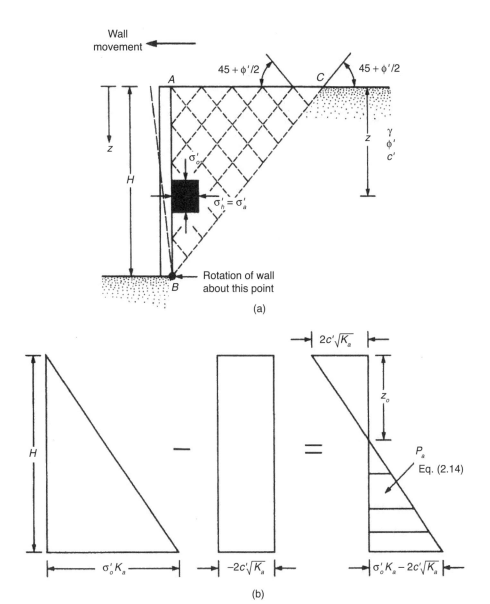

FIGURE 2.3 Rankine active pressure.

$$P_a = \frac{1}{2} K_a \gamma H^2 - 2c'H \sqrt{K_a} \qquad (2.13)$$

After the occurrence of the tensile crack:

$$P_a = \frac{1}{2}(H - z_o)\left(K_a\gamma H - 2c'\sqrt{K_a}\right) \qquad (2.14)$$

For granular soil with $c' = 0$, the magnitude of z_o is 0, so

$$P_a = \frac{1}{2} K_a \gamma H^2 \qquad (2.15)$$

For saturated cohesive soils (undrained condition), $\phi = 0$ and $c = c_u$; hence, $K_a = 1$. Thus

$$z_o = \frac{2c_u}{\gamma} \qquad (2.16)$$

$$P_a = \frac{1}{2} \gamma H^2 - 2c_u H \quad \text{(before occurrence of tensile crack)} \qquad (2.17)$$

$$P_a = \frac{1}{2}\left(H - \frac{2c_u}{\gamma}\right)(\gamma H - 2c_u) \quad \text{(after occurrence of tensile crack)} \qquad (2.18)$$

where c_u = undrained cohesion.

Example 1

For a 6-m-high retaining wall with a vertical back and a horizontal backfill of c'-ϕ' soil, $\gamma = 17$ kN/m³, $\phi' = 25°$, and $c' = 10$ kN/m². Determine:

    a. Depth of the tensile crack
    b. P_a before the occurrence of the tensile crack
    c. P_a after the occurrence of the tensile crack

Solution

Part a

$$K_a = \tan^2\left(45 - \frac{\phi'}{2}\right) = \tan^2\left(45 - \frac{25}{2}\right) = 0.406$$

From Equation 2.12:

$$z_o = \frac{2c'}{\gamma\sqrt{K_a}} = \frac{(2)(10)}{(17)\left(\sqrt{0.406}\right)} \approx \mathbf{1.85\ m}$$

Part b
From Equation 2.13:

$$P_a \text{ (before crack)} = \frac{1}{2} K_a \gamma H^2 - 2c'H\sqrt{K_a}$$

$$= \left(\frac{1}{2}\right)(0.406)(17)(6)^2 - (2)(10)(6)\left(\sqrt{0.406}\right)$$

$$= 47.8 \text{ kN/m}$$

Part c
From Equation 2.14:

$$P_a \text{ (after crack)} = \frac{1}{2}(H - z_o)\left(K_a \gamma H - 2c'\sqrt{K_a}\right)$$

$$= \left(\frac{1}{2}\right)(6 - 1.85)\left[(0.406)(17)(6) - (2)(10)\left(\sqrt{0.406}\right)\right]$$

$$= 59.5 \text{ kN/m}$$

2.4 Rankine Active Pressure with Inclined Backfill

Figure 2.4 shows a frictionless retaining wall with a vertical back and an inclined backfill. The backfill is inclined at an angle α with the horizontal. If the backfill is a *granular soil* ($c' = 0$), the magnitude of σ'_a at any depth z can be expressed as

$$\sigma'_a = \gamma z K_a \tag{2.19}$$

where

$$K_a = \cos\alpha \; \frac{\cos\alpha - \sqrt{\cos^2\alpha - \cos^2\phi'}}{\cos\alpha + \sqrt{\cos^2\alpha - \cos^2\phi'}} \tag{2.20}$$

The direction of σ'_a will be inclined at an angle α with the horizontal.
    The total force per unit length of the wall is

$$P_a = \frac{1}{2} K_a \gamma H^2 \tag{2.21}$$

Table 2.1 gives the variation of K_a with and α and ϕ'.

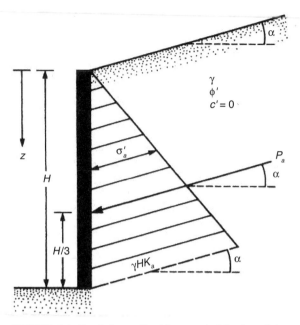

FIGURE 2.4 Retaining wall with a vertical back and in-clined granular backfill.

If the backfill (Figure 2.4) is a *cohesive soil with* $\phi' \neq 0$ *and* $c' \neq 0$, then the Rankine active pressure at any depth z can be given as (Mazindrani and Ganjali 1997)

$$\sigma_a' = \gamma z K_a = \gamma z K_a' \cos \alpha \qquad (2.22)$$

where

$$K_a' = \frac{1}{\cos^2 \phi'} \left\{ 2\cos^2 \alpha + 2\left(\frac{c'}{\gamma z}\right) \cos \phi' \sin \phi' - \sqrt{\left[\begin{array}{c} 4\cos^2 \alpha(\cos^2 \alpha - \cos^2 \phi') \\ + 4\left(\frac{c'}{\gamma z}\right)^2 \cos^2 \phi' \\ + 8\left(\frac{c'}{\gamma z}\right) \cos^2 \alpha \sin \phi' \cos \phi' \end{array} \right]} \right\} - 1 \qquad (2.23)$$

Values of K_a' are given in Table 2.2. For a problem of this type, the depth of tensile crack is given as:

TABLE 2.1 Values of K_a (Equation 2.20)

α (deg)	ϕ' (deg)						
	28	30	32	34	36	38	40
0	0.361	0.333	0.307	0.283	0.260	0.238	0.217
5	0.366	0.337	0.311	0.286	0.262	0.240	0.219
10	0.380	0.350	0.321	0.294	0.270	0.246	0.225
15	0.409	0.373	0.341	0.311	0.283	0.258	0.235
20	0.461	0.414	0.374	0.338	0.306	0.277	0.250
25	0.573	0.494	0.434	0.385	0.343	0.307	0.275

TABLE 2.2 Values of K_a' (Equation 2.23)

ϕ' (deg)	α (deg)	$\dfrac{c'}{\gamma z}$			
		0.025	0.050	0.100	0.500
15	0	0.550	0.512	0.435	−0.179
	5	0.566	0.525	0.445	−0.184
	10	0.621	0.571	0.477	−0.186
	15	0.776	0.683	0.546	−0.196
20	0	0.455	0.420	0.350	−0.210
	5	0.465	0.429	0.357	−0.212
	10	0.497	0.456	0.377	−0.218
	15	0.567	0.514	0.417	−0.229
25	0	0.374	0.342	0.278	−0.231
	5	0.381	0.348	0.283	−0.233
	10	0.402	0.366	0.296	−0.239
	15	0.443	0.401	0.321	−0.250
30	0	0.305	0.276	0.218	−0.244
	5	0.309	0.280	0.221	−0.246
	10	0.323	0.292	0.230	−0.252
	15	0.350	0.315	0.246	−0.263

$$z_c = \frac{2c'}{\gamma} \sqrt{\frac{1 + \sin \phi'}{1 - \sin \phi'}} \tag{2.24}$$

For this case, the active pressure is inclined at an angle with the horizontal (as shown in Figure 2.4).

Chu (1991) provided a more generalized case for Rankine active pressure for a *frictionless* retaining wall with an inclined back face and inclined *granular* backfill ($c' = 0$), as shown in Figure 2.5. For this case, active pressure at any depth z can be given by the expression

$$\sigma_a' = \frac{\gamma z \cos \alpha \sqrt{1 + \sin^2 \phi' - 2 \sin \phi' \cos \psi_a}}{\cos \alpha + \sqrt{\sin^2 \phi' - \sin^2 \alpha}} \tag{2.25}$$

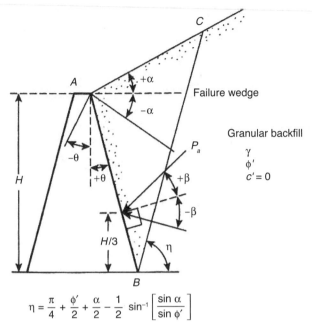

$$\eta = \frac{\pi}{4} + \frac{\phi'}{2} + \frac{\alpha}{2} - \frac{1}{2} \ \sin^{-1} \left[\frac{\sin \alpha}{\sin \phi'} \right]$$

FIGURE 2.5 Generalized case of Rankine active pressure with a granular backfill.

where

$$\psi_a \ = \ \sin^{-1} \left(\frac{\sin \alpha}{\sin \phi'} \right) \ - \ \alpha \ + \ 2\theta \tag{2.26}$$

The pressure σ_a' will be inclined at an angle β with the plane drawn at a right angle to the back face of the wall, and

$$\beta \ = \ \tan^{-1} \left(\frac{\sin \phi' \sin \psi_a}{1 \ - \ \sin \phi' \cos \psi_a} \right) \tag{2.27}$$

The active force P_a for unit length of the wall can then be calculated as

$$P_a \ = \ \frac{1}{2} \gamma H^2 K_a \tag{2.28}$$

where K_a = Rankine active earth pressure coefficient for the generalized case, or

$$K_a \ = \ \frac{\cos(\alpha \ - \ \theta) \ \sqrt{1 \ + \ \sin^2 \phi' \ - \ 2 \sin \phi' \cos \psi_a}}{\cos^2 \theta \left(\cos \alpha \ + \ \sqrt{\sin^2 \phi' \ - \ \sin^2 \alpha} \right)} \tag{2.29}$$

Example 2

For a frictionless retaining wall with a vertical backfill, $H = 6$ m, $\alpha = 5°$, $\gamma = 16$ kN/m³, $c' = 9.6$ kN/m², and $\phi' = 20°$. Determine the active force per unit length of the wall after the occurrence of the tensile crack and the location of the resultant P_a.

Solution

From Equation 2.24:

$$z_c = \frac{2c'}{\gamma}\sqrt{\frac{1 + \sin\phi'}{1 - \sin\phi'}} - \frac{(2)(9.6)}{16}\sqrt{\frac{1 + \sin 20}{1 - \sin 20}} = 1.71 \text{ m}$$

$$\frac{c'}{\gamma z} = \frac{9.6}{(16)(6)} = 0.1$$

From Table 2.2, for $\phi' = 20°$, $\alpha = 5°$, and $c'/\gamma z = 0.1$, the value of $K_a' = 0.357$.
At $z = 6$ m,

$$\sigma_a' = \gamma z K_a \cos\alpha = (16)(6)(0.357)(\cos 5) = 34.14 \text{ kN/m}^2$$

$$P_a = \frac{1}{2}(H - z_o)(\sigma_a') = \frac{1}{2}(6 - 1.71)(34.14) = \textbf{73.23 kN/m}$$

The resultant P_a will act at a distance of $(6 - 1.71)/3 = \textbf{1.43 m above the bottom of the wall.}$

2.5 Coulomb's Active Pressure

Figure 2.6 shows a retaining wall of height H with an inclined back face and a *granular* ($c' = 0$) inclined backfill. The angle of friction between the backfill soil and the back face of the wall is δ'. If it is assumed that the failure surface is a plane as shown by the line *BC*, then the active force per unit length of the wall is (Coulomb 1776)

$$P_a = \frac{1}{2}K_a\gamma H^2 \tag{2.30}$$

where K_a = Coulomb's active earth pressure coefficient, or

$$K_a = \frac{\cos^2(\phi' - \theta)}{\cos^2\theta\cos(\delta' + \theta)\left[1 + \sqrt{\dfrac{\sin(\delta' + \phi')\sin(\phi' - \alpha)}{\cos(\delta' + \theta)\cos(\theta - \alpha)}}\right]} \tag{2.31}$$

where θ = inclination of the back face of the wall with the vertical and α = inclination of the backfill with the horizontal.

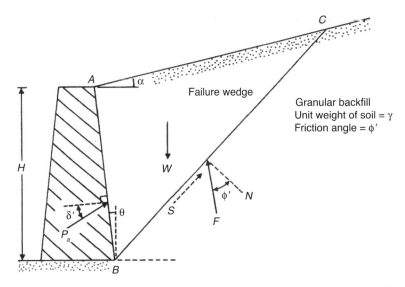

FIGURE 2.6 Coulomb's active earth pressure. (Note: *BC* is the failure plane, *W* = weight of the wedge *ABC*, *S* and *N* = shear and normal forces on plane *BC*, and *F* = resultant of *S* and *N*.)

Table 2.3 gives the variation of K_a with ϕ' and δ' for $\theta = 0°$ and $\alpha = 0°$. Tables 2.4 and 2.5 give the variation of K_a with α, ϕ', and θ for $\delta' = \frac{2}{3}\phi'$ and $\frac{1}{2}\phi'$. The active force P_a acts at a distance of $H/3$ above the bottom of the wall and is inclined at an angle δ' with the normal drawn to the back face of the wall.

2.6 Active Earth Pressure with Earthquake Forces

Coulomb's active earth pressure theory can be extended to take into account earthquake forces. Figure 2.7 shows a retaining wall with a *granular* backfill. *ABC* is the failure wedge. The forces per unit length of the wall that need to be considered for equilibrium of wedge *ABC* are

- Weight of the wedge W
- Horizontal inertia force $k_h W$

TABLE 2.3 Values of K_a (Equation 2.31) for $\theta = 0°$ and $\alpha = 0°$

ϕ' (deg)	δ' (deg)					
	0	5	10	15	20	25
28	0.3610	0.3448	0.3330	0.3251	0.3203	0.3186
30	0.3333	0.3189	0.3085	0.3014	0.2973	0.2956
32	0.3073	0.2945	0.2853	0.2791	0.2755	0.2745
34	0.2827	0.2714	0.2633	0.2579	0.2549	0.2542
36	0.2596	0.2497	0.2426	0.2379	0.2354	0.2350
38	0.2379	0.2292	0.2230	0.2190	0.2169	0.2167
40	0.2174	0.2089	0.2045	0.2011	0.1994	0.1995
42	0.1982	0.1916	0.1870	0.1341	0.1828	0.1831

TABLE 2.4 Values of K_a (Equation 2.31) ($\delta' = \frac{2}{3}\phi'$)

α (deg)	φ' (deg)	θ (deg)					
		0	5	10	15	20	25
0	28	0.3213	0.3588	0.4007	0.4481	0.5026	0.5662
	29	0.3091	0.3467	0.3886	0.4362	0.4908	0.5547
	30	0.2973	0.3349	0.3769	0.4245	0.4794	0.5435
	31	0.2860	0.3235	0.3655	0.4133	0.4682	0.5326
	32	0.2750	0.3125	0.3545	0.4023	0.4574	0.5220
	33	0.2645	0.3019	0.3439	0.3917	0.4469	0.5117
	34	0.2543	0.2916	0.3335	0.3813	0.4367	0.5017
	35	0.2444	0.2816	0.3235	0.3713	0.4267	0.4919
	36	0.2349	0.2719	0.3137	0.3615	0.4170	0.4824
	37	0.2257	0.2626	0.3042	0.3520	0.4075	0.4732
	38	0.2168	0.2535	0.2950	0.3427	0.3983	0.4641
	39	0.2082	0.2447	0.2861	0.3337	0.3894	0.4553
	40	0.1998	0.2361	0.2774	0.3249	0.3806	0.4468
	41	0.1918	0.2278	0.2689	0.3164	0.3721	0.4384
	42	0.1840	0.2197	0.2606	0.3080	0.3637	0.4302
5	28	0.3431	0.3845	0.4311	0.4843	0.5461	0.6190
	29	0.3295	0.3709	0.4175	0.4707	0.5325	0.6056
	30	0.3165	0.3578	0.4043	0.4575	0.5194	0.5926
	31	0.3039	0.3451	0.3916	0.4447	0.5067	0.5800
	32	0.2919	0.3329	0.3792	0.4324	0.4943	0.5677
	33	0.2803	0.3211	0.3673	0.4204	0.4823	0.5558
	34	0.2691	0.3097	0.3558	0.4088	0.4707	0.5443
	35	0.2583	0.2987	0.3446	0.3975	0.4594	0:5330
	36	0.2479	0.2881	0.3338	0.3866	0.4484	0.5221
	37	0.2379	0.2778	0.3233	0.3759	0.4377	0.5115
	38	0.2282	0.2679	0.3131	0.3656	0.4273	0.5012
	39	0.2188	0.2582	0.3033	0.3556	0.4172	0.4911
	40	0.2098	0.2489	0.2937	0.3458	0.4074	0.4813
	41	0.2011	0.2398	0.2844	0.3363	0.3978	0.4718
	42	0.1927	0.2311	0.2753	0.3271	0.3884	0.4625
10	28	0.3702	0.4164	0.4686	0.5287	0.5992	0.6834
	29	0.3548	0.4007	0.4528	0.5128	0.5831	0.6672
	30	0.3400	0.3857	0.4376	0.4974	0.5676	0.6516
	31	0.3259	0.3713	0.4230	0.4826	0.5526	0.6365
	32	0.3123	0.3575	0.4089	0.4683	0.5382	0.6219
	33	0.2993	0.3442	0.3953	0.4545	0.5242	0.6078
	34	0.2868	0.3314	0.3822	0.4412	0.5107	0.5942
	35	0.2748	0.3190	0.3696	0.4283	0.4976	0.5810
	36	0.2633	0.3072	0.3574	0.4158	0.4849	0.5682
	37	0.2522	0.2957	0.3456	0.4037	0.4726	0.5558
	38	0.2415	0.2846	0.3342	0.3920	0.4607	0.5437
	39	0.2313	0.2740	0.3231	0.3807	0.4491	0.5321
	40	0.2214	0.2636	0.3125	0.3697	0.4379	0.5207
	41	0.2119	0.2537	0.3021	0.3590	0.4270	0.5097
	42	0.2027	0.2441	0.2921	0.3487	0.4164	0.4990
15	28	0.4065	0.4585	0.5179	0.5868	0.6685	0.7670
	29	0.3881	0.4397	0.4987	0.5672	0.6483	0.7463
	30	0.3707	0.4219	0.4804	0.5484	0.6291	0.7265
	31	0.3541	0.4049	0.4629	0.5305	0.6106	0.7076
	32	0.3384	0.3887	0.4462	0.5133	0.5930	0.6895
	33	0.3234	0.3732	0.4303	0.4969	0.5761	0.6721

TABLE 2.4   Values of K_a (Equation 2.31) ($\delta' = \frac{2}{3}\phi'$) (continued)

α (deg)	φ' (deg)	θ (deg)					
		0	5	10	15	20	25
	34	0.3091	0.3583	0.4150	0.4811	0.5598	0.6554
	35	0.2954	0.3442	0.4003	0.4659	0.5442	0.6393
	36	0.2823	0.3306	0.3862	0.4513	0.5291	0.6238
	37	0.2698	0.3175	0.3726	0.4373	0.5146	0.6089
	38	0.2578	0.3050	0.3595	0.4237	0.5006	0.5945
	39	0.2463	0.2929	0.3470	0.4106	0.4871	0.5805
	40	0.2353	0.2813	0.3348	0.3980	0.4740	0.5671
	41	0.2247	0.2702	0.3231	0.3858	0.4613	0.5541
	42	0.2146	0.2594	0.3118	0.3740	0.4491	0.5415
20	28	0.4602	0.5205	0.5900	0.6714	0.7689	0.8880
	29	0.4364	0.4958	0.5642	0.6445	0.7406	0.8581
	30	0.4142	0.4728	0.5403	0.6195	0.7144	0.8303
	31	0.3935	0.4513	0.5179	0.5961	0.6898	0.8043
	32	0.3742	0.4311	0.4968	0.5741	0.6666	0.7799
	33	0.3559	0.4121	0.4769	0.5532	0.6448	0.7569
	34	0.3388	0.3941	0.4581	0.5335	0.6241	0.7351
	35	0.3225	0.3771	0.4402	0.5148	0.6044	0.7144
	36	0.3071	0.3609	0.4233	0.4969	0.5856	0.6947
	37	0.2925	0.3455	0.4071	0.4799	0.5677	0.6759
	38	0.2787	0.3308	0.3916	0.4636	0.5506	0.6579
	39	0.2654	0.3168	0.3768	0.4480	0.5342	0.6407
	40	0.2529	0.3034	0.3626	0.4331	0.5185	0.6242
	41	0.2408	0.2906	0.3490	0.4187	0.5033	0.6083
	42	0.2294	0.2784	0.3360	0.4049	0.4888	0.5930

TABLE 2.5   Values of K_a (Equation 2.31) ($\delta' = \phi'/2$)

α (deg)	φ' (deg)	θ (deg)					
		0	5	10	15	20	25
0	28	0.3264	0.3629	0.4034	0.4490	0.5011	0.5616
	29	0.3137	0.3502	0.3907	0.4363	0.4886	0.5492
	30	0.3014	0.3379	0.3784	0.4241	0.4764	0.5371
	31	0.2896	0.3260	0.3665	0.4121	0.4645	0.5253
	32	0.2782	0.3145	0.3549	0.4005	0.4529	0.5137
	33	0.2671	0.3033	0.3436	0.3892	0.4415	0.5025
	34	0.2564	0.2925	0.3327	0.3782	0.4305	0.4915
	35	0.2461	0.2820	0.3221	0.3675	0.4197	0.4807
	36	0.2362	0.2718	0.3118	0.3571	0.4092	0.4702
	37	0.2265	0.2620	0.3017	0.3469	0.3990	0.4599
	38	0.2172	0.2524	0.2920	0.3370	0.3890	0.4498
	39	0.2081	0.2431	0.2825	0.3273	0.3792	0.4400
	40	0.1994	0.2341	0.2732	0.3179	0.3696	0.4304
	41	0.1909	0.2253	0.2642	0.3087	0.3602	0.4209
	42	0.1828	0.2168	0.2554	0.2997	0.3511	0.4117
5	28	0.3477	0.3879	0.4327	0.4837	0.5425	0.6115
	29	0.3337	0.3737	0.4185	0.4694	0.5282	0.5972
	30	0.3202	0.3601	0.4048	0.4556	0.5144	0.5833
	31	0.3072	0.3470	0.3915	0.4422	0.5009	0.5698

TABLE 2.5 Values of K_a (Equation 2.31) ($\delta' = \phi'/2$) (continued)

α (deg)	φ' (deg)	θ (deg)					
		0	5	10	15	20	25
	32	0.2946	0.3342	0.3787	0.4292	0.4878	0.5566
	33	0.2825	0.3219	0.3662	0.4166	0.4750	0.5437
	34	0.2709	0.3101	0.3541	0.4043	0.4626	0.5312
	35	0.2596	0.2986	0.3424	0.3924	0.4505	0.5190
	36	0.2488	0.2874	0.3310	0.3808	0.4387	0.5070
	37	0.2383	0.2767	0.3199	0.3695	0.4272	0.4954
	38	0.2282	0.2662	0.3092	0.3585	0.4160	0.4840
	39	0.2185	0.2561	0.2988	0.3478	0.4050	0.4729
	40	0.2090	0.2463	0.2887	0.3374	0.3944	0.4620
	41	0.1999	0.2368	0.2788	0.3273	0.3840	0.4514
	42	0.1911	0.2276	0.2693	0.3174	0.3738	0.4410
10	28	0.3743	0.4187	0.4688	0.5261	0.5928	0.6719
	29	0.3584	0.4026	0.4525	0.5096	0.5761	0.6549
	30	0.3432	0.3872	0.4368	0.4936	0.5599	0.6385
	31	0.3286	0.3723	0.4217	0.4782	0.5442	0.6225
	32	0.3145	0.3580	0.4071	0.4633	0.5290	0.6071
	33	0.3011	0.3442	0.3930	0.4489	0.5143	0.5920
	34	0.2881	0.3309	0.3793	0.4350	0.5000	0.5775
	35	0.2757	0.3181	0.3662	0.4215	0.4862	0.5633
	36	0.2637	0.3058	0.3534	0.4084	0.4727	0.5495
	37	0.2522	0.2938	0.3411	0.3957	0.4597	0.5361
	38	0.2412	0.2823	0.3292	0.3833	0.4470	0.5230
	39	0.2305	0.2712	0.3176	0.3714	0.4346	0.5103
	40	0.2202	0.2604	0.3064	0.3597	0.4226	0.4979
	41	0.2103	0.2500	0.2956	0.3484	0.4109	0.4858
	42	0.2007	0.2400	0.2850	0.3375	0.3995	0.4740
15	28	0.4095	0.4594	0.5159	0.5812	0.6579	0.7498
	29	0.3908	0.4402	0.4964	0.5611	0.6373	0.7284
	30	0.3730	0.4220	0.4777	0.5419	0.6175	0.7080
	31	0.3560	0.4046	0.4598	0.5235	0.5985	0.6884
	32	0.3398	0.3880	0.4427	0.5059	0.5803	0.6695
	33	0.3244	0.3721	0.4262	0.4889	0.5627	0.6513
	34	0.3097	0.3568	0.4105	0.4726	0.5458	0.6338
	35	0.2956	0.3422	0.3953	0.4569	0.5295	0.6168
	36	0.2821	0.3282	0.3807	0.4417	0.5138	0.6004
	37	0.2692	0.3147	0.3667	0.4271	0.4985	0.5846
	38	0.2569	0.3017	0.3531	0.4130	0.4838	0.5692
	39	0.2450	0.2893	0.3401	0.3993	0.4695	0.5543
	40	0.2336	0.2773	0.3275	0.3861	0.4557	0.5399
	41	0.2227	0.2657	0.3153	0.3733	0.4423	0.5258
	42	0.2122	0.2546	0.3035	0.3609	0.4293	0.5122
20	28	0.4614	0.5188	0.5844	0.6608	0.7514	0.8613
	29	0.4374	0.4940	0.5586	0.6339	0.7232	0.8313
	30	0.4150	0.4708	0.5345	0.6087	0.6968	0.8034
	31	0.3941	0.4491	0.5119	0.5851	0.6720	0.7772
	32	0.3744	0.4286	0.4906	0.5628	0.6486	0.7524
	33	0.3559	0.4093	0.4704	0.5417	0.6264	0.7289
	34	0.3384	0.3910	0.4513	0.5216	0.6052	0.7066
	35	0.3218	0.3736	0.4331	0.5025	0.5851	0.6853
	36	0.3061	0.3571	0.4157	0.4842	0.5658	0.6649
	37	0.2911	0.3413	0.3991	0.4668	0.5474	0.6453

TABLE 2.5 Values of K_a (Equation 2.31) ($\delta' = \phi'/2$) (continued)

α (deg)	φ' (deg)	θ (deg)					
		0	5	10	15	20	25
	38	0.2769	0.3263	0.3833	0.4500	0.5297	0.6266
	39	0.2633	0.3120	0.3681	0.4340	0.5127	0.6085
	40	0.2504	0.2982	0.3535	0.4185	0.4963	0.5912
	41	0.2381	0.2851	0.3395	0.4037	0.4805	0.5744
	42	0.2263	0.2725	0.3261	0.3894	0.4653	0.5582

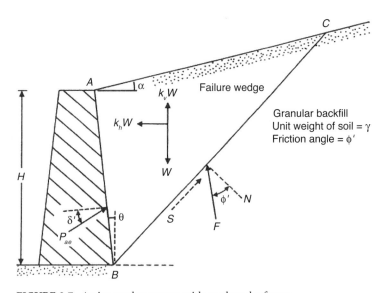

FIGURE 2.7 Active earth pressure with earthquake forces.

- Vertical inertia force $k_v W$
- Active force per unit length of the wall P_{ae}
- Resultant F of the normal and shear forces along the failure surface BC

Note that

$$k_h = \frac{\text{Horizontal component of earthquake acceleration}}{g} \tag{2.32}$$

$$k_v = \frac{\text{Vertical component of earthquake acceleration}}{g} \tag{2.33}$$

where g = acceleration due to gravity.

For this case, the active force per unit length of the wall P_{ae} can be given as

$$P_{ae} = \frac{1}{2}\gamma H^2 (1 - k_v) K_{ae} \tag{2.34}$$

where

$$K_{ae} = \frac{\cos^2(\phi' - \theta - \bar{\beta})}{\cos^2\theta \cos\bar{\beta} \cos(\delta' + \theta + \bar{\beta})} \\ \left\{ 1 + \left[\frac{\sin(\delta' + \phi')\sin(\phi' - \alpha - \bar{\beta})}{\cos(\delta' + \theta + \bar{\beta})\cos(\theta - \alpha)} \right]^{1/2} \right\}^2 \tag{2.35}$$

and

$$\bar{\beta} = \tan^{-1}\left(\frac{k_h}{1 - k_v} \right) \tag{2.36}$$

Equations 2.34 and 2.35 generally are referred to as the *Mononobe-Okabe equations* (Mononobe 1929; Okabe 1926). The variation of K_{ae} with $\theta = 0°$ and $k_v = 0$ is given in Table 2.6. The active force P_{ae} will be inclined at an angle δ' with the normal drawn to the back face of the wall. Figure 2.8 shows the variation of $k_{ae}\cos\delta'$ with k_h and ϕ' for $k_v = 0$, $\alpha = 0$, $\theta = 0$, and $\delta' = \phi'/2$.

It is important to note from the term $\sin(\phi' - \alpha - \bar{\beta})$ that if $\phi' - \alpha - \bar{\beta}$ is less than 0, no real solution of K_{ae} is possible. Hence, for stability:

$$\alpha \le \phi' - \bar{\beta} \tag{2.37}$$

Seed and Whitman (1970) have shown that Equation 2.34 can be rewritten as

$$P_{ae} = \frac{1}{2}\gamma H^2 (1 - k_v) K_a(\theta^*, \alpha^*) \left[\frac{\cos^2(\theta + \bar{\beta})}{\cos\bar{\beta}\cos^2\theta} \right] \tag{2.38}$$

where

$$\theta^* = \theta + \bar{\beta} \tag{2.39}$$

and

$$\alpha^* = \alpha + \bar{\beta} \tag{2.40}$$

$K_a(\theta^*, \alpha^*)$ = static active earth pressure coefficient K_a (see Tables 2.4 and 2.5) for a retaining wall with its back face inclined at an angle θ^* with the vertical and with a backfill inclined at an angle α^* with the horizontal.

TABLE 2.6   Values of K_{ae} (Equation 2.35) with $\theta = 0°$ and $k_v = 0$

k_h	δ' (deg)	α (deg)	ϕ' (deg) 28	30	35	40	45
0.1	0	0	0.427	0.397	0.328	0.268	0.217
0.2			0.508	0.473	0.396	0.382	0.270
0.3			0.611	0.569	0.478	0.400	0.334
0.4			0.753	0.697	0.581	0.488	0.409
0.5			1.005	0.890	0.716	0.596	0.500
0.1	0	5	0.457	0.423	0.347	0.282	0.227
0.2			0.554	0.514	0.424	0.349	0.285
0.3			0.690	0.635	0.522	0.431	0.356
0.4			0.942	0.825	0.653	0.535	0.442
0.5			—	—	0.855	0.673	0.551
0.1	0	10	0.497	0.457	0.371	0.299	0.238
0.2			0.623	0.570	0.461	0.375	0.303
0.3			0.856	0.748	0.585	0.472	0.383
0.4			—	—	0.780	0.604	0.486
0.5			—	—	—	0.809	0.624
0.1	$\phi'/2$	0	0.396	0.368	0.306	0.253	0.207
0.2			0.485	0.452	0.380	0.319	0.267
0.3			0.604	0.563	0.474	0.402	0.340
0.4			0.778	0.718	0.599	0.508	0.433
0.5			1.115	0.972	0.774	0.648	0.552
0.1	$\phi'/2$	5	0.428	0.396	0.326	0.268	0.218
0.2			0.537	0.497	0.412	0.342	0.283
0.3			0.699	0.640	0.526	0.438	0.367
0.4			1.025	0.881	0.690	0.568	0.475
0.5			—	—	0.962	0.752	0.620
0.1	$\phi'/2$	10	0.472	0.433	0.352	0.285	0.230
0.2			0.616	0.562	0.454	0.371	0.303
0.3			0.908	0.780	0.602	0.487	0.400
0.4			—	—	0.857	0.656	0.531
0.5			—	—	—	0.944	0.722
0.1	$\tfrac{2}{3}\phi'$	0	0.393	0.366	0.306	0.256	0.212
0.2			0.486	0.454	0.384	0.326	0.276
0.3			0.612	0.572	0.486	0.416	0.357
0.4			0.801	0.740	0.622	0.533	0.462
0.5			1.177	1.023	0.819	0.693	0.600
0.1	$\tfrac{2}{3}\phi'$	5	0.427	0.395	0.327	0.271	0.224
0.2			0.541	0.501	0.418	0.350	0.294
0.3			0.714	0.655	0.541	0.455	0.386
0.4			1.073	0.921	0.722	0.600	0.509
0.5			—	—	1.034	0.812	0.679
0.1	$\tfrac{2}{3}\phi'$	10	0.472	0.434	0.354	0.290	0.237
0.2			0.625	0.570	0.463	0.381	0.317
0.3			0.942	0.807	0.624	0.509	0.423
0.4			—	—	0.909	0.699	0.573
0.5			—	—	—	1.037	0.800

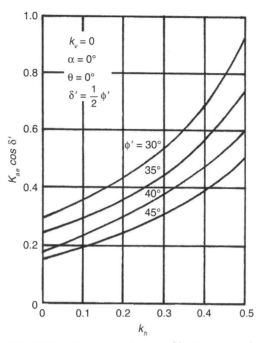

FIGURE 2.8 Variation of $K_{ae} \cos \delta'$ with k_h and ϕ'.

2.6.1 Location of the Resultant Force P_{ae}

Seed and Whitman (1970) proposed a simple procedure to determine the location of the line of action of the resultant P_{ae}. Their method is as follows:

1. Let

$$P_{ae} = P_a + \Delta P_{ae} \qquad (2.41)$$

where P_a = Coulomb's active force as determined from Equation 2.30 and ΔP_{ae} = additional active force caused by the earthquake effect.
2. Calculate P_a (Equation 2.30).
3. Calculate P_{ae} (Equation 2.34 or 2.38).
4. Calculate $\Delta P_{ae} = P_{ae} - P_a$.
5. According to Figure 2.9, P_a will act at a distance of $H/3$ from the base of the wall. Also, ΔP_{ae} will act at a distance of $0.6H$ from the base of the wall.
6. Calculate the location of P_{ae} as

$$\bar{z} = \frac{P_a\left(\dfrac{H}{3}\right) + \Delta P_{ae}(0.6H)}{P_{ae}} \qquad (2.42)$$

where $\bar{z}$ = distance of the line of action of P_{ae} from the base of the wall.

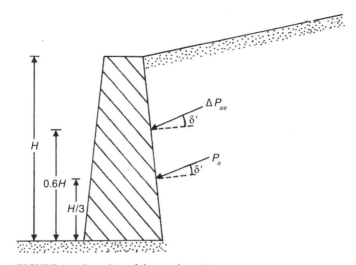

FIGURE 2.9 Location of the resultant P_{ae}.

Note that the line of action of P_{ae} will be inclined at an angle of δ' to the normal drawn to the back face of the retaining wall.

Example 3

For a retaining wall, $H = 5$ m, $\gamma = 15$ kN/m^3, $\phi' = 30°$, $\delta' = 15°$, $\theta = 5°$, $\alpha = 5°$, $k_v = 0$, and $k_h = 0.18$. Determine P_{ae} and $\bar{z}$.

Solution

$$\bar{\beta} = \tan^{-1}\left(\frac{k_h}{1-k_v}\right) = \tan^{-1}\left(\frac{0.18}{1-0}\right) = 10.2° \approx 10°$$

$$\theta^* = \theta + \bar{\beta} = 5 + 10 = 15°$$

$$\alpha^* = \alpha + \bar{\beta} = 5 + 10 = 15°$$

$$\frac{\delta'}{\phi'} = \frac{15}{30} = 0.5$$

From Table 2.5 for $\alpha^* = 15°$, $\theta^* = 15°$, $\phi' = 30°$, and $\delta'/\phi' = 0.5$, the magnitude of K_a is 0.5419. From Equation 2.38:

$$P_{ae} = \frac{1}{2}\gamma H^2 K_a(\theta^*, \alpha^*)(1-k_v)\left[\frac{\cos^2(\theta+\bar{\beta})}{\cos\bar{\beta}\cos^2\theta}\right]$$

$$= \left(\frac{1}{2}\right)(15)(5)^2(0.5419)(1-0)\left[\frac{\cos^2(15)}{\cos 10 \cos^2 5}\right] = \textbf{97 kN/m}$$

Determination of $\bar{z}$
From Equation 2.30:

$$P_a = \frac{1}{2} K_a \gamma H^2$$

For $\phi' = 30°$, $\delta'/\phi' = 0.5$, $\theta = 5°$, and $\alpha = 5°$, the magnitude of K_a is 0.3601 (Table 2.5).

$$P_a = \left(\frac{1}{2}\right)(0.3601)(15)(5)^2 = 67.42 \text{ kN/m}$$

$$\Delta P_{ae} = P_{ae} - P_a = 97 - 67.52 = 29.48 \text{ kN/m}$$

$$\bar{z} = \frac{P_a\left(\dfrac{H}{3}\right) + \Delta P_{ae}(0.6H)}{P_{ae}}$$

$$= \frac{(67.52)\left(\dfrac{5}{3}\right) + (29.48)(0.6 \times 5)}{97} = 2.07 \text{ m}$$

2.7 Rankine Passive Pressure

Figure 2.10a shows a *frictionless* retaining wall with a vertical back face and a c'-ϕ' soil backfill. If the wall is pushed into the soil mass, a triangular soil mass *ABC* will fail. The plane *BC* will make an angle $45 - \phi'/2$ with the horizontal. At this point, the effective horizontal pressure at a depth z is the Rankine passive earth pressure and can be given as

$$\sigma_p' = \sigma_o' K_p + 2c'\sqrt{K_p} \tag{2.43}$$

where σ_p' = vertical effective stress (= γz in Figure 2.10a) and K_p = Rankine passive earth pressure coefficient

$$= \tan^2\left(45 + \frac{\phi'}{2}\right) \tag{2.44}$$

Figure 2.10b shows the variation of σ_p' with depth. The force per unit length of the wall P_p can be obtained by calculating the area of the pressure distribution diagram, or

$$P_p = \frac{1}{2}\gamma H^2 K_p + 2c'H\sqrt{K_p} \tag{2.45}$$

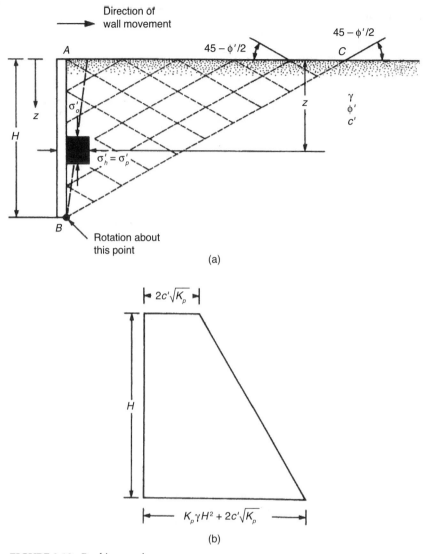

FIGURE 2.10 Rankine passive pressure.

The location of the line of action $\bar{z}$ above the bottom of the wall can be obtained by taking the moment of the pressure diagram about the bottom of the wall, or

$$\bar{z} = \frac{\left(\dfrac{1}{2} \gamma H^2 K_p \right) \left(\dfrac{H}{3} \right) + \left(2c'H \sqrt{K_p} \right) \left(\dfrac{H}{2} \right)}{P_p}$$

(2.46)

$$= \frac{\dfrac{1}{6} \gamma H^3 K_p + c'H^2 \sqrt{K_p}}{P_p}$$

2.8 Rankine Passive Pressure with Inclined Backfill

Chu (1991) developed a general expression for Rankine passive earth pressure for a frictionless retaining wall with an inclined back and a *granular sloping backfill* ($c' = 0$), as shown in Figure 2.11. The following are the relationships in reference to Figure 2.11.

σ'_p = pressure at any depth z

$$= \frac{\gamma z \cos \alpha \sqrt{1 + \sin^2 \phi' + 2 \sin \phi' \cos \psi_p}}{\cos \alpha - \sqrt{\sin^2 \phi' - \sin^2 \alpha}} \tag{2.47}$$

where

$$\psi_p = \sin^{-1} \left(\frac{\sin \alpha}{\sin \phi'} \right) + \alpha - 2\theta \tag{2.48}$$

The inclination β of σ'_p as shown in Figure 2.11 is

$$\beta = \tan^{-1} \left(\frac{\sin \phi' \sin \psi_p}{1 + \sin \phi' \cos \psi_p} \right) \tag{2.49}$$

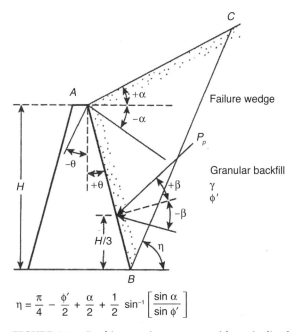

$$\eta = \frac{\pi}{4} - \frac{\phi'}{2} + \frac{\alpha}{2} + \frac{1}{2} \sin^{-1} \left[\frac{\sin \alpha}{\sin \phi'} \right]$$

FIGURE 2.11 Rankine passive pressure with an inclined granular backfill.

The passive force per unit length of the wall is

$$P_p = \frac{1}{2}\gamma H^2 K_p \qquad (2.50)$$

where

$$K_p = \frac{\cos(\alpha - \theta)\sqrt{1 + \sin^2\phi' + 2\sin\phi'\cos\psi_p}}{\cos^2\theta\left(\cos\alpha - \sqrt{\sin^2\phi' - \sin^2\alpha}\right)} \qquad (2.51)$$

As a special case, if $\theta = 0$,

$$K_p = \cos\alpha\,\frac{\cos\alpha + \sqrt{\cos^2\alpha - \cos^2\phi'}}{\cos\alpha - \sqrt{\cos^2\alpha - \cos^2\phi'}} \qquad (2.52)$$

$$\sigma'_a = \gamma z K_p \qquad (2.53)$$

and

$$P_p = \frac{1}{2}K_p\gamma H^2 \qquad (2.54)$$

The variation of K_p with ϕ' and α as given by Equation 2.52 is given in Table 2.7.

Backfill of c'-ϕ' Soil. If the backfill of a frictionless retaining wall with a vertical back face ($\theta = 0$) is a c'-ϕ' soil (see Figure 2.4), then the Rankine passive pressure at any depth z can be expressed as (Mazindrani and Ganjali 1997)

$$\sigma'_p = \gamma z K_p = \gamma z K'_p \cos\alpha \qquad (2.55)$$

TABLE 2.7 Passive Earth Pressure Coefficient K_p (Equation 2.52)

α (deg)	28	30	32	34	36	38	40
0	2.770	3.000	3.255	3.537	3.852	4.204	4.599
5	2.715	2.943	3.196	3.476	3.788	4.136	4.527
10	2.551	2.775	3.022	3.295	3.598	3.937	4.316
15	2.284	2.502	2.740	3.003	3.293	3.615	3.977
20	1.918	2.132	2.362	2.612	2.886	3.189	3.526
25	1.434	1.664	1.894	2.135	2.394	2.676	2.987

Column header: ϕ' (deg)

TABLE 2.8 Values of K'_p (Equation 2.56)

ϕ' (deg)	α (deg)	$\dfrac{c'}{\gamma z}$			
		0.025	0.050	0.100	0.500
15	0	1.764	1.829	1.959	3.002
	5	1.716	1.783	1.917	2.971
	10	1.564	1.641	1.788	2.880
	15	1.251	1.370	1.561	2.732
20	0	2.111	2.182	2.325	3.468
	5	2.067	2.140	2.285	3.435
	10	1.932	2.010	2.162	3.339
	15	1.696	1.786	1.956	3.183
25	0	2.542	2.621	2.778	4.034
	5	2.499	2.578	2.737	3.999
	10	2.368	2.450	2.614	3.895
	15	2.147	2.236	2.409	3.726
30	0	3.087	3.173	3.346	4.732
	5	3.042	3.129	3.303	4.674
	10	2.907	2.996	3.174	4.579
	15	2.684	2.777	2.961	4.394

where

$$K'_p = \frac{1}{\cos^2 \phi'} \left\{ 2\cos^2\alpha + 2\left(\frac{c'}{\gamma z}\right)\cos\phi'\sin\phi' + \sqrt{\left[4\cos^2\alpha(\cos^2\alpha - \cos^2\phi') + 4\left(\frac{c'}{\gamma z}\right)^2\cos^2\phi' + 8\left(\frac{c'}{\gamma z}\right)\cos^2\alpha\sin\phi'\cos\phi' \right]} - 1 \right\} \tag{2.56}$$

The variation of K'_p with ϕ', α, and $c'/\gamma z$ is given in Table 2.8.

2.9 Coulomb's Passive Pressure

Figure 2.12 shows a retaining wall with an inclined back face (similar to Figure 2.6) with an inclined granular backfill ($c' = 0$). The angle of friction between the wall and granular backfill is δ'. The failure wedge in the soil in the passive case is *ABC*. *BC* is assumed to be a plane. This

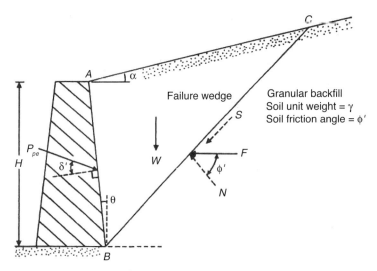

FIGURE 2.12 Coulomb's passive pressure.

is Coulomb's passive case. *Coulomb's passive earth pressure* per unit length of the wall thus can be given as

$$P_p = \frac{1}{2}\gamma H^2 K_p \qquad (2.57)$$

where K_p = Coulomb's passive earth pressure coefficient, or

$$K_p = \frac{\cos^2(\phi' + \theta)}{\cos^2\theta\cos(\delta' - \theta)\left[1 - \sqrt{\dfrac{\sin(\phi' + \delta')\sin(\phi' + \alpha)}{\cos(\delta' - \theta)\cos(\alpha - \theta)}}\right]^2} \qquad (2.58)$$

The variation of K_p with ϕ' and δ' (for $\theta = 0°$ and $\alpha = 0°$) is given in Table 2.9. It can be seen from this table that for a given value of ϕ', the value of K_p increases with the wall friction. Note that the resultant passive force P_p will act at a distance $H/3$ from the bottom of the wall and will be inclined at an angle δ' to the normal drawn to the back face of the wall.

2.10 Passive Pressure with Curved Failure Surface (Granular Soil Backfill)

The assumption of plane failure surface in the backfill (as described in Section 2.5) gives fairly good results for calculation of active earth pressure. However, this assumption may grossly

TABLE 2.9 Values of K_p (Equation 2.58) for $\theta = 0°$ and $\alpha = 0°$

ϕ' (deg)	δ' (deg)				
	0	5	10	15	20
15	1.698	1.900	2.130	2.405	2.735
20	2.040	2.313	2.636	3.030	3.525
25	2.464	2.830	3.286	3.855	4.597
30	3.000	3.506	4.143	4.977	6.105
35	3.690	4.390	5.310	6.854	8.324
40	4.600	5.590	6.946	8.870	11.772

overestimate the actual passive earth pressure, particularly when $\delta' > \phi'/2$. This is on the unsafe side for design considerations.

Figure 2.13 shows a curved failure surface in a *granular soil backfill* ($c' = 0$) for passive pressure consideration. The curved surface defined by BC is usually taken as an arc of a logarithmic spiral. CD is a plane. Several solutions have been proposed by various investigators to obtain the passive pressure coefficient K_p using a failure surface such as that shown in Figure 2.13. Some of these solutions are summarized below.

Terzaghi and Peck's Solution. Based on the trial wedge solution suggested by Terzaghi and Peck (1967):

$$K_p = \frac{P_p}{0.5\gamma H^2} \tag{2.59}$$

The variation of K_p with δ' for $\theta = 0$ (vertical back face) and $\alpha = 0$ (horizontal backfill) is shown in Figure 2.14.

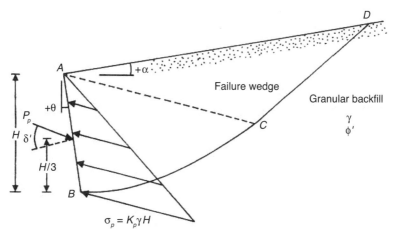

FIGURE 2.13 Curved failure surface for passive pressure determination.

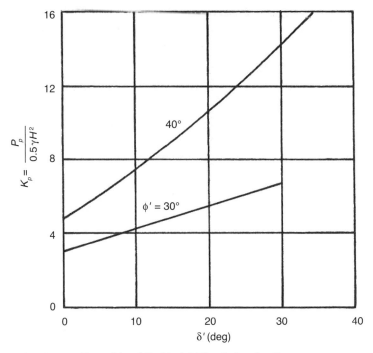

FIGURE 2.14 Terzaghi and Peck's (1967) solution for K_p.

Shields and Tolunay's Solution. Shields and Tolunay (1973) used the method of slices and obtained the variation of K_p for θ and $\alpha = 0$. The variation of $K_p = P_p/0.5\gamma H^2$ with ϕ' and δ'/ϕ' based on this solution is shown in Figure 2.15.

Zhu and Qian's Solution. Zhu and Qian (2000) used the method of triangular slices (such as in zone *ABC* in Figure 2.13) to obtain the variation of K_p. According to this analysis (for $\alpha = 0$)

$$K_p = \frac{P_p}{0.5\gamma H^2} = K_{p(\delta'=0)}R \tag{2.60}$$

where K_p = passive earth pressure coefficient for given values of θ, δ', and ϕ'; $K_{p(\delta'=0)}$ = passive earth pressure coefficient for given values of θ, ϕ' with $\delta' = 0$; and R = a modification factor which is a function of ϕ', θ, and δ'/ϕ'.

The variations of $K_{p(\delta'=0)}$ are given in Table 2.10, and the interpolated values of R are given in Table 2.11.

Caquot and Kerisel's Solution. According to Caquot and Kerisel's (1948) solution for $\alpha = 0$ and $\theta \neq 0$

$$K_p = \frac{P_p}{0.5\gamma \left(\dfrac{H}{\cos\theta}\right)^2} = K_{p(\delta'/\theta'=1)}(R') \tag{2.61}$$

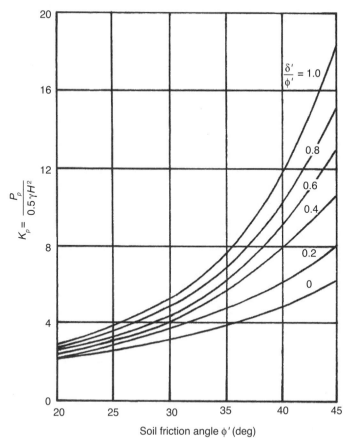

FIGURE 2.15 K_p based on Shields and Tolunay's (1973) analysis. (Note: $\theta = 0$, $\alpha = 0$.)

TABLE 2.10 Variation of $K_{p(\delta'=0)}$ (Equation 2.60)

ϕ' (deg)	θ (deg)						
	30	25	20	15	10	5	0
20	1.70	1.69	1.72	1.77	1.83	1.92	2.04
21	1.74	1.73	1.76	1.81	1.89	1.99	2.12
22	1.77	1.77	1.80	1.87	1.95	2.06	2.20
23	1.81	1.81	1.85	1.92	2.01	2.13	2.28
24	1.84	1.85	1.90	1.97	2.07	2.21	2.37
25	1.88	1.89	1.95	2.03	2.14	2.28	2.46
26	1.91	1.93	1.99	2.09	2.21	2.36	2.56
27	1.95	1.98	2.05	2.15	2.28	2.45	2.66
28	1.99	2.02	2.10	2.21	2.35	2.54	2.77
29	2.03	2.07	2.15	2.27	2.43	2.63	2.88
30	2.07	2.11	2.21	2.34	2.51	2.73	3.00
31	2.11	2.16	2.27	2.41	2.60	2.83	3.12
32	2.15	2.21	2.33	2.48	2.68	2.93	3.25
33	2.20	2.26	2.39	2.56	2.77	3.04	3.39

TABLE 2.10 Variation of $K_{p(\delta'=0)}$ (Equation 2.60) (continued)

ϕ' (deg)	\multicolumn{7}{c}{θ (deg)}
	30	25	20	15	10	5	0
34	2.24	2.32	2.45	2.64	2.87	3.16	3.53
35	2.29	2.37	2.52	2.72	2.97	3.28	3.68
36	2.33	2.43	2.59	2.80	3.07	3.41	3.84
37	2.38	2.49	2.66	2.89	3.18	3.55	4.01
38	2.43	2.55	2.73	2.98	3.29	3.69	4.19
39	2.48	2.61	2.81	3.07	3.41	3.84	4.38
40	2.53	2.67	2.89	3.17	3.53	4.00	4.59
41	2.59	2.74	2.97	3.27	3.66	4.16	4.80
42	2.64	2.80	3.05	3.38	3.80	4.34	5.03
43	2.70	2.88	3.14	3.49	3.94	4.52	5.27
44	2.76	2.94	3.23	3.61	4.09	4.72	5.53
45	2.82	3.02	3.32	3.73	4.25	4.92	5.80

TABLE 2.11 Variation of R (Equation 2.60)

θ (deg)	$\dfrac{\delta'}{\phi'}$	\multicolumn{4}{c}{R for ϕ' (deg)}
		30	35	40	45
0	0.2	1.2	1.28	1.35	1.45
	0.4	1.4	1.6	1.8	2.2
	0.6	1.65	1.95	2.4	3.2
	0.8	1.95	2.4	3.15	4.45
	1.0	2.2	2.85	3.95	6.1
5	0.2	1.2	1.25	1.32	1.4
	0.4	1.4	1.6	1.8	2.1
	0.6	1.6	1.9	2.35	3.0
	0.8	1.9	2.35	3.05	4.3
	1.0	2.15	2.8	3.8	5.7
10	0.2	1.15	1.2	1.3	1.4
	0.4	1.35	1.5	1.7	2.0
	0.6	1.6	1.85	2.25	2.9
	0.8	1.8	2.25	2.9	4.0
	1.0	2.05	2.65	3.6	5.3
15	0.2	1.15	1.2	1.3	1.35
	0.4	1.35	1.5	1.65	1.95
	0.6	1.55	1.8	2.2	2.7
	0.8	1.8	2.2	2.8	3.8
	1.0	2.0	2.6	3.4	4.95
20	0.2	1.15	1.2	1.3	1.35
	0.4	1.35	1.45	1.65	1.9
	0.6	1.5	1.8	2.1	2.6
	0.8	1.8	2.1	2.5	3.55
	1.0	1.9	2.4	3.2	4.8

where $K_{p(\delta'/\phi'=1)}$ = passive earth pressure coefficient with $\delta' = \phi'$, and R' = a reduction factor for actual δ' (which is a function of ϕ' and δ'/ϕ').

The variation of $K_{p(\delta'/\phi'=1)}$ with ϕ' and θ is shown in Figure 2.16. Table 2.12 gives the values of R' as a function of ϕ' and δ'/ϕ'.

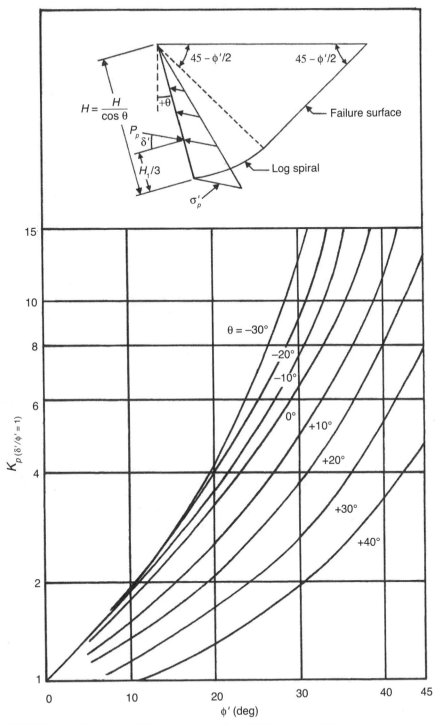

FIGURE 2.16 Variation of $K_{p(\delta'/\phi'=1)}$ with ϕ' and θ (Equation 2.61).

TABLE 2.12 Reduction Factor R' for Use in Equations 2.61 and 2.62

ϕ' (deg)	$\dfrac{\delta'}{\phi'}$							
	0.7	0.6	0.5	0.4	0.3	0.2	0.1	0
10	0.978	0.962	0.946	0.929	0.912	0.898	0.881	0.864
15	0.961	0.934	0.907	0.881	0.854	0.830	0.803	0.775
20	0.939	0.901	0.862	0.824	0.787	0.752	0.716	0.678
25	0.912	0.860	0.808	0.759	0.711	0.666	0.620	0.574
30	0.878	0.811	0.746	0.686	0.627	0.574	0.520	0.467
35	0.836	0.752	0.674	0.603	0.636	0.475	0.417	0.362
40	0.783	0.682	0.592	0.512	0.439	0.375	0.316	0.262
45	0.718	0.600	0.500	0.414	0.339	0.276	0.221	0.174

If $\theta = 0$ and $\alpha \neq 0$, the passive earth pressure coefficient can be expressed as:

$$K_p = \frac{P_p}{0.5\gamma H^2} = K_{p(\delta'/\theta'=1)}(R') \tag{2.62}$$

The variation of K_p with ϕ' and $\dfrac{\alpha}{\phi'}$ is shown in Figure 2.17. The reduction factor R' shown in Table 2.12 also can be used in Equation 2.62.

Example 4

For a retaining wall with a granular soil backfill as shown in Figure 2.13, $H = 4$ m, $\theta = 0$, $\alpha = 0$, $\gamma = 16$ kN/m³, $\phi' = 30°$, and $\delta' = 15°$. Estimate P_p by:

    a. Terzaghi and Peck's method
    b. Shields and Tolunay's method
    c. Zhu and Qian's method
    d. Caquot and Kerisel's method

Solution
Part a

$$P_p = \frac{1}{2} K_p \gamma H^2$$

From Figure 2.14 for $\phi' = 30°$ and $\delta' = 15°$, the value of K_p is about 4.6. Hence:

$$P_p = \left(\frac{1}{2}\right)(4.6)(16)(4)^2 = \textbf{588.8 kN/m}$$

Part b
$\delta'/\phi' = 15/30 = 0.5$. For $\phi' = 30°$ and $\delta'/\phi' = 0.5$, the value of K_p (Figure 2.15) is 4.13, so

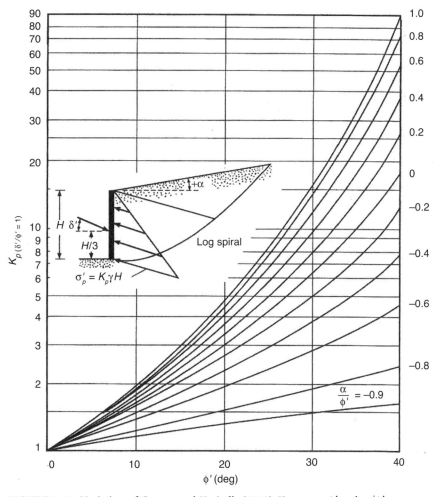

FIGURE 2.17 Variation of Caquot and Kerisel's (1948) $K_{p(\delta'/\phi'=1)}$ ϕ' and α/ϕ'.

$$P_p = \left(\frac{1}{2}\right) K_p \gamma H^2 = \left(\frac{1}{2}\right)(4.13)(16)(4)^2 = \mathbf{528.6\ kN/m}$$

Part c

$$P_p = \frac{1}{2}\gamma H^2 K_{p(\delta'=0)} R$$

For $\phi' = 30°$ and $\theta = 0$, the value of $K_{p(\delta'=0)}$ (Table 2.10) is 3. Again, from Table 2.11, for $\theta = 0$, $\delta'/\phi' = 0.5$, and $\phi' = 30°$, the value of R is about 1.5, so

$$P_p = \left(\frac{1}{2}\right)(16)(4)^2(3)(1.5) = \mathbf{576\ kN/m}$$

Part d

$$P_p = \left(\frac{1}{2}\right)(\gamma)\left(\frac{H}{\cos\theta}\right)^2 K_{p(\delta'/\phi'=1)}(R')$$

With $\theta = 0$ and $\phi' = 30°$, the value of $K_{p(\delta'/\phi'=1)}$ is about 6.5 (Figure 2.16). Again, from Table 2.12, for $\phi' = 30°$ and $\delta'/\phi' = 0.5$, the value of $R' = 0.746$. Hence:

$$P_p = \left(\frac{1}{2}\right)(16)\left(\frac{4}{\cos 0}\right)^2 (6.5)(0.746) = \textbf{620.7 kN/m}$$

2.11 Passive Pressure under Earthquake Conditions (Granular Backfill)

Figure 2.18 shows a retaining wall with a granular soil as the backfill material. If the wall is pushed toward the soil mass, it is assumed that, at a certain stage, failure in the soil will occur along a plane BC. At failure, the force P_{pe} per unit length of the retaining wall is the *dynamic passive force*. The force per unit length of the wall that needs to be considered for equilibrium of the soil wedge is shown in Figure 2.18. The notations $W, \phi', \delta', \gamma, F, k_h$, and k_v have the same meaning as described in Figure 2.7 (Section 2.6). Using the basic assumptions for the soil given in Section 2.6, the passive force P_{pe} also may be derived as (Kapila 1962)

$$P_{pe} = \frac{1}{2}\gamma H^2(1 - k_v)K_{pe} \tag{2.63}$$

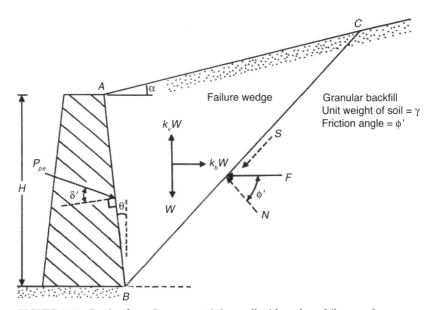

FIGURE 2.18 Passive force P_{pe} on a retaining wall with a plane failure surface.

where

$$K_{pe} = \cfrac{\cos^2(\phi' + \theta - \bar{\beta})}{\cos\bar{\beta}\,\cos^2\theta\,\cos(\delta' - \theta + \bar{\beta})}$$

$$\left\{ 1 - \left[\frac{\sin(\phi' + \delta')\sin(\phi' + \alpha - \bar{\beta})}{\cos(\alpha - \theta)\cos(\delta' - \theta + \bar{\beta})} \right]^{\frac{1}{2}} \right\}^2$$

(2.64)

and $\bar{\beta} = \tan^{-1}(k_h/1 - k_v)$.

Figure 2.19 shows the variation of K_{pe} for various values of soil friction angle ϕ' and k_h (with $k_v = \alpha = \theta = \delta' = 0$). From the figure, it can be seen that, with other parameters remaining the same, the magnitude of K_{pe} increases with the increase in soil friction angle ϕ'.

The relationship for passive earth pressure on a retaining wall with a *granular horizontal backfill* and vertical back face under earthquake conditions was evaluated by Subba Rao and Choudhury (2005) using the pseudo-static approach to the method of limit equilibrium. The

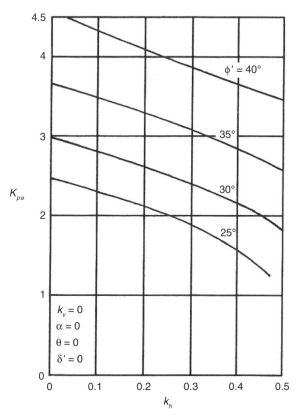

FIGURE 2.19 Variation of K_{pe} with soil friction angle and k_h.

curved failure surface in soil assumed in the analysis was similar to that shown in Figure 2.13 (with $\theta = 0$, vertical back face and $\alpha = 0$, horizontal backfill). Based on this analysis, the passive force P_{pe} can be expressed as

$$P_{pe} = \left[\frac{1}{2} \gamma H^2 K_{p\gamma(e)} \right] \frac{1}{\cos \delta'} \tag{2.65}$$

where $K_{p\gamma(e)}$ = passive earth pressure coefficient in the normal direction to the wall.

$K_{p\gamma(e)}$ is a function of k_h and k_v. The variations of $K_{p\gamma(e)}$ for $\delta'/\phi' = 0.5$ and 1 are shown in Figures 2.20 and 2.21. The passive pressure P_{pe} will be inclined at an angle δ' to the back face of the wall and will act at a distance of $H/3$ above the bottom of the wall.

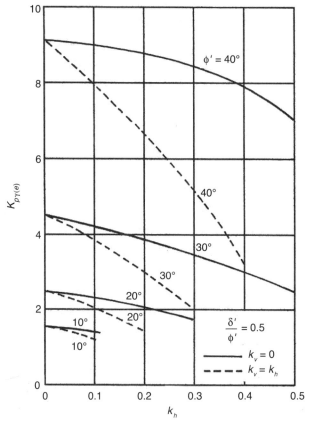

FIGURE 2.20 Variation of $K_{p\gamma(e)}$ (Equation 2.65) for $\delta'/\phi' = 0.5$.

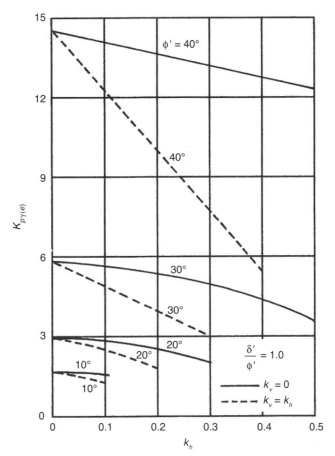

FIGURE 2.21 Variation of $K_{p\gamma(e)}$ (Equation 2.65) for $\delta'/\phi' = 1.0$.

References

Caquot, A. and Kerisel, J. (1948). *Tables for Calculation of Passive Pressure, Active Pressure and Bearing Capacity of Foundations,* Gauthier-Villars, Paris.

Chu, S.C. (1991). Rankine analysis of active and passive pressures on dry sand. *Soils Found.,* 31(4):115–120.

Coulomb, C.A. (1776). Essai sur une application des règles de maximis et minimis a quelques problèmes de statique, relatifs a l'architecture. *Mem. R. Sci. Paris,* 3:38.

Jaky, J. (1944). The coefficient of earth pressure at rest. *J. Soc. Hung. Archit. Eng.,* 7:355–358.

Kapila, J.P. (1962). Earthquake resistant design of retaining walls. *Proceedings, 2nd Earthquake Symposium,* University of Roorkee, Roorkee, India.

Massarsch, K.R. (1979). Lateral earth pressure in normally consolidated clay. *Proceedings, 7th European Conference on Soil Mechanics and Foundation Engineering,* Vol. 2, Brighton, England, 245–250.

Mayne, P.W. and Kulhawy, F.H. (1982). K_o-OCR relationships in soil. *J. Geotech. Eng. Div. ASCE,* 108(6):851–872.

Mazindrani, Z.H. and Ganjali, M.H. (1997). Lateral earth pressure problem of cohesive backfill with inclined surface. *J. Geotech. Geoenviron. Eng.*, 123(2):110–112.

Mononobe, N. (1929). Earthquake-proof construction of masonry dams. *Proceedings, World Engineering Conference*, Vol. 9, Tokyo, October–November, 274–280.

Okabe, S. (1926). General theory of earth pressure. *J. Jpn. Soc. Civ. Eng.*, 12(1).

Rankine, W.M.J. (1857). On stability on loose earth. *Philos. Trans. R. Soc. London Part I*, pp. 9–27.

Seed, H.B. and Whitman, R.V. (1970). Design of earth retaining structures for dynamic loads. *Proceedings, Specialty Conference on Lateral Stresses in the Ground and Design of Earth Retaining Structures*, American Society of Civil Engineers, 103–147.

Shields, D.H. and Tolunay, A.Z. (1973). Passive pressure coefficients by method of slices. *J. Soil Mech. Found. Div. ASCE*, 99(SM12):1043–1053.

Subba Rao, K.S. and Choudhury, D. (2005). Seismic passive earth pressures in soil. *J. Geotech. Geoenviron. Eng.*, 131(1):131–135.

Terzaghi, K. and Peck, R.B. (1967). *Soil Mechanics in Engineering Practice*, 2nd edition, John Wiley & Sons, New York.

Zhu, D.Y. and Qian, Q. (2000). Determination of passive earth pressure coefficient by the method of triangular slices. *Can. Geotech. J.*, 37(2):485–491.

3

Design of
Shallow Foundations

by
Nagaratnam Sivakugan
James Cook University, Townsville, Australia

Marcus Pacheco
Universidade do Estado do Rio de Janeiro, Brazil

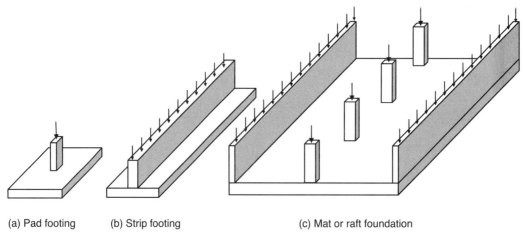

(a) Pad footing   (b) Strip footing   (c) Mat or raft foundation

FIGURE 3.1 Types of shallow foundations.

3.1 Introduction

A foundation is a structural element that is expected to transfer a load from a structure to the ground safely. The two major classes of foundations are *shallow foundations* and *deep foundations*. A shallow foundation transfers the entire load at a relatively shallow depth. A common understanding is that the depth of a shallow foundation (D_f) must be less than the breadth (B). Breadth is the shorter of the two plan dimensions. Shallow foundations include pad footings, strip (or wall) footings, combined footings, and mat foundations, shown in Figure 3.1. Deep foundations have a greater depth than breadth and include piles, pile groups, and piers, which are discussed in Chapter 4. A typical building can apply 10–15 kPa per floor, depending on the column spacing, type of structure, and number of floors.

Shallow foundations generally are designed to satisfy two criteria: *bearing capacity* and *settlement*. The bearing capacity criterion ensures that there is adequate safety against possible bearing capacity failure within the underlying soil. This is done through provision of an adequate factor of safety of about 3. In other words, shallow foundations are designed to carry a working load of one-third of the failure load. For raft foundations, a safety factor of 1.7–2.5 is recommended (Bowles 1996). The settlement criterion ensures that settlement is within acceptable limits. For example, pad and strip footings in granular soils generally are designed to settle less than 25 mm.

3.2 Stresses beneath Loaded Areas

In particular for computing settlement of footings, it is necessary to be able to estimate the stress increase at a specific depth due to the foundation loading. The theories developed for computing settlement often assume the soil to be a homogeneous, isotropic, weightless elastic continuum.

3.2.1 Point and Line Loads

Boussinesq (1885) showed that in a homogeneous, isotropic elastic half-space, the vertical stress increase ($\Delta\sigma_v$) at a point within the medium, due to a point load (Q) applied at the surface (see Figure 3.2), is given by

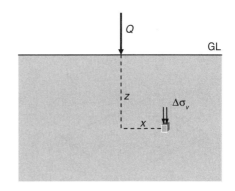

$$\Delta\sigma_v = \frac{3Q}{2\pi z^2}\left[\frac{1}{1 + (x/z)^2}\right]^{5/2} \quad (3.1)$$

FIGURE 3.2 Stress increase beneath a point or line load.

where z and x are the vertical and horizontal distance, respectively, to the point of interest from the applied load.

Westergaard (1938) did similar research, assuming the soil to be reinforced by closely spaced rigid sheets of infinitesimal thicknesses, and proposed a slightly different equation:

$$\Delta\sigma_v = \frac{Q}{2\pi z^2}\frac{\sqrt{\dfrac{1 - 2v}{2 - 2v}}}{\left[\left(\dfrac{1 - 2v}{2 - 2v}\right) + \left(\dfrac{x}{z}\right)^2\right]^{3/2}} \quad (3.2)$$

Westergaard's equation models anisotropic sedimentary clays with several thin seams of sand lenses interbedded with the clays. The stresses computed from the Boussinesq equation generally are greater than those computed from the Westergaard equation. As it is conservative and simpler, the Boussinesq equation is more popular and will be used throughout this section.

If the point load is replaced by an infinitely long line load in Figure 3.2, the vertical stress increase $\Delta\sigma_v$ is given by:

$$\Delta\sigma_v = \frac{2Q}{\pi z}\left[\frac{1}{1 + (x/z)^2}\right]^2 \quad (3.3)$$

3.2.2 Uniform Rectangular Loads

The vertical stress increase at a depth z beneath the corner of a uniform rectangular load (see Figure 3.3a) can be obtained by breaking the rectangular load into an infinite number of point loads ($dq = Q\,dx\,dy$) and integrating over the entire area. The vertical stress increase is given by

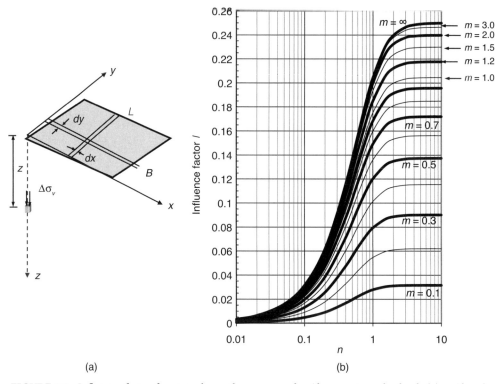

(a)   (b)

FIGURE 3.3 Influence factor for stress beneath a corner of uniform rectangular load: (a) uniformly loaded rectangle and (b) chart.

$$\Delta\sigma_v = Iq \qquad (3.4)$$

where q is the applied pressure and the influence factor I is given by:

$$I = \frac{1}{4\pi} \left[\left(\frac{2mn\sqrt{m^2 + n^2 + 1}}{m^2 + n^2 + m^2n^2 + 1} \right) \left(\frac{m^2 + n^2 + 2}{m^2 + n^2 + 1} \right) + \tan^{-1}\left(\frac{2mn\sqrt{m^2 + n^2 + 1}}{m^2 + n^2 - m^2n^2 + 1} \right) \right] \qquad (3.5)$$

Here $m = B/z$ and $n = L/z$. Variation of I with m and n is shown in Figure 3.3b. Using the equation or Figure 3.3b, the vertical stress increase at any point within the soil, under a uniformly loaded rectangular footing, can be found. This will require breaking up the loaded area into four rectangles and applying the principle of superposition. This can be extended to T-shaped or L-shaped areas as well.

At a depth z, $\Delta\sigma_v$ is the maximum directly below the center and decays with horizontal distance. Very often, the value of $\Delta\sigma_v$ is estimated by assuming that the soil pressure applied at the footing level is distributed through a rectangular prism, with slopes of 2 (vertical):1 (horizontal) in both directions, as shown in Figure 3.4. Assuming the 2:1 spread in the load, the vertical stress at depth z below the footing becomes:

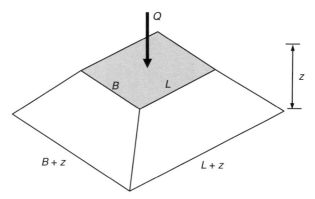

FIGURE 3.4 Average vertical stress increase with 2:1 distribution.

$$\Delta\sigma_v = \frac{Q}{(B + z)(L + z)} \tag{3.6}$$

In the case of strip footings, Equation 3.6 becomes:

$$\Delta\sigma_v = \frac{Q}{B + z} \tag{3.7}$$

3.2.3 Newmark's Chart for Uniformly Loaded Irregular Areas

The vertical stress increase at depth z below the center of a uniformly loaded circular footing of radius r is given by:

$$\Delta\sigma_v = \left\{ 1 - \frac{1}{[(r/z)^2 + 1]^{3/2}} \right\} q \tag{3.8}$$

The values of r/z for $\Delta\sigma_v = 0.1q$, $0.2q...1.0q$ are given in Table 3.1. Newmark (1942) developed the influence chart shown in Figure 3.5 using the values given in Table 3.1. Each block in the chart contributes an equal amount of vertical stress increase at any point directly below the center. This chart can be used to determine the vertical stress increase at depth z directly below any point (X) within or outside a uniformly loaded irregular area.

The following steps are required for computing $\Delta\sigma_v$ at depth z below P:

1. Redraw (better to use tracing paper) the plan of the loaded area to a scale where z is equal to the scale length given in the diagram.
2. Place the plan on top of the influence chart such that the point of interest P on the plan coincides with the center of the chart.

TABLE 3.1 Influence Circle Radii for Newmark's Chart

$\Delta\sigma_v/q$	0.1	0.2	0.3	0.4	0.5	0.6	0.7	0.8	0.9	1.0
r/z	0.270	0.401	0.518	0.637	0.766	0.918	1.110	1.387	1.908	∞

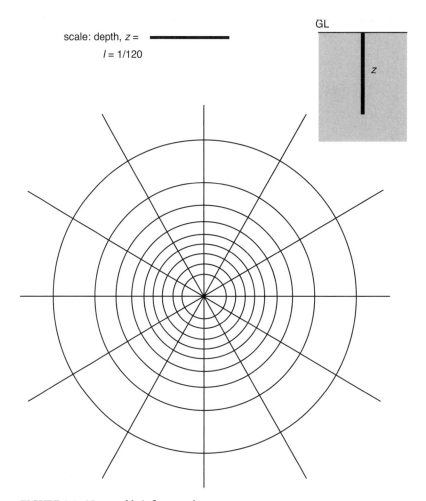

FIGURE 3.5 Newmark's influence chart.

3. Count the number of blocks (say, n) covered by the loaded area (include fractions of the blocks).
4. Compute $\Delta\sigma_v$ as $\Delta\sigma_v = Inq$, where I is the influence value for Newmark's chart. For the one in Figure 3.5, where there are 200 blocks, $I = 1/120 = 0.00833$.

3.3 Bearing Capacity of Shallow Foundations

Several researchers have studied bearing capacity of shallow foundations, analytically and using model tests in laboratories. Let's look at some historical developments and three of the major bearing capacity equations with corresponding correction factors.

Typical pressure-settlement plots in different types of soils are shown in Figure 3.6. Three different failure mechanisms, namely *general shear*, *local shear*, and *punching shear*, were recognized by researchers. General shear failure is the most common mode of failure, and it occurs in firm ground, including dense granular soils and stiff clays, where the failure load is well defined (see Figure 3.6a). Here, the shear resistance is fully developed along the entire

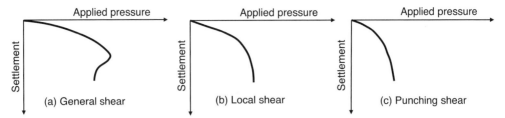

FIGURE 3.6 Failure modes of a shallow foundation.

failure surface that extends to the ground level, and a clearly formed heave appears at the ground level near the footing. The other extreme is punching shear failure, which occurs in weak, compressible soils such as very loose sands, where the failure surface does not extend to the ground level and the failure load is not well defined, with no noticeable heave at the ground level (Figure 3.6c). In between these two modes, there is local shear failure (Figure 3.6b), which occurs in soils of intermediate compressibility such as medium-dense sands, where only slight heave occurs at the ground level near the footing.

In reality, the ground conditions are always improved through compaction before placing the footing. For shallow foundations in granular soils with $D_r > 70\%$ and stiff clays, the failure will occur in the general shear mode (Vesic 1973). Therefore, it is reasonable to assume that the general shear failure mode applies in most situations.

From bearing capacity considerations, the allowable bearing capacity (q_{all}) is defined as

$$q_{all} = \frac{q_{ult}}{F} \tag{3.9}$$

where q_{ult} is the ultimate bearing capacity, which is the average contact pressure at the soil-footing interface when the bearing capacity failure occurs, and F is the factor of safety, which typically is taken as 3 for the bearing capacity of shallow foundations.

3.3.1 Historical Developments

Prandtl (1921) modeled a narrow metal tool bearing against the surface of a block of smooth softer metal, which later was extended by Reissner (1924) to include a bearing area located *below* the surface of the softer metal. The Prandtl-Reissner plastic limit equilibrium plane strain analysis of a hard object penetrating a softer material later was extended by Terzaghi (1943) to develop the first rational bearing capacity equation for strip footings embedded in soils. Terzaghi assumed the soil to be a semi-infinite, isotropic, homogeneous, weightless, rigid plastic material; the footing to be rigid; and the base of the footing to be sufficiently rough to ensure there is no separation between the footing and the underlying soil. It also was assumed that the failure occurs in the general shear mode (Figure 3.7).

3.3.2 Terzaghi's Bearing Capacity Equation

Assuming that the bearing capacity failure occurs in the general shear mode, Terzaghi expressed his first bearing capacity equation for a strip footing as:

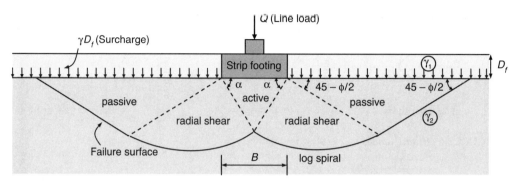

FIGURE 3.7 Assumed failure surfaces within the soil during bearing capacity failure.

$$q_{ult} = cN_c + \gamma_1 D_f N_q + 0.5B\gamma_2 N_\gamma \tag{3.10}$$

Here, c is the cohesion and γ_1 and γ_2 are the unit weights of the soil above and below, respectively, the footing level. N_c, N_q, and N_γ are the bearing capacity factors, which are functions of the friction angle. The ultimate bearing capacity is derived from three distinct components. The first term in Equation 3.10 reflects the contribution of cohesion to the ultimate bearing capacity, and the second term reflects the frictional contribution of the overburden pressure or surcharge. The last term reflects the frictional contribution of the self-weight of the soil in the failure zone.

For *square* and *circular footings*, the ultimate bearing capacities are given by Equations 3.11 and 3.12, respectively:

$$q_{ult} = 1.2cN_c + \gamma_1 D_f N_q + 0.4B\gamma_2 N_\gamma \tag{3.11}$$

$$q_{ult} = 1.2cN_c + \gamma_1 D_f N_q + 0.3B\gamma_2 N_\gamma \tag{3.12}$$

It must be remembered that the bearing capacity factors in Equations 3.11 and 3.12 are still for strip footings. For local shear failure, where the failure surface is not fully developed and thus the friction and cohesion are not fully mobilized, Terzaghi reduced the values of the friction angle and cohesion by one-third to:

$$\phi' = \tan^{-1}(0.67\phi) \tag{3.13}$$

$$c' = 0.67c \tag{3.14}$$

Terzaghi neglected the shear resistance provided by the overburden soil, which was treated as a surcharge (see Figure 3.7). Also, he assumed that $\alpha = \phi$ in Figure 3.7. Subsequent studies by several others show that $\alpha = 45 + \phi/2$ (Vesic 1973), which makes the bearing capacity factors different than what were originally proposed by Terzaghi. With $\alpha = 45 + \phi/2$, the bearing capacity factors N_q and N_c become:

$$N_q = e^{\pi \tan \phi} \tan^2 \left(45 + \frac{\phi}{2} \right) \tag{3.15}$$

TABLE 3.2 Expressions for N_γ

Expression	Reference
$(N_q - 1) \tan (1.4\phi)$	Meyerhof (1963)
$1.5 (N_q - 1) \tan \phi$	Hansen (1970)
$2.0 (N_q - 1) \tan \phi$	European Committee for Standardisation (1995)
$2.0 (N_q + 1)$	Vesic (1973)
$1.1 (N_q - 1) \tan(1.3\phi)$	Spangler and Handy (1982)
$0.1054 \exp(9.6\phi)$[a]	Davis and Booker (1971)
$0.0663 \exp(9.3\phi)$[b]	Davis and Booker (1971)

[a] Rough footing with ϕ in radians.
[b] Smooth footing with ϕ in radians.

$$N_c = (N_q - 1) \cot \phi \qquad (3.16)$$

The above expression for N_c is the same as the one originally proposed by Prandtl (1921), and the expression for N_q is the same as the one given by Reissner (1924). While there is general consensus about Equations 3.15 and 3.16, various expressions for N_γ have been proposed in the literature, the most frequently used of which are those proposed by Meyerhof (1963) and Hansen (1970). Some of these different expressions for N_γ are presented in Table 3.2.

For undrained loading in clays, when $\phi_u = 0$, it can be shown that $N_q = 1$, $N_\gamma = 0$, and $N_c = 2 + \pi$ (= 5.14). Skempton (1951) studied the variation of N_c with shape and the depth of the foundation. He showed that for a strip footing, it varies from $2 + \pi$ at the surface to 7.5 at a depth greater than $5B$, and for a square footing, it varies between 2π at the surface and 9.0 at a depth greater than $5B$. Therefore, for pile foundations, it generally is assumed that $N_c = 9$.

Most of the bearing capacity theories (e.g., Prandtl, Terzaghi) assume that the footing-soil interface is rough. Concrete footings are made by pouring concrete directly on the ground, and therefore the soil-footing interface is rough. Schultze and Horn (1967) noted that from the way concrete footings are cast in place, there is adequate friction at the base, which mobilizes friction angles equal to or greater than ϕ. Even the bottom of a metal storage tank is not smooth, since the base is always treated with paint or asphalt to resist corrosion (Bowles 1996). Therefore, the assumption of a rough base is more realistic than a smooth one. Based on experimental studies, Vesic (1975) stated that foundation roughness has little effect on the ultimate bearing capacity, provided the footing load is vertical.

Meyerhof's N_γ, used predominantly in North America, and Hansen's, used in Europe, appear to be the most popular of the above. The values of N_γ proposed by Meyerhof (1963), Hansen (1970), Vesic (1973), and in *Eurocode 7* (European Committee for Standardisation 1995) are shown in Figure 3.8, along with the values of N_q and N_c. For $\phi < 30°$, Meyerhof's and Hansen's values are essentially the same. For $\phi > 30°$, Meyerhof's values are larger, the difference increasing with ϕ. The Indian standard recommends Vesic's N_γ factor (Raj 1995). The *Canadian Foundation Engineering Manual* recommends Hansen's N_γ factor (Canadian Geotechnical Society 1992).

3.3.3 Meyerhof's Bearing Capacity Equation

In spite of the various improvements that were made to the theoretical developments proposed by Terzaghi, his original form of the bearing capacity equation is still being used because of its

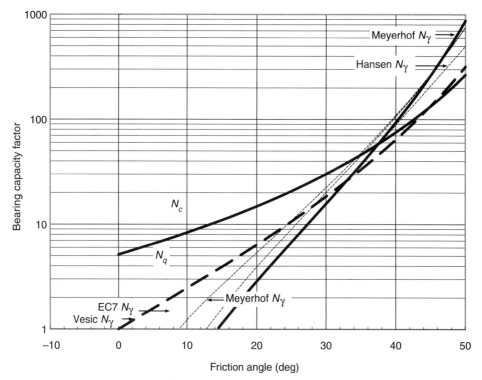

FIGURE 3.8 Bearing capacity factors.

simplicity and practicality. Terzaghi neglected the shear resistance within the overburden soil (i.e., above the footing level), which was included in the modifications made by Meyerhof (1951) that are discussed here. Meyerhof's (1963) modifications, which are being adapted worldwide, are summarized here. Meyerhof (1963) proposed the general bearing capacity equation of a rectangular footing as

$$q_{\text{ult}} = s_c d_c i_c c N_c + s_q d_q i_q \gamma_1 D_f N_q + s_\gamma d_\gamma i_\gamma 0.5 B \gamma_2 N_\gamma \qquad (3.17)$$

where N_c, N_q, and N_γ are the bearing capacity factors of a *strip* footing. The shape of the footing is accounted for through the shape factors s_c, s_q, and s_γ. The depth of the footing is taken into account through the depth factors d_c, d_q, and d_γ. The inclination factors i_c, i_q, and i_γ account for the inclination in the applied load. These factors are summarized below.

Shape factors (Meyerhof 1963):

$$s_c = 1 + 0.2 \, \frac{B}{L} \, \tan^2\left(45 + \frac{\phi}{2}\right) \qquad (3.18)$$

$$s_q = s_\gamma = 1 + 0.1 \, \frac{B}{L} \, \tan^2\left(45 + \frac{\phi}{2}\right) \qquad \text{for } \phi \geq 10° \qquad (3.19)$$

$$s_q = s_\gamma = 1 \qquad \text{for } \phi = 0 \qquad (3.20)$$

Depth factors (Meyerhof 1963):

$$d_c = 1 + 0.2 \, \frac{D_f}{B} \, \tan\left(45 + \frac{\phi}{2}\right) \tag{3.21}$$

$$d_q = d_\gamma = 1 + 0.1 \, \frac{D_f}{B} \, \tan\left(45 + \frac{\phi}{2}\right) \quad \text{for } \phi \geq 10° \tag{3.22}$$

$$d_q = d_\gamma = 1 \quad \text{for } \phi = 0 \tag{3.23}$$

Inclination factors (Meyerhof 1963; Hanna and Meyerhof 1981):

$$i_c = i_q = \left(1 - \frac{\alpha°}{90}\right)^2 \tag{3.24}$$

$$i_\gamma = \left(1 - \frac{\alpha}{\phi}\right)^2 \quad \text{for } \phi \geq 10° \tag{3.25}$$

$$i_\gamma = 1 \quad \text{for } \phi = 0 \tag{3.26}$$

In Equations 3.24 and 3.25, α is the inclination (in degrees) of the footing load to the vertical. It should be noted that in spite of the load being inclined, the ultimate bearing capacity computed from Equation 3.17 gives its vertical component.

3.3.3.1 Plane Strain Correction

It has been reported by several researchers that the friction angle obtained from a plane strain compression test is greater than that obtained from a triaxial compression test by about 4–9° in dense sands and 2–4° in loose sands (Ladd et al. 1977). A conservative estimate of the plane strain friction angle may be obtained from the triaxial friction angle by (Lade and Lee 1976):

$$\phi_{ps} = 1.5\phi_{tx} - 17° \quad \text{for } \phi_{tx} > 34° \tag{3.27}$$

$$\phi_{ps} = \phi_{tx} \quad \text{for } \phi_{tx} \leq 34° \tag{3.28}$$

Allen et al. (2004) related the peak friction angles from direct shear and plane strain compression tests through the following equation:

$$\phi_{ps} = \tan^{-1}(1.2 \tan \phi_{ds}) \tag{3.29}$$

The soil element beneath the centerline of a strip footing is subjected to plane strain loading, and therefore, the plane strain friction angle must be used in calculating its bearing capacity. The plane strain friction angle can be obtained from a plane strain compression test. The loading condition of a soil element along the vertical centerline of a square or circular footing more closely resembles axisymmetric loading than plane strain loading, thus requiring a triaxial friction angle, which can be determined from a consolidated drained or undrained triaxial compression test.

On the basis of the suggestions made by Bishop (1961) and Bjerrum and Kummeneje (1961) that the plane strain friction angle is 10% greater than that from a triaxial compression test, Meyerhof proposed the corrected friction angle for use with rectangular footings as:

$$\phi_{\text{rectangular}} = \left(1.1 - 0.1\,\frac{B}{L}\right)\phi_{\text{tx}} \qquad (3.30)$$

The above equation simply enables interpolation between ϕ_{tx} (for $B/L = 1$) and ϕ_{ps} (for $B/L = 0$). The friction angles available in most geotechnical designs are derived from triaxial tests in the laboratory or *in situ* penetration tests.

3.3.3.2 Eccentric Loading

When the footing is applied with some eccentricity, the ultimate bearing capacity is reduced. Meyerhof (1963) suggested the effective footing breadth (B') and length (L') as:

$$B' = B - 2e_B \qquad (3.31)$$

$$L' = L - 2e_L \qquad (3.32)$$

where e_B and e_L are the eccentricities along the breadth and length, respectively, as shown in Figure 3.9.

For footings with eccentricities, B' and L' should be used in computing the ultimate bearing capacity (Equation 3.17) and shape factors (Equations 3.18 and 3.19). In computing the depth factors (Equations 3.21 and 3.22), B should be used. The unhatched area ($A' = B' \times L'$) in Figure 3.9 is the effective area which contributes to the bearing capacity, and therefore, the ultimate footing load is computed by multiplying the ultimate bearing capacity by this area A'. It should be noted that when the hatched area is disregarded, the load is applied at the center of the remaining area.

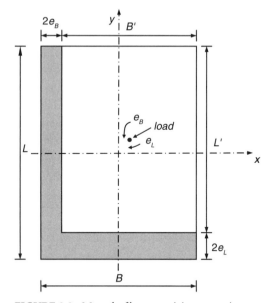

FIGURE 3.9 Meyerhof's eccentricity correction.

3.3.4 Hansen's Bearing Capacity Equation

Based on theoretical and experimental work, Hansen (1970) and Vesic (1973, 1975) proposed the following bearing capacity equation for drained and undrained conditions:

$$q_{\text{ult}} = s_c d_c i_c b_c g_c c N_c + s_q d_q i_q b_q g_q \gamma D_f N_q + s_\gamma d_\gamma i_\gamma b_\gamma g_\gamma 0.5 B \gamma N_\gamma \qquad (3.33)$$

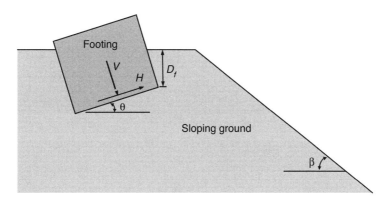

FIGURE 3.10   Base and ground inclination.

In addition to the shape (*s*), depth (*d*), and inclination (*i*) factors, they included base inclination (*b*) and ground inclination (*g*) factors. Base inclination factors account for any inclination in the base of the footing. This may become necessary when the footing is required to carry an inclined load. The ground inclination factors account for the reduction in bearing capacity when the footing is located on sloping ground, as shown in Figure 3.10. The equations to compute these factors are summarized below.

Shape factors (Hansen 1970):

$$s_c = 1 + \left(\frac{B}{L} \right) \left(\frac{N_q}{N_c} \right) \tag{3.34}$$

$$s_q = 1 + \left(\frac{B}{L} \right) \tan \phi \tag{3.35}$$

$$s_\gamma = 1 - 0.4 \left(\frac{B}{L} \right) \tag{3.36}$$

Depth factors (Hansen 1970):

$$d_c = 1 + 0.4 \, \frac{D_f}{B} \tag{3.37}$$

$$d_q = 1 + 2 \, \frac{D_f}{B} \, \tan \phi \, (1 - \sin \phi)^2 \tag{3.38}$$

$$d_\gamma = 1 \tag{3.39}$$

When $D_f > B$, the factor D_f/B should be replaced by $\tan^{-1} (D_f/B)$.

Load inclination factors (Hansen 1970):

$$i_c = 0.5 + 0.5\sqrt{\left(1 - \frac{H}{cBL}\right)} \qquad \text{for } \phi = 0 \qquad (3.40)$$

$$i_c = i_q - \frac{1 - i_q}{N_q - 1} \qquad \text{for } \phi > 0 \qquad (3.41)$$

$$i_q = \left(1 - \frac{0.5H}{V + cBL \cot \phi}\right)^5 \qquad (3.42)$$

$$i_\gamma = \left[1 - \frac{\left(0.7 - \dfrac{\theta^\circ}{450}\right)H}{V + cBL \cot \phi}\right]^5 \qquad (3.43)$$

The cohesion mobilized at the footing-soil contact area must be used for c in Equations 3.40, 3.42, and 3.43. The U.S. Army (1993) recommends using adhesion or a reduced value of cohesion.

Base inclination factors (Hansen 1970):

$$b_c = 1 - \frac{\theta^\circ}{147} \qquad (3.44)$$

$$b_q = \exp(-0.0349\theta^\circ \tan \phi) \qquad (3.45)$$

$$b_\gamma = \exp(-0.0471\theta^\circ \tan \phi) \qquad (3.46)$$

Ground inclination factors (Hansen 1970):

$$g_c = 1 - \frac{\beta^\circ}{147} \qquad (3.47)$$

$$g_q = g_\gamma = (1 - 0.5 \tan \beta)^5 \qquad (3.48)$$

3.3.5 Vesic's Bearing Capacity Equation

Vesic's bearing capacity equation is the same as Hansen's, but with slight differences in the bearing capacity factor N_γ and the last three inclination factors (i, b, and g), which are less conservative.

Shape factors (Vesic 1975):

$$s_c = 1 + \left(\frac{B}{L} \right) \left(\frac{N_q}{N_c} \right) \tag{3.49}$$

$$s_q = 1 + \left(\frac{B}{L} \right) \tan \phi \tag{3.50}$$

$$s_\gamma = 1 - 0.4 \left(\frac{B}{L} \right) \tag{3.51}$$

Depth factors (Vesic 1975):

$$d_c = 1 + 0.4 \, \frac{D_f}{B} \qquad \text{for } \phi = 0 \tag{3.52}$$

$$d_c = d_q - \frac{1 - d_q}{N_q - 1} \qquad \text{for } \phi > 0 \tag{3.53}$$

$$d_q = 1 + 2 \, \frac{D_f}{B} \, \tan \phi (1 - \sin \phi)^2 \tag{3.54}$$

$$d_\gamma = 1 \tag{3.55}$$

When $D_f > B$, the factor D_f/B should be replaced by $\tan^{-1}(D_f/B)$.

Load inclination factors (Vesic 1975):
If V and H are the components of the load perpendicular and parallel to the base of the footing, the load inclination factors i_c, i_q, and i_γ are given by:

$$i_c = 1 - \frac{mH}{AcN_c} \qquad \text{for } \phi = 0 \tag{3.56}$$

$$i_c = i_q - \frac{1 - i_q}{N_q - 1} \qquad \text{for } \phi > 0 \tag{3.57}$$

$$i_q = \left(1 - \frac{H}{V + cBL \cot \phi} \right)^m \tag{3.58}$$

$$i_\gamma = \left(1 - \frac{H}{V + cBL \cot \phi} \right)^{m+1} \tag{3.59}$$

where

$$m = \frac{2 + B/L}{1 + B/L}$$

if the load is inclined in the direction parallel to the breadth and

$$m = \frac{2 + L/B}{1 + L/B}$$

if the load is inclined in the direction parallel to the length. The cohesion mobilized at the footing-soil contact area must be used for c in Equations 3.56, 3.58, and 3.59. The U.S. Army (1993) recommends using adhesion or a reduced value of cohesion.

Base inclination factors (Vesic 1975):

$$b_c = 1 - \frac{\phi^\circ}{147} \qquad \text{for } \phi = 0 \tag{3.60}$$

$$b_c = b_q - \frac{1 - b_q}{N_q - 1} \qquad \text{for } \phi > 0 \tag{3.61}$$

$$b_q = b_\gamma = \left(1 - \frac{\theta^\circ \tan \phi}{57}\right)^2 \tag{3.62}$$

where θ is the inclination (in degrees) of the base of the footing to horizontal (see Figure 3.10).

Ground inclination factors (Vesic 1975):

$$g_c = 1 - \frac{\beta^\circ}{147} \qquad \text{for } \phi = 0 \tag{3.63}$$

$$g_c = g_q - \frac{1 - g_q}{N_q - 1} \qquad \text{for } \phi > 0 \tag{3.64}$$

$$g_q = g_\gamma = (1 - \tan \beta)^2 \tag{3.65}$$

where β is the inclination of the slope in degrees, $\beta < \phi$, and $\theta + \beta \le 90^\circ$ (see Figure 3.10). On a sloping ground, when $\phi = 0$, $N_g = -2 \sin \beta$.

It should be noted that the ultimate bearing capacity equation for clays under undrained conditions ($\phi_u = 0$) sometimes is given in the literature slightly differently as (Aysen 2002; Bowles 1988)

$$q_{ult} = (1 + s_c + d_c - i_c - b_c - g_c)c_u N_c + \gamma D_f \tag{3.66}$$

and consequently the reported correction factors for Equation 3.32 are slightly different (U.S. Army 1993; Cernica 1995; Coduto 2001; McCarthy 2007; European Committee for Standardisation 1995).

3.3.6 Gross and Net Pressures and Bearing Capacities

The ultimate bearing capacities computed using Equations 3.10–3.12, 3.17, 3.33, and 3.66 are all *gross* ultimate bearing capacities. There already is an overburden pressure of γD_f acting at the foundation level. The net ultimate bearing capacity is the maximum additional soil pressure that can be sustained before failure. Therefore, net ultimate bearing capacity is obtained by subtracting the overburden pressure from the gross ultimate bearing capacity. Similarly, the net applied pressure is the additional pressure applied at the foundation level in excess of the existing overburden pressure. The safety factor with respect to bearing capacity failure is therefore defined in terms of the net values as:

$$F = \frac{q_{ult,net}}{q_{applied,net}} = \frac{q_{ult,gross} - \gamma D_f}{q_{applied,gross} - \gamma D_f} \tag{3.67}$$

In most spread footing designs, the gross pressures are significantly larger than the overburden pressures. Only in problems that involve removal of large overburden pressures, such as foundations for basements, can gross and net pressures be significantly different. In clays under undrained conditions ($\phi_u = 0$), $N_c = 5.14$, $N_q = 1$, and $N_\gamma = 0$. Therefore, the net ultimate bearing capacity of a shallow foundation can be written as:

$$q_{ult,net} = 5.14 c_u \left(1 + 0.2 \frac{D_f}{B} \right) \left(1 + 0.2 \frac{B}{L} \right) \tag{3.68}$$

3.3.7 Effects of the Water Table

When computing the ultimate bearing capacity in terms of effective stress parameters, it is necessary to use the correct unit weights, depending on the location of the water table. If the water table lies at or above ground level, γ' must be used in both terms in the bearing capacity equation (Equation 3.10). If the water table lies at the footing level, γ_m must be used in the second term and γ' in the third term in the bearing capacity equation. It can be seen from Figure 3.7 that the failure zone within the soil is confined to a depth of B below the footing width. Therefore, if the water table lies at B or more below the footing, the bulk unit weight (γ_m) must be used in both terms in the bearing capacity equation. Terzaghi and Peck (1967) stated that the friction angle is reduced by 1–2° when a sand is saturated. Therefore, if a future rise in the water table is expected, the friction angle may be reduced slightly in computing the ultimate bearing capacity.

3.3.8 Presumptive Bearing Pressures

Presumptive bearing pressures are very approximate and conservative safe bearing pressures that can be assumed in preliminary designs. They are given in building codes and geotechnical textbooks (see U.S. Army 1993; Bowles 1988). Here, the specified values do not reflect the site

TABLE 3.3 Presumed Bearing Capacity Values

Soil Type	Bearing Capacity (kPa)
Rocks	
Hard and sound igneous and gneissic rock	10,000
Hard limestone/sandstone	4,000
Schist/slate	3,000
Hard shale/mudstone or soft sandstone	2,000
Soft shale/mudstone	600–1,000
Hard sound chalk or soft limestone	600
Granular soils	
Dense gravel or sand/gravel	>600
Medium-dense gravel or sand/gravel	200–600
Loose gravel or sand/gravel	<200
Dense sand	>300
Medium-dense sand	100–300
Loose sand	<100
Cohesive soils	
Very stiff clays	300–600
Stiff clays	150–300
Firm clays	75–150
Soft clays and silts	<75

After BS8004:1986 (British Standards Institution 1986) and Canadian Geotechnical Society (1992).

or geologic conditions, shear strength parameters, or the foundation dimensions. Some typical values are given in Table 3.3.

3.4 Pressure Distribution beneath Eccentrically Loaded Footings

The pressure distribution beneath a *flexible* footing often is assumed to be uniform if the load is applied at the center. This is not the case when the load is applied with some eccentricity in one or both directions. Eccentricity can be introduced through moments and/or lateral loads such as wind loads. It can reduce the ultimate bearing capacity, and with the reduced effective area, the allowable load on the footing is reduced even further.

In a strip footing, when the line load is applied with an eccentricity of e, as shown in Figure 3.11a, the soil pressure at any point beneath the footing is given by

$$q(x) = \frac{Q}{B}\left(1 + \frac{12ex}{B^2}\right) \tag{3.69}$$

where x is the horizontal distance from the centerline. The maximum and minimum values of the soil pressure, which occur at the two edges of the strip footing, at $x = 0.5B$ and $x = -0.5B$, respectively, are given by:

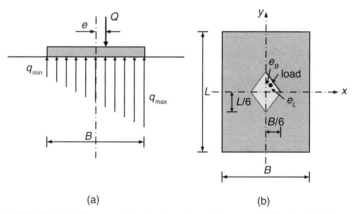

FIGURE 3.11 Pressure distribution beneath eccentrically loaded footings: (a) strip footing with one-way eccentricity and (b) rectangular footing with two-way eccentricity.

$$q_{max} = \frac{Q}{B}\left(1 + \frac{6e}{B}\right) \tag{3.70}$$

$$q_{min} = \frac{Q}{B}\left(1 - \frac{6e}{B}\right) \tag{3.71}$$

It can be seen from Equation 3.71 that the soil pressure beneath the footing will be compressive at all points provided $e < B/6$. Since there cannot be tensile normal stress between the foundation and the soil, when e exceeds $B/6$, one edge of the footing will lift off the ground, reducing the contact area, resulting in redistribution of the contact pressure. It is therefore desirable to limit the eccentricity to a maximum of $B/6$.

In a rectangular footing with eccentricities of e_B and e_L in the direction of breadth and length, respectively, the contact pressure at any point beneath the footing is given by:

$$q(x, y) = \frac{Q}{BL}\left(1 + \frac{12e_B}{B^2}x + \frac{12e_L}{L^2}y\right) \tag{3.72}$$

Here, the origin is at the center of the footing and the x- and y-axes are in the direction of breadth and length, respectively (see Figure 3.11b). The lightly shaded area at the center of Figure 3.11b, a rhombus, is known as the kern. Provided the foundation load acts within this area, the contact stresses are compressive at all points beneath the footing.

3.5 Settlement of Shallow Foundations in Cohesive Soils

When foundations are subjected to vertical loads, there will be settlement. Depending on whether the underlying soils are cohesive or granular, the settlement pattern can be quite different. In saturated cohesive soils, the settlements consist of three components: *immediate settlement* (s_i), *consolidation settlement* (s_c), and *secondary compression* (s_s). Immediate settlement occurs immediately after the load is applied and is instantaneous. Consolidation settlement occurs due to the expulsion of water from the soil and dissipation of excess pore water pressure. This can take place over a period of several years. Secondary compression settlement, also known as creep, occurs after the consolidation is completed. Therefore, there will be no *excess* pore water pressure during the secondary compression stage.

3.5.1 Immediate Settlement

Immediate settlement, also known as distortion settlement, initial settlement, or elastic settlement, occurs immediately upon the application of the load, due to lateral distortion of the soil beneath the footing. In clays, where drainage is poor, it is reasonable to assume that immediate settlement takes place under undrained conditions where there is no volume change (i.e., $v = 0.5$). The average immediate settlement under a flexible footing generally is estimated using the theory of elasticity, using the following equation, originally proposed by Janbu et al. (1956):

$$s_i = \frac{qB}{E_u} \mu_0 \mu_1 \tag{3.73}$$

The values of μ_1 and μ_2, originally suggested by Janbu et al. (1956), were modified later by Christian and Carrier (1978), based on the work by Burland (1970) and Giroud (1972). The values of μ_0 and μ_1, assuming $v = 0.5$, are given in Figure 3.12. Obtaining a reliable estimate of the undrained Young's modulus (E_u) of clays through laboratory or *in situ* tests is quite difficult. It can be estimated using Figure 3.13, proposed by Duncan and Buchignani (1976) and the U.S. Army (1994). E_u/c_u can vary from 100 for very soft clays to 1500 for very stiff clays. Typical values of the elastic modulus for different types of clays are given in Table 3.4. Immediate settlement generally is a small fraction of the total settlement, and therefore a rough estimate often is adequate.

TABLE 3.4 Typical Values of Elastic Modulus for Clays

Clay	E (MPa)
Very soft clay	0.5–5
Soft clay	5–20
Medium clay	20–50
Stiff clay, silty clay	50–100
Sandy clay	25–200
Clay shale	100–200

After U.S. Army (1994).

3.5.2 Consolidation Settlement

Consolidation is a time-dependent process in saturated clays, where the foundation load is gradually transferred from the pore water to the soil skeleton. Immediately after loading, the entire applied normal stress is carried by the water in the voids, in the form of excess pore water pressure. With time, the pore water drains out into the more porous granular soils at the boundaries, thus dissipating the excess pore water pressure and increasing the effective stresses. Depending on the thickness of the clay layer, and its consolidation characteristics, this process can take from a few days to several years.

Consolidation settlement generally is computed assuming one-dimensional consolidation, and then a correction factor is applied for three-dimensional effects (Skempton and Bjerrum 1957). In one-dimensional consolidation, the normal strains and drainage are assumed to take place only in the vertical direction. This situation arises when the applied pressure at the ground level is uniform and is of a very large lateral extent, as shown in Figure 3.14.

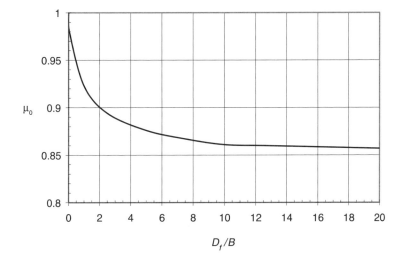

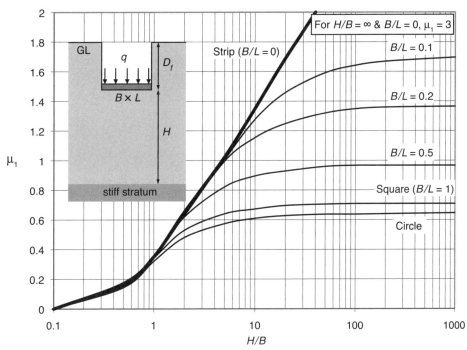

FIGURE 3.12 Values of μ_0 (top) and μ_1 (bottom) for immediate settlement computation (after Christian and Carrier 1978).

In a clay layer with an initial thickness of H and a void ratio of e_0, the final consolidation settlement s_c due to the applied pressure q can be estimated from

$$s_c = \frac{\Delta e}{1 + e_0} H \qquad (3.74)$$

where Δe is the change in the void ratio due to the applied pressure q. H and e_0 can be obtained from the soil data, and Δe has to be computed as follows.

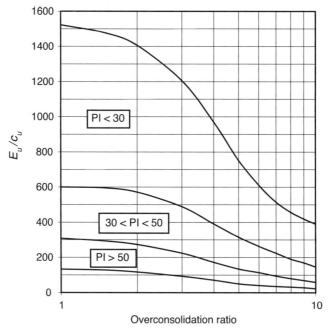

FIGURE 3.13 E_u/c_u values (after Duncan and Buchignani 1976; U.S. Army 1994).

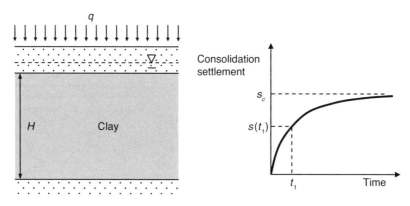

FIGURE 3.14 One-dimensional consolidation settlement within a clay layer.

Three different cases, as shown in Figure 3.15, are discussed here. Point I corresponds to the initial state of the clay, where the void ratio and the vertical stress are e_0 and σ'_{vo}, respectively. With the vertical stress increase of $\Delta\sigma_v$, consolidation takes place, and the void ratio decreases by Δe. Point F corresponds to the final state, at the end of consolidation. Point P corresponds to the preconsolidation pressure (σ'_p) on the virgin consolidation line.

   Case I. If the clay is normally consolidated, Δe can be computed from:

$$\Delta e = C_c \log \left(\frac{\sigma'_{vo} + \Delta\sigma_v}{\sigma'_{vo}} \right) \qquad (3.75)$$

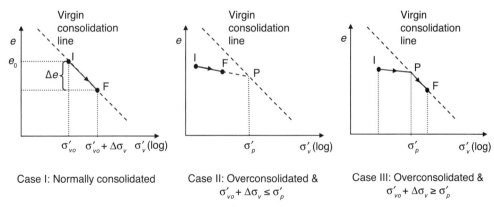

FIGURE 3.15 Δe calculations from e vs. log σ'_v plot.

Case II. If the clay is overconsolidated and $\sigma'_{vo} + \Delta\sigma_v \leq \sigma'_p$ (i.e., the clay remains overconsolidated at the end of consolidation), Δe can be computed from:

$$\Delta e = C_r \log \left(\frac{\sigma'_{vo} + \Delta\sigma_v}{\sigma'_{vo}} \right) \tag{3.76}$$

Case III. If the clay is overconsolidated and $\sigma'_{vo} + \Delta\sigma_v \geq \sigma'_p$ (i.e., the clay becomes normally consolidated at the end of consolidation), Δe can be computed from:

$$\Delta e = C_r \log \left(\frac{\sigma'_p}{\sigma'_{vo}} \right) + C_c \log \left(\frac{\sigma'_{vo} + \Delta\sigma_v}{\sigma'_p} \right) \tag{3.77}$$

In one-dimensional consolidation, assuming the pressure at the ground level is applied over a large lateral extent, $\Delta\sigma_v = q$ at any depth. In the case of footings where the loading is not one-dimensional, $\Delta\sigma_v$ can be significantly less than the footing pressure q and can be estimated using the methods discussed in Section 3.2.

Another but less desirable method to compute the consolidation settlement is to use the coefficient of volume compressibility (m_v). The final consolidation settlement can be written as:

$$s_c = m_v q H \tag{3.78}$$

The main problem with this apparently simple method is that m_v is stress dependent, and therefore a value appropriate to the stress level must be used. The consolidation settlement $s(t_1)$ at a specific time t_1 can be determined from the U_{avg}-T plot in Figure 1.17.

3.5.3 Secondary Compression Settlement

Secondary compression settlement takes place at constant effective stress, when there is no more dissipation of excess pore water pressure. For simplicity, it is assumed to start occurring

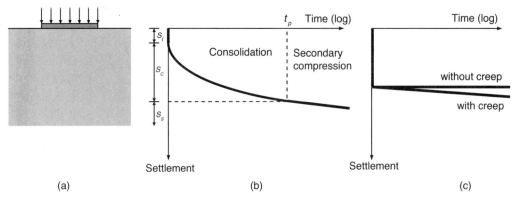

FIGURE 3.16 Settlement in soils: (a) footing under pressure, (b) settlement in cohesive soils, and (c) settlement in granular soils.

when the primary consolidation is completed at time t_p (see Figure 3.16), and the settlement increases linearly with the logarithm of time. Secondary compression settlement can be estimated using the following equation:

$$s_s = C_a \left(\frac{H}{1 + e_p} \right) \log \left(\frac{t}{t_p} \right) \qquad \text{for } t > t_p \qquad (3.79)$$

Here, e_p is the void ratio at the end of primary consolidation and C_α is the coefficient of secondary compression or the secondary compression index, which can be determined from a consolidation test or estimated empirically. Assuming that the void ratio decreases linearly with the logarithm of time, C_α is defined as:

$$C_\alpha = \frac{\Delta e}{\Delta \log t} \qquad (3.80)$$

Mesri and Godlewski (1977) reported that C_α/C_c is a constant for a specific soil and suggested typical values. In the absence of consolidation test data, C_α can be assumed to be 0.03–0.08 times C_c. While the upper end of the range applies to organic and highly plastic clays, the lower end of the range is suitable for inorganic clays. Secondary compression settlement can be quite significant in organic clays, especially in peat.

3.6 Settlement of Shallow Foundations in Granular Soils

Settlement of footings in granular soils is instantaneous, with some possibility for long-term creep. There are more than 40 different settlement prediction methods, but the quality of the predictions is still very poor, as demonstrated at the Settlement 94 settlement prediction symposium in Texas in 1994 (Briaud and Gibbens 1994).

The five most important factors that govern the settlement of a footing are the applied pressure, soil stiffness, footing breadth, footing depth, and footing shape. Soil stiffness often is quantified indirectly through penetration resistance such as the N-value or blow count from a standard penetration test or through tip resistance q_c from a cone penetration test. Das and Sivakugan (2007) summarized the empirical correlations relating soil stiffness to penetration resistance.

3.6.1 Terzaghi and Peck Method

Terzaghi and Peck (1967) proposed the first rational method for predicting settlement of a shallow foundation in granular soils. They related the settlement of a square footing of width B (in meters) to that of a 300-mm square plate, obtained from a plate loading test, through the following expression:

$$\delta_{\text{footing}} = \delta_{\text{plate}} \left(\frac{2B}{B + 0.3} \right)^2 \left(1 - \frac{1}{4} \frac{D_f}{B} \right) \tag{3.81}$$

The last term in Equation 3.81 accounts for the reduction in settlement with the increase in footing depth. Leonards (1986) suggested replacing ¼ by ⅓, based on additional load test data. The values of δ_{plate} can be obtained from Figure 3.17, which summarizes the plate loading test data given by Terzaghi and Peck (1967). This method originally was proposed for square footings, but can be applied to rectangular and strip footings with caution. The deeper influence zone and increase in the stresses within the soil mass in the case of rectangular or strip footings are compensated for by the increase in the soil stiffness.

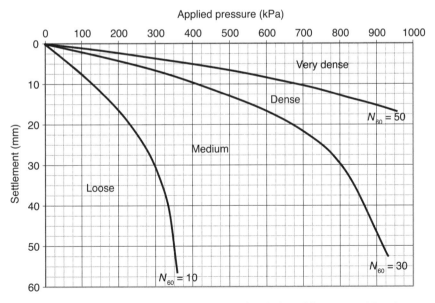

FIGURE 3.17 Settlement of 300-mm × 300-mm plate (adapted from Terzaghi et al. 1996; load test data from late Professor G.A. Leonards).

3.6.2 Schmertmann et al. Method

Based on the theory of elasticity, Schmertmann (1970) proposed that the vertical normal strain (ε_z) at a depth z below the footing is given by

$$\varepsilon_z = \frac{q}{E_z} I_z \tag{3.82}$$

where E_z and I_z are Young's modulus and the strain influence factor, respectively, at depth z. Based on some finite element studies and load tests on model footings, Schmertmann proposed the influence factor as shown in Figure 3.18a, which is known as the $2B$-0.6 distribution. The influence factor increases linearly from 0 at the footing level to 0.6 at a depth of $0.5B$ below the footing and then decreases linearly to 0 at a depth of $2B$ below the footing. Integrating the above equation and dividing the granular soil beneath the footing into sublayers of constant Young's modulus, the vertical settlement can be expressed as

$$s = q_{net} C_1 C_2 \sum_{z=0}^{z=2B} \frac{I_z dz}{E_z} \tag{3.83}$$

where C_1 and C_2 are the correction factors to account for the embedment and strain relief due to the removal of overburden and the time dependence of settlement, respectively, and q_{net} is the net applied pressure at the footing level. C_1 and C_2 are given by

$$C_1 = 1 - 0.5 \left(\frac{\sigma'_{vo}}{q_{net}} \right) \geq 0.5 \tag{3.84}$$

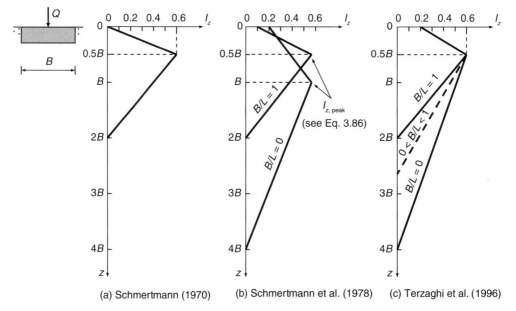

(a) Schmertmann (1970)　　(b) Schmertmann et al. (1978)　　(c) Terzaghi et al. (1996)

FIGURE 3.18 Schmertmann et al.'s influence factors.

$$C_2 = 1 + 0.2 \log \left(\frac{t}{0.1} \right) \qquad (3.85)$$

where σ'_{vo} is the effective *in situ* overburden stress at the footing level, and t is the time since loading (in years). Leonards (1986), Holtz (1991), and Terzaghi et al. (1996) suggest that C_2 = 1, disregarding the time-dependent settlements in granular soils. They suggest that the time-dependent settlements in the footings studied by Schmertmann probably are due to the thin layers of clays and silts interbedded within the sands in Florida, from where most of Schmertmann's load test data come. Schmertmann (1970) recommended that Young's modulus be derived from the static cone resistance as $E = 2q_c$. Leonards (1986) suggested that E $(kg/cm^2) = 8N_{60}$ for normally consolidated sands, where N_{60} is the blow count from a standard penetration test, not corrected for overburden (1 kg/cm^2 = 98.1 kPa).

Schmertmann's (1970) original method does not take the footing shape into account. Realizing the need to account for the footing shape, Schmertmann et al. (1978) made some modifications to the original method. The modified influence factor diagram is shown in Figure 3.18b, where the strain influence factor extends to a depth of 2B for square footings and 4B for strip footings, peaking at depths of 0.5B and B, respectively. The peak value of the influence factor is given by

$$I_{z,\text{peak}} = 0.5 + 0.1 \sqrt{\frac{q_{\text{net}}}{\sigma'_{vo}}} \qquad (3.86)$$

where σ'_{vo} is the original overburden pressure at a depth of 0.5B below the footing for square footings and B below the footing for strip footings, where the peak values occur. The equations for computing the settlement and the correction factors remain the same. Schmertmann et al. (1978) suggested that $E = 2.5q_c$ for axisymmetric loading and $E = 3.5q_c$ for plane strain loading, based on the observation by Lee (1970) that Young's modulus is about 40% greater for plane strain loading compared to axisymmetric loading. They suggested that for rectangular footings, the settlement be calculated separately for $B/L = 0$ and 1 and interpolated on the basis of B/L.

Terzaghi et al. (1996) suggested a simpler influence factor diagram, shown in Figure 3.18c, with the influence factors starting and peaking at the same points but extending to depths of 2B and 4B for square and strip footings. For rectangular footings, they suggested an interpolation function to estimate the depth of influence z_I (see Figure 3.18c) as:

$$z_I = 2B \left(1 + \log \frac{L}{B} \right) \qquad (3.87)$$

Terzaghi et al. (1996) suggest taking $E = 3.5q_c$ for axisymmetric loading and increasing it by 40% for plane strain loading and suggest the following expression for E of a rectangular footing:

$$E_{\text{rectangular footing}} = 3.5 \left(1 + 0.4 \log \frac{L}{B} \right) q_c \qquad (3.88)$$

These modifications give more realistic and less conservative estimates of settlements. Nevertheless, the above values of $E = (3.5\text{–}4.9)q_c$ are significantly larger than what is recommended in the literature.

3.6.3 Burland and Burbidge Method

Burland et al. (1977) collated more than 200 settlement records of shallow foundations of buildings, tanks, and embankments on granular soils and plotted settlement per unit pressure against the footing breadth, as shown in Figure 3.19, defining the upper limits for possible settlement that can be expected. It is a good practice to use this figure to check whether the settlement predicted by a specific method falls within the bounds. They suggested that the "probable" settlement is about 50% of the upper limit shown in the figure and that in most cases the maximum settlement will be unlikely to exceed 75% of the upper limit.

Burland and Burbidge (1985) reviewed the above settlement records and proposed an indirect and empirical method for estimating settlement of shallow foundations in granular soils, based on N-values from standard penetration tests that are not corrected for overburden pressure. The influence depth (z_I) was defined as

$$z_I = B^{0.7} \tag{3.89}$$

where z_I and B are in meters. They expressed the compressibility of the soil by a compressibility index (I_c), which is similar to the coefficient of volume compressibility (m_v) used in consolidation of saturated clays. For normally consolidated granular soils, I_c was related to the average blow count within the influence depth $\overline{N}_{60}$ by

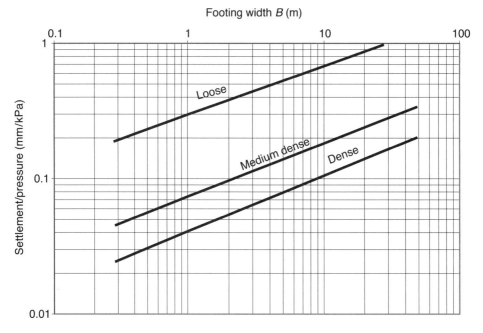

FIGURE 3.19 Upper limits of settlement per unit pressure (after Burland et al. 1977).

$$I_c = \frac{1.71}{N_{60}^{1.4}} \tag{3.90}$$

where I_c is in MPa^{-1}. For overconsolidated granular soils, I_c is one-third of what is given in Equation 3.90. Burland and Burbidge (1985) suggested that the settlement can be estimated from:

$$s = qI_c z_I \tag{3.91}$$

It should be noted that Equation 3.91 is similar in form to Equation 1.36, which is used for estimating consolidation settlement in clays. In normally consolidated granular soils, Equation 3.91 becomes:

$$s = q \frac{1.71}{N_{60}^{1.4}} B^{0.7} \tag{3.92}$$

In overconsolidated granular soils, if the preconsolidation pressure (σ'_p) can be estimated, Equation 3.91 becomes:

$$s = \frac{1}{3} q \frac{1.71}{N_{60}^{1.4}} B^{0.7} \quad \text{if } q \le \sigma'_p \tag{3.93}$$

$$s = \left(q - \frac{2}{3} \sigma'_p \right) \frac{1.71}{N_{60}^{1.4}} B^{0.7} \quad \text{if } q \ge \sigma'_p \tag{3.94}$$

For fine sands and silty sands below the water table, where $N_{60} > 15$, driving of the split-spoon sampler can dilate the sands, which can produce negative pore water pressure that would increase the effective stresses and hence overestimate the blow count. Here, Terzaghi's correction given below should be applied:

$$N_{60,\text{corrected}} = 15 + 0.5(N_{60} - 15) \tag{3.95}$$

In gravel or sandy gravel, N should be increased by 25% using Equation 3.96:

$$N_{60,\text{corrected}} = 1.25N_{60} \tag{3.96}$$

The settlements estimated above apply to square footings. For rectangular or strip footings, settlement has to be multiplied by the following factor (f_s):

$$f_s = \left(\frac{1.25L/B}{0.25 + L/B} \right)^2 \tag{3.97}$$

The settlement estimated above implies that there is granular soil to a depth of at least z_I. If the thickness (H_s) of the granular layer below the footing is less than the influence depth, the settlement has to be multiplied by the following reduction factor (f_l):

$$f_l = \frac{H_s}{z_I}\left(2 - \frac{H_s}{z_I}\right) \tag{3.98}$$

Burland and Burbidge (1985) noted some time-dependent settlement of footings and suggested a multiplication factor (f_t) given by

$$f_t = 1 + R_3 + R_t \log\frac{t}{3} \tag{3.99}$$

where R_3 takes into consideration the time-dependent settlement during the first three years of loading, and the last component accounts for the time-dependent settlement that takes place after the first three years at a slower rate. Suggested values for R_3 and R_t are 0.3–0.7 and 0.2–0.8, respectively. The lower end of the range is applicable for static loads and the upper end for fluctuating loads such as bridges, silos, and tall chimneys.

3.6.4 Accuracy and Reliability of the Settlement Estimates and Allowable Pressures

Das and Sivakugan (2007) reviewed the different settlement prediction methods and discussed the current state-of-the-art. The three methods discussed above in detail are the most popular for estimating settlement of shallow foundations in granular soils. It is well known that these methods overestimate settlement in general and thus are conservative. Sivakugan et al. (1998) studied 79 settlement records where the footing width was less than 6 m and concluded that the settlement predictions by Terzaghi and Peck (1967) and Schmertmann (1970) overestimate settlement by about 2.18 and 3.39 times, respectively.

Tan and Duncan (1991) introduced two parameters—accuracy and reliability—to quantify the quality of settlement predictions and applied these to 12 different methods using a large database of settlement records. *Accuracy* was defined as the average ratio of the predicted settlement to the measured settlement. *Reliability* is the probability that the predicted settlement is greater than the measured settlement. Therefore, an ideal settlement prediction method will have an accuracy of 1 and reliability approaching 100%. There often is a trade-off between accuracy and reliability. The Terzaghi and Peck (1967) method has high reliability but poor accuracy, which shows that the estimates are conservative. On the other hand, the Burland and Burbidge (1985) method has good accuracy but poor reliability, which shows that the predictions are more realistic, but it does not always overestimate like the Schmertmann et al. (1978) method or Terzaghi and Peck (1967) method and is less conservative.

It is widely documented in the literature that the design of shallow foundations in granular soils is almost always governed more by settlement considerations than bearing capacity. Therefore, more care is required in the settlement computations. The allowable bearing capacity values, on the basis of limiting settlement to 25 mm, estimated by the Burland and Burbidge (1985) and Terzaghi and Peck (1967) methods are shown in Figure 3.20. The Burland and Burbidge (1985) charts were developed for square footings on normally consolidated sands, with no consideration given to time-dependent settlement. If the sand is overconsolidated with a preconsolidation pressure of σ_p' and $q < \sigma_p'$, the allowable pressure from Figure 3.20 should be multiplied by 3. If the sand is overconsolidated and $q > \sigma_p'$, add $0.67\sigma_p'$ to the value obtained from Figure 3.20. For any other value of limiting settlement, the

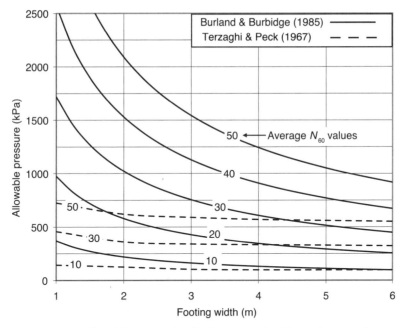

FIGURE 3.20 Allowable pressure for footings on sands with maximum settlement of 25 mm.

allowable pressure from Figure 3.20 must be adjusted proportionally. To limit the probability that the actual settlement will exceed 25 mm, it may be necessary to limit the maximum settlement to 16 mm and reduce the allowable soil pressure proportionally (Terzaghi et al. 1996). It can be seen in Figure 3.20 that the Burland and Burbidge (1985) method gives significantly smaller settlements and higher allowable pressures compared to the more conservative Terzaghi and Peck (1967) method.

Meyerhof (1956, 1974) suggested an expression for allowable pressure that would limit settlement to 25 mm, which again underestimates the allowable pressure significantly. Bowles (1996) suggested increasing this value by 50%, whereby the modified Meyerhof equation becomes

$$q_{\text{allowable}} = 1.25 N_{60} \left(\frac{B + 0.3}{B} \right)^2 \left(1 + \frac{1}{3} \frac{D_f}{B} \right) \tag{3.100}$$

where $D_f/B \leq 1$ and $B > 1.2$ m. Equation 3.100 gives a slightly higher allowable pressure than the Terzaghi and Peck (1967) values, but significantly less than the Burland and Burbidge (1985) values shown in Figure 3.20.

3.6.5 Probabilistic Approach

The magnitude of settlement can have a different meaning depending on which method was used in the settlement computations. Sivakugan and Johnson (2004) proposed a probabilistic design chart, based on several settlement records reported in the literature, to quantify the probability that the settlement predicted by a certain method will exceed a specific limiting value in the field. Three separate charts for the Terzaghi and Peck, Schmertmann et al., and Burland and Burbidge methods are given in Figure 3.21. For example, if the settlement predicted by the Schmertmann et al. method is 20 mm, the probability that the actual settlement will exceed 25 mm is 0.2.

3.7 Raft Foundations

A raft foundation, also known as a mat foundation, is a large, thick concrete slab that supports all or some of the columns and/or walls of a structure. A raft also can support an entire structure, such as a silo, storage tank, chimney, tower, and foundation machinery. A hollow raft can reduce the heavy self-weight of a large slab and yet provide enough structural stiffness. A widely accepted practical criterion is to use a raft when more than 50% of the building plan projection is covered by footings. Other purposes of rafts include increasing the foundation area and thus increasing the foundation bearing capacity whenever possible, bridging over small compressible pockets to minimize differential settlements, resisting hydrostatic uplift, facilitating basement waterproofing, and redistributing horizontal soil and water thrust through the structure or support peripheral columns and walls. Reduced raft thicknesses can be achieved efficiently by introducing structural stiffeners such as plates thickened under columns, a two-way beam and slab, and a flat plate with pedestals or basement walls incorporated into the raft (Teng 1975; Bowles 1996). Thinner slabs also can be designed to resist high uplift pressures by introducing vertical prestressed anchors or tension piles (Danziger et al. 2006). Compared to footings, a raft spreads the structural load over a larger area in the soil and reduces the bearing pressure. Because of the high stiffness of the thick concrete slab, a raft can reduce differential settlement. Differential settlement also can be minimized by simultaneously taking into account the slab stiffness and the stiffness of the superstructure.

3.7.1 Structural Design Methods for Rafts

The methods for raft foundation design are classified as *rigid* and *flexible*. The rigid method (also known as the conventional method) is still widely used in practice because of its simplicity. It is also used to check or validate results obtained by more sophisticated flexible methods.

The *rigid method* assumes that a thick slab is infinitely rigid when compared to the soil, and hence the flexural deflections are negligible and do not influence the contact pressure, which is assumed to vary linearly as a result of simultaneous rigid body translation and rotation of the raft. Closed-form solutions to estimate the contact pressure underneath a rigid eccentrically loaded circular raft can be found elsewhere (e.g., Teng 1975). For a rigid rectangular raft with area $B \times L$, the contact pressure q at any point, with coordinates x and y with respect to a Cartesian coordinate system passing through the centroid of the raft area (see Figure 3.22), is given by

$$q(x, y) = \frac{Q_t}{B \times L} \pm \frac{M_x}{I_x} y \pm \frac{M_y}{I_y} x \qquad (3.101)$$

where

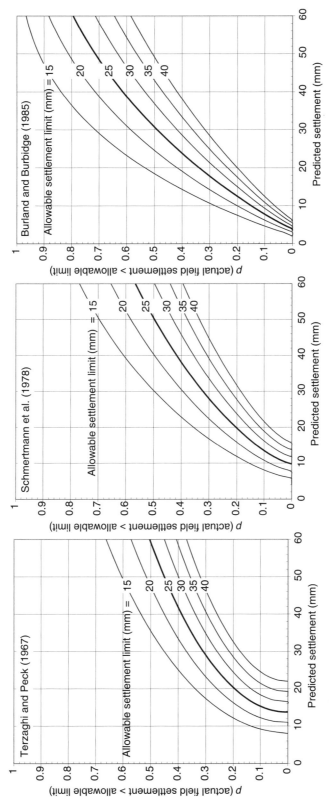

FIGURE 3.21 Probabilistic design charts (after Sivakugan and Johnson 2004).

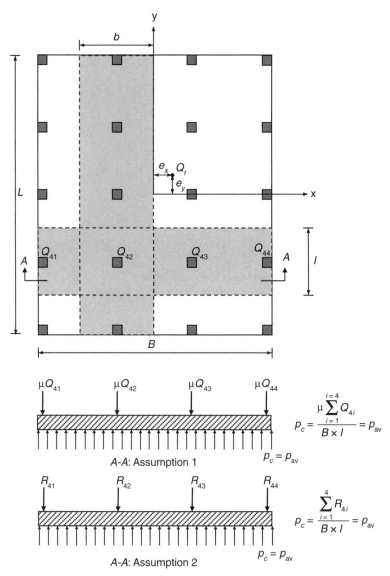

FIGURE 3.22 Assumptions for the rigid method of raft design.

$$Q_t = \sum_{i=1}^{n} Q_i = \text{total load on the raft (sum of all column loads)}$$

$$M_x = Q_t e_y = \text{moment of the column loads about the } x\text{-axis}$$

$$M_y = Q_t e_x = \text{moment of the column loads about the } y\text{-axis}$$

$$e_x, e_y = \text{eccentricities about the } x\text{- and } y\text{-axes, respectively}$$

$$I_x = BL^3/12 = \text{moment of inertia about the } x\text{-axis}$$

$$I_y = BL^3/12 = \text{moment of inertia about the } y\text{-axis}$$

The contact pressure distribution given by Equation 3.101 is used to estimate raft settlement, bearing capacity, bending moment, and shear forces. Static equilibrium in the vertical direction causes the resultant of column loads Q_t to be equal and opposite to the resultant load obtained from integration of the reactive contact pressure in Equation 3.101. For simplicity, the rigid method assumes that the raft is analyzed by tributary areas in each of two perpendicular directions, similar to the structural design of two-way flat slabs, as shown by the shaded areas in Figure 3.22. To calculate bending moments and shear forces, each of two perpendicular bands is assumed to represent independent continuous beams under a constant average upward pressure q_{av} estimated by Equation 3.101. This simplification violates equilibrium because bending moments and shear forces at the common edge between adjacent bands are neglected. Therefore, the contact pressure q_c obtained by dividing the sum of the column loads in each band by the total area of the band is not equal to q_{av} computed from Equation 3.101. Hence, one of the following two assumptions is made in practice to estimate bending moments and shear forces in the assumed independent beams shown in Figure 3.22: (1) the actual column loads Q_{ij} are multiplied by an adjustment factor μ to make the contact pressure q_c equal to q_{av} (Das 1984) or (2) the columns are assumed to be rigid supports whose reactions R_{ij} are calculated assuming uniform contact pressure q_{av}. The above simplifications produce adjusted loads (μQ_{ij} with assumption 1) or rigid support reactions (R_{ij} with assumption 2) that are not equal to the corresponding column loads. This error is tolerated in practice provided the raft is regarded as rigid, which satisfies the following requirements (American Concrete Institute 1988):

1. The column loads and column spaces do not vary from each other by more than 20%.
2. The spacing (l) between column loads is such that

$$b \leq \frac{1.75}{\sqrt[4]{\dfrac{k_s l}{4 E_f I_f}}} \tag{3.102}$$

where l is the width of the band, E_f is the modulus of elasticity of the foundation material, I_f is the moment of inertia of the cross section of the equivalent continuous beam, and k_s is the coefficient of the subgrade reaction defined as

$$k_s = \frac{q}{\delta} \tag{3.103}$$

where δ is the settlement produced by a gross bearing pressure q. k_s is measured in pressure per unit of length, sometimes referred as force per cubic length, which should not be confused with the soil unit weight. If the above requirements are not met, the raft should be designed as a flexible raft. In addition to calculation of bending moments and shear forces, the punching shear under each column also must be checked.

Flexible methods are based on analytical linear elastic solutions (Milović 1992; Hemsley 1998) and numerical solutions such as the method of finite differences and method of finite elements, where the stiffness of both the soil and structural members can be taken into account. Early *flexible numerical methods* were based on the numerical solution of the fourth-order differential equation governing the flexural behavior of a plate by the method of finite differences. The raft is treated as a linear elastic structural element whose soil reaction is replaced by an infinite number of independent linear elastic springs, following the Winkler hypothesis. The soil elastic constant is given by the coefficient of the subgrade reaction k_s defined by Equation 3.103.

Let's consider an infinitely long beam of width b (m) and thickness h (m) resting on the ground and subjected to a point load where the soil reaction is $q^\star$ (kN/m) at distance x from the origin. From the principles of engineering mechanics, it can be shown that the bending moment at distance x is

$$M = E_{raft} I_{raft} \frac{d^2 z}{dx^2} \qquad (3.104)$$

The shear force at distance x is

$$V = \frac{dM}{dx} = E_{raft} I_{raft} \frac{d^3 z}{dx^3} \qquad (3.105)$$

The soil reaction at distance x is

$$q^\star = \frac{dV}{dx} = E_{raft} I_{raft} \frac{d^4 z}{dx^4} = -zk' \qquad (3.106)$$

Here, I_{raft} is the moment of inertia of the cross section of the beam about the bending axis, given by $bh^3/12$, and k' is the subgrade reaction of the Winkler beam (in kN/m^2), which is related to k_s by:

$$k' = k_s b \qquad (3.107)$$

Therefore, Equation 3.106 becomes:

$$E_{raft} I_{raft} \frac{d^4 z}{dx^4} = -zk_s b \qquad (3.108)$$

Equation 3.108 can be solved with appropriate boundary conditions, and the deflections of the Winkler beam on elastic springs can be obtained.

Approximate methods to estimate k_s as a function of the soil and foundation material elastic constants can be found elsewhere (e.g., Teng 1975; Bowles 1996; Das 2007; Coduto 2001; Lopes 2000). A major limitation of the early flexible methods comes from the unrealistic estimates of the coefficient of the subgrade reaction. The difficulty in estimating k_s comes from the fact that it is not a fundamental soil property, and its magnitude depends on factors such as the (1) width of the loaded area, (2) shape of the loaded area, (3) depth of the loaded area, (4) location on the raft for which settlement is being considered, and (5) time. Thus, considerable judgment and personal experience are required to select appropriate k_s values for design purposes. A practical way to estimate k_s roughly is to calculate the settlement δ by any method outlined in Sections 3.5 and 3.6 and back calculate k_s by Equation 3.103.

The modulus of the subgrade reaction can be obtained from a plate load test, typically using a 300-mm square plate. Typical values of $k_{0.3}$ (from a 300-mm plate) are given in Table 3.5. In granular soils, the value of k_s for a $B \times L$ rectangular footing is given by:

TABLE 3.5 Typical Values of Modulus of Subgrade Reaction
($k_{0.3}$) of a 300-mm Plate

Soil	$k_{0.3}$ (MN/m³)		
Granular soils	Loose	Medium	Dense
Dry or moist	10–25	25–125	125–250
Saturated	10–15	25–40	125–150
Cohesive soils	Stiff	Very stiff	Hard
	10–20	20–40	40+

$$k_s = k_{0.3} \left(\frac{B + 0.3}{2B} \right)^2 \left(\frac{1 + 0.5B/L}{1.5} \right) \tag{3.109}$$

In cohesive soils:

$$k_s = k_{0.3} \left(\frac{0.3}{B} \right) \left(\frac{1 + 0.5B/L}{1.5} \right) \tag{3.110}$$

Vesic (1961) suggested that k_s can be estimated from Equation 111 by:

$$k_s = 0.65 \frac{E_s}{B(1 - v^2)} \sqrt{\frac{E_s B^4}{E_F I_F}} \tag{3.111}$$

Here, E_s = Young's modulus of the soil, E_F = Young's modulus of the footing, I_F = moment of inertia of the foundation's cross section, and v = Poisson's ratio of the soil.

Flexible methods based on the coefficient of the subgrade reaction generally are not suitable for reliable estimates of total settlement, although such methods may provide acceptable estimates of differential settlement. Since bending moments and shear forces generally are not very sensitive to variations in k_s, flexible methods are still widely used for structural raft design.

Due to the increasing popularity of efficient, user-friendly computational codes for geotechnical design based on finite elements and advanced versions of the finite difference method, it was possible to overcome some of the limitations of the rigid method and early flexible methods, such as equilibrium and compatibility requirements, irregular soil layers, nonlinear inelastic soil response, three-dimensional modeling, soil-structure interaction, coupled flow-deformation analyses, and dynamic loading. Three-dimensional numerical modeling of a soil-structure interaction problem is still a time-consuming task, and hence numerical modeling commonly is limited to simpler *two-dimensional* analyses whenever feasible. The finite element method and advanced versions of the finite difference method allow different possibilities for modeling the soil stiffness, including linear elastic analysis, hyperbolic models, and elastic-plastic and viscous-plastic models (e.g., Duncan and Chang 1970; Chen and Saleeb 1982; Desai and Siriwardena 1984). Although more refined elastic-plastic and viscous-plastic models represent a better idealization of the soil response, simpler models such as linear elastic analyses limited by a simple failure criterion (e.g., Mohr-Coulomb), hyperbolic models, and the Cam-Clay elastic-plastic model probably are the most widely used in practice due to their simplicity, ease in estimating input parameters, computational speed, and numerical stability.

3.7.2 Bearing Capacity and Settlement of Rafts

Raft bearing capacity and settlement calculations generally follow the same conventional methods outlined previously for pad footings. Although numerical analyses are gaining increased acceptance, conventional methods still are widely used in practice, either in preliminary estimates or to cross-check numerical calculations. Numerical methods rely on input soil constants estimated mostly by laboratory tests. Apart from the differences between field and laboratory soil responses, actual stress path dependency in the field may not be adequately reproduced by simple paths simulated in the laboratory. This is especially true for the simpler numerical models more commonly used in practice, as they rely on a reduced number of soil constants, which may not thoroughly simulate the complex soil behavior. Hence, some adjustments in the input soil constants may be needed in numerical analyses, and conventional methods may be very helpful in validating numerical results.

An increased raft area (and a corresponding decreased contact pressure) generally can increase the raft bearing capacity. However, increasing the raft area may not prevent bearing capacity failure in cases of low-strength soils underneath a raft, including cases where the raft bears on firm deposits underlain by soft sediments. For clays under undrained conditions (ϕ_u = 0), N_γ = 0, and hence the ultimate bearing capacity becomes independent of the foundation width B. Increasing the raft area also may be ineffective in reducing total settlement because a larger foundation width also will encompass a deeper volume of the deformable soil mass. Therefore, increasing the raft area to reduce total settlement or to increase bearing capacity in weak soils may be costly and inefficient. Total settlement can be minimized by a larger foundation area in cases where the compressible soil stratum is at a relatively shallow depth (hence part of the contact pressure can be distributed to a more resistant soil layer beneath the weak soil) or when the soil stiffness increases significantly with depth.

An efficient way to increase bearing capacity and decrease total settlement is to design *floating* (or *compensated*) *rafts* whereby the total weight of construction is compensated for by previous excavation of the same or a slightly higher total weight of soil and water (Golder 1975; Zeevaert 1983). This leads to a higher gross ultimate bearing capacity as a result of the deeper raft depth D_f and lower total settlement because the soil elements beneath the raft become overconsolidated. Fully compensated floating rafts in normally consolidated deposits settle less than noncompensated or partially compensated rafts, where loading reaches the virgin compression and produces undesired long-term deformations.

Excavation unloading causes bottom heave that may progress to bottom failure. Zeevaert (1983) classifies bottom heave as P-heave (*plastic heave*, ultimately leading to bottom failure), E-heave (*elastic heave*, caused by nearly instantaneous elastic unloading and further upward relief by seepage pressure), S-heave (*swelling heave*, time-dependent upward displacement at constant total stress), and D-heave (*driving heave*, nearly instantaneous upward movement caused by the soil displaced by pile driving in the case of piled rafts).

Figure 3.23 shows the unloading path OBC of a fully unloaded normally consolidated soil element located right beneath the bottom of the excavation. Excavation unloading causes an upward displacement corresponding to ABC, where AB is nearly instantaneous E-heave and BC is time-dependent S-heave. Time-dependent S-heave depends on the amount of time the excavation remains open, and hence construction reloading should proceed as quickly as

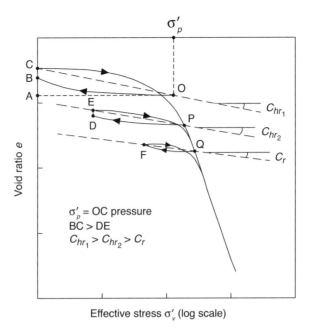

FIGURE 3.23 Estimation of E-heave and S-heave.

possible to minimize BC. For settlement calculations, S-heave should be measured soon after the bottom of the excavation is reached and added to settlements estimated by the recompression index C_r (FQ). For preliminary estimates of settlements produced by E- and S-heave recompression, the recompression index C_{hr1} (CO) may be selected for soil elements close to the excavation bottom. C_{hr1} depends on the duration the sample is exposed to relaxation until it is recompressed in the laboratory. Hence, C_{hr1} is higher than the recompression index C_r (FQ), selected for deeper soil elements not affected by S-heave. Intermediate values C_{hr2} (EP) between C_{hr1} and C_r may be selected at intermediate depths, where unloading may produce smaller S-heave (DE). D-heave is minimized in piled rafts by an alternating driving program, where a sufficient distance is allowed between piles during driving. Next, intermediate piles are driven between two previously driven piles only after a prescribed time has elapsed. D-heave also is minimized by driving from the center toward the edges of the raft. Piles can minimize E- and S-heave as long as they are driven from the surface, before excavation commences. This allows mobilization of negative skin friction along the pile right from the early stages of excavation. Upon construction reloading, rafts with full friction piles are efficient for controlling differential settlements and building tilting induced by eccentric loading, uneven primary and secondary consolidation settlements, neighboring construction, dewatering, and deformation of wall supports. Design criteria for piled rafts can be found elsewhere (e.g., Zeevaert 1983; Franke et al. 2000; Katzenbach et al. 2000; Poulos 2000).

Compensated nonpiled rafts are efficient for ensuring adequate bearing capacity and tolerable settlements only when not influenced by unforeseen features in the compressible strata and from adjacent existing or new buildings. Massad (2005) explains the excessive tilting of some buildings in the city of Santos, Brazil, as due to local overconsolidation produced by mobile sand dunes. Figure 3.24 shows possible types of damage caused by adjacent construction (Teixeira 2003). In Figure 3.24a, buildings A and B are constructed at approximately the

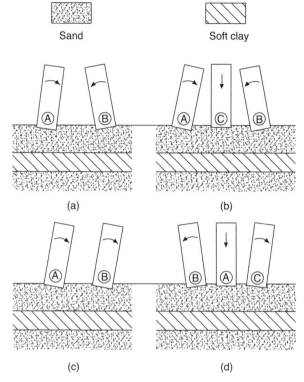

Sand

Soft clay

(a)

(b)

(c)

(d)

FIGURE 3.24   Influence of adjacent buildings (Teixeira 2003).

same time. In Figure 3.24b, buildings A and B are constructed at approximately the same time and building C shortly thereafter. In Figure 3.24c, building B is constructed long after building A. In Figure 3.24d, buildings B and C are constructed nearly simultaneously after existing building C. The interaction with adjacent construction always should be carefully investigated in foundation design. Excavation and subsequent building construction always must rely on rigorous and permanent monitoring of the new building and all adjacent buildings for mutual settlement and tilting.

3.8 Shallow Foundations under Tensile Loading

Tensioned foundations are common in civil engineering applications such as transmission towers, harbors, basement slabs under pressure, industrial equipment, etc. Procedures for the design of tensioned foundations are discussed in this section, including specific recommendations for the more common transmission tower foundations. Starting with a distinction between shallow and deep modes of failure, this section presents the most common failure mechanisms for shallow failure under tension and procedures for calculation of the foundation tensile capacity under vertical and inclined loading. Emphasis is given to the influence of the strength of the compacted backfill compared to the strength of the natural soil. The design considerations presented here are the results of three decades of research carried out at the Federal University of Rio de Janeiro since the 1970s, based on many full-scale tensile tests on different types of transmission tower foundations in several soil formations throughout Brazil (Barata et al. 1978, 1979, 1985; Danziger 1983; Danziger et al. 1989; Pereira Pinto 1985; Ruffier dos Santos 1985, 1999; Garcia 2005, Danziger et al. 2006). The practical recommendations are based on the second author's experience in the design and construction of foundations for extra-high-voltage transmission lines over the last 30 years across Brazil, including very long transmission systems in the Amazon region and the Itaipu 750-kV transmission system. The criteria for predicting tensioned foundation capacity as discussed in this session are based mainly on the comprehensive work developed at the University of Grenoble (e.g., Martin 1966; Biarez and Barraud 1968; Martin 1973), due to its wide applicability for different types of soils, failure modes, load inclinations, and embedment depths and its good agreement with several full-scale tests on different types of foundations in a wide variety of soils.

In tensile foundations, shear strains are more pronounced than volumetric strains in contributing to displacement. In foundations under compressive loads, especially in weak soils, volumetric strains are predominant in contributing to settlement. As a result, tensioned foundations generally produce smaller displacements compared to foundations under the same compressive load in the same type of soil. Therefore, the design of foundations under tensile loads is conceived under limit equilibrium criteria in most cases, in contrast to compressed foundations, where consideration of limit equilibrium and settlement is important. Further discussion on prediction of displacement of tensioned foundations is provided by Trautmann and Kulhawy (1988) and Sakai and Tanaka (2007). Finite element analyses also are useful for predicting displacement of tensioned foundations, although more accurate three-dimensional simulations may be time consuming for design purposes. In this section, the design recommendations are restricted to limit equilibrium analyses.

3.8.1 Tensile Loads and Failure Modes

Tensioned foundations can be subjected to *permanent* as well as *transient loading*. In the case of transmission lines, permanent loading is caused by *angle* and *anchor* loading in the towers. Angle loading occurs when there is a change in the direction of the transmission line at the tower. Anchor loading occurs on one side of the first and last tower in a row of towers (called *end-of-line* or *anchor* towers), resulting in unbalanced forces at the sides of the towers, produced by different cable tension and construction load. For design purposes, temperature variation in the conductors also may be regarded as permanent loading. Transient loading occurs due to wind load (usually the dominant design load) and sudden mechanical failure of the conductors.

Self-supported transmission towers (Figure 3.25a) can apply alternate concentric compression/tension loads (Figures 3.26a and 3.26b) or eccentric loads (Figure 3.26c) to the foundation. Guyed towers (Figure 3.25b) transmit concentric orthogonal tension loads to the inclined guy foundation and compressive eccentric loads to the central mast foundation. For the typical design and inclination of a tower guy (about 30–35° to vertical), the effect of load inclination should be accounted for in the foundation design, as the ultimate tensile capacity is dependent on the load/plate inclination.

The foundation design loads usually are provided by the tower manufacturer. The foundation loads are calculated under different load hypotheses. For self-supported towers (Figure 3.25a), the design loads are given by superposition of the vertical (tension/compression) and two mutually perpendicular horizontal loads that act transversely and along the transmission line. The foundation designer takes into account the most unfavorable load hypothesis for

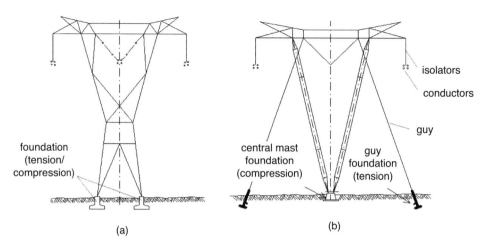

FIGURE 3.25 Most common types of towers: (a) self-supported tower and (b) guyed tower.

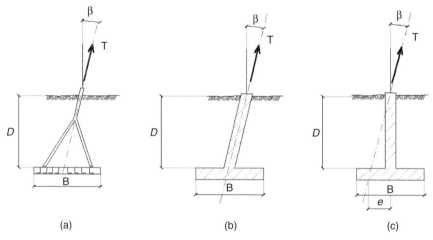

FIGURE 3.26 Common foundations for self-supported towers: (a) steel grillage, (b) footing with inclined pedestal, and (c) footing with vertical pedestal.

each foundation element, making a clear distinction between permanent and transient loading. A safety factor of 3 for permanent loads and 2 for transient loads, with respect to the theoretical ultimate tensile capacity, generally is recommended for tensioned foundations. Intermediate values may be used for simultaneous permanent and transient loading.

Steel grillage foundations (Figures 3.26a) or footings with inclined pedestals (Figure 3.26b) for self-supported transmission towers are subjected to a resultant tension/compression load that is approximately in the same direction as the tower leg, thus transmitting mostly concentric loading to the foundation. Moreover, the usual slope of a typical self-supported tower leg is small. Thus, for practical purposes, the tensile capacity of steel grillage foundations or footings with inclined pedestals for self-supported towers is calculated for vertical loading only, neglecting the secondary effects of load inclination and minor eccentricities. In contrast, self-supported towers on footings with a vertical pedestal (Figure 3.26c) introduce eccentricities in two orthogonal directions, parallel and perpendicular to the direction of the transmission line. The behavior of tensioned foundations under an eccentric oblique load was studied by Meyerhof (1973a, 1973b). A simplified procedure for practical design is to determine the equivalent reduced foundation dimensions to account for the double eccentricity, similar to the design of compressive eccentric loads discussed in Section 3.3. Horizontal loading and foundation eccentricity may play a dominant role in the design of piled transmission tower foundations on weak soils. There are situations in practice where foundation pedestals need to be high, such as in cases of significant seasonal variation in the flooding level of rivers or inundated areas crossed by transmission lines. High pedestals are much easier to build vertically than inclined. The overturning moments generated in such cases may be very high, and the corresponding footing dimensions would be very large. Hence, the use of prestressed anchors at the foundation corners is generally a cost-effective way to absorb high overturning moments (Danziger et al. 2006).

Vertical or nearly vertical tensioned plates can fail in shallow and deep modes of failure (Martin 1966; Biarez and Barraud 1968; Meyerhof and Adams 1968; Martin 1973), as shown in Figure 3.27a for firm soils and Figure 3.27b for weak soils. In the shallow mode (Figure 3.27a.1 and b.1), the failure surface reaches the ground level, and all applied tensile load is resisted by the plate. Thus, the shaft (or pedestal) transmits the applied tensile load directly from the structural member to the plate. In the deep mode (Figure 3.27a.2 and b.3), the tensile load is shared by the plate and the shaft where the failure surface around the plate does not reach the ground level. Therefore, the applied tensile load is not entirely transmitted to the

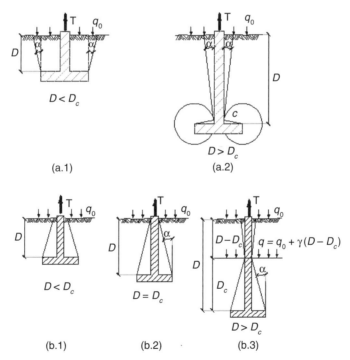

FIGURE 3.27 Shallow and deep failure modes: (a) firm soils where $\alpha < 0$ and (b) weak soils where $\alpha > 0$. (a.1) Shallow mode in firm soils, (a.2) deep mode in firm soils, (b.1) shallow mode in weak soils, (b.2) intermediate mode in weak soils, and (b.3) deep mode in weak soils (Biarez and Barraud 1968; Martin and Cochard 1973).

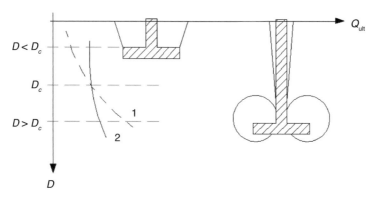

FIGURE 3.28 Determination of the critical depth (Biarez and Barraud 1968; Martin and Cochard 1973).

plate. The ultimate tensile load Q_{ult} obtained as a function of the plate depth in the shallow and deep modes is shown qualitatively in Figure 3.28. The dashed and solid lines represent the shallow and deep modes, respectively (Biarez and Barraud 1968; Martin and Cochard 1973). The two curves intersect at the critical depth D_c, where the failure mode changes from shallow to deep or vice versa. To determine whether failure will be in the shallow or deep mode, the calculations for both modes should be performed and the one that corresponds to the smaller tensile resistance chosen. However, full-scale load tests indicate that the critical depth usually is less than two to three times the diameter of a circular plate or the width of a square plate. Therefore, for typical depths and dimensions of ordinary shallow foundations used in transmission towers, failure would be in the shallow mode. Thus, the behavior of foundations under tensile loading discussed in this section is limited to the shallow mode.

The shapes of the failure surface in the shallow and deep modes are dictated by the type of soil in which the foundation is placed (Biarez and Barraud 1968) and by the inclination of the tensile load (Martin and Cochard 1973). The simplified shallow mode shown in Figure 3.29 was developed for homogeneous firm soils under vertical loading. The actual curvilinear failure surface observed in tensile tests is replaced by an equivalent simplified conical surface

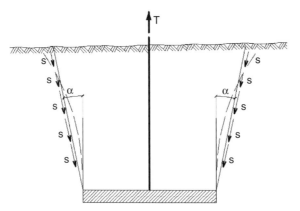

FIGURE 3.29 Observed (dashed line) and simplified (solid line) failure modes in firm soils (Biarez and Barraud 1968).

with a slope α, as indicated in Figures 3.27 and 3.29. The shape of the failure surface (and hence slope α) depends on the type of soil and the friction angle φ, as shown in Figure 3.27. For shallow plates, Biarez and Barraud (1968) and Martin and Cochard (1973) conceived three cases of distinct failure modes, depending on the soil type:

1. Granular soils (dense or loose), where the failure surface develops *outward* with an average inclination $\alpha = -\phi$
2. Firm clayey soils with $\phi > 15°$, where the failure surface develops *outward* with an average inclination $\alpha = -\phi/4$
3. Soft clayey soils with $\phi < 15°$, where the failure surface develops *inward* with an average inclination $\alpha = \tan^{-1}(0.2)$

The convention used here is that $\alpha < 0$ for a failure surface that propagates outward from the plate and $\alpha > 0$ when this surface propagates inward. The above failure modes have been observed in model tests in homogeneous soils in the laboratory and often have been confirmed by full-scale tests. The shallow mode for weak soils such as saturated soft clays (case 3) generally is of little importance in practice, since in this case the weak soil above the foundation almost always is replaced by more resistant, preselected compacted backfills, as in Figure 3.30. In situations such as Figure 3.30a, the undrained tensile capacity of the plate is estimated simply as $c_u p_b D$, where c_u = undrained shear strength of the clay, p_b = perimeter of the plate, and D = plate depth. In situations such as Figure 3.30b, the tensile capacity is estimated either as in case 1 or case 2.

The failure modes shown in Figures 3.27 and 3.29 are applicable to homogeneous soils. Sakai and Tanaka (2007) investigated the tensile capacity of layered soils. To account for the inhomogeneity introduced by the compacted backfill, the tensile capacity is controlled by the weaker of the two materials: backfill or surrounding natural soil. If the backfill is weaker than the natural soil, the failure takes place at the vertical interface ($\alpha = 0$). If the natural soil is weaker, the failure takes places within the natural soil, with the conical failure surface propagating outward from the plate ($\alpha = -\phi/4$ or $\alpha = -\phi$).

The effect of load inclination β (with respect to the vertical direction) in homogeneous soils is shown in Figure 3.31a for the shallow mode of failure and in Figure 3.31b for the deep mode (Martin and Cochard 1973). In the shallow mode, however, depending on the load inclination and the relative strength of the compacted backfill with respect to the natural soil, the actual shallow failure mode is likely to depart from the idealized modes shown in Figure

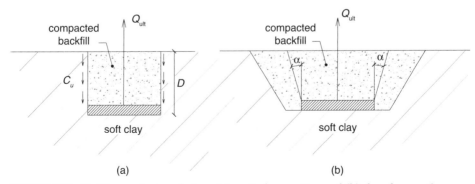

FIGURE 3.30 Uplift capacity in soft clays: (a) vertical excavation and (b) sloped excavation.

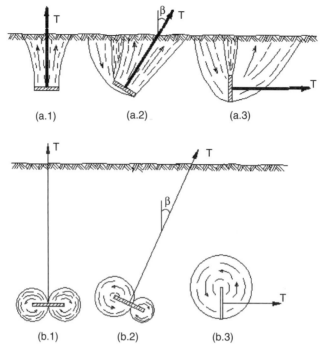

FIGURE 3.31 Failure modes for inclined load in firm soils: (a.1) vertical load in shallow mode, (a.2) inclined load in shallow mode, (a.3) horizontal load in shallow mode, (b.1) vertical load in deep mode, (b.2) inclined load in deep mode, and (b.3) horizontal load in deep mode.

3.31a and produce distinct failure angles α_L and α_R (at the left and right edges of the plate), as in Figure 3.32. The failure modes for shallow inclined plates at moderate load inclination ($\beta < 30°$), as in the case of guyed transmission towers shown in Figure 3.25b, are similar to the ones for horizontal plates under uplift loading (Martin and Cochard 1973). For steeper inclinations ($\beta > 30°$), the failure modes change as the angle β increases (Figure 3.31b). In all the models discussed below, it is assumed that the load is acting normal to the plate.

3.8.2 Tensile Capacity Equations in Homogeneous Soils: Grenoble Model (Martin and Cochard 1973)

3.8.2.1 Moderately Inclined Plates ($\beta < 30°$), Including Horizontal Plates ($\beta = 0°$)

As with most methods discussed in the literature, the uplift capacity Q_{ult} of plates installed at a shallow depth can be expressed by tensile capacity factors, similar to the bearing capacity formulae, as (Biarez and Barraud 1968; Martin and Cochard 1973)

$$Q_{ult} = p_b \frac{D}{\cos \beta} \left[cM_c + \frac{D}{\cos \beta} (M_\phi + M_\gamma) + q_0 M_q \right] \tag{3.112}$$

$$+ \gamma S_b D + W \cos \beta$$

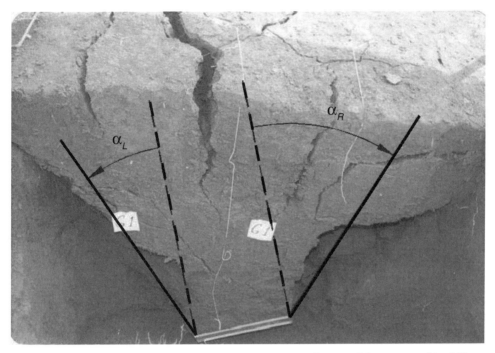

FIGURE 3.32 Tensile tests on inclined grillages showing the influence of the compacted backfill.

where D = depth, p_b = plate perimeter, S_b = plate area, c = cohesion, γ = unit weight of the soil, W = foundation self-weight, and q_0 = external surcharge acting at the ground level. M_c, $(M_\phi + M_\gamma)$, and M_q are dimensionless tensile capacity factors dependent on the soil type and friction angle ϕ, calculated by the set of formulae given in Appendix A. The term

$$\left[cM_c + \frac{D}{\cos\beta}(M_\phi + M_\gamma) + q_0 M_q \right]$$

in Equation 3.112 accounts for the average shear stress acting on the failure surface and is usually the dominant term in the foundation tensile capacity. In the absence of external surcharge at the soil surface (the most common situation in practice), the term $q_0 M_q$ vanishes. The term W, the foundation self-weight, is negligible in the case of steel grillage foundations. The term $\gamma S_b D$ accounts for the weight of the soil above the plate.

3.8.2.2 Steeply Inclined Plates

The following applies to a steeply inclined ($\beta > 30°$) shallow rectangular plate under a concentric load acting normal to the plate:

$$Q_{ult} = BL(cN_c + 0.5B\gamma N_\phi + q_0 N_q) + W\cos\beta \qquad (3.113)$$

where B is the width and L is the length of a rectangular plate. In the case of a circular plate, the plate area is calculated assuming an equivalent radius $R_e = (B + L)/\pi$. The term $(cN_c + 0.5B\gamma N_\phi + q_0 N_q)$ accounts for the average shear stress acting on the failure surface and usually is the dominant term in the foundation tensile capacity. The tensile capacity factors N_c, N_ϕ, and N_q are given by the set of formulae in Appendix B. For load inclinations close to the limit $\beta = 30°$, it is advisable to calculate the tensile capacity separately by Equations 3.112 and 3.113 and choose the smaller value. The tensile capacity factors applicable to Equations 3.112 and 3.113 are obtained easily with spreadsheets or programmable hand calculators.

Example 1. Determine the uplift capacity of a 2.50-m × 2.50-m horizontal square grillage embedded $D = 2.30$ m. The soil strength parameters are $c = 15$ kPa and $\phi = 25°$, and the unit weight is $\gamma = 16$ kN/m³. The backfill is assumed to be stronger than the natural soil.

In this example, the strength parameters of the more resistant backfill are not needed in the calculations, since failure is expected to develop through the natural soil ($\alpha = -\phi/4$). The uplift capacity factors applicable to Equation 3.12 are obtained from Appendix A as $M_c = 0.83$ and $M_\phi + M_\gamma = 0.22$. Neglecting the weight of the grillage and assuming no surcharge at the soil surface ($W = 0$ and $q_0 = 0$), the uplift capacity is calculated from Equation 3.12 taking $\beta = 0$ (vertical loading) as:

$$Q_{ult} = 4 \times 2.50 \times 2.30[15.0 \times 0.83 + 16.0 \times 2.30 \times 0.22] + 2.50^2 \times 2.30 \times 16.0$$

$$Q_{ult} = 472.6 + 230.0 = 702.6 \text{ kN}$$

Example 2. For the same horizontal grillage as in example 1, now the backfill is less resistant than the natural soil. The backfill strength parameters are $c = 15$ kPa and $\phi = 25°$. The backfill unit weight is $\gamma = 16$ kN/m³.

In this example, the strength parameters of the more resistant natural soil are not needed in the calculations, since failure is expected to develop at the interface of the backfill with the natural soil ($\alpha = 0$), controlled by the strength of the backfill. From Appendix A, the tensile capacity factors are calculated as $M_c = 0.66$, $M_\phi + M_\gamma = 0.19$, and $M_q = 0.31$. Therefore, from Equation 3.12:

$$Q_{ult} = 4 \times 2.50 \times 2.30[15.0 \times 0.66 + 16.0 \times 2.30 \times 0.19] + 2.50^2 \times 2.30 \times 16.0$$

$$Q_{ult} = 388.5 + 230.0 = 618.5 \text{ kN}$$

The above examples show that for the same strength parameters in both cases, the uplift capacity in example 1 (failure through the less resistant natural soil) is about 13% higher than in example 2 (failure through the interface). This illustrates the need for adequate compaction of the backfill, as a poorly compacted backfill significantly decreases the overall tensile capacity.

Example 3. Determine the tensile capacity of a rectangular plate ($B = 0.5$ m and $L = 1.35$ m), inclined $\beta = 33.5°$, embedded $D = 1.28$ m, and loaded normally to the plate. The soil strength parameters are $c = 9$ kPa and $\phi = 23°$, and the soil unit weight is $\gamma = 13.4$ kN/m³.

Assuming initially that the plate is at moderate inclination, it follows from Appendix A (for $\phi = 23°$ and $\alpha = -23/4 = -5.75°$) that $M_c = 0.897$, $M_\phi + M_\gamma = 0.21$, and $M_q = 0.267$. Neglecting the plate self-weight and assuming no surcharge at the soil surface ($q_0 = 0$ and $W = 0$), then:

$$Q_{ult} = 2x(0.5 + 1.35)x \frac{1.28}{\cos 33.5°} x \left(9x0.897 + 18.4x \frac{1.28}{\cos 33.5°} x0.21 \right)$$

$$+ 0.5x1.35x1.28x18.4$$

$$Q_{ult} = 79.5 + 15.9 = 95.4 \text{ kN}$$

Example 4. Assuming in example 3 that the plate inclination is steep, it follows from Appendix B (for $\phi = 23°$) that $N_c = 9.523$, $N_\phi = 13.563$, and $N_q = 4.042$. Therefore, taking $q_0 = 0$ and $W = 0$:

$$Q_{ult} = 0.5x1.35x(9x9.523 + 0.5x18.4x0.5x13.563) = 99.96 \text{ kN}$$

Comparing examples 3 and 4, the smaller value $Q_{ult} = 95.4$ kN (plate at moderate inclination) should be taken as the plate tensile capacity.

Figure 3.32 shows that the failure surface propagates through the natural soil and compacted backfill in the case of an inclined foundation. Thus, good engineering judgment is required to select the strength parameters to be used in the design of inclined foundations in practice, due to the influence of the compacted backfill. For safe design, however, it is recommended that the strength parameters from the smaller values corresponding to either the natural soil or the compacted backfill be chosen. As with vertical tensioned foundations, proper backfill compaction is essential for adequate foundation performance.

Appendix A

Tensile Capacity Factors for Shallow Plates at Moderate Load Inclination (β < 30°) or Horizontal Plates (β = 0): Grenoble Model (Martin and Cochard 1973)

$$Q_{ult} = p_b \frac{D}{\cos \beta} \left[cM_c + \frac{\gamma D}{\cos \beta} (M_\phi + M_\gamma) + q_0 M_q \right]$$

$$+ \gamma S_b D + W \cos \beta$$

M_c, $(M_\phi + M_\gamma)$, and M_q are dimensionless tensile capacity factors dependent on the friction angle ϕ and calculated by the following set of formulae:

$$M_c = M_{c0} \left(1 - \frac{\tan \alpha}{2} \frac{D}{R} \frac{1}{\cos \beta} \right)$$

$$M_{c0} = - \frac{\tan \alpha}{\tan \phi} + \frac{f}{H} \cos \phi \left(1 + \frac{\tan \alpha}{\tan \phi} \right)$$

$$\frac{f}{H} = \tan\left(\frac{\pi}{4} + \frac{\phi}{2}\right)\frac{\cos n - \sin\phi\cos m}{\cos n + \sin\phi\cos m}$$

$$m = -\frac{\pi}{4} + \frac{\phi}{2} + \alpha$$

$$\sin n = \sin\phi\sin m$$

$$M_\phi + M_\gamma = (M_{\phi 0} + M_{\gamma 0})\left(1 - \frac{\tan\alpha}{3}\frac{D}{R}\frac{1}{\cos\beta}\right)$$

$$M_{\phi 0} + M_{\gamma 0} = \frac{\sin\phi\cos(\phi + 2\alpha)}{2\cos^2\alpha}$$

$$M_q = M_{q 0}\left(1 - \frac{\tan\alpha}{2}\frac{D}{R}\frac{1}{\cos\beta}\right)$$

$$M_{q 0} = M_{c 0}\tan\phi + \tan\alpha$$

β is the load inclination to the vertical (which is zero for horizontal plates) and R is the radius of a circular plate or the equivalent radius of a rectangular plate with dimensions $B \times L$, calculated as $R = (B+L)/\pi$, except in the case of saturated clays, where $R = (B+L)/4$. D is the plate depth, p_b is the plate perimeter, S_b is the plate area, c is the soil cohesion, γ is the unit weight of the soil, W is the self-weight of the foundation element, and q_0 is the external surcharge acting at the ground level.

Appendix B

Tensile Capacity Factors for Shallow Plates at Steep Load Inclination ($\beta > 30°$): Grenoble Model (Martin and Cochard 1973)

$$Q_{\text{ult}} + BL(cN_c + 0.5B\gamma N_\phi + q_0 N_q) + W\cos\beta$$

N_c, N_ϕ, and N_q are dimensionless tensile capacity factors dependent on the friction angle ϕ and calculated by the following set of formulae:

$$N_\phi = A_\phi + B_\phi\left(\frac{D}{B} - \frac{1}{2}\sin\beta\right) + \left(C_\phi + \frac{B}{L}E_\phi\right)\left(\frac{D}{B} - \frac{1}{2}\sin\beta\right)^2$$

$$N_q = B_q + \left(C_q + \frac{B}{L}E_q\right)\left(\frac{D}{B} - \frac{1}{2}\sin\beta\right)$$

$$N_c = N_q \cot \phi$$

$$A_\phi = b_o - p_o$$

$$B_\phi = 2(b_o - p_o)$$

$$C_\phi = 2 \sin \phi \left(\frac{b_o b_\phi}{l_2} + \frac{p_o p_\phi}{l_1} \right)$$

$$E_\phi = 2 \sin \phi (b_o b_\phi + p_o p_\phi)$$

$$B_q = 0.5 B_\phi$$

$$C_q = 2 \sin \phi \left(\frac{b_o b_\phi}{l_2} + \frac{p_o p_\phi}{l_1} \right)$$

$$E_q = 2 \sin \phi (b_o b_\phi + p_o p_\phi)$$

$$b_o = \sin \beta \, \exp - (0.6 - 1.7\beta)\phi$$

$$p_o = \sin(\beta - \phi) \, \exp - \left(2.45 + \frac{1}{\beta} - 0.8\beta \right)\phi$$

$$b_\phi = \cos \beta$$

$$p_\phi = 1.1 \cos \phi$$

$$b_o = \frac{1 + \sin \phi}{1 - \sin \phi} \, \frac{\sqrt{1 - \sin^2 \phi \sin^2 \beta} - \sin \phi \cos \beta}{\sqrt{1 - \sin^2 \phi \sin^2 \beta} + \sin \phi \cos \beta}$$

$$p_o = \frac{1 - \sin \phi}{1 + \sin \phi} \, \exp - [(\pi - 2\beta) \tan \phi]$$

$$b_\phi = \frac{\cos \phi}{1 - \sin \phi} \, \frac{\sqrt{2 - \sin^2 \phi \left(\cos \frac{\phi}{2} - \sin \frac{\phi}{2} \right)^2} - \sin \phi \left(\cos \frac{\phi}{2} - \sin \frac{\phi}{2} \right)}{\sqrt{2 - \sin^2 \phi \left(\cos \frac{\phi}{2} - \sin \frac{\phi}{2} \right)^2} + \sin \phi \left(\cos \frac{\phi}{2} - \sin \frac{\phi}{2} \right)}$$

$$p_\phi = \frac{\cos\phi}{1 + \sin\phi} \exp - \left[\left(\frac{\pi}{2} - \phi \right) \tan\phi \right]$$

$$l_1 = \tan\left(\frac{\pi}{4} - \frac{\phi}{2} \right) \exp - [(\pi - 2\beta) \tan\phi]$$

$$l_2 = \tan\left(\frac{\pi}{4} + \frac{\phi}{2} \right) \frac{\sin\left[\left(\frac{\pi}{4} + \frac{\beta}{2} \right) - \left(\frac{\phi}{2} - \frac{n}{2} \right) \right]}{\sin\left[\left(\frac{\pi}{4} + \frac{\beta}{2} \right) + \left(\frac{\phi}{2} - \frac{n}{2} \right) \right]}$$

$$\sin n = \sin\beta \sin\phi$$

β (>30°) is the load inclination to the vertical. B is the width and L is the length of a rectangular plate. In the case of a circular plate, the plate area is calculated assuming an equivalent radius $R = (B + L)/\pi$. W is the self-weight of the foundation and q_0 is the external surcharge acting at the ground level.

References

Allen, T.M., Bathurst, R.J., Holtz, R.D., Lee, W.F., and Walters, D. (2004). New method for prediction of loads in steel reinforced soil walls. *J. Geotech. Geoenviron. Eng.*, 130(11):1109–1120.

American Concrete Institute (1988). Suggested analysis and design procedures for combined footings and mats. Report by ACI Committee 336. *J. ACI*, May–June:304–324.

Aysen, A. (2002). *Soil Mechanics: Basic Concepts and Engineering Applications*, A.A. Balkema, 459 pp.

Barata, F.E., Pacheco, M.P., and Danziger, F.A.B. (1978). Uplift tests on drilled piers and footings built in residual soil. *Proceedings 6th Brazilian Conference on Soil Mechanics and Foundation Engineering*, Vol. 3, 1–37.

Barata, F.E., Pacheco, M.P., Danziger, F.A.B., and Pereira Pinto, C. (1979). Foundations under pulling loads in residual soil—analysis and application of the results of load tests. *Proceedings 6th Pan-American Conference on Soil Mechanics and Foundation Engineering*, Lima.

Barata, F.E., Danziger, F.A.B., and Pereira Pinto, C. (1985). Behaviour of inclined plates in residual soil submitted to uplift load. *Proceedings 11th International Conference on Soil Mechanics and Foundation Engineering*, San Francisco, 2163–2166.

Biarez, J. and Barraud, Y. (1968). *Adaptation des Fondations de Pylônes au Terrain par les Méthodes de la Méchanique des Sols*, Rapport 22-06 de la CIGRÉ, Paris.

Bishop, A.W. (1961). Discussion on soil properties and their measurement. *Proceedings of International Conference on Soil Mechanics*, Vol. 3, 97.

Bjerrum, L. and Kummeneje, O. (1961). *Shearing Resistance of Sand Samples with Circular and Rectangular Cross Sections*, Norwegian Geotechnical Institute Publication No. 44, 1.

Boussinesq, J. (1885). *Application des Potentials a L'etude de L'equilibre et du Mouvement des Solides Elastiques*, Gauthier-Villars, Paris.

Bowles, J.E. (1988). *Foundation Analysis and Design*, 4th edition, McGraw-Hill, 1004 pp.

Bowles, J.E. (1996). *Foundation Analysis and Design,* 5th edition, McGraw-Hill, 1024 pp.

Briaud, J.-L. and Gibbens, R.M. (1994). *Predicted and Measured Behaviour of Five Spread Footings on Sand,* Geotechnical Special Publication 41, ASCE, Texas.

British Standards Institution (1986). *BS8004:1986. British Standard Code of Practice for Foundations,* BSI, London.

Burland, J.B. (1970). Discussion, Session A. *Proceedings of Conference on In Situ Investigations in Soils and Rocks,* British Geotechnical Society, London, 61–62.

Burland, J.B. and Burbidge, M.C. (1985). Settlement of foundations on sand and gravel. *Proc. Inst. Civ. Eng.,* I(78):1325–1381.

Burland, J.B., Broms, B.B., and De Mello, V.F.B. (1977). Behaviour of foundations and structures. *Proceedings of 9th International Conference on Soil Mechanics and Foundation Engineering,* Tokyo, Vol. 2, 495–538.

Canadian Geotechnical Society (1992). *Canadian Foundation Engineering Manual,* 3rd edition.

Cernica, J.N. (1995). *Geotechnical Engineering: Foundation Design,* John Wiley, 486 pp.

Chen, W.F. and Saleeb, A.F. (1982). *Constitutive Equations for Engineering Materials,* Vol. 1: Elasticity and Modeling, John Wiley, New York.

Christian, J.T. and Carrier III, W.D. (1978). Janbu, Bjerrum and Kjaernsli's chart reinterpreted. *Can. Geotech. J.,* 15:123–128.

Coduto, D.P. (2001). *Foundation Design: Principles and Practices,* 2nd edition, Prentice Hall, 883 pp.

Danziger, F.A.B. (1983). *Ultimate Capacity of Foundations under Vertical Tension Load* (in Portuguese), M.S. thesis, COPPE, Federal University of Rio de Janeiro, Brazil.

Danziger, F.A.B., Pereira Pinto, C., and Danziger, B.R. (1985). Uplift load tests on grillages for guyed towers in Itaipu transmission system. *Proceedings 11th International Conference on Soil Mechanics and Foundation Engineering,* San Francisco, 1193–1196.

Danziger, F.A.B., Danziger, B.R., and Pacheco, M.P. (2006). The simultaneous use of piles and prestressed anchors in foundation design. *Eng. Geol.,* 87(3–4):163–177.

Das, B.M. (1984). *Principles of Foundation Engineering,* Brooks/Cole, Belmont, CA.

Das, B.M. (2007). *Principles of Foundation Engineering,* 6th edition, Thomson, Canada.

Das, B.M. and Sivakugan, N. (2007). Settlements of shallow foundations on granular soils—an overview. *Int. J. Geotech. Eng.,* 1(1):19–29.

Davis, E.H. and Booker, J.R. (1971). The bearing capacity of strip footings from the standpoint of plasticity theory. *Proceedings of the 1st ANZ Conference on Geomechanics,* Melbourne, Vol. 1, 276–282.

De Beer, E.E. (1970). Experimental determination of the shape factors and bearing capacity factors in sand. *Geotechnique,* 20(4):387–411.

Desai, C.S. and Siriwardena, H.J. (1984). *Constitutive Laws for Engineering Materials,* Prentice Hall, Englewood Cliffs, NJ.

Duncan, J.M. and Buchignani, A.L. (1976). *An Engineering Manual for Settlement Studies,* Geotechnical Engineering Report, Department of Civil Engineering, University of California, Berkeley, 94 pp.

Duncan, J.M. and Chang, C.Y. (1970). Nonlinear analysis of stress and strains in soils. *J. Soil Mech. Found. Div. ASCE,* 96(SM5):1629–1653.

European Committee for Standardisation (1995). *Eurocode 7: Geotechnical Design—Part 1: General Rules,* ENV 1997-1, Brussels.

Franke, E., Mossallamy, Y.E., and Wittmann, P. (2000). Calculation methods for raft foundations in Germany. *Design Applications of Raft Foundations,* J.S. Hemsley, Ed., Thomas Telford, London, 283–322.

Garcia, O.C. (2005). *Influence of Backfill Quality on the Uplift Capacity of Foundations under Uplift Load,* M.S. thesis, COPPE, Federal University of Rio de Janeiro, Brazil.

Giroud, J.-P. (1972). Settlement of rectangular foundation on soil layer. *J. Soil Mech. Found. Div. ASCE,* 98(SM1):149–154.

Golder, H.Q. (1975). Floating foundations. *Foundation Engineering Handbook,* H. Winterkorn and H.-Y. Fang, Eds., Van Nostrand Reinhold, New York, 537–555.

Hanna, A.M. and Meyerhof, G.G. (1981). Experimental evaluation of bearing capacity of footings subjected to inclined loads. *Can. Geotech. J.,* 18(4):599–603.

Hansen, J.B. (1970). *A Revised and Extended Formula for Bearing Capacity,* Bulletin 28, Danish Geotechnical Society, Copenhagen.

Hemsley, J.A. (1998). *Elastic Analysis of Raft Foundations,* Thomas Telford, London.

Holtz, R.D. (1991). Stress distribution and settlement of shallow foundations. *Foundation Engineering Handbook,* 2nd edition, H.-Y. Fang, Ed., Van Nostrand Reinhold, New York, 166–222.

Janbu, N., Bjerrum, L., and Kjaernsli, B. (1956). *Veiledning ved losning av fundamenteringsoppgaver,* Norwegian Geotechnical Institute Publication 16, Oslo, 30–32.

Katzenbach, R., Arslan, U., and Moormann, C. (2000). Piled raft foundation projects in Germany (2000). *Design Applications of Raft Foundations,* J.A. Hemsley, Ed., Thomas Telford, London, 323–391.

Ladd, C.C., Foote, R., Ishihara, K., Schlosser, F., and Poulos, H.G. (1977). Stress deformation and strength characteristics. State-of-the-art report. *Proceedings of the 9th ICSMFE,* Tokyo, Vol. 2, 421–494.

Lade, P.V. and Lee, K.L. (1976). *Engineering Properties of Soils,* Report UCLA-ENG-7652, 145 pp.

Lee, K.L. (1970). Comparison of plane strain and triaxial tests on sand. *J. Soil Mech. Found. Div. ASCE,* 96(SM3):901–923, Paper 7276.

Leonards, G.A. (1986). *Advanced Foundation Engineering—CE683,* Lecture Notes, Purdue University.

Lopes, F.R. (2000). Design of raft foundations on Winkler springs. *Design Applications of Raft Foundations,* J.A. Hemsley, Ed., Thomas Telford, London, 127–154.

Martin, D. (1966), *Étude à la Rupture de Différents Ancrages Solicitées Verticalement,* Thèse de docteur-ingénieur, Faculté des Sciences de Grenoble, France.

Martin, D. (1973). Calcul de pieux et fondations a dalle des pylônes de transport d'énergie électrique, étude théorique e d'éssais en laboratoire et in-situ. *Proceedings, Institut Technique du Batiment et des Traveaux Public,* Supplément au No. 307–308.

Martin, D. and Cochard, A. (1973). *Design of Anchor Plates,* Rapport 22–10 de la CIGRÉ, Paris.

Massad, F. (2005). Marine soft clays of Santos, Brazil: building settlements and geological history. *Proceedings 18th International Conference on Soil Mechanics and Geotechnical Engineering,* Vol. 2, 405–408.

McCarthy, D.F. (2007). *Essentials of Soil Mechanics and Foundations,* 7th edition, Pearson Prentice Hall, 850 pp.

Mesri, G. and Godlewski, P.M. (1977). Time- and stress-compressibility interrelationship. *J. Geotech. Eng.,* 103(5):417–430.

Meyerhof, G.G. (1951). The ultimate bearing capacity of foundations. *Geotechnique,* 2(4):301–331.

Meyerhof, G.G. (1956). Penetration tests and bearing capacity of cohesionless soils. *J. Soil Mech. Found. Div. ASCE,* 82(SM1).

Meyerhof, G.G. (1963). Some recent research on bearing capacity of foundations. *Can. Geotech. J.,* 1(1):16–26.

Meyerhof, G.G. (1973a). The uplift capacity of foundations under oblique loads. *Can. Geotech. J.,* 10:64–70.

Meyerhof, G.G. (1973b). Uplift resistance of inclined anchors and piles. *Proceedings 8th International Conference on Soil Mechanics and Foundation Engineering,* Moscow, Vol. 2, 167–172.

Meyerhof, G.G. (1974). Penetration testing outside Europe: general report. *Proceedings of the European Symposium on Penetration Testing,* Stockholm, 2.1, 40–48.

Meyerhof, G.G. and Adams, J.I. (1968). The ultimate uplift capacity of foundations. *Can. Geotech. J.*, 5(4):225–244.

Milović, D. (1992). *Stresses and Displacements for Shallow Foundations*, Elsevier, Amsterdam.

Newmark, N.M. (1942). *Influence Charts for Computation of Stresses in Elastic Soils*, Bulletin No. 338, University of Illinois Experiment Station.

Pereira Pinto, C. (1985). *Behavior of Anchors for Guyed Towers in Residual Soil* (in Portuguese), M.S. thesis, COPPE, Federal University of Rio de Janeiro, Brazil.

Poulos, H.G. (2000). Practical design procedures for piled raft foundations. *Design Applications of Raft Foundations*, J.A. Hemsley, Ed., Thomas Telford, London, 425–467.

Prandtl, L. (1921). Uber die eindringungsfestigkeit plastischer baustoffe und die festigkeit von schneiden. *Z. Angew. Math. Mech.*, 1(1):15–20.

Raj, P.P. (1995). *Geotechnical Engineering*, Tata-McGraw-Hill.

Reissner, H. (1924). Zum erddruck-problem. *Proceedings of First International Congress on Applied Mechanics*, Delft, 295–311.

Robertson, P.K. and Campanella R.G. (1983). Interpretation of cone penetration test. Part II. Clay. *Can. Geotech. J.*, 20(4):734–745.

Ruffier dos Santos, A.P. (1985). *Analysis of Foundations under Pulling Loads by the Finite Element Method* (in Portuguese), M.S. thesis, COPPE, Federal University of Rio de Janeiro, Brazil.

Ruffier dos Santos, A.P. (1999). *Ultimate Capacity of Foundations under Uplift Load in Slopes*, Ph.D. thesis, COPPE, Federal University of Rio de Janeiro, Brazil.

Sakai, T. and Tanaka, T. (2007). Experimental and numerical study of uplift behavior of shallow circular anchor in two-layered sand. *J. Geotech. Geoenviron. Eng.*, 133(4):469–477.

Schmertmann, J.H. (1970). Static cone to compute static settlement over sand. *J. Soil Mech. Found. Div. ASCE*, 96(SM3):1011–1043.

Schmertmann, J.H., Hartman, J.P., and Brown, P.R. (1978). Improved strain influence factor diagrams. *J. Geotech. Eng. Div. ASCE*, 104(GT8):1131–1135.

Schultze, E. and Horn, A. (1967). The base friction for horizontally loaded footings in sand and gravel. *Geotechnique*, 17(4):329–347.

Sivakugan, N. and Johnson, K. (2004). Settlement predictions in granular soils: a probabilistic approach. *Geotechnique*, 54(7):499–502.

Sivakugan, N., Eckersley, J.D., and Li, H. (1998). Settlement predictions using neural networks. *Australian Civil Engineering Transactions*, CE40, The Institution of Engineers, Australia, 49–52.

Skempton, A.W. (1951). The bearing capacity of clays. *Proceedings of Building Research Congress*, London, Vol. 1, 180–189.

Skempton, A.W. and Bjerrum, L. (1957). A contribution to settlement analysis of foundations on clay. *Geotechnique*, 7:168–178.

Spangler, M.G. and Handy, R.L. (1982). *Soil Engineering*, 4th edition, Harper & Row, New York, 819 pp.

Tan, C.K. and Duncan, J.M. (1991). Settlement of footings on sands: accuracy and reliability. *Proceedings of Geotechnical Engineering Congress 1991*, Geotechnical Special Publication No. 27, ASCE, Colorado, 1, 446–455.

Teixeira, A.H. (2003). A method for analysis of time-settlement curves for buildings at the beach city of Santos (in Portuguese). *Proc. Workshop on Past, Present and Future of Buildings at the Beach City of Santos*, Brazilian Society for Soil Mechanics and Geotechnical Engineering, 25–38.

Teng, W.C. (1975). Mat foundations. *Foundation Engineering Handbook*, H. Winterkorn and H.-Y. Fang, Eds., Van Nostrand Reinhold, New York, 528–536.

Terzaghi, K. (1943).*Theoretical Soil Mechanics*, John Wiley, New York.

Terzaghi, K. and Peck, R.B. (1967). *Soil Mechanics in Engineering Practice*, 2nd edition, John Wiley & Sons, New York.

Terzaghi, K., Peck, R.B., and Mesri, G. (1996). *Soil Mechanics in Engineering Practice,* 3rd edition, John Wiley & Sons.

Trautmann, C.H. and Kulhawy, F.H. (1988). Uplift load-displacement behavior of spread foundations. *J. Geotech. Eng.,* 114(2):168–184.

U.S. Army (1993). *Bearing Capacity of Soils,* Technical Engineering and Design Guides, ASCE, 142 pp.

U.S. Army (1994). *Settlement Analysis,* Technical Engineering and Design Guides, ASCE, 136 pp.

Vesic, A.S. (1961). Bending of beams resting on isotropic solid. *J. Eng. Mech. Div. ASCE,* 87(EM2):35–53.

Vesic, A.S. (1973). Analysis of ultimate loads of shallow foundations. *J. Soil Mech. Found. Div. ASCE,* 99(SM1):45–73.

Vesic, A.S. (1975). Bearing capacity of shallow foundations. *Foundation Engineering Handbook,* H.F. Winterkorn and H.-Y. Fang, Eds., Van Nostrand Reinhold, chap. 3.

Westergaard, H.M. (1938). A problem of elasticity suggested by a problem in soil mechanics: soft material reinforced by numerous strong horizontal sheets. *Contributions to the Mechanics of Solids,* MacMillan, New York.

Zeevaert, L. (1983). *Foundation Engineering for Difficult Subsoil Conditions,* Van Nostrand Reinhold, New York.

4

Foundation-Soil Interaction

by
Priti Maheshwari
Indian Institute of Technology Roorkee, India

4.1 Introduction

A system is defined as a collection of entities or processes that act and interact together toward accomplishment of a logical end. This logical end is the production of an output that corresponds to an external input. Therefore, an interaction problem is quite important for the analysis of any kind of system and especially for the systems related to applied mathematics and engineering. In the case of foundation engineering, the system under consideration has three components: the structure, the structural foundation, and the supporting soil/rock media. The external input is the various loading conditions for which the response of the system is to be studied. Conventional analysis and design methods treat structure as independent of foundation as well as the supporting soil. However, in reality, the structure, structural foundation, and supporting soil/rock media act as one integral compatible unit; therefore, analysis of the soil-foundation interaction problem is quite essential to study the response of a system, in the form of deformations and stresses, under external loading conditions. The superstructure remains in firm contact with the structural foundation, and the foundation is in contact with the supporting soil media. Forces transferred from the superstructure to the foundation govern the settlements of the foundation and the supporting soil media. These settlements, in turn, govern the stresses in the foundation as well as in the superstructure. Therefore, the behavior of the supporting soil media is a function of the stresses transferred to it, and the behavior of the foundation is a function of the settlement or deformational characteristics of the soil media. This interdependence of the behavior of the foundation and the supporting soil gives rise to the foundation-soil interaction problem. In the solution of the foundation-soil interaction problem, the whole system is first represented by a mathematical model comprised of the deformational characteristics of the supporting soil medium, interface conditions, and the flexibility of the foundation. Various research workers have developed several constitutive models to represent the supporting soil medium based on its type and deformational characteristics. Various parameters of the models represent the system characteristics (i.e., the characteristics of the structure–supporting foundation–soil system). These parameters can be physically interpretable parameters or sometimes fitting parameters.

In this chapter, various aspects related to the foundation-soil interaction problem are addressed. First of all, various constitutive models (lumped parameter as well as distributed) are presented for the idealization of various types of soil media, including linear elastic, nonlinear elastic, elastoplastic, and viscoelastic characteristics. Methods adopted for estimation of the parameters associated with these models are discussed, and typical representative values of the parameters are reported. Application of the foundation-soil interaction to the problems of shallow footings such as isolated footings, strip footings, combined footings, and raft foundations are discussed. Various research workers have contributed by means of different methods of analysis; however, some typical applications also are discussed in detail. The last section of this chapter deals with the application of the interaction to pile foundations under axial loads, lateral loads, and moments. Although the main focus is the foundation-soil interaction problems, the stiffness of the structure or the manner in which the structural stiffness is transmitted to the foundation has quite a significant influence on the response of the foundation-soil system. Therefore, a complete analysis and design procedure should consider the interaction between all three components. In view of this, a few typical studies are discussed which deal with structure-foundation-soil interaction problems.

4.2 Modeling of the Ground (Soil Mass) and Constitutive Equations

The mechanics of the interaction between a foundation and the subsoil must take into account the effects of the complex states of stress, strain, and environment on the mechanical behavior of different classes of materials. This requires that the different variables involved be related by means of fundamental equations, including equilibrium equations, kinematic equations, compatibility equations, constitutive equations, and a set of boundary conditions. The mechanics of the interaction between a foundation and the soil is governed by the mechanical response of the compressible subsoil. Soil behaves elastically or nearly so under small stresses. The strain remains constant as long as the stress is fixed and disappears immediately upon removal of the load. However, the inelastic strain does not disappear after removal of the stress, representing the plastic behavior of soil.

In cohesive soils (composed of clay minerals), the strength of the films of adsorbed water surrounding the grains accounts for the resistance of soil to deformation (Šuklje 1969; Findley et al. 1976). These soils exhibit elastic action upon loading; then a slow and continuous increase in strain at a decreasing rate is observed. A continuously decreasing strain follows as an initial elastic recovery upon the removal of stress. This type of response is said to be viscoelastic behavior. The time-dependent behavior of such soils must be expressed by a constitutive equation which includes time as a variable in addition to the stress and strain variables. Viladkar (1989) has summarized various constitutive laws that represent the behavior of soils.

Due to the inherent complexity in the behavior of the soil mass, various models have been developed for the response of foundation-soil interaction problems. Generally, the response of these models is represented by the surface deflection caused by an external system of forces. The response represents the displacement characteristics of the upper boundary of the soil which is in contact with the foundation (i.e., soil-foundation interface). The displacement characteristics form a major portion of the information necessary in foundation-soil interaction analysis.

Two approaches have been adopted for modeling the soil mass: the discrete approach and the continuum approach. Various models used in these approaches are presented in this section.

4.2.1 Discrete Approach

In the discrete approach, the soil mass is replaced by a finite number of equivalent springs, which results in the simplest model using this approach. The response of the model can be studied only at a finite number of points where the springs have been connected to the foundation. To make the model more realistic, sometimes these springs are employed in combination with a shear layer or dashpots.

The discrete approach, because of its simplicity, has been widely adopted for analyzing various foundation-soil interaction problems. Kerr (1964), Šuklje (1969), Findley et al. (1976), and Selvadurai (1979) have summarized various fundamental models developed by employing this approach. These fundamental models have been further extended by various research workers. Some of the discrete models adopted for modeling the ground are presented below.

4.2.1.1 Winkler Model

Winkler (1867) proposed a model of soil media which assumes that the deflection w of the soil medium at any point on the surface is directly proportional to the stress p applied at that point and independent of stresses applied at other locations; that is,

$$p(x, y) = kw(x, y) \tag{4.1}$$

where k is the modulus of subgrade reaction in units of stress per unit length. Winkler's idealization of the soil mass is comprised of a system of mutually independent springs that have a spring constant k. An important feature of this model is that displacement occurs only under the loaded area. The surface displacements of the Winkler model are shown in Figure 4.1 for various types of loading. The Winkler model cannot distinguish between an infinitely rigid load and a uniform flexible load (Figures 4.1c and 4.1d).

4.2.1.2 Filonenko-Borodich Model

Filonenko-Borodich (1940, 1945) proposed a model to eliminate the inherent deficiency of the Winkler model in depicting the continuous behavior of real soil masses. This model provides continuity between the individual spring elements in the Winkler model by connecting them to a thin elastic membrane under a constant tension T (Figure 4.2). The equilibrium of the

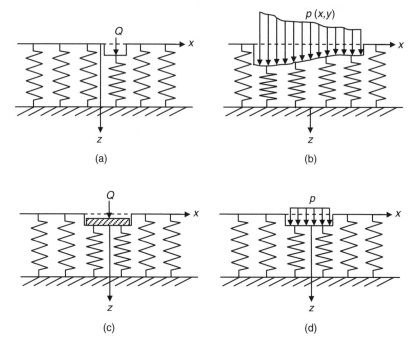

FIGURE 4.1 Surface displacements of the Winkler model due to (a) a concentrated load, (b) a nonuniform load, (c) a rigid load, and (d) a uniform flexible load.

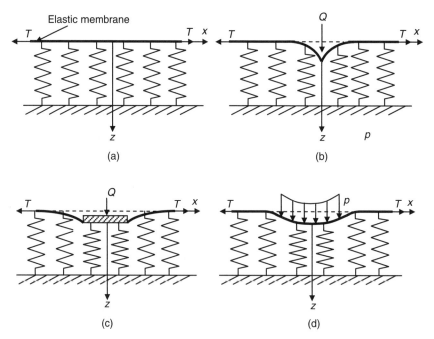

FIGURE 4.2 Surface displacements of the Filonenko-Borodich model: (a) basic model, (b) concentrated load, (c) rigid load, and (d) uniform flexible load.

membrane-spring system yields the surface deflection of the soil medium due to a pressure p as

$$p(x, y) = kw(x, y) - T\nabla^2 w(x, y) \tag{4.2}$$

where

$$\nabla^2 = \frac{\partial^2}{\partial x^2} + \frac{\partial^2}{\partial y^2}$$

is Laplace's differential operator in rectangular Cartesian coordinates. The two elastic constants k and T characterize the soil model. Typical surface deflection profiles due to concentrated, flexible, and rigid external loads are depicted in Figure 4.2.

4.2.1.3 Hetényi Model

Hetényi (1946) proposed a model in which the interaction between the independent spring elements was established by incorporating an imaginary elastic plate (in three-dimensional problems) or an elastic beam (in two-dimensional problems). The surface deflection due to a pressure p is given by

$$p(x, y) = kw(x, y) - D\nabla^4 w(x, y) \tag{4.3}$$

where

$$D = \frac{E_p h^3}{12(1 - \nu_p^2)}$$

is the flexural rigidity of the plate, h is the thickness of the plate, and E_p and ν_p are the elastic constants for the plate material.

4.2.1.4 Pasternak Model

Pasternak (1954) presented a model that assumes shear interaction between the spring elements; this was accomplished by connecting these spring elements to a layer of incompressible vertical elements deforming only in transverse shear (Figure 4.3). A free body diagram of an element of the shear layer is depicted in Figure 4.3. Force equilibrium in the z direction yields the relation

$$p(x, y) = kw(x, y) - G\nabla^2 w(x, y) \tag{4.4}$$

where G is the shear modulus of the shear layer, which is considered to be isotropic in the x, y plane.

Equation 4.4 coincides with Equation 4.2 if T is replaced by G. Thus, the surface deflection profiles for this model and the Filonenko-Borodich model are quite similar. The Filonenko-

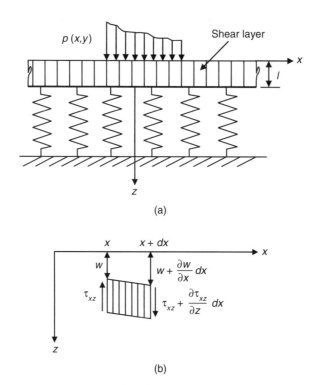

(a)

(b)

FIGURE 4.3 Pasternak model: (a) basic model and (b) stresses in the shear layer.

Borodich (1940, 1945), Hetényi (1946), and Pasternak (1954) models reduce to the Winkler (1867) model as the respective parameters T, D, and G tend to zero.

4.2.1.5 Kelvin-Voigt Model

The Kelvin-Voigt model is constructed by a combination of a Hookean spring element in series with a Kelvin model (Figure 4.4). Various research workers have employed this model to explain the phenomenon of primary compression, consolidation, and secondary compression of clayey soils and have developed theories (Merchant 1939; Taylor and Merchant 1940; Gibson and Lo 1961).

The constitutive relation for this model is

$$\varepsilon = \frac{\sigma}{k_1} + \frac{\sigma}{k_2}\left(1 - e^{-\frac{k_2}{\eta_2}t}\right) \qquad (4.5)$$

where σ is the total applied stress and ε is the total strain. k_1 and k_2 are spring constants and η_2 is the dashpot constant (coefficient of viscosity), as shown in Figure 4.4.

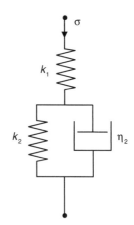

FIGURE 4.4 Kelvin-Voigt model

This model shows an instantaneous strain of σ/k_1 at time $t = 0$. The strain described by Equation 4.5 increases at a decreasing rate and asymptotically approaches a value of $\sigma/(k_1 + k_2)$ when time t tends to infinity. Under applied stress, the viscous element undergoes strain at a decreasing rate, thus transferring a greater and greater portion of the applied load to the Hookean spring element. Finally, the entire applied stress is carried by the Hookean elements of the model.

4.2.1.6 Burger's Model

Burger's model is used for soils that exhibit creep behavior and is composed of a Maxwell model connected in series with a Kelvin model (Figure 4.5). The constitutive equation for this model can be derived by considering the strain response under the constant stress of each of the elements coupled in series. The total strain at any time t will be the sum of the strains in the three elements of Burger's model (viz., the Kelvin model and the spring and dashpot in the Maxwell model). This yields a constitutive equation of Burger's model as

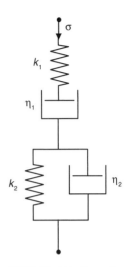

FIGURE 4.5 Burger's model.

$$\sigma + \left(\frac{\eta_1}{k_1} + \frac{\eta_1}{k_2} + \frac{\eta_2}{k_2} \right) \dot{\sigma} + \frac{\eta_1 \eta_2}{k_1 k_2} \ddot{\sigma} = \eta_1 \dot{\varepsilon} + \frac{\eta_1 \eta_2}{k_2} \ddot{\varepsilon} \qquad (4.6)$$

where k_1 and k_2 are spring constants and η_1 and η_2 are dashpot constants, as shown in Figure 4.5. This model finds wide application in the study of the time-dependent behavior of soils, underground tunnels, and excavations.

4.2.1.7 Generalized Maxwell Model

The Maxwell model is represented by a viscous damper and an elastic spring connected in series. Several Maxwell models in series or parallel result in the generalized Maxwell model, as presented in Figure 4.6. Maxwell models connected in series (Figure 4.6a) result in the following constitutive equation:

$$\dot{\varepsilon} = \dot{\sigma} \sum_{i=1}^{n} \frac{1}{k_i} + \sigma \sum_{i=1}^{n} \frac{1}{\eta_i} \qquad (4.7)$$

where $\dot{\sigma}$ and $\dot{\varepsilon}$ are the applied stress rate and strain rate, respectively; k_i and η_i are the spring constant (modulus of subgrade reaction) and dashpot constant (coefficient of viscosity), respectively, for the ith Maxwell body; and n is the total number of Maxwell bodies connected in series. The above equation is equivalent to the stress-strain rate relation for a single Maxwell model and describes the same mechanical behavior.

Several Maxwell models connected in parallel (Figure 4.6b) represent instantaneous elasticity, delayed elasticity with various retardation times, stress relaxation with various relax-

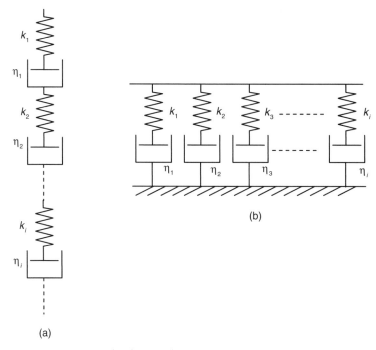

FIGURE 4.6 Generalized Maxwell model (a) in series and (b) in parallel.

ation times, and also viscous flow. This generalized Maxwell model (Figure 4.6b) is convenient for predicting the stress associated with a prescribed strain variation, as the same prescribed strain is applied to each individual element and the resulting stress is the sum of the individual contributions. The ith element would yield the stress-strain relation as

$$\sigma_i = \frac{D}{\dfrac{D}{k_i} + \dfrac{1}{\eta_i}} \, \varepsilon \tag{4.8}$$

where D is the differential operator with respect to time (i.e., $D = d/dt$).

Upon summing both sides of Equation 4.8 and simplifying, the generalized constitutive relation takes the following form:

$$\left[\left(\frac{D}{k_1} + \frac{1}{\eta_1}\right)\left(\frac{D}{k_2} + \frac{1}{\eta_2}\right)\left(\frac{D}{k_3} + \frac{1}{\eta_3}\right) \ldots\right]\sigma$$

$$= \begin{bmatrix} D\left(\dfrac{D}{k_2} + \dfrac{1}{\eta_2}\right)\left(\dfrac{D}{k_3} + \dfrac{1}{\eta_3}\right)\ldots \\[2ex] + D\left(\dfrac{D}{k_1} + \dfrac{1}{\eta_1}\right)\left(\dfrac{D}{k_3} + \dfrac{1}{\eta_3}\right)\ldots + \ldots \end{bmatrix}\varepsilon \tag{4.9}$$

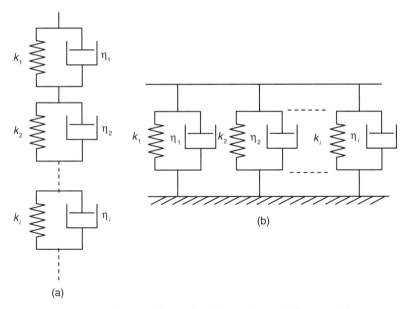

FIGURE 4.7 Generalized Kelvin model (a) in series and (b) in parallel.

4.2.1.8 Generalized Kelvin Model

A purely viscous damper and a purely elastic spring connected in parallel form the basic unit of the Kelvin model. Several Kelvin models connected in series or parallel (Figure 4.7) result in the generalized Kelvin model. The strain contribution of the ith element of the generalized Kelvin model resulting in a series combination of several Kelvin models is

$$\varepsilon_i = \frac{1}{D\eta_i + k_i}\,\sigma \tag{4.10}$$

where D is the differential operator with respect to time (i.e., $D = d/dt$).

Summing up both sides of Equation 4.10 and on further simplification, the open form of the constitutive equation can be obtained as:

$$[(D\eta_1 + k_1)(D\eta_2 + k_2)(D\eta_3 + k_3)\ldots]\varepsilon$$
$$= [(D\eta_2 + k_2)(D\eta_3 + k_3)\ldots \tag{4.11}$$
$$+ (D\eta_1 + k_1)(D\eta_3 + k_3)\ldots + \ldots]\sigma$$

If several Kelvin models are connected in parallel, they do not exhibit any different behavior than an equivalent Kelvin model. The constitutive equation for n Kelvin models connected in series is

$$\sigma = \varepsilon \sum_{i=1}^{n} k_i + \dot{\varepsilon} \sum_{i=1}^{n} \eta_i \tag{4.12}$$

4.2.2 Continuum Approach

As mentioned above, in the discrete approach, the soil is replaced by distinct spring elements or sometimes spring elements in combination with dashpots. In the case of soil media, surface deflections occur not only immediately under the loaded region but also within certain limited zones outside the loaded region. To account for this continuous behavior, soil often is treated as infinitely divisible media, which leads to the idea of an infinitesimal volume. This infinitesimal volume is treated as a particle of the continuum. The distribution of the continuum is considered to be continuous without any gaps or voids. Various models in the form of constitutive relations that are employed in this approach are presented below for the analysis of soil-foundation interaction problems.

4.2.2.1 Elastic Half-Space Approach

In this approach, soil media are modeled as three-dimensional continuous elastic solids or elastic continua. Generally, the distribution of displacements and stresses in such media remains continuous under external loading systems. Boussinesq (1878, 1885) analyzed the problem of a semi-infinite homogeneous isotropic linear elastic solid subjected to a concentrated load acting normal to the plane boundary, and this analysis initiated the continuum representation of soil media.

In the most general three-dimensional form, the stresses and strains in linear elasticity are related as

$$\{\sigma\} = [D]\{\varepsilon\} \tag{4.13}$$

where the matrix $[D]$ is known as the elastic constitutive matrix and is comprised of elements in terms of the elastic properties of the soil. These elements can be expressed in terms of several different parameters, such as:

1. *Modulus of elasticity E*—Relates axial strain to axial stress in a simple tension or compression test
2. *Poisson's ratio* ν—Relates axial strain to transverse normal strain in a simple tension or compression test
3. *Shear modulus G*—Relates shear stress to shear strain
4. *Bulk modulus K*—Relates volumetric strain ε_{vol} to octahedral normal stress
5. *Lame's constants* λ and μ—Relate stresses and strains as:

$$\sigma_x = \lambda \varepsilon_{vol} + 2\mu\varepsilon_x \tag{4.14}$$

Similar equations can be written for σ_y and σ_z as well as

$$\tau_{xy} = \mu\gamma_{xy} \tag{4.15}$$

with similar equations for other shear stresses.

If the constitutive relationships and the strain-deformation relations are known, surface displacement profiles of an elastic half-space can be obtained for various loading conditions. Davis and Selvadurai (1996) have summarized some special problems that hold a fundamental position in relation to the elastic solutions (Boussinesq's problem [1878], Flamant's problem [1892], Kelvin's problem [Thompson 1848], Cerrutti's problem [1884], Mindlin's problem [1936], etc.).

The displacement in the z direction, $w(r, z)$ in an isotropic elastic half-space due to the action of a concentrated force Q (Figure 4.8) on its boundary as per Boussinesq (1885) is

$$w(r, z) = \frac{Q}{4\pi GR} \left[2(1 - v) + \frac{z^2}{R^2} \right] \tag{4.16}$$

where G and v are the shear modulus and Poisson's ratio of the elastic material and $R^2 = r^2 + z^2$. As per Equation 4.16, surface deflection becomes zero as r tends to infinity (Figure 4.8). Surface deflection at the boundary due to a uniform load p distributed over a radius a is calculated as

$$w(a, 0) = \frac{2(1 - v)pa}{\pi G} \tag{4.17}$$

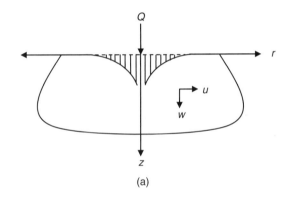

(a)

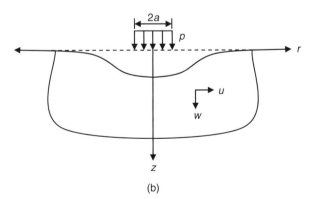

(b)

FIGURE 4.8 Typical surface displacement profiles of an elastic half-space subjected to (a) a concentrated load Q and (b) a uniform load p of radius a.

Boussinesq's solution has been used extensively to determine the deflection profile for other loadings (such as line, triangular, rectangular, etc.) by employing the principle of superposition.

The cross-anisotropic relation of Equation 4.13 also can be expressed in terms of a strain-stress matrix. For a three-dimensional situation, this can be presented as

$$\varepsilon_x = \frac{\sigma_x}{E_h} - \nu_h \frac{\sigma_y}{E_h} - \nu_v \frac{\sigma_z}{E_v}; \qquad \gamma_{xy} = \frac{2\tau_{xy}(1 + \nu_h)}{E_h}$$

$$\varepsilon_y = -\nu_h \frac{\sigma_x}{E_h} + \frac{\sigma_y}{E_h} - \nu_v \frac{\sigma_z}{E_v}; \qquad \gamma_{yz} = \frac{2\tau_{yz}(1 + \nu_v)}{E_v} \qquad (4.18)$$

$$\varepsilon_z = \nu_v \frac{\sigma_x}{E_v} - \nu_v \frac{\sigma_y}{E_v} + \frac{\sigma_z}{E_v}; \qquad \gamma_{zx} = \frac{2\tau_{zx}(1 + \nu_v)}{E_v}$$

where E_h and E_v can be interpreted as the modulus of elasticity for loading in the horizontal plane and along the vertical axis, respectively. Poisson's ratio relating the loading along one horizontal axis to strains along the other horizontal axis is ν_h. The relation between extensional strains in the horizontal plane and vertical loadings or between vertical extensional strains and horizontal loadings is controlled by the other Poisson's ratio ν_v.

4.2.2.2 Nonlinear Elastic Half-Space Approach

The relations between stresses and strains for soils are much more complex than the simple linearly elastic relations described in Section 4.2.2.1. In order to represent foundation-soil interaction problems more realistically therefore, some form of nonlinear relations must be used, as given below.

4.2.2.2.1 Bilinear Models

The simplest type of nonlinear relation is the bilinear one, illustrated in Figure 4.9. The material has the initial modulus E_1 until the modulus reduces, after which the modulus is changed to E_2. Before change of modulus, therefore, the incremental stress-strain relation can be written as

$$\{\Delta\sigma\} = [D_1]\{\Delta\varepsilon\} \qquad (4.19)$$

and after change of modulus can be written as

$$\{\Delta\sigma\} = [D_2]\{\Delta\varepsilon\} \qquad (4.20)$$

where $[D_1]$ and $[D_2]$ are elasticity matrices before and after change of modulus, respectively. The drawback of this method is that the bulk and shear moduli are reduced equally. The material becomes compressible just as it becomes highly deformable after change of moduli values and often gives unreliable results. It is, therefore, much better to reduce the shear modulus and keep the bulk modulus constant.

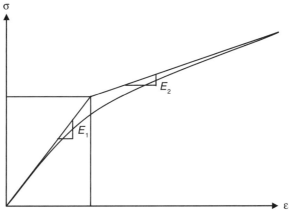

FIGURE 4.9 Bilinear model.

4.2.2.2.2 *Quasi-linear Model*

A nonlinear stress-strain curve can be divided into a number of linear curves, leading to the so-called multilinear, piecewise linear, or quasi-linear models. In the initial stages involving nonlinear analyses, the piecewise linear approach (Figure 4.10) involves interpolation on the basis of a set of data points $(\sigma_i, \varepsilon_i)$ on the given stress-strain curve. The tangent modulus E_t is defined as the slope of the chord between two computed points. The constitutive equations can be written in incremental form as

$$\{d\sigma\}_m = [D_t]_m \{d\varepsilon\}_m \tag{4.21}$$

where m denotes the mth increment of stress $\{d\sigma\}$ and strain $\{d\varepsilon\}$, and $[D_t]_m$ denotes the tangent constitutive matrix corresponding to the mth increment (Figure 4.10).

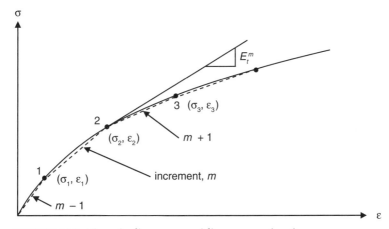

FIGURE 4.10 Piecewise linear or quasi-linear approximation.

4.2.2.2.3 Hyperbolic Model

Kondner (1963) and Kondner and Zelasko (1963) have shown that nonlinear stress-strain curves for both clay and sand may be approximated with a high degree of accuracy by a hyperbola (Figure 4.11) of the form

$$\frac{\varepsilon}{\sigma_1 - \sigma_3} = a + b\varepsilon \tag{4.22}$$

where ε is the axial strain and a and b are constants of the hyperbola.

The plot $\varepsilon/(\sigma_1 - \sigma_3)$ vs. ε gives a straight line, where a is the intercept on the y-axis and b is the slope of the line (Figure 4.11b). The reciprocal of b represents the ultimate compressive strength of the soil, which is larger than the failure compressive strength. This is expected because the hyperbola remains below the asymptote at all values of strain. The ratio R_f of

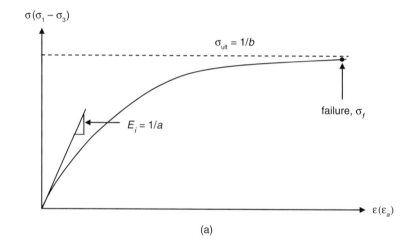

(a)

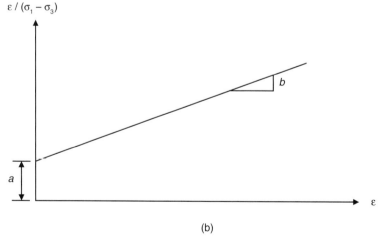

(b)

FIGURE 4.11 Hyperbolic model: (a) hyperbolic simulation of stress-strain curve and (b) transformed hyperbola.

compressive strength $(\sigma_1 - \sigma_3)_f$ to the ultimate compressive value σ_u varies from 0.75 to 1.0 for different soils independent of the confining pressure (Kondner 1963). The inverse of a represents the initial tangent modulus E_i.

Duncan and Chang (1970) have stated Kondner's expression in terms of the shear strength defined by the Mohr-Coulomb failure criterion and initial tangent modulus as

$$(\sigma_1 - \sigma_3) = \left[1 - \frac{R_f(1 - \sin\phi)(\sigma_1 - \sigma_3)}{2c\cos\phi + 2\sigma_3\sin\phi} \right] E_i \qquad (4.23)$$

where c is cohesion, ϕ is the angle of internal friction, E_i is the initial tangent modulus, and $R_f = (\sigma_1 - \sigma_3)_f/(\sigma_1 - \sigma_3)_{ult}$. The material tangent modulus E_t can therefore be written as:

$$E_t = \left[1 - \frac{R_f(1 - \sin\phi)(\sigma_1 - \sigma_3)}{2c\cos\phi + 2\sigma_3\sin\phi} \right]^2 E_i \qquad (4.24)$$

By employing the relation between the initial tangent modulus and the confining pressure σ_3 as given by Janbu (1963), the above expression takes the form

$$E_t = \left[1 - \frac{R_f(1 - \sin\phi)(\sigma_1 - \sigma_3)}{2c\cos\phi + 2\sigma_3\sin\phi} \right]^2 Kp_a \left(\frac{\sigma_3}{p_a} \right)^n \qquad (4.25)$$

where K and n are experimentally determined parameters. p_a is the atmospheric pressure and was introduced to make K a dimensionless number.

A similar relation for the tangent Poisson's ratio was developed by Kulhawy et al. (1969) based on the hyperbolic concept as

$$\nu_t = \frac{G - F\log(\sigma_3/p_a)}{(1 - A)^2} \qquad (4.26)$$

where

$$A = \frac{(\sigma_1 - \sigma_3)d}{Kp_a(\sigma_3/p_a)^n \left[1 - \frac{R_f(1 - \sin\phi)(\sigma_1 - \sigma_3)}{2c\cos\phi + 2\sigma_3\sin\phi} \right]} \qquad (4.27)$$

where G, F, and d are the material parameters.

All the parameters can be obtained from laboratory triaxial compression tests conducted for a given stress path. However, the hyperbolic model can yield satisfactory results only in cases of geotechnical problems that involve monotonic loadings. For problems that involve loading and unloading and various stress paths in soil, the results from hyperbolic simulation may not be reliable. One of the major limitations is that the hyperbolic model includes only one stress path, whereas loading and/or unloading can cause a wide range of stress paths. The hyperbolic model also is not able to account for the second-order dilatancy effects. Expression 4.26 loses significance as soon as $v_t > 0.5$. Hence the hyperbolic model of a given curve for a specific stress path should be used with care and essentially for cases that involve monotonic loading.

4.2.2.2.4 Parabolic Model

Hansen (1963) proposed two additional functional representations of stress-strain relationships:

$$(\sigma_1 - \sigma_3) = \left(\frac{\varepsilon}{a + b} \right)^{1/2} \tag{4.28}$$

$$(\sigma_1 - \sigma_3) = \frac{\varepsilon^{1/2}}{a + b} \tag{4.29}$$

Equation 4.28 accounts for the possibility of parabolic variation of stress-strain curves at small strains. Equation 4.29 is an alternative form to account for the parabolic variation and possesses the property of giving a maximum value of $(\sigma_1 - \sigma_3)$ for finite strain; that is, it is suitable when the curve shows a decrease after the peak stress.

4.2.2.3 Elastoplastic Half-Space Approach

The behavior of most geological media is quite different from that of metals, and their strength is dependent on the hydrostatic stress. Under fully or partially drained conditions, the strength of soil often increases with mean pressure and exhibits frictional characteristics. There are certain exceptions, such as the undrained behavior of clays, which can be similar to the behavior of metals. In view of this, true representation of the characteristics of soils cannot be accomplished with the help of the above-mentioned constitutive relations. The constitutive relations that arise from plasticity theory must be used. These usually are incremental in nature; that is, stresses and strains are related entirely by their incremental or differential behavior. It is not possible to relate total stress to total strain directly without knowledge of the loading path. The essential features of plasticity theory are (1) a yield function that separates the elastic and plastic states of soil, (2) a plastic potential function that defines the direction of plastic straining when yielding occurs, (3) a hardening/softening law that describes the dependence of the yield function on plastic strains, and (4) some assumed elastic behavior of the yield surface. Clearly, all four of these assumptions have to be checked against experimental evidence before satisfactory performance of the model can be expected. There are many yield criteria available for representation of soil behavior. Viladkar et al. (1995) have presented the convenient forms of these criteria for use in the elastoplastic analysis of geological materials like soils and rocks.

4.2.2.3.1 Mohr-Coulomb Model

It has long been noted that the Mohr envelope to a series of Mohr's circles of stress usually is curved (Figure 4.12), and therefore a general expression for the yield surface can be written as

$$F = |\tau| - f(\sigma_n') = 0 \tag{4.30}$$

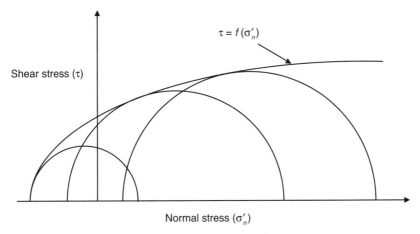

FIGURE 4.12 Envelope to Mohr-Coulomb circles of stress.

where $|\tau|$ and σ'_n represent the absolute value of shearing stress and the effective normal stress on the failure plane, respectively. $f(\sigma'_n)$ is a function chosen to represent the nonlinearity of the Mohr envelope. The linear form of Equation 4.30 is commonly known as the Mohr-Coulomb yield criterion, which is a generalization of the Coulomb failure law and can be written as

$$\tau - \sigma'_n \tan\phi - c = 0 \qquad (4.31)$$

where c and ϕ denote cohesion and the angle of internal friction, respectively. Graphically, Equation 4.31 represents a straight line tangent to the largest principal stress circle, as shown in Figure 4.13, and was first presented by Mohr. By inspection of Figure 4.13, the linearized equation (Equation 4.31) can be written in terms of major and minor principal stresses as:

$$(\sigma'_1 - \sigma'_3) = 2c \cos\phi + (\sigma'_1 + \sigma'_3) \sin\phi \qquad (4.32)$$

4.2.2.3.2 Drucker-Prager Model

The Mohr-Coulomb yield surface exhibits singularities at the corners of the hexagon in the principal stress space whenever the stresses are represented by one of the ridges of the yield surface and is not suitable for use as a plastic potential. To avoid such singularities, Drucker and Prager (1952) approximated the angular yield surface by using a right circular cone, which is given by

$$F = 3\alpha\sigma'_m + \beta\bar{\sigma} - K = 0 \qquad (4.33)$$

where

$$\alpha = \frac{\sin\phi}{\sqrt{3 + \sin^2\phi}}, \quad \beta = \sqrt{3}, \quad K = \frac{3c \cos\phi}{\sqrt{3 + \sin^2\phi}} \qquad (4.34)$$

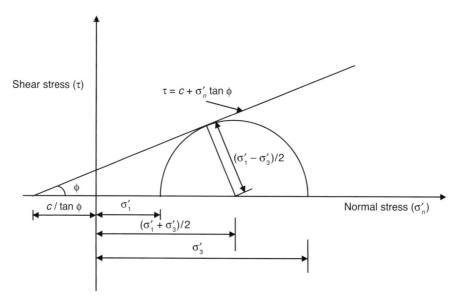

FIGURE 4.13 Mohr-Coulomb failure envelope.

It can be shown that the Drucker-Prager yield criterion will always give a lower bound to the Mohr-Coulomb representation. In terms of the invariants of the stress tensor, the Drucker-Prager yield criterion can be written as

$$f = \sqrt{J_{2D}} + \alpha J_1 - k \tag{4.35}$$

where α and k are positive material parameters, J_1 is the first invariant of the stress tensor, and J_{2D} is the second invariant of the deviatoric stress tensor. Equation 4.35 represents a straight line on a J_1 vs. $(J_{2D})^{1/2}$ plot (Figure 4.14). In three-dimensional principal stress space, the criterion plots as a right circular cone. When the state of stress reaches the failure surface (Equation 4.35), the material undergoes plastic deformations. The material can undergo plastic deformations while the stress point is moving on the failure surface.

The two material parameters α and k can be determined from the slope and the intercept of the failure envelope plotted on the J_1-$(J_{2D})^{1/2}$ space (Figure 4.14). In order to establish a failure envelope for a material, it is necessary to perform laboratory tests such as conventional triaxial, true triaxial, or plane strain tests up to the failure. The values of α and k can be expressed in terms of cohesion c and angle of internal friction ϕ. However, the values of c and ϕ determined by conventional triaxial compression tests are different from those determined under plane strain conditions. The values of α and k can be expressed as follows.

Conventional triaxial compression:

$$\alpha = \frac{2 \sin \phi}{\sqrt{3}\,(3 - \sin \phi)}, \qquad k = \frac{6c \cos \phi}{\sqrt{3}\,(3 - \sin \phi)} \tag{4.36a}$$

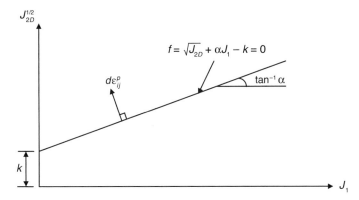

FIGURE 4.14 Drucker-Prager yield criterion in terms of stress invariants.

Plane strain condition:

$$\alpha = \frac{\tan\phi}{\sqrt{9 + 12\tan^2\phi}}, \qquad k = \frac{3c}{\sqrt{9 + 12\tan^2\phi}} \qquad (4.36b)$$

4.2.2.3.3 Critical State Model

Frictional criteria like the Mohr-Coulomb and Drucker-Prager yield criteria do not represent soil behavior adequately. The few drawbacks are prediction of unreasonably large dilation with associated flow rules and the occurrence of yielding well below the Mohr-Coulomb failure envelope. If the soil sample is loaded, the frictional yield criteria would predict a reversible linear stress-strain behavior, but the observed stress-strain response actually would show deviation from linearity and permanent strain after removal of the load. These drawbacks can be avoided by using strain-dependent cap models.

Drucker et al. (1957) were the first to suggest that soil can be treated as a work-hardening material which would eventually reach a perfectly plastic state. The proposed yield surface consisted of a Drucker-Prager yield surface with a spherical end cap, the position and size of which depended upon the volumetric strain. Roscoe et al. (1958) proposed a model which also distinguished between yielding and ultimate failure by introducing the concept of a critical state line in conjunction with the strain-dependent yield surface, which was called the Cam-Clay model. This was improved upon by Roscoe and Burland (1968), who proposed an elliptical shape for the strain-dependent yield surface, which became known as the modified Cam-Clay model. Originally, the theory was developed for a triaxial stress condition, but Roscoe and Burland (1968) extended it to a plane strain situation, still using the material parameters determined from triaxial tests.

An elliptical yield surface of this type, which is a function of the first two stress invariants, is shown in Figure 4.15. The normality principle assuming an associated flow rule applies to the elliptical surface, and since the surface is completely smooth, the direction of viscoplastic straining is uniquely defined. At the intersection of the critical state line and the ellipse, the normal to the yield surface is vertical; therefore, the failure state is reached. The expression for the yield surface can be written as

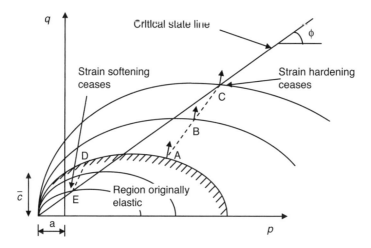

FIGURE 4.15 Graphical representation of critical state yield surface in the space of two stress invariants p and q.

$$F_c = p^2 - p_o p + \frac{q^2}{M^2} \tag{4.37a}$$

where p is the mean pressure $= J_1/3$, p_o is the initial mean pressure, M is the slope of the critical state line, and q is the deviatoric stress $= \sqrt{3}(J_2')^{1/2}$.

The hardening rule is defined as a function of the plastic volumetric strain ε_{ii}^p as

$$\varepsilon_{ii}^p = \frac{\lambda - K}{1 + e_o} \log_{10} \frac{p_o'}{p_o} \tag{4.37b}$$

where e_o is the initial void ratio; λ and K are the compression and swelling indices, respectively; and p_o' is the preconsolidation pressure.

The generalization of the above model was given by Zienkiewicz et al. (1975) with the help of a third stress invariant in terms of θ. This model is an elliptical model whose section in the π-plane is similar to that for the Mohr-Coulomb criterion. Figure 4.15 leads to a surface in which various parameters are expressed as

$$a = \frac{\bar{c}}{\tan \phi} \tag{4.38a}$$

where

$$\tan \phi = \frac{3 \sin \phi}{\left(\sqrt{3} \cos \theta - \sin \theta \sin \phi \right)} \tag{4.38b}$$

and

$$\bar{c} = \frac{3c \cos \phi}{\left(\sqrt{3} \cos \theta - \sin \theta \sin \phi \right)} \tag{4.38c}$$

Thus, the equation for the yield surface becomes

$$F = \frac{3q^2}{(p_{co} \tan \phi)^2} + \frac{(p - p_{co} + a)^2}{p_{co}^2} \tag{4.39}$$

where $2p_{co}$ is the major axis of the ellipse and p_{co} is the initial preconsolidation pressure. Subsequent hardening is related to change of volumetric stress by means of a consolidation test; thus:

$$\Delta p_{co} = F(\varepsilon_v^p) = p_{co} \exp(-\chi \varepsilon_v^p) \tag{4.40a}$$

where χ is a constant given by

$$\chi = \frac{1 + e_o}{\lambda - k} \tag{4.40b}$$

where e_o is the initial void ratio, and λ and k are the compression and swelling indices, respectively, determined from odometer tests.

4.2.2.4 Viscoelastic Half-Space Approach

Soil is a three-phase system comprised of solid, liquid, and gaseous materials. Therefore, soil resists the effects of external forces in a manner different from simple solid continua. In noncohesive soils, the external force is resisted by intergranular friction at the contact surfaces. In cohesive soils, composed of clay minerals, the strength of the films of adsorbed water surrounding the grains accounts for the resistance of the soil to deformation (Šuklje 1969).

Soils exhibit elasticity as well as creep under constant stress. Creep occurs at a rate that either remains constant or varies with time. Stress relaxation under constant applied strain also is observed in soils. This behavior of soil can be described by viscoelastic models comprised of rheological elements, namely a Hookean elastic body, Newtonian viscous liquid, Saint Venant plastic body, and Pascal's liquid. Rheological models are constructed in an intuitive way, and the corresponding relationships between stresses and strains are deduced and compared with experimental observations. This comparison controls the applicability of the assumed rheological models. Some of these rheological models were discussed in Sections 4.2.1.5–4.2.1.8. Constitutive relations as presented in Sections 4.2.1.5–4.2.1.8 which correspond to the various models can be directly employed to represent soils that exhibit viscoelastic behavior.

4.3 Estimation of Model Parameters

Before an analysis of any situation involving the stressing of soil can be undertaken, it is necessary to determine the constitutive equation of state for the soil and the constants in the equation that describe its behavior. Various constitutive relations that represent soil behavior were discussed in the previous section. This section includes the methodologies used to estimate the constants or parameters of the different models (modulus of subgrade reaction, elastic constants, shear strength parameters, coefficient of viscosity, etc.).

4.3.1 Modulus of Subgrade Reaction

If a foundation of width B undergoes settlement Δs due to a load Δq per unit area, the modulus of subgrade reaction k_s is defined as (Figure 4.16):

$$k_s = \frac{\Delta q}{\Delta s} \tag{4.41}$$

k_s (in kN/m^3) describes the constant of the Winkler model or two-parameter models (Filonenko-Borodich model, Hetényi model, Pasternak model, etc.).

   In practical situations, the fundamental assumptions of modeled soil behavior may not be completely satisfied, and therefore the value of the modulus of subgrade reaction is not a unique property of the given soil medium. The modulus of subgrade reaction is determined from plate loading tests and is affected by factors such as the size, shape, and embedded depth of the plate. Terzaghi (1955), Teng (1962), Selvadurai (1979), Bowles (1996), and Das (1999) have presented methods for evaluation of the modulus of subgrade reaction in a comprehensive manner. Terzaghi (1955) proposed that k_s for footings of width B could be obtained from plate load test data using the following equations:

For footings on stiff clay $$k_s = k_{s1} \frac{B_1}{B} \tag{4.42a}$$

For footings on sand $$k_s = k_{s1} \left(\frac{B + B_1}{2B} \right)^2 \tag{4.42b}$$

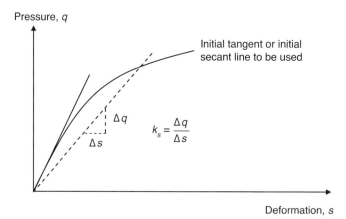

FIGURE 4.16 Determination of modulus of subgrade reaction.

where B_1 is the dimension of the square plate used in the plate load test to produce k_{s1}. For a rectangular footing of width B and length mB resting on stiff clay or medium-dense sand, the modulus of subgrade reaction is obtained as

$$k_s = k_{s1} \frac{m + 0.5}{1.5m} \tag{4.43}$$

where k_{s1} is the value of the modulus of subgrade reaction obtained from a plate load test using a 0.3×0.3 m or other size plate.

Considering the average values of stress and strain beneath a rigid plate resting at a depth D below the surface of a granular soil medium, it was shown by Terzaghi (1955), Teng (1962), and Bowles (1996) that the modulus of subgrade reaction k_s at depth D is related to the modulus of subgrade reaction k_{s1} of the plate located at the surface of the granular soil medium as:

$$k_s = k_{s1} \left(1 + 2 \frac{D}{B} \right) \tag{4.44}$$

The effects of size, shape, and depth of embedment of the footing can be combined to obtain the modulus of subgrade reaction by employing Equations 4.42–4.44.

Vesic (1961) proposed a relationship between the modulus of subgrade reaction and the stress-strain modulus E_s. For all practical purposes, this relationship reduces to

$$k_s = \frac{E_s}{B(1 - v_s^2)} \tag{4.45}$$

where v_s is Poisson's ratio of the soil.

Biot (1937) compared the solutions using both the Winkler model and the elastic continuum model for a particular soil-foundation interaction problem and expressed the modulus of subgrade reaction k_s in terms of the elastic constants of the soil medium E_s and v_s. This correlation was obtained by comparing the maximum bending moment of an infinite beam subjected to a concentrated force P using both soil models. The following expression was obtained as a measure of k in terms of elastic constants of the soil medium and the properties of the infinite beam:

$$k_s = \frac{1.23 E_s}{(1 - v_s^2)b} \left[\frac{E_s b^4}{16 C(1 - v_s^2) E_b I} \right]^{0.11} \tag{4.46}$$

where b is the width of the beam, $E_b I$ is the flexural rigidity of an infinite beam, and C is a dimensionless parameter ($C = 1.0$ for uniform pressure distribution across the width of the beam and $1.0 < C < 1.13$ for uniform deflection across the width of the beam). This technique for obtaining the modulus of subgrade reaction was substantiated by means of experimental studies and has been used extensively by various research workers.

TABLE 4.1 Range of Modulus of Subgrade Reaction k_{s1}

Soil Type		k_{s1} (MN/m³)
Sand (dry or moist)	Loose	8–25
	Medium	25–125
	Dense	125–375
Sand (saturated)	Loose	10–15
	Medium	35–40
	Dense	130–150
Clay	Stiff	12–25
	Very stiff	25–30
	Hard	>50

After Das (1999).

Selvadurai (1979), Bowles (1996), and Das (1999) have presented typical ranges of values for the modulus of subgrade reaction k_{s1} and k_s for various types of soils. The range for the modulus of subgrade reaction k_{s1} adapted from Das (1999) is presented in Table 4.1.

Daloglu and Vallabhan (2000) developed a method for evaluation of an equivalent modulus of subgrade reaction to be used in the Winkler model using nondimensional parameters for the analysis of a slab on a layered soil medium. The results from the study have been compared by means of numerical examples with those obtained from the modified Vlazov model (Vlazov and Leontiev 1966) and by using the value of k_s suggested by Biot (1937) and Vesic (1961). It was concluded that if a constant value of the modulus of subgrade reaction is used for a uniformly distributed load, the displacements would be uniform and there would be no bending moment and shear force in the slab. It was recommended that higher values of k_s closer to the edges of the slab have to be used for realistic results. The value of the modulus of subgrade reaction was observed to be dependent on the depth of the soil layer. Plots have been provided for nondimensional values of the modulus of subgrade reaction k_s for different nondimensional depths of the soil layer, from which an equivalent value of k_s can be computed when the complete geometry and properties of the overall system are known.

4.3.2 Elastic Constants

The elastic constants are the modulus of elasticity E_s and Poisson's ratio ν_s, which characterize the isotropic elastic continuum model. According to their definitions, these constants are assumed to be independent of test procedure or size of the sample used. Several factors, such as levels of applied isotropic and deviatoric stresses, stress history, type and rate of application of load, sample disturbance, and influence of physical properties (moisture content, void ratio, etc.), affect the measured values of elastic constants as far as the elastic behavior of the soil medium is concerned.

4.3.2.1 Poisson's Ratio

Poisson's ratio for a soil is evaluated from the ratio of radial strain to axial strain during a triaxial compression test. As mentioned above, the test procedure plays an important role in its determination. Bowles (1996) and Das (1999) have presented ranges of values for Poisson's ratio for various types of soil. Typical ranges adapted from Das (1999) are presented in Table 4.2.

TABLE 4.2   Elastic Parameters of Various Soils

Soil Type	Modulus of Elasticity E_s (MN/m²)	Poisson's Ratio v_s
Loose sand	10.35–24.15	0.20–0.40
Medium-dense sand	17.25–27.60	0.25–0.40
Dense sand	34.50–55.20	0.30–0.45
Silty sand	10.35–17.25	0.20–0.40
Sand and gravel	69.00–172.50	0.15–0.35
Soft clay	4.1–20.7	
Medium clay	20.7–41.4	0.20–0.50
Stiff clay	41.4–96.6	

After Das (1999).

4.3.2.2 Modulus of Elasticity

The modulus of elasticity E_s of the soil medium often is determined from unconfined compression, triaxial compression, or odometer tests. Field tests such as plate loading tests and pressuremeter tests also may be used for determination of the *in situ* modulus of elasticity of the soil. Some typical values of the modulus of elasticity for various types of soils adapted from Das (1999) are presented in Table 4.2.

4.3.3 Constants That Describe Two-Parameter Elastic Models of Soil Behavior

The material constants in this category include the modulus of subgrade reaction k_s and the parameter G_p. They can be determined by the expressions

$$k_s = \frac{E_s}{H(1 + v_s)(1 - 2v_s)}; \quad G_p = \frac{E_s H}{6(1 + v_s)} \quad (4.47)$$

where H is the thickness of the soil layer, and the values of E_s and v_s can be determined as discussed in Section 4.3.2.

Similar expressions can be obtained for multilayer soil media. However, these have been found to be quite complicated (Vlazov-Leontiev 1966; Rao et al. 1971).

4.3.4 Constants for Viscoelastic Half-Space Models

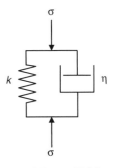

FIGURE 4.17   Kelvin model.

The method of estimating the constants that describe the behavior of soil in the constitutive relations of viscoelastic half-space models is described in this section with the help of a representative model in the form of a Kelvin model (Figure 4.17). A Kelvin model is used to represent the saturated soil mass in the drained condition and consists of a spring and a dashpot in parallel (Figure 4.17) such that the strains experienced by the two components under constant applied stress are the same, whereas the stresses shared are different. The material constants (i.e., spring and dashpot constants) can be obtained using consolidation test data or triaxial compression test data.

4.3.4.1 Determination of Material Constants Using Consolidation Test Data

Viladkar et al. (1992, 1993) developed the procedure for determination of the material constants of a Kelvin model by employing consolidation test data. The rheological equation for a Kelvin model in a uniaxial stress situation is

$$\sigma = k\varepsilon + \eta\dot{\varepsilon} \tag{4.48}$$

where σ is the total applied stress, ε is the total strain, and $\dot{\varepsilon}$ is the strain rate. k and η are the spring constant and dashpot constant, respectively, and are determined using consolidation test data.

The solution to the above differential equation can be obtained, with the help of appropriate boundary and initial conditions, at any time T as:

$$\varepsilon^T = \frac{\sigma}{k}\left[1 - e^{-(k/\eta)T}\right] \tag{4.49}$$

The steady state is reached at time $T = \infty$, and at this state:

$$\varepsilon^T = \varepsilon^\infty = \frac{\sigma}{k} \tag{4.50}$$

From this equation, the spring constant k can be approximated as

$$k = \frac{\sigma}{\varepsilon^\infty} \tag{4.51}$$

where ε^∞ is the final strain at the end of the steady state.

Taking the natural logarithm of Equation 4.49 and rearranging it, the dashpot constant at any time T can be expressed as:

$$\eta^T = \frac{-kT}{\ln\left(1 - \dfrac{\varepsilon^T}{\varepsilon^\infty}\right)} \tag{4.52}$$

The stress in the Kelvin model splits into its deviatoric and hydrostatic stress components; the spring and dashpot constants (k' and η' for deviatoric and k'' and η'' for hydrostatic) can be determined for the two situations using Equations 4.51 and 4.52 if the strains under the two stress conditions are known at any time T and correspond to the steady state. The hydrostatic (or volumetric) and deviatoric strains occur simultaneously in a saturated soil mass subjected to a three-dimensional stress situation under the fully drained condition. In such a situation, if the vertical component of strain $\Delta\varepsilon_1$ at any given time during deformation can be evaluated, then it can be expressed as its hydrostatic component $\Delta\varepsilon_1^h$ as

$$\Delta\varepsilon_1^h = \frac{1}{3}\Delta\varepsilon_v \tag{4.53}$$

and the deviatoric component $\Delta\varepsilon_1^d$ can be expressed as

$$\Delta\varepsilon_1^d = \Delta\varepsilon_1 - \frac{1}{3}\Delta\varepsilon_v \tag{4.54}$$

where $\Delta\varepsilon_v$ is the volumetric strain due to hydrostatic stress (Šuklje 1969). The methods for evaluation of the above strains and determination of the rheological constants on the basis of the strains evaluated are discussed in the following sections.

4.3.4.1.1 Hydrostatic Stress Condition

Application of stress at the surface causes an excess pore water pressure u to develop at points in the underlying saturated soil, where u is expressed in terms of the pore pressure coefficients A and B and the incremental principal stresses $\Delta\sigma_1$, $\Delta\sigma_2$, and $\Delta\sigma_3$. The state of stress can be separated into its hydrostatic and deviatoric components as

Total stress = Hydrostatic stress + Deviatoric stress

$$
\begin{bmatrix}
\Delta\sigma_1 & 0 & 0 \\
0 & \Delta\sigma_2 & 0 \\
0 & 0 & \Delta\sigma_3
\end{bmatrix}
$$

$$
=
\begin{bmatrix}
\Delta\sigma_v & 0 & 0 \\
0 & \Delta\sigma_v & 0 \\
0 & 0 & \Delta\sigma_v
\end{bmatrix}
\tag{4.55}
$$

$$
+
\begin{bmatrix}
(\Delta\sigma_1 - \Delta\sigma_v) & 0 & 0 \\
0 & (\Delta\sigma_2 - \Delta\sigma_v) & 0 \\
0 & 0 & (\Delta\sigma_3 - \Delta\sigma_v)
\end{bmatrix}
$$

where

$$\Delta\sigma_v = \frac{1}{3}(\Delta\sigma_1 + \Delta\sigma_2 + \Delta\sigma_3) \tag{4.56}$$

The volumetric strain $\Delta\varepsilon_v$ due to hydrostatic stress can be obtained as

$$\Delta\varepsilon_v = \frac{\Delta\sigma_v}{K} = m_{vi}\Delta\sigma_v = 1.5 m_v \Delta\sigma_v \tag{4.57}$$

where K is the bulk modulus, m_{vi} is the coefficient of volume compressibility determined from a triaxial isotropic consolidation test for a three-dimensional stress situation, and m_v is the coefficient of volume compressibility for one-dimensional consolidation. The relationship between the two coefficients is given by Head (1984). The vertical component of volumetric strain due to the hydrostatic stress condition when the steady state is reached is given by

$$\Delta\varepsilon_1^{h\infty} \;=\; \frac{1}{3}\,\Delta\varepsilon_v^{\infty} \;=\; \frac{1}{3}\,(1.5\,m_v\Delta\sigma_v) \;=\; \frac{1}{2}\,m_v\Delta\sigma_v \qquad (4.58)$$

Therefore, the spring constant k'' under the hydrostatic condition is given by

$$k'' \;=\; \frac{\Delta\sigma_v}{\Delta\varepsilon_1^{h\infty}} \qquad (4.59a)$$

and the dashpot viscosity coefficient at any time T, using Equation 4.52, is given by

$$\eta''^{\,T} \;=\; \frac{-k''\,T}{\ln\!\left(1 \,-\, \dfrac{\Delta\varepsilon_1^{hT}}{\Delta\varepsilon_1^{h\infty}}\right)} \qquad (4.59b)$$

where ε_1^{hT} is the hydrostatic strain at any time T during consolidation and can be expressed as

$$\Delta\varepsilon_1^{hT} \;=\; U_h\,\Delta\varepsilon_1^{h\infty} \qquad (4.60)$$

where U_h is the degree of hydrostatic consolidation. Therefore, Equation 4.59b takes the following form:

$$\eta''^{\,T} \;=\; \frac{-k''\,T}{\ln\,(1 \,-\, U_h)} \qquad (4.61)$$

The time T required to reach a certain percentage of consolidation can be determined from Terzaghi's one-dimensional consolidation theory.

4.3.4.1.2 Deviatoric Stress Condition

The deviatoric strain at infinite time when the steady state condition is reached is given by

$$\Delta\varepsilon_1^{d\infty} \;=\; \Delta\varepsilon_1^{\infty} \,-\, \Delta\varepsilon_1^{h\infty} \qquad (4.62a)$$

where $\Delta\varepsilon_1^{h\infty}$ is given by Equation 4.58 and the strain due to the applied stress increment tensor (Equation 4.55) $\Delta\varepsilon_1^{\infty}$ is given by

$$\Delta\varepsilon_1^{\infty} \;=\; \frac{\Delta s}{H} \;=\; \frac{m_v\Delta u H}{H} \;=\; m_v[\Delta\sigma_3 + A(\Delta\sigma_1 - \Delta\sigma_3)] \qquad (4.62b)$$

where Δs is the vertical compression of a soil layer of thickness H caused by the increase in pore water pressure Δu given by Skempton and Bjerrum (1957).

The spring constant k' under the deviatoric condition is therefore given by

$$k' = \frac{\Delta\sigma_1 - \Delta\sigma_v}{\Delta\varepsilon_1^{d\infty}} \tag{4.63}$$

and the dashpot viscosity coefficient at any time T is given by employing Equation 4.52:

$$\eta'^T = \frac{-k'T}{\ln(1 - U_d)} \tag{4.64}$$

This equation is analogous to Equation 4.61, where U_d is the degree of deviatoric consolidation expressed as

$$U_d = \frac{\Delta\varepsilon_1^{dT}}{\Delta\varepsilon_1^{d\infty}} \tag{4.65}$$

where $\Delta\varepsilon_1^{dT}$ is the deviatoric strain at any time T.

Since $\Delta\varepsilon_1^{dT}$ and $\Delta\varepsilon_1^{hT}$ correspond to the same time T, the value of U_d can be taken as that of U_h. The degree of consolidation U that corresponds to any time T during the consolidation period can easily be obtained. The dashpot viscosity coefficients η'' and η' (Equations 4.61 and 4.64) are both functions of load and time and will vary accordingly.

4.3.4.2 Determination of Material Constants Using Triaxial Test Data

Sharma (1989) considered a nonlinear Kelvin model consisting of a Hookean element with a spring constant k and a dashpot with a constant η, both connected in parallel, and proposed a methodology for the determination of the rheological constants from triaxial tests. The same methodology is presented here.

The model considered is similar to the earlier model (Figure 4.17) and follows the same constitutive relationship (Equation 4.48). The rheological constants can be defined as follows. The spring constant k is

$$k = \frac{1}{a + b\varepsilon} \tag{4.66}$$

and at any time T the dashpot constant is

$$\eta = \eta_o T^{1-N} \tag{4.67}$$

where a, b, η_o, and N are constants. Thus, the governing differential equation for the proposed model becomes:

$$\sigma = \frac{1}{a + b\varepsilon} \varepsilon + \eta_o T^{1-N} \frac{d\varepsilon}{dT} \tag{4.68}$$

The solution to the above equation can be written as

$$\frac{T^N}{N\eta_o} = \frac{b}{f}(\varepsilon - \varepsilon_o) - \frac{a}{f^2} \ln\left(\frac{a\sigma + f\varepsilon}{a\sigma + f\varepsilon_o}\right) \tag{4.69}$$

where

$$f = b\sigma - 1 \tag{4.70a}$$

and ε_o is the initial strain at time $T = 0$. If this initial strain is zero, Equation 4.69 can be simplified as:

$$\frac{T^N}{N\eta_o} = \frac{b\varepsilon}{f} - \frac{a}{f^2} \ln\left(1 + \frac{f\varepsilon}{a\sigma}\right) \tag{4.70b}$$

Equation 4.70 contains four constants: a, b, η_o, and N. These constants can be determined separately under both hydrostatic and deviatoric stress conditions on the basis of triaxial tests which can be conducted on soil samples. Subsequently, spring and dashpot constants due to these stress conditions can be evaluated.

To determine the above constants, it is essential to know the experimental strain-time curve under the constant stress condition. It is easier to obtain this curve for the hydrostatic stress situation, but it is difficult to obtain it directly for the deviatoric stress condition. The deviatoric strains can be computed from strains for the total stress situation by subtracting the hydrostatic strains. Triaxial tests can be conducted on identical soil samples under total and hydrostatic stress situations to obtain the axial strain vs. time curve. The data obtained from these tests can be analyzed to estimate the spring and dashpot constants for the hydrostatic and deviatoric stress conditions.

The axial strains due to the total stress condition are calculated by dividing the observed axial displacements by the original length of the soil sample. The volumetric strains are calculated by dividing the volume change (which can be observed in the form of drained water from the soil sample in a burette) by the original volume of the sample. The strain vs. time plot can then be obtained for the hydrostatic stress condition, taking one-third of the volumetric strain as axial strain. The deviatoric strains are computed by subtracting one-third of the volumetric strains from the axial strains obtained under the total stress condition, and the strain vs. time curve for the deviatoric stress condition also can be plotted. It can be observed that axial strain vs. time curves for both the hydrostatic and deviatoric stress conditions tend to become asymptotic when the rate of strain can be assumed to be zero.

When the strain rate tends to become zero, the strain becomes the final strain ε_f, and Equation 4.68 takes the form

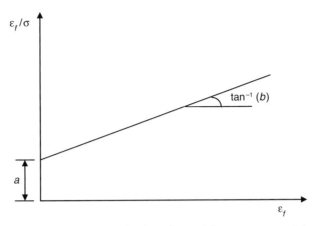

FIGURE 4.18a Determination of material constants a and b for nonlinear Kelvin model.

$$\sigma = \frac{\varepsilon_f}{a + b\varepsilon_f} \quad \text{or} \quad \frac{\varepsilon_f}{\sigma} = a + b\varepsilon_f \quad (4.71)$$

The final strain can be obtained from the strain-time curves. The above equation suggests that if the ratio ε_f/s is plotted against ε_f, a linear relationship would be obtained such that a is the intercept on the ε_f/σ axis and b is the slope of the straight line (Figure 4.18a).

The constants η_o and N can be computed using Equations 4.70a and 4.70b as follows. Equation 4.71 can be rewritten as:

$$\frac{(b\sigma - 1)}{a\sigma} = -\frac{1}{\varepsilon_f} \quad \text{or} \quad \frac{f}{a\sigma} = -\frac{1}{\varepsilon_f} \quad (4.72)$$

Substituting the value of f and the above expression in Equation 4.70b and rearranging the terms gives

$$F(T) = \frac{T^N}{N\eta_o} = -\frac{a}{(1 - b\sigma)^2} \ln\left(1 - \frac{\varepsilon}{\varepsilon_f}\right) - \frac{b\varepsilon}{(1 - b\sigma)} \quad (4.73)$$

where $F(T)$ is a function of time T. If $F(T)$ is plotted against time on a logarithmic plot (Figure 4.18b), the constants N and η_o can be computed.

4.4 Application to Shallow Foundations

In addition to the conventional methods of analyzing the soil-foundation interaction phenomenon, various research workers have proposed different methods that employ various

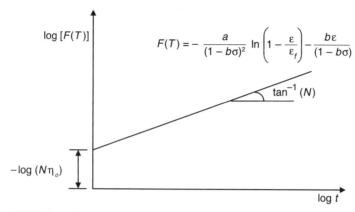

FIGURE 4.18b   Determination of material constants N and η_o for non-linear Kelvin model.

constitutive models for the analysis of shallow foundations. Some typical studies are presented below that pertain to the analysis of various types of shallow foundations on different types of soil which take into consideration the interaction between the soil and the foundation.

4.4.1 Strip Footings

Strip footings are shallow footings subjected to a uniformly distributed load. Usually these are analyzed under a plain strain condition. Khadilkar and Varma (1977) addressed the interference effect of two adjacent strip footings resting on cohesionless soil by employing the finite element method and by invoking the nonlinear stress-dependent and inelastic soil behavior. Gazetas (1980) presented an analytical-numerical formulation for dynamic and static analysis of strip foundations on an elastic isotropic medium consisting of heterogeneous layers. The main emphasis was on the dynamic aspect of the analysis. Small and Booker (1984) analyzed a horizontally layered elastic material using an exact finite-layer flexibility matrix. This method is useful in overcoming the difficulty which can arise due to incompressible behavior in undrained conditions. Li and Dempsey (1988b) proposed a solution for a rigid strip footing on an elastic layer. Azam et al. (1991) investigated the performance of strip footings on homogeneous soil and also a stratified deposit containing two soil layers, both with and without a continuous void, using the finite element method. Maheshwari and Madhav (2006) presented an elastic approach for the analysis of strip footings on layered soil and investigated the effect of a thin and a very stiff soil layer sandwiched between two soil layers on deformation and stress distribution. Maheshwari and Viladkar (2007) extended this study to understand the influence of relative thickness and modular ratio on the response of the strip footing.

The interference phenomenon is quite common in the case of shallow footings and can only be dealt with by considering the soil-footing interaction. Khadilkar and Varma (1977) analyzed the problem of the interference of two strip footings resting on cohesionless soil using the finite element method by considering nonlinear stress-dependent and inelastic soil behavior. The stress deformation study first was conducted for an isolated footing and subsequently was extended to the interfering footings at various spacings for rigid and flexible foundations. A quadrilateral finite element composed of four constant strain triangles was adopted for the discretization. The stress-strain behavior of sands was approximated by using the hyperbolic model presented by Duncan and Chang (1970) (Equation 4.23). The model parameters suggested by Duncan and Chang (1970) for dense silica sand at a relative density of 100% were used, with $\phi = 36.5°$, $R_f = 0.91$, $K = 2000$, and $n = 0.54$. The unit weight of sand was taken as 17 kN/m³. First, the initial stresses corresponding to at rest conditions were introduced in the soil. An incremental procedure was adopted to invoke the nonlinear stress-dependent and inelastic behavior of the soil in the analysis. In this procedure, the stress components of the elements are accumulated at the end of each load step, and the tangent moduli for the successive load increments are computed from the resulting principal stresses after ascertaining the strength criterion based on the Mohr-Coulomb failure hypothesis. The inelasticity of the soil behavior was taken into account in the analysis by adopting the appropriate unload and reload moduli for elements where the major principal stress value σ_1 decreases for the progressive load increments on the footing. The modulus was calculated from Equation 4.74 until the element developed a value of σ_1 which exceeded the corresponding value prior to unloading:

$$E_{ur} = K_{ur} \left(\frac{\sigma_3}{p_a} \right)^n \tag{4.74}$$

where the parameter K_{ur} was assumed to have a value of 2120. The load intensity on the footing was incrementally increased to 14 t/m², and this value was found to exceed the ultimate bearing capacity of an isolated footing on this sand.

The influence of friction between the rigid footing and the soil was considered by employing special joint elements in plane strain in the nonlinear analysis. The normal and shear stresses at the footing-soil contact interface were computed after each load increment and a friction rule was applied by prescribing a coefficient of sliding friction $U_f = 0.5$. The load-settlement characteristics were obtained from the analysis for the case of isolated and interfering footings for both rigid and flexible strip footings. It was found that interfering footings in certain cases indicate an increase in bearing capacity governed by the settlement criteria. However, at smaller spacings, the interference causes greater differential settlement. The settlement pattern of interfering footings is indicated in Figure 4.19, and numerical values for some typical cases are given in Tables 4.3a and 4.3b.

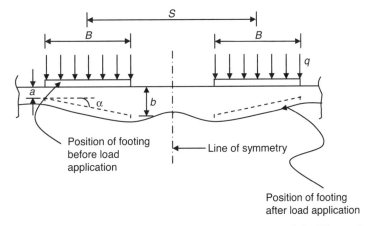

FIGURE 4.19 Settlement pattern of interfering footings (Khadilkar and Varma 1977).

TABLE 4.3a Settlement Pattern for Interfering Rigid Footings with Rough Interface

Loading intensity q (t/m^2)	Settlement (mm) $S = 2B$		Tilt α (radians)	Settlement (mm) $S = 3B$		Tilt α (radians)
	a	b		a	b	
4.0	2.7	3.4	0.0035	3.4	3.3	−0.0005
6.0	5.1	4.9	−0.0010	6.5	5.4	−0.0050
8.0	8.3	8.0	−0.0015	8.8	7.9	−0.0045
10.0	11.9	11.5	−0.0020	11.8	10.3	−0.0075
12.0	13.6	15.3	0.0085	14.7	11.7	−0.0150

Based on the results of Khadilkar and Varma (1977).

TABLE 4.3b Settlement Pattern for Interfering Footings for Coefficient of Sliding Friction = 0.5

Loading intensity q (t/m^2)	Settlement (mm) $S = 2B$		Tilt α (radians)	Settlement (mm) $S = 3B$		Tilt α (radians)
	a	b		a	b	
4.0	2.2	2.7	0.0025	5.3	5.0	−0.0015
6.0	5.0	6.3	0.0065	6.5	5.8	−0.0035
8.0	5.6	6.9	0.0065	8.1	6.5	−0.0008
10.0	6.5	10.0	0.0175	9.0	−1.1	−0.0505
12.0	9.2	13.2	0.0200	10.0	−0.9	−0.0545

Based on the results of Khadilkar and Varma (1977).

The horizontal stress components in the soil continuum below the closer vicinity of the footing on the interfering side are higher, and this resulted in larger soil moduli due to confinement. Therefore, the interfering rigid footings tilt away from each other during the initial stages of loading ($a < b$, Figure 4.19). As loading progresses, these increased soil moduli in this region build up greater vertical stress components, which for smaller spacings (2B and below) are large enough to cause many soil elements to fail, resulting in the footings tilting toward each other (positive α).

For greater spacings (i.e., 3B and above), the vertical stress components in the region, as mentioned above, were not found to build to such levels as to cause failure while the increased horizontal stress component prevailed. Therefore, the footings tilt progressively away from each other (Table 4.3a) as the load is increased incrementally.

The results for the influence of friction at the contact surface are presented in Table 4.3b. It was observed that the interfering footings for U_f = 0.5 yield greater total and differential settlements than rigid rough interfering footings.

The finite element analysis was further extended to obtain the displacement and stress patterns for some other cases of interfering footings of larger widths. It was noted that there is qualitative agreement in the settlement and tilt patterns with values obtained for smaller footing widths. However, for the same spacing of interfering footings, the magnitude of tilts associated with wider footings was found to be smaller compared with footings of smaller widths for an applied load intensity. This study clearly brought out the interesting behavior of two strip footings at various spacings as influenced by the assumed constitutive response of the soil and was found to be helpful in better understanding the problem of interference.

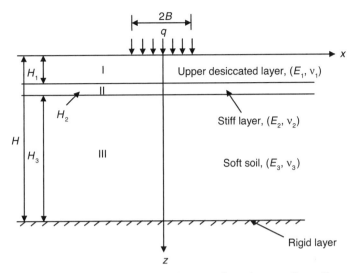

FIGURE 4.20 Strip footing resting on three-layer soil medium (Maheshwari and Madhav 2006).

Maheshwari and Madhav (2006) analyzed a strip footing resting on a three-layer soil medium by employing the theory of elasticity approach. The main purpose of this investigation was to evaluate and quantify the effect of the thin but very strong and stiff layer on the distribution of stresses on the soil and the settlement of the lower normally consolidated alluvial deposit. The soil deposit was modeled as depicted in Figure 4.20. The second layer (II) was considered to be the stiffest layer and the third layer (III) was the softest layer (i.e., $E_1 < E_2$ and $E_3 < (E_1$ and $E_2)$. The governing differential equations for this model were derived from the theory of elasticity approach as

$$G\nabla^2 u + \frac{G}{(1 - 2v)} \frac{\partial}{\partial x}\left(\frac{\partial u}{\partial x} + \frac{\partial w}{\partial z}\right) = 0$$

$$G\nabla^2 w + \frac{G}{(1 - 2v)} \frac{\partial}{\partial z}\left(\frac{\partial u}{\partial x} + \frac{\partial w}{\partial z}\right) = 0$$

(4.75)

and the stresses can be expressed in terms of displacements as

$$\sigma_x = \frac{E}{(1 + v)(1 - 2v)}\left[(1 - v)\frac{\partial u}{\partial x} + v\frac{\partial w}{\partial z}\right]$$

$$\sigma_z = \frac{E}{(1 + v)(1 - 2v)}\left[v\frac{\partial u}{\partial x} + (1 - v)\frac{\partial w}{\partial z}\right]$$

(4.76)

$$\tau_{xz} = G\left(\frac{\partial w}{\partial x} + \frac{\partial u}{\partial z}\right)$$

where E and ν are the elastic modulus and Poisson's ratio, respectively; σ_x and σ_z are the normal stresses in the x and z directions, respectively; τ_{xz} is the shear stress; G represents the shear modulus; and u and w are the independent displacements in the x and z directions, respectively.

The governing equations were solved with the help of appropriate boundary and continuity conditions as follows.

Stress conditions:

$$\sigma_z = \frac{E_1}{(1 + \nu_1)(1 - 2\nu_1)} \left[(1 - \nu_1) \frac{\partial w}{\partial z} + \nu_1 \frac{\partial u}{\partial x} \right] = q \tag{4.77}$$

$$\text{for } x \leq B; \; z = 0$$

$$\sigma_z = \frac{E_1}{(1 + \nu_1)(1 - 2\nu_1)} \left[(1 - \nu_1) \frac{\partial w}{\partial z} + \nu_1 \frac{\partial u}{\partial x} \right] = 0 \tag{4.78}$$

$$\text{for } x > B; \; z = 0$$

$$\tau_{xz} = G \left(\frac{\partial w}{\partial x} + \frac{\partial u}{\partial z} \right) = 0 \quad \text{for all } x; \; z = 0 \tag{4.79}$$

Displacement boundary conditions:

$$u = 0 \quad \text{for } x \leq B; \; z = 0 \tag{4.80a}$$

and

$$u, w = 0 \quad \text{for all } x; \; z = H \tag{4.80b}$$

Continuity conditions at the interface where $z = H_1$ in terms of displacements:

$$\frac{E_1}{(1 + \nu_1)(1 - 2\nu_1)} \left[(1 - \nu_1) \frac{\partial w}{\partial z} + \nu_1 \frac{\partial u}{\partial x} \right]$$

$$= \frac{E_2}{(1 + \nu_2)(1 - 2\nu_2)} \left[(1 - \nu_2) \frac{\partial w}{\partial z} + \nu_2 \frac{\partial u}{\partial x} \right] \tag{4.81}$$

and

$$G_1 \left(\frac{\partial w}{\partial x} + \frac{\partial u}{\partial z} \right) = G_2 \left(\frac{\partial w}{\partial x} + \frac{\partial u}{\partial z} \right) \tag{4.82}$$

Similarly, at the interface where $z = H_1 + H_2$:

$$\frac{E_2}{(1 + \nu_2)(1 - 2\nu_2)} \left[(1 - \nu_2) \frac{\partial w}{\partial z} + \nu_2 \frac{\partial u}{\partial x} \right]$$

$$= \frac{E_3}{(1 + \nu_3)(1 - 2\nu_3)} \left[(1 - \nu_3) \frac{\partial w}{\partial z} + \nu_3 \frac{\partial u}{\partial x} \right] \tag{4.83}$$

and

$$G_2 \left(\frac{\partial w}{\partial x} + \frac{\partial u}{\partial z} \right) = G_3 \left(\frac{\partial w}{\partial x} + \frac{\partial u}{\partial z} \right) \tag{4.84}$$

where the various terms are as defined in Figure 4.20.

Equation 4.75 along with above-mentioned boundary conditions were expressed in finite difference form and solved by employing the Gauss-Siedel iterative technique to obtain the horizontal and vertical displacements of the footing at various nodes inside the soil medium. Once the displacements were evaluated, Equation 4.76 was used to evaluate the respective stresses. For the sake of simplicity, Poisson's ratio was kept constant at 0.3 for all three soil layers. A detailed parametric study was conducted to study the influence of the presence of a thin but very stiff soil layer sandwiched between two relatively softer soil layers. The thin but very stiff middle layer was found to act as a plate, and it redistributed the stresses uniformly on the very soft lower soil layer. The stresses on the lower soft soil layer were found to reduce to a large extent. The effect of the position of the middle stiff layer also was studied. The closer the middle stiff layer was to the ground surface, the less the displacement. The effect of variation of the modular ratio of the third and second layers (i.e., E_3/E_2) was not very significant in the stress redistribution, but it helped in the reduction of stress on the lower soft soil layer.

Maheshwari and Viladkar (2007) extended the above-mentioned analysis (Figure 4.20) to study the influence of the relative thickness and relative modular ratio of adjacent soil layers on the resulting vertical displacements and vertical stress redistribution. A detailed parametric study was carried out for this purpose, and relevant parameters were adopted for both conventional and industrial structures such as silos, chimneys, cooling towers, overhead tanks, etc. The input parameters for this study are given in Table 4.4.

For typical industrial structures such as silos, chimneys, etc., an increase in the normalized thickness of the upper soil layer H_1/H could be of help in reducing the vertical displacement along the thickness of the middle soil layer by about 75% (Figure 4.21). The corresponding reduction in vertical stress at the center of the footing could be of the order of about 17% (Figure 4.22). This also was found to be true in the case of conventional structures. The variation in the normalized thickness of the middle soil layer H_2/H was found to affect the vertical displacement along the thickness of the upper and the middle soil layers significantly. However, its effect along the thickness of the lower layer was negligible.

Figures 4.23 and 4.24 show the typical effect of variation of E_1 and modular ratios E_2/E_1 and E_3/E_2 on vertical displacement and vertical stress distribution along the soil interfaces for various parametric values listed in the plot. The maximum vertical stress occurs at the center

TABLE 4.4 Range of Values of Various Parameters Considered for Parametric Study (Maheshwari and Viladkar 2007)

Parameter	Symbol	Range of Values Conventional Structures	Range of Values Industrial Structures	Units
Applied load intensity	q	0.15	0.30	MN/m
Half-width of loaded region	B	1.0	4.0	m
Thickness of soil stratum	H	16	40	m
Elastic modulus of upper soil layer	E_1	30–120	80	MPa
Poisson's ratio	ν	0.3	0.3	—
Relative thickness of upper soil layer	H_1/H	0.05–0.3	0.05–0.3	—
Relative thickness of middle soil layer	H_2/H	0.05–0.3	0.05–0.3	—
Modular ratio with respect to upper and middle soil layer	E_2/E_1	0.5–4	0.25–2	—
Modular ratio with respect to middle and lower soil layer	E_3/E_2	0.5–2	0.25–1.25	—

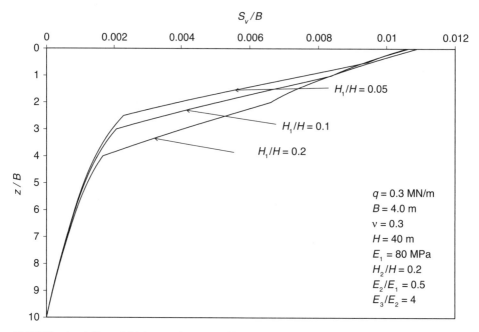

FIGURE 4.21 Effect of thickness of upper soil layer on vertical displacement along central axis of footing (Maheshwari and Viladkar 2007).

of the footing, gradually reducing with distance from the central axis and vanishing at the boundary. Further, an increase in the modular ratio E_2/E_1 was found to be of help in reducing the vertical displacement below the center of the footing. A reduction in vertical displacement was observed with an increase in the modular ratio E_3/E_2. At the layer interfaces, this reduction was found to be 65–70% for both conventional as well as industrial structures. The

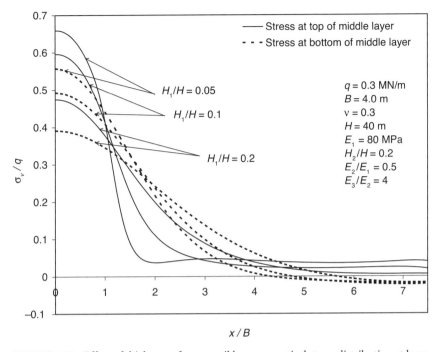

FIGURE 4.22 Effect of thickness of upper soil layer on vertical stress distribution at layer interfaces (Maheshwari and Viladkar 2007).

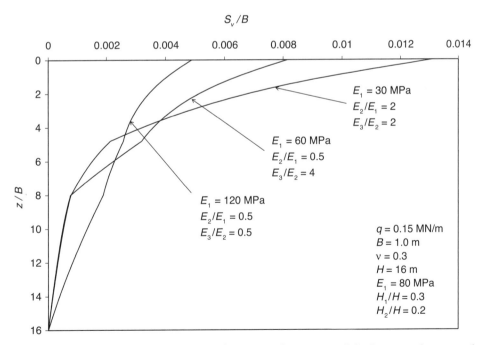

FIGURE 4.23 Effect of variation of E_1, E_2/E_1, and E_3/E_2 on vertical displacement along central axis of footing (Maheshwari and Viladkar 2007).

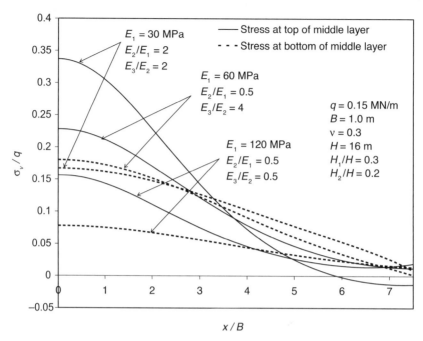

FIGURE 4.24 Effect of variation of E_1, E_2/E_1, and E_3/E_2 on vertical stress distribution at layer interfaces (Maheshwari and Viladkar 2007).

corresponding effect on the vertical stress at the bottom of the middle layer was found to be more pronounced compared to that on top of the middle layer.

4.4.2 Isolated Footings

Isolated footings usually are subjected to various types of loading, such as concentrated, triangular, rectangular, etc. Boussinesq's solution (Equations 4.16 and 4.17) has been used very widely to determine the deflection and stress profiles for such types of loading by means of the principle of superposition. Typical distribution of the vertical stress beneath the point load is depicted in Figure 4.25.

The soil-foundation interaction of isolated footings has been analyzed by various research workers for different shapes of footings resting on sand, clay, or a layered soil stratum by employing various approaches as presented above. Desai and Reese (1970) analyzed circular footings resting on single-layer and two-layer cohesive subsoils by the finite element method. The suggested method used the nonlinear stress-strain relationship obtained from triaxial tests. An excellent correlation was obtained between the results from laboratory experiments and the finite element method. Borodachev (1976), Li and Dempsey (1988a), and Dempsey and Li (1989) analyzed a rectangular footing by employing the elastic half-space approach. Viladkar et al. (1992) presented a three-dimensional viscoelastic finite element formulation to analyze a square footing. The approach considered the stress-strain-time response of the supporting soil medium (represented as a Kelvin model). Further, a methodology for evaluating time-dependent viscoelastic constants for the soil mass also was presented. The behavior of a square footing was compared with that predicted by the Skempton and Bjerrum theory and the results were found to be in good agreement. Papadopoulos (1992) presented a new

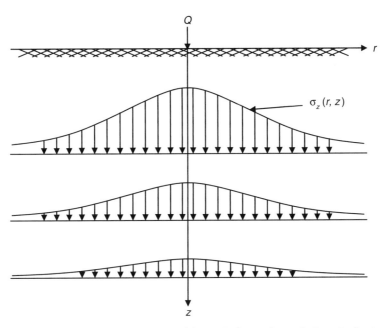

FIGURE 4.25 Typical distribution of the vertical stress beneath the point load.

method for the estimation of settlements of shallow foundations in cohesionless soils. The method included a simplified model for the distribution of applied stresses with depth, incorporated certain experimental results concerning the stress-strain relation in cohesionless soils, and proposed a formula for selection of the effective depth below which deformations can be considered negligible. It was found that the final expressions for the calculation of settlement, although they have the form of elastic-type solutions, were, in effect, nonlinear functions of the foundation width and of the applied loading. A rectangular footing was analyzed and the influence of certain basic factors, such as the effect of stress history and the width of the foundation, on the settlement of cohesionless soils was examined. It was concluded that the trends from known statistical correlations were predicted satisfactorily by the proposed method. Back calculations of settlement using data from published cases also showed generally satisfactory approximation.

Mayne and Poulos (1999) proposed a methodology for obtaining approximate displacement influence factors for calculating the magnitude of drained and undrained settlements of shallow foundations by simple numerical integration of elastic stress distribution in a spreadsheet. Influence factors for circular foundations resting on soils with homogeneous (constant modulus with depth) to Gibson-type (linearly increasing modulus) profiles with finite layer thicknesses were obtained by summing the unit strains from incremental vertical and radial stress changes. The influence of foundation rigidity and embedment was addressed by approximate modifier terms obtained from prior finite element studies. An equivalent circular foundation was used to approximate other geometric areas. Results were compared with closed-form analytical and rigorous numerical solutions, where available. A new solution for Gibson soil of finite thickness was presented.

Conniff and Kiousis (2007) introduced a novel elastoplastic three-degree-of-freedom medium which can be employed to model foundation settlements under combined loadings. A soil-structure interaction problem can then be solved by replacing the soil mass with this three-degree-of-freedom elastoplastic medium, thus significantly reducing the size of the problem. The model was developed by extending the classical plasticity concepts to the force deformation level. Its ability to predict foundation deformations was evaluated using finite element solutions of a typical shallow foundation (rectangular and circular) problem and was found to be reasonably accurate as well as to save a significant amount of time.

Some of the typical studies are discussed below in detail in order to understand the interaction behavior of isolated footings of different shapes resting on various type of soils.

Desai and Reese (1970) investigated the behavior of rigid circular footings on a single layer of clay and on two layers of clay using the finite element method. The proposed method employed nonlinear stress-strain curves which were obtained from triaxial tests. Two plate loading tests also were conducted to predict the load displacement characteristic of a steel footing with a diameter of 3 in. and thickness of 0.5 in., and the results were compared with those obtained from the finite element method.

For the finite element analysis, material properties were obtained from the actual test behavior and were specified in terms of deformation moduli. These moduli depend on the state of stress for a nonlinear material. Quadrilateral elements were adopted for discretizing the soil-footing system. The stiffness matrix of the entire continuum was obtained from individual element stiffnesses by means of the direct stiffness method. The Gaussian elimination technique was used to solve the resulting simultaneous equations for unknown displacements. Subsequently, the stresses and strains were computed from the known displacements.

Stress-strain curves for clays were obtained from undrained triaxial tests for different confining pressures. The incremental load method was employed to compute the constitutive matrix corresponding to the ith state of stress, as given below.

In triaxial test conditions, the generalized constitutive relation yields

$$\varepsilon_{zz} = \frac{1}{E}(\sigma_{zz} - 2\nu\sigma_{rr}) = \frac{\sigma_{zz} - \sigma_{rr}}{E} + \frac{\sigma_{rr}(1 - 2\nu)}{E} \quad (4.85)$$

where σ_{zz} is the axial stress, σ_{rr} is the confining pressure, and ε_{zz} is the axial strain. For a constant confining pressure σ_{rr}, differentiation of Equation 4.85 would yield the deformation modulus E as:

$$E = \frac{d(\sigma_{zz} - \sigma_{rr})}{d\varepsilon_{zz}} \quad (4.86)$$

For a load increment, the deformation modulus E can be computed from

$$E = \frac{(\sigma_{zz} - \sigma_{rr})_i - (\sigma_{zz}\sigma_{rr})_{i-1}}{(\varepsilon_{zz})_i - (\varepsilon_{zz})_{i-1}} \quad (4.87)$$

where i indicates the current total load after the ith increment and $i - 1$ indicates the total load after the $(i - 1)$th increment.

Poisson's ratio of the soils was assumed to be constant at 0.485 during the deformation process. To start the computations, values of the initial modulus E_o were computed from the initial slopes of the stress-strain curves. With the onset of the deformation process, the deformation modulus E was modified as per Equation 4.87.

A nonlinear stress-strain relationship was used for the analysis of a rigid circular footing resting on layered soil media and showed that the finite element method can be adopted for the solution of very complex problems of foundation engineering that involve nonhomogeneous materials, arbitrary boundary conditions, and nonlinear stress-strain behavior. The proposed method finds application in the design and bearing capacity computations of foundations in layered cohesive soils.

Dempsey and Li (1989) attempted to address the soil-foundation interaction problem of a rigid rectangular footing ($a \times b$) on an elastic layer under symmetric loading. The soil was modeled by using the elastic half-space approach. The contact between the soil and the footing was assumed to be frictionless. The soil layer was assumed to rest on a rigid base. Burmister's (1956) point load solution was employed and extended for this study. The contact pressure at the edges and corners of a rigid footing is singular; therefore, these singularities were treated by discretizing the contact region as per Gauss-Chebyshev quadrature.

A detailed parametric study was conducted to investigate the influence of Poisson's ratio (0.1–0.5), normalized depth ($d/a = 0.3$ to infinity), and aspect ratio of the footing ($r = b/a$ = 1 to infinity) on displacement and the contact pressure at the center of a footing. Figures 4.26 and 4.27 show, respectively, typical results for depicting the variation in contact pressure and displacement at the center of a footing with depth. The corresponding Poisson's ratio was 0.3 and the results adopted an aspect ratio r that varied from 1 to infinity (strip footing).

The results from a parametric study in the form of footing indentations and center contact pressures were found to be independent of Poisson's ratio for an elastic half-space (i.e., normalized depth, d/a tending to infinity). The intensity of the pressure singularities reduces

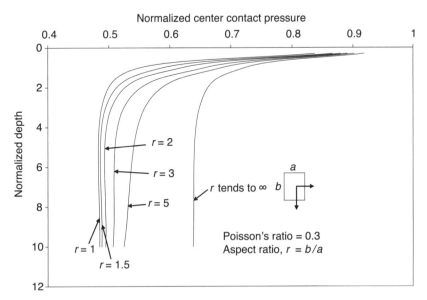

FIGURE 4.26 Effect of aspect ratio on normalized contact pressure at the center of a rectangular footing (based on the results of Dempsey and Li 1989).

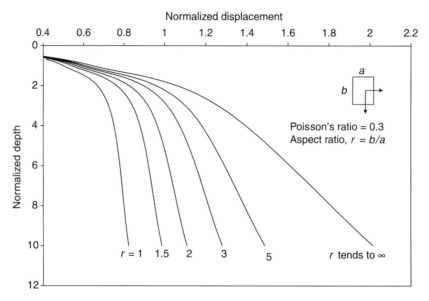

FIGURE 4.27   Effect of aspect ratio on normalized displacement of a rectangular footing (based on the results of Dempsey and Li 1989).

near the edge of the footing with the reduction in soil layer depth. This causes the pressure to increase in the center; therefore, to get a general idea of the pressure distribution, the pressure was normalized as the ratio of the center pressure to the average pressure. In general, for all values of Poisson's ratio, this ratio was found to be less than one, indicating that the center pressure is less than the average pressure (Figure 4.26).

When the thickness of the soil layer is less, the indentations for an incompressible material (Poisson's ratio = 0.5) were found to be much smaller than those for other Poisson's ratios and also decreased with increasing aspect ratio. This was observed because of the greater increase in stiffness due to bonded contact at the rigid base compared to the reduction in stiffness due to a higher aspect ratio. The elastic solutions thus obtained and presented (e.g., Figures 4.26 and 4.27) were found to be useful in the design and analysis of rectangular footings.

Viladkar et al. (1992) presented a three-dimensional viscoelastic finite element formulation to study the behavior of a soil-footing system that takes into account the stress-strain-time response of the supporting soil medium. A square footing resting on stratified soil was analyzed and the footing was assumed to behave in a linearly elastic manner. The soil supporting the footing was modeled as a linear viscoelastic half-space. The problem considered was treated as quasi-static and isothermal due to the presence of clay soil. The shear strength of the soil was not exceeded, and the assumption of infinitesimal strain theory was considered as valid.

The saturated soil mass was considered to be in the drained condition and was modeled as a Kelvin body (Figure 4.17). Constitutive relations were developed for the soil mass, modeled as a viscoelastic body, for uniaxial and hydrostatic stress conditions. Analysis of the soil-footing interaction problem was carried out by employing the finite element method. The equations for element stiffness matrices were derived and evaluated using the Gauss integration technique. The viscoelastic finite element formulation for saturated clay soils in the drained condition and a conventional elastic finite element formulation for a soil-foundation system were integrated to obtain the solution. The material constants were determined using consolidation test data as discussed in Section 4.3.4.1.

The problem of a square footing 2 m × 2 m in plan and 0.75 m thick subjected to a uniform pressure intensity of 150 kN/m^2 was analyzed in order to establish the proposed three-dimensional finite element formulation. The footing was resting on a clay layer underlain by sand layers. Various properties of different soil layers are depicted in Figure 4.28. The behavior of the footing was studied with respect to settlement of the footing, variation of the stress below the footing with time, computational time, and effect of the length of the time step.

Due to symmetry, a quarter of the soil-footing system was considered for analysis. A rigid footing was discretized as a single eight-node three-dimensional isoparametric element, and the soil medium was discretized into a number of similar elements. The side boundaries of the soil system were considered at a distance of more than five times the width of the footing to nullify their influence. Rheological constants were obtained for each of the elements for hydrostatic and deviatoric conditions. The time-settlement curves of the footing were obtained from viscoelastic finite element analysis of the soil-footing system. Two sets of results were obtained: one corresponding to an average value of the dashpot constants and the other corresponding to time-variant dashpot constants. Typical results in the form of settlement of the footing with time are presented in Figure 4.29 for two different values of the time step: Δt = 0.06 and 0.24 yr. The curves, with average values of dashpot constants as well as with time-variant dashpot constants, indicate that settlement increases at a faster rate initially and

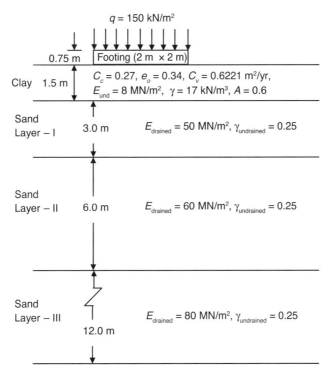

FIGURE 4.28 Square footing on stratified soil medium (as considered by Viladkar et al. 1992).

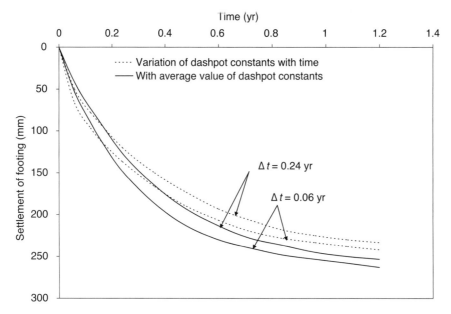

FIGURE 4.29 Settlement of footing with time for different time steps (based on the results of Viladkar et al. 1992).

becomes asymptotic with time. At a time corresponding to about 96% of consolidation and with a time step of 0.06 yr, the settlement of footing was observed to be 263.0 and 242.0 mm for average and time-variant dashpot constants, respectively. These values were found to match well with those obtained from the Skempton and Bjerrum (1957) method. The differences in the results were attributed to the proposed method and corresponding modeling of the soil-footing system.

The settlement of the footing at a time step Δt of 0.06 yr was observed to be same at a time instant of 0.162 yr for both cases of the dashpot constants. Before the time instant of 0.162 yr, the values of the dashpot constants, considered to vary with time, were less and later were found to be larger than the average values of dashpot constants. Therefore, the results indicated that the dashpot constants influence settlement. A similar observation was found for a different time step of 0.24 yr (Figure 4.29).

The settlement predicted by the proposed approach at time $T = 0$ was zero. This was due to the fact that the footing was placed on clay which was assumed to be fully saturated. However, the elastic settlement of the sand layer would appear at time $T = 0$. Due to formulation of the problem, the solution directly moved into the time domain. For the first increment of time, the displacements of the nodes in clay were considered to be zero, as the initial condition of no settlement was assumed. Therefore, elastic settlement of sand was absent at time $T = 0$. However, the analysis was able to take care of this aspect and it appeared in the displacement reached at the end of each time increment.

The increase in stress with time in the clay element just below the footing was found to be negligible, and it was suggested that it can be neglected for all practical purposes.

The consolidation period usually is considered to be 90–95% of the time for total settlement to occur; therefore, the total time for settlement to occur was considered to be the time required for about 96% consolidation, which was 1.2 yr for the soil-footing system studied.

The effect of the length of the time step also was studied by varying the time steps from 0.03 yr to 0.24 yr, and it was proposed that a suitable time step should maintain a balance between accuracy and economy in the time required to analyze the interaction problem.

Keeping in mind the aspects of accuracy and economy, it was concluded that an analysis with average values of the dashpot constants and a time increment of one-tenth of the time required for 95% consolidation of the clay layer would yield satisfactory results.

4.4.3 Combined Footings

Combined footings are used in situations where the spacing of the structural columns is so close that there would be interference from the pressure bulbs below the isolated column footings. These footings transfer the load from the superstructure to the soil near the ground surface and usually are analyzed in the form of beams. Many research workers have proposed different approaches, including analytical, numerical, and seminumerical methods, to model soil media and simulate the interaction between the soil and combined footings.

The assumption of the elastic nature of soil media is not unrealistic because most structures are subjected to working loads that cause stresses in the soil mass which remain within the elastic limit. Therefore, many mathematical analyses employ a beams on elastic foundation approach for simulation of the soil-footing interaction problem. However, the nonlinearity of soils can come into play due to the stress concentration that may occur along the edges of the foundations; in such cases, soil media can be modeled with the help of any of the models discussed in Section 4.2 to address the nonlinear or time-dependent behavior. Wang et al. (2005) presented a state-of-the-art review on the analysis of beams and plates on elastic foundations and summarized the most commonly adopted soil models and the literature pertaining to their application.

Hetényi (1946) reported the solutions for finite and infinite beams resting on a Winkler foundation for a number of cases of practical interest. The method of superposition can be used to solve problems of more general loading conditions by combining the solutions for the basic cases. Other approaches also are employed to analyze a beam resting on a Winkler foundation, including iterative methods (Gazis 1958; Yankelevsky et al. 1989), matrix methods (Iyenger 1965; Lee and Yang 1993; Bowles 1996), Green's function approach (Gao et al. 1999; Guo and Weitsman 2002), and finite element methods (Kaschiev and Mikhajlov 1995; Noorzaei et al. 1995; Erguven and Gedikli 2003). Yin (2000) developed a Timoshenko model for the analysis of the interaction action between a reinforced structure and the soil. Maheshwari (2004, 2008) modified Hetényi's model for the analysis of a combined footing resting on reinforced earth beds. Various research workers have analyzed a beam resting on a two-parameter model, including Selvadurai (1979), Zhaohua and Cook (1983), and Morfidis (2002) among others. A combined footing modeled as a beam can be analyzed by employing the elastic continuum approach (Cheung and Zeinkiewicz 1965; Cheung and Nag 1968). A few typical studies are discussed in detail below.

Hetényi (1946) presented an analysis of various types of beams on an elastic foundation for different loading conditions. In the case where the soil medium is modeled as a Winkler foundation, the differential equation for the deflection curve of such a beam of unit width is

$$EI \ \frac{d^4y}{dx^4} = -k_s y + q \tag{4.88}$$

where EI is the flexural rigidity of the beam, q is the external distributed loading, y is the deflection of the beam, and x is the dimension along the length of the beam. Along the unloaded parts of the beam, where no distributed load is acting, the general solution to the above equation becomes

$$y = e^{\lambda x}(C_1 \cos \lambda x + C_2 \sin \lambda x) + e^{-\lambda x}(C_3 \cos \lambda x + C_4 \sin \lambda x) \qquad (4.89)$$

where the constants C_1 to C_4 can be evaluated by means of appropriate boundary conditions and

$$\lambda = \sqrt[4]{\frac{k_s}{4EI}} \qquad (4.90)$$

The term $1/\lambda$ is referred to as the characteristic length. As can be observed, as the length of the beam increases, the end effects become less pronounced, and therefore length becomes an important parameter influencing the response of the beam. Beams are classified in terms of length as:

Short beams	For $\lambda l < \pi/4$
Beams of medium length	For $\pi/4 < \lambda l < \pi$
Long beams	For $\lambda l > \pi$

where l is the length of the beam.

A closed-form solution for various types of loadings, such as a uniformly distributed patch strip load, symmetrically placed concentrated load, etc., can be obtained (Hetényi 1946). For some problems, difficulty arises in determining the integration constants; this can be avoided by using the method of superposition, which also can be employed when the beam is subjected to axial forces or twisting moments in addition to lateral loads.

Bowles (1996) presented the general formulation for the finite element analysis of beams on an elastic foundation. Yin (2000) suggested a method for obtaining closed-form solutions for a reinforced Timoshenko beam on an elastic foundation subjected to any vertical pressure loading. A particular solution was then obtained for uniform pressure loading. The choice of a Timoshenko beam was based on the fact that the Winkler model (based on pure bending beam theory) and Pasternak model (based on pure shearing of the beam) take an extreme point of view on the deformation behavior of a beam. The basic equations for a Timoshenko beam on an elastic foundation can be reduced to

$$D\,\frac{d^4 w}{dx^4} - \frac{k_s D}{C}\,\frac{d^2 w}{dx^2} + k_s w = q - \frac{D}{C}\,\frac{d^2 q}{dx^2} \qquad (4.91)$$

where q is the pressure loading and may be a function of x also. C is the shear stiffness, and D is the bending stiffness, which is expressed as

$$D = EI_c + E_g(y_g - y_c)^2 \qquad (4.92)$$

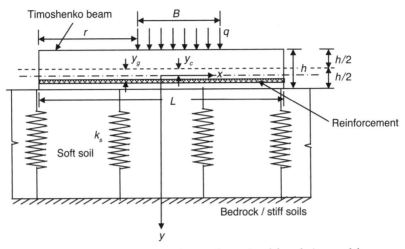

FIGURE 4.30 Schematic diagram of a one-dimensional foundation model as proposed by Yin (2000).

where y_c is the location of the neutral line, E is Young's modulus of the beam, E_g is the tensile stiffness of the reinforcement, and y_g is the location of the reinforcement, as shown in Figure 4.30.

A finite beam of length L subjected to any form of pressure loading $q = f(x)$ was considered and the general procedure for obtaining the closed-form solution was presented. Further, the solution was obtained for a particular case of uniform vertical pressure loading. A detailed parametric study was carried out to consider the influence of the shear stiffness of the reinforcement, tensile modulus of the reinforcement, and location of the pressure loading. The width of the loaded region B was taken as 0.1 m. For the undrained condition, the soft clay was assumed to have an elastic modulus of 2000 kPa and Poisson's ratio of 0.5. The spring constant was then evaluated using the expression

$$k_s = \frac{E_{\text{clay}}}{B(1 - v^2)}$$

and k_s was calculated as 26,666.7 kN/m³. Because the granular soil was placed on soft ground, it was difficult to compact the granular soil to a high density to achieve a high stiffness; therefore, accordingly, the elastic modulus for the granular soil was taken as 50,000 kPa. For the parametric study, the tensile modulus of the reinforcement was varied as 5000, 10,000, and 50,000 kN/m. The length and thickness of the beam were taken as 3 and 0.64 m, respectively. The location of the reinforcement y_g was 0.24 m (Figure 4.30). A uniform loading intensity of 1000 kPa was applied to the Timoshenko beam.

To study the influence of the shear stiffness of the reinforcement on the response of a foundation beam, two cases were considered. In one case, the shear stiffness was assumed to be fully mobilized and in the other to be zero. The maximum settlement in the absence of shear stiffness of the reinforcement layer was found to be larger than the settlement obtained with shear stiffness. However, an opposite trend was observed for maximum mobilized tension of the reinforcement. The difference was found to be 17.6% for maximum settlement and 9.03% for maximum tension. In both these cases, the location of the loaded region (i.e., the distance r) was taken as 1.45 m (see Figure 4.30).

To study the influence of the tensile modulus of the reinforcement layer E_g in relation to full shear stiffness C, three cases ($E_g = 5000$, 10,000 and 50,000 kN/m) were analyzed. In all three cases, the location of the loaded region r was taken as 1 m. Settlement and rotation (radians) were found to decrease with an increase in E_g, while bending moment, tension in the reinforcement, and shear force were found to increase with an increase in E_g. The decrease in maximum settlement was from 3.69 mm to 3.02 mm and the increase in maximum tension was from 10.75 kN/m to 35.35 kN/m for a corresponding decrease in tensile modulus from 50,000 kN/m to 5000 kN/m.

In comparing the results of these five cases, it was observed that the location of the vertical pressure loading r affects the shape of the profiles for deflection, rotation angle, and mobilized tension in the reinforcement. However, a negligible effect was observed as far as magnitude of maximum settlement and tension was concerned.

The main limitations of this model are the linear behavior of the beam and springs, nonoccurrence of sliding between granular soils and the reinforcement, and lack of consideration of large deformation/deflection.

Maheshwari (2004) proposed a generalized procedure for the analysis of beams on reinforced earth beds. In Hetényi's model, the interaction between the independent spring elements is established by including an imaginary elastic beam. However, Maheshwari (2004) considered this imaginary beam as a physical entity (reinforcement has some finite bending stiffness). Further, a finite length of the foundation and the reinforcing beam was considered along with the weight of the soil lying above the reinforcing beam (Figure 4.31). The reinforcing beam was assumed to be smooth. The governing differential equations for the proposed model were obtained as

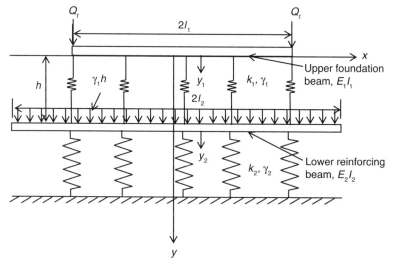

FIGURE 4.31 Proposed model for the analysis of combined footing on reinforced earth beds (Maheshwari 2004).

$$\frac{d^4 y_2}{dx^4} = \frac{E_1 I_1}{k_1} \frac{d^8 y_1}{dx^8} + \frac{d^4 y_1}{dx^4}, \qquad 0 \le x \le l_1 \qquad (4.93)$$

$$\frac{d^8 y_1}{dx^8} + \frac{k_1}{E_1 I_1 E_2 I_2} \left(E_1 I_1 + \frac{k_2}{k_1} E_1 I_1 + E_2 I_2 \right) \frac{d^4 y_1}{dx^4}$$

$$(4.94)$$

$$+ \frac{k_1 k_2}{E_1 I_1 E_2 I_2} y_1 = \frac{\gamma_1 h k_1}{E_1 I_1 E_2 I_2}, \qquad 0 \le x \le l_1$$

$$\frac{d^4 y_2}{dx^4} + \frac{k_2}{E_2 I_2} y_2 = \frac{\gamma_1 h}{E_2 I_2}, \qquad l_1 \le x \le l_2 \qquad (4.95)$$

where y_1 and y_2 are the deflections of the upper and the lower beams, respectively (Figure 4.31); $E_1 I_1$ and $E_2 I_2$ are their flexural rigidities; l_1 and l_2 are their lengths; h is the depth of placement of the reinforcing beam; γ_1 is the unit weight of the upper soil layer; and k_1 and k_2 are the stiffness of the two soil layers, as shown in Figure 4.31. x is any location along the length of the beam. The above equations were converted into nondimensional form, and it was possible to obtain their closed-form solution as

$$y_1' = e^{\lambda_1 z}(C_1 \cos \lambda_1 z + C_2 \sin \lambda_1 z)$$

$$+ e^{-\lambda_1 z}(C_3 \cos \lambda_1 z + C_4 \sin \lambda_1 z)$$

$$(4.96a)$$

$$+ e^{\lambda_2 z}(C_5 \cos \lambda_2 z + C_6 \sin \lambda_2 z)$$

$$+ e^{-\lambda_2 z}(C_7 \cos \lambda_2 z + C_8 \sin \lambda_2 z) + \gamma_1' h' r, \qquad 0 \le z \le z_1$$

and

$$y_2' = \frac{1}{R} \left(\frac{R}{r} \right)^{3/4} \left(\frac{d^4 y_1'}{dz^4} + y_1' \right), \qquad 0 \le z \le z_1$$

$$= e^{\lambda_3 z}(C_9 \cos \lambda_3 z + C_{10} \sin \lambda_3 z)$$

$$(4.96b)$$

$$+ e^{-\lambda_3 z}(C_{11} \cos \lambda_3 z + C_{12} \sin \lambda_3 z)$$

$$+ y_1' h' = \left(\frac{R}{r} \right)^{-1/4}, \qquad z_1 \le z \le z_2$$

where various nondimensional terms are defined as

$$y_1' = \frac{y_1 E_1 I_1}{Q_t R_1^3}, \qquad y_2' = \frac{y_2 E_2 I_2}{Q_t R_2^3}$$

$$R = \frac{E_1 I_1}{E_2 I_2}, \qquad r = \frac{k_1}{k_2}$$

$$\gamma_1' = \frac{\gamma_1 R_1^2}{Q_t}, \qquad h' = \frac{h}{R_1}$$

$$z = \frac{x}{R_1}, \qquad z_1 = \frac{l_1}{R_1}, \qquad z_2 = \frac{l_2}{R_1}$$

$$\lambda_1 = \sqrt[4]{\frac{\alpha + \beta}{4}}, \qquad \lambda_2 = \sqrt[4]{\frac{\alpha - \beta}{4}}$$

where

$$\alpha = \frac{A}{2}, \qquad \beta = \sqrt{\frac{A^2}{4} - B}$$

and

$$A = \left(1 + \frac{R}{r} + R\right), \qquad B = \frac{R}{r}$$

and

$$\lambda_3 = \sqrt{\frac{a}{2}}$$

where

$$a^2 = \frac{R}{r}$$

External loads were treated as shear and accommodated via boundary conditions. The 12 constants of integration in the above equations were obtained with the help of appropriate boundary and continuity conditions, which are presented below in nondimensional form.

Boundary Conditions
For the upper beam, at $z = 0$:

$$\frac{dy_1'}{dz} = 0 \quad \text{and} \quad \frac{d^3 y_1'}{dz^3} = 0 \qquad (4.97a)$$

and at $z = z_1$:

$$\frac{d^2 y_1'}{dz^2} = 0 \quad \text{and} \quad \frac{d^3 y_1'}{dz^3} = -1 \qquad (4.97b)$$

For the lower beam, at $z = 0$:

$$\frac{dy_2'}{dz} = 0 \quad \text{and} \quad \frac{d^3 y_2'}{dz^3} = 0 \qquad (4.97c)$$

and at $z = z_2$:

$$\frac{d^2 y_2'}{dz^2} = 0 \quad \text{and} \quad \frac{d^3 y_2'}{dz^3} = 0 \qquad (4.97d)$$

Continuity Conditions
At $z = z_1$:

$$y_2'\Big|_{z_1 - \varepsilon} = y_2'\Big|_{z_1 + \varepsilon}, \quad \frac{dy_2'}{dz}\Big|_{z_1 - \varepsilon} = \frac{dy_2'}{dz}\Big|_{z_1 + \varepsilon} \qquad (4.97e)$$

and at $z = z_1$:

$$\frac{d^2 y_2'}{dz^2}\Big|_{z_1 - \varepsilon} = \frac{d^2 y_2'}{dz}\Big|_{z_1 + \varepsilon}, \quad \frac{d^3 y_2'}{dz^3}\Big|_{z_1 - \varepsilon} = \frac{d^3 y_2'}{dz^3}\Big|_{z_1 + \varepsilon} \qquad (4.97f)$$

where ε tends to zero.

These equations were solved to obtain the constants C_1 to C_{12} using a Cholesky decomposition scheme. Using the constants C_1 to C_{12} in the appropriate expressions, deflections, bending moments, and shear forces of the upper and lower beam can be obtained.

This model was further modified by Maheshwari (2008) to address the separation of the combined footing from the ground surface. This separation was assumed to occur only due to the application of external loads. The weight of the foundation beam also was considered while modeling the soil-foundation system. Separation from the ground surface was taken into consideration by invoking the condition that the contact pressure must be zero for negative deflection values of the foundation beam. The governing differential equations were solved by employing a finite difference scheme with the help of appropriate boundary, continuity, and interface conditions. The proposed model was first validated by comparing the results of a particular case (without considering the separation between the beam and the

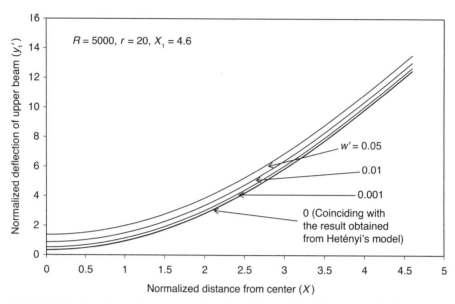

FIGURE 4.32 Variation of deflection with weight of the combined footing (Maheshwari 2008).

ground surface) to those from Hetényi's model. The normalized deflection profiles of the upper beam (idealization for a combined footing) for various values of the normalized weight of the beam w' are depicted in Figure 4.32. It can be observed that the deflection profile for $w' = 0$ coincides with the results obtained from Hetényi's model. The results from this study were compared with those obtained by Maheshwari (2004) in which the beam was assumed to have perfect contact with the ground surface (Figure 4.33). In the case of separation between the ground surface and the upper beam (tensionless foundation), the negative

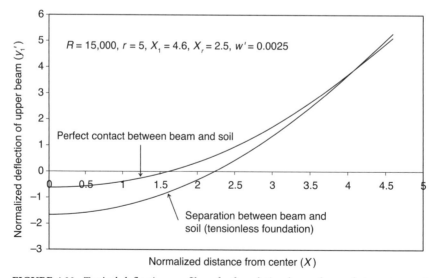

FIGURE 4.33 Typical deflection profiles of a foundation beam for perfect contact and separation between the soil and the beam (Maheshwari 2008).

deflection (lift up) was found to be greater. The maximum lift up at the center of the beam was found to increase by 167% as the soil is unable to take any tension (Figure 4.33). It further became obvious that a tensionless foundation affects negative deflection more than positive deflection of a foundation beam.

A detailed parametric study was conducted to determine the influence of the length ratio of beams and the relative stiffness of the soil layers. The response of the soil-footing system was found to be the same for a length ratio of beams greater than 1.5. It was observed that the relative stiffness of the soil layers influences the response of the model significantly. The amount of lift up of the beam was found to be less for higher values of relative stiffness of the soil layers, as shown in Figure 4.34. As the relative stiffness of the soil layers is reduced from 45 to 30, settlement of the footing was found to be reduced by 58%.

Various research workers have proposed many approaches which employ methods other than the beams on elastic foundation approach. The constitutive relations representing the behavior of soil in these studies form various nonlinear/viscoelastic or elastoplastic models (Section 4.2).

The settlement, contact pressure, and bending moment in the elastic combined footing are affected by the structural stiffness, type of connection between the columns and the combined footing, and compressibility of the subsoil. In view of this, Noorzaei et al. (1995) analyzed a soil-structure interaction problem of a plane frame-combined footing-soil system, taking into account the elastoplastic behavior of the underlying compressible soil stratum and its strain-hardening characteristics. The methodology makes use of an incremental elastoplastic stress-strain relationship and the associated flow rule. This elastoplastic behavior was modeled with the help of two Drucker-Prager yield criteria (DPYC). For the first yield criterion (DPYC-1), the yield locus passes through the outer apices of the Mohr-Coulomb hexagon (Equation 4.36a), and for the other yield criterion (DPYC-2), the yield locus passes through the inner apices of the Mohr-Coulomb hexagon (Equation 4.36b). A coupled finite-infinite element model was used in the analysis. Three-node isoparametric beam bending elements with three

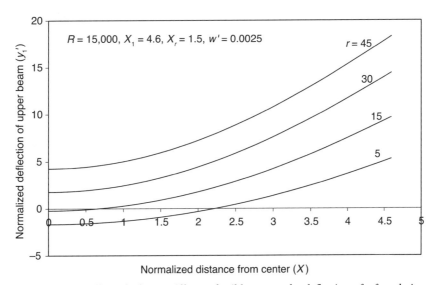

FIGURE 4.34 Effect of relative stiffness of soil layers on the deflection of a foundation beam (Maheshwari 2008).

degrees of freedom (u, v, ϕ) were used to discretize the combined footing. Three different types of analyses were carried out: elastoplastic interactive analysis with no hardening, elastoplastic analysis with strain hardening, and linear interactive analysis. The interactive behavior with the strain-hardening characteristics was compared with that of an elastic–perfectly plastic response and also with the linear elastic response. Some of the properties of the soil were obtained on the basis of unsaturated undrained triaxial tests. The shear strength parameters c and ϕ were obtained as 25 kN/m² and 29°. The initial tangent modulus E_i and Poisson's ratio were found to be 7500 kN/m² and 0.343 and the strain-hardening parameter H to be 500 kN/m². The load intensity on the foundation beam q was 40 kN/m² and the corresponding collapse loads from the analysis were $1.6q$ and $2.2q$ for rigid plastic analysis and strain-hardening analysis, respectively, for DPYC-1. However, for DPYC-2, they were $1.5q$ and $2.2q$, respectively.

It was observed from the deformation profiles that at low values of the load factor (collapse load/applied load), the total vertical settlements obtained from elastoplastic analysis are almost the same as those from linear interactive analysis. However, a significant departure was observed at higher values of the load factor. This is due to the fact that most of the soil elements start yielding at higher load factors and there is a progressive spread of the plastic zone. Vertical settlements at the center and at the edge of the foundation beam for the maximum load factor were compared for the rigid plastic and the strain-hardening behavior of the soil mass. For DPYC-1, a reduction of around 7 and 5.7% in vertical settlement along the length of the beam was observed in elastic–rigid plastic analysis and elastic-plastic analysis with strain hardening, respectively. However, the corresponding reduction for DPYC-2 was around 6 and 2.6%.

It was further observed that all the interactive analyses give essentially the same contact pressure up to a load factor of 2.0. However, the contact pressure obtained with DPYC-1 was found to be greater compared to the other criterion (DPYC-2) and when the subsoil is considered as behaving in a linear elastic manner. The spread of the plastic zone for the elastic–rigid plastic and elastic-plastic analysis with strain-hardening behavior of the soil mass was obtained. It was observed that the plastic zone initially spreads vertically and then toward the center of the soil-foundation system, causing more redistribution of forces and moments in the frame members and the foundation beam.

4.4.4 Raft Foundations

A raft or mat foundation is used where the base soil has a low bearing capacity and/or the column loads are so large that more than 50% of the area is covered by conventional spread footings. These foundations are modeled as plates and can be analyzed by employing a plates on elastic foundation approach. However, as in the case of combined footings, the interaction problem also can be analyzed by taking the nonlinear or time-dependent behavior of the soil into consideration.

There are two approaches to the plate theory of elastic foundations to describe the bending action of a plate: the thin plate and the thick plate approach. For a thin plate, Kirchhoff (1850) assumed that the plane cross sections normal to the undeformed mid-surface would remain normal to the deformed mid-surfaces (Timoshenko and Goodier 1970). The deflection of a plate w subjected to an applied load $p(x, y)$ and reaction $q(x, y)$ is given by

$$D\nabla^4 w(x, y) = p(x, y) - q(x, y) \qquad (4.98)$$

where D is the flexural rigidity of the plate.

For a thick plate, the Reissner (1947) plate theory, which can take into account the shear effect, is more appropriate. The general governing differential equation for the deflection of the plate is

$$D\nabla^4 w(x, y) = p(x, y) - q(x, y)$$

$$- \frac{h^2}{10} \left(\frac{2 - \nu_p}{1 - \nu_p} \right) \nabla^2 \left[p(x, y) - q(x, y) \right] \tag{4.99}$$

where ν_p and h are Poisson's ratio and the thickness of the plate, respectively.

Wang et al. (2005) have reviewed the state-of-the-art for plates on an elastic foundation. Various analytical methods have been developed for the analysis of thin plates resting on an elastic foundation. In these analyses, the soil has been modeled either as a Winkler foundation or as a two-parameter foundation. Due to complexity, the analytical solutions could be obtained only for foundations of special shapes, such as an infinite plate or a circular plate. Kerr (1964) emphasized the proper mathematical formulation of the physical problems under consideration, and in view of this, a critical study of a number of foundation models was presented. The differential equations for a circular plate resting on various types of foundation models, such as the Pasternak model, Reissner model, and viscoelastic Pasternak model, were obtained. Appropriate boundary conditions also were discussed, along with initial conditions for time-dependent behavior of the soil (viscoelastic model).

Selvadurai (1979) has presented and summarized various analytical solutions for infinite and circular plates on elastic foundations. Analysis of circular plates has drawn more attention from research workers compared to rectangular foundations. The analytical treatment of a rectangular plate problem is much more complicated than the axisymmetric case of a circular plate. General analysis of the rectangular plate problem involves two spatial variables, and the boundary conditions have to be specified on all four edges of the plate. Selvadurai (1979) considered rectangular plates with free edges and symmetrical loading in both the x and y directions and obtained an analytical solution. The soil medium was represented as either a Winkler model or as a two-parameter elastic model.

In addition to the analytical methods for the solution of the soil-raft interaction problem, various numerical and seminumerical methods also are employed. These include the substructure method (Hain and Lee 1974), finite difference method (Chakravorty and Ghosh 1975; Selvadurai 1979; Vallabhan and Das 1991), finite strip method (Booker and Small 1986; Rajapakse 1988; Chow et al. 1989), finite element method (Cheung and Zienkiewicz 1965; Yang 1972; Bowles 1996; Wang and Cheung 2001), and boundary element method (Beskos 1997; Providakis and Beskos 1999).

The raft and the supporting soil are two of the elements in a three-element system. The stiffness of the third element, the structure, may have a significant influence on the distribution of loads and moments transmitted to the raft, resulting in an effect on the differential settlement pattern and the moments and shears induced in the raft. Therefore, the structure, foundation, and supporting soil should be analyzed as a system comprised of three compatible and interacting elements. Hain and Lee (1974) proposed a substructure approach for the analysis of such a three-element system. The supporting soil was modeled as a Winkler model or a layer of linear elastic material, and a detailed study was carried out to understand the influence of the stiffness of the frame structure and the choice of the soil model on the settlements and moments induced in the raft. The substructure method facilitates the inclusion of the contribution of the stiffness of the structure and the supporting soil in the stiffness matrix of the raft. This approach is discussed next.

The free body equilibrium equation for the structure and the raft was written in matrix form as

$$[K]\{U\} = \{P\} \tag{4.100}$$

where $[K]$ is the structure-raft stiffness matrix (superstructure), $\{U\}$ is the nodal displacements, and $\{P\}$ is the external nodal forces. This equation system can be partitioned into a two-equation system as

$$\begin{bmatrix} K_{bb} & K_{bi} \\ K_{ib} & K_{ii} \end{bmatrix} \begin{Bmatrix} U_b \\ U_i \end{Bmatrix} = \begin{Bmatrix} P_b \\ P_i \end{Bmatrix} \tag{4.101}$$

where the set of nodal displacements is divided into $\{U_b\}$ for boundary displacements common to the superstructure and the supporting soil and $\{U_i\}$ for interior displacements of the superstructure. The corresponding set of external forces is $\{P_b\}$ and $\{P_i\}$, respectively.

From the partial inversion of Equation 4.101, the free body equilibrium equation in matrix form for the boundary nodes of the superstructure was obtained:

$$[K_{bb} - K_{bi}K_{ii}^{-1}K_{ib}]\{U_b\} = \{P_b\} - [K_{bi}K_{ii}^{-1}]\{P_i\} \tag{4.102}$$

Employing one of the mathematical models for the analysis of an elastic foundation, the equilibrium equation for the supporting medium was

$$[K_s]\{\delta\} = \{F\} \tag{4.103}$$

where $[K_s]$ is the supporting medium stiffness matrix, $\{\delta\}$ is the supporting medium nodal displacements, and $\{F\}$ is the corresponding external nodal forces.

Similarly,

$$\{\delta\} = \begin{Bmatrix} U_b \\ \delta_s \end{Bmatrix} \tag{4.104}$$

and

$$\{F\} = \begin{Bmatrix} F_b \\ F_s \end{Bmatrix} \tag{4.105}$$

where $\{U_b\}$ is the vector comprised of nodal displacements along the soil-raft interface, $\{F_b\}$ is the corresponding external forces, $\{\delta_s\}$ is the remaining supporting medium nodal displacements, and $\{F_s\}$ is the corresponding external forces.

The partitioned form of Equation 4.103 was then obtained as:

$$
\begin{bmatrix} K_{sbb} & K_{sbs} \\ K_{ssb} & K_{sss} \end{bmatrix} \begin{Bmatrix} U_b \\ \delta_s \end{Bmatrix} = \begin{Bmatrix} F_b \\ F_s \end{Bmatrix} \tag{4.106}
$$

The full set of equations for the raft, which includes the contribution of stiffness from both the structure as well as the supporting soil below, was obtained by combining Equations 4.102 and 4.106:

$$
\begin{bmatrix} K_{sbb} + K_{bb} - K_{bi} K_{ii}^{-1} K_{ib} & K_{sbs} \\ K_{ssb} & K_{sss} \end{bmatrix} \begin{Bmatrix} U_b \\ \delta_s \end{Bmatrix}
$$
$$
= \begin{Bmatrix} F_b + P_b - K_{bi} K_{ii}^{-1} P_i \\ F_s \end{Bmatrix} \tag{4.107}
$$

The above equation was solved for $\{\delta\}$ using the available techniques (Gauss and Cholesky). After evaluating $\{\delta\}$, the actual column loads were extracted from the vector S_b, which is calculated by extracting U_b from $\{\delta\}$:

$$
\{S_b\} = [K_{bb} - K_{bi} K_{ii}^{-1} K_{ib}] \{U_b\} + [K_{bi} K_{ii}^{-1}] \{P_i\} \tag{4.108}
$$

Similarly, the foundation reaction forces were calculated by back substitution of $\{\delta\}$ into Equation 4.103, and the superstructure member forces and stresses were calculated.

The influence of the stiffness of the structure on the performance of a flexible rectangular raft was studied by analyzing two multibay structures for a range of raft and soil stiffnesses. Results were presented for both a Winkler soil model and a linear elastic soil model. For the Winkler model, the pressure p at any point along the contact surface was expressed as

$$
p = kw \tag{4.109}
$$

where k is the modulus of subgrade reaction and w is the settlement of the point.

The term relative flexibility λL was defined while analyzing the raft to express the flexibility of the raft compared with the compressibility of the soil:

$$
\lambda L = \sqrt[4]{\frac{kL^4}{4E_R I_R}} \tag{4.110}
$$

where $E_R I_R$ is the flexural rigidity of the raft per unit width and the plan dimensions of the raft are $B \times L$ ($B < L$).

Typically, diagrams for variation of column load with relative flexibility and bending moment are plotted by employing the above-mentioned method of analysis, representing the soil by a Winkler model. An increase in relative flexibility was found to result in a transfer of load to the interior columns due to the convex settlement profile of the raft. The actual magnitude of the load redistribution was found to be a function of the absolute stiffness of the structure as well as of the soil.

The same analysis was carried out by representing the soil as a linear elastic model. In this case, the soil layer was considered to be of infinite lateral extent and of finite or infinite thickness. One of the simplest cases was where the layer was considered to be isotropic and homogeneous with an elastic modulus E_s and Poisson's ratio ν_s. From these analyses, a relative flexibility parameter F analogous to λL for the Winkler model was derived as

$$F = \frac{\pi L^4 E_s (1 - \nu_R^2)}{16(1 - \nu_s^2) E_R I_R B} \qquad (4.111)$$

where ν_R is Poisson's ratio of the raft material.

The influence of raft flexibility on the distribution of column load, total and differential settlements, and maximum moments was determined. The settlement profile was observed to be concave with the linear elastic model, thus leading to a transfer of load to the edge and corner columns. This was found to be in contrast to the predictions based on the Winkler model. The same contrast was observed in the case of bending moments.

A relationship was established between the relative flexibility parameters for the Winkler model and the linear elastic model as

$$(\lambda L)^4 = 1.55 F \qquad (4.112)$$

which was found to be independent of the absolute values of E_s and k.

From the pattern of results obtained from the above analysis, it was clear that the predicted behavior of the structure-raft-soil system depends on the choice of the soil model. By selecting "equivalent" values of the subgrade modulus k and elastic parameters E_s and ν_s, it was possible to obtain identical values of raft settlement at a specified point on the raft and for a specific raft-soil flexibility. Even under these special circumstances, the settlement profiles and the associated column loads and raft moments were quite different, except for very large raft flexibility.

Results for the linear elastic model were found to be consistent with the commonly observed concave settlement profile, and therefore it was concluded that the linear elastic model should be used in preference to the Winkler model.

Vallabhan and Das (1991) proposed a mathematical model for the analysis of an axisymmetric circular tank foundation in which two parameters were used to represent the elastic foundation. The elastic properties of the soil stratum were assumed to be constant or to vary linearly in the vertical direction, and the soil stratum was assumed to rest on a rigid surface. The modified Vlazov model used to represent the soil employed a parameter λ to characterize the distribution of the vertical displacement in the elastic foundation. Figure 4.35 shows a circular plate of radius R resting on an elastic foundation of uniform thickness H with a rigid base. The system was axisymmetric in geometry, loading, and boundary conditions. The governing equations for this interaction problem were derived by using the minimum potential energy theorem. The potential energy function was a function of flexural rigidity of the plate,

$$D = \frac{E_p h^3}{12(1 - \nu_p^2)}$$

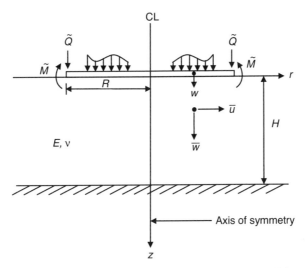

FIGURE 4.35 Circular plate on an elastic foundation (Vallabhan and Das 1991).

lateral displacement of the circular plate w; displacements along the r and z axes in the soil continuum $\bar{u}$ and $\bar{w}$; components of stress at a point in the soil continuum σ_r, σ_θ, σ_z, and σ_{rz}; modulus of elasticity of the plate E_p; Poisson's ratio of the plate v_p; lateral loading on the plate $q(r)$; prescribed bending moment $\tilde{M}$ and shear force $\tilde{Q}$ on the plate; and thickness of the plate h.

The constitutive equations for the elastic foundation were given by

$$\begin{Bmatrix} \sigma_r \\ \sigma_\theta \\ \sigma_z \\ \tau_{rz} \end{Bmatrix} = \frac{E(1-v)}{(1+v)(1-2v)}$$

$$\begin{bmatrix} 1 & v/(1-v) & v/(1-v) & 0 \\ v/(1-v) & 1 & v/(1-v) & 0 \\ v/(1-v) & v/(1-v) & 1 & 0 \\ 0 & 0 & 0 & (1-2v)/[2(1-v)] \end{bmatrix} \quad (4.113)$$

$$\begin{Bmatrix} \partial\bar{u}/\partial r \\ \bar{u}/r \\ \partial\bar{w}/\partial z \\ \partial\bar{u}/\partial z + \partial\bar{w}/\partial r \end{Bmatrix}$$

where E and v are the elastic modulus and Poisson's ratio of the elastic foundation. Vlazov assumptions were applied: the vertical displacement $\bar{w}(r,z) = w(r)\phi(z)$ such that $\phi(0) = 1$ and $\phi(H) = 0$ and the horizontal displacement $\bar{u}(r,z) = 0$ everywhere in the elastic foundation, where $\phi(z)$ is a function that describes the decay of the vertical displacement $\overline{w}(r, z)$ in the direction of the z axis.

By employing the constitutive equations and assumptions, the potential function was modified and the field equation and boundary conditions were obtained. The field equation was solved by employing the finite difference method. As the field equation was a quasi-linear fourth-order differential equation, the finite difference equations were expressed by a qui-diagonal matrix equation with five coefficients using the central difference. The appropriate boundary conditions also were expressed in finite difference form, and subsequently the equations were solved with the help of special numerical techniques.

Vallabhan and Das (1991) first published the data on elastic settlements of circular plates on an elastic continuum with a finite thickness for the general loading condition. The proposed model was employed for a numerical example. To establish the accuracy of the solution, results were compared with those obtained from a finite element model in which rectangular finite elements were employed. The numerical displacements obtained from these examples using the proposed model were compared with the corresponding finite element displacements, and a fairly good correlation was observed. Any variation of material properties with depth easily can be incorporated into the proposed model by modifying the parameters of the model.

Wang and Cheung (2001) proposed a finite element method for the analysis of plates on a cross-anisotropic foundation. This problem can be solved in two ways. The first is to divide both the plate and the foundation into a number of elements. Since the foundation is considered to be a half-space, its boundary is far away from the plate, and thus the foundation has to be divided into a large number of elements. This results in a large amount of data and computing time. In the second way of solving the problem, the foundation reaction is treated as a force external to the plate, and therefore only the plate needs to be divided into a number of elements.

Wang and Cheung (2001) adopted the latter way for the analysis of a plate resting on a cross-anisotropic foundation using the isoparametric element method. The plate was modeled as a Reissner-Midlin plate. The rotations of a point on the plate were chosen as independent variables, and therefore it was possible to consider the shear deformations at the same time. The proposed method, therefore, finds application for thick and thin plates of different shapes. The surface displacement due to a point force acting on a cross-anisotropic foundation was taken as the fundamental solution. Displacement fields were uniquely specified by an independent variation of the deflection w and two independent variations θ_x and θ_y of the two angles defining the direction of the line originally normal to the mid-surface of the plate. The load vector and geometry for the isoparametric plate bending element were expressed for each node on the plate. The element stiffness matrix relating nodal forces to the corresponding nodal displacements was derived for each element on the plate, and the integration was carried out in a natural coordinates system by employing the Gauss integration method. The equilibrium equation for the plate acted upon by the forces from the upper structure F and the foundation forces Q finally was expressed as

$$(K + K_f)\delta = F \qquad\qquad (4.114)$$

where K and K_f are the global stiffness matrix for the plate and the foundation, respectively, and δ is the displacement vector.

Interaction between the plate and the foundation was considered to determine the stiffness matrix for the foundation K_f. This matrix represents the relationship between the force Q and the displacement δ. By considering the fundamental solution for the surface displacement due to a point load acting on a cross-anisotropic foundation, this relationship was established.

To exhibit the effectiveness of the proposed isoparametric finite element method, two numerical examples were analyzed. For the first example, a square plate was considered on an isotropic elastic foundation. Due to the symmetry of the problem, a quarter of the plate was divided into equal-size elements by using different meshes (2×2, 4×4, 6×6, and 6×8). Results from the study were found to be in good agreement with those obtained from the spline method and the displacement method. At the boundary of the plate, especially at the corners of the plate, a stress concentration was observed. However, no stress concentration was observed in the internal part of the plate. This agreed well with the normally accepted stress concentration profile. The second example considered a square plate resting on a cross-anisotropic soil medium. A detailed parametric study was carried out to investigate the influence of cross-anisotropy on the displacement and contact pressure, and corresponding charts were developed.

The method also lifted the restriction that contact pressure must be assumed to be uniformly distributed around each nodal point. Due to eight-node elements, the model was found to be useful for a plate with different boundaries and applicable in all cases. As the rotation angles of the plate were assumed to be independent of deflection, the method was found to be applicable to both thin and thick plates.

4.5 Application to Pile Foundations

The mechanisms that resist applied loads in pile foundations are quite different compared to those for shallow foundations. Vertical compressive loads are resisted by a combination of skin friction and end bearing, vertical uplift loads by a combination of dead weight and skin friction, horizontal loads by lateral earth pressures, and moments by converting them to axial compression and uplift (Figure 4.36). Therefore, treatment of the soil-pile interaction problem is somewhat different, although the same models discussed in previous sections can be adopted for representation of the soil media. Analyses focus on the interaction of piles with soil media

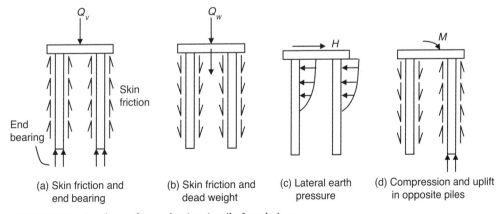

FIGURE 4.36 Load transfer mechanism in pile foundations.

primarily under vertical and lateral loads. Various research workers have proposed different theories for analysis of the soil-pile interaction problem.

Analytical approaches to the soil-pile interaction problem for laterally loaded piles have been developed in two separate directions. The first uses the conceptual model of treating the soil restraint as discrete springs. The model is improved by allowing the spring stiffness to vary along the length of the pile (Matlock and Reese 1960; Sogge 1981). This approach has two main limitations. First, difficulties exist in choosing appropriate p-y curves for a given combination of pile size and soil type. Second, replacement of the soil continuum by discrete springs precludes extension of the analysis to pile groups since interaction between neighboring piles may not be taken into account. The second development in solutions for laterally loaded piles has made use of methods in which the soil is modeled by employing the continuum approach. Methods such as boundary element and finite element analysis are adopted for this approach. The characteristics of the soil medium can be represented by various constitutive models, some of which are those by Banerjee and Davies (1978), Poulos and Davis (1980), Randolph (1981), Chen and Poulos (1993), Poulos (1999), and Xu and Poulos (2000).

For axially loaded piles and pile groups, different studies have been conducted employing various constitutive models as discussed in previous sections. Some of these are by Randolph and Wroth (1978), Banerjee and Davies (1978), Hain and Lee (1978), Poulos and Davis (1980), Nogami and Chen (1984), Chin and Poulos (1991), Poulos (1999), and Xu and Poulos (2000). Guo (2000) presented closed-form solutions for the radial consolidation of the soil around a driven pile, assuming that the soil skeleton deforms viscoelastically. Subsequently, the load-settlement response of the pile also was predicted.

Matlock and Reese (1960) presented a general method for determining moments and displacements of a vertical pile embedded in a granular soil and subjected to lateral load and moment at the ground surface. A Winkler model was employed to represent the soil medium (i.e., the soil was replaced by a series of infinitely close independent elastic springs). The response of a pile of length L under a lateral load Q_g and moment M_g was obtained in the form of its deflection, slope of the deflected shape, moment, and shear force at any depth z below the ground surface as follows:

Pile deflection
$$x_z(z) = A_x \frac{Q_g T^3}{E_p I_p} + B_x \frac{M_g T^2}{E_p I_p}$$

Slope of the deflected shape
$$\theta_z(z) = A_\theta \frac{Q_g T^2}{E_p I_p} + B_\theta \frac{M_g T}{E_p I_p}$$

Moment of the pile
$$M_z(z) = A_m Q_g T + B_m M_g$$

Shear force on the pile
$$V_z(z) = A_v Q_g + B_v \frac{M_g}{T}$$

where E_p is the modulus of elasticity of the pile material and I_p is the moment of inertia of the pile section. A_x, B_x, A_θ, B_θ, A_m, B_m, A_v, and B_v are coefficients, and T is the characteristic length of the soil-pile system and is defined as

$$T = \sqrt[5]{\frac{E_p I_p}{n_h}}$$

where n_h is the constant of the modulus of the horizontal subgrade reaction. Das (1999) presented representative values of n_h as shown in Table 4.5.

Randolph (1981) developed simple algebraic expressions which allowed the behavior of flexible piles under lateral loading to be calculated in terms of fundamental soil properties. These expressions were based on the results of a parametric study conducted using the finite element method and treating the soil as an elastic continuum with a linearly varying soil modulus. The expressions enabled immediate estimates of the active length of the pile, ground-level

TABLE 4.5 Representative Values of n_h

Soil Type		n_h (MN/m^3)
Dry or moist sand	Loose	1.8–2.2
	Medium	5.5–7.0
	Dense	15.0–18.0
Submerged sand	Loose	1.0–1.4
	Medium	3.5–4.5
	Dense	9.0–12.0

After Das (1999).

deformations, and maximum bending moment down the pile. In addition, the patterns of soil movement around a laterally loaded pile obtained from finite element analysis were used to develop expressions for interaction factors between neighboring piles, by which the solutions for single piles may be extended to deal with pile groups.

First of all, a detailed parametric study was performed for piles in homogeneous soil, characterized by a shear modulus G and Poisson's ratio ν, and also in soil with stiffness proportional to depth z. The latter type of soil was characterized by a parameter m to indicate the rate of increase in shear modulus with depth as

$$G = mz = mr_o \frac{z}{r_o} \qquad (4.115)$$

For a particular pile radius r_o, it was found more convenient to use the term mr_o (i.e., the rate of increase of the shear modulus with each pile radius), which has the same dimensions as the shear modulus. For this parametric study, the range of stiffness ratios was taken as

$$100 \le \frac{E_p}{G} \le 10^6 \qquad \text{and} \qquad 400 \le \frac{E_p}{mr_o} \le 4 \times 10^6$$

where E_p is the effective Young's modulus of the pile and is defined as

$$E_p = \frac{(EI)_p}{(\pi r_o^4/4)}$$

where $(EI)_p$ is the bending rigidity of the pile.

Further, the pile length l was varied to correlate the critical pile length, or the critical slenderness ratio $(l/r_o)_c$, with the stiffness ratio.

It was found that the influence of variation in Poisson's ratio on the deformation of a laterally loaded pile could be adequately represented by considering a parameter G^*, given by

$$G^* = G \left(1 + \frac{3v}{4} \right)$$

for homogeneous soil and

$$m^* = m \left(1 + \frac{3v}{4} \right)$$

for soil with stiffness proportional to depth z.

The results from a finite element study were presented in the form of algebraic expressions of the same form as those obtained by Hetényi (1946), from which the lateral response of single piles may be readily calculated. To model pile deformations at ground level using finite element analyses, the expressions proposed for homogeneous soil conditions were

$$u = 0.25 \frac{H}{G^* r_o} \left(\frac{E_p}{G^*} \right)^{-1/7} + 0.27 \frac{M}{G^* r_o^2} \left(\frac{E_p}{G^*} \right)^{-3/7}$$

$$\theta = 0.27 \frac{H}{G^* r_o^2} \left(\frac{E_p}{G^*} \right)^{-3/7} + 0.8 \frac{M}{G^* r_o^3} \left(\frac{E_p}{G^*} \right)^{-5/7}$$

(4.116)

where H is the applied lateral load, M is the moment applied to the pile at ground level, and u and θ are the lateral deflection and rotation of the pile at ground level, respectively. The results from the above equations were compared with those obtained from finite element analyses and other results available in the literature and were found to be in good agreement, with only about a 10% difference observed.

For the case of soil with stiffness proportional to depth, the corresponding expressions were obtained as

$$u = 0.54 \frac{H}{m^* r_o^2} \left(\frac{E_p}{m^* r_o} \right)^{-3/9} + 0.60 \frac{M}{m^* r_o^3} \left(\frac{E_p}{m^* r_o} \right)^{-5/9}$$

$$\theta = 0.60 \frac{H}{m^* r_o^3} \left(\frac{E_p}{m^* r_o} \right)^{-5/9} + 1.13 \frac{M}{m^* r_o^4} \left(\frac{E_p}{m^* r_o} \right)^{-7/9}$$

(4.117)

The results in this case also were found to be in good agreement with those obtained from previous studies. The critical slenderness ratio was correlated with the stiffness ratio by noting the effect of pile length on ground-level deformations and was given as

$$\left(\frac{l}{r_o} \right)_c = 2 \left(\frac{E_p}{G^*} \right)^{2/7}$$

(4.118)

for homogeneous soil and

$$\left(\frac{l}{r_o}\right)_c = 2\left(\frac{E_p}{m^* r_o}\right)^{2/9} \tag{4.119}$$

for soil with stiffness proportional to depth.

To consider the more general variation of soil stiffness with depth, Equations 4.116 and 4.117 were combined, as were Equations 4.118 and 4.119. To accomplish this, a characteristic modulus G_c was defined, which was the average value of G^* over the active length of the pile (i.e., over depths less than l_c). Subsequently, this concept of a characteristic modulus was used to combine Equations 4.116 and 4.117 to obtain the general expressions for ground-level deformations of a laterally loaded pile as

$$u = \frac{(E_p/G_c)^{1/7}}{\rho_c G_c}\left[0.27H\left(\frac{l_c}{2}\right)^{-1} + 0.3M\left(\frac{l_c}{2}\right)^{-2}\right]$$

$$\tag{4.120}$$

$$\theta = \frac{(E_p/G_c)^{1/7}}{\rho_c G_c}\left[0.3H\left(\frac{l_c}{2}\right)^{-2} + 0.8(\rho_c)^{1/2}M\left(\frac{l_c}{2}\right)^{-3}\right]$$

where the parameter ρ_c, which reflects the relative homogeneity of the soil deposit, was defined as:

$$\rho_c = \frac{G^*_{z=l_c/4}}{G_c} \tag{4.121}$$

The solutions for the lateral response of single piles were extended to deal with groups of closely spaced piles by using interaction factors. Approximate expressions for these interaction factors were presented as a function of the parameters ρ_c, E_p, G_c, and r_o and the direction of loading and spacing between the piles.

The expressions developed were found to be quite accurate within the framework of an elastic soil response and simple enough to be of practical use in estimating the response of laterally loaded piles and pile groups.

Pile groups always have a pile cap on top which is designed as a raft, and the stiffness of the structure influences the distribution of loads and moments transmitted to the raft. Therefore, the complete system, which consists of a structure–raft–supporting soil–pile group, should be analyzed. In view of this, Hain and Lee (1978) developed an analysis to predict the behavior of a raft-pile foundation system in which the interaction of the raft, supporting soil, and pile group was taken into consideration. The raft was considered to be a flexible elastic plate, with variable stiffness and any geometric shape, supported by a random group of identical compressible friction piles, and the supporting soil was represented as an elastic homogeneous or nonhomogeneous material. The raft was modeled by using rectangular plate bending finite elements, and the pile group–supporting soil system was modeled using Mindlin's equation for a deep homogeneous soil mass. Further, modified linear elastic analysis was employed for a layer of finite thickness, and a finite element analysis was used for nonhomogeneous soil situations. The analysis of a pile-reinforced continuum, apart from the analysis of a single pile, required consideration of four interactions between a pile and/or the surface of the continuum. These interactions were pile-to-pile interaction, surface-to-pile interaction, pile-to-surface interaction, and surface-to-surface interaction. Interaction factors were determined for a soil modulus increasing linearly with depth. It was found that the increase in value of the modulus with depth caused a significant reduction in the interaction between adjacent piles. The pile-pile interaction factor was defined as the ratio of the additional displacement due to unit load on an adjacent pile to the displacement of the pile due to unit load. Two pile-surface interaction factors were defined as

$$\beta_p = \frac{\text{Additional displacement of a pile due to unit surface pressure}}{\text{Displacement of a pile due to unit load}}$$

$$\beta_s = \frac{\text{Additional displacement of the surface due to unit pile load}}{\text{Displacement of the surface due to unit load}}$$

where β_p is the pile–soil surface interaction factor and β_s is the soil surface–pile interaction factor. The surface interaction effect can be evaluated using various available methods for a homogeneous or nonhomogeneous continuum.

After establishing the basic interaction between the supporting soil and the pile group, stiffness equations for the supporting soil–pile group system were derived. Compatibility and equilibrium equations were satisfied between this system and the raft, resulting in a set of stiffness equations representing the raft–supporting soil–pile group system. In obtaining this, two assumptions were made: (1) vertical forces were transmitted only from the raft to the head of a pile and (2) each pile occupied the whole of the "constant pressure" area around a particular node.

The reduction in settlement was found to be more effective with an increase in pile stiffness and length. Differential settlement was found to increase with increasing raft flexibility, but a corresponding reduction in maximum bending moments induced in the raft was observed. It was evident that the ultimate load capacity of the individual piles affected the performance of the system. Two raft-pile systems were reanalyzed by this approach, and encouraging agreement between measured and predicted settlements and pile loads was found.

Chin and Poulos (1991) employed a hybrid approach for the analysis of axially loaded vertical piles and pile groups embedded in layered soil. In this approach, a single pile response was represented by load-transfer (t-z) curves, whereas pile-soil-pile interaction was obtained accurately using the available analytical solutions for a two-layer system and in an appropriate manner for a Gibson soil. The schematic diagram in Figure 4.37 depicts the composition of an axially loaded pile group. The pile group problem was decomposed into two systems: (1) group piles acted upon by external applied loads $\{Q\}$ and the pile-soil interaction forces $\{P_p\}$ acting on the piles and (2) an extended layered soil continuum acted upon by the pile-soil interaction forces $\{P_s\}$ at an imaginary pile-soil interface. The group piles were divided into a number of elastic discrete bar elements with an axial mode of deformation. This approach was different than the continuum approach in the manner in which the soil flexibility coefficients were evaluated. This hybrid approach modeled the single pile response using load-

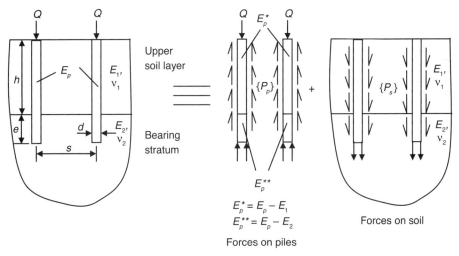

FIGURE 4.37 Composition of an axially loaded pile group (Chin and Poulos 1991).

transfer curves, whereas the pile-soil-pile interaction was obtained using the fundamental point load solutions for a layered soil. The hyperbolic shear stress–strain model was used to adequately represent the nonlinear behavior of the soil. The results from this approach were compared with those obtained from the continuum approach for layered soil and were found to be in good agreement. The influence of the socketing length e/d, pile–soil stiffness ratio E_p/E_1, and soil stiffness ratio E_2/E_1 on the response of the pile was studied, and the results were compared with those obtained from the more rigorous elastic continuum method.

The elastic interaction factor for two equally loaded single piles socketed into the lower bearing layer was obtained, and again the results were in good agreement with the continuum solutions. The interaction effect was found to be more significant for a less compressible pile and for decreasing pile spacing. Further, the results also were compared with field measurements, and it was found that the computed solutions generally tend to overpredict settlements measured at the end of construction. An advantage of this analysis was reduced computation time to form the single-pile flexibility coefficients.

An advantage of the finite element method is its ability to deal with complex configurations of structures and soil media. However, the proper location of the finite outer boundary often is crucial in obtaining an accurate solution, and selection of the minimum distance from the region of interest often depends on engineering experience and judgment. This problem is solved by using a combination of infinite and finite elements. Infinite elements are used to simulate the far-field behavior of the soil medium, while the standard finite elements are used to model the pile and the neighboring soil. Such a technique for the analysis and better understanding of pile-soil interaction under lateral loading has been adopted by Chen and Poulos (1993). Piles within a group may suffer some reduction in capacity due to interaction effects; therefore, a single isolated pile was analyzed first and then the analysis was extended to the pile group. This necessitated a modification to the elastic approach to consider the local yield of the soil, which required knowledge of the ultimate soil resistance p_u.

The computer program AVPULL was developed to implement the proposed approach. This program was verified by the analysis of a hollow cylinder subjected to internal pressure in an isotropic elastic mass. The same problem also was analyzed by the conventional finite element method to illustrate the advantage of the combined infinite and finite element method over the conventional finite element method. For the case of finite and infinite elements, displacements in both the near and far field were found to be in good agreement with the analytical solution, while reasonable agreement was found for stresses.

After verification of the computer program, the problem of piles was analyzed as a plane strain situation using eight-node isoparametric finite elements to model piles and the neighboring soil (near-field), while mapped infinite elements were used to model the soil in the far-field region. The pile was assumed to be rigid and square in shape. The soil was purely cohesive undrained clay and was modeled as an elastoplastic material obeying the Tresca yield criterion. A Goodman-type interface element was used to model the separation and slip between the soil and the pile, and it was assumed to follow the same constitutive relation as the soil. The properties of the interface element were assumed on basis of the data available in the literature.

The stress acting on the perimeter of the pile at each level was found to change from its initial uniform state to a nonuniform state. The average stress p (per unit length of pile) over the pile width w perpendicular to the direction of pile movement was calculated from the sum of stresses acting in the direction of pile movement. Thus, at a particular depth, a p-y curve of the pile (Figure 4.38) was constructed and the ultimate soil pressure was obtained.

First a single isolated pile 0.5 m wide was analyzed, and initial stress conditions representing a depth of six times the pile width were selected. Due to symmetry, only half of the problem was analyzed, and symmetry was represented with the help of rollers. The boundary conditions for the other three boundaries were automatically satisfied with the use of mapped infinite elements for modeling the far-field behavior of the soil. A uniform displacement y was applied to the pile in the direction of pile movement in an incremental fashion. For each increment, a single point on the p-y curve (Figure 4.38) for the pile was obtained by means of computation of the shear and normal stresses at the interface. The soil was assumed to have zero tensile strength and was allowed to separate from the pile once it was required to transmit any tensile stress. The parameters chosen for this analysis were as follows:

Elastic modulus of the soil	$E_s = 2667.0$ kN/m^2
Undrained shear strength of the soil	$c_u = 7.62$ kN/m^2
Poisson's ratio	$v_s = 0.495$
Coefficient of earth pressure at rest	$k_o = 1.0$
Shear and normal stiffness of the interface element	k_s and $k_n = 1570.0$ kN/m^2/m

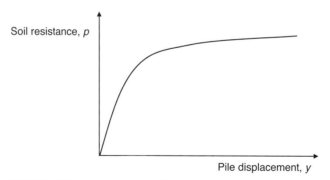

FIGURE 4.38 p-y curve for a pile.

Pile adhesion ratio	$f_c = c_a/c_u = 1.0$
Width of the pile	$w = 0.5$ m

The normalized ultimate soil resistance p_u/c_u was found to be equal to 11.7, which agreed well with the analytical predictions, although separation was not permitted in the analytical study. This verified the proposed method and generated confidence for the analysis of pile groups. The deformed mesh at failure clearly depicted the development of slip between the soil and the sides of the pile. When the pile displaced forward, the neighboring soil was found to move around the pile from the front toward the back. The displacement of the soil was found to decrease with increasing distance away from the pile. A parametric study was carried out to understand the influence of the elastic modulus of the soil, elastic stiffness of the interface, and pile adhesion on the normalized ultimate soil resistance. The elastic modulus of the soil was found to have an insignificant effect on the value of the ultimate soil resistance p_u, although there was a tendency for p_u to be slightly higher for stiffer soils. The elastic stiffness of the interface element was found to have a significant bearing on the ultimate soil resistance. The normalized ultimate soil resistance was found to increase with the increase in stiffness. This increase was found to be around 33%, which corresponds to an increase in stiffness of the interface element from 157 to 15,700 kN/m^2/m. The effect of pile adhesion was accounted for in the form of the ratio f_c (= c_a/c_u). The ultimate soil resistance was found to reduce significantly (about 24–25%) as the ratio f_c tends to zero (i.e., for a perfectly smooth pile). However, little reduction (about 6%) was observed for a pile of intermediate roughness. It was concluded from this parametric study that for a rigid pile in a purely cohesive soil, the ultimate soil resistance is governed primarily by the properties of the interface between the soil and the pile and pile adhesion.

The analysis was further extended to pile groups, and four cases were considered, as listed in Table 4.6. In each case, it was assumed that the piles were connected by a rigid cap and were sufficiently rigid to displace equally at all levels. The *p-y* curves for the pile group were obtained in a manner similar to that for a single pile. It was concluded from the analysis of these four cases that no significant group effect on piles within a group needs to be considered when piles are in one row (either infinitely long or a row of limited length). However, a substantial group effect was noticed when piles were in parallel rows, resulting in a major reduction in the ultimate lateral resistance of all piles in the group.

The viscoelastic behavior of the neighboring soil becomes an important aspect for many soft clays when analyzing a pile and a pile group, and its effect therefore should be accounted for in the analysis. Many studies have been conducted on this aspect; one of the most recent studies is by Guo (2000), who presented closed-form solutions for the radial consolidation of soil around a driven pile, assuming that the soil skeleton deforms viscoelastically. To predict

TABLE 4.6 Configurations of a Pile Group (Chen and Poulos 1993)

Case a	An infinitely long row of piles (different spacings were considered)
Case b	Two infinitely long rows of piles ($S_h/w = S_v/w = 3$; $S_h/w = 6$, $S_v/w = 3$; $S_h/w = 3$, $S_v/w = 6$)
Case c	A three-pile group (different spacings were considered)
Case d	A 3×2 pile group

Note: S_h = spacing of piles in the direction perpendicular to the direction of loading; S_v = spacing of piles in the direction parallel to the direction of loading; w = width of pile.

the load-settlement response, variation of pile-soil stiffness with dissipation of pore pressure must be quantified. For this purpose, logarithmic variation of the initial pore pressure with radius was assumed. This initial pore pressure was generated due to the expansion of a cylindrical cavity in an ideal elastic–perfectly plastic soil. In the process of reconsolidation, the soil was assumed to be a viscoelastic medium and was described by a model as depicted in Figure 4.4 (Merchant 1939). Volumetric strain was first generated and expressed as the sole variable of excess pore water pressure for a plane strain condition. Governing equations were established for radial consolidation of a viscoelastic medium, and a general solution was obtained by employing logarithmic variation of initial pore pressure for radial consolidation. It was concluded that a viscoelastic solution can be obtained from the available elastic solutions using the principle of correspondence. The viscosity of a soil was found to significantly increase consolidation time and therefore pile-head settlement. However, a negligible effect on soil strength or pile capacity was observed. The overall pile response measured from three case studies was used to back analyze the time-dependent variation of shear modulus and strength. This study showed that the variation in the normalized soil-pile interaction stiffness (shear modulus of the soil) due to reconsolidation is consistent with the variation in pore pressure dissipation on the pile-soil interface and the increase in soil strength. Therefore, radial consolidation theory was found to be quite accurate for predicting the time-dependent properties following pile installation.

References

Azam, G., Hsieh, C.W., and Wang, M.C. (1991). Performance of strip footing on stratified soil deposit with void. *J. Geotech. Eng. Div. ASCE*, 117(5):753–772.

Banerjee, P.K. and Davies, T.G. (1978). The behaviour of axially and laterally loaded single piles embedded in nonhomogeneous soils. *Geotechnique*, 28(3):309–326.

Beskos, D.E. (1997). Boundary element methods in dynamic analysis. II. 1986–1996. *Appl. Mech. Rev.*, 50(3):149–197.

Biot, M.A. (1937). Bending of an infinite beam on an elastic foundation. *J. Appl. Mech.*, 4:A1–A7.

Booker, J.R. and Small, J.C. (1986). The behaviour of an impermeable flexible raft on a deep layer of consolidating soil. *Int. J. Numer. Anal. Methods Geomech.*, 10:311–327.

Borodachev, N.M. (1976). Contact problem for a stamp with a rectangular base. *J. Appl. Math. Mech.*, 40:554–560.

Boussinesq, J. (1878). Équilibre d'élasticité d'un solide isotrope sans pesanteur, supportant différents poids. *C.R. Acad. Sci. Paris*, 86:1260–1263.

Boussinesq, J. (1885). *Application des Potentiels a l'Etude de l' Equilibre et du Mouvement des Solides Elastique*, Gauthier-Villars, Paris.

Bowles, J.E. (1996). *Foundation Analysis and Design*, 5th edition, McGraw-Hill, Singapore.

Burmister, D.M. (1956). Stress and displacement characteristics of a two-layer rigid base soil system: influence diagrams and practical applications. *Proc. Highw. Res. Board*, 35:773–814.

Cerrutti, V. (1884). Sulla deformazione di uno strato isotropo indefinite limitato da due piani paealleli. *Atti Accad. Naz. Lincei Rend.*, 4(1):521–522.

Chakravorty, A.K. and Ghosh, A. (1975). Finite difference solution for circular plates on elastic foundations. *Int. J. Numer. Methods Eng.*, 9:73–84.

Chen, L. and Poulos, H.G. (1993). Analysis of pile-soil interaction under lateral loading using infinite and finite elements. *Comput. Geotech.*, 15:189–220.

Cheung, Y.K. and Nag, D.K. (1968). Plates and beams on elastic foundations: linear and nonlinear behaviour. *Geotechnique*, 18:250–260.

Cheung, Y.K. and Zienkiewicz, O.C. (1965). Plates and tanks on elastic foundation—an application of the finite element method. *Int. J. Solids Struct.*, 1:451–461.

Chin, J.T. and Poulos, H.G. (1991). Axially loaded vertical piles and pile groups in layered soil. *Int. J. Numer. Anal. Methods Geomech.*, 15:497–511.

Chow, Y.K., Swaddiwudhipong, S., and Phoon, K.F. (1989). Finite strip analysis of strip footings: horizontal loading. *Comput. Geotech.*, 8(1):65–86.

Conniff, D.E. and Kiousis, P.D. (2007). Elastoplastic medium for foundation settlements and monotonic soil–structure interaction under combined loadings. *Int. J. Numer. Anal. Methods Geomech.*, 31:789–807.

Daloglu, A.T. and Vallabhan, C.V.G. (2000). Values of *k* for slab on Winkler foundation. *J. Geotech. Geoenviron. Eng.*, 126(5):463–471.

Das, B.M. (1999). *Principles of Foundation Engineering*, 4th edition, PWS Publishing, Pacific Grove, CA.

Davis, R.O. and Selvadurai, A.P.S. (1996). *Elasticity and Geomechanics*, Cambridge University Press, Cambridge.

Dempsey, J.P. and Li, H. (1989). A rigid rectangular footing on an elastic layer. *Geotechnique*, 39(1):147–152.

Desai, C.S. and Abel, J.F. (1972). *Introduction to the Finite Element Method*, CBS Publishers and Distributors, India.

Desai, C.S. and Reese, L.C. (1970). Analysis of circular footings on layered soils. *J. Soil Mech. Found. Eng.*, 96(SM4):1289–1310.

Drucker, D.C. and Prager, W. (1952). Solid mechanics and plastic analysis for limit design. *Q. Appl. Math.*, 10(2):157–165.

Drucker, D.C., Gibson, R.E., and Henkel, D.J. (1957). Soil mechanics and work hardening theories of plasticity. *J. Soil Mech. Found. Eng.*, 122:338–346.

Duncan, J.M. and Chang, C.Y. (1970). Nonlinear analysis of stress and strain in soils. *J. Soil Mech. Found. Eng.*, 96(SM5):1629–1651.

Erguven, M.E. and Gedikli, A. (2003). A mixed finite element formulation for Timoshenko beam on Winkler foundation. *Computat. Mech.*, 31(3–4):229–237.

Filonenko-Borodich, M.M. (1940). Some approximate theories of the elastic foundation (in Russian). *Uch. Zap. Mosk. Gos. Univ. Mekh.*, 46:3–18.

Filonenko-Borodich, M.M. (1945). A very simple model of an elastic foundation capable of spreading the load (in Russian). *Sb. Tr. Mosk. Elektro. Inst. Inzh. Trans.*, 53.

Findley, W.N., Lai, J.S., and Onaran, K. (1976). *Creep and Relaxation of Nonlinear Viscoelastic Materials*, North-Holland, New York.

Flamant, A.A. (1892). Sur la repartition des pressions dans un solide rectangulaire chargé transversalement. *C.R. Acad. Sci.*, 114:1465–1468.

Gao, J., Selvarathinam, A., and Weitsman, Y.J. (1999). Analysis of adhesively jointed composite beams. *J. Sandwich Struct. Mater.*, 1:323–339.

Gazetas, G. (1980). Static and dynamic displacements of foundations on heterogeneous multilayered soils. *Geotechnique*, 30(2):159–177.

Gazis, D.C. (1958). Analysis of finite beams on elastic foundation. *J. Struct. Eng.*, Paper No. 1722, 84.

Gibson, R.E. and Lo, K.Y. (1961). *A Theory of Consolidation for Soils Exhibiting Secondary Compression*, Publication No. 41, Norwegian Geotechnical Institute; *Acta Polytech. Scand.*, 296/191:10.

Guo, W.D. (2000). Visco-elastic consolidation subsequent to pile installation. *Comput. Geotech.*, 26:113–144.

Guo, Y.J. and Weitsman, Y.J. (2002). Solution method for beams on nonuniform elastic foundations. *J. Eng. Mech.*, 128(5):592–594.

Hain S.J. and Lee, I.K. (1974). Rational analysis of raft foundation. *J. Geotech. Eng.*, 100(GT7): 843–860.

Hain, S.J. and Lee, I.K. (1978). The analysis of flexible raft-pile systems. *Geotechnique*, 28(1):65–83.

Hansen, J.B. (1963). Discussion on hyperbolic stress-strain response of cohesive soils. *J. Soil Mech. Found. Eng.*, 89(SM4):241–242.

Head, K.H. (1984). *Manual of Soil Laboratory Testing*, Vol. 3, Pentech Press, London.

Hetényi, M. (1946). *Beams on Elastic Foundation*, University of Michigan Press, Ann Arbor.

Iyengar, K.T.S.R. (1965). Matrix analysis of finite beams on elastic foundations. *J. Inst. Eng. India*, 45:837–855.

Janbu, N. (1963). Soil compressibility as determined by oedometer and triaxial tests. *Proceedings of the European Conference of Soil Mechanics and Foundation Engineering*, Wiesbaden, 19–25.

Kaschiev, M.S. and Mikhajlov, K. (1995). Beam resting on a tensionless Winkler foundation. *Comput. Struct.*, 55(2):261–264.

Kerr, A.D. (1964). Elastic and viscoelastic foundation models. *J. Appl. Mech.*, 31(3):491–498.

Khadilkar, B.S. and Varma, B.S. (1977). Analysis of interference of strip footings by FEM. *Proceedings of the 9th International Conference on Soil Mechanics and Foundation Engineering*, Vol. 1, 597–600.

Kirchhoff, G. (1850). Uber das Gleichgewicht und die bewegung einer elastischen Scheibe. *J. Reine Angew. Math.*, 40:51–88.

Kondner, R.L. (1963). Hyperbolic stress-strain response: cohesive soils. *J. Soil Mech. Found. Eng.*, 89(SM1):115–143.

Kondner, R.L. and Zelasko, J.S. (1963). A hyperbolic stress-strain formulation of sand. *Proceedings of the 2nd Pan American Conference on Soil Mechanics and Foundation Engineering*, Vol. 1, 289–324.

Kulhawy, F.H., Duncan, J.M., and Seed, H.B. (1969). *Finite Element Analysis of Stresses and Movements in Embankments during Construction*, Report No. TE 69-4, Office of Research Services, University of California, Berkeley.

Lee, S.Y. and Yang, C.C. (1993). Non-conservative instability of a Timoshenko beam resting on Winkler elastic foundation. *J. Sound Vib.*, 162(1):177–184.

Li, H. and Dempsey, J.P. (1988a). Unbonded contact of a square plate on an elastic half-space or a Winkler foundation. *J. Appl. Mech.*, 55:430–436.

Li, H. and Dempsey, J.P. (1988b). Unbonded contact of a finite Timoshenko beam on an elastic layer. *J. Eng. Mech.*, 114:1265–1284.

Maheshwari, P. (2004). *Response of Reinforced Granular Bed Soft Soil System to Static and Moving Loads*, Ph.D. thesis, Department of Civil Engineering, Indian Institute of Technology, Kanpur.

Maheshwari, P. (2008). Analysis of beams on tensionless reinforced granular fill-soft soil system. *Int. J. Numer. Anal. Methods Geomech.*, 32(12):1479–1494.

Maheshwari, P. and Madhav, M.R. (2006). Analysis of a rigid footing lying on three layered soil using the finite difference method. *Geotech. Geol. Eng.*, 24(4):851–869.

Maheshwari, P. and Viladkar, M.N. (2007). Strip footings on a three layer soil system: theory of elasticity approach. *Int. J. Geotech. Eng.*, 1(1):47–59.

Matlock, H. and Reese, L.C. (1960). Generalized solutions for laterally loaded piles. *J. Soil Mech. Found. Div. ASCE*, 86(SM5):63–91.

Mayne, P.W. and Poulos, H.G. (1999). Approximate displacement influence factors for elastic shallow foundations. *J. Geotech. Geoenviron. Eng.*, 125(6):453–460.

Merchant, W. (1939). *Some Theoretical Considerations on the One-Dimensional Consolidation of Clay*, M.S. thesis, Massachusetts Institute of Technology.

Mindlin, R.D. (1936). Force at a point in the interior of a semi-infinite solid. *Physics*, 7:195–202.

Morfidis, K. (2002). Formulation of a generalized beam element on a two parameter elastic foundation with semi-rigid connections and rigid offsets. *Comput. Struct.*, 80(25):1919–1934.

Nogami, T. and Chen, H.L. (1984). Simplified approach for axial pile group response analysis. *J. Geotech. Eng.*, 110(GT9):1239–1255.

Noorzaei, J., Viladkar, M.N., and Godbole, P.N. (1995). Influence of strain hardening on soil-structure interaction of framed structures. *Comput. Struct.*, 55(5):789–795.

Papadopoulos, B.P. (1992). Settlements of shallow foundations on cohensionless soils. *J. Geotech. Eng.*, 118(3):377–393.

Pasternak, P.L. (1954). On a new method of analysis of an elastic foundation by means of two foundation constants (in Russian). *Gosudarstvennoe Izdatelstro Liberaturi po Stroitelstvui Arkhitekture*, Moscow.

Poulos, H.G. (1999). Approximate computer analysis of pile groups subjected to loads and ground movement. *Int. J. Numer. Anal. Methods Geomech.*, 23:1021–1041.

Poulos, H.G. and Davis, E.H. (1980). *Pile Foundation Analysis and Design,* John Wiley & Sons, New York.

Providakis, C.P. and Beskos, D.E. (1999). Dynamic analysis of plates by boundary elements. *Appl. Mech. Rev.*, 52(7):213–236.

Rajapakse, R.K.N.D. (1988). The interaction between circular elastic plate and a transversely isotropic half space. *Int. J. Numer. Anal. Methods Geomech.*, 12:419–436.

Randolph, M.F. (1981). The response of flexible piles to lateral loading. *Geotechnique*, 31(2): 247–259.

Randolph, M.F. and Wroth, C.P. (1978). Analysis of deformation of vertically loaded piles. *J. Geotech. Eng.*, 104(GT12):1465–1488.

Rao, N.S.V. Kameswara, Das, Y.C., and Anandakrishnan, M. (1971). Variational approach to beams on elastic foundations. *J. Eng. Mech.*, 97(EM2):271–294.

Reissner, E. (1947). On the bending of elastic plates. *Q. Appl. Math.*, 5:55–68.

Roscoe, K.H. and Burland, J.B. (1968). *On the Generalized Stress-Strain Behaviour of Wet Clay in Engineering Plasticity,* J. Heyman and F.A. Leckie, Eds., University Press, Cambridge, 535–609.

Roscoe, K.H., Schofield, A.N., and Wroth, C.P. (1958). On yielding in soils. *Geotechnique*, 8:22–53.

Selvadurai, A.P.S. (1979). *Elastic Analysis of Soil-Foundation Interaction,* Elsevier Scientific Publishing, Netherlands.

Sharma, R.P. (1989). *Interaction of Frame Structures on Visco-elastic Foundations,* Ph.D. thesis, Department of Civil Engineering, University of Roorkee, Roorkee, India.

Skempton, A.W. and Bjerrum, L. (1957). A contribution to settlement analysis of foundation in clay. *Geotechnique*, 7:168–178.

Small, J.C. and Booker, J.R. (1984). Finite layer analysis of layered elastic materials using a flexibility approach. 1. Strip loadings. *Int. J. Numer. Methods Eng.*, 20:1025–1037.

Sogge, R.L. (1981). Laterally loaded pile design. *J. Geotech. Eng.*, 107:1179–1199.

Šuklje, L. (1969). *Rheological Aspects of Soil Mechanics,* John Wiley & Sons, Great Britain.

Taylor, D.W. and Merchant, W. (1940). A theory of clay consolidation accounting for secondary compression. *J. Math. Phys.*, 19:167–185.

Teng, W.C. (1962). *Foundation Design,* Prentice Hall, Englewood Cliffs, NJ.

Terzaghi, K. (1955). Evaluation of coefficient of subgrade reaction. *Geotechnique*, 5(4):297–326.

Thompson, W. (Lord Kelvin) (1848). On the equations of equilibrium of an elastic solid. *Cambr. Dubl. Math.*, 3:87–89.

Timoshenko, S. and Goodier, J.N. (1970). *Theory of Elasticity,* McGraw-Hill, New York.

Vallabhan, C.V.G. and Das, Y.C. (1991). Analysis of circular tank foundations. *J. Eng. Mech.*, 117(4):789–797.

Vesic, A.S. (1961). Bending of beams resting on isotropic elastic solid. *J. Eng. Mech.*, 87(EM2): 35–53.

Viladkar, M.N. (1989). Special Lecture Notes on Constitutive Laws for Soils (unpublished).

Viladkar, M.N., Sharma, R.P., and Ranjan, G. (1992). Visco-elastic finite element formulation for isolated foundations on clays. *Comput. Struct.*, 43(2):313–324.

Viladkar, M.N., Ranjan, G., and Sharma, R.P. (1993). Soil-structure interaction in the time domain. *Comput. Struct.*, 46(3):429–442.

Viladkar, M.N., Noorzaei, J., and Godbole, P.N. (1995). Convenient forms of yield criteria in elasto-plastic analysis of geological materials. *Comput. Struct.*, 54(2):327–337.

Vlazov, V.Z. and Leontiev, U.N. (1966). *Beams, Plates and Shells on Elastic Foundations* (translated from Russian), Israel Program for Scientific Translations, Jerusalem.

Wang, Y.H. and Cheung, Y.K. (2001). Plate on cross-anisotropic foundation analyzed by the finite element method. *Comput. Geotech.*, 28:37–54.

Wang, Y.H., Tham, L.G., and Cheung, Y.K. (2005). Beams and plates on elastic foundations: a review. *Prog. Struct. Eng. Mater.*, 7(4):174–182.

Winkler, E. (1867). *In Die Lehre von der Elastizitat und Festigkeit,* Domonicus, Prague.

Xu, K.J. and Poulos, H.G. (2000). General elastic analysis of piles and pile groups. *Int. J. Numer. Anal. Methods Geomech.*, 24:1109–1138.

Yang, T.Y. (1972). A finite element analysis of plates on a two-parameter foundation model. *Comput. Struct.*, 2:593–614.

Yankelevsky, D.Z., Eisenberge, M., and Adin, M.A. (1989). Analysis of beams on nonlinear Winkler foundation. *Comput. Struct.*, 31(2):287–292.

Yin, J.-H. (2000). Closed-form solution for reinforced Timoshenko beam on elastic foundation. *J. Eng. Mech.*, 126(8):868–874.

Zhaohua, F. and Cook, R.D. (1983). Beam element on two-parameter elastic foundations. *J. Eng. Mech.*, 109(6):1390–1402.

Zienkiewicz, O.C., Humpheson, C., and Lewis, R.W. (1975). Associated and non-associated visco-plasticity and plasticity in soil mechanics. *Geotechnique*, 25(4):671–689.

5

Design of Pile Foundations

by
Sanjeev Kumar
Southern Illinois University Carbondale, Carbondale, Illinois

5.1 Introduction

The load on any structure, irrespective of its size, shape, type, and function, has to be transferred to soil or rock unless the structure is floating in space or water. The structural element that transfers a structural load to the ground is called a foundation. For any project that requires foundation design and construction, the first and obvious question to be answered is whether a shallow or a deep foundation is needed. As the names suggest, a foundation that transfers the structural load to the ground at a shallow depth is called a shallow foundation, and a foundation that transfers the load at deeper depths is called a deep foundation. Selection of the type of foundation generally is based on many factors, including but not limited to the magnitude and type of the design load, strength and compressibility of site soils, project performance criteria, availability of foundation construction materials, and foundation cost.

Design and construction of shallow foundations generally are cheaper as long as anticipated settlements are within the acceptable limits and the stresses in the soil mass are less than the soil strength. Therefore, on many projects, if the soil strength and structural load combination is such that shallow foundations bearing on the existing soils are not practical, ground improvements in conjunction with shallow foundations are evaluated before selecting a deep foundation system. The engineer also should understand that use of deep foundations is not a panacea for all subsurface conditions. There are many subsurface conditions where construction of pile foundations is impractical and cost prohibitive. Some of the most common practical situations where use of deep foundations may be more economical or may be required are

1. Heavy column loads (vertical, uplift, or horizontal) and moments
2. Soft soil or unsuitable fill near the ground surface
3. Expansive (or collapsible) soils near the ground surface
4. Foundations for offshore towers, transmission towers, etc.
5. Foundations for structures where there is significant erosion or scour potential

Sometimes deep foundations also are used to stabilize slopes and site soils. Deep foundations used for these purposes generally experience limited vertical loads but may be subjected to significant lateral loads. Discussion of the design, construction, and testing of these types of deep foundations is beyond the scope of this chapter.

5.2 Foundation Support Cost Index

Economic evaluation of the potential foundation types suitable for a particular project is an essential part of any foundation design and construction project. For subsurface conditions at a particular site, several foundation alternatives may satisfy project requirements; however, only one foundation type may be the most economical. One of the ways various foundation alternatives can be compared is by the foundation support cost (FSC) index, which is defined as the ratio of the total cost of an installed foundation alternative to the allowable load it is designed to support:

$$\text{FSC index} = \frac{\text{Total cost of installed foundation alternative}}{\text{Allowable load supported by the foundation alternative}} \quad (5.1)$$

It is important to note that the total foundation cost must include all costs associated with the foundation design, construction, and testing (e.g., need for excavation and retention system, any environmental restrictions, type and cost of foundation testing program, need for and type and size of pile cap, need for and cost of predrilling, etc.).

Komurka (2004) has provided a detailed study that describes the use of the FSC concept. For large projects, it is highly recommended that the FSC index for various foundation alternatives be calculated to select a particular type of foundation system. Within a particular type of foundation alternative (e.g., pile foundation alternative), the FSC index can be developed for various types and sizes of piles in order to select the most economical.

With the development of new pile design and testing methods and new equipment for installation of piles, great opportunities are now available for optimizing pile sizes and types, which in turn would result in installation of efficient and cost-effective pile foundation systems without compromising safety or service life of the project. Table 5.1 provides cost-saving recommendations for pile foundation systems.

5.3 Types of Deep Foundations

Many different types of deep foundations are available. However, deep foundations can be broadly divided into the categories shown in Figure 5.1. Selection of a particular type of deep foundation is based on many factors, but constructability and cost normally control selection of a deep foundation. Basic technical information about commonly used piles is presented in Table 5.2.

5.3.1 Classification of Pile Foundations

Based on various variables, deep foundations can be classified as listed in Table 5.3. Figures 5.2–5.6 are photographs of various types of piles. Figure 5.7 shows a steel casing for construction of drilled shafts, and Figure 5.8 shows construction of a geopier.

TABLE 5.1 Cost-Saving Recommendations for Pile Foundation Systems (Hannigan et al. 2006)

Factor	Inadequacy of Older Methods	Cost-Saving Recommendation	Remarks
A. Design structural load capacity of piles	Allowable pile material stresses may not address site-specific considerations	1. Use realistic allowable stresses for pile materials in conjunction with adequate construction control procedures (i.e., load testing, dynamic pile monitoring, and wave equation) 2. Determine potential pile types and carry candidate pile types forward in the design process 3. Optimize pile size for loads	1. Rational consideration of factors A and B may decrease cost of a foundation by 25% or more 2. Significant cost savings can be achieved by optimization of pile type and section for the structural loads with consideration of pile driveability requirements
B. Design geotechnical capacity of soil and rock to carry load transferred by piles	1. Inadequate subsurface explorations and laboratory testing 2. Rules of thumb and prescription values used in lieu of static design may result in overly conservative designs 3. High potential for change orders and claims	1. Perform thorough subsurface exploration, including *in situ* and laboratory testing, to determine design parameters 2. Use rational and practical methods of design 3. Perform wave equation driveability analysis 4. Use design-stage pile load testing on large pile-driving projects to determine load capacities (load tests during design stage)	1. Reduction of safety factor can be justified because some of the uncertainties about load-carrying capacities of piles are reduced 2. Rational pile design generally will lead to shorter pile lengths and/or smaller number of piles
C. Alternate foundation design	Alternate foundation designs are rarely used even when possibilities of cost savings exist by allowing alternates in contract documents	For major projects, consider inclusion of alternate foundation designs in the contract documents if estimated costs of feasible foundation alternatives are within 15% of each other	Alternative designs often generate more competition, which can lead to lower costs

D. Plans and specifications	1. Unrealistic specifications 2. Uncertainties due to inadequate subsurface exploration force contractors to inflate bid prices	1. Prepare detailed contract documents based on thorough subsurface exploration, understanding of contractor's difficulties, and knowledge of pile techniques and equipment 2. Provide subsurface information to the contractor	1. Lower bid prices will result if the contractor is provided with all the available subsurface information 2. Potential for contract claims is reduced with realistic specifications
E. Construction determination of pile load capacity during installation	Often-used dynamic formulas such as Engineering News are unreliable	1. Eliminate use of dynamic formulas for construction control as experience is gained with wave equation analysis 2. Use wave equation analysis coupled with dynamic monitoring for construction control and load capacity evaluation 3. Use pile load tests on projects to substantiate capacity predictions by wave equation and dynamic monitoring	1. Reduced factor of safety may allow shorter pile lengths and/or smaller number of piles 2. Pile damage due to excessive driving can be eliminated by using dynamic monitoring equipment 3. Increased confidence and lower risk result from improved construction control

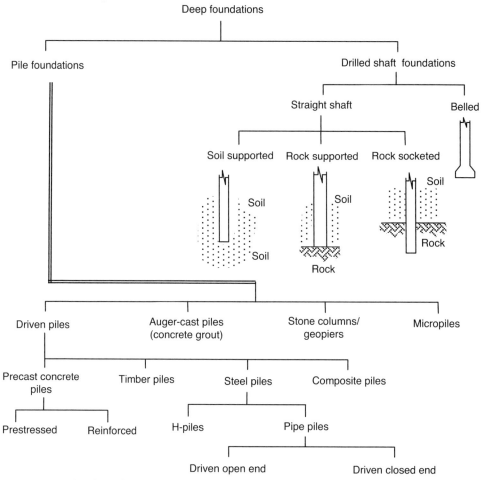

FIGURE 5.1 Flowchart showing various types of deep foundations.

5.4 Allowable Stress and Load and Resistance Factor Design of Deep Foundations

Allowable stress design (ASD) of pile foundations has been in use in geotechnical engineering practice for over a century. It is based on the simple concept that the allowable load (Q_{all}) that can be transferred to a pile is equal to the ultimate load (Q_{ult}) divided by a factor of safety (FS):

$$Q_{all} = \frac{Q_{ult}}{FS} \tag{5.2}$$

The ultimate load may be envisioned as the load that will cause failure (yield) stresses in either the pile material or the surrounding soils without considering deformations in the pile material or settlements in the surrounding soil. The factors of safety generally used range between 2 and 3.5, depending primarily on reliability of the design method and construction control method.

The load and resistance factor design (LRFD) approach currently is being used worldwide in structural design practice, whereas use of LRFD in geotechnical engineering practice is still limited. However, use of LRFD in foundation design is being adopted at a very rapid pace. In October 2007, the Federal Highway Administration decided to use the LRFD approach to designing foundations for any bridge design that it supports financially. For any engineer involved in the design of pile foundations, it is extremely important to understand the LRFD methodology. Misinterpretation or incorrect application of the LRFD procedure can result in unsafe or impractical design. Equation 5.3 forms the basis of the LRFD methodology:

$$\sum \gamma_i Q_i \leq \phi R_n \qquad (5.3)$$

On the left-hand side of the above equation, Q_i refers to the effect of all loads or forces and γ_i is the load factor (multiplier) which accounts for the variability of loads, lack of accuracy in the analysis, and the probability of the simultaneous occurrence of different loads (AASHTO 2007). Subscript *i* refers to the force type (e.g., dead load, live load, snow load, and so on). The left-hand side of Equation 5.3 also is referred to as *factored load*. On the right-hand side of Equation 5.3, R_n refers to nominal resistance, which is the maximum resistance available, and ϕ is a resistance factor (multiplier) which accounts for variability in material properties, structural dimensions, and workmanship and uncertainty in the prediction of resistance (AASHTO 2007). The right-hand side of Equation 5.3 also is referred to as *factored resistance* (R_r); that is, $\phi R_n = R_r$.

The primary difference between the ASD and LRFD methodologies is the way of accounting for uncertainties. In the ASD method, uncertainties are blended into a single factor of safety, whereas in the LRFD method, uncertainties are assigned to load and resistance separately. In order to compare the LRFD method with the ASD method, the LRFD load and resistance factors can be viewed as partial factors of safety and the combined effect of load and resistance factors is similar to the effect of the factor of safety in the ASD method. Comparison of Equations 5.2 and 5.3 suggests that the factor of safety is equivalent to the ratio of the load factor to the resistance factor.

Take a closer look at Equation 5.3. In order to account for the variability of loads, lack of accuracy in analysis, and probability of the simultaneous occurrence of different loads, it makes sense to increase the calculated loads. Therefore, for most design conditions (except when the load effects tend to resist failure), the load factor γ_i is equal to or greater than 1.0. On the other hand, in order to account for variability in material properties, structural dimensions, and workmanship and uncertainty in the prediction of resistance, it makes sense to reduce the calculated maximum resistance. Therefore, for most design conditions, the resistance factor ϕ is less than or equal to 1.0. Table 5.4 presents some commonly used load and resistance factors from AASHTO (2007).

Assuming that each soil layer has fairly uniform soil properties and the soil properties are known with reasonable accuracy, calculation of ultimate load Q_{ult} that a pile can resist (ASD method) and the nominal resistance R_n of a pile (LRFD method) is essentially the same, and the same basic equations are used to calculate Q_{ult} or R_n. When soil properties measured or estimated show some scatter, mean soil properties are used with the LRFD method.

Although calculations for estimating Q_{ult} and R_n are the same, it is extremely important for structural and geotechnical engineers to communicate clearly whether the structure under consideration is being designed using the ASD or LRFD method. Otherwise, the recommended capacities may either have too small a factor of safety or too great a factor of safety.

TABLE 5.2 Technical Information about Commonly Used Piles

Pile Type	Typical Cross Section	Typical Lengths	Typical Axial Loads
Timber piles	12- to 20-in. (300- to 500-mm) butt diameter 5- to 10-in. (120- to 230-mm) toe diameter	15–120 ft (5–35 m)	20–100 kips (100–500 kN)
Steel H-piles	Various sections ranging from HP 8 × 36 (HP 200 × 53) through HP 14 × 117 (HP 360 × 174)	15–150 ft (5–45 m)	125–550 kips (600–2500 kN)
Steel pipe piles (open or closed end)	8–48 in. (200–1200 mm) Larger sections also are available	15–150 ft (5–45 m)	125–550 kips (600–2500 kN) Capacities above 3000 kips (13,000 kN) could be obtained with steel H-pile and concrete as core)
Precast concrete piles	10–36 in. (250–900 mm) square 10–24 in. (250–600 mm) circular	30–50 ft (10–15 m)	90–225 kips (400–1000 kN)
Prestressed concrete	10–36 in. (250–900 mm) square 10–24 in. (250–600 mm) circular	50–150 ft (15–45 m)	90–1000 kips (400–4500 kN)
Auger-cast or continuous-flight auger piles	16- to 30-in. (400- to 760-mm) diameter	15–100 ft (5–30 m)	60–200 kips (250–875 kN)
Micropiles	4- to 8-in. (100- to 200-mm) diameter	40–100 ft (12–25 m)	70–250 kips (300–1100 kN) Generally installed in 5-ft sections

TABLE 5.2 Technical Information about Commonly Used Piles (continued)

Advantages	Disadvantages	Remarks
Low initial cost, easy to handle, resistance to decay if fully submerged	Prone to damage due to driving stresses, difficult to splice, prone to decay if not completely submerged	Displacement pile, good for granular material
Easy to handle, relatively high capacity, easy to splice, can penetrate through stiff soils and light obstructions, also can penetrate through soft rock or weathered rock with toe protection, small soil displacement	Possibility of damage during driving due to hard major obstructions such as boulders, vulnerable to corrosion	Good end-bearing pile, low-displacement pile, increase the pile size or reduce the allowable load if installation is in a corrosive environment
Easy to handle, relatively high capacity, easy to splice, open-end piles can penetrate through stiff soils and light obstructions, open-end pipes with cutting shoe also can penetrate through soft rock or weathered rock, small soil displacement with open-end pipes, closed-end pipes are easy to inspect and clean after installation	Possibility of damage during driving due to hard major obstructions such as boulders, vulnerable to corrosion, large ground vibrations when installed closed ended	Displacement pile if installed closed ended, high bending resistance
Resistance to corrosion, easy to manufacture	Possibility of damage during transportation and installation, difficult to splice, low lateral and uplift load capacity, large ground vibrations during driving	High-displacement pile suitable for granular soils, possibility of significant tensile stress during driving to rock
Resistance to corrosion, easy to manufacture, relatively high load capacity	Possibility of damage during transportation and installation, difficult to splice, low lateral and uplift load capacity, large ground vibrations during driving	High-displacement pile suitable for granular soils, possibility of significant tensile stress during driving to rock
Minimum vibrations during installation, cost effective, high skin resistance	Need for significant quality control, needs extensive subsurface exploration, no indirect correlations to estimate capacity based on measurements during installation, difficult to install reinforcing cage	Techniques are available to verify workmanship, excessive auger cuttings, not suitable for highly compressible material such as peat
Installation under low headroom and limited access conditions, low vibrations and noise, small amount of soil, suitable for installation in soils that contain boulders	Must be used in groups, relatively expansive	Suitable for foundation underpinning, suitable for most subsurface conditions

TABLE 5.3 Classification of Pile Foundations

Basis of Classification	Classification
Pile material	Precast reinforced concrete, prestressed concrete, steel, timber, composite, gravel, or stone
Method of installation	Driven, cast-in-place, bored, jetted
Load transfer mechanism	End bearing, friction piles, combination of end-bearing and friction piles
Soil displacement during pile installation	Nondisplacement, low or partial displacement, high or full displacement
Mode of loading	Axially loaded, transverse or laterally loaded, moment resisting
Shape	Square (solid or hollow), octagonal (solid or hollow), circular (solid or hollow), fluted, H, pipe, others

FIGURE 5.2 Stack of steel micropiles.

FIGURE 5.3 Closed end of a steel pipe pile.

FIGURE 5.4 Stack of steel H-piles.

FIGURE 5.5 Reinforced concrete fluted piles.

FIGURE 5.6 Reinforced concrete square and circular piles.

FIGURE 5.7   Steel casing for construction of drilled shafts.

FIGURE 5.8   Construction of a geopier.

5.5 Axial Capacity of Piles in Compression

Axial capacity of piles primarily depends on how and where the applied loads are transferred into the ground. Based on the location of the load transfer in deep foundations, they can be classified as follows:

1. *End- or point-bearing piles*—The load is primarily distributed at the tip or base of the pile.
2. *Frictional piles*—The load is distributed primarily along the length of the pile through friction between the pile material and the surrounding soil.
3. *Combination of friction and end bearing*—The load is distributed both through friction along the length of the pile and at the tip or base of the pile.

Figure 5.9 shows types of deep foundations based on the location of load transfer.

TABLE 5.4 Commonly Used Load and Resistance Factors (AASHTO 2007)

Load factors		
For permanent structures	Dead load	1.25–1.50
	Live load	1.30–1.75
	Seismic	1.0
Resistance factors for single-pile foundations		
Axial compression	Clay and mixed soils	
	α-Method	0.35
	β-Method	0.25
	λ-Method	0.40
	Sand	
	Nordlund method	0.45
	Meyerhof method	0.35
	End bearing on rock	0.45
Uplift resistance	Nordlund method	
	α-Method	0.35
	β-Method	0.25
	λ-Method	0.40
	Meyerhof method	0.25
	Load test	0.60

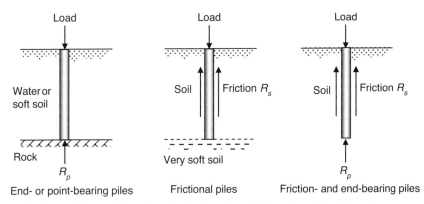

FIGURE 5.9 Types of deep foundations based on the location of load transfer.

In general, the ultimate load-carrying capacity of a pile or shaft can be calculated as

$$Q_{ult} = R_s + R_p \tag{5.4}$$

where R_s = load resisted due to friction and R_p = load resisted at the pile tip or point.

5.5.1 Load Transfer Mechanism in Pile Foundations

As discussed above, any load applied to a pile is resisted by the skin resistance and the resistance at the tip of the pile. In order to understand the load transfer mechanism, refer to Figure 5.10. Consider that a pile is installed at a site, and the pile is capable of transferring load through skin

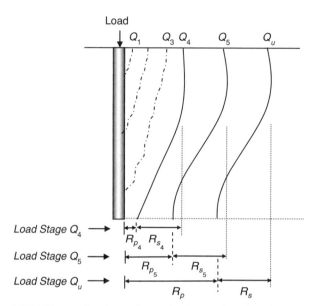

FIGURE 5.10 Load transfer mechanism in deep foundations.

friction and point. Also assume that when load is applied at the top of the pile, we have a mechanism of measuring the magnitude of the load that is transferred through skin friction and point separately.

If a very small amount of load (say Q_1) is applied at the top of the pile, all of the load may be resisted by the skin friction near the top of the pile, and the tip of the pile may not experience the application of load Q_1. At this stage, total load resisted by the pile can be calculated as:

$$Q_1 = R_{s_1}$$

If the load is gradually increased, skin friction along more and more of the length of the pile would resist the load, and a stage will come when the tip of the pile also will start contributing to resisting the applied load. This is shown by the curve for load Q_4 in Figure 5.10. At this stage, total load resisted by the pile can be calculated as

$$Q_4 = R_{s_4} + R_{p_4}$$

where R_{s_4} = magnitude of the load resisted by the skin friction and R_{p_4} = magnitude of the load resisted by the pile tip.

If the load on the pile is increased further, the magnitude of the load resisted by skin friction and point would increase, and a stage will come when all the skin resistance is mobilized (i.e., the skin resistance reaches its maximum value). In other words, any additional load on the pile will be resisted by the pile tip. This stage is shown by the curve for load Q_5 in Figure 5.10. At this stage, total load resisted by the pile can be calculated as

$$Q_5 = R_{s_5} + R_{p_5}$$

where R_{s_5} = maximum skin friction capacity of the pile and R_{p_5} = magnitude of the load resisted by the pile tip.

Further increase in the applied load will be resisted by the pile tip; that is, there will be no increase in the skin resistance since it has reached its maximum value. Ultimately, a stage will come when the point resistance also reaches its maximum value. The total load at this stage has fully mobilized its skin friction and point capacity, which means any further increase in the load will cause pile failure. This stage is shown by the curve for load Q_u in Figure 5.10. At this stage, total load resisted by the pile can be calculated as:

$$Q_u = R_s + R_p$$

The above equation is the same as Equation 5.4. Note that R_s is the same as R_{s_5}.

5.5.2 Pile Settlement and Resistance Mobilization

It is very important to understand that movements required to completely mobilize R_s and R_p are significantly different. Therefore, Equation 5.4 should be used with great care. Calculation of R_s and R_p should be consistent with the amount of deformation required to mobilize them, which in turn depends on the amount of maximum acceptable settlement in the pile or shaft:

Movement required to mobilize R_s ≈ 0.2–0.3 in., irrespective of the pile diameter or length

Movement required to mobilize R_p ≈ 10–25% of the pile diameter or width (10% for driven piles and 25% for drilled piles)

For a 15-in.-diameter driven pile, the approximate amount of movement (or settlement) required to mobilize R_p is 10% of 15 in. (i.e., 1.5 in.). Now, if 1.5 in. of settlement in the pile foundation is acceptable, the load-carrying capacity of the pile can be calculated by adding R_s and R_p according to Equation 5.4, because this movement is large enough to fully mobilize both the skin resistance and point resistance. However, if settlement of 1.5 in. is not acceptable, R_p would not be fully mobilized and the point capacity available would be less than R_p. In other words, if only ½ in. of settlement is acceptable, full skin resistance R_s would be available since it would be fully mobilized; however, full point capacity R_p would not be available because ½ in. of movement is not sufficient for the full point capacity to be mobilized. Therefore, the smaller value of R_p, consistent with the amount of expected settlement, should be used. It is important to note that Q_{ult} (or R_n for the LRFD method) is based on limiting strength without considering the amount of deformation or settlement. Therefore, Q_{ult} is calculated by first calculating R_s and R_p separately, assuming that the movement is significant enough to mobilize both R_s and R_p, and then adding R_s and R_p.

5.6 Ultimate Static Capacity of Single Piles in Cohesionless Soils

Over the years, many methods have been developed to estimate the ultimate load-carrying capacity of single piles. It is very important for designers to understand the applicability of a particular method to the project being designed and assumptions and limitations of the method being used. Only selected methods are discussed in this chapter.

5.6.1 Point Capacity

From the design of shallow foundations, the ultimate bearing capacity of shallow foundations can be calculated as

$$q_u = cN_cF_c + q'N_qF_q + \tfrac{1}{2}\gamma BN_\gamma F_\gamma \tag{5.5}$$

where the F factors depend on the shape and depth of the foundation.

If we incorporate the effect of shape and depth in determination of the N factors, the equation for bearing capacity of shallow foundations may be modified for deep foundations as:

$$q_u = cN_c^* + q'N_q^* + \tfrac{1}{2}\gamma BN_\gamma^* \tag{5.6}$$

For deep foundations, the third term in the above equation generally is small because of the small diameter or width of the piles. Therefore, for deep foundations, the equation to calculate ultimate bearing pressure at the tip or point of the pile can be reduced to Equation 5.7:

$$q_u \text{ or } q_p = cN_c^* + q'N_q^* \tag{5.7}$$

The capacity of deep foundations generally is expressed in terms of load they can carry. Therefore, the above equation can be modified to obtain the point capacity by multiplying the pressure by the point area of the pile:

$$R_p = q_uA_p \Rightarrow A_p \times (cN_c^* + q'N_q^*) \tag{5.8}$$

where R_p = point capacity of the pile, A_p = point or tip area of the pile (refer to Section 5.12 for additional discussion), q' = effective overburden pressure, c = soil cohesion near the pile tip, and N_c^* and N_q^* = bearing capacity factors for deep foundations which are related to the length and diameter of piles and the angle of internal friction of soils.

Bearing capacity factor N_c^* is commonly taken as 9. Several recommendations for bearing capacity factor N_q^* are available. Figure 5.11 shows the range of values for N_q^* recommended by various researchers. Note the wide range of recommended values.

5.6.2 Skin Friction Capacity

Frictional capacity of a single pile can be calculated by considering the frictional resistance between the pile material and the soil surrounding the pile. In order to understand the basic equation used to calculate the frictional capacity of a pile, let's first consider a small portion of the pile ΔL (refer to Figure 5.12).

If p is the perimeter of the pile and f is the unit frictional resistance, then the frictional capacity offered by a small portion of the pile can be calculated as:

$$\Delta R_s = p\Delta Lf \tag{5.9}$$

The frictional capacity of the entire pile length can then be calculated as:

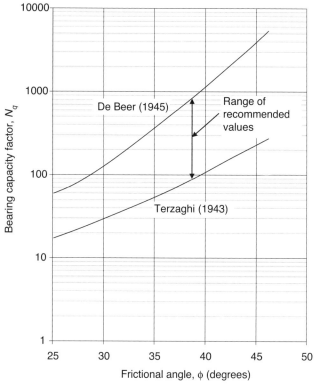

FIGURE 5.11   Range of theoretical values for N_q^* recommended by various researchers (data from Vesic 1963).

$$R_s = \Sigma p \Delta L f \tag{5.10}$$

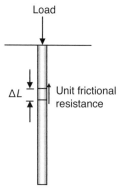

FIGURE 5.12   Estimation of frictional capacity.

Note that the unit frictional resistance will depend on several factors, including the pile material, cohesion in the soil surrounding the pile, and angle of internal friction of the soil surrounding the pile.

Let's first review the basic principle of frictional resistance. Refer to Figure 5.13a, which shows a massless block resting on another surface. Let's assume that the friction angle between the block material and the surface on which the block is resting is equal to δ. If a pressure σ is applied on the block and horizontal force P is applied in an attempt to move the block, frictional resisting force will develop at the contact, as shown in Figure 5.13a. The maximum magnitude of this frictional force (or resistance) can be calculated by:

$$F_r = \mu \times \sigma \times \text{Area} \tag{5.11}$$

The discussion presented above also is true if the whole setup is turned 90°, as shown in Figure 5.13b.

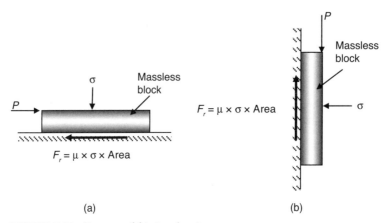

FIGURE 5.13 Concept of frictional resistance.

The basic principle of frictional resistance presented above now can be extended to estimate the frictional capacity of piles in sand. For the case of a pile embedded in sand, let's first estimate the frictional capacity of a small portion of the pile ΔL at a depth z from the ground surface. Refer to Figure 5.14. The frictional capacity of the small portion of the pile can be calculated from

$$\Delta R_s = \underbrace{K \times (q')_z}_{\sigma} \times \underset{\mu}{\underset{\uparrow}{\tan \delta}} \times \underbrace{p \times \Delta L}_{\text{Area}} \qquad (5.12)$$

where ΔR_s = frictional capacity offered by pile length ΔL, $(q')_z$ = effective vertical pressure (or overburden pressure) at depth z, K = coefficient to convert vertical pressure to lateral pressure, p = perimeter of the pile, and δ = frictional angle between the pile material and soil, generally taken as between 0.5 and 0.8 of the friction angle of soil ϕ. Note that the format of Equation 5.12 is the same as that of Equation 5.11. The term $[K \times (q')_z \tan \delta]$ is commonly referred to as unit frictional resistance. For cohesive soils, the unit friction is related to cohesion c, as discussed in subsequent sections.

Since the vertical pressure (and, in turn, the horizontal pressure) will be different at different depths, the skin friction capacity of the pile can be calculated by dividing the pile into smaller sections, calculating the capacity of each section using Equation 5.9, and then taking the sum of the capacity of each pile section; that is:

$$R_s = \Sigma \Delta R_s \qquad (5.13)$$

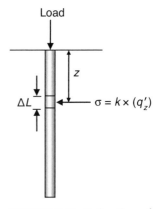

FIGURE 5.14 Estimation of frictional capacity.

Field studies have shown that the unit frictional resistance of piles embedded in cohesionless soils increases with depth. However, beyond a certain depth, the unit frictional resistance remains more or less constant, as illustrated in Figure 5.15. This depth, beyond which the unit

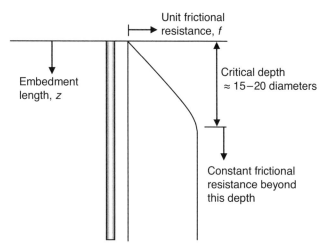

FIGURE 5.15 Concept of critical depth.

frictional resistance does not increase, is called the critical depth and has been observed to vary between 15 to 20 times the pile diameter.

A critical look at Equation 5.12 reveals that one of the most important parameters that can affect the skin friction capacity of piles in cohesionless soils is coefficient K. Several studies have shown that the value of K varies between 0.5 and 1.5 depending on several factors, including pile installation technique used, roughness of the pile surface, type of soil, etc. Although the value of coefficient K varies with depth, it is common practice to consider the value of K to be constant unless there is a significant change in the type and density of sand. The value of K is related to Rankine's coefficient of lateral earth pressure (K_0), and the following vales are commonly used in practice to estimate the skin friction capacity of piles:

$K = K_0$	Bored or jetted piles
$K \cong 1.4K_0$	Low-displacement driven piles
$K \cong 1.8K_0$	High-displacement driven piles

where $K_0 = 1 - \sin \phi$ for sands.

5.6.3 Meyerhof Method

The Meyerhof method of estimating single-pile capacity is primarily based on the analyses of numerous pile load tests in a variety of cohesionless soils. This method is quick and simple for preliminary estimates of pile capacities based on the results of standard penetration tests (SPTs). Because of the wide-scale use of SPTs for subsurface exploration, this method is widely used for preliminary estimates of pile capacities. However, the method should be used with caution because of the nonreproducibility of SPT N-values.

5.6.3.1 Point Capacity (Meyerhof Method)

Meyerhof (1976) proposed that N_c^* and N_q^* may be estimated from Figure 5.16. For piles installed in sand, cohesion c is 0 and Equation 5.8 can be rewritten as:

$$R_p = q_u A_p \Rightarrow A_p \times (q'N_q^*) \tag{5.14}$$

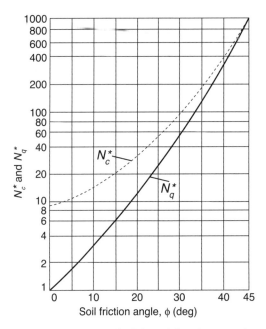

FIGURE 5.16 Meyerhof (1976) bearing capacity factors N_c^* and N_q^* (adapted from Das 1999).

Equation 5.14 shows that as the length of a pile in sand increases, the point capacity R_p also increases because the overburden pressure q' increases. However, Meyerhof (1976) observed that point capacity increases with the depth of embedment but reaches a limiting value after the ratio of the embedment length of the pile L_b in the bearing stratum (the soil stratum in which the pile tip is located) to the diameter of the pile D reaches a critical value, as illustrated in Figure 5.17.

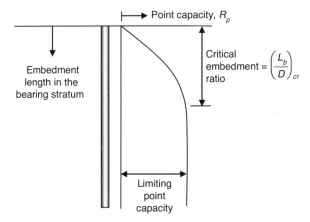

FIGURE 5.17 Increase in the point capacity with depth of embedment in the bearing stratum.

Based on field observations, Meyerhof (1976) suggested that the limiting point capacity can be calculated as:

$$(R_p)_{\lim} = A_p (1000 \times N_q^* \tan \phi) \qquad \text{in lb, area of the pile } A_p \text{ in ft}^2$$

$$(R_p)_{\lim} = A_p (50 \times N_q^* \tan \phi) \qquad \text{in kN, area of the pile } A_p \text{ in m}^2$$

(5.15)

Meyerhof also suggested that for piles embedded at least 10 pile diameters in the sand or gravel-bearing stratum, the point capacity can be approximated using SPT data as

$$R_p = A_p \left[800(N_{cor}) \left(\frac{L}{D} \right) \right] \leq A_p \left[8000(N_{cor}) \right]$$

$$\text{in lb, area of the pile } A_p \text{ in ft}^2$$

(5.16)

$$R_p = A_p \left[40(N_{cor}) \left(\frac{L}{D} \right) \right] \leq A_p \left[400(N_{cor}) \right]$$

$$\text{in kN, area of the pile } A_p \text{ in m}^2$$

where N_{cor} is the average of corrected SPT N-values between 10 pile diameters above and 3 pile diameters below the pile tip. It is recommended that N_{cor} be taken as $(N_1)_{60}$ (i.e., N-values corrected for overburden and 60% hammer efficiency).

For open-end piles in cohesionless soils, Tomlinson (1994) recommended that the static pile capacity be calculated using a limiting value of 105 ksf for the unit toe resistance regardless of the pile size or soil density because higher toe resistance does not develop due to yielding of soil plug rather than bearing capacity failure of the soil below the plug (Hannigan et al. 2006).

5.6.3.2 Skin Friction Capacity (Meyerhof Method)

Meyerhof suggested that skin friction capacity of piles embedded in sand or gravel can be approximated using SPT data as follows.

High-displacement driven piles:

$$R_s = \sum 40(N_{cor})pL \leq 2000pL \qquad \text{in lb, } p \text{ and } L \text{ in ft}$$

$$R_s = \sum 2(N_{cor})pL \leq 100pL \qquad \text{in kN, } p \text{ and } L \text{ in m}$$

(5.17)

Low-displacement driven piles:

$$R_s = \sum 40(N_{cor})pL \leq 2000pL \qquad \text{in lb, } p \text{ and } L \text{ in ft}$$

$$R_s = \sum 2(N_{cor})pL \leq 100pL \qquad \text{in kN, } p \text{ and } L \text{ in m}$$

(5.18)

where N_{cor} is the average of corrected SPT N-values along the embedded length of the pile. It is typical to divide the soil profile into 10- to 20-ft- (3- to 6-m-) thick sublayers and skin friction capacity is estimated using Equation 5.17 or 5.18. It is recommended that N_{cor} be taken as $(N_1)_{60}$ (i.e., N-values corrected for overburden and 60% hammer efficiency).

5.6.4 Nordlund Method

The Nordlund method is a semiempirical method which is based on results of several pile load tests on various pile types (steel H-piles, timber piles, steel pipe piles, Raymond step-taper piles, etc.) ranging in size from 10 to 20 in. (250 to 500 mm) embedded in cohesionless soils. This method considers the increased skin friction of tapered piles and includes the effects of volume of soil displaced and friction angle between the soils and pile material. Figure 5.18 presents various variables considered by Nordlund (1963).

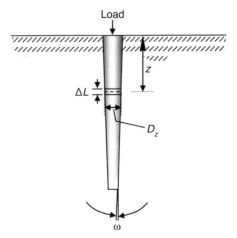

FIGURE 5.18   Variables considered by Nordlund (1963).

5.6.4.1 Point Capacity (Nordlund Method)

Nordlund (1963) proposed that the point capacity of a pile (shown in Figure 5.18) can be estimated by

$$R_p = \alpha q' N_q^* A_p \qquad (5.19)$$

where α = a dimensionless factor that depends on the friction angle of the soil and L/D ratio of the pile, N_q^* = a bearing capacity factor, q' = effective overburden pressure at the pile base not to exceed 3 ksf (150 kPa), and A_p = cross-sectional area of the pile base. Factors α and N_q^* can be obtained from Figures 5.19 and 5.20, respectively.

5.6.4.2 Skin Friction Capacity (Nordlund Method)

Nordlund proposed that the ultimate skin friction capacity of a pile (shown in Figure 5.18) can be calculated by

$$R_s = \sum_{z=0}^{z=L} K_\delta C_F (q')_z \frac{\sin(\delta + \omega)}{\cos \omega} D_z \Delta L \qquad (5.20)$$

where α = friction angle between the soil and pile material, ϕ = friction angle of the soil, ω = pile taper angle with vertical, z = depth from the ground line, L = length of the pile, ΔL = pile length increment, D_z = pile diameter at depth z, K_δ = coefficient of lateral earth pressure at depth z (at the center of the pile length increment) based on the angle of pile taper and

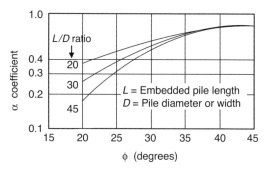

FIGURE 5.19 Dimensionless factor α for Nordlund method (adapted from Hannigan et al. 2006).

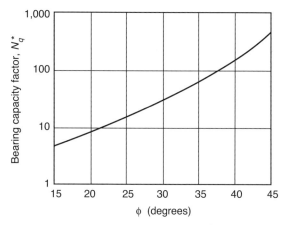

FIGURE 5.20 Bearing capacity factor N_q^* for Nordlund method (adapted from Hannigan et al. 2006).

displaced volume V, C_F = correction factor for K_δ when $\delta \neq \phi$, and $(q')_z$ = effective overburden pressure at depth z.

In order to estimate the skin friction capacity of piles, the displaced volume of soil is calculated using Figure 5.21, which presents the relationship between δ/ϕ and the volume of soil displaced for various types of piles proposed by Nordlund (1979). The coefficient of lateral earth pressure K_δ is then obtained from Figures 5.22–5.25 based on the pile taper angle and displaced volume of soil. A correction factor C_F is estimated using Figure 5.26 based on the frictional angle of the soil and δ/ϕ.

5.6.5 Effective Stress Method

The effective stress method can be used to estimate capacities of piles installed in cohesionless, cohesive, or layered soils. Effective stress soil parameters are used to calculate the pile capacities.

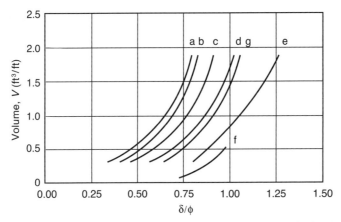

FIGURE 5.21 Displaced volume of soil for Nordlund method: (a) closed-end pipe and nontapered portion of monotube piles, (b) timber piles, (c) precast concrete piles, (d) Raymond step-taper piles, (e) Raymond uniform-taper piles, (f) H-piles, and (g) tapered portion of monotube piles (adapted from Hannigan et al. 2006).

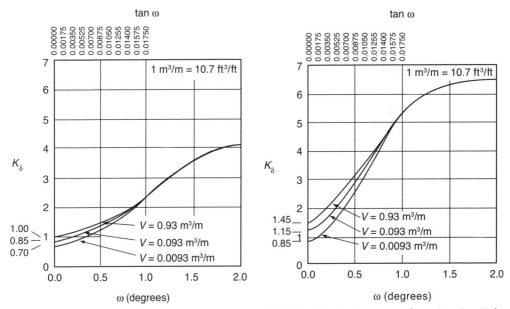

FIGURE 5.22 Design curves for estimating K_δ by Nordlund method where $\phi = 25°$ (adapted from Hannigan et al. 2006).

FIGURE 5.23 Design curves for estimating K_δ by Nordlund method where $\phi = 30°$ (adapted from Hannigan et al. 2006).

5.6.5.1 Point Capacity (Effective Stress Method)

Fellenius (1991) suggested that the point capacity of single piles installed in cohesionless or cohesive soils using effective stress soil parameters can be estimated by

$$R_p = A_p \times (q'N_t) \tag{5.21}$$

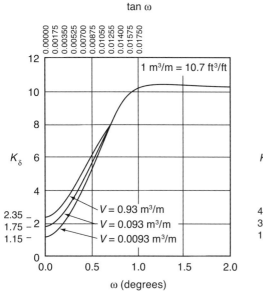

FIGURE 5.24 Design curves for estimating K_δ by Nordlund method where $\phi = 35°$ (adapted from Hannigan et al. 2006).

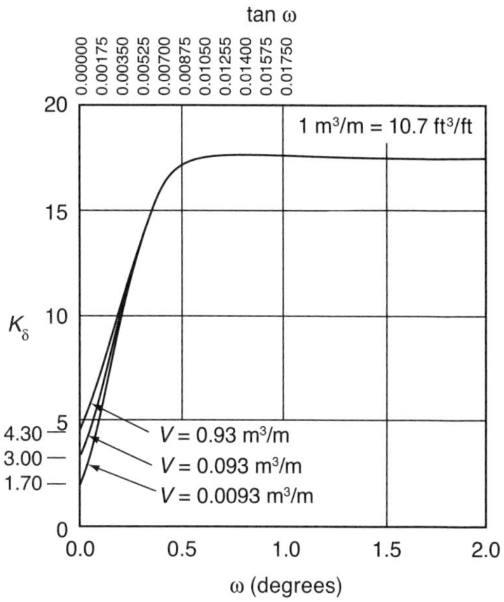

FIGURE 5.25 Design curves for estimating K_δ by Nordlund method where $\phi = 40°$ (adapted from Hannigan et al. 2006).

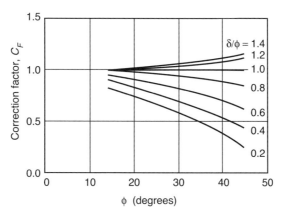

FIGURE 5.26 Correction factor C_F for Nordlund method (adapted from Hannigan et al. 2006).

where q' = effective overburden pressure at the pile tip and N_t = bearing capacity coefficient. Note that the format of Equation 5.21 is the same as Equation 5.14. Recommended values of N_t are given in Table 5.5 and Figure 5.27. For piles tips installed in clay, Fellenius (1991) recommends an N_t of 3.

5.6.5.2 Skin Friction Capacity (Effective Stress Method)

Fellenius (1991) suggested that the skin friction capacity of single piles installed in cohesionless or cohesive soils using effective stress soil parameters can be estimated by

TABLE 5.5 Recommended Range of N_t

Soil Type	Effective Soil Friction Angle (ϕ)	Bearing Capacity Coefficient (N_t)
Clay	25–30	3–30
Silt	28–34	20–40
Sand	32–40	30–150
Gravel	35–45	60–300

Based on Fellenius (1991); adapted from Hannigan et al. (2006).

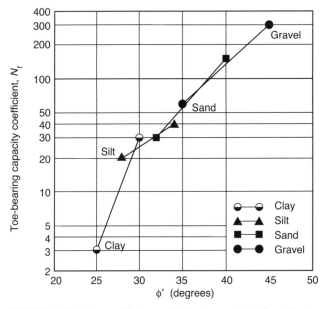

FIGURE 5.27 N_t vs. effective soil friction angle ϕ for effective stress method (based on Fellenius 1991; adapted from Hannigan et al. 2006).

$$R_s = \sum p \times \Delta L \times \beta \times (q')_z \tag{5.22}$$

where $(q')_z$ = effective overburden pressure at the center of depth increment and β = Bjerrum-Burland beta coefficient. Note that the format of Equation 5.22 is the same as Equation 5.12 if $\beta = K \tan \phi$. Recommended values of β are presented in Table 5.6 and Figure 5.28. Alternatively, β can be estimated as $\beta = K \tan \phi$.

5.7 Ultimate Static Capacity of Single Piles in Cohesive Soils

5.7.1 Point Capacity of Piles in Clay

As discussed earlier, the general equation to estimate point capacity of piles bearing on soil is

TABLE 5.6   Recommended Range of β

Soil Type	Effective Soil Friction Angle (φ)	Beta Coefficient (β)
Clay	25–30	0.23–0.40
Silt	28–34	0.27–0.50
Sand	32–40	0.30–0.60
Gravel	35–45	0.35–0.80

Based on Fellenius (1991); adapted from Hannigan et al. (2006).

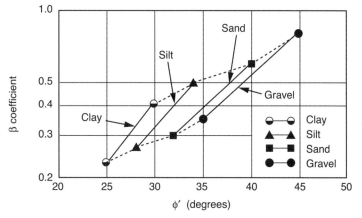

FIGURE 5.28   β vs. effective soil friction angle φ for effective stress method (based on Fellenius 1991; adapted from Hannigan et al. 2006).

$$R_p = q_u A_p \implies A_p \times (cN_c^* + q'N_q^*)$$

which is the same as Equation 5.18.

For clays under undrained condition, the angle of internal friction of soil φ is zero. For φ = 0, N_q^* from Figure 5.16 is equal to 1.0, which makes the second term q (i.e., γz). This is the weight of overburden, which generally is assumed to be balanced by the weight of the pile, and therefore this term is neglected. The bearing capacity factor N_c^* is taken as 9 for φ = 0. Therefore, the point capacity of piles embedded in clay can be calculated from Equation 5.23:

$$R_p = A_p \times (c \times 9) \implies 9c \times A_p \qquad (5.23)$$

5.7.2 Frictional Capacity of Piles in Clay

The basic equation for estimating the skin friction capacity of piles (Equation 5.10) is applicable to piles embedded in both sand and clays. However, determination of the unit friction factor f is significantly different from that presented for sands:

$$R_s = \Sigma p \Delta L f$$

Although several methods of estimating the unit frictional resistance are available in the literature, the three most commonly used methods are

1. λ-method
2. α-method
3. β-method (effective stress method)

5.7.2.1 λ-Method

Based on the results of pile load tests, Vijayvergiya and Focht (1972) proposed a method to estimate the skin friction capacity of piles embedded in overconsolidated clays. This method is commonly known as the λ-method. According to this method, the skin friction capacity of piles in clays can be estimated by

$$R_s = p \times L \times \lambda(\sigma_0' + 2\bar{c}_u) \tag{5.24}$$

where p = perimeter of the pile, L = length of the pile, λ = a coefficient that is based on the embedment length of the pile and can be obtained from Figure 5.29 (note that the embedment length is in meters), σ_0' = mean effective vertical stress, and C_u = mean undrained shear strength.

For a layered soil profile, the mean values of undrained shear strength c_u and effective vertical stress σ_0' can be calculated from Equations 5.25 and 5.26, respectively:

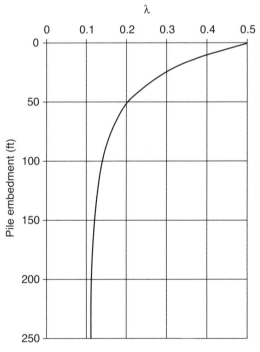

FIGURE 5.29 Relationship between pile embedment length and λ (data from Vijayvergiya and Focht 1972).

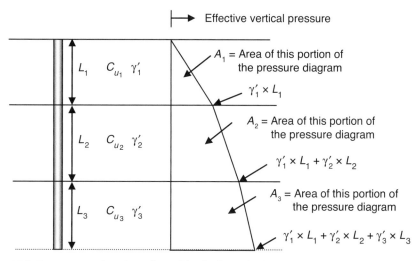

FIGURE 5.30 Explanation of variables for λ-method.

$$\bar{c}_u = \frac{(c_{u_1}L_1 + c_{u_2}L_2 + c_{u_3}L_3 + \ldots)}{L_1 + L_2 + L_3 + \ldots} \tag{5.25}$$

$$\sigma'_0 = \frac{(A_1 + A_2 + A_3 + \ldots)}{L_1 + L_2 + L_3 + \ldots} \tag{5.26}$$

The variables used in the above equations are explained in Figure 5.30. Note that only one value of λ based on the pile embedment length is used in Equation 5.24.

5.7.2.2 α-Method

According to the α-method, the skin friction capacity of a portion of a pile ΔL at a depth z can be calculated using

$$\Delta R_s = p \times \Delta L \times \alpha \times c_u \tag{5.27}$$

where c_u = undrained cohesion of the soil at a depth z and α = an empirical adhesion factor.

The adhesion factor α may be estimated from Figure 5.31. The skin friction capacity of the entire pile can be calculated by summing the capacities of various portions of the pile using Equation 5.28:

$$R_s = \sum p \times \Delta L \times \alpha \times c_u \tag{5.28}$$

It is important to note that the value of α depends on many factors, including strength of the clay, pile dimensions, roughness of the pile, method of pile installation used, and time after

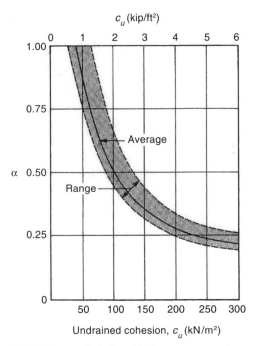

FIGURE 5.31 Relationship between α and c_u (adapted from Das 1999).

installation. Figure 5.31 shows that the adhesion factor decreases sharply with the unconfined compression strength of the clay. Tomlinson (1980) presented the variation in pile adhesion (αc_u) with the undrained shear strength of clay as shown in Figure 5.32.

5.7.2.3 β-Method (Effective Stress Method)

Unlike the λ-method and the α-method, which are based on undrained parameters, the β-method is based on the effective stress or drained soil parameters. This method was proposed by Burland (1973) and makes the following assumptions:

1. The effective or drained cohesion adjacent to the pile is zero.
2. The effective horizontal pressure on the pile surface after installation of the pile is approximately equal to the pressure before pile installation (i.e., lateral earth pressure coefficient is approximately equal to K_0).
3. The excess pore water pressure generated due to pile installation near the pile surface dissipates during the period between pile driving and loading.

The procedure to estimate skin friction capacity of piles in clay is the same as presented earlier in Section 5.6.5.2). By making the above assumption, the skin friction capacity of a portion of a pile ΔL at a depth z can be calculated using the following equation (which is the same as Equation 5.22):

$$R_s = \sum p \times \Delta L \times \beta \times (q')_z$$

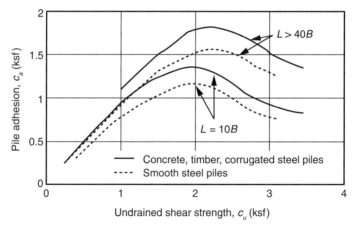

FIGURE 5.32 Pile adhesion in clays (based on Tomlinson 1980; adapted from Hannigan et al. 2006).

Recommended values of β are given in Table 5.6 and Figure 5.28. Alternatively, β can be estimated as $\beta = K \tan \phi$, where ϕ = drained friction angle of remolded clay near the pile surface and K = coefficient of lateral earth pressure, which can be estimated as $K = 1 - \sin \phi$ for normally consolidated clays and $K = (1 - \sin \phi) \times \sqrt{\text{overconsolidation ratio}}$ for overconsolidated clays.

5.8 Design Capacity of Single Piles

In accordance with allowable stress design, it is common practice to calculate the design capacity (allowable capacity) of a single pile by applying a factor of safety to the ultimate static load determined as per Sections 5.6 and 5.7. The purpose of the factor of safety is to incorporate the effects of various factors including but not limited to variability of the soil and rock, lack of confidence in developing input parameters such as soil and rock properties, construction control during pile installation, and limitations of the method used for estimating ultimate pile capacity. In general, a factor of safety between 2 and 4 is used, depending on the level of confidence in these factors. Design and allowable capacity of piles can be calculated by:

$$Q_{\text{allowable}} = \frac{Q_{\text{ult}}}{\text{FS}} \qquad (5.29)$$

Confidence in factors related to soil and rock profile and properties can be enhanced by implementing quality subsurface exploration and field and laboratory testing programs. Therefore, it makes sense to relate the factor of safety to the level of confidence in pile installation and testing. Hannigan et al. (2006) recommended the factors of safety in Table 5.7, which are based on the construction control method selected and associated level of field observations.

TABLE 5.7 Recommended Factors of Safety Based on the Construction Control Method Selected (Hannigan et al. 2006)

Construction Control Method	Factor of Safety
Static load test (ASTM D-1143) with wave equation analysis	2.00
Dynamic testing (ASTM D-4945) with wave equation analysis	2.25
Indicator piles with wave equation analysis	2.50
Wave equation analysis	2.75
Gates dynamic formula	3.50

Note that the recommended factor of safety when static load tests are planned is almost half of that recommended for use with the Gates dynamic formula. More detailed field observation and a testing program result in higher confidence and hence a lower factor of safety (i.e., higher pile capacity). Therefore, the design engineer must consider the advantages and disadvantages of using a particular design and construction control method and the impacts on the project cost.

5.9 Effect of Pile Driving on Pile Capacity

Method of installation of piles and soil type have a significant effect on the long-term capacity of piles. Pile driving can cause substantial disturbance and remolding of soils around a pile. In addition, substantial change in pore water pressure occurs in soils around the pile. Based on field measurements, Poulos and Davis (1980) presented results which show that the pore water pressure near a pile can be as high as two times the effective overburden pressure but drop sharply within a distance of 5–7.5 pile diameters.

In cohesionless soils, driving of displacement piles also can cause a significant increase in the relative density of loose and medium-dense sand. The zone of densification may extend 3–5 pile diameters around a pile, as shown in Figure 5.33a. Densification of cohesionless soils

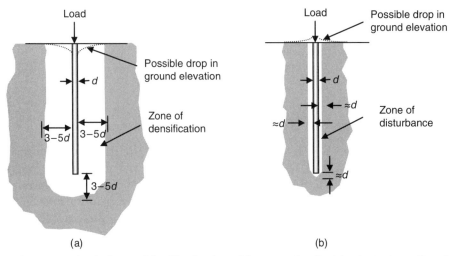

FIGURE 5.33 Typical zone of densification/remolding around a pile: (a) cohesionless soils and (b) cohesive soils.

TABLE 5.8 Recommended Values of Soil Setup Factor

Soil Type	Range of Soil Setup Factor	Recommended Soil Setup Factor
Clay	1.2–5.5	2.0
Silt-clay	1.0–2.0	1.0
Silt	1.5–5.0	1.5
Sand-clay	1.0–6.0	1.5
Sand-silt	1.2–2.0	1.2
Fine sand	1.2–2.0	1.2
Sand	0.8–2.0	1.0
Sand-gravel	1.2–2.0	1.0

Based on Rausche et al. (1996); adapted from Hannigan et al. (2006).

may cause a drop in the ground around a pile. Since pile capacity depends on the relative density of the soil around a pile, an increase in the relative density due to pile driving generally results in an increase in pile capacity. For piles driven into soft or normally consolidated saturated cohesive soils, remolding of soils occur within a distance of approximately 1 pile diameter. Radial compression of cohesive soils may cause ground heave, as shown in Figure 5.33b. The soil around the pile goes through a recovery phase after disturbance during pile driving. The magnitude of recovery and the time it takes to recover cause a change in pile capacity.

The change in pore water pressure during and after pile driving can significantly affect the short-term and long-term pile capacities. The time required for a pile to reach its long-term capacity depends on how fast the excess pore water dissipates. Field measurements have shown that the capacity of piles driven in saturated clays, silts, and fine sands increases with time after their installation. This increase in pile capacity is caused by a phenomenon known as soil setup. On the other hand, the capacity of piles driven into dense saturated sands may decrease with time due to the development of negative pore water pressures during and immediately after pile driving. This is known as soil relaxation.

Table 5.8 presents the recommended values of the soil setup factor, which is defined as the ratio of long-term pile capacity divided by the capacity of the pile at the end of driving. A relaxation factor, which is defined similar to the setup factor, in the range of 0.5–0.9 has been reported in the literature. If the capacity of a pile driven into soils where soil relaxation is possible needs to be verified, it is recommended that a static pile load test or a restrike test be delayed for a week after pile driving.

5.10 Ultimate Load-Carrying Capacity and Resistance to Driving

The long-term ultimate load-carrying capacity of piles installed in soils depends on the resistance provided by soils. Therefore, only soil layers that are expected to provide resistance throughout the life of the project should be considered for determination of ultimate load-carrying capacity. However, the effects of soil layers present during pile installation should be considered to determine the resistance to pile driving. As an example, consider the soil profile shown in Figure 5.34 in which soils to a depth of z have the potential to be scoured. These soils may not be available to provide resistance throughout the life of the project but will be present during pile installation. Therefore, the resistance from soils present within the potential scour

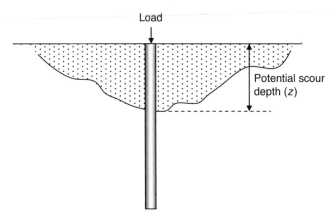

FIGURE 5.34 Soil profile with scour potential.

zone should be ignored for determination of long-term ultimate pile load-carrying capacity but should be included in determining resistance to pile driving.

Similarly, consider the soil profile shown in Figure 5.35. Due to the new fill, the soft clay layer has the potential for compression under the weight of the new fill. Therefore, the soft clay layer and the layers above should not be included in determining the long-term ultimate pile load-carrying capacity but should be included to determine resistance to pile driving. In fact, the soft clay layers and the layers above it may impart significant additional load on the piles due to down-drag forces. This phenomenon is commonly known as negative skin friction. Particular attention should be given to this phenomenon when interpreting the results of pile load tests.

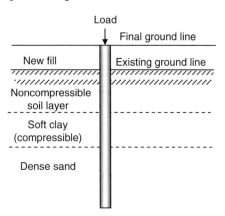

FIGURE 5.35 Soil profile with compressible soft clay layer.

5.11 Capacity of Piles Bearing on Rock

If rock is within 150 ft below the ground surface and soils above the rock do not have sufficient load-carrying capacity, piles are commonly driven or augered to bedrock. Pile foundations bearing on rock generally are designed to carry large loads. Because of the significant difference in the stiffness of the bedrock and the overlying soil, only the end-bearing or point capacity of piles is calculated.

The point capacity of piles bearing on bedrock should be calculated in two steps: (1) capacity based on the strength of rock and (2) capacity based on the yield strength of the pile material. The lower value of the capacity calculated from step 1 and step 2 should be selected as the point capacity of the pile. Unless a pile is bearing on soft rock such as shale or other poor quality rocks (rock quality designation less than 50), the capacity calculated from the strength of the rock is higher than that calculated from the yield strength of the pile material. Therefore, in most cases, calculation of the capacity of pile bearing on rock based on the properties of the pile material is sufficient.

The most common types of piles which are driven to rock include steel H-piles, steel pipe piles, and prestressed concrete piles. When piles are driven to rock, the exact area of the pile tip in contact with the bedrock is not known with reasonable certainty. In addition, the quality of the rock below the pile tip and the depth of penetration of the pile tip into the bedrock bring additional uncertainty to the performance of piles bearing on rock. Therefore, it is important to perform field observations during pile installation and pile load tests to verify the load-carrying capacity of piles bearing on bedrock.

5.11.1 Capacity Based on Strength of Bedrock

The ultimate capacity of a pile based on the strength of rock can be calculated by

$$Q_u = R_p = A_p q_u (N_\phi + 1) \tag{5.30}$$

where q_u = unconfined compression strength of the bedrock, $N_\phi = \tan^2(45 + \phi/2)$, ϕ = the drained angle of internal friction, and A_p = point area at the tip of the pile, which may be taken as equal to the actual area of the pile.

For steel H-piles or pipe piles, if a driving shoe is used at the tip of the pile or if the tip of the pile has the potential to become plugged, the point area through which the load is transferred to the rock may be higher than the actual area of the pile. Therefore:

$$Q_{allowable} = \frac{A_p q_u (N_\phi + 1)}{FS} \tag{5.31}$$

The unconfined compression strength on rock generally is obtained by performing unconfined compression strength tests on a small-diameter and intact sample of bedrock in the laboratory. Bedrock generally has irregularities and fractures which may or may not show up in small-diameter samples. Studies have shown that the unconfined compression strength of rock decreases as the sample diameter increases. The strength from a 2-in.-diameter sample may be four to five times greater than that obtained from a large-diameter sample or from field tests on bedrock. Therefore, the unconfined compression strength of bedrock for design purposes is generally taken as one-fourth to one-fifth of the strength measured in the laboratory as given by:

$$Q_{(u)design} = \frac{q_{u(lab)}}{4 \text{ to } 5} \tag{5.32}$$

It is important to note that the number 4 or 5 in Equation 5.32 is not a factor of safety. Instead, it is applied to consider the scaling effect in measuring the unconfined compression strength of the bedrock.

Equation 5.31 can be rewritten as:

TABLE 5.9 Typical Values of Unconfined Compression Strength and Effective Angle of Internal Friction of Rocks

Rock Type	Compressive Strength, q_u (psi)	Internal Friction Angle ϕ (degrees)
Basalt	28,000–67,000	40–50
Granite	10,000–38,000	35–50
Quartzite	16,000–44,800	35–50
Limestone	2,450–28,400	30–45
Marble	7,900–27,000	25–30
Sandstone	4,900–20,000	25–45
Slate	6,950–31,000	5–30
Shale	500–6,500	5–20

$$Q_{\text{allowable}} = \frac{A_p(N_\phi + 1)}{\text{FS}} \times \frac{q_{u(\text{lab})}}{4 \text{ to } 5} \tag{5.33}$$

Typical values of unconfined compression strength of common types of rocks from laboratory samples and typical values of the effective angle of internal friction of rocks are given in Table 5.9.

5.11.2 Capacity Based on Yield Strength of the Pile Material

If a pile is driven to a sound rock, which has sufficient capacity, the ultimate design load based on the yield strength of the pile material can be calculated by

$$Q_u = \sigma_y \times A_p \tag{5.34}$$

$$Q_{\text{allowable}} = \frac{\sigma_y \times A_p}{\text{FS}} \tag{5.35}$$

where σ_y = design yield strength of the pile material (for steel piles, the design yield strength of steel is generally taken as one-third to one-half of the actual yield strength reported by the manufacturer, but this reduction is not a factor of safety), A_p = actual area at the pile (note that for a steel H-pile or steel pipe pile, A_p is the area of the steel only since the yield strength of the pile material is used in Equation 5.35), and FS = an acceptable factor of safety.

5.12 Special Considerations for Calculation of A_p

As discussed earlier, the area of the tip of the pile is needed to calculate the point capacity of a pile. For almost all types of piles except steel pipe piles driven open ended and steel H-piles, the area of the pile tip is clearly defined and easy to calculate (i.e., full base area). However, for steel pipe piles driven open ended and steel H-piles, calculation of the area of the pile tip is more complex and depends on the formation of a competent soil plug. In the case of piles embedded in soil where a competent soil plug forms, the pile tip area should be taken as the full base area (i.e., the area of the steel and soil plug), as shown in Figure 5.36.

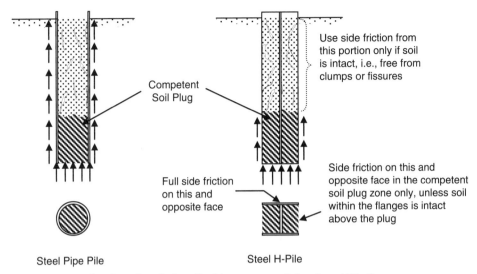

FIGURE 5.36 Plugging of steel pipe pile driven open ended and steel H-pile.

When a steel pipe pile is driven open ended, soil enters the pipe and starts formation of the plug. After penetrating a certain distance into the soil, the soil inside the pipe starts behaving as a part of the pile and starts moving with the pile. Formation of a competent soil plug depends on several factors, including but not limited to the size of the pile, method of installation of the pile, soil type and density or consistency, and penetration depth. An ideal and most desirable situation is that no soil plug forms under the dynamic load of pile driving, but a competent plug forms after driving. This can be achieved to a certain extent by carefully selecting the characteristics of the pile-driving hammer and controlling acceleration of the pile during driving.

According to Paikowsky and Whitman (1990), formation of a competent soil plug may be assumed in steel pipe piles if the penetration-to-diameter ratio is greater than 25–35 for sands and 10–20 for clays. For steel H-piles, the penetration-to-diameter ratio required for formation of the soil plug is smaller because of the much smaller space between the flanges.

For most piles embedded in soil, penetration is generally greater than 25–35 times the diameter or width of the pile. Therefore, assumption of the presence of a competent soil plug is reasonable. However, steel pipe piles and steel H-piles often are driven to bedrock. Due to the significant difference in the stiffness of soil and bedrock, load transfer at the point primarily occurs through the actual area of the steel. Therefore, for piles driven to bedrock, the actual area of the steel, without any soil plug, should be used for calculation of point capacity.

5.13 Special Considerations for Calculation of Perimeter

The perimeter of the pile is needed to calculate the frictional capacity of a pile. For almost all types of piles except steel pipe piles driven open ended and steel H-piles, calculation of the perimeter of the pile is straightforward. However, for steel pipe piles driven open ended and steel H-piles, an effective perimeter depends on many factors.

For steel pipe pile when a competent soil plug forms near the pile tip, resistance due to friction between the outside surface of the pile over the embedded length and the surrounding soil is available to resist the load. Therefore, the outside perimeter of the pile should be taken into account in calculation of the frictional capacity of the pile. However, if a competent soil plug does not form, resistance due to friction between both the outside and inside surfaces of the pile over the embedded length and the soil may be considered in estimating the frictional capacity of the pile (in this case, the point capacity of the pile will be minimal). If the soil inside the pipe has the potential to develop fissures and/or clumps (as shown in Figure 5.36), resistance from that portion of the pile should be ignored.

Estimation of the skin frictional capacity of steel H-piles is more complex than other piles. If the soil within the flanges of a steel H-pile is intact throughout the embedded length of the pile, the perimeter of the box as shown in Figure 5.36 can be used for calculation of skin friction capacity. However, it is important to understand that frictional resistance along the two flanges will develop due to friction between steel and soil, whereas on the other two faces it will be due to friction between soil and soil. In most practical situations, skin friction capacity can be calculated by considering friction between steel and soil along all four faces. If the soil within the flanges of a steel H-pile has the potential to develop fissures and/or clumps (e.g., stiff clays), frictional resistance from the faces where the contact is soil to soil should be calculated from the zone of the competent soil plug only, as shown in Figure 5.36.

As discussed in Section 5.8, for most piles embedded in soil, penetration generally is greater than 25–35 times the diameter or width of the pile, and assumption of the presence of a competent soil plug is reasonable. Therefore, for pipe piles and H-piles embedded in soil, it is reasonable to calculate the perimeter by assuming the piles to be fully plugged. If the piles are driven to bedrock, it is common practice to ignore the frictional resistance of the piles because of the significant difference in the stiffness of soil and bedrock.

5.14 Maximum Stresses in Driven Piles

In order for piles to perform as designed and intended, it is important that stresses in piles remain within structural limits during installation and service life. Therefore, maximum allowable material stresses should be within the limits given in Table 5.10.

5.15 Uplift Capacity of Single Piles

Because of seismic and other dynamic loads of considerable magnitude, the penetration depth of a pile foundation may be controlled by its uplift capacity. It is obvious that piles derive resistance to uplift loads from friction between the pile material and the surrounding soil. For large-diameter piles (e.g., concrete-filled pipe piles and drilled shafts), the weight of the pile itself also provides significant resistance against uplifting.

Based on information available in the literature, the uplift capacity of a single pile generally ranges from about 70–100% of the skin friction capacity in compression. Therefore, it is common practice to take the allowable uplift capacity of a single pile as one-third of the skin friction capacity in compression unless the uplift capacity of a pile is verified in the field by performing an uplift pile load test. When a field test is performed, the uplift capacity of a single pile can be taken as one-half of the failure load determined from the uplift load test.

Where there is the potential for loss of contact between the soil and the pile near the ground surface (e.g., due to desiccation or application of cyclic loads), resistance from the soil down to an appropriate depth should be ignored. Also, when the effects of down drag due to

TABLE 5.10 Maximum Allowable Stresses

Pile Type	Stresses during Driving	Stresses during Service Life
Timber piles	3 × allowable working stress (compression and tension)	0.8–1.2 ksi (5.5–8.3 MPa) Southern pine = 1.2 ksi (8.3 MPa) Douglas fir = 1.2 ksi (8.3 MPa) Red oak = 1.1 ksi (7.6 MPa) Eastern hemlock = 0.8 ksi (5.5 MPa)
Steel H-piles	$0.9f_y$	$0.25f_y$ (can be increased to $0.33f_y$ if damage to the pile is unlikely and pile load tests will be performed to verify the design)
Steel pipe piles (open or closed end)	$0.9f_y$	$0.25f_y$ (can be increased to $0.33f_y$ if damage to the pile is unlikely and pile load tests will be performed to verify the design)
		For concrete-filled pipe piles, maximum allowable stresses may be taken as $0.25f_y + 0.40f_c'$
Precast reinforced concrete piles	$0.85f_c'$ (compression) $0.9f_y$ (tension)	$0.33f_c'$
Precast prestressed concrete	$0.85f_c' - f_{pe}$ (compression) $0.25 \times \sqrt{f_c'} + f_{pe}$ (tension, SI units) $3 \times \sqrt{f_c'} + f_{pe}$ (tension, U.S. units)	$0.33f_c' - 0.27f_{pe}$

Modified from Hannigan et al. (2006).

f_y = yield strength of steel, f_c' = unconfined compression strength of concrete at 28 days of curing, f_{pe} = effective prestress after losses (for prestressed concrete piles).

the presence of a compressible soil layer are considered in the calculation of the skin friction capacity, the uplift capacity of the pile should be adjusted appropriately.

5.16 Lateral Capacity of Single Piles

Piles that sustain lateral loads of significant magnitude are used in offshore structures, water-front structures, bridges, buildings, industrial plants, locks and dams, and retaining walls. Piles used to stabilize slopes also are subjected to lateral loading. The lateral loads on piles are derived from earth pressures, wind pressures, current forces from flowing water, earthquakes, impact loads from barges and other vessels, and moving vehicles. Even if the above loads are not present, lateral load on piles can result from eccentric application of vertical load (Kumar 1993). In many cases, the design of piles is controlled by their lateral capacity instead of their axial capacity.

Lateral loads on piles are accommodated either by designing vertical piles to resist the lateral load or by using battered piles. However, use of battered piles to resist lateral loads in seismically active areas is not recommended because battered piles, due to their higher stiffness, can cause punching through pile caps. Analysis and design of laterally loaded piles must consider structural capacity of piles and deformations in surrounding soils. Therefore, it is important that an engineer working on lateral load design have appropriate knowledge of both geotechnical and structural engineering.

The basic concept behind the analysis of laterally loaded piles is that a horizontal load on a pile induces resistive pressures by the soil opposite to the direction of the load. The magnitude and distribution of the soil resistance along the length of the pile depend on three closely related factors: (1) soil stiffness, (2) pile stiffness, and (3) pile head fixity.

The stiffness of a pile is a relative parameter that depends on stiffness of the pile material, stiffness of the surrounding soil, and pile length. Piles subjected to lateral loads are commonly classified as *short* or *rigid* piles and *long* or *flexible* piles depending on the stiffness of pile material, stiffness of the soil, and pile length. The differentiation originates from the fact that a long, flexible pile tends to bend under lateral loads, while the short rigid pile tends to rotate or translate; hence the terms flexible and rigid (Kumar and Cisco 2006).

Another important parameter that affects the design and analysis of piles under lateral load is pile head fixity. In general, pile head fixity refers to whether or not the pile head is allowed to rotate. A pile is considered free headed if it is free to rotate at its top, as shown in Figure 5.37a. Under this condition, movement of both piles, due to a lateral load, will result in lateral and rotational movement of the pile head. A pile is considered to be fixed headed if it is not allowed to rotate at its top. A pile head embedded in a pile cap generally is considered to be fixed, since it is not free to rotate. Under this condition, piles will translate without rotation of the pile head, as shown in Figure 5.37b.

Pile stiffness and pile head fixity are very important interrelated factors in the design of laterally loaded piles. Deflection of a free-head pile can be reduced by as much as one-half for a given lateral load with the introduction of a pile cap. Therefore, the stiffness of the pile is greater for a fixed-head pile than for a pile with a top that is free to rotate. The fixity issue is actually more complex in most cases since some piles are considered to have some degree of fixity when the pile head is not completely restrained from rotation.

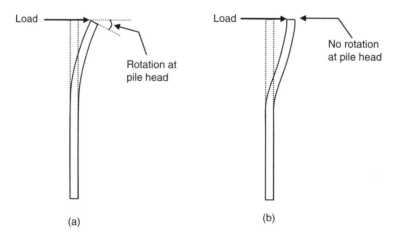

FIGURE 5.37 Rotation at the pile head: (a) free-head pile and (b) fixed-head pile.

Generally, the ultimate lateral capacity of piles is determined based on the following two criteria:

1. Ultimate soil resistance
2. Acceptable lateral deflection of single piles

The first approach estimates the ultimate load in terms of the shear failure of the soil (i.e., ultimate lateral load capacity based on ultimate soil resistance), and the second and most widely used approach estimates the ultimate capacity in terms of acceptable lateral deflections.

5.16.1 Ultimate Lateral Capacity of Single Piles Based on Ultimate Soil Resistance

There are two commonly used methods for estimating the ultimate lateral capacity of a single pile under lateral loading in terms of ultimate lateral resistance of the soil against the pile: Brinch Hansen's method (1961) and Broms' method (1964a, 1964b). Both methods utilize soil resistance and use equations of static equilibrium to solve for the ultimate lateral load, from which the allowable lateral load can be determined. Broms' method is applicable for both long and short piles, while Brinch Hansen's method is applicable to short piles.

Lateral load design of most of the piles used in practice is based on acceptable deflection rather than ultimate soil resistance. Therefore, further discussion is limited to designing laterally loaded piles based on acceptable lateral deflection.

5.16.2 Lateral Load Capacity of Single Piles Based on Acceptable Lateral Deflection

An acceptable lateral load against an acceptable deflection is most often determined from a load-deflection curve developed through lateral pile load analysis. Due to the complexity and trial-and-error nature of analysis, computer programs are used to accomplish some or all of this task. The most widely used method for determining allowable deflections at working loads is the *p-y* curve method (Matlock 1970; Reese et al. 1974; Reese and Welch 1975; Bhushan et al. 1979), which is an extension of the subgrade reaction method (Reese and Matlock 1956; Matlock and Reese 1961, 1962). Other methods include the elastic continuum approach (Poulos 1971a, 1971b), the strain wedge model approach (Norris 1986; Ashour et al. 1998), and the $K_{h\max}$ approach (Kumar 1993; Prakash and Kumar 1996). The *p-y* curve method is the most commonly used method in practice. Therefore, only this method is discussed herein. Since the *p-y* curve method is based on the modulus of subgrade reaction approach, a brief discussion of the modulus of subgrade reaction is presented first.

5.16.2.1 Modulus of Subgrade Reaction

The subgrade reaction method is primarily based on the Winkler (1867) soil model in which the soil along a pile is replaced by a series of infinitely closely spaced independent and elastic springs. The stiffness of the springs may be defined by the modulus of subgrade reaction k_h using

$$k_h = -\frac{p}{y} \tag{5.36}$$

where p is the soil reaction per unit length of the pile and y is the lateral deformation of the pile. The negative sign indicates that the deflection is opposite to the soil reaction. The modulus is the secant modulus obtained from the p-y relationship. For overconsolidated clays, the modulus of subgrade reaction typically is assumed to be constant with depth, whereas for sands and normally consolidated clays, the modulus of subgrade reaction usually is assumed to increase linearly with depth, and k_h can be related to depth x by

$$k_h = n_h x \qquad (5.37)$$

where n_h is the coefficient of the horizontal subgrade reaction and x is the depth from the ground surface to any point along the pile length.

The behavior of the pile is assumed to be elastic, which results in the following fourth-order differential equation:

$$E_p I_p \frac{d^4 y}{dx^4} + Q \frac{d^2 y}{dx^2} + E_s y = 0 \qquad (5.38)$$

where E_p = modulus of elasticity of the pile, I_p = moment of inertia of the pile section, E_s = modulus of elasticity of the soil, Q = lateral load at the ground surface, and y = deflection at the ground surface.

The derivation of this equation for a beam on an elastic foundation was given by Hetenyi (1946). Solutions for this equation to determine deflection and maximum moments in a free-head pile, as a function of depth, in the form of nondimensional coefficients and charts were developed by Reese and Matlock (1956) and Matlock and Reese (1961, 1962). Solution of the differential equation yields a set of curves similar to those shown in Figure 5.38.

The p-y curve method also is rooted in the Winkler model and separates the soil into a series of infinitely closely spaced independent and elastic springs, which in this case are defined by p-y curves: p is the unit pressure in terms of load per unit length of the pile due to the lateral load applied to the pile at a particular deflection y. In other words, p-y curves are modeled to represent the soil reaction with deflection at any point along the pile. The nonlinear differential equations are solved using the finite difference numerical technique. For each set of applied boundary loads (shear and moment), an iterative solution is needed which satisfies the static equilibrium and achieves compatibility between force and deflection in every element. Because of the iterative nature of analysis, it is necessary to use a computer program. The computer programs most commonly used to analyze single piles under lateral loads are LPILE (Reese et al. 2004) and FLPIER, developed by the University of Florida in collaboration with the Florida Department of Transportation. Further discussion in this chapter is limited to the LPILE computer program.

The most accurate p-y curve would be developed from the results of a full-scale load test, which is expensive and impractical to carry out at every site. Instead, typical p-y curves have been developed by the author's researchers using the results of full-scale field load tests and laboratory tests, which model the behavior of soils of varying types and conditions. The shape of the p-y curve depends on the type of soil and soil properties. A p-y curve also depends on many other factors, such as depth, location of the groundwater table, pile width, loading conditions, etc. For analysis using LPILE, the typical p-y curves generated by the program from input parameters supplied by the user can be used or the user has the option to input p-y curves. LPILE Version 5.0 has p-y curves for the following soils built into the program: soft

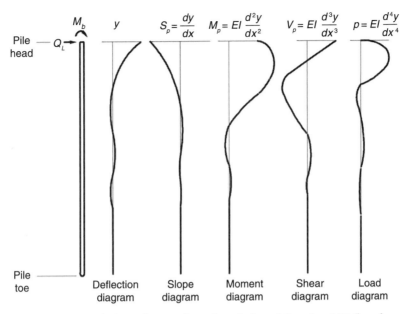

FIGURE 5.38 Typical set of curves from the solution of Equation 5.38 (based on Reese and Matlock 1956; adapted from Hannigan et al. 2006).

clay, stiff clay without free water, stiff clay with free water, stiff clay without free water using initial k, sand, API sand, liquefied sand, silt, string rock (vuggy limestone), and weak rock. The reader should refer to the LPILE technical and user manuals for detailed information about these p-y curves. To develop p-y curves for clays, soil strain parameter E_{50} and soil modulus parameter k are needed, whereas to develop p-y curves for sands, soil modulus parameter k is needed. The strain parameter E_{50} represents the axial strain at which 50% of the undrained strength is developed in a laboratory compression test. Representative soil parameters for clays and sands commonly used are given in Tables 5.11 and 5.12, respectively.

Figures 5.39 and 5.40 show p-y curves for clays and sand, respectively, developed using LPILE 5.0. The unit weight of the soil for all curves was selected as 110 lb/ft^3; undrained cohesion for soft and stiff clays was taken as 1000 and 2000 lb/ft^2, respectively; and the angle of internal friction for loose and dense sand was taken as 30 and 38°, respectively.

In order to perform analysis of a single pile subjected to lateral load, all data are input using various menu screens. Figure 5.41 shows the various menus to input pile properties, soil

TABLE 5.11 Representative Values of Soil Strain Parameter E_{50} and Soil Modulus Parameter k for Clays

Soil Consistency	Average Undrained Shear Strength (lb/ft^2)	Soil Strain Parameter (E_{50})	k (lb/in.^{3}) Static Loading	k (lb/in.^{3}) Cyclic Loading
Soft	250–500	0.02	30	
Medium	500–1000	0.01	100	
Stiff	1000–2000	0.007	500	200
Very stiff	2000–4000	0.005	1000	400
Hard	4000–8000	0.004	2000	1000

TABLE 5.12　Representative Values of Soil Modulus Parameter k (lb/in.^{3}): Static and Cyclic Loading

	Loose	Medium	Dense
Sand below groundwater table	20	60	125
Sand above groundwater table	25	90	225

properties, and loading conditions and the control options for terminating the iteration process. Several different types of analysis can be performed (see Figure 5.42); for example, Type 1 analysis consists of computing the response using elastic pile stiffness, and Type 3 analysis consists of computing the response using nonlinear pile stiffness. The user can control the output by specifying the information that needs to be included in the output file (Figure 5.43). Also, depending on the project requirements, various response relationships can be viewed graphically.

Figures 5.44 and 5.45 show typical responses of a steel H-pile under free-head and fixed-head conditions, respectively. Comparison of Figures 5.44 and 5.45 shows that the deflections in the free-head pile are significantly greater compared to the fixed-head pile. Also, for a free-head pile, the maximum bending moment in the pile occurs at some depth below the top of the pile, whereas for a fixed-head pile, the maximum bending moment generally occurs at the top of the pile.

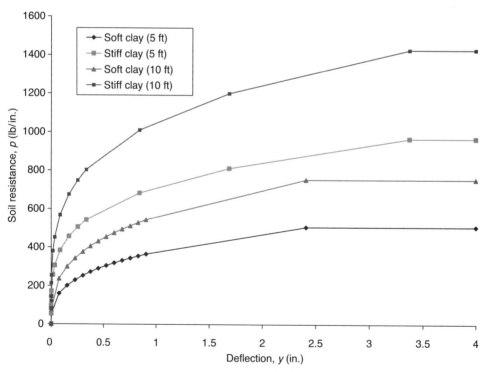

FIGURE 5.39　*p-y* curves for soft and stiff clay developed using LPILE 5.0.

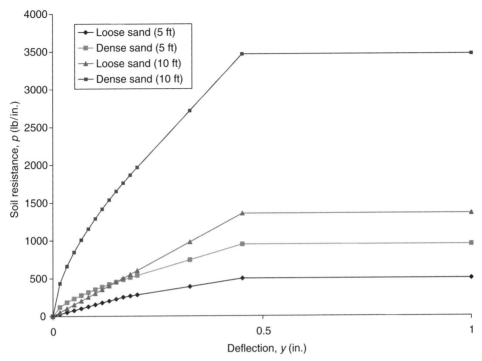

FIGURE 5.40 *p-y* curves for loose and dense sand developed using LPILE 5.

5.17 Design of Pile Groups

So far, the discussion of pile foundation design has been limited to the design of single piles. However, a pile foundation is rarely constructed using a single pile. Instead, a pile foundation generally is constructed by installing a cluster of piles, known as a pile group, as shown in Figure 5.46. Piles in a pile group generally are connected though a pile cap. The design of pile groups subjected to static loads depends on several factors, such as pile material, pile spacing, pile length, pile head fixity, and soil conditions. Most of the difficulty in the design of pile groups comes from the interaction of piles because piles are typically spaced close to one another, and thus interaction between the pile and adjacent soil becomes very important. Although the most accurate and appropriate method for analyzing pile groups is to account for pile-to-pile and pile-to-soil interaction simultaneously for each pile in a group, the most commonly used method is to estimate the group response based on the response of a single pile in a group and then apply factors to account for various effects. Because of the complex nature of pile-soil-pile interaction, computer programs such as GROUP (Reese et al. 2003) and FLPIER (Hoit and McVay 1996) are commonly used to analyze and design pile groups.

Figure 5.47a shows the typical shape of the stress zone along a single pile subjected to an axial compression load, Figure 5.47b shows overlapping of the stress zones of piles installed close to each other, and Figure 5.47c shows the typical shape and size of a stress zone for a group of piles installed close to each other. It is important to understand that if the piles are spaced sufficiently far away from each other such that their stress zones do not overlap, the

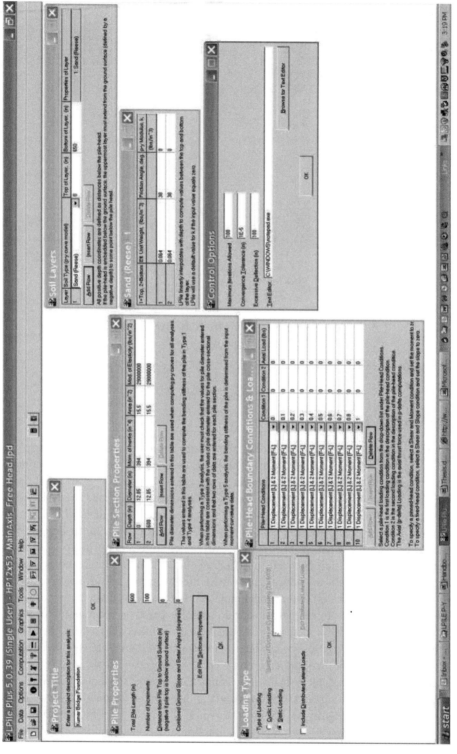

FIGURE 5.41   Various drop-down menus in LPILE 5.0 for data input.

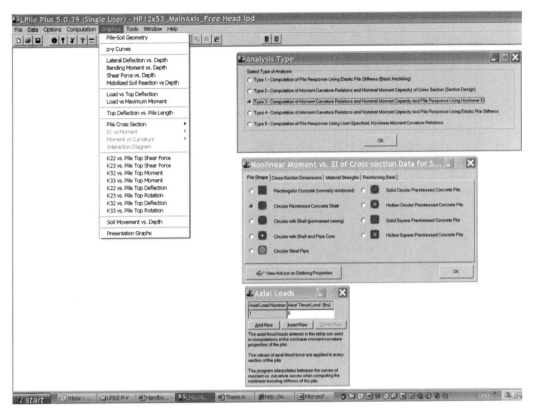

FIGURE 5.42 Drop-down menus in LPILE 5.0 for selecting type of analysis and graphics.

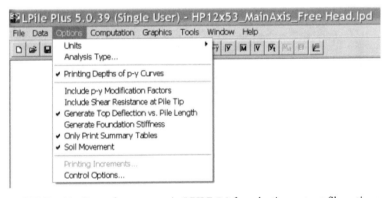

FIGURE 5.43 Drop-down menu in LPILE 5.0 for selecting output file options.

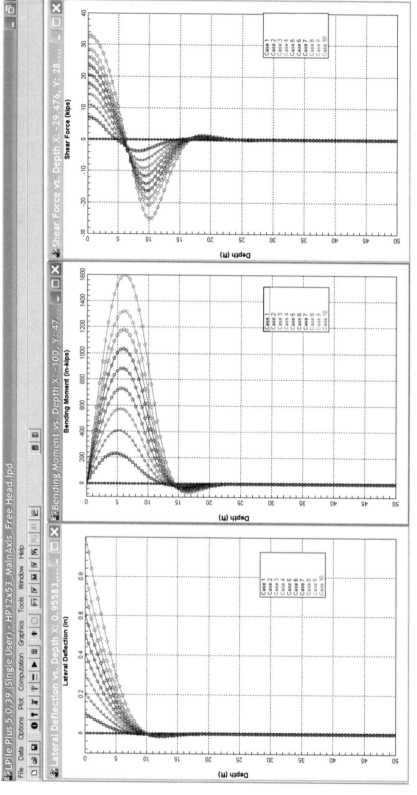

FIGURE 5.44 Typical responses of a steel H-pile under free-head condition developed using LPILE 5.0.

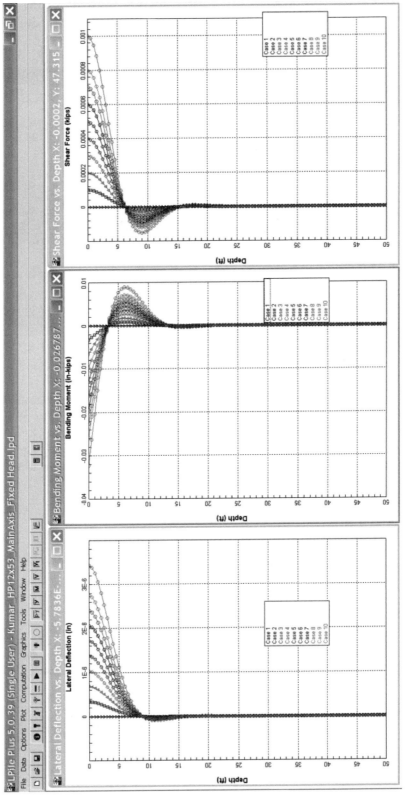

FIGURE 5.45 Typical responses of a steel H-pile under fixed-head condition developed using LPILE 5.0.

FIGURE 5.46 A group of steel H-piles for construction of a bridge pier (before construction of the pile cap).

shape and size of the heavily stressed zone for a pile group will be different. In practice, however, the piles in most pile groups are installed close to each other and their stress zone is similar to Figure 5.47c.

Overlapping of stress zones in a pile group generally reduces the ultimate capacity of the pile group unless the method of installation of piles changes the characteristics of the soils surrounding the piles (e.g., densification due to pile driving; see Section 5.9). The increased stress as shown in Figure 5.47c could increase the settlement of the pile group significantly. Figure 5.48 shows a typical load-deflection curve of a pile group compared to the load-

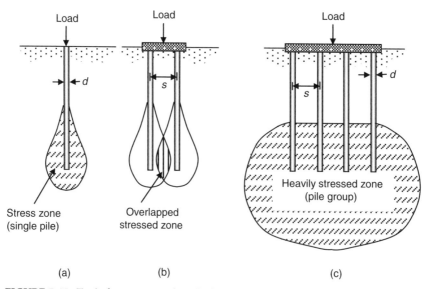

FIGURE 5.47 Typical stress zones for pile foundations under axial compression load.

deflection curve obtained by adding the individual ultimate capacity of all piles in a
group. Reduction of the group capacity at
any deflection is the result of a group effect.
The group capacity often is assessed in terms
of group efficiency η, which usually is defined by

$$\eta = \frac{Q_{ug}}{NQ_u} \qquad (5.39)$$

where Q_{ug} = ultimate capacity of the group,
Q_u = ultimate capacity of a single pile in the
pile group, and N = number of piles in the
group.

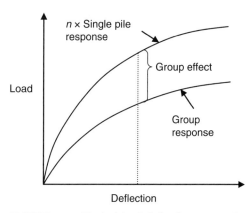

FIGURE 5.48 Typical load-deflection curve of a
pile group compared to that obtained by adding the
individual ultimate capacity of all piles in a group.

5.17.1 Capacity of Pile Groups Subjected to Axial Compression Loads

5.17.1.1 Pile Groups in Cohesive Soils

When a pile is driven in cohesive soils, an increase in the pore water pressure occurs in the soils
surrounding the pile. Excess pore water pressure can cause uplift in other adjacent piles during
driving. Increase in pore water pressure also can cause significant reduction in group efficiency
(as low as 0.4). However, this reduction is short term. After dissipation of the excess pore water
pressure, group efficiency is likely to increase. The rate of dissipation of the excess pore water
pressure depends on properties of the soil and size of the pile group. Typically, the excess pore
water pressure dissipates within 1–6 weeks, but may take up to a year for large pile groups. In
order to minimize installation problems and ground heave, it is recommended that piles in
cohesive soils be driven at a minimum center-to-center spacing of 3 pile diameters or 3 ft.

Published literature suggests that the efficiency of a pile group installed in cohesive soils is
influenced by the type of contact between the pile cap and the ground. When the contact
between the pile cap and the ground is firm, the piles and the soil within a pile group act as
a unit. Hannigan et al. (2006) recommend the group efficiencies given in Table 5.13 for
estimating the group capacity of piles driven into cohesive soils.

For piles installed in saturated clay or cohesionless soils underlain by weak cohesive soils,
the ultimate bearing capacity also should be checked, assuming block failure, as shown in
Figure 5.49. The lesser of the ultimate bearing capacity calculated by using group efficiency
factors and that calculated assuming block failure should be used as the ultimate bearing
capacity of the pile group.

5.17.1.2 Ultimate Group Capacity against Block Failure

The ultimate group capacity of a pile group installed in cohesive soils assuming block failure
can be calculated as follows.

Average skin friction resistance of the block:

$$R_{sg} = 2(B + L) \times Z \times S_{u1} \qquad (5.40)$$

TABLE 5.13 Group Efficiencies for Estimating Ultimate Capacity of Pile Group in Cohesive Soils

Undrained Shear Strength of Clay	Firm Contact between Pile Cap and Ground	Group Efficiency[a]	Remarks
Less than 2 ksf	NO	$\eta = 0.7$ for $s/d = 3$ $\eta = 1.0$ for $s/d = 6$	Use linear interpolation for s/d values between 3 and 6; also check ultimate group capacity against block failure
Less than 2 ksf	YES	$\eta = 1.0$ for $s/d \geq 3$	Also check ultimate group capacity against block failure
Greater than or equal to 2 ksf	Not applicable	$\eta = 1.0$ for $s/d \geq 3$	Also check ultimate group capacity against block failure

[a] s = center-to-center spacing between the piles, and d = pile diameter or width.

Point resistance of the block:

$$R_{pg} = B \times L \times S_{u2} \times N_c \quad (5.41)$$

Ultimate capacity of the pile group:

$$Q_{ug} = R_{sg} + R_{pg} \quad (5.42)$$

where R_{sg} = ultimate skin resistance of the block, R_{pg} = ultimate point resistance of the block, Q_{ug} = ultimate capacity of the pile group against block failure, B = width of the pile group (block), L = length of the pile group (block), Z = embedment depth of piles in the block, S_{u1} = weighted average of

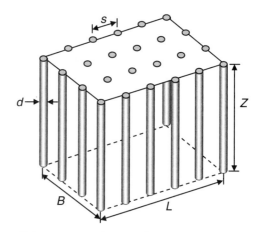

FIGURE 5.49 Pile group for block failure.

undrained shear strength of clays along the embedment depth, S_{u2} = average undrained shear strength of clays below the point of the pile group (block) to a depth of $2B$, and N_c = a bearing capacity factor.

The value of N_c for rectangular blocks generally is taken as 9. However, for large groups or for relatively short pile embedment depths, N_c can be estimated by:

$$N_c = 5 \times \left(1 + 0.2 \frac{Z}{B}\right) \times \left(1 + 0.2 \frac{B}{L}\right) \leq 9 \quad (5.43)$$

As discussed earlier, due to the development of excess pore water pressures, the group efficiency factors may be as low as 0.4 immediately after pile installation. Therefore, if a pile group in clay is expected to experience full group load shortly after construction, group capacity may range between 0.4 and 0.8 times the long-term ultimate group capacity. In that case, skin friction resistance R_{sg} of the pile group (block) should be calculated using the remolded value of the undrained shear strength S_{u1}.

5.17.1.3 Pile Groups in Cohesionless Soils

As discussed earlier, when a pile is driven in cohesionless soils, densification of soils around the pile occurs at a distance ranging from 3 to 5 pile diameters (refer to Figure 5.33). It is common practice to estimate the ultimate capacity of piles using the soil parameters before taking any densification into consideration. Therefore, when piles are installed at a center-to-center spacing of less than 3 pile diameters, overlapping of the zone of densification could result in an ultimate group capacity greater than the sum of the ultimate capacities of the individual piles in the pile group. When piles are installed at a center-to-center spacing greater than 3 pile diameters, piles generally act as individual piles. Therefore, for all practical purposes, when piles are installed in cohesionless soils, the ultimate capacity of a pile group can be taken as the sum of the ultimate capacities of the individual piles irrespective of the pile spacing (i.e., group efficiency of 1.0). However, the following should be taken into consideration:

- Sequence of pile installation, especially for displacement piles, is very important in order to avoid overdensification of soils. Therefore, it is recommended that pile installation start from the center of the group and move outward.
- Although some codes allow installation of piles at a center-to-center spacing less than 3 piles diameters, it is recommended that driven piles not be installed at a spacing less than 3 pile diameters to optimize group efficiency and minimize pile installation problems.
- If jetting or predrilling is used for pile installation, densification of surrounding soils is not likely to take place. Therefore, group efficiency is likely to be less than 1.0. It is recommended that jetting or predrilling be avoided for pile installations in cohesionless soils. For drilled piles, such as auger-cast or continuous-flight auger piles, a group efficiency of 0.7 is recommended when piles are installed at a center-to-center spacing less than 3 pile diameters.
- If the pile group is installed in a stratum that is underlain by a weak layer of cohesive soils, the ultimate group capacity should be checked assuming block failure. The lesser of the ultimate bearing capacity calculated using group efficiency factors and that calculated assuming block failure should be taken as the ultimate bearing capacity of the pile group. The procedure discussed in Section 5.17.1.2 for calculating the ultimate group capacity assuming block failure can be used by replacing the undrained shear strength with unit frictional resistance.

5.17.2 Capacity of Pile Groups Subjected to Lateral Loads

Pile-soil-pile interaction in a pile group subjected to lateral loads is much more prevalent compared to that in pile groups subjected to vertical compression loads. Lateral deflection of a pile in a pile group can be two to three times greater than that of an isolated single pile subjected to the same lateral load. There are many factors that contribute to the behavior of pile groups under lateral loads. Some of the most important factors are the shadowing effect, center-to-center spacing of piles, and the size of the pile group.

Figure 5.50 shows the typical stress zones of an isolated single pile and a pile in a pile group subjected to lateral load. As shown in the figure, the stress zones of piles in a pile group overlap, thus reducing soil resistance due to overstressing of the soil in front of the pile. Therefore, if the piles are closely spaced, overlapping of the stress zones could reduce soil resistance

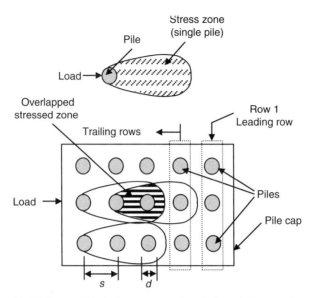

FIGURE 5.50 Typical stress zones for pile foundations under lateral load.

significantly. It is obvious that the piles in the front row (row 1 or the leading row) draw their resistance from the less stressed zone compared to those in the trailing rows. Reduction of soil resistance for the trailing rows is known as the shadow effect. If the piles are spaced such that their stress zones do not overlap, the lateral capacity of a pile group can be taken as the sum of the lateral capacity of each pile in the pile group.

Rollins et al. (1998) reported that pile-soil-pile interaction becomes insignificant when center-to-center spacing between piles, in the direction of loading, is greater than 6 pile diameters. However, Prakash (1962) has shown that the shadowing effect begins to occur if the center-to-center spacing is less than 8 pile diameters in the direction of loading. Prakash (1962), Cox et al. (1984), Brown et al. (1987, 1988), and Lieng (1988) have shown that the shadowing effect perpendicular to the direction of loading becomes insignificant when center-to-center spacing is greater than 3 pile diameters. However, AASHTO (2007) recommends considering some interaction when pile spacing perpendicular to the direction of loading is less than 5 pile diameters. It is important to note that most pile groups in practice are constructed by installing piles at a center-to-center spacing of 2.5 to 4 pile diameters. Therefore, the shadowing effect has a significant influence on the lateral capacity of pile groups.

One of the most common approaches to estimate the response of a pile group subjected to lateral loads is based on the response of a single pile in the same group. This approach consists of predicting or measuring the response of a single pile and then modifying the response by using group efficiency factors, interaction factors, or response modifiers. Most of the factors and modifiers have been developed based on results of both full-scale and laboratory-scale lateral load tests on various pile groups subjected to lateral loading. Some of the methods used to obtain the pile group response from a single-pile response when piles in a group are closely spaced are

- *p*-multiplier method
- Soil modifier method
- Group reduction (efficiency) factor method

- Modulus of subgrade reaction multiplier method
- Equivalent pier method

The soil modifier method is similar to and may be considered a variation of the p-multiplier method. The least common of these methods is the equivalent pier method. The most commonly used method in practice is the p-multiplier method. Therefore, only this method is discussed further.

5.17.2.1 *p*-Multiplier Method

The basic concept behind the p-multiplier method is that the soil resistance around a pile in a pile group is reduced due to the shadowing effect, while the pile stiffness is unchanged. Brown et al. (1987) suggested that the reduced soil resistance can be modeled by modifying the p-y curve such that the p-value is reduced at each deflection by a factor commonly known as the p-multiplier; that is, the p-multiplier is a factor applied to account for pile-soil-pile action. The p-multiplier reduces the soil resistance of every pile in a particular row by the same amount.

5.17.2.2 Published *p*-Multipliers

Several studies have been conducted to estimate the value of the p-multiplier for different rows of piles. Table 5.14 presents some of the published p-multipliers which were developed based on results of many lateral load tests on varying pile types, soil conditions, group configurations, and center-to-center spacing of piles. Brown et al. (1987) and Rollins et al. (1998)

TABLE 5.14 Published p-Multipliers

Soil Type	Test Type	s/d	p-Multipliers for Row 1, 2, and 3+	Deflection (in.)	Reference
Stiff clay	Field test	3	0.70, 0.50, 0.40	2	Brown et al. (1987)
	Field test	3	0.70, 0.60, 0.50	1.2	Brown et al. (1987)
Medium clay	Model test Cyclic load	3	0.60, 0.45, 0.40	2.4	Moss (1997)
Clayey silt	Field test	3	0.60, 0.40, 0.40	1.0–2.4	Rollins et al. (1998)
Very dense sand	Field test	3	0.80, 0.40, 0.30	1	Brown et al. (1988)
Medium-dense sand	Centrifuge model	3	0.80, 0.40, 0.30	3	McVay et al. (1995)
	Centrifuge model	5	1.0, 0.85, 0.70	3	McVay et al. (1995)
Loose medium sand	Centrifuge model	3	0.65, 0.45, 0.35	3	McVay et al. (1995)
	Centrifuge model	5	1.0, 0.85, 0.70	3	McVay et al. (1995)
Loose fine sand	Field test	3	0.80, 0.70, 0.30	1–3	Ruesta and Townsend (1997)
All soils	Based on laboratory tests conducted by Cox et al. (1984), Schmidt (1981, 1985), and Lieng (1988)	Leading row: $p_m = 0.7309\ (s/d)^{0.2579}$ for $s/d < 3.37$ $p_m = 1.0$ for $s/d \geq 3.37$ Trailing row: $p_m = 0.5791\ (s/d)^{0.3251}$ for $s/d < 5.37$ $p_m = 1.0$ for $s/d \geq 5.37$			Reese et al. (2003)

Adapted from Hannigan et al. (2006) with additional information from Reese et al. (2003).

TABLE 5.15 p-Multipliers from AASHTO (2007)

Center-to-center spacing of piles (in the direction of loading)	p-Multipliers		
	Row 1	Row 2	Row 3+
3 × pile diameter	0.7	0.5	0.35
5 × pile diameter	1.0	0.85	0.7

developed p-multipliers based on a full-scale experimental study, whereas Cox et al. (1984), Schmidt (1981, 1985), and Lieng (1988) developed p-multipliers based on laboratory tests. Based on an analysis of the data of Cox et al. (1984), Dunnavant and O'Neill (1986) related p-multipliers to pile spacing. Experimental studies have shown that there are several factors that can affect the group capacity; however, the position of a particular pile in a pile group has received the most attention.

Most p-multipliers published in the literature are based on the relative position of the piles in the direction of the loading. There is general consensus that the piles in the leading row sustain the maximum load (i.e., have minimum reduction in their load-carrying capacity). Therefore, the p-multiplier of the piles in the leading row is highest and in some cases close to 1. Reese et al. (2003) plotted the data from Cox et al. (1984) and Schmidt (1981, 1985) and concluded that if the center-to-center spacing of the piles in a group is greater than 3.37, the p-multiplier for the leading row may be taken as 1. For spacing less than 3.37, the p-multiplier reduces almost linearly to a value of 0.7 for piles spaced next to each other. For trailing rows, Reese et al. (2003) plotted the data from Cox et al. (1984), Schmidt (1981, 1985), and Lieng (1988) and concluded that if the center-to-center spacing of the piles in a group is greater than 5.37 pile diameters, the p-multiplier may be taken as 1.0. For spacing less than 5.37 pile diameters, the p-multiplier reduces almost linearly to a value of 0.58 for piles spaced next to each other. Table 5.15, which is taken from AASHTO (2007), shows the average of p-multipliers proposed by various researchers.

5.17.2.3 Procedure to Develop Load-Deflection Response of Pile Group Using p-Multipliers

The following is a step-by-step procedure to develop the lateral load-deflection response of a pile group:

1. Develop the site-specific p-y curves for a single pile. This may be done most accurately by using an instrumented lateral load test on a test pile at the site. However, for projects where no pile load test is performed, the p-y curve can be developed from known soil properties. Computer programs are commonly used to develop the p-y curves.
2. The p-values from the p-y curves for every single pile are multiplied by the p-multiplier specific to each row to develop a modified p-y curve for any pile in a particular row. Computer programs such as LPILE can perform this task.
3. Develop a lateral load-deflection curve for a single pile that represents a particular row in a pile group using the modified p-y curves for that pile. A computer program is usually used to develop lateral load-deflection curves.
4. At each deflection, multiply the value of the load by the number of piles in a specific row to obtain the lateral load-deflection response of that row.

5. Repeat the above procedure for other rows in the pile group.
6. At each deflection, add the load values of each row to obtain the lateral load-deflection response of the pile group.
7. The group lateral capacity with respect to an acceptable deflection is determined by selecting the load for that deflection.

5.17.3 Capacity of Pile Groups Subjected to Axial Uplift Loads

On most projects, the uplift capacity of pile groups does not control the pile group design. However, there are many situations where the uplift capacity of a pile group may control the pile design, particularly the embedment length of piles in a pile group. According to AASHTO (2002, 2007), the uplift capacity of pile groups can be taken as the lesser of:

1. The design uplift capacity of a single pile times the number of piles in a pile group
2. The uplift resistance of the pile group considered as a block

There are several approaches to estimating the uplift capacity of a pile group considered as a block. AASHTO (2007), based on the recommendations of Tomlinson (1987), recommends calculating the uplift capacity of a pile group using the soil block as shown in Figure 5.51a for cohesive soils and Figure 5.51b for cohesionless soils. It is recommended that the side slope of the block for cohesionless soil be 1 horizontal to 4 vertical (1H:4V). In addition, it is recommended that the buoyant (effective) weight of the soils below the groundwater table be used to calculate the uplift capacity. For cohesive soils, the uplift capacity of the pile group can be calculated by (refer to Figure 5.51a):

$$Q_{ulg} = 2 \times (B + L) \times Z \times c_{u1} + W_g \qquad (5.44)$$

where Q_{ulg} = ultimate uplift capacity of the pile group, B = width of the pile group (block), L = length of the pile group (block), Z = embedment depth of piles in the block, c_{u1} = weighted average of undrained shear strength of clays along the embedment depth, and W_g = effective weight of the pile group including the weight of the pile cap.

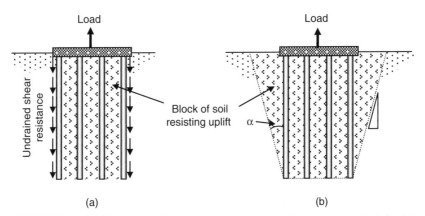

FIGURE 5.51 Uplift capacity of a pile group considering the group as a block for (a) cohesive soils and (b) cohesionless soils (modified from AASHTO 2007).

AASHTO (2002) recommends that the uplift capacity of a block can be estimated as the lesser of:

1. Two-thirds the effective weight of the pile group and the soil contained within a block defined by the perimeter of the pile group and the embedded length of the piles
2. Two-thirds the effective weight of the pile group and the soil contained within a block defined by the perimeter of the pile group and the embedded pile length plus one-half the total soil shear resistance on the peripheral surface of the pile group

For pile groups in cohesionless soils (refer to Figure 5.51b), it is the author's opinion that instead of using the slope of the side block recommended by AASHTO (2007), an angle α equal to half of the friction angle of the soil could be used.

5.18 Settlement of Pile Foundations

No foundation design is complete unless both the load-carrying capacity and anticipated settlements have been estimated. In most practical situations, settlement of deep foundations is less than ½ in. and therefore is not of major concern, unless the structure is sensitive to settlements. However, certain conditions can produce settlements greater than ½ in., which requires special attention. Some of these conditions include but are not limited to:

- Large pile groups designed to resist the load primarily through end bearing
- Presence of a highly compressible stratum near the toe of the pile
- Significant increase in the down drag on the pile

Since piles will actually be subjected to allowable loads only, it is common practice to estimate settlement of piles under allowable loads.

5.18.1 Settlement of Pile Groups in Cohesionless Soils

Pile groups installed in cohesionless soils experience only immediate or elastic settlements unless they are underlain by compressible soils. Immediate settlements in most cases are small and of no major concern. When cohesionless soils are not underlain by a compressible soil layer, immediate settlements can be estimated using the following equations recommended by Meyerhof (1976).

For clean sand:

$$s = \frac{4q_0 \sqrt{B} \, I_z}{(\overline{N}_1)_{60}} \tag{5.45}$$

For silty sand:

$$s = \frac{8q_0 \sqrt{B} \, I_z}{(\overline{N}_1)_{60}} \tag{5.46}$$

where

s = immediate settlement due to compression of soil (in.)

q_0 = allowable pressure (ksf); that is, $q_0 = \dfrac{\text{Allowable group load in kips}}{B \times L}$

B = width of the pile group (ft)

L = length of the pile group (ft)

I_z = influence factor which is taken as $I_z = \left(1 - \dfrac{Z}{8B}\right) \geq 0.5$

Z = embedment depth of piles in the group (ft)

$(\overline{N}_1)_{60}$ = average corrected N-values to a depth B below the tip of the pile group

Settlement estimated using the above equations does not include elastic compression of the pile material. Therefore, settlement due to the pile material should be added to calculate the total expected settlement. Elastic compression of piles can be calculated by

$$s_e = \frac{Q_{ag} Z}{A E_p} \tag{5.47}$$

where Q_{ag} = allowable axial load on the pile group, Z = length of the piles, A = cross-sectional area of the pile, and E_p = modulus of elasticity of the pile material.

The modulus of elasticity of steel is 29×10^6 psi and the modulus of elasticity of concrete can be calculated by

$$E_p \text{ (psi)} = 57{,}000 \sqrt{f_c'} \tag{5.48}$$

where f_c' is the compressive strength of concrete (psi) after 28 days of curing.

5.18.2 Settlement of Pile Groups in Cohesive Soils

Pile groups installed in cohesive soils or groups underlain by cohesive soils within a depth B below the pile group tip may experience both immediate settlements and long-term consolidation settlements. However, in normally consolidated or slightly overconsolidated soils, consolidation settlement generally is the major source of settlement. The method of calculating consolidation settlement of pile groups is the same as for shallow foundations with the exception of estimating the new applied load ΔP at various depths in the layer that is expected to undergo consolidation settlement.

For pile groups in clay, an equivalent footing method proposed by Terzaghi and Peck (1967) generally is used to estimate the new applied load ΔP. Figure 5.52a shows the concept of equivalent footing. As shown in Figure 5.52a, the load on the pile group is assumed to be transferred through an equivalent footing of size $B \times L$ bearing at a depth of $Z/3$ above the tip of the pile group. The pressure is assumed to spread out at a rate of 2 vertical to 1 horizontal ($2V{:}1H$). AASHTO (2007), based on the work of Duncan and Buchignani (1976), recommends that if the pile group tip is embedded in a firm soil layer, as shown in Figure 5.52b, the location of an equivalent footing be taken at a depth of $Z_b/3$ above the pile group tip, where Z_b is the embedment depth in the firm soil layer. Consolidation settlement is then calculated as follows.

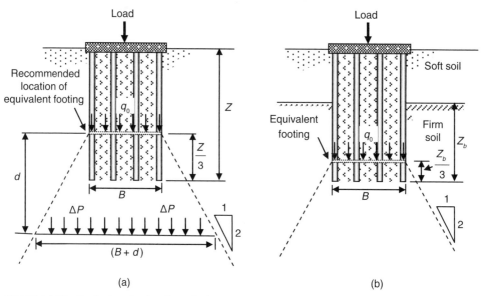

FIGURE 5.52 Equivalent footing for calculating consolidation settlement of a pile group in clay.

Bearing pressure at the location of an equivalent footing:

$$q_0 = \frac{Q_{\text{allowable}}}{B \times L} \qquad (5.49)$$

Pressure at depth d below the bearing elevation of the equivalent footing:

$$\Delta P = \frac{Q_{\text{allowable}}}{(B + d) \times (L + d)} \qquad (5.50)$$

Consolidation settlement due to a pile group can be calculated as follows. For normally consolidated soil, if $\overline{P}_0 = P_c$ and $\overline{P}_0 + \Delta P > P_c$:

$$s_c = \frac{C_c H}{1 + e_0} \log \frac{\overline{P}_0 + \Delta P}{\overline{P}_0} \qquad (5.51)$$

For highly overconsolidated soil, if $\overline{P}_0 < P_c$ and $\overline{P}_0 + \Delta P \leq P_c$:

$$s_c = \frac{C_r H}{1 + e_0} \log \frac{\overline{P}_0 + \Delta P}{\overline{P}_0} \qquad (5.52)$$

For slightly overconsolidated soil, if $\overline{P}_0 < P_c$ and $\overline{P}_0 + \Delta P > P_c$:

$$s_c = \frac{C_r H}{1 + e_0} \log \frac{P_c}{\overline{P}_0} + \frac{C_c H}{1 + e_0} \log \frac{\overline{P}_0 + \Delta P}{P_c} \qquad (5.53)$$

In the above equations, $Q_{allowable}$ = allowable capacity of the pile group, B = width of the pile group, L = length of the pile group, Z = embedment depth of piles in the block, q_0 = allowable pressure at the bearing elevation of the equivalent footing, d = depth below the bearing elevation of the equivalent footing at which an increase in pressure needs to be calculated, ΔP = increase in pressure due to the pile group at depth d, s_c = consolidation settlement, H = thickness of the clay layer in which consolidation settlement needs to be calculated, C_c = compression index, C_r = recompression index, e_0 = initial void ratio, P_c = preconsolidation pressure, and $\bar{P}_0$ = effective overburden pressure before application of the new load.

The reader should refer to the discussion on settlement of shallow foundations for additional information on consolidation settlement. The location of the equivalent footing discussed above is for a homogeneous clay layer above and below the tip of the pile group. Hannigan et al. (2006), based on the recommendations of Cheney and Chassie (2002), suggest that the location of the equivalent footing for various subsurface conditions may be taken as shown in Figure 5.53.

Example 1: Point Capacity of a Pile in Sand

Problem Statement

A 16-in.-diameter × 50-ft-long precast concrete pile is driven in sand that has a moist unit weight of 110 pcf and an angle of internal friction of 30°. The compressive strength of the concrete used to manufacture the pile is 5000 psi. Calculate the point capacity of the pile.

Solution

For sand, $c = 0$.

Meyerhof Method

From Figure 5.16, $\phi = 30°$ and $N_q^* = 58$. Therefore:

$$R_p = A_p \times (cN_c^* + q'N_q^*) \Rightarrow A_p \times q'N_q^*$$

$$R_p = \pi/4 \left(\frac{16 \text{ in.}}{12}\right)^2 \times [(110 \text{ pcf} \times 50 \text{ ft}) \times 58] \Rightarrow 445,408 \text{ lb} \Rightarrow 445 \text{ kips}$$

The limiting value of R_p is

$$(R_p)_{lim} = A_p (1000 \times N_q^* \tan \phi) \qquad \text{in lb, area of the pile } A_p \text{ in ft}^2$$

$$(R_p)_{lim} = \pi/4 \left(\frac{16 \text{ in.}}{12}\right)^2 \times 1000 \times 58 \times \tan 30° \Rightarrow 46,756 \text{ lb} \Rightarrow 46.75 \text{ kips}$$

Since the limiting value of the point capacity is smaller than the point capacity calculated earlier, the limiting value of 46.75 kips should be used as the point capacity:

$$R_p = 46 \text{ kips}$$

(a)
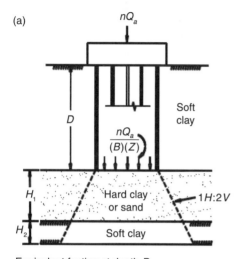

Equivalent footing at depth D
Settlement of pile group = Compression of layers H_1 and H_2 under pressure distribution shown

(b)
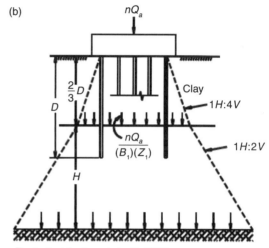

Equivalent footing at depth $2/3\,D$
Settlement of pile group = Compression of layer H under pressure distribution shown

(c)
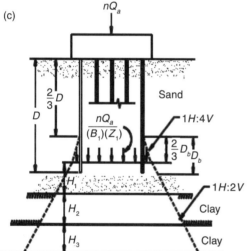

Equivalent footing at depth $8/9D$
Settlement of pile group = Compression of layers H_1, H_2, and H_3 under pressure distribution shown nQ_a is limited by bearing capacity of clay layers

(d)
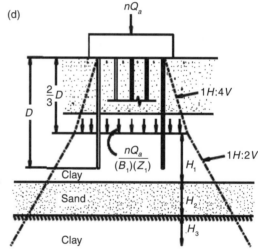

Equivalent footing at depth $2/3\,D$
Settlement of pile group = Compression of layers H_1, H_2, and H_3 under pressure distribution shown

Notes: (1) Plan area of perimeter of pile group = $(B)(Z)$
      (2) Plan area $(B_1)(Z_1)$ = projection of area $(B)(Z)$ at depth based on shown pressure distribution
      (3) For relatively rigid pile cap, pressure distribution is assumed to vary with depth as above
      (4) For flexible slab or group of small separate caps, compute pressures by elastic solutions

FIGURE 5.53 Location of equivalent footing and pressure distribution for various subsurface conditions (adapted from Hannigan et al. 2006).

Nordlund Method

$$R_p = \alpha q' N_q^* A_p$$

$$\frac{L}{D} = \frac{50 \times 12}{16} = 37.5$$

From Figure 5.19, $\alpha = 0.5$. From Figure 5.20, $N_q^* = 30$. Therefore:

$$R_p = 0.5 \times 110 \times 50 \times 30 \times \pi \left(\frac{16 \text{ in.}}{12}\right)^2 = 115{,}191 \text{ lb}$$

$$R_p = 115 \text{ kips}$$

Effective Stress Method

$$R_p = A_p \times (q' N_t)$$

From Figure 5.27, $N_t = 25$. Therefore:

$$R_p = \frac{\pi}{4} \left(\frac{16 \text{ in.}}{12}\right)^2 \times 110 \times 50 \times 25 = 191{,}986 \text{ lb}$$

$$R_p = 192 \text{ kips}$$

Example 2: Skin Friction Capacity of a Pile in Sand

Problem Statement

A 16-in.-diameter × 50-ft-long precast concrete pile is driven in sand that has a moist unit weight of 110 pcf, a corrected N-value of 12, and an angle of internal friction of 30°. The compressive strength of the concrete used to manufacture the pile is 5000 psi. Calculate the skin friction capacity of the pile. Also calculate the total pile capacity assuming that the acceptable pile movement is large enough to mobilize both the skin friction and point capacity.

Solution

Meyerhof Method

A precast concrete pile is a high-displacement pile. Therefore:

$$R_s = \sum 40(N_{\text{cor}}) pL \leq 2000 pL \qquad \text{in lb, } p \text{ and } L \text{ in ft}$$

$$R_s = 40 \times 12 \times \left(\pi \times \frac{16 \text{ in.}}{12}\right) \times 50 = 100{,}531 \text{ lb}$$

The limiting value of R_s is

$$R_s = 2000 \times \left(\pi \times \frac{16 \text{ in.}}{12} \right) \times 50 = 418,879 \text{ lb}$$

$$R_s = 100,531 \text{ lb} = 100.5 \text{ kips}$$

Nordlund Method

$$R_s = \sum_{z=0}^{z=L} K_\delta C_F (q')_z \frac{\sin(\delta + \omega)}{\cos \omega} D_z \Delta L$$

Let's assume $\delta = \frac{2}{3}\phi$ (i.e., $\delta = \frac{2}{3} \times 30° \Rightarrow 20°$). The pile taper angle $\omega = 0$. From Figure 5.21, the displaced volume $V = 0.70$ ft^3/ft. From Figure 5.23, $K_\delta = 1.10$. From Figure 5.26, $C_F = 0.88$. $(q')_z$ at the middle of the soil layer $= 110 \times 25 = 2750$ psf. Therefore:

$$R_s = 1.10 \times 0.88 \times 2750 \times \frac{\sin(20 + 0)}{\cos 0} \times \left(\frac{16 \text{ in.}}{12} \right) \times 50 = 60,697 \text{ lb}$$

$$R_s = 60.7 \text{ kips}$$

Effective Stress Method

Since it is a driven pile, assume $K \approx 1.0$. The critical depth beyond which the unit frictional resistance can be assumed to be constant can be taken as 15 times the pile diameter; that is:

$$L_{cr} = 15 \times \frac{16 \text{ in.}}{12} \Rightarrow 20 \text{ ft}$$

At a depth of 20 ft, the effective overburden pressure σ'_v can be calculated as:

$$(q')_z = 110 \text{ pcf} \times 20 \text{ ft} \Rightarrow 2200 \text{ psf}$$

Let's assume $\delta = \frac{2}{3}\phi$ (i.e., $\delta = \frac{2}{3} \times 30° \Rightarrow 20°$). The unit frictional resistance at a depth of 20 ft (i.e., the critical depth) can be calculated as:

$$f = K(q')_z \tan \delta \Rightarrow 1.0 \times 2200 \text{ psf} \times \tan 20°$$

$$f = 800.7 \text{ psf}$$

The unit frictional resistance diagram along with critical depth are shown in Figure 5.54.

Now the skin friction capacity of the pile can be calculated by considering two sections: one from the ground surface to a depth of 20 ft (i.e., the critical depth) and the other from 20 to 50 ft since the unit frictional resistance within this section is constant.

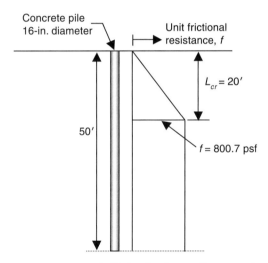

FIGURE 5.54 Unit frictional resistance diagram and critical depth for Example 2.

From 0.0 to 20 ft, the average unit friction can be used; that is:

$$f_{av} = \left(\frac{0.0 + 800.7 \text{ psf}}{2} \right) \Rightarrow 400.35 \text{ psf}$$

The skin friction capacity offered by the pile between depths of 0–20 ft can be calculated as:

$$R_s = 400.35 \text{ psf} \times \left(\pi \times \frac{16 \text{ in.}}{12} \times 20 \text{ ft} \right) \Rightarrow 33{,}541.0 \text{ lb} \Rightarrow 33.5 \text{ kips}$$

The skin friction capacity offered by the pile between depths of 20–50 ft can be calculated as (note that the unit friction is constant within this section):

$$R_s = 800.735 \text{ psf} \times \left(\pi \times \frac{16 \text{ in.}}{12} \times 30 \text{ ft} \right) \Rightarrow 100{,}622.7 \text{ lb} \Rightarrow 100.622 \text{ kips}$$

The total skin capacity can be calculated by adding the skin capacity of each section; that is:

$$R_s = 100.6 \text{ kips} + 33.5 \text{ kips} \Rightarrow 134.1 \text{ kips}$$

Example 3: Capacity of a Pile in Clay

Problem Statement

A 16-in.-diameter × 50-ft-long precast concrete pile is driven in the soil stratigraphy shown in Figure 5.55. All soil layers are clay. The soil properties of each clay layer are shown in the figure. The effective angle of internal friction of all clay layers is 25°. The compressive strength of the concrete used to manufacture the pile is 5000 psi. Calculate the point and skin friction capacities of the pile using the λ-method, α-method, and β-method. Also calculate the total pile capacity assuming that the acceptable pile movement is large enough to mobilize both the skin friction and point capacity.

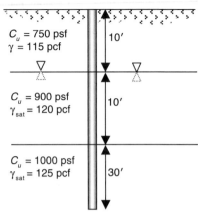

| Layers 1 and 2 | Normally consolidated |
| Layer 3 | Overconsolidated with an overconsolidation ratio of 2 |

Solution

Point Capacity

FIGURE 5.55 Soil properties of clay layers for Example 3.

Note that the pile tip is in layer 3. Therefore, the soil properties of layer 3 should be used for calculation of the point capacity of the pile.

For clay in an undrained condition, $\phi = 0$. From Figure 5.16, $N_q^* = 0$ and $N_c^* = 9$. Therefore, the point capacity can be calculated as:

$$R_p = 9C \times A_p = \frac{\pi}{4} \times \left(\frac{16 \text{ in.}}{12}\right)^2 \times (9 \times 1000 \text{ psf})$$

$$R_p = 12{,}566.4 \text{ lb} \Rightarrow 12.56 \text{ kips}$$

Skin Friction Capacity

λ-Method

According to the λ-method, the skin friction capacity can be calculated by

$$R_s = p \times L \times \lambda(\sigma_0' + 2\overline{C}_u)$$

where

$$\overline{C}_u = \frac{(C_{u_1}L_1 + C_{u_2}L_2 + C_{u_3}L_3 + \ldots)}{L_1 + L_2 + L_3 + \ldots}$$

$$\sigma_0' = \frac{(A_1 + A_2 + A_3 + \ldots)}{L_1 + L_2 + L_3 + \ldots}$$

To calculate A_1, A_2, and A_3, let's first calculate the effective overburden pressure at various depths:

Effective overburden pressure at a depth of 10 ft = 115 × 10 = 1150 psf
Effective overburden pressure at a depth of 20 ft = 1150 + (120 − 62.4) × 10 = 1726 psf
Effective overburden pressure at a depth of 50 ft = 1726 + (125 − 612.4) × 30 = 3604 psf

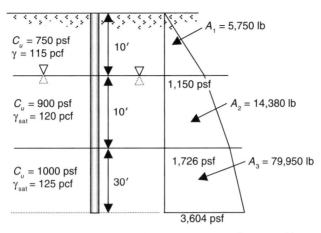

FIGURE 5.56 Effective overburden pressure at the top and bottom of each layer calculated using the λ-method for Example 3.

Now A_1, A_2, and A_3 can be calculated as follows:

$$A_1 = \frac{0 + 1150}{2} \times 10 \text{ ft} \Rightarrow 5750$$

$$A_2 = \frac{1150 + 1726}{2} \times 10 \text{ ft} \Rightarrow 14{,}380$$

$$A_3 = \frac{1726 + 3604}{2} \times 30 \text{ ft} \Rightarrow 79{,}950$$

The values of the overburden pressures and A_1, A_2, and A_3 are shown in Figure 5.56. $\overline{C}_u$ and σ'_0 now can be calculated as follows:

$$\overline{C}_u = \frac{750 \text{ psf} \times 10 \text{ ft} + 900 \text{ psf} \times 10 \text{ ft} + 1000 \text{ psf} \times 30 \text{ ft}}{50 \text{ ft}} \Rightarrow 930 \text{ psf}$$

$$\sigma'_0 = \frac{5750 + 14{,}380 + 79{,}950}{50 \text{ ft}} \Rightarrow 2001.6 \text{ psf}$$

From Figure 5.29, $\lambda \cong 0.21$. Therefore:

$$R_s = \pi \times \frac{16 \text{ in.}}{12} \times 50 \text{ ft} \times 0.21 \times (2001.6 \text{ psf} + 2 \times 930 \text{ psf})$$

$$R_s = 169{,}842 \text{ lb} \Rightarrow 169.8 \text{ kip}$$

α-Method

According to the α-method, the skin friction capacity of a pile can be calculated by

$$R_s = \sum p \times \Delta L \times \alpha \times c_u$$

From Figure 5.31, values of α can be estimated based on the undrained cohesion of each soil layer. The values are as follows:

Layer 1	$C_u = 750$ psf	$\alpha = 1.0$
Layer 2	$C_u = 900$ psf	$\alpha = 0.9$
Layer 3	$C_u = 1000$ psf	$\alpha = 0.8$

$$R_s = 1.0 \times 750 \text{ psf} \times \pi \times \frac{16 \text{ in.}}{12} \times 10 \text{ ft}$$

$$+ 0.9 \times 900 \text{ psf} \times \pi \times \frac{16 \text{ in.}}{12} \times 10 \text{ ft}$$

$$+ 0.83 \times 1000 \text{ psf} \times \pi \times \frac{16 \text{ in.}}{12} \times 30 \text{ ft}$$

$$R_s = 169,646.00 \text{ lb} \Rightarrow 169.64 \text{ kips}$$

β-Method (Effective Stress Method)

The skin friction capacity of a pile using the β-method can be calculated by the following equation:

$$R_s = \sum p \times \Delta L \times K \tan \phi_R \times (q')_z$$

Layers 1 and 2 are normally consolidated. Therefore:

$$K = 1 - \sin \phi_R \Rightarrow 1 - \sin 25° \Rightarrow 0.577$$

Layer 3 is overconsolidated. Therefore:

$$K = (1 - \sin \phi_R) \times \sqrt{\text{OCR}} \Rightarrow (1 - \sin 25°) \times \sqrt{2} \Rightarrow 0.816$$

where OCR is the overconsolidation ratio.

The effective overburden pressure at the top and bottom of each layer was calculated earlier for the λ-method and is shown in Figure 5.56. Now the skin friction capacity offered by each of the soil layers can be estimated using the β-method as follows:

$$R_{s_1} = \pi \times \frac{16 \text{ in.}}{12} \times 10 \text{ ft} \times 0.577 \times \tan 25° \times \frac{0.0 + 1150}{2} = 6480.4 \text{ psf}$$

$$R_{s_2} = \pi \times \frac{16 \text{ in.}}{12} \times 10 \text{ ft} \times 0.577 \times \tan 25° \times \frac{1150 + 1726}{2} = 16,206.7 \text{ psf}$$

$$R_{s_3} = \pi \times \frac{16 \text{ in.}}{12} \times 30 \text{ ft} \times 0.816 \times \tan 25° \times \frac{1726 + 3604}{2} = 127,429.4 \text{ psf}$$

Total skin resistance can be calculated as:

$$R_s = R_{s_1} + R_{s_2} + R_{s_3} = 6480.4 + 16{,}206.7 + 127{,}429.4$$

$$R_s = 150{,}116.5 \text{ lb} \Rightarrow 150.1 \text{ kips}$$

To summarize, the skin friction capacity using all three methods is

λ-method	$R_s = 169.8$ kips
α-method	$R_s = 169.6$ kips
β-method	$R_s = 150.1$ kips

Since all three methods resulted in reasonably close values of R_s, the skin friction capacity for the pile can be taken as the average of the capacities from all three methods; that is:

$$R_s = \frac{169.8 + 169.6 + 150.1}{3} \Rightarrow 163.2 \text{ kips}$$

Total Pile Capacity

The point capacity of this pile is 12.6 kips. Since movements in the pile are sufficient to mobilize skin friction and point resistances are acceptable, the total pile capacity can be calculated as:

$$Q_{\text{ult}} = R_s + R_p \Rightarrow 163.2 + 12.6 \Rightarrow 175.8 \text{ kips}$$

If a factor of safety of 2 is acceptable, the allowable pile capacity can be calculated as:

$$R_{\text{allowable}} = \frac{Q_{\text{ult}}}{\text{FS}} \Rightarrow \frac{175.8 \text{ kips}}{2} \Rightarrow 87.9 \text{ kips}$$

The recommend pile capacity is 85 kips.

Example 4: Capacity of a Pile End Bearing on Rock

Problem Statement

A steel H-pile of size 10×57 is driven to limestone bedrock at a depth of 60 ft through loose to medium-dense, medium sand. The yield strength of the pile material reported by the manufacturer is 50 ksi. The average unconfined compression strength of the bedrock measured from 2-in.-diameter rock cores in the laboratory is 20,000 psi, and the effective angle of friction is 35°. Calculate the allowable capacity of the pile assuming a factor of safety of 2. Also calculate the expected settlement. The modulus of elasticity of steel is 29,000 ksi.

Solution

The properties of a steel H-pile section of size 10 × 57 from standard tables are as follows:

Flange width = 10 in.
Depth of pile section = 10 in. (refer to Figure 5.57)
Actual area of the pile = 16.8 in.^{2}
(Note that this is the area of the steel only)

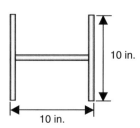

FIGURE 5.57 Depth of the pile section in Example 4.

Capacity Based on Strength of Bedrock

The design unconfined compression strength of the bedrock is

$$q_{u(design)} = \frac{q_{u(lab)}}{4 \text{ to } 5}$$

$$q_{u(design)} = \frac{20{,}000}{4} \Rightarrow 5000 \text{ psi}$$

The ultimate pile capacity can be calculated as follows:

$$Q_u = R_p = A_p q_u (N_\phi + 1)$$

$$Q_u = 16.8 \text{ in.}^2 \times 5000 \text{ psi} \times \left[\tan^2 \left(45 + \frac{35}{2} \right) + 1 \right] \Rightarrow 393{,}974.5 \text{ lb}$$

$$Q_u = 393.9 \text{ kips}$$

$$Q_{allowable} = \frac{393.9}{2} \Rightarrow 196.9 \text{ kips}$$

Capacity Based on Yield Strength of the Pile

The design yield strength of the pile material is 16.67 ksi (taken as one-third of the yield strength). The ultimate capacity of the pile can be calculated as follows:

$$Q_u = \sigma_y \times A_p$$

$$Q_u = 16.67 \text{ ksi} \times 16.8 \text{ in.}^2 \Rightarrow 280 \text{ kips}$$

$$Q_{allowable} = \frac{280}{2} \Rightarrow 140 \text{ kips}$$

Since the capacity based on the yield strength of the pile material is lower than that calculated based on the strength of the bedrock, the capacity based on the yield strength of the pile material should be used as the design capacity. Therefore:

$$Q_{allowable} = 140 \text{ kips}$$

Since the pile is supported on rock, all the load is assumed to be transferred through the pile tip. Therefore, settlement in the pile can be estimated by using the basic equation from the mechanics of deformable bodies:

$$S = \frac{PL}{AE} = \frac{140 \text{ kips} \times (60 \times 12) \text{ in.}}{16.8 \text{ in.}^2 \times 29,000 \text{ ksi}} = 0.207 \text{ in.}$$

References

AASHTO (2002). *Standard Specifications for Highway Bridges—Division 1 and 2,* American Association of State Highway and Transportation Officials, Washington, D.C.

AASHTO (2007). *AASHTO LRFD Bridge Design Specifications,* American Association of State Highway and Transportation Officials, Washington, D.C.

Ashour, M., Norris, G.M., and Pilling, P. (1998). Lateral loading of a pile in layered soil using the strain wedge model. *J. Geotech. Geoenviron. Eng.,* 124(4):303–315.

Bhushan, K., Haley, S.C., and Fong, P.T. (1979). Lateral load tests on drilled piers in stiff clay. *J. Geotech. Eng. Div. ASCE,* 105(GT8):969–985.

Brinch Hansen, J. (1961). The ultimate resistance of rigid piles against transversal forces. *Dan. Geotech. Inst. (Geotek. Inst.) Bull.,* 12:5–9.

Broms, B. (1964a). The lateral resistance of piles in cohesive soils. *J. Soil Mech. Found. Div. ASCE,* 90(SM2):27–63.

Broms, B. (1964b). The lateral resistance of piles in cohesionless soils. *J. Soil Mech. Found. Div. ASCE,* 90(SM3):123–156.

Brown, D.A., Reese, L.C., and O'Neill, M.W. (1987). Cyclic lateral loading of a large-scale pile group. *J. Geotech. Eng.,* 113(11):1326–1343.

Brown, D.A., Clark, M., and Reese, L.C. (1988). Lateral load behavior of pile group in sand. *J. Geotech. Eng.,* 114(11):1261–1276.

Burland, J. (1973). Shaft friction of piles in clay—a simple fundamental approach. *Ground Eng.,* 6(3).

Cheney, R.S. and Chassie, R.G. (2002). *Soils and Foundation Workshop Manual,* Report No. HI-88-009, Federal Highway Administration, U.S. Department of Transportation, Washington, D.C.

Cisco, R. (2004). *Parametric Study on Various Methods of Analyzing Pile Groups Subjected to Static Lateral Loads,* Master's thesis, Southern Illinois University, Carbondale.

Cox, W.R., Dixon, D.A., and Murphy, B.S. (1984). Lateral load tests of 5.4 mm diameter piles in very soft clay in side-by-side and in-line groups. *Laterally Loaded Deep Foundations: Analysis and Performance,* ASTM, West Conshohocken, PA.

Das, B.M. (1999). *Principals of Foundation Engineering,* 4th edition, Brooks/Cole Publishing, Pacific Grove, CA.

De Beer, E.E. (1945). Etude des fondations sur pilotis et des fondations directes. *Ann. Trav. Publics Belg.,* 46:1–78.

Duncan, J.M. and Buchignani, A.L. (1976). *An Engineering Manual for Settlement Studies,* Department of Civil Engineering, University of California, Berkeley.

Dunnavant, T.W. and O'Neill, M.W. (1986). Evaluation of design-oriented methods for analysis of vertical pile groups subjected to lateral load. *Numerical Methods in Offshore Piling,* Institut Francais du Petrole, Labortoire Central des Ponts et Chausses.

Fellenius, B.H. (1991). *Pile Foundations, Foundation Engineering Handbook,* 2nd edition, H.S. Fang, Ed., Van Nostrand Reinhold, New York.

Hannigan, P.J., Goble, G.G., Thendean, G., Likins, G.E., and Rausche, F. (2006). *Design and Construction of Driven Pile Foundations,* NHI-05-42, Federal Highway Administration, U.S. Department of Transportation, Washington, D.C.

Hetenyi, M. (1946). *Beams on Elastic Foundation,* University of Michigan Press, Ann Arbor.

Hoit, M.L. and McVay, M.C. (1996). *FLPIER User's Manual,* University of Florida, Gainesville.

Komurka, V.E. (2004). Incorporating set-up and support cost distributions into driven pile design. *Current Practices and Future Trends in Deep Foundations,* Geotechnical Special Publication No. 125, American Society of Civil Engineers, 16–49.

Kumar, S. (1993). *Non-linear Load Deflection Prediction of Single Piles in Sand Using a Sub-grade Reaction Approach,* M.S. thesis, University of Missouri, Rolla.

Kumar, S. and Cisco, R. (2006). State-of-the-art of analysis of pile groups subjected to lateral loads. *Proceedings of International Conference on New Developments in Geoenvironmental and Geotechnical Engineering,* Korean Institute of Construction Technology Education, Incheon, Korea, November 9–11.

Lieng, F.T. (1998). *Behavior of Laterally Loaded Piles in Sand—Large Scale Model Tests,* Ph.D. thesis, Department of Civil Engineering, Norwegian Institute of Technology.

Matlock, H. (1970). Correlation for design of laterally loaded piles in soft clay. *Proceedings Offshore Technology Conference,* Paper OTC 1204, Houston, TX.

Matlock, H. and Reese, L.C. (1961). Foundation analysis of offshore pile supported structures. *Proceedings Fifth International Conference on Soil Mechanics and Foundation Engineering,* Paris, 2, 91–97.

Matlock, H. and Reese, L.C. (1962). Generalized solutions for laterally loaded piles. *Trans. Am. Soc. Civ. Eng.,* 127(Part 1):1220–1247.

McVay, M., Casper, R., and Shang, T.I. (1995). Lateral response of three-row groups in loose to dense sands at 3D and 5D pile spacing. *J. Geotech. Geoenviron. Eng.,* 121(5):436–441.

Meyerhof, G.G. (1976). Bearing capacity and settlement of pile foundations. *J. Geotech. Eng. Div. ASCE,* 102(3):195–228.

Moss, R.E.S. (1997). *Cyclic Lateral Loading of Model Pile Groups in Clay Sand, Phase 2B,* Master's thesis research, Utah State University.

Nordlund, R.L. (1963). Bearing capacity of piles in cohesionless soils. *J. Soil Mech. Found. Div. ASCE,* SM3.

Nordlund, R.L. (1979). Point Bearing and Shaft Friction of Piles in Sand, 5th Annual Short Course on the Fundamentals of Deep Foundation Design, Rolla, Missouri.

Norris, G.M. (1986). Theoretically based BEF laterally loaded pile analysis. *Proceedings 3rd International Conference on Numerical Methods in Offshore Piling,* Nantes, France, 361–386.

Paikowsky, S.G. and Whitman, R.V. (1990). The effects of plugging on pile performance and design. *Can. Geotech. J.,* 27(4):429–440.

Poulos, H.G. (1971a). Behavior of laterally loaded piles. II. Single piles. *J. Soil Mech. Found. Div. ASCE,* 97(SM5):711-731.

Poulos, H.G. (1971b). Behavior of laterally loaded piles. II. Pile groups. *J. Soil Mech. Found. Div. ASCE,* 97(SM5):733–751.

Poulos, H.G. and Davis, E.H. (1980). *Pile Foundation Analysis and Design,* John Wiley and Sons, New York.

Prakash, S. (1962). *Behavior of Driven Pile Groups under Lateral Loads,* Ph.D. thesis, University of Illinois, Urbana.

Prakash, S. and Kumar, S. (1996). Nonlinear lateral pile deflection prediction in sands. *J. Geotech. Eng.,* 122(2):130–138.

Rausche, F., Thendean, G., Abou-matar, H., Likins, G., and Goble, G. (1996). *Determination of Pile Driveability and Capacity from Penetration Tests,* FHWA Contract No. DTFH61-91-C-00047,

Final Report, Federal Highway Administration, U.S. Department of Transportation (from Hannigan et al. 2006).

Reese, L.C. and Matlock, H. (1956). Non-dimensional solutions for laterally loaded piles with soil modulus assumed proportional to depth. *Proceedings 8th Texas Conference on Soil Mechanics and Foundation Engineering,* Austin, 1–41.

Reese, L.C. and Welch, R.C. (1975). Lateral loading of deep foundations in stiff clay. *J. Geotech. Eng. Div. ASCE,* 101(GT7):633–649.

Reese, L.C., Cox, W.R., and Koop, F.D. (1974). Analysis of laterally loaded piles in sand. *Proceedings Offshore Technology Conference,* Houston, TX, Paper OTC 2080, 473–483.

Reese, L.C., Wang, S.T., and Vasquez, L. (2003). *Computer Program GROUP, Version 6.0—Analysis of a Group of Piles Subjected to Axial and Lateral Loads, Technical Manual,* Ensoft, Inc.

Reese, L.C., Wang, S.T., Isenhower, W., and Arrellaga, J.A. (2004). *Computer Program LPILE Plus, Version 5.0—A Program for the Analysis of Piles and Drilled Shafts under Lateral Loads, Technical Manual,* Ensoft, Inc.

Rollins, K.M., Peterson, K.T., and Weaver, T.J. (1998). Lateral load behavior of full-scale pile group in clay. *J. Geotech. Geoenviron. Eng.,* 124(6):468–478.

Ruesta, P.F. and Townsend, F.C. (1997). Evaluation of laterally loaded pile group at Roosevelt Bridge, *J. Geotech. Geoenviron. Eng.,* 123(12):1153–1161.

Schmidt, H.G. (1981). Group action of laterally loaded bored piles. *Proceedings, 10th International Conference on Soil Mechanics and Foundations Engineering,* Stockholm, 833–837.

Schmidt, H.G. (1985). Horizontal load tests on large diameter bored piles. *Proceedings, 11th International Conference on Soil Mechanics and Foundations Engineering,* San Francisco, 1569–1573.

Terzaghi, K. (1943). *Theoretical Soil Mechanics,* John Wiley, New York.

Terzaghi, K. and Peck, R.B. (1967). *Soil Mechanics in Engineering Practice,* John Wiley and Sons, New York.

Tomlinson, M.J. (1980). *Foundation Design and Construction,* 4th edition, Pitman Advanced Publishing Program.

Tomlinson, M.J. (1987). *Pile Design and Construction Practice,* Viewpoint, London, 415.

Tomlinson, M.J. (1994). *Pile Design and Construction Practice,* 4th edition, E&F Spon, London, 411.

Vesic, A.S. (1963). Bearing capacity of deep foundations in sand. *Highw. Res. Rec.,* 39:112–153.

Vijayvergiya, V.N. and Focht, Jr., J.A. (1972). A new way to predict capacity of piles in clay. *4th Offshore Technology Conference,* Houston, OTC Paper 1718.

Walsh, K.D., Fréchette, D.N., Houston, W.N., and Houston, S.L. (2000). *State of the Practice for Design of Groups of Laterally Loaded Drilled Shafts,* Transportation Research Record 1736, Paper 00-1306, U.S. Department of Transportation, Washington, D.C.

Winkler, E. (1867). *Die Lehre von Elasticitaet und Festigkeit,* Verlag H. Dominious, Prague.

6
Retaining Walls

by

Aniruddha Sengupta
Indian Institute of Technology, Kharagpur, India

6.1 Introduction

A retaining wall is a **structure** whose primary purpose is to provide lateral support to soil and rock. Some of the common types of retaining walls are gravity walls, cantilever walls, counterfort walls, diaphragm walls, crib walls, gabion walls, bored pile (contiguous and secant) walls, sheet pile walls, and mechanically stabilized walls.

A gravity retaining wall (Figure 6.1a) is built of plain concrete or stone **masonry**. The stability of a gravity retaining wall depends on its own weight and the weight of the soil resting on it. It is considered to be a **rigid** structure. Sometimes a minimum amount of steel **reinforcement** also is used in the construction of a gravity retaining wall to minimize the size of the wall section. This type of wall is referred to as a semigravity wall (Figure 6.1b).

A **cantilever** retaining wall (Figure 6.1c) is built of reinforced concrete. It consists of a thin stem and a base slab. The stem of a cantilever retaining wall is provided with reinforcement at the back. It also is provided with temperature reinforcement near the exposed front face to control cracking that might occur due to temperature changes.

A counterfort retaining wall (Figure 6.1d) is similar to a cantilever wall. In this type of wall, thin vertical concrete slabs known as **counterforts** are placed at regular intervals to tie the wall and the base slab together. The purpose of the counterforts is to reduce the shear and the bending moments.

A diaphragm wall (Figure 6.2) is a thin retaining structure which is constructed using the **slurry** trench technique. This technique involves excavating a narrow trench that is kept full

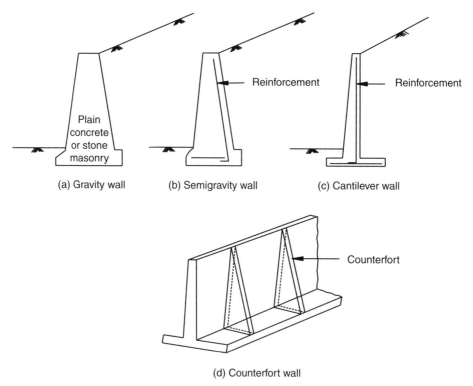

(a) Gravity wall   (b) Semigravity wall   (c) Cantilever wall

(d) Counterfort wall

FIGURE 6.1 Different types of walls.

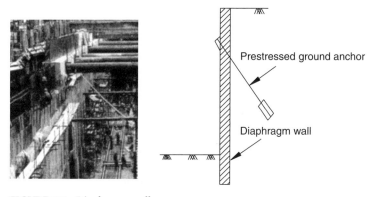

FIGURE 6.2 Diaphragm wall.

of clay and bentonite slurry. The slurry exerts hydraulic pressure against the trench walls and acts as shoring to prevent collapse. A diaphragm wall is constructed by excavating the trench in discontinuous sections. Once the excavation of a panel is complete, a steel reinforcement cage is placed in the center of the panel. Concrete is tremied in one continuous operation. The finished wall may be cantilever, anchored, or **propped** for lateral support.

A crib wall (Figure 6.3) consists of **interlocking** concrete/wooden members that form cells. These are then filled with compacted soil or boulders.

A gabion wall (Figure 6.4) is similar to a crib wall. It is constructed of gabions, which are double-twisted wire mesh containers of variable size that are uniformly partitioned into internal cells, interconnected with other similar units, and filled with stones.

In a contiguous bored pile wall (Figure 6.5a), reinforced concrete piles are installed at center-to-center spacing of generally 150 mm greater than their diameter, thus leaving gaps in the structural wall. This option usually is suitable where the retained soil is firm to stiff and where the groundwater table is below the level of the maximum excavation. A secant bored pile wall (Figure 6.5b) is similar to a contiguous bored pile wall, but the gap between piles is filled by secant piles made of unreinforced cement/bentonite mix for the hard/soft wall and weak concrete for the hard/firm wall. This type of wall is constructed by installing the primary piles, and then the secondary piles are formed in reinforced concrete, cutting into the primary piles. In secant bored pile walls, the ingress of water to any subsequent excavation is substantially reduced.

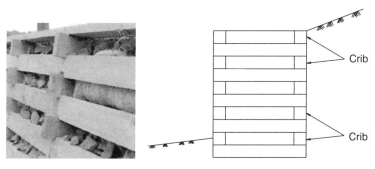

FIGURE 6.3 Crib wall.

FIGURE 6.4 Gabion wall.

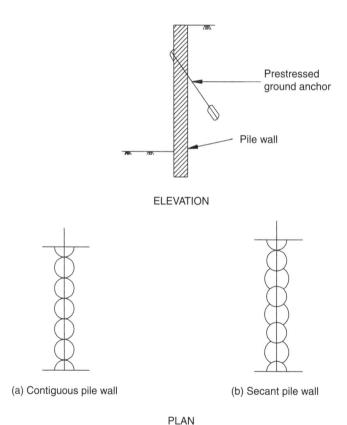

FIGURE 6.5 Pile wall: (a) contiguous and (b) secant.

A sheet pile wall (Figure 6.6) consists of interlocking members that are driven into place. Usually the sheet piles are steel sections which come in different shapes and sizes, with interlocking joints that enable the individual segments to be connected to form a solid wall. This type of flexible wall often is used for waterfront construction.

A mechanically stabilized wall (Figure 6.7) is the most modern type of wall. In this type of wall, the thin facing skin is held in position by a large number of thin reinforcing strips tied

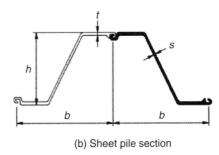

(a) Sheet pile wall   (b) Sheet pile section

FIGURE 6.6 Sheet pile wall.

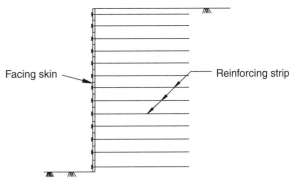

FIGURE 6.7 Mechanically stabilized wall.

to it and running through the backfill material. The backfill soil is held in position by the mechanical friction between the reinforcing strips and the backfill soil.

6.2 Initial Proportioning of Retaining Walls

Over the years, some guidelines have evolved regarding the initial trial dimensions of gravity and cantilever retaining walls which have been found to give satisfactory outcomes (the general proportions of various retaining wall components are shown in Figure 6.8). These guidelines are based on the total height of the wall H, which must be fixed in relation to the height of the soil to be retained. The top width of the stem of a retaining wall should not be less than 300 mm for proper placement of concrete. The increase in the width of the stem typically is between 20 and 60 mm per meter height of the stem. The depth D_f to the bottom of

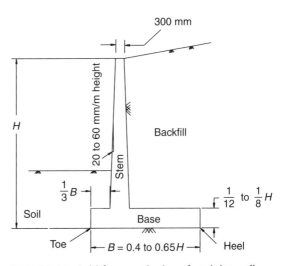

FIGURE 6.8 Initial proportioning of retaining wall.

the base slab is fixed based on the theories for shallow foundations. However, it should not be less than 600 mm. The thickness of the base slab typically is between $1/12$ and $1/8 H$. The width of the base slab B is 0.4–$0.7H$. The smaller B-to-H ratio is for firm soil and when the retaining soil is horizontal. The ratio increases with decreasing strength of the foundation soil and increasing slope of the backfill. The projection of the toe from the stem is $0.1H$ for a cantilever wall and 0.12–$0.17H$ for a gravity wall.

6.3 Lateral Earth Pressure Theories

Lateral earth pressures that act on a retaining wall play a pivotal role in the design and stability calculations of a wall. The lateral earth pressure acting on the back of a wall is the **driving force** that can cause instability, such as sliding and rotation, of the wall. Thus, determination of the lateral earth pressures acting on a wall is important.

There are two classical earth pressure theories: (1) Coulomb's (1776) earth pressure theory and (2) Rankine's (1857) earth pressure theory. Both theories propose to estimate the magnitudes of two lateral earth pressures: active earth pressure and passive earth pressure.

When a rigid wall, such as a counterfort wall, does not move even after the backfill soil is placed, the lateral pressure P exerted by the backfill on the wall is termed at-rest pressure and is expressed as

$$P = \frac{1}{2}\gamma H^2 K_o \qquad (6.1)$$

where γ = unit weight of the backfill soil, H = height of the retaining wall, and K_o = coefficient of earth pressure at rest.

The coefficient of earth pressure at rest K_o can be obtained from the theory of elasticity as

$$K_o = \frac{\nu}{1-\nu} \qquad (6.2)$$

where ν = **Poisson's ratio** of the backfill soil. Typical values of Poisson's ratio for different soils are given in Table 6.1.

A good approximation for K_o is given by Jaky (1944), according to whom

$$K_o = 1 - \sin\phi \qquad (6.3)$$

where ϕ = **angle of internal friction** of the backfill soil. Typical values of the friction angle for different soils are given in Table 6.2. Table 6.3 gives the value of K_o for different types of backfill soil.

If the lateral pressures acting on a wall are such that the wall rotates about the toe and moves away

TABLE 6.1 Typical Range of Poisson's Ratio (ν) for Different Soils

Type of Backfill Soil	ν
Loose sand	0.2–0.35
Dense sand	0.3–0.4
Sandy soil	0.15–0.25
Silt	0.3–0.35
Unsaturated clay	0.35–0.4
Saturated clay	0.5
Clay with sand and silt	0.3–0.42

TABLE 6.2 Typical Range of Friction Angle (ϕ) for Different Soils

Type of Backfill Soil	ϕ (deg)
Sand and gravel	30–40
Silty sand	20–30
Compacted clay	20–30
Soft clay	30–15

from the backfill soil, as may be the case in a cantilever retaining wall, the lateral earth pressure gradually reduces to a minimum after a particular displacement. This lateral pressure is termed the active earth pressure P_a. If, on the other hand, the lateral pressures acting on a wall are such that the wall moves into the backfill soil, the lateral earth pressure gradually reaches a maximum possible value after a certain displacement. This maximum possible value of lateral earth pressure is called the passive earth pressure

TABLE 6.3 Coefficient of Earth Pressure at Rest (K_o) for Different Soils

Type of Backfill Soil	K_o
Dry loose sand (void ratio, $e = 0.8$)	0.64
Dry dense sand (void ratio, $e = 0.6$)	0.49
Loose saturated sand	0.46
Dense saturated sand	0.36
Low-plastic compacted clay	0.42
High-plastic compacted clay	0.60
Organic silty clay	0.57

P_p. This type of situation may arise if the anchor forces are high enough to move the anchored retaining wall toward the backfill. The movement of the wall required to mobilize the passive pressure is far greater than that required to mobilize the active pressure. Table 6.4 gives the movement of the wall X in terms of wall height H required to mobilize the active and passive conditions (Department of the Navy 1982).

6.3.1 Coulomb's Earth Pressure Theory

As per Coulomb's earth pressure theory for cohesionless soil, the active earth pressure P_a acting on a wall is given by

$$P_a = \frac{1}{2} \gamma H^2 K_a \qquad (6.4)$$

where K_a = the active earth pressure coefficient and is given by

$$K_a = \frac{\cos^2(\phi - \alpha)}{\cos^2\alpha \cos(\alpha + \delta)\left[1 + \sqrt{\dfrac{\sin(\phi + \delta)\sin(\phi - i)}{\cos(\alpha + \delta)\cos(\alpha - i)}}\right]^2} \qquad (6.5)$$

where α = inclination (with respect to the vertical axis) of the back face of the wall, δ = friction between the wall and the backfill soil, and i = slope of the backfill soil.

Typical values of **wall friction** for different backfill soils are given in Table 6.5. If no information is known regarding the wall friction, two-thirds of ϕ can be used as an estimate.

TABLE 6.4 Movement (X) of Wall Required to Activate Active and Passive Conditions

Type of Backfill Soil	X/H for Active State	X/H for Passive State
Dense sand	0.0005	0.0002
Loose sand	0.002	0.006
Soft clay	0.02	0.04
Stiff clay	0.01	0.02

The passive pressure acting on a wall with cohesionless backfill is given by

$$P_p = \frac{1}{2} \gamma H^2 K_p \qquad (6.6)$$

where K_p = the passive earth pressure coefficient, given by

$$K_p = \frac{\cos^2(\phi + \alpha)}{\cos^2 \alpha \cos(\alpha - \delta) \left[1 - \sqrt{\dfrac{\sin(\phi + \delta)\sin(\phi + i)}{\cos(\alpha - \delta)\cos(\alpha - i)}} \right]^2} \qquad (6.7)$$

TABLE 6.5 Wall Friction Angle (δ) for Different Backfill Soils

Type of Backfill Soil	δ (deg)
Coarse sand	20–28
Fine sand	15–25
Silty clay	12–16
Stiff clay	15–20
Gravel	27–30

Coulomb's theory assumes that the backfill soil is isotropic, homogeneous, and cohesionless. The rupture surface is planer. The failure wedge can be treated as a rigid body.

6.3.2 Rankine's Earth Pressure Theory

Rankine, in his earth pressure theory, assumed that the wall is vertical and smooth or frictionless. The rupture surface is planer. The backfill soil is cohesionless. According to Rankine, the active earth pressure is given by

$$P_a = \frac{1}{2} \gamma H^2 K_A \qquad (6.8)$$

where K_A is the active earth pressure coefficient, given by

$$K_A = \cos i \, \frac{\cos i - \sqrt{\cos^2 i - \cos^2 \phi}}{\cos i + \sqrt{\cos^2 i - \cos^2 \phi}} \qquad (6.9)$$

The passive earth pressure is given by

$$P_p = \frac{1}{2} \gamma H^2 K_p \qquad (6.10)$$

where K_p is the passive earth pressure coefficient, expressed as

$$K_p = \cos i \, \frac{\cos i + \sqrt{\cos^2 i - \cos^2 \phi}}{\cos i - \sqrt{\cos^2 i - \cos^2 \phi}} \qquad (6.11)$$

If the backfill soil is horizontal (that is, $i = 0$), Rankine's above expressions for the active and passive earth pressure coefficients reduce to

$$K_A \ = \ \frac{1 \ - \ \sin \phi}{1 \ + \ \sin \phi} \ = \ \tan^2 \left(45 \ - \ \frac{\phi}{2} \right) \ = \ \frac{1}{N_\phi} \qquad (6.12)$$

and

$$K_P \ = \ \frac{1 \ + \ \sin \phi}{1 \ - \ \sin \phi} \ = \ \tan^2 \left(45 \ + \ \frac{\phi}{2} \right) \ = \ N_\phi \qquad (6.13)$$

Thus, under this condition, K_A and K_P are reciprocals of each other.

If the wall is vertical and smooth, and the backfill soil is horizontal (that is, $i = \gamma = 0$ and $\alpha = 90°$), Coulomb's equations for active and passive pressures also reduce to the above forms of Rankine's equations.

6.3.3 Earth Pressure Theory for Clayey Soil

The active earth pressure for a clayey soil is given by

$$P_a \ = \ \frac{1}{2} \gamma H^2 \ \frac{1}{N_\phi} \ - \ 2c \ \frac{H}{\sqrt{N_\phi}} \qquad (6.14)$$

where N_ϕ is given by $N_\phi = \tan^2 (45 + \phi/2)$ and $c = $ **cohesion** of the soil.

For soft soil, $\phi = 0$ and $N_\phi = 1$. Therefore,

$$P_a \ = \ \frac{1}{2} \gamma H^2 \ - \ 2cH \qquad (6.15)$$

The expression for the passive earth pressure in clayey soil is given by

$$P_p \ = \ \gamma H N_\phi \ + \ 2c \sqrt{N_\phi} \qquad (6.16)$$

6.3.4 Pressures Due to Surcharge Load and Groundwater

When calculating total lateral pressures acting on a wall, lateral pressures due to **surcharge load** on the ground and due to the steady groundwater table need to be accounted for.

6.3.5 Earth Pressures Acting on a Wall in a Braced Excavation

Vertical or near-vertical cuts often are required in the construction of foundations for high-rise buildings and underground transportation facilities and in laying underground cables and water and sewer lines. The vertical faces of a cut are protected by temporary bracing systems to avoid failure. First, vertical steel or timber beams, called soldier beams, are driven into the ground. After excavation is started, horizontal timber planks or steel plates called lagging are placed between the soldier beams. After excavation reaches a desired depth, horizontal steel beams called wales and struts are installed to support the side walls. Instead of soldier beams, interlocking sheet pile walls often are utilized as side walls. In contrast to the ordinary retaining

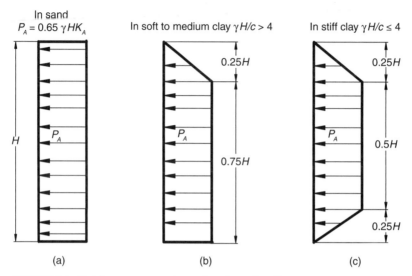

FIGURE 6.9 Peck's pressure envelopes for a braced wall.

walls discussed above, braced walls show different yielding behavior, in which the lateral deformation gradually increases with depth. As a result, the lateral earth pressures acting on braced walls also are different. Figure 6.9 shows the earth pressure envelopes for braced walls in sand and clay as proposed by Peck (1943, 1969). Peck suggested using $P_A = 0.65\gamma HK_A$ for sand, whichever is the higher of $P_A = \gamma H[1 - (4c/\gamma H)]$ or $0.3\gamma H$ to calculate earth pressure envelopes for soft to medium clay, and about $0.3\gamma H$ for stiff clay. When both sand and clay are encountered in an excavation, Peck proposed using the equivalent (weighted average) value of cohesion c and the unit weight of the soil γ for calculation of earth pressures.

6.3.6 Earth Pressures Acting on a Wall during an Earthquake

During an earthquake, there is an increase in the lateral pressure exerted by backfill. This increase depends on many factors, including intensity and type of the earthquake, natural frequency of the wall, nature of the backfill, etc. Total lateral earth pressure (static plus dynamic) in the active condition is computed by the Mononobe-Coulomb formula (Seed and Whitman 1970; Fang and Chen 1995) as

$$P_{ae} = \frac{1}{2}\gamma H^2 K_{ae} \tag{6.17}$$

with

$$K_{ae} = \frac{\cos^2(\phi - \theta - \alpha)}{\cos\theta \cos^2\alpha \cos(\delta + \alpha + \theta)\left[1 + \sqrt{n}\right]^2} \tag{6.18}$$

where

$$n = \frac{\sin(\phi + \delta) \sin(\phi - \theta - i)}{\cos(\delta + \alpha + \theta) \cos(i - \alpha)}$$

$\theta = \tan^{-1} \beta$

β = horizontal earthquake acceleration/gravity acceleration

The corresponding expression for K_{ae} for the Mononobe-Rankine formula is given by:

$$K_{ae} = \cos i \ \frac{\sqrt{\begin{array}{l} [\cos(i - \theta) - \sqrt{\cos^2(i + \theta) - \cos^2 \phi}]^2 \\ + [\sin(i + \theta) - \sin(i - \theta)]^2 \end{array}}}{\cos \theta [\cos(i + \theta) + \sqrt{\cos^2(i + \theta) - \cos^2 \phi}]} \qquad (6.19)$$

The dynamic increment of the pressure is obtained by subtracting the static earth pressure from the total earth pressure. The dynamic pressure acts at $\frac{2}{3}H$ for walls with a back slope less than or equal to $1(H){:}3(V)$. For walls with a back slope greater than $1(H){:}3(V)$, the dynamic pressure is applied at $0.58H$. Distribution for this point of application increases uniformly from zero at the plane of analysis to $6P_{ae}/5H$ at $H/3$, where P_{ae} is the horizontal component of the dynamic pressure, and then remains constant up to the surface of the backfill. If the retaining wall is holding back water on the upstream side, as in seawalls, the **hydrodynamic pressure** also needs to be included to account for the wave action during an earthquake event.

6.4 Forces Acting on a Retaining Wall

The forces acting on a gravity wall and a cantilever retaining wall are shown in Figures 6.10a and 6.10b, respectively. The resistive force acting on a wall consists of a net vertical force acting

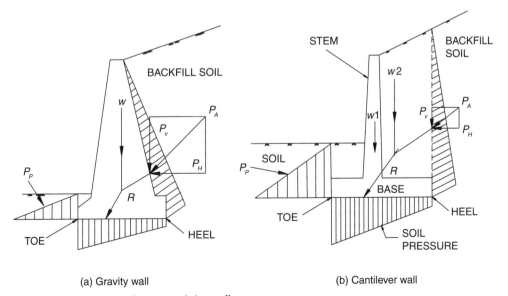

(a) Gravity wall (b) Cantilever wall

FIGURE 6.10 Forces acting on retaining walls.

on the wall (sum of the self-weight of the wall W, the weight of the backfill soil, and the surcharge load, minus the **uplift pressures** acting below the wall). The driving force acting on the wall is calculated as the summation of the net lateral earth pressures (active pressure minus passive pressure), lateral pressures due to the groundwater, and lateral pressures due to the surcharge load. For an earthquake condition, the inertial load acting horizontally through the centroid also needs to be included as a driving force.

6.5 Stability Checks of a Retaining Wall

The following stability checks are necessary for a retaining wall.

6.5.1 Overturning about the Toe

The factor of safety F_o against overturning of a wall about its toe is expressed as

$$F_o = \frac{\sum M_R}{\sum M_O} \tag{6.20}$$

where $\sum M_R$ = sum of the moments of forces resisting overturning and $\sum M_O$ = sum of the moments of forces overturning about the toe. A factor of safety of 2 usually is required against overturning.

6.5.2 No Tension at the Base

The eccentricity e of the resultant force acting on the base slab of a retaining wall is calculated as

$$e = \frac{B}{2} - \frac{\sum M}{\sum V} \tag{6.21}$$

where B = width of the base slab of a retaining wall, $\sum M = \sum M_R - \sum M_O$ = sum of the moments due to all the forces acting on the retaining wall, and $\sum V$ = sum of all the vertical forces acting on the wall.

For no tensile soil pressure to develop at the base, eccentricity e should be less than or equal to $B/6$. When this condition is satisfied, the criterion for overturning is automatically satisfied. If $e > B/6$, there will be tension at the heel of the base slab, and a redistribution of soil pressure takes place to keep it compressive throughout.

6.5.3 Allowable Maximum Pressure on the Foundation Soil

The maximum pressure acting at the base slab of a retaining wall is given by:

$$P_{max} = \frac{\sum V}{B} \left(1 + \frac{6e}{B} \right) \tag{6.22}$$

P_{max} should not exceed the design allowable soil pressure obtained from the **bearing capacity** of the foundation soil and settlement, considering the eccentricity of the resultant load.

The ultimate bearing capacity q_u of a shallow strip footing carrying an eccentric load (Meyerhof 1963) is given by

$$q_u = c_f N_c d_c i_c + q N_q d_q i_q + \frac{1}{2} \gamma_f B' N_\gamma d_\gamma i_\gamma \qquad (6.23)$$

where

$$q = \gamma_f D_f \qquad (6.24)$$

$$B' = B - 2e \qquad (6.25)$$

In the above equations, B is the width of the bottom slab of a wall, c_f is the cohesion of the foundation soil, γ_f is the unit weight of the foundation soil, ϕ_f is the frictional strength of the foundation soil, D_f is the depth of embedment of the wall, and N_c, N_q, and N_γ are bearing capacity factors (Vesic 1973, 1974) as given in Table 6.6.

d_c, d_q, and d_γ are depth factors (Hansen 1970), given by

TABLE 6.6 Bearing Capacity Factors

ϕ	N_c	N_q	N_γ
0	5.14	1.00	0.00
5	6.49	1.57	0.45
10	8.35	2.47	1.22
15	10.98	3.94	2.65
20	14.83	6.40	5.39
25	20.72	10.66	10.88
30	30.14	18.40	22.40
35	46.12	33.30	48.03
40	75.31	64.20	109.41
45	133.88	134.88	271.76
50	266.89	319.07	762.89

$$d_c = 1 + 0.4 \left(\frac{D_f}{B'} \right) \qquad (6.26)$$

$$d_q = 1 + 2 \tan \phi_f (1 - \sin \phi_f)^2 \frac{D_f}{B'} \qquad (6.27)$$

$$d_\gamma = 1 \qquad (6.28)$$

i_c, i_q, and i_γ are load inclination factors (Hanna and Meyerhof 1981), given by

$$i_c = i_q = \left(1 - \frac{\varphi}{90} \right)^2 \qquad (6.29)$$

$$i_\gamma = \left(1 - \frac{\varphi}{\phi_f} \right)^2 \qquad (6.30)$$

where φ is the inclination of the load, given by

$$\varphi = \tan^{-1} \left(\frac{\sum H}{\sum V} \right) \qquad (6.31)$$

where $\sum H$ = sum of the horizontal forces acting on a wall. A factor of safety of 3 usually is required against bearing capacity failure.

6.5.4 Sliding Stability

The sliding stability along the base of the wall as well as the deep-seated shear failure need to be checked. The factor of safety for sliding stability of the wall along the base is calculated as

$$F_{\text{sliding}} = \frac{\left(\sum V\right)\tan\delta + BC_a}{\sum H} \tag{6.32}$$

where C_a = **adhesion** between the base slab and foundation soil. For sandy soil and gravel, adhesion is zero. For clayey soil, adhesion depends on its consistency. Typically, it may be assumed as one-half of cohesion c of the foundation soil. The typical ranges for adhesion of clays with respect to cohesion are given in Table 6.7.

If subsurface investigation reveals the existence of a continuous weak soil layer in the foundation, the sliding stability of the wall along that weak layer needs to be checked as well. The typical value for the factor of safety against sliding stability is 1.5 for normal conditions and 1.1 for an earthquake condition.

TABLE 6.7 Adhesion Factor (C_a/c) for Different Backfill Soils

Type of Clayey Soil	C_a/c
Stiff to hard clay	0.25–0.3
Stiff clay	0.3–0.4
Medium-stiff clay	0.4–0.7
Soft to very soft clay	1.0

6.5.5 Other Checks

If seepage pressure may develop in the backfill and in the foundation of a retaining wall, it is necessary to check for maximum upward gradient and the factor of safety against **piping** and **bottom heaving**.

In a sheet pile wall, maximum interlock tensile force needs to be checked to prevent rupture at the interlocks.

6.6 Stability Analysis of Rigid Retaining Walls

6.6.1 Gravity Wall

For the calculations in this section:

Unit weight of concrete	$\gamma_c = 23.56$ kN/m^3
Unit weight of the backfill	$\gamma = 18$ kN/m^3
Strength of the backfill	$c = 0$, $\phi = 30°$
Wall friction with the backfill	$\delta = 0$
Slope of the backfill	$i = 0$
Unit weight of the foundation soil	$\gamma_f = 20$ kN/m^3
Strength of the foundation soil	$c_f = 100$ kN/m^2, $\phi_f = 20°$
Total height of the wall	$H' = 5 + 0.7 = 5.7$ m

Since $i = \delta = 0$, Equation 6.12 is used to calculate K_A:

$$K_A = \tan^2 \left(45 - \frac{30}{2} \right) = \frac{1}{3}$$

The active earth pressure acting in the horizontal direction is computed from Equation 6.8 as:

$$P_A = \left(\frac{1}{2} \right) (18)\,(5.7)^2 \left(\frac{1}{3} \right) = 97.47 \text{ kN/m} = \sum H$$

6.6.1.1 Factor of Safety against Overturning

Calculate the sum of the moments (about the toe) of forces resisting overturning $\sum M_R$, as given in Table 6.8.

TABLE 6.8 Calculation of Resisting Moments

Area (Refer to Figure 6.11)	Weight (kN/m)	Moment Arm from Toe (m)	Moment about Toe (kN-m/m)
1	$(4)(0.7)(23.56) = 65.97$	$0.5(4) = 2.0$	131.94
2	$0.5(1.25)(5)(23.56) = 73.625$	$0.5 + \tfrac{2}{3}(1.25) = 1.33$	97.92
3	$(0.5)(5)(23.56) = 58.9$	$0.5 + 1.25 + \tfrac{1}{2}(0.5) = 2.0$	117.8
4	$0.5(1.25)(5)(23.56) = 73.625$	$0.5 + 1.25 + 0.5 + \tfrac{1}{3}(1.25) = 2.67$	196.58
5	$0.5(1.25)(5)(18) = 56.25$	$0.5 + 1.25 + 0.5 + \tfrac{2}{3}(1.25) = 3.08$	173.25
6	$(0.5)(5)(18) = 45.0$	$0.5 + 3 + \tfrac{1}{2}(0.5) = 3.75$	168.75
	$\sum V = 373.37$		$\sum M_R = 886.24$

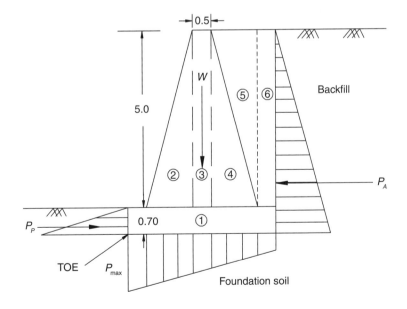

FIGURE 6.11 Stability analysis of gravity wall.

The sum of the overturning moment about the toe is

$$\sum M_O = P_A \frac{H'}{3} = (97.47)\left(\frac{5.7}{3}\right) = 185.19$$

Per Equation 6.20, the factor of safety against overturning is

$$F_o = \frac{886.24}{185.19} = 4.78$$

6.6.1.2 Factor of Safety against Sliding

Assuming friction between the wall and foundation soil, $\delta = \frac{2}{3}\phi_f$ and $C_a = \frac{1}{2}c_f$, and neglecting the passive earth pressure acting on the left side of the bottom slab of the wall, the factor of safety against sliding along the base is calculated from Equation 6.23 as:

$$F_{sliding} = \frac{(373.37)\tan\left(\frac{2}{3}\,20\right) + (4)\left(\frac{1}{2}\,100\right)}{(97.47)} = 2.96$$

6.6.1.3 Maximum Pressure Acting on the Foundation Soil

The width of the bottom slab B is

$$B = 0.5 + 1.25 + 0.5 + 1.25 + 0.5 = 4.0 \text{ m}$$

The eccentricity of the resultant force acting on the base slab is calculated from Equation 6.21 as:

$$e = \frac{4}{2} - \frac{(886.24 - 185.19)}{(373.37)} = 0.122 < \frac{B}{6}$$

The maximum pressure acting on the foundation soil is computed from Equation 6.22 as:

$$P_{max} = \frac{373.37}{4}\left(1 + \frac{6(0.122)}{4}\right) = 110.42 \text{ kN/m}^2$$

The above pressure must be less than or equal to the allowable foundation soil-bearing pressure.

From Table 6.6, for $\phi_f = 20°$, the bearing capacity factors are $N_c = 14.83$, $N_q = 6.40$, and $N_\gamma = 5.39$.

From Equation 6.24, the overburden load q is

$$q = (20)(0.7) = 14 \text{ kN/m}^2$$

From Equation 6.25,

$$B' = 4 - 2(0.122) = 3.756 \text{ m}$$

The depth factors are calculated as follows. From Equation 6.26:

$$d_c = 1 + 0.4 \left(\frac{0.7}{3.756} \right) = 1.075$$

From Equation 6.27:

$$d_q = 1 + 2 \tan 20(1 - \sin 20)^2 \frac{0.7}{3.756} = 1.12$$

From Equation 6.28:

$$d_\gamma = 1$$

The load inclination factors are calculated as follows. From Equation 6.31, the load inclination φ is

$$\varphi = \tan^{-1} \left(\frac{97.47}{373.37} \right) = 14.63°$$

From Equation 6.29:

$$i_c = i_q = \left(1 - \frac{14.63}{90} \right)^2 = 0.70$$

From Equation 6.30:

$$i_\gamma = \left(1 - \frac{14.63}{20.0} \right)^2 = 0.072$$

From Equation 6.23, the ultimate bearing capacity q_u is given by:

$$q_u = (100)(14.83)(1.075)(0.7) + (14)(6.4)(1.12)(0.7)$$

$$+ \frac{1}{2}(20)(3.756)(5.39)(1)(0.072)$$

$$q_u = 1200.78 \text{ kN/m}^2$$

The factor of safety against bearing capacity failure is

$$F_{bearing} = \frac{q_u}{P_{max}} = \frac{1200.78}{110.42} = 10.87$$

6.6.2 Cantilever Rigid Wall

For the calculations in this section:

Unit weight of concrete	$\gamma_c = 23.56 \text{ kN/m}^3$
Unit weight of the backfill	$\gamma = 18 \text{ kN/m}^3$
Strength of the backfill	$c = 0, \phi = 30°$
Wall friction with the backfill	$\delta = 0$
Slope of the backfill	$i = 10°$
Unit weight of the foundation soil	$\gamma_f = 20 \text{ kN/m}^3$
Strength of the foundation soil	$c_f = 100 \text{ kPa}, \phi_f = 20°$
Total height of the wall	$H' = 6.0 \text{ m}$

Referring to Figure 6.12:

$$H' = 2.6 \tan 10° + 6 = 6.46 \text{ m}$$

For $\phi = 30°$ and $i = 10°$, per Equation 6.9:

$$K_A = \cos 10 \, \frac{\cos 10 - \sqrt{\cos^2 10 - \cos^2 30}}{\cos 10 + \sqrt{\cos^2 10 - \cos^2 30}} = 0.350$$

Per Equation 6.8, the active earth pressure acting on the wall is

$$P_a = \frac{1}{2}(18)(6.46)^2(0.35) = 131.45 \text{ kN/m}$$

$$P_h = \sum H = P_a \cos i = 131.45 \cos 10° = 129.45 \text{ kN/m}$$

$$P_v = P_a \sin i = 131.45 \sin 10° = 22.83 \text{ kN/m}$$

6.6.2.1 Factor of Safety against Overturning

Calculate the sum of the moments (about the toe) of forces resisting overturning $\sum M_R$, as given in Table 6.9.

The sum of the overturning moment about the toe is

$$\sum M_O = P_h \frac{H'}{3} = (129.45)\left(\frac{6.46}{3}\right) = 278.75$$

Per Equation 6.20, the factor of safety against overturning is

$$F_o = \frac{2185.12}{278.75} = 7.84$$

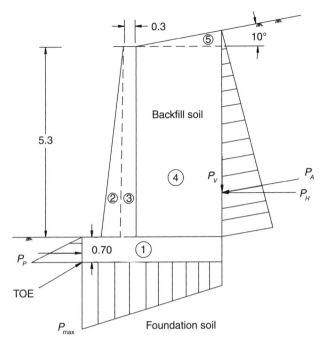

FIGURE 6.12 Stability analysis of cantilever wall.

TABLE 6.9 Calculation of Resisting Moments

Area (Refer to Figure 6.12)	Weight (kN/m)	Moment Arm from Toe (m)	Moment about Toe (kN-m/m)
1	$(4)(0.7)(23.56) = 659.68$	$0.5(4) = 2.0$	1319.36
2	$0.5(0.4)(5.3)(23.56) = 24.97$	$0.7 + ⅔(0.4) = 0.97$	24.44
3	$(0.3)(5.3)(23.56) = 37.46$	$0.7 + 0.4 + ½(0.3) = 1.25$	46.83
4	$(2.6)(5.3)(18) = 284.04$	$0.7 + 0.7 + ½(2.6) = 2.7$	669.71
5	$0.5(2.6)(0.46)(18) = 10.76$	$0.7 + 0.7 + ⅔(2.6) = 3.13$	33.68
6	$P_v = 22.83$	4.0	91.32
	$\sum V = 1003.74$		$\sum M_R = 2185.12$

6.6.2.2 Factor of Safety against Sliding

Assuming the friction between the wall and foundation soil $\delta = ⅔\phi_f$ and $C_a = ½c_f$, and neglecting the passive earth pressure acting on the left side of the bottom slab of the wall, the factor of safety against sliding along the base is calculated from Equation 6.23 as:

$$F_{\text{sliding}} = \frac{(1003.74) \tan\left(\dfrac{2}{3}\,20\right) + (4)\left(\dfrac{1}{2}\,100\right)}{(129.45)} = 3.38$$

6.6.2.3 Maximum Pressure Acting on the Foundation Soil

The width of the bottom slab B is

$$B = 0.7 + 0.7 + 2.6 = 4.0 \text{ m}$$

The eccentricity of the resultant force acting on the base slab is calculated from Equation 6.21 as:

$$e = \frac{4}{2} - \frac{(2185.12 - 278.75)}{(1003.74)} = 0.101 < \frac{B}{6}$$

The maximum pressure acting on the foundation soil is computed from Equation 6.22 as:

$$P_{max} = \frac{1003.74}{4} \left[1 + \frac{6(0.101)}{4} \right] = 288.95 \text{ kN/m}$$

The above pressure must be less than or equal to the allowable foundation soil-bearing pressure.

From Table 6.6, for $\phi_f = 20°$, the bearing capacity factors are $N_c = 14.83$, $N_q = 6.40$, and $N_\gamma = 5.39$.

From Equation 6.24, the overburden load q is

$$q = (20)(0.7) = 14 \text{ kN/m}^2$$

From Equation 6.25,

$$B' = 4 - 2(0.101) = 3.798 \text{ m}$$

The depth factors are calculated as follows. From Equation 6.26:

$$d_c = 1 + 0.4 \left(\frac{0.7}{3.798} \right) = 1.074$$

From Equation 6.27:

$$d_q = 1 + 2 \tan 20 (1 - \sin 20)^2 \frac{0.7}{3.798} = 1.06$$

From Equation 6.28:

$$d_\gamma = 1$$

The load inclination factors are calculated as follows. From Equation 6.31, load inclination φ is

$$\phi \;=\; \tan^{-1}\left(\frac{129.45}{1003.74}\right) \;=\; 7.35°$$

From Equation 6.29:

$$i_c \;=\; i_q \;=\; \left(1 - \frac{7.35}{90}\right)^2 \;=\; 0.84$$

From Equation 6.30:

$$i_\gamma \;=\; \left(1 - \frac{7.35}{20.0}\right)^2 \;=\; 0.4$$

From Equation 6.23, the ultimate bearing capacity q_u is given by:

$$q_u \;=\; (100)(14.83)(1.074)(0.84) \;+\; (14)(6.4)(1.06)(0.84)$$

$$+ \; \frac{1}{2}\,(20)(3.798)(5.39)(1)(0.4)$$

$$q_u \;=\; 1499.57 \ \ \text{kN/m}^2$$

The factor of safety against bearing capacity failure is

$$F_{\text{bearing}} \;=\; \frac{q_u}{P_{\text{max}}} \;=\; \frac{1499.57}{288.95} \;=\; 5.19$$

6.7 Stability Analysis of Cantilever Sheet Pile Wall

6.7.1 In Sandy Soils

In a cantilever sheet pile wall, the depth of embedment into the ground and the maximum moment acting on the wall usually are determined. The earth pressures acting on a cantilever sheet pile wall are shown by the dashed lines in Figure 6.13. The resultant earth pressure acting on the wall is shown by the solid line. In the figure, D is the minimum depth of embedment that corresponds to a factor of safety equal to 1. O is the point below dredge line where the active earth pressure is equal to the passive earth pressure. O' is the point of rotation of the sheet pile wall.

6.7.1.1 Forces Acting on the Wall

The forces acting on the wall are as follows:

1. Active earth pressure acting from the top of the backfill to the point of rotation O' behind the wall

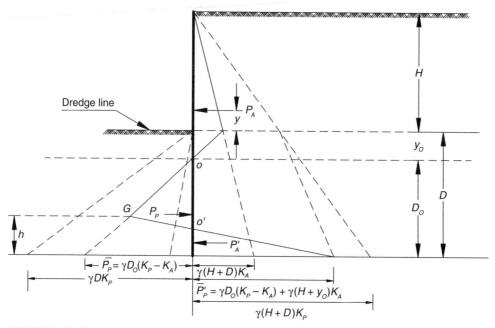

FIGURE 6.13 Pressures on a cantilever sheet pile wall in sandy soil.

2. Passive earth pressure acting in front of the wall from the dredge line up to the point of rotation
3. Passive earth pressure acting behind the wall between the point of rotation and the bottom of the wall
4. Active earth pressure acting in front of the wall between the point of rotation and the bottom of the wall

6.7.1.2 Location of Point *O*

The depth y_o to point *O* from the dredge line is determined by equating the active and the passive earth pressures acting at point *O* by

$$\gamma(H + y_o)K_A = \gamma y_o K_P$$

or

$$y_o = \frac{\gamma H K_A}{\gamma(K_P - K_A)} \qquad (6.33)$$

Let

$$\overline{P}_P = \gamma D_o(K_P - K_A)$$

and

$$\overline{P}_P' = \gamma D_o(K_P - K_A) + \gamma(H + y_o)K_A$$

Then h is calculated from

$$P_A - \frac{1}{2}\bar{P}_P D_o + \frac{1}{2}h(\bar{P}_P + \bar{P}_P') = 0$$

as

$$h = \frac{\bar{P}_P D_o - 2P_A}{\bar{P}_P + \bar{P}_P'} \tag{6.34}$$

The depth of embedment in sandy soil is calculated by taking the moment of all forces about the bottom of the wall and equating it to zero:

$$P_A(D_o + y') - \frac{1}{2}\bar{P}_P D_o \frac{D_o}{3} + \frac{1}{2}(\bar{P}_P + \bar{P}_P')h\frac{h}{3} = 0 \tag{6.35}$$

The above equation is solved for D_o by trial and error. The depth D is then calculated as:

$$D = D_o + y_o \tag{6.36}$$

The minimum depth D thus obtained typically is increased by 20–40% in the design.

6.7.2 In Clayey Soils

The pressure distribution on a sheet pile wall for this case is shown in Figure 6.14. The forces acting on the wall are as follows:

1. Active earth pressure acting behind the wall is per Equation 6.14. At the surface of the backfill:

$$\bar{P} = -2c\sqrt{K_A}$$

If $\phi = 0$,

$$\bar{P} = -2c \quad \text{(tensile)}$$

At the dredge level:

$$\bar{P}_A = \gamma H K_A - 2c\sqrt{K_A}$$

If $\phi = 0$,

$$\bar{P}_A = \gamma H - 2c$$

The depth to zero active earth pressure is given by:

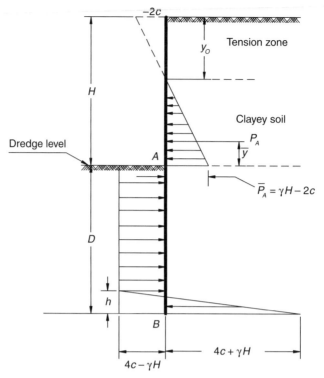

FIGURE 6.14 Pressures on a cantilever sheet pile wall in clayey soil.

$$y_o = \frac{2c}{\gamma \sqrt{K_A}} \tag{6.37}$$

2. The passive pressure acting in front of the wall per Equation 6.16 is

$$\overline{P}_P = \gamma D K_P + 2c \sqrt{K_P}$$

For $\phi = 0$, the pressure at the dredge line is given by:

$$\overline{P}_P = 2c$$

Therefore, the resultant pressure acting at the dredge line is

$$\overline{P}_P - \overline{P}_A = 4c - \gamma H$$

The resultant pressure acting at any depth z below the dredge line is

$$P_P - P_A = (\gamma z + 2c) - [\gamma(H + z) - 2c] = 4c - \gamma H$$

If passive pressure is developed behind the wall at the bottom of the wall,

$$P_P - P_A = [\gamma(H + D) + 2c] - (\gamma D - 2c) = 4c + \gamma H$$

$\sum H = 0$ yields

$$P_A - (4c - \gamma H)D + \frac{1}{2}(4c - \gamma D + 4c + \gamma D)h = 0$$

or

$$h = \frac{(4c - \gamma H) - P_A}{4c} \qquad (6.38)$$

Another equilibrium equation $\sum M = 0$ about the bottom of the wall yields

$$P_A(\bar{y} + D) + (4ch)\left(\frac{h}{3}\right) - (4c - \gamma H)D\frac{D}{2} = 0$$

or

$$(4c - \gamma H)D^2 - 2P_A D + \frac{P_A(12c\bar{y} + P_A)}{(2c + \gamma H)} = 0 \qquad (6.39)$$

The depth of embedment D is obtained by solving the above quadratic equation. The D thus obtained typically is increased by 20–40% in the design. Alternatively, a factor of safety could be applied to the values of c and ϕ.

6.8 Stability Analysis of Anchored Sheet Pile Wall

The stability of an anchored sheet pile wall may be analyzed by three methods:

1. *Free earth method*—The end of the sheet pile embedded in the ground is considered to be simply supported.
2. *Fixed earth method*—The end of the sheet pile embedded in the ground is considered to be fixed.
3. *Equivalent beam method*—The sheet pile wall is analyzed as a beam with net lateral earth pressures acting like surcharge loads. The beam is considered to be simply supported at the anchor and fixed at the embedded end with a reactive force R acting. Since the moment at the point of inflection is zero, the whole sheet pile is analyzed as two beams.

6.8.1 Anchored Sheet Pile Wall in Sandy Soil by Free Earth Method

The lateral pressures acting on a sheet pile wall (shown in Figure 6.15) are as follows:

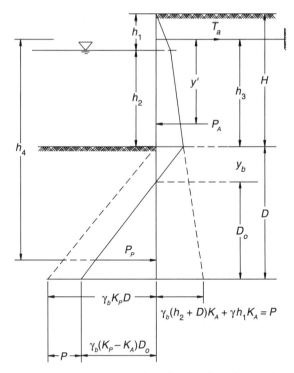

FIGURE 6.15 Pressure on anchored sheet pile wall in sandy soil.

1. Active earth pressure P_A due to the backfill soil acting at a distance y' from the anchor
2. Passive pressure due to the soil in front of the wall

$$P_P = \frac{1}{2} \gamma_b (K_P - K_A) D_o^2$$

acting at

$$h_4 = h_3 + y_b + \frac{2}{3} D_o$$

where γ_b is the buoyant unit weight of the soil. The distance y_b can be calculated as:

$$y_b = \frac{\bar{P}_A}{y_b (K_P - K_A)} \tag{6.40}$$

3. Tensile force T_a in the anchor rod

For equilibrium, $\sum M = 0$ about the anchor rod. Therefore,

$$P_A y' = P_P h_4$$

or

$$\left[\frac{\gamma_b(K_P - K_A)}{3} \right] D_o^3 + \frac{\gamma_b(K_P - K_A)}{2} (h_3 + y_b) D_o^2 - P_A y' = 0 \qquad (6.41)$$

D_o is obtained by solving the above quadratic equation. The minimum depth of embedment is $D = D_o + y_b$. The depth is increased by 20–40% in the design.

The tensile force in the anchor rod is calculated as

$$T_a = P_A - P_P \qquad (6.42)$$

6.8.2 Anchored Sheet Pile Wall in Clayey Soil below Dredge Line

The pressure distribution on a sheet pile wall for this case is shown in Figure 6.16. The surcharge load at the dredge line due to the backfill is

$$q = \gamma h_1 + \gamma_b h_2$$

The active earth pressure force due to the sandy backfill is given by P_A acting at a distance $\bar{y}$ from the anchor rod.

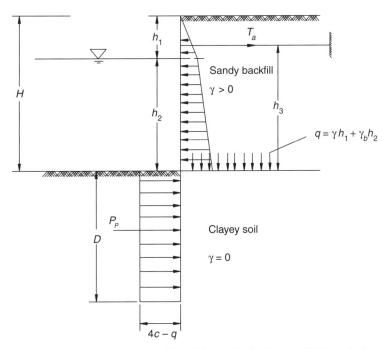

FIGURE 6.16 Pressures on anchored sheet pile in clayey soil below dredge line.

The active pressure at the dredge line is given by:

$$\bar{P}_A = q - 2c$$

The passive pressure at the dredge line is given by:

$$\bar{P}_P = 2c$$

The resultant pressure at the dredge line is

$$\bar{P}_P - \bar{P}_A = 4c - q$$

The resultant pressure acting on the wall remains constant with depth in clayey soil.
Taking the moment of all forces about the anchor rod,

$$P_A \bar{y} - D(4c - q) \left(h_3 + \frac{D}{2} \right) = 0 \qquad (6.43)$$

The depth of embedment D is obtained by solving the above quadratic equation. The depth of embedment should be increased by 20–40% for the design.
The anchor force is obtained, as before, by

$$T_a = P_A - P_P$$

Because it is flexible, the anchored sheet pile wall yields and redistributes lateral earth pressures acting on it. This tends to reduce the maximum bending moment as calculated by the free earth method. The maximum design moment acting on a wall computed by the free earth method can be reduced by a procedure suggested by Rowe (1952, 1957).

6.8.3 Anchored Sheet Pile Wall in Sandy Soil by Equivalent Beam Method

Figure 6.17 shows the lateral pressures acting on a wall per the fixed earth pressure/equivalent beam methods. The following pressures are acting on the wall:

1. Active pressure acting at the top of the wall:

$$P_A = qK_A = 20(0.28) = 5.6 \text{ kN/m}^2$$

2. Active earth pressure acting at B:

$$P_B = \gamma z K_A + qK_A = 18(3)(0.28) + 20(0.28) = 20.72 \text{ kN/m}^2$$

3. Active earth pressure acting just above the dredge line:

$$P_{c1} = P_B + \gamma_{sub} z K_A = 20.72 + 8(10)(0.28) = 43.12 \text{ kN/m}^2$$

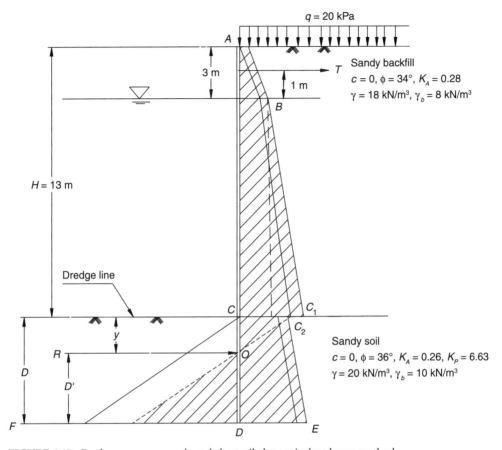

FIGURE 6.17 Earth pressure on anchored sheet pile by equivalent beam method.

4. Active earth pressure acting just below the dredge line:

$$P_{c2} = \gamma z K_A' + q K_A' = 18(3)(0.26) + 8(10)(0.26) + 20(0.26)$$
$$= 40.0 \text{ kN/m}^2$$

5. Active earth pressure acting at a depth D from the dredge line:

$$P_E = P_{c2} + \gamma_{sub}' D K_A' = 40.0 + 10(0.26)D = 40 + 2.6D$$

6. Passive pressure acting at a depth D from the dredge line:

$$P_F = \gamma_{sub}' D K_P' = 10(6.63)D = 66.3D$$

The location of the point of zero pressure O is

$$y = \frac{P_{c2}}{\gamma_{sub}'(K_P' - K_A')} = \frac{40}{10(6.63 - 0.26)} = 0.628 \text{ m}$$

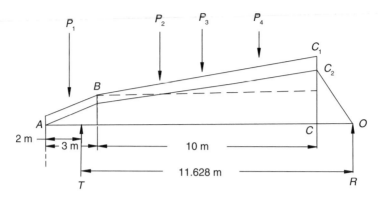

FIGURE 6.18 Equivalent beam.

Assume that the **point of contraflexure** is at the point of zero pressure. Then, the equivalent beam is as shown in Figure 6.18.

The forces acting on the beam are as follows:

1. $P_1 = \dfrac{(P_A + P_B)(3)}{2} = \dfrac{(5.6 + 20.72)(3)}{2} = 22.68$ kN/m

    acting at $\dfrac{5.6(3)\left(\dfrac{3}{2}\right) + \left(\dfrac{1}{2}\right)(3)(20.72 - 5.6)\left(\dfrac{3}{3}\right)}{\left(\dfrac{1}{2}\right)(5.6 + 20.72)(3)}$

    $= 1.213$ m from the anchor

2. $P_2 = 20.72(10) = 207.2$ kN/m acting at $\dfrac{(13 - 3)}{2} + 1$

    $= 6$ m from the anchor

3. $P_3 = \dfrac{(43.12 - 20.72)(10)}{2} = 112$ kN/m acting at $\dfrac{2}{3}(10) + 1$

    $= 7.67$ m from the anchor

4. $P_4 = \dfrac{40(0.628)}{2} = 12.56$ kN/m acting at $11 + \dfrac{0.628}{3}$

    $= 11.21$ m from the anchor

5. Force R acting at a distance 11.628 m from the anchor

$\sum M = 0$ about the anchor yields:

$$R(11.628) + 22.68(1.213) - 207.2(6) - 112(7.67) - 12.56(11.21) = 0$$

$$R = 190.53 \ \text{kN/m}$$

Therefore, the anchor force T is

$$T = P_1 + P_2 + P_3 + P_4 - R$$

$$= 22.68 + 207.2 + 112 + 12.56 - 190.53 = 163.9 \ \text{kN/m}$$

Taking the moment about the bottom of the sheet pile wall,

$$\sum M = \frac{\gamma'_{sub}(K'_P - K'_A)D'^2}{2}\left(\frac{D'}{3}\right) - RD' = 0$$

$$D' = \sqrt{\frac{6(190.53)}{10(6.63 - 0.26)}} = 4.24 \ \text{m}$$

Therefore,

$$D = D' + y = 4.24 + 0.628 = 4.9 \ \text{m}$$

The depth of embedment is 6 m.

6.9 Anchorage Systems for Sheet Pile Walls

The major components of an anchorage system for a sheet pile wall consist of tie-rods, wales, and anchors.

6.9.1 Tie-Rods

Tie-rods usually are round structural steel bars with upset threaded ends. They are subjected to tensions most of the time. Usually turnbuckles are provided in every tie-rod to take up slack that might develop due to consolidation of the recent backfill. The pull on a tie-rod theoretically is calculated as

$$P = \frac{Tl}{\cos\theta} \tag{6.44}$$

where T = anchor force per meter width of the wall, l = center-to-center distance between rods, and θ = angle of inclination of the tie-rod with horizontal.

The design value of pull (P_{design}) is obtained by increasing the theoretical value of the tension in the tie-rod by 30% and 50–100% at **splices** and connections to account for the increase in force due to accidental overloading. The design area of the rod is obtained from

$$A = \frac{P_{\text{design}}}{\sigma_{\text{all}}} \tag{6.45}$$

where σ_{all} is the allowable stress in a steel bar.

6.9.2 Wales

The horizontal reaction from an anchored sheet pile wall is transferred to the tie-rods by a **flexural member** known as a wale. It usually consists of two structural channel sections placed with their webs back-to-back in the horizontal position, as shown in Figure 6.19. The best location for the wales is on the outer face of the sheet pile wall, where the piles will bear against the wales. For design purposes, wales are considered to be between a continuous beam and a single-span simply supported beam. The maximum moment for a continuous beam is calculated as

$$M_{\text{max}} = \frac{1}{10} P_{\text{design}} l^2 \tag{6.46}$$

and for a simply supported beam is calculated as

$$M_{\text{max}} = \frac{1}{8} P_{\text{design}} l^2 \tag{6.47}$$

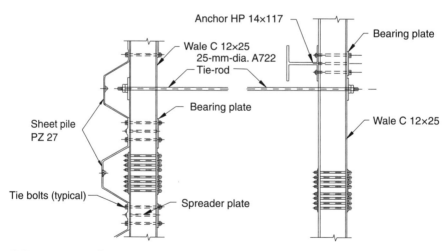

FIGURE 6.19 Anchorage system.

The **section modulus** of wales S is given by

$$S = \frac{M_{max}}{\sigma_{all}} \tag{6.48}$$

where σ_{all} is the allowable bending stress in steel.

Initially wales are tack-welded to the sheet piles and later connected to them by plates and bolts. The **pullout force** in a bolt is calculated as

$$P_{bolt} = P_{design} w F_s \tag{6.49}$$

where w is the width of a single sheet pile and F_s is a factor of safety, typically between 1.2 and 1.5.

6.9.3 Anchors

The general types of anchor used in sheet pile walls are

1. Short sheet piles
2. Vertical anchor piles
3. Tiebacks
4. Anchor beams supported by batter (tension and compression) piles
5. Anchor plates and beams (deadman)

6.9.3.1 Short Sheet Piles as Anchor

Short steel sheet piles (Figure 6.20) driven in the form of a continuous wall often are used as anchors. The resistance is derived from passive pressure developed in front of the anchor wall as the tie-rods pull against it. The anchor wall is analyzed by methods discussed in Section 6.8. Full passive pressure is developed only when the active and passive failure zones do not intersect. The tie-rod connection to the anchor wall ideally should be located at the place where the resultant earth pressure is acting.

6.9.3.2 Vertical Anchor Piles

Vertical anchor piles also are used as anchors for anchored sheet pile walls, as shown in Figure 6.21a. The piles should be designed for the lateral load in the form of an anchor force.

6.9.3.3 Tiebacks

Grouted tiebacks (Figure 6.21b) are constructed by drilling steel rods through the retaining wall into the soil or bedrock on the other side. Grout is then pumped under pressure into the tieback anchor holes so that the rods can utilize soil resistance to prevent tieback pullout and wall destabilization. The ultimate resistance P_u offered by a tieback in sandy soil is given by

$$P_u = \pi\, dl\sigma'_v K \tan \phi \tag{6.50}$$

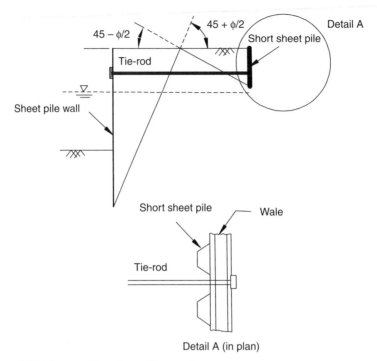

FIGURE 6.20 Short sheet pile as anchor.

where K = earth pressure coefficient, σ'_v = average effective vertical stress, d = diameter of the grouted bulb, and l = length of the grouted portion of the tieback.

The earth pressure coefficient K can be assumed to be at rest. The lower limit of K can be taken to be equal to Rankine's active earth pressure coefficient.

The ultimate resistance in clays is given by

$$P_u \; = \; \pi d l c_a \tag{6.51}$$

The value of adhesion c_a often is assumed to be two-thirds of the undrained cohesion of clays. The ultimate resistance obtained from above expressions is multiplied by a factor of safety of 1.5–2 to obtain the allowable resistance of a tieback.

6.9.3.4 Anchor Beams Supported by Batter Piles

Anchor beams supported by batter piles often are used to anchor sheet pile walls, especially where the subsoil is rock or good enough to support the pile loads. The anchor beam with batter piles should be located away from the active zone behind the retaining wall, as shown in Figure 6.21c.

6.9.3.5 Anchor Plates and Beams (Deadman)

The ultimate resistance P_u of a continuous ($l/h > 5$) anchor plate or deadman (Figure 6.21d) with length l and height h in sandy soil located at or near the ground surface ($H/h \le 1.5$) is given by Teng (1962) as:

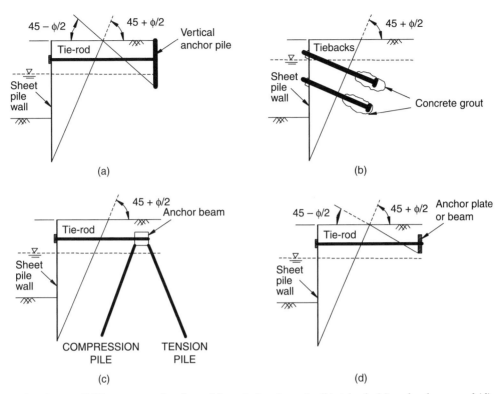

FIGURE 6.21 Different types of anchors: (a) vertical anchor pile, (b) tieback, (c) anchor beam, and (d) anchor plate.

$$P_u = P_P - P_A \qquad (6.52)$$

The pressure distribution on both sides of the anchor plate in sandy soil is shown in Figure 6.22. The allowable resistance is calculated by dividing P_u by a factor of safety of 2. If the anchor plate or deadman is located near the ground surface but is short in length, the resistance along the curved sliding surfaces at the edges should be considered. The expression for the ultimate capacity of a short anchor plate or deadman in sandy soil is given by

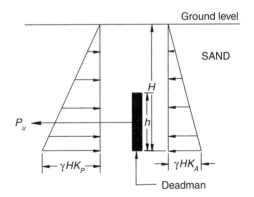

FIGURE 6.22 Pressure distribution on a deadman.

$$P_u \leq l(P_P - P_A) + \frac{1}{3} K_o \gamma \left(\sqrt{K_P} - \sqrt{K_A} \right) H^3 \tan \phi \qquad (6.53)$$

where l = length of the anchor plate, H = height of the anchor plate, γ = unit weight of the soil, ϕ = angle of internal friction, and K_o = coefficient of earth pressure at rest (typically taken as 0.4).

For cohesive soils, the value of P_u is given by

$$P_u \leq l(P_P - P_A) + 2cH^2 \qquad (6.54)$$

where c = cohesion of the soil.

6.10 Design Example of an Anchorage System

6.10.1 Tie-Rods

For the anchor sheet pile wall shown in Figure 6.17, the anchor force was computed as $T = 163.9$ kN/m. Using a spacing of tie-rods of $l = 3$ m and assuming a level tie-rod (that is, $\phi = 0$), pull on the tie-rod is computed from Equation 6.44 as:

$$P = \frac{163.9(3)}{1} = 491.7 \text{ kN per tie-rod}$$

The above value is increased by 30% for the design. Then $P_{design} = 639.2$ kN. The required cross-sectional area of the tie-rod is then obtained from Equation 6.45.

6.10.2 Wales

To calculate the maximum moment M_{max}, the average between simple and continuous supports is used (Equations 6.46 and 6.47):

$$M_{max} = \frac{1}{9} P_{design} l^2 = \frac{1}{9} (639.2)(3)^2 = 639.2 \text{ kN-m}$$

The section modulus of wales can be calculated as

$$S_{xx} = \frac{M_{max}}{\sigma_{all}}$$

where σ_{all} is the allowable stress in steel (in bending).

6.10.3 Anchor Wall

Find the location of the anchor wall (see Figure 6.23):

$$Y = 6 \tan 27 + 13 \tan 28 = 10 \text{ m}$$

$$Z = 6 \cot 36 + 13 \cot 34 - Y = 17.5 \text{ m}$$

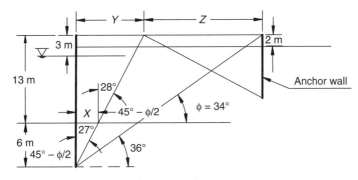

FIGURE 6.23 Location of anchor wall.

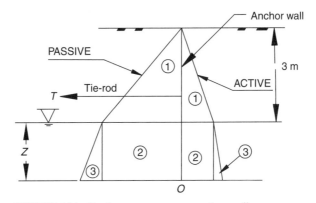

FIGURE 6.24 Earth pressures on an anchor wall.

Find the anchor force and depth of embedment of the anchor wall for the earth pressure distribution on the anchor wall (Figure 6.24):

$$P_{P1} - P_{A1} = \frac{1}{2}(18)(5.72 - 0.28)(3)^2 = 440.64 \text{ kN}$$

$$P_{P2} - P_{A2} = 3(18)(5.72 - 0.28)Z = 293.76Z$$

$$P_{P3} - P_{A3} = \frac{1}{2}(8)(5.72 - 0.28)Z^2 = 21.76Z^2$$

$$\sum H = 0, (P_{P1} - P_{A1}) + (P_{P2} - P_{A2}) + (P_{P3} - P_{A3}) - T = 0$$

or

$$440.64 + 293.76Z + 21.76Z^2 = T$$

$$\sum M = 0, \; (P_{P1} - P_{A1}) \left(Z + \frac{3}{3} \right) + (P_{P2} - P_{A2}) \frac{Z}{2}$$

$$+ (P_{P3} - P_{A3}) \frac{Z}{3} - T(Z + 1) = 0$$

or

$$440.64(Z + 1) + 293.76 \frac{Z^2}{2} + 21.76 \frac{Z^3}{2} - T(Z + 1) = 0$$

Combine the above two equations and solve for Z and T.

Use $Z = 1$ m; that is, the anchor wall should be driven up to 4 m. The corresponding anchor force is $T = 756.2$ kN/m. Therefore, the factor of safety is

$$\frac{756.2}{163.9} = 4.6$$

6.11 Design Example of a Braced Wall System

Figure 6.25 shows a typical braced wall system used for stabilizing a near-vertical cut in soil. The design calculations for different components of the wall system are as follows:

1. *Calculate the earth pressure diagram*—Unit weight of the soil $\gamma = 18$ kN/m³, cohesion of the soil $c = 50$ kN/m², and total depth of excavation = 8 m:

$$\frac{\gamma H}{c} = \frac{18(8)}{50} = 2.9 < 4$$

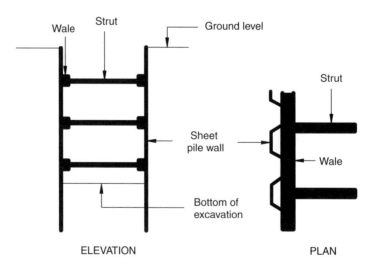

FIGURE 6.25 Typical braced wall system.

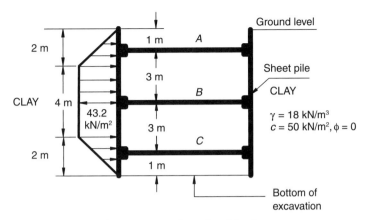

FIGURE 6.26 Earth pressures on braced wall in clay.

$$P_A = 0.3\gamma H = 0.3(18)(8) = 43.2 \text{ kN/m}^2$$

The earth pressure acting on a braced wall is shown in Figure 6.26.
2. *Determine the strut loads—*

$$\sum M_B = 0, \; F_A(3) = \frac{1}{2}(43.2)(2)\left(\frac{2}{3}+2\right) + (2)(43.2)\left(\frac{2}{2}\right)$$

$$F_A = 67.2 \text{ kN/m}$$

$$\sum V = 0, \; F_{B1} = \frac{1}{2}(2)(43.2) + (43.2)(2) - F_A$$

$$F_{B1} = 62.4 \text{ kN/m}$$

From symmetry:

$$F_{B2} = 62.4 \text{ kN/m}$$

and

$$F_C = 67.2 \text{ kN/m}$$

3. *Calculate the strut loads—*Struts are designed as a horizontal column subjected to bending. They should have a minimum spacing of 2.75 m to create working space. Assume the horizontal spacing $s = 3$ m center to center:

$$P_A = F_A(s) = 67.2(3) = 201.6 \text{ kN}$$

$$P_B = (F_{B1} + F_{B2})(s) = (62.4 + 62.4)(3) = 374.4 \text{ kN}$$

$$P_C = F_C(3) = (67.2)(3) = 201.6 \text{ kN}$$

4. *Locate the point of contraflexure (x)*—The point of contraflexure is the point where the shear force is zero and the bending moment is maximum:

$$x = \frac{F_{B1}}{43.2} = 1.44 \text{ m}$$

5. *Calculate the maximum moment M_{max} in the sheet pile wall and find the appropriate section*—

$$M_{max} = M_E = (F_{B1})(1.44)(43.2)(1.44)\left(\frac{1.44}{2}\right)$$

$$= 45.07 \text{ kN-m/m of wall}$$

The section modulus of the wall S is then calculated as

$$S = \frac{M_{max}}{\sigma_{all}}$$

where σ_{all} = allowable stress in steel in bending. Choose the appropriate section of sheet pile based on the section modulus thus obtained.

6. *Determine the section modulus of wales*—Wales are treated as continuous horizontal members pinned at the struts. The maximum moment in wales will occur at B and is given by:

$$M_{max} = \frac{(F_{B1} + F_{B2})}{8} S^2 = \frac{(62.4 + 62.4)}{8} (3)^2$$

$$= 140.4 \text{ kN-m/m of wall}$$

The section modulus of wales S is

$$S = \frac{M_{max}}{\sigma_{all}}$$

Find the appropriate section of wales based on the above value of the section modulus.

6.12 Mechanically Stabilized Retaining Walls

A mechanically stabilized earth retaining wall is a flexible wall composed of three elements: (1) wall facing, (2) soil reinforcement such as strip- or grid-type reinforcement, and (3) compacted backfill.

The wall facing element usually is a precast concrete member. Other types of facing elements are steel plates, wooden planks, and concrete interlocking panels.

Mainly two types of soil reinforcement are used in mechanically stabilized retaining wall construction. Wide and thin metallic strips, tied to the face elements and placed at regular horizontal and vertical spacing, often are utilized to reinforce the backfill soil. Alternatively, **geogrids** composed of high-strength polymers are placed horizontally at regular intervals between the compacted backfill to reinforce it. The geogrids often are attached to the wall facing elements.

Since drainage is not usually provided in mechanically stabilized retaining walls, the backfill should be a permeable granular material to prevent buildup of pore water pressure. Lightweight compaction equipment is used to compact the backfill placed on top of each layer of reinforcement strips or geogrid.

The soil reinforcement and the compacted backfill derive frictional resistance and interlocking resistance between each other. When the mechanically stabilized soil mass is subjected to shear stresses, it tends to transfer them to the reinforcement elements. Also, reinforcement elements tend to redistribute stresses away and prevent development of localization of stresses.

For a mechanically stabilized retaining wall, both external and internal stability must be checked. Analysis of external stability is similar to that for a gravity wall. The factor of safety against sliding (both along the base and overall), bearing capacity failure, and overturning should be checked. The resultant of the vertical forces should lie within the middle third of the base of the reinforced soil mass.

Analysis of internal stability includes determination of maximum tie force, horizontal and vertical spacing of ties, effective tie length, and a factor of safety against tie breaking or tie pullout for a given area of ties.

The lateral earth pressure in a mechanically stabilized retaining wall is determined from Rankine's active earth pressure theory (Equations 6.8 and 6.12). The tie force T at any depth Z is then determined as

$$T = (\gamma Z K_A)(S_v S_h) \tag{6.55}$$

where γ = unit weight of the backfill soil, S_v = vertical spacing of the ties, and S_h = horizontal spacing of the ties.

The maximum tie force T_{max} will develop at the bottommost ties and is given by

$$T_{max} = (\gamma H K_A)(S_v S_h) \tag{6.56}$$

where H = height of the wall.

The factor of safety against tie break is then calculated as

$$F_s = \frac{wt\sigma_y}{\gamma H K_A S_v S_h} \tag{6.57}$$

where w = width of the tie, t = thickness of the tie, and σ_y = yield stress of the material.

The factor of safety against tie pullout is given by

$$F_p = \frac{4 l_{eff} w \tan \phi_i}{3 K_A S_v S_h} \tag{6.58}$$

where l_{eff} = effective length of a tie (length of a tie outside the active earth pressure zone of the wall) and ϕ_i = soil-tie friction angle.

For a mechanically stabilized retaining wall with geogrid reinforcements, the factor of safety against break at any depth z is

$$F_s = \frac{\sigma_{all}}{\gamma z K_A S_v} \tag{6.59}$$

The factor of safety against pullout at any depth z is given by

$$F_p = \frac{2 l_e \tan \phi_i}{S_v K_A} \tag{6.60}$$

$$l_e = \frac{H - z}{\tan \left(45 + \dfrac{\phi}{2} \right)} \tag{6.61}$$

where H = total height of the wall.

When geotextile is used as reinforcement, the facing of the wall is formed by lapping the geotextile sheet over a lap length l_p. This lap length is determined as:

$$l_p = \frac{F_p S_v K_A}{4 \tan \phi_i} \tag{6.62}$$

The minimum lap length is 1 m. In the absence of data, the interface friction angle between geotextile and soil may be assumed to be two-thirds of the soil friction angle.

6.13 Failure of Retaining Walls

Several failures of retaining walls have been reported in the literature (Sengupta and Venkateshwarlu 2002). Some of the common causes are (1) long-term increase in pressures in the backfill (e.g., Euston Station Wall in London, Mill Lane Wall in London, Railroad Wall and U.S. Public Road Walls in the U.S.), (2) cyclic freeze-thaw pressures (e.g., Water Street Wall in Wisconsin), (3) high water pressure behind the wall (e.g., Lingfield Railway Bridge in London, Highway Fill Wall in Greece, and Development Wall in India), and (4) compaction pressure (e.g., Eisenhower Lock in New York and Bund Wall in London).

Most of the failures have been observed when the backfill material is clay. Clean, free-draining, granular sand or gravel usually is recommended as backfill material. In clayey backfill, swelling pressures, high pore water pressures, and ice-related forces may substantially increase the thrust on the wall and should be carefully considered in the design.

As discussed above, compaction-induced excessive pressures also can damage the wall. Small vibrator plate (hand-operated) compactors can be used effectively to densify granular backfill adjacent to the wall. They do not induce high lateral loads because of their light weight.

Failure of the excavation behind a wall also can occur if the cut slope is too steep and an adequate factor of safety is not maintained.

Failure of retaining walls also occurs when the foundation is not competent and there is excessive settlement (see Figure 6.27). Rapid failure of a retaining wall occurs when the wall is supported on soft clay and there is undrained shear failure (deep-seated rotational failure) beneath the foundation (see Figure 6.28).

Rupture of ties (see Figure 6.29) in mechanically stabilized retaining walls and rupture and slippage at the interlock between sheet piles in sheet pile walls also are known to cause failure.

If water seepage force exists behind the retaining wall, precautions should be taken against the development of high seepage gradient behind and below the wall, which can induce instability, such as blowout of

FIGURE 6.27 Cracks on retaining wall due to excessive settlement of foundation.

FIGURE 6.28 Deep-seated rotational failure of wall in soft clay.

FIGURE 6.29 Failure of mechanically stabilized retaining wall due to rupture of ties.

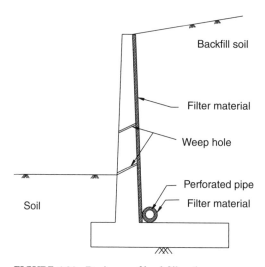

FIGURE 6.30 Drainage of backfill soil.

the foundation soil. The common practice of draining the backfill soil includes construction of **weep holes** in the wall at different elevations and/or providing horizontal drains (such as perforated drain pipes encased in **filter material**) behind the retaining wall (see Figure 6.30).

Defining Terms

Adhesion: The tendency of certain dissimilar molecules (like those of a wall and soil) to cling together due to attractive forces.

Angle of internal friction: The friction (expressed in degrees) resisting motion between elements of a solid material (e.g., rock and soil) while it undergoes deformation due to shearing. It is a part of the shear strength of a material that is dependent on normal stress.

Bearing capacity: Capacity of a structure to carry load without shear failure and excessive settlement.

Bottom heaving: Instability due to heaving of the bottom of an excavation.

Cantilever: A structure supported at one end only. The free end can deform but cannot support moment.

Cohesion: The part of shear strength that is independent of the normal effective stress in mass movements.

Counterfort: Similar to a buttress but located at the toe instead of the heel of a wall. The fin of a counterfort is designed as a tensile member.

Driving force: A force, usually earth pressure and water pressure, acting to move a wall.

Filter material: A material that prevents movement of fines (erosion), piping, and clogging of drains.

Flexural member: A structural component that is designed to take stress due to bending.

Geogrids: A range of polymeric products in the form of mesh used to reinforce soils.

Grout: A construction material used to embed anchors in the ground, connect sections of precast concrete, fill voids, and seal joints. Grout generally is composed of a mixture of water, cement, sand, and sometimes fine gravel.

Hydrodynamic pressure: Pressure due to wave action, etc. in the sea.

Interlocking: Sheet piles come with male and female connections. They are joined by slipping the male end into the female end (grooves) to form a continuous wall. An interlock may fail if hoop stress at the joint is excessive.

Masonry: The building of structures from individual units laid in and bound together by mortar. The common materials of masonry construction are brick, stone, etc.

Piping: Loss of fine material with flow, resulting in failure. This kind of failure happens if high seepage gradient exists in the foundation.

Point of contraflexure: Location at which no bending occurs. In a bending moment diagram, it is the point at which the bending moment curve intersects with the zero line.

Poisson's ratio: Ratio of the contraction or transverse strain (normal to the applied load) to the extension or axial strain (in the direction of the applied load).

Propped: A structural engineering term that means supported by a member.

Pullout force: Force required to pull out an anchor from the ground.

Reinforcement: Steel bars commonly used in concrete and masonry structures to increase tensile strength of the structure.

Rigid: Inflexible or stiff.

Section modulus: The section modulus of a beam is the ratio of the second moment of area to the distance of the extreme compressive fiber from the neutral axis in a typical cross section of a beam. It is directly related to the strength of the beam.

Slurry: A thick suspension of solids (clay, cement, or bentonite) in fluid (water).

Splices: Joints between two anchors or tie bars.

Structure: Usually refers to any large, man-made object, such as a wall, building, dam, etc.

Surcharge load: External load due to equipment, snow, etc.

Uplift pressure: Upward-acting pressure below a foundation due to upward water forces and soil reactions.

Wall friction: Friction between the backfill soil and the retaining wall.

Weep hole: A drain drilled into a concrete wall.

References

Coulomb, C.A. (1776). Essai sur une application des regeles des maximis et minimis a quelques problemes de statique. *Mem. Acad. R. Sci. Paris,* 7:38.

Das, B.M. (1999). *Principles of Foundation Engineering,* 4th edition, Brooks/Cole, New York.

Department of the Navy (1982). *Design Manual—Soil Mechanics, Foundations and Earth Structures,* NAVFAC DM-7, Washington, D.C.

Fang, Y.-S. and Chen, T.-J. (1995). Modification of Mononobe-Okabe theory, technical notes. *Geotechnique,* 45(1):165–167.

Hanna, A.M. and Meyerhof, G.G. (1981). Experimental evaluation of bearing capacity footings subjected to inclined loads. *Can. Geotech. J.,* 18(4):599–603.

Hansen, J.B. (1953). *Earth Pressure Calculation,* Danish Technical Press, Institute of Danish Civil Engineering, Copenhagen.

Hansen, J.B. (1970). A revised and extended formula for bearing capacity. *Dan. Geotech. Inst. Bull.,* p. 28.

Institution of Structural Engineers (1951). *Earth Retaining Structures,* Civil Engineering Code of Practice No. 2, London.

Jaky, J. (1944). The coefficient of earth pressure at rest. *J. Soc. Hung. Archit. Eng.,* 78(22):355–358.

Juran, I. (1982). *Design of Reinforced Earth Structures,* Report No. GE-82/02, Louisiana State University, Baton Rouge.

Juran, I. and Schlosser, F. (1978). Theoretical analysis of failure in reinforced earth structures. *Symposium on Earth Reinforcement,* ASCE, 528–555.

Lee, K.L., Adams, B.D., and Vagneron, J.J. (1973). Reinforced earth retaining walls. *J. Soil Mech. Found. Div. ASCE,* 99(10):745–763.

Meyerhof, G.G. (1963). Some recent research on the bearing capacity of foundations. *Can. Geotech. J.,* 1(1):16–26.

Okabe, S. (1924). General theory on earth pressures and seismic stability of retaining wall and dam. *J. Jpn. Soc. Civ. Eng.,* 10(5):1277–1323.

Peck, R.B. (1943). Earth pressure measurements in open cuts, Chicago (IL) subway, *Trans. Am. Soc. Civ. Eng.,* 108:1008–1058.

Peck, R.B. (1969). Deep excavation and tunneling in soft ground. *Proceedings of the 7th International Conference on Soil Mechanics and Foundation Engineering,* Mexico City, 225–290.

Rankine, W.J.M. (1857). On the stability of loose earth. *Trans. R. Soc. London,* p. 147.

Rowe, P.W. (1952). Anchored sheet pile walls. *Proc. Inst. Civ. Eng.,* 1(1):27–70.

Rowe, P.W. (1955). A theoretical and experimental analysis of sheet pile walls. *Proc. Inst. Civ. Eng.,* 1(1).

Rowe, P.W. (1957). Sheet pile walls in clay. *Proc. Inst. Civ. Eng.,* 7:629–654.

Schlosser, F. and Long, N. (1974). Recent results in French research on reinforced earth. *J. Constr. Div. ASCE,* 100(3):113–237.

Seed, H.B. and Whitman, R.V. (1970). Design of retaining structures for dynamic loads. *Proceedings of the ASCE Specialty Conference on Lateral Stresses in the Ground and Design of Earth Retaining Structures,* Cornell University, Ithaca, NY, 103–147.

Sengupta, A. and Venkateshwarlu, G. (2002). Lateral earth pressures in clayey backfill. *Indian Geotech. J.,* 32(2):65–85.

Taylor, D.W. (1948). *Fundamentals of Soil Mechanics,* John Wiley and Sons, New York.

Teng, W.C. (1962). *Foundation Design,* Prentice Hall, Englewood Cliffs, NJ.

Terzaghi, K. (1941). General wedge theory of earth pressure. *Trans. Am. Soc. Civ. Eng.,* 106:68–97.

Terzaghi, K. (1943). *Theoretical Soil Mechanics,* John Wiley and Sons, New York.

Terzaghi, K. (1954). Anchored bulkheads. *Trans. Am. Soc. Civ. Eng.,* 119:1243–1324.

Terzaghi, K. and Peck, R. (1967). *Soil Mechanics in Engineering Practice,* John Wiley and Sons, New York.

Tschebotarioff, G.P. (1951). *Soil Mechanics, Foundations and Earth Structures,* McGraw-Hill, New York.

Tschebotarioff, G.P. (1973). *Foundations, Retaining and Earth Structures,* 2nd edition, McGraw-Hill, New York.

USACE (1989). *Engineering and Design: Retaining and Flood Walls,* EM 1110-2-2502, Department of the Army, Washington, D.C.

U.S. Steel (1969). *Steel Sheet Piling, Design Manual,* ADUSS25-3848-01, U.S. Steel Corporation, Pittsburgh.

Vesic, A.S. (1973). Analysis of ultimate loads of shallow foundations. *J. Soil Mech. Found. Div. ASCE,* 96(2):561–584.

Vesic, A.S. (1974). Bearing capacity of shallow foundations. *Foundation Engineering Handbook,* Van Nostrand Reinhold, New York, 751.

7

Slope Stability

by
Khaled Sobhan
Florida Atlantic University, Boca Raton, Florida

7.1 Introduction

Rational analysis of natural and man-made slopes for the assessment of stability and the forensic geotechnical interpretation of landslides dates back more than 75 years. Modern-day geotechnical engineers can justifiably claim significant advancement in the analysis of the deformation and stability of slopes, especially during the last 40 years with the advent of powerful computing tools and increased use of the finite element methods. Yet, our ability to predict the forces of nature that govern the instabilities of slopes and their occasional catastrophic failure remain, at best, inadequate and uncertain. The reason is quite simple. We are attempting to define phenomena in nature with mathematical formulations, but due to variability in site conditions and soil properties, this process rarely will be accurate or accomplished with a high degree of precision. This was eloquently stated, though with some degree of frustration, by Ralph Peck after a catastrophic 1965 landslide near Seattle, Washington (Peck 1967):

We simply do not understand the reasons for the rapid development of the slide in what was expected to be a period of grace during which remedial measures could be carried out. Hence we have lost much of our confidence in our ability to predict the behavior of a natural hillside or in the results of our remedial measures. On this project, it is evident that nature was able to outwit us, and we fear that she can and will do so on similar occasions in the future. This, I submit, is the present state of the art.

Stability analysis of existing natural slopes and the safe design of man-made slopes, which include cuts, excavations, and embankments, historically have driven the field of soil mechanics through some important advancements. These include better understanding of the short-term and long-term stability of slopes and the significance of total and effective stress analysis of slopes, as well as the use of drained and undrained shear strength in field applications. In a recent state-of-the-art paper, Duncan (1996) stated: "The first prerequisite for performing effective slope stability analysis is to formulate the right problem, and to formulate it correctly." Indeed, reliable analysis of slopes is strongly dependent on understanding and identification of field drainage conditions, choice of correct shear strength (drained or undrained), and employing the appropriate analysis technique (total or effective stress analysis). By definition, when a slope fails or there is movement in a slope, it is called a landslide. Investigations of numerous failed slopes and landslides across the world have enriched our experience to further refine our analysis and computational techniques, increased our confidence in designing safer slopes, and improved our understanding of appropriate remedial measures or slope stabilization methods. The objectives of this chapter are to discuss some of the most critical soil mechanics concepts relevant to stability of slopes and to develop a practical reference guide which contains various slope stability analysis methods and slope stabilization techniques that can be readily used by students, educators, and engineering professionals.

7.2 Goals of Slope Stability Analysis

It is important to establish the primary goals of slope stability analysis. Engineers generally are concerned with the safe and economical design of embankments, landfills, excavations, cuts, and earthen dams. These slopes constitute man-made construction which results in disturbance of the natural site conditions and probably natural stability, by either removal of stresses or application of new loads. In addition, many engineering projects may interfere with natural slopes, such as hillsides, by adding a surcharge load such as a building on top of the slope. Moreover, the stability of a natural slope may be crucial for the safety of people and for structures built near the bottom of the slope. Although this chapter primarily is concerned with the engineering stability analysis of slopes, it is crucial to point out that many interrelated factors, including environmental, geological, economical, societal, and in numerous cases legal parameters, may be associated with the movement and/or failure of slopes. An excellent review of some of these factors was presented by Varnes (1978). The goals of slope stability analysis can be summarized as follows:

1. Assessment of the structural stability of natural and man-made slopes based on geotechnical investigations, historical data, and a sound mechanistic approach complemented by empirical observations and experience

2. Analysis of landslides to understand failure mechanisms, verify the accuracy of stability analysis techniques, and assess the potential for future landslides
3. Development of strategies for safely redesigning failed slopes and planning preventive remedial measures
4. Evaluation of the effects of seismic loading and environmental conditions on slopes and embankments

7.3 Slope Movements and Landslides

Slopes consist of geologic materials such as rocks, cobbles, boulders, soils, artificial fills, and combinations of these materials. In general, the visible movement of the slope-forming materials in the downward and outward directions, including their movement within the slope, is termed a *landslide* (although all movement does not involve slides), during which shear failure may occur along a specific surface or simultaneously along a combination of surfaces, called *slip surfaces*. Some of the main components of a landslide are shown in Figure 7.1 and defined in Table 7.1.

7.3.1 Types of Slope Movements

Slope movements can be divided into the following six groups (Varnes 1978; Cornforth 2005): falls, topples, slides, spreads, flows, and complex landslides. These groups are briefly described below and shown schematically in Figure 7.2.

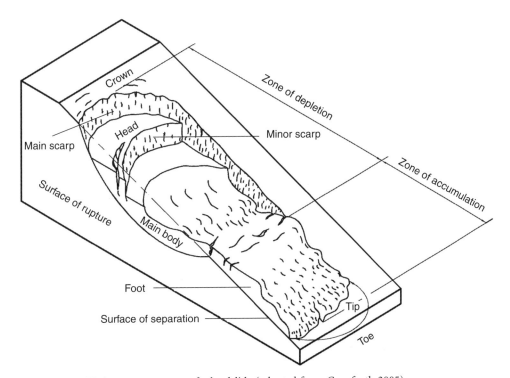

FIGURE 7.1 Various components of a landslide (adapted from Cornforth 2005).

TABLE 7.1 Definition of Landslide Components

Component	Definition
Crown	Practically undisplaced zone above the main scarp
Main scarp	Steep surface on undisturbed ground at upper edge of the landslide caused by movement of displaced material
Head	Upper parts of the landslide between displaced material and main scarp
Minor scarp	A steep surface on the displaced material produced by differential movement
Main body	Part of the displaced material that overlies the surface of rupture
Foot	The portion of the landslide that has moved beyond the toe of the surface of rupture
Tip	The point on the toe farthest from the top of the landslide
Toe	The lower curved margin of the displaced material
Surface of rupture	Surface that forms the lower boundary of the displaced material below the original ground surface
Surface of separation	Part of the original ground surface now overlain by the foot of the landslide
Zone of depletion	Area of landslide within which displaced material lies below the original ground surface
Zone of accumulation	Area of landslide within which displaced material lies above the original ground surface

After Varnes (1978).

Falls consist of a detached mass of any size that initially belonged to a slope undergoing rapid free-fall due to gravity. They may be accompanied by leaping, bounding, or rolling (Figure 7.2a).

Topples are created by the forward rotation of a unit or units about a pivot point under the action of gravity or forces exerted by adjacent units. They may be called tilting without collapse (Figure 7.2b).

Slides are shear strains and displacements along one or several surfaces. Movement may be progressive, originating from a local shear failure and ultimately becoming a defined surface of rupture (Figure 7.2c). Slides can be divided into rotational and translational slides. *Rotational slides* are slightly deformed slumps along a surface of rupture which is curved concavely upward. Slump movements occur along these internal slip surfaces. Slumps in combination with other movements constitute the majority of slope problems encountered in the engineering profession (Varnes 1978). Some of the common types of slumps are shown in Figure 7.3. *Translational slides* involve outward and downward movement of mass on a relatively planar surface with minor rotation. A system that moves as a single unit is sometimes called a block slide.

Spreads are lateral spreading or extension due to shear failure or tensile fracture along nearly horizontal soil layers (Figure 7.2d).

Flows resemble a viscous fluid and are created by the distribution of velocities and displacements within the moving mass. They have short-lived and practically invisible slip surfaces (Figure 7.2e).

Complex landslides occur when a slope undergoes a combination of multiple types of movements within its various parts or at different times during its development.

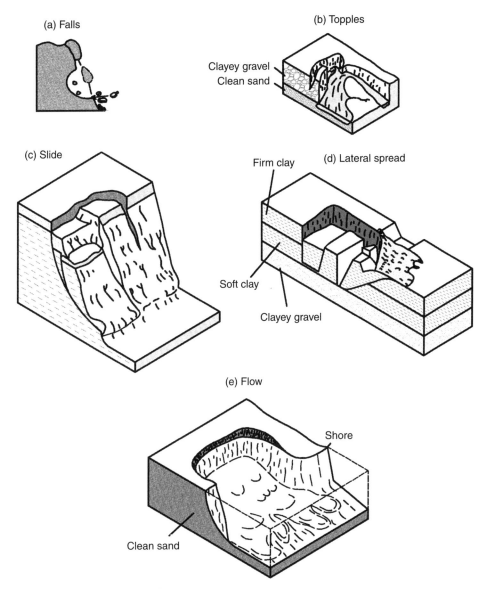

FIGURE 7.2 Various types of slope movement.

7.3.2 Factors Contributing to Slope Movement

Slope movement often is a complex process that involves a continuous series of events, from cause to effect. It usually is very difficult to identify a single definitive cause that initiated a particular slope movement. Most frequently, a combination of geologic, topographic, climatic, human, and other factors contribute simultaneously to the triggering of a movement. All sliding type of slope movements (failures) generally are associated with an increase in shear stresses and/or a decrease in the shear strength of the slope material. Since the focus of this chapter is sliding-type failures, it is important to identify the principal factors which contribute to (1) an increase in shear stresses and (2) a reduction in shear strength within a slope. These factors are summarized in Tables 7.2 and 7.3.

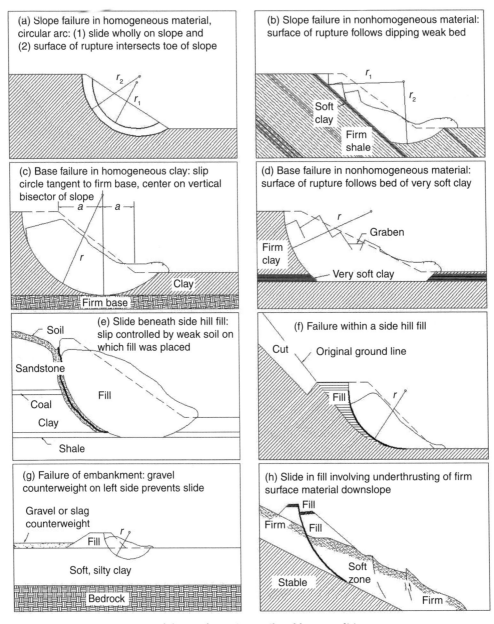

FIGURE 7.3 Common types of slumps for various soil and base conditions.

7.4 Soil Mechanics Principles for Slope Analysis

7.4.1 Introduction

Most sliding type of slope failures and their stability analyses involve determination of the shear stresses required for static equilibrium of the potential sliding mass and the available shear strength, which provide a factor of safety. Definition, significance, and calculation of the factor of safety (denoted by F) are a major focus of this chapter and are discussed later. The

TABLE 7.2 Factors Contributing to Increased Shear Stresses in Slopes

Factor	Description of Mechanism
Removal of lateral support	*Erosion*: (1) By streams, (2) by glaciers, (3) by waves and marine currents, and (4) by weathering, wetting and drying, and frost action
	Slope movement: (1) Previous rockfall and slides, (2) subsidence, and (3) large-scale faulting that creates new slopes
	Human agents: (1) Cuts, quarries, pits, and canals; (2) removal of retaining walls and sheet piling; and (3) creation of lakes and reservoirs and alteration of their levels
Surcharge loading	*Natural causes*: (1) Weight of rain, snow, and water from springs; (2) materials accumulation due to past landslides; (3) avalanches and debris flow from collapse of accumulated volcanic materials; (4) vegetation; and (5) seepage pressure of percolating water
	Human agents: (1) Construction of fill, (2) stockpiles of rocks and waste piles, (3) weight of buildings/structures, and (4) water leakage from sewers, pipelines, and reservoirs
Transitory earth stresses	*Natural causes*: (1) Earthquakes, (2) thunder, and (3) adjacent slope failure
	Human agents: (1) Blasting, (2) machinery, and (3) traffic
Removal of underlying support	*Natural causes*: (1) Undercutting of banks by rivers or waves; (2) subareal weathering, wetting and drying, and frost action; (3) subterranean erosion; and (4) failure in underlying materials
	Human agents: Mining, excavation, and other similar actions
Lateral pressure	*Natural causes*: (1) Accumulated water in cracks, (2) freezing of water in cracks, (3) swelling of clays, and (4) mobilization of residual stresses
Volcanic processes	*Natural causes*: Stress fields in crater walls are modified due to expansion or compression of magma chambers, changes in lava levels, and tremors

After Highway Research Board (1958).

choice of drained or undrained shear strength in the analysis of slopes requires an understanding of the field drainage conditions and is intimately connected with the concept of short-term (end-of-construction) and long-term stability analysis. These fundamental soil mechanics principles relevant to slope analysis are described in this section.

7.4.2 Concept of Total and Effective Stresses

Consider a point M located at a depth h below the ground surface, as shown in Figure 7.4. The soil saturated unit weight is γ, and the groundwater table is located on the surface.

Total stress σ_t at point M is equal to the sum of two components: (1) the effective stress σ', which is the sum of all interparticle contact forces divided by the total contact area, and (2) the pore water pressure u at point M. Therefore:

$$\sigma_t = \sigma' + u \qquad (7.1)$$

As shown in Figure 7.4, the total stress at point M is γh, and the pore water pressure is $\gamma_w h$, where γ_w is the unit weight of water. This implies that

TABLE 7.3 Factors Contributing to a Reduction in Shear Strength of Slopes

Factor	Description of Mechanism
Inherent material characteristics	*Composition and texture*: Organic materials, sedimentary clays and shales, certain decomposed rocks, sensitive clays, loose sands, etc. are weak in terms of shear strength
	Structure and slope geometry: (1) Faults, joints, bedding planes, and other discontinuities; (2) massive beds over highly plastic zones; (3) alternating beds of permeable and highly impermeable materials; and (4) certain slope orientations
Weathering and physiochemical reactions	(1) Softening of fissured clays, (2) disintegration of rocks due to thermal or frost action, (2) decrease of clay cohesion due to water absorption and subsequent swelling, (3) saturation-induced increase in compressibility, (4) changes in clay physical properties due to exchangeable ions, (5) drying and shrinkage cracks in clays with subsequent water infiltration, and (6) loss of cementation due to solution
Saturation	Intergranular effective stresses reduced due to saturation caused by (1) natural phenomena such as rain and snowmelt or (2) human action such as diversion of streams, blockage of drainage, irrigation and ponding, and deforestation
Changes in structure	Fissuring and fracturing of soils and rocks, progressive creep, and disturbance in saturated loose sands and sensitive clays

After Highway Research Board (1958).

$$\sigma' = h(\gamma - \gamma_w) \tag{7.2}$$

In other words, effective stress is equal to total stress minus pore water pressure. Effective stress is a calculated quantity and cannot be measured. The concept of effective stress was introduced by Karl Terzaghi, who stated: "All the measurable effects of a change of stress, such as compression, distortion, and a change of shearing resistance, are *exclusively* due to changes in the effective stresses σ'_1, σ'_2, and σ'_3. Hence, every investigation of the stability of a saturated body of soil requires the knowledge of both the total and the neutral stresses" (Terzaghi 1936b). The shear strength of all types of soils under any condition of drainage (drained or undrained) is dependent on the effective stress.

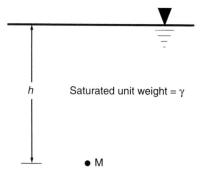

FIGURE 7.4 Total and effective stresses in saturated soil.

Let us now evaluate the total and effective stress conditions within a soil mass due to application of a fill load, similar to the construction of an embankment slope. This scenario is illustrated in Figure 7.5. Let us suppose

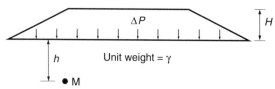

FIGURE 7.5 Embankment with height H.

that the fill load applies a uniform pressure of ΔP on the surface, which is transmitted to point M. Since the ground is saturated, right after construction there is an instantaneous rise in pore water pressure by ΔP, which is called the excess pore water pressure and is denoted by u_e. Accordingly, at time $T = 0$,

Total stress: $$\sigma_{t(T=0)} = \gamma h + \Delta P \tag{7.3}$$

Pore pressure: $$u_{(T=0)} = \gamma_w h + u_e = \gamma_w h + \Delta P \tag{7.4}$$

Effective stress: $$\sigma'_{(T=0)} = \sigma_{t(T=0)} - u_{(T=0)} = h(\gamma - \gamma_w) = \sigma' \tag{7.5}$$

Therefore, there is no change in effective stress right after the application of the external load. This is called the undrained condition and often is referred to as the end-of-construction condition in slope stability analysis (discussed later).

A long time after the application of the fill load, the excess pore water pressure is dissipated ($u_e = 0$), and the external pressure ΔP is supported by intergranular contact stresses, with a net increase in effective stress. Accordingly, at time $T = \alpha$,

Total stress: $$\sigma_{t(T=\alpha)} = \gamma h + \Delta P \tag{7.6}$$

Pore pressure: $$u_{(T=\alpha)} = \gamma_w h + u_e = \gamma_w h \tag{7.7}$$

Effective stress: $$\sigma'_{(T=\alpha)} = \sigma_{t(T=\alpha)} - u_{(T=\alpha)} = h(\gamma - \gamma_w) + \Delta P \tag{7.8}$$

$$= \sigma' + \Delta P$$

This is called the drained or long-term condition for slope stability analysis. The concept of the drained and undrained conditions and their significance in slope stability analyses are discussed in the following sections.

7.4.3 Drained and Undrained Conditions

7.4.3.1 Definition

In rudimentary terms, the drained and undrained conditions are related to the ability or inability of water to drain from or into the soil when equilibrium stress conditions are altered. More specifically, changes in stress due to loading or unloading tend to change the pore water pressure, and two drainage conditions may arise:

1. *Drained condition*—During the length of time soil is undergoing some changes in stresses, water is able to freely move in or out of the pores such that there is essentially no change in the pore water pressure. How quickly drainage can occur depends on the soil permeability characteristics.
2. *Undrained condition*—There is no flow of water in or out of the pores during the length of time the soil is subjected to some changes in stresses. Since water is incompressible, the changes in soil stresses will cause the pore water pressure to undergo appropriate changes.

These principles can be explained with respect to the example shown in Figure 7.4. During rapid construction of the fill layer, the foundation soil (assuming it is a clay layer with low permeability) will not have sufficient time to drain the pore water freely into the surrounding medium in response to the stress changes, which will tend to decrease the void volumes. As a result, the pore pressure will rise and the foundation will represent an undrained soil.

If the fill load is left in place for a long time after the completion of construction, the state of the foundation soil will transform from an undrained to a drained condition. This is because of the fact that sufficient time will be available during application of the constant fill load for the water to flow out of the soil mass.

7.4.3.2 Identification of Drained or Undrained Condition

It follows from the above discussion that the length of time (after changes in loading) plays the most important role in defining the drained or undrained condition in a soil mass. Inherent soil characteristics will dictate the length of time required for a soil mass to transform from an undrained condition to a drained condition. Terzaghi's theory of consolidation provides a sound approach for estimating the degree of drainage during construction or loading in terms of the dimensionless time factor T:

$$T = C_v \frac{t}{H_{dr}^2} \tag{7.9}$$

where C_v is the coefficient of consolidation (ft^2/year or m^2/year), t is the loading or construction time (years), and H_{dr} is the length of the drainage path or the maximum distance the water particles have to travel to flow out of the soil mass. The average degree of consolidation U (compression at any time t divided by the compression at the end of consolidation) is considered to be in excess of 99% at $T = 3.0$ (Lambe and Whitman 1979). Accordingly, the soil can be regarded as drained if T exceeds 3.0 and undrained if T is less than 0.01, while an intermediate value of T ($0.01 < T < 3.0$) suggests that both drained and undrained conditions should be considered in the analysis (Duncan 1996). A practical measure of the real time required for the degree of consolidation to reach 99% is denoted by t_{99} and can be estimated from Equation 7.9 by setting the time factor $T = 4$ as follows (Duncan and Wright 2005):

$$t_{99} = 4 \frac{H_{dr}^2}{C_v} \tag{7.10}$$

Equation 7.10 provides a logical quantitative basis for calculating the length of real time required (after loading is initiated) for the soil mass to transform from an undrained to a drained condition. If critical conditions are expected before this time is reached, an undrained analysis should be performed.

Typical values of C_v are presented in Table 7.4, which shows that the normal range is between 0.4×10^{-4} and 10×10^{-4} cm^2/s, while in some soils it can reach as high as 60×10^{-4} cm^2/s. Theoretical values of t_{99} for different values of C_v and practical ranges for drainage path lengths (H_{dr}) are shown in Figure 7.6. It has been found that sands and gravels reach drainage equilibrium fairly quickly compared to clays, which may require tens or even hundreds of years. Therefore, it is quite logical to perform drained analysis in sands and undrained analysis in clays for practical applications.

TABLE 7.4 Typical Values of C_v for Various Soils

Soil Type	C_v ($\times 10^{-4}$ cm^2/s)
Boston blue clay (CL)	40 ± 20
Organic silt (OH)	2–10
Glacial lake clays (CL)	6.5–10.7
Chicago silty clay (CL)	8.5
Swedish medium-sensitive clays (CL-CH)	0.4–3.0
San Francisco Bay mud (CL)	2–4
Mexico City clay (MH)	0.9–1.5

After Holtz and Kovacs (1981).

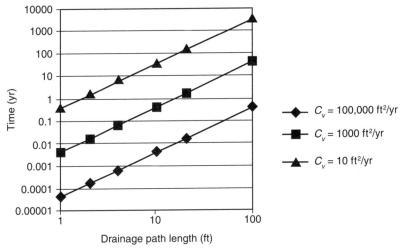

FIGURE 7.6 Theoretical values of t_{99}.

7.4.3.3 Shear Strength

A comprehensive discussion of shear strength was presented in Chapter 1. An important point to emphasize here is the fact that shear strength of a soil is governed by effective stress whether failure occurs under undrained or drained conditions. However, depending on the time-dependent drainage conditions that exist in the field, either a total stress analysis or an effective stress analysis can be performed for slope stability calculations (discussed later). The shear strength is expressed by the Mohr-Coulomb failure criterion as

$$\tau_{ff} = c' + \sigma' \tan \phi' \tag{7.11}$$

where τ_{ff} is the shear stress on the failure plane at failure, σ' is the effective normal stress on the failure plane at failure, c' is the effective stress cohesion, and ϕ' is the effective stress angle of internal friction. The interrelationships between the failure envelopes in triaxial tests under unconsolidated undrained, consolidated undrained, and consolidated drained conditions are compared in Figure 7.7 for a silty clay soil. The close agreement between the effective stress failure envelopes for both the consolidated drained and consolidated undrained cases is noteworthy. In general, the relative comparisons between effective and total stress analysis shown in Figure 7.7 should be of practical interests to engineers.

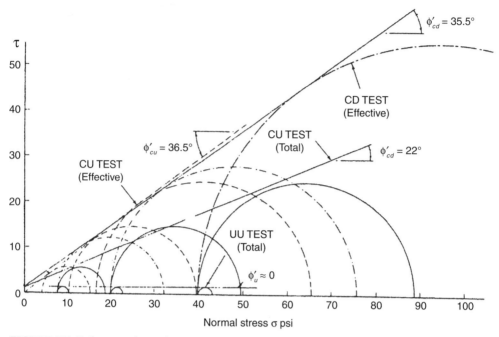

FIGURE 7.7 Failure envelopes for a silty clay soil (CD = consolidated drained, CU = consolidated undrained, and UU = unconsolidated undrained) in triaxial tests (adapted from Bishop and Bjerrum 1960).

7.4.3.4 Rotation of Principal Stress Direction

Rotation of the principal stress direction along the potential slip plane within a slope affects the stability analysis, since it can reduce the shear strength of certain clays. The nature of this rotation is illustrated schematically in Figure 7.8. The natural state of stress in the ground is shown at point A, where the major principal effective stress σ_1' and minor principal effective stress σ_3' ($= k_0 \, \sigma_1'$, where k_0 is the coefficient of lateral earth pressure at rest) are aligned in the vertical and horizontal directions, respectively. After the slope is constructed by excavation, the stress conditions as well as the orientation of the principal planes change along an assumed slip surface ABC. Near the top of the slope at A, the principal stresses are aligned in the same direction as the natural ground. At the toe

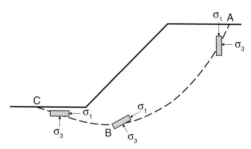

FIGURE 7.8 Rotation of principal stresses on the potential slip plane.

of the slope at point C, the principal stresses have rotated through a 90° angle, with the major principal stress acting in the horizontal direction. At an interior point B, the stresses have rotated through an intermediate angle. Due to inherent material anisotropy, the shear strengths at A, B, and C will be different, and therefore, the factor of safety calculations according to most limit equilibrium methods, which assume uniform shear strength along the slip plane, will be in error.

7.5 Essential Concepts for Slope Analysis

7.5.1 Introduction

Engineers deal with the stability conditions of primarily two types of slopes: (1) natural slopes such as hillsides and (2) man-made or engineered slopes such as fills or embankments and excavations or cuts. Various causes and factors, both natural and man-made, contribute to critical stress conditions within a slope such that failure or movement is initiated. Stability of sliding type of movement almost always is expressed in terms of a factor of safety F, which is calculated on the basis of shear strength, using the concepts of limit equilibrium analysis, a practice which has been employed for at least three-quarters of a century across the world. Principles of the limit equilibrium method, factor of safety, and various slope analysis strategies under different critical conditions are discussed in this section.

7.5.2 Limit Equilibrium Analysis

Principles of limit equilibrium methods are employed both during the design phase of a slope and during the forensic back analysis of a slope that has failed. Designing a slope requires computation of a factor of safety in terms of shear stresses and available strength. Conversely, when a slope has failed, it is implicitly assumed that the factor of safety is unity, and limit equilibrium analysis is performed to estimate the average shear strength that existed along the failure plane at failure. Numerous techniques for calculating the factor of safety have been developed based on limit equilibrium analysis. However, all limit equilibrium analysis techniques consist of the following general steps (Morgenstern and Sangrey 1978):

1. The shape and mechanism of a sliding surface are hypothesized, and in most cases, a circular slip surface is assumed, as shown by surface ABC in Figure 7.9.
2. The shearing resistance along the slip plane required to maintain the static equilibrium of the sliding mass is determined from the principles of statics. This just-stable condition of the slope is called a stage of limiting equilibrium.

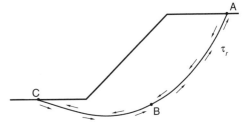

FIGURE 7.9 A circular slip surface in limit equilibrium analysis.

3. The available shear strength along the slip plane is divided by the required limiting shearing resistance calculated in step 2 to determine a factor of safety F.
4. The slip mechanism with the minimum F, called the critical slip surface, is obtained by iterative procedures.

7.5.3 Factor of Safety

It is evident from the above discussion that the limit equilibrium procedure of slope stability analysis involves calculation of a factor of safety F, defined as

$$F = \frac{s}{\tau_r} \tag{7.12}$$

where s is the available shear strength on the failure plane and τ_r is the shear stress on the failure plane required to maintain a just-stable equilibrium condition. It is implicitly assumed that F will be constant at every point on the failure surface ABC. This also implies that a constant proportion of the shear strength is mobilized at every point on the failure surface. Expressing the shear strength s in terms of effective stresses from Equation 7.11, the required shear stress τ_r is given by

$$\tau_r = \frac{c'}{F} + \frac{\sigma' \tan \phi'}{F} = c_d' + \sigma' \tan \phi_d' \tag{7.13}$$

where $c_d' = c'/F$ and $\tan \phi_d' = \tan \phi'/F$ are called the developed shear strength parameters (in terms of effective stress), which are actually mobilized at every point on the slip surface to resist the sliding of the slope mass. In other words, the factor of safety refers to the factor by which available shear strength parameters must be reduced to achieve the limiting equilibrium condition of the slope.

For short-term or end-of-construction stability analysis of slopes constructed over normally consolidated silts and clays, it is appropriate and convenient to perform a total stress analysis using the so-called $\phi_u = 0$ concept (Skempton 1948). Shear strength in this case is expressed in terms of the undrained shear strength c_u, which is one-half of the compressive strength in an unconfined or undrained triaxial test. In terms of total stresses, the shear strength s is equal to c_u, and Equation 7.13 takes the following form:

$$\tau_r = \frac{c_u}{F} \tag{7.14}$$

It follows from basic mechanics that the static equilibrium condition refers to the equilibrium of forces in both the horizontal and vertical directions and the equilibrium of moments about any point. Most limit equilibrium methods satisfy only some of these conditions, and very few satisfy all of the required conditions for static equilibrium. The reason is simple. In most of the available methods of slope analysis, the number of unknowns exceeds the number of equilibrium equations, and therefore, simplifying assumptions need to be made regarding the magnitude and location of unknown forces to satisfy static equilibrium. Since assumptions vary from one procedure to another, the mathematical formulations also vary, resulting in different values of the calculated factor of safety for the same slope. Some of these details are discussed later in this chapter.

7.5.4 Critical Stability Conditions in Slopes

Design and stability checks of engineered and natural slopes often depend on correctly identifying the drainage conditions and, based on that, choosing an appropriate analysis strategy. Engineers have to select between total and effective stress analysis, choose either drained or undrained shear strength, and perform analysis for either long-term or short-term stability conditions. Depending on the complexity of the site conditions and the type of structure being constructed, any particular project may include a combination of some of these schemes. For example, let us consider the case of a granular embankment constructed over a clay foundation.

Due to its high permeability, sand is expected to be in a drained condition even during the construction process and certainly at the end of construction and thereafter. Accordingly, both the short-term and long-term analysis of the embankment should be carried out under drained conditions using the effective stress approach (using drained shear strength). On the other hand, due to the relatively low permeability of the clay foundation, it will be in an undrained condition both during construction and at the end of construction. Therefore, in the short term, the analysis of the clay foundation should be carried out under undrained conditions using a total stress approach (using undrained shear strength). However, any long-term analysis of the clay should be carried out in terms of drained or effective stress conditions, since all excess pore water pressure is expected to be dissipated after sufficient time has elapsed. It follows from the above example that it will be appropriate (in a short-term analysis) to use an effective stress analysis approach for the sand embankment, while simultaneously performing an undrained total stress analysis on the clay foundation.

So, how do we determine which condition—the end of construction (short term) or the long term—would be most critical for a slope? The answer depends on the permeability of the soil (which governs the dissipation of excess pore water pressure) and on the type of construction (a slope made by embankment fill or excavation). The following examples illustrate these cases (after Bishop and Bjerrum 1960).

7.5.4.1 End-of-Construction Stability

Let us consider various phases during the life of an embankment slope built on a clay foundation. The excess pore water pressure, shear stress, and factor of safety all will change with time, as shown in Figure 7.10, and will reach equilibrium under the applied stress after sufficient time has elapsed.

Analysis of Figure 7.10 reveals that the most critical condition is short-term stability at the end of construction, when the factor of safety reaches its minimum value. This can be explained as follows. During the construction phase, the embankment load will increase the shear stress τ along a potential slip plane in the clay foundation. The excess pore water pressure Δu due to the applied stress at an element P can be calculated from Equation 7.15 (Skempton 1954) as

$$\Delta u = B[\Delta\sigma_3 + A(\Delta\sigma_1 - \Delta\sigma_3)] \tag{7.15}$$

where $\Delta\sigma_1$ = change in major principal stress, $\Delta\sigma_3$ = change in minor principal stress, A = an empirical parameter related to excess pore water pressure developed due to changes in shear stress, and B = an empirical coefficient related to soil compressibility and the degree of saturation. Since $B = 1$ for a saturated soil and A is positive for a normally consolidated or lightly overconsolidated clay, there will be a positive excess pore water pressure, which will reach its maximum value at the end of construction. Since shear strength (which depends on effective stress) will concurrently decrease with the rise in pore water pressure, the factor of safety will reach its minimum value at the end of construction.

The excess pore water pressure will gradually dissipate with time and eventually reach an equilibrium condition that corresponds to the groundwater level. Since the embankment height and, therefore, the shear stress remain constant, the increase in shear strength brought about by the decrease in excess pore water pressure will gradually improve the factor of safety as time progresses. Therefore, in this case, it is only necessary to perform short-term stability

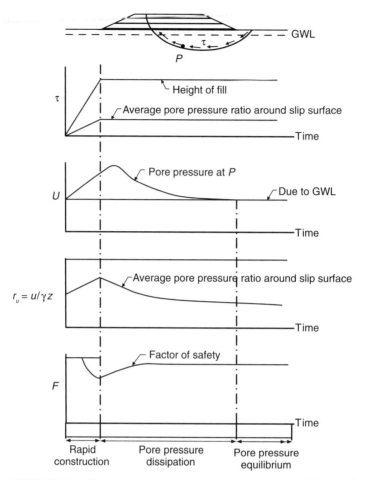

FIGURE 7.10 Changes in pore water pressure, shear stress, and factor of safety during various phases of the life of an embankment constructed over a clay foundation (adapted from Bishop and Bjerrum 1960).

analysis because the slope is at its most critical or vulnerable condition at the end of construction.

7.5.4.2 Long-Term Stability

When a slope is constructed by excavation in a clay soil, the mechanics of critical stability analysis are reversed. In this case, the pore water pressure in the clay decreases due to the removal of the excavated material. This is explained by assuming $B = 1$ and rearranging Equation 7.15 in the following form:

$$\Delta u = [\Delta\sigma_1 + \Delta\sigma_3)/2] + (A - 1/2)(\Delta\sigma_1 - \Delta\sigma_3) \qquad (7.16)$$

The first term in Equation 7.16 is the mean principal stress, which will be reduced due to excavation. This in turn will cause a decrease in the pore water pressure Δu. The second term, which is the shear stress, will also reduce the pore water pressure unless A is greater than ½. It should be noted that the value of parameter A is generally less than ½ for lightly to heavily

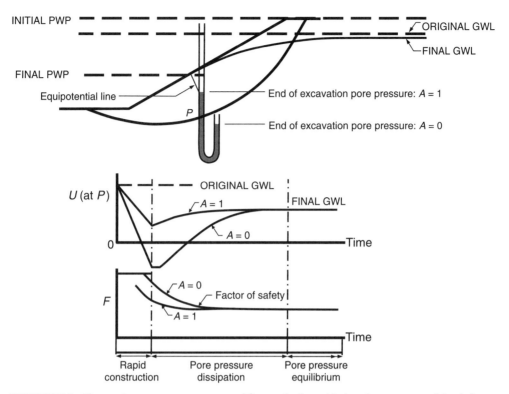

FIGURE 7.11 Changes in pore water pressure and factor of safety with time for an excavated (cut) slope in clay (adapted from Bishop and Bjerrum 1960).

overconsolidated clays and typically more than 1 for normally consolidated clays (Lambe and Whitman 1979). In any case, there will be a net decrease in pore water pressure, or in other words, there will be an increase in negative excess pore water pressure.

Figure 7.11 shows the variation in pore water pressure and the factor of safety with time during and after construction of an excavation slope in soils where $A = 0$ and $A = 1$. During excavation, the average shear stress along the potential failure surface increases (the factor of safety decreases). After completion of the excavation, the shear stress remains constant. However, as shown in Figure 7.11, the negative excess pore water pressure will dissipate with time and eventually will reach an equilibrium condition with the groundwater level. This implies an increase in positive pore water pressure and a simultaneous decrease in effective stress, shear strength, and the factor of safety. If the excavation geometry and the applied stresses do not change any further, the factor of safety will attain its minimum value after the pore pressure reaches equilibrium state with the groundwater level. Therefore, in the case of slopes constructed by excavation in saturated clay, only long-term stability conditions need to be evaluated.

7.5.5 Recommended Factor of Safety

Table 7.5 provides the minimum required factor of safety for slopes of earthen and rock-fill dams and embankments. Various analysis and drainage conditions are incorporated in the recommended values. The values are based on historical past performance data and experi-

TABLE 7.5   Minimum Design Factor of Safety (Corps of Engineers 2003)

Analysis or Drainage Condition	Minimum Required Factor of Safety	Notes
End-of-construction stability	1.3	Higher values may be needed if embankment is constructed on soft soils and is greater than 50 ft in height
Long-term stability	1.5	Steady seepage
Rapid drawdown	1.1^a–1.3^b	[a] When drawdown occurs from maximum surcharge pool
		[b] When drawdown occurs from maximum storage pool

ence. Although some form of risk or reliability analysis, along with knowledge of the probable economic consequence in case of failure, may impact the values listed in Table 7.5, slope stability analysis in practice continues to be dictated largely by accumulated past experience. Therefore, Table 7.5 provides initial guidelines which can be modified according to project-specific criteria.

7.5.6 Total or Effective Stress: Theory vs. Practice

Since failure is governed by effective stress, in principle, analyses at all times and for all conditions (drained or undrained) should be carried out using the effective stress analysis approach. In theory, it is entirely possible to do so for end-of-construction, long-term stability, and any intermediate analysis. From practical considerations, it is more convenient to use a total stress analysis approach under undrained conditions for analysis of end-of-construction stability. This is because effective stress analysis involves estimation or actual measurements of the field excess pore water pressure, a task which is often difficult or yields inaccurate results. Table 7.6 can be used as a practical guide for the selection of analysis strategies under various commonly encountered conditions in the field.

TABLE 7.6   Guidelines for the Choice of Total or Effective Stress Stability Analysis

Stability Condition	Drainage Condition	Analysis Strategy
Short term or end of construction	Undrained, saturated	Total stress analysis: $\phi_u = 0$, $s = c_u$
Short term or end of construction	Undrained, partially saturated	Total stress analysis or effective stress analysis: c_u, ϕ_u from unconsolidated undrained tests or c, ϕ with estimated pore pressures
Long term	Drained, saturated	Effective stress analysis: c, ϕ from consolidated drained tests, pore pressure calculated from equilibrium groundwater level

After Lambe and Whitman (1979).

7.6 Analysis of Slope Stability

7.6.1 Introduction

Systematic study of slope failures and stability analysis dates back at least 75 years, with many crucial developments in concepts, methods, and procedures taking place during a 30-year span from the mid-1930s to the late 1960s (Terzaghi 1936a; Fellenius 1936; Taylor 1937; Janbu 1954a, 1954b; Bishop 1955; Skempton 1948, 1954, 1964; Bishop and Bjerrum 1960; Morgenstern and Price 1965, 1967; Peck 1967; Spencer 1967; Whitman and Bailey 1967). These early pioneering works formed the basis for further development and refinement in slope stability analysis procedures and practice that continued during the following 35 years up to the present day, especially in the area of computational studies, including three-dimensional and finite element analysis aided by the advent of powerful digital computers. Excellent summaries of these developments are available in the recent literature (Duncan 1996; Abramson et al. 2002; Duncan and Wright 2005).

It is beyond the scope of this chapter to provide extensive coverage of the numerous techniques and procedures that are available for slope stability analysis. The objective here is to describe several well-known procedures which are most widely used in practice. Numerical examples are provided wherever necessary to illustrate a procedure. Relative comparisons are made between various methods by outlining their respective features and comparing the computed factor of safety for each technique.

Only sliding type of slope movements or failures are considered in this chapter. Slopes are analyzed under two major categories:

1. *Single free body or block procedures* — The slope mass is analyzed as a single body or multiple blocks with a planar or circular slip surface.
2. *Method of slices* — The slope mass is divided into discrete vertical slices or elements, and the equilibrium condition of each slice is analyzed. Both circular (most common) and noncircular slip surfaces can be considered.

In the analysis procedures presented here, a two-dimensional cross section of the slope (plane strain condition) is used, assuming that (1) the slope extends to infinity along a direction perpendicular to the cross section and (2) the failure occurs along the entire length of the slip surface, which is also infinitely long perpendicular to the cross section of interest.

7.6.2 Single Free Body and Block Procedures

7.6.2.1 Infinite Slopes

Slopes can be considered as infinite in the case of large landslides, where the solid mass is moving approximately parallel to the ground surface or the face of the slope. A planar slip surface is assumed. The slope extends infinitely in the lateral and longitudinal directions, and the length of the slide is very long relative to the depth or height of the sliding surface. These conditions are presented schematically in Figure 7.12a, and the free body diagram of a sliding block PQRS is presented in Figure 7.12b. Since the slope is infinite, any two planes perpendicular to the slope (such as PS and QR) will have equal, opposite, and collinear forces P_{PS} and P_{QR}

acting on them and will cancel each
other. Employing the concepts of limit
equilibrium, a factor of safety can now
be computed for the infinite slope, considering interaction with the groundwater table and conducting an effective
stress analysis.

7.6.2.1.1 Effective Stress (c′-φ′) Analysis

As shown in Figure 7.12a, the slip surface is located at a depth z below the
surface of the slope. Let us suppose that
the groundwater table is located at a
height h from the slip surface, and a
steady seepage condition exists parallel
to the slope. Let the slope angle be β
and the width and length of the sliding
block PQRS be b and l, respectively.
Figure 7.12b shows the forces acting on
the free body diagram, which include
the weight of the block W, the normal
force N, the shear forces S on the sliding plane, and the pore water force U
perpendicular to the sliding surface. If
the soil unit weight is γ, then the weight
of block $W = \gamma z b$. Summing the forces
parallel and perpendicular to the slip
plane:

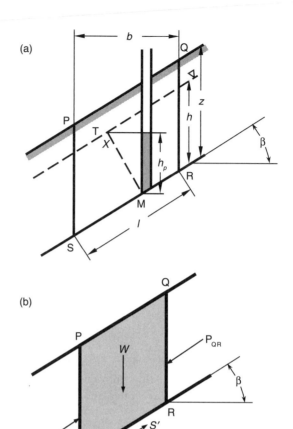

FIGURE 7.12 Analysis of an infinite slope.

$$N + U = W \cos \beta = \gamma z b \cos \beta \qquad (7.17)$$

$$S = W \sin \beta = \gamma z b \sin \beta \qquad (7.18)$$

The pore water pressure u acting along the base can be obtained by knowing the piezometric head h_p at point M located at the middle of the base RS. Since the groundwater table acts
as a flow line, the line MT normal to the water surface is an equipotential line. Therefore, u
can be expressed as:

$$u = \gamma_w h_p = \gamma_w h \cos^2 \beta \qquad (7.19)$$

Accordingly, the pore water force U is given by:

$$U = ul = \gamma_w h l \cos^2 \beta = \gamma_w h b \cos \beta \qquad (7.20)$$

It follows from Equation 7.17 that:

$$N = \gamma z b \cos\beta - \gamma_w h b \cos\beta = b \cos\beta (\gamma z - \gamma_w h) \qquad (7.21)$$

The factor of safety F can be computed using Equation 7.12:

$$F = \frac{s}{\tau_r} = \frac{l(c' + \sigma' \tan\phi')}{l\tau_r} = \frac{lc' + N \tan\phi'}{S}$$

$$= \frac{bc' + b \cos\beta (\gamma z - \gamma_w h) \tan\phi'}{\gamma z b \sin\beta} \qquad (7.22)$$

Equation 7.22 can be written in a simplified form as follows:

$$F = \frac{c' \sec\beta + (\gamma z - \gamma_w h) \tan\phi'}{\gamma z \sin\beta} \qquad (7.23)$$

Equation 7.23 represents a general case where the groundwater table is located between the slope surface and the potential slip surface in an infinite slope. Some special cases are discussed below.

Case 1: Submerged Slope. In this case, the groundwater table is at the slope surface such that $h = z$. Accordingly, F can be calculated from Equation 7.23 as

$$F = \frac{c' \sec\beta + \gamma' \tan\phi'}{\gamma \tan\beta} \qquad (7.24)$$

where $\gamma' = \gamma - \gamma_w$. For sands or normally consolidated clays, $c' = 0$. Therefore:

$$F = \frac{\gamma' \tan\phi'}{\gamma \tan\beta} \qquad (7.25)$$

Case 2: Dry Slope. In this case, the groundwater table is located below the slip surface such that $h = 0$. Therefore, F can be computed from Equation 7.23 as follows:

$$F = \frac{c' \sec\beta + \gamma z \tan\phi'}{\gamma z \tan\beta} \qquad (7.26)$$

If $c' = 0$, Equation 7.26 is further simplified to:

$$F = \frac{\tan\phi'}{\tan\beta} \qquad (7.27)$$

Since $F = 1.0$ at the limiting equilibrium condition, it follows from Equation 7.27 that the shear strength parameter on the slip surface is equal to the slope angle; that is, $\phi' = \beta$. Cornforth

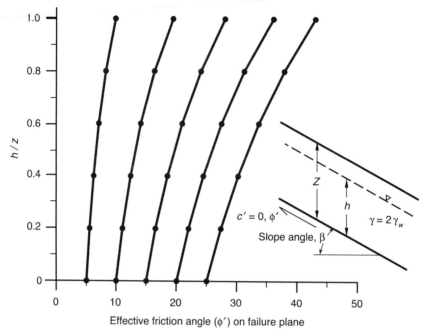

FIGURE 7.13 Prediction of effective friction angle on the slip surface of infinite slopes assuming $\gamma = \gamma_w$ (adapted from Cornforth 2005).

(2005) developed simple design charts based on Equation 7.23 for $c' = 0$ soils at the limiting equilibrium condition ($F = 1.0$). These charts, shown in Figure 7.13, allow prediction of the effective friction angle ϕ' on the slip surface for different slope angles and groundwater levels (in terms of h/z ratios) on the verge of slope failure. The chart is useful for back analysis of failed slopes.

7.6.2.1.2 Total Stress ($\phi_u = 0$) Analysis

The factor of safety can be determined using a total stress analysis from Equations 7.14 and 7.18 as follows:

$$F = \frac{c_u}{\tau_r} = \frac{l c_u}{l \tau_r} = \frac{b c_u}{S \cos \beta} = \frac{b c_u}{\gamma z b \sin \beta \cos \beta} \tag{7.28}$$

Therefore:

$$F = \frac{c_u}{\gamma z \sin \beta \cos \beta} \tag{7.29}$$

7.6.2.2 Circular Slip Surface

Single free body analysis also can be carried out assuming a circular slip surface, as shown in Figure 7.14. The method is known as the Swedish circle method and employs a total stress ($\phi_u = 0$) analysis in cohesive soil (Fellenius 1922). An alternate definition of the factor of safety based on moment equilibrium is used in this approach.

As shown in Figure 7.14, the slope is assumed to fail along a circular slip surface ABC. The weight of the sliding mass W acts through its center of gravity and is responsible for the driving (overturning) moment about the center of rotation O given by

$$M_d = Wd \qquad (7.30)$$

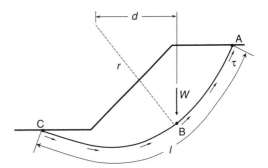

FIGURE 7.14 Single free body analysis with circular slip surface.

where d is the moment arm. The resistive moment is provided by the shear force acting along the slip plane. If $s = c_u$ is the uniform undrained shear strength acting along the slip surface of length l, then the resistive moment M_r is given by

$$M_r = c_u l r \qquad (7.31)$$

where r is the radius of the circular arc. The factor of safety F is given by:

$$F = \frac{M_r}{M_d} = \frac{c_u l r}{Wd} \qquad (7.32)$$

7.6.2.3 Analysis of Sliding Block Failures

Sliding block failures take place when the slope mass is underlain by a relatively thin weak stratum, as shown schematically in Figure 7.15. The failure surface is denoted by three planar slip surfaces AB, BC, and CD, and the landslide is divided into two wedges (ABF and CDE) and a central block (BCEF) by drawing two imaginary vertical lines BF and CE. The active wedge ABF applies a driving force to the central block, which is resisted by the shear strength along the bottom plane BC and the passive wedge CDE.

Abramson et al. (2002) proposed a simple procedure based on the Rankine earth pressure theories for calculation of the factor of safety. Although the procedure is iterative, hand

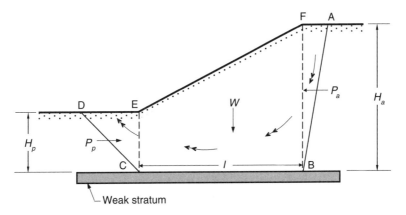

FIGURE 7.15 Analysis of sliding block failure.

calculations can be performed with reasonable accuracy. The procedure is outlined in the following paragraphs.

The factor of safety F is calculated from the ratio of the sum of horizontal resistive forces to the sum of horizontal driving forces. Referring to Figure 7.15, F is given by

$$F = \frac{P_p + sl}{P_a} \tag{7.33}$$

where P_a = Rankine active force applied by wedge ABF, P_p = Rankine passive force applied by wedge CDE, s = shear strength along the interface BC, and l = length of the base BC (area of the base per unit width).

If W is the weight of the central block and u is the pore water pressure, then the effective normal force N' at the base is given by

$$N' = W \cos \alpha - ul \tag{7.34}$$

where α is the inclination of the base BC. The shear strength sl is given by

$$sl = c'l + (W \cos \alpha - ul) \tan \phi' \tag{7.35}$$

where c' and ϕ' are the effective shear strength parameters at the base BC. Substituting Equation 7.35 into Equation 7.33, the factor of safety becomes

$$F = \frac{P_p + c'l + (W \cos \alpha - ul) \tan \phi'}{P_a} \tag{7.36}$$

The Rankine active and passive forces are calculated in a fashion similar to the forces on an earth retaining structure, assuming the vertical lines BF and CE to be the virtual "retaining walls." The active and passive forces are given by

$$P_a = \frac{1}{2} K_a \gamma H^2 - 2c'H \sqrt{K_a} \tag{7.37}$$

$$P_p = \frac{1}{2} K_p \gamma H^2 + 2c'H \sqrt{K_p} \tag{7.38}$$

where H is the height of the "retaining wall" and K_a and K_p are active and passive earth pressure coefficients, given as follows:

$$K_a = \tan^2 \left(45° - \frac{\phi'}{2} \right)$$
$$\tag{7.39}$$
$$K_p = \tan^2 \left(45° + \frac{\phi'}{2} \right)$$

7.6.3 Method of Slices

7.6.3.1 Introduction

In the method of slices, the slope mass above the assumed failure surface, which can be either circular or noncircular, is divided into a number of vertical slices, and the mechanics of limiting equilibrium are considered for each of the individual slices. Contributions from all slices are summed together to determine the total applied shear stress and the available shear strength along the failure surface. Equation 7.12 is then used to determine the factor of safety. This process of discretization has a huge advantage (over single-body procedures) when nonhomogeneous soil conditions are encountered in practice, with spatial variations in soil properties that result in unknown distribution of stresses along the slip surface. In addition, complex slope geometries, unusual seepage patterns, noncircular slip surfaces, and various boundary conditions can be analyzed using the method of slices, but usually with the aid of a powerful computer. The number of slices used in practice and considered suitable for hand calculations generally is between 8 and 12 slices, depending on the complexity of the soil profile.

7.6.3.2 Location of Critical Slip Surface

In limit equilibrium analysis, a number of trial slip surfaces (circular and noncircular) are assumed, and the calculations are repeated a sufficient number of times to determine the minimum factor of safety and the corresponding critical failure surface. Three common methods of searching for the critical failure circle are shown in Figure 7.16 and described below (after Corps of Engineers 2003):

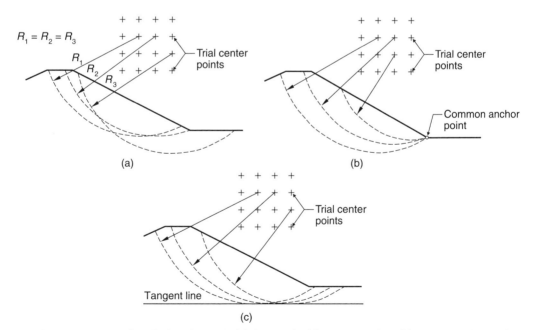

FIGURE 7.16 Procedure for locating critical failure circle: (a) constant radius, (b) common point, and (c) fixed tangent.

1. *Constant radius method*—The radius R is held constant while the location of the center is varied until the minimum factor of safety is obtained (Figure 7.16a).
2. *Common point method*—All circles are passed through a common point such as the toe of the slope, while both the centers and radii are varied until the minimum factor of safety is obtained (Figure 7.16b).
3. *Fixed tangent method*—All circles are made tangent to a fixed line, while both the centers and radii are varied until the minimum factor of safety is obtained (Figure 7.16c).

7.6.3.3 System of Forces and Equilibrium Analysis

Figure 7.17a is a schematic of a slope mass subdivided into n slices. The free body of an interior slice ABCD, with all possible forces acting on the slice, is shown in Figure 7.17b. Various components of the free body diagram are presented in Table 7.7.

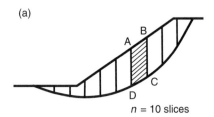

(a)

Calculation of the factor of safety using limit equilibrium concepts involves analysis of force and moment equilibrium of n number of slices. Referring to Figure 7.17b and Table 7.7, the total number of unknowns and the total number of equations involved in the equilibrium analysis of the system of slices can be determined. The types and number of unknown variables and the number of available equations are listed in Table 7.8. There are $6n - 2$ unknowns and only $4n$ equations, which makes the system statically indeterminate. If the location of the normal force N is assumed to be in the middle of the base (a common assumption), then the number of unknowns is reduced to $5n - 2$. This will require an additional $n - 2$ assumptions to transform the problem into a statically determinate system.

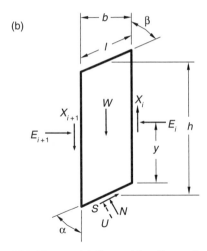

(b)

Various methods of slope stability analysis use different sets of assumptions. Some of the common methods and their assumptions are listed in Table 7.9, which shows that the methods differ not only in their assumptions but also in the manner in which

FIGURE 7.17 (a) Slope with n slices and (b) free body diagram of a typical slice ABCD.

the equilibrium conditions are satisfied. Among the methods listed, Spencer's, Morgenstern and Price's, and Sarma's are called "complete" equilibrium methods because they fully satisfy static equilibrium. All other methods listed in Table 7.9 only partially fulfill the conditions of static equilibrium. Some of these methods, which are commonly used in practice, are described in more detail in the following sections.

7.6.3.4 Ordinary Method of Slices

Description. The ordinary method of slices was developed by Fellenius (1936) and is also known as the Swedish circle method. It is considered to be one of the simplest methods suitable for hand calculations. Typical slice geometry and the free body diagram are shown in Figure 7.18. In this method, the interslice force E is neglected. This provides $n - 1$ assumptions, although only a total of $n - 2$ assumptions is needed for static equilibrium (Section 7.6.3.3).

TABLE 7.7 Parameters Associated with the Free Body of an Individual Slice

Slice Geometry		Slice Forces	
h	average height of slice	W	weight of slice
b	width of slice	N	normal force at slice base
l	length of slice base	S	shear force at slice base
α	inclination of slice base	E	interslice normal force
β	inclination of slice top	X	interslice shear force
y	location of interslice normal force	U	pore water force at slice base

TABLE 7.8 Number of Known Equations and Unknown Variables for n Slices

Known Equations (Total Number = $4n$)	
Source of Equation for Each Slice	Total Number for n Slices
Force equilibrium in horizontal direction	n
Force equilibrium in vertical direction	n
Moment equilibrium	n
Relationship between normal stress and shear strength at the slice base given by Mohr-Coulomb criterion	n

Unknown Variables (Total Number = $6n - 2$)	
Variables	Total Number for n Slices
Factor of safety (F)	1
Normal force at slice base (N)	n
Location of normal force N	n
Shear force at slice base (S)	n
Interslice normal force (E)	$n - 1$
Interslice shear force (X)	$n - 1$
Location of interslice normal force	$n - 1$

Hence, the system of slices is overdetermined, and in general it is not possible to completely satisfy statics. The factor of safety is obtained by considering the moment equilibrium about the center of the critical slip circle. Accordingly, the system of unknown variables and the available number of equations for this method are given in Table 7.10.

Mathematical Formulation. The procedure is similar to the moment equilibrium method described earlier for single free body analysis for circular slip surfaces (Section 7.6.2.2). Referring to Figure 7.18, the sum of the driving moments (M_d) about the center of the circular slip surface is given by

$$M_d = \sum_{i=1}^{n} W_i d_i \tag{7.40}$$

where W_i and d_i are the weight and moment arm of the ith slice, respectively, and n is the total number of slices. It should be noted that the slices which produce counterclockwise moments will actually reduce the overturning moments and help improve the factor of safety.

If r is the radius of the circle and α_i is the base inclination of the ith slice, then, using $d_i = r \sin \alpha_i$, it follows from Equation 7.40 that:

TABLE 7.9 Commonly Used Slope Stability Analysis Methods and Their Assumptions

Procedure	Assumptions and Characteristics	Equilibrium Conditions Satisfied		
		$\sum F_x = 0$	$\sum F_y = 0$	$\sum M = 0$
Ordinary method of slices (Fellenius 1936)	Circular slip surfaces only; interslice forces are zero	No	No	Yes
Simplified Bishop (1955) method	Circular slip surfaces only; interslice shear force is zero	No	Yes	Yes
Corps of Engineers (1970, 2003) modified Swedish method	Slip surfaces of any shape; interslice force is parallel to the ground surface or inclined at an angle equal to slope of a line connecting the crest and the toe (called average embankment slope)	Yes	Yes	No
Lowe and Karafiath's (1960) method	Slip surfaces of any shape; interslice force is inclined at an angle of $(\frac{1}{2}\alpha + \beta)$	Yes	Yes	No
Janbu's (1954a, 1954b) simplified method	Slip surfaces of any shape; interslice shear force is zero	Yes	Yes	No
Spencer's (1967, 1973) method	Slip surfaces of any shape; interslice forces are parallel with unknown inclination	Yes	Yes	Yes
Morgenstern-Price (1965) method	Slip surfaces of any shape; interslice shear forces are related to the interslice normal forces by $X = \lambda f(x)E$, where λ is an unknown scaling factor and $f(x)$ is an assumed function with prescribed values at slice boundaries	Yes	Yes	Yes
Sarma's (1973) method	Slip surfaces of any shape; interslice shear force is related to the interslice shear strength by $X = \lambda f(x)S_v$, where λ is an unknown scaling factor, $f(x)$ is an assumed function with prescribed values at the slice boundaries, and S_v is the available shear force depending on c' and ϕ' along the slice boundaries	Yes	Yes	Yes

Adapted from Abramson et al. (2002).

$$M_d = r \sum_{i=1}^{n} W_i \sin \alpha_i \qquad (7.41)$$

The resistive moment M_r is provided by the shear forces S_i generated at the bottom of the individual slices. Shear force S_i is related to the shear stress τ_r, shear strength s_i, and factor of safety F through Equation 7.12 as follows:

$$S_i = l_i \tau_{r_i} = \frac{l_i s_i}{F} \qquad (7.42)$$

The resistive moment is given by:

$$M_r = \sum_{i=1}^{n} r S_i \qquad (7.43)$$

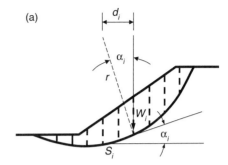

(a)

Substituting S_i from Equation 7.42 into Equation 7.43 and equating the driving and resistive moments:

$$r \sum W_i \sin \alpha_i = \frac{r}{F} \sum l_i s_i \qquad (7.44)$$

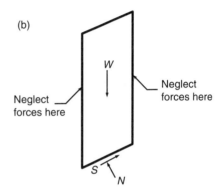

(b)

Rearranging Equation 7.44, the factor of safety is given by:

$$F = \frac{\sum l_i s_i}{\sum W_i \sin \alpha_i} \qquad (7.45)$$

The quantity $l_i s_i$ can be expressed in terms of the effective stress parameters c' and ϕ' as

$$l_i s_i = c_i' l_i + N_i' \tan \phi_i' \qquad (7.46)$$

FIGURE 7.18 (a) Typical slice geometry and (b) free body diagram for the ordinary method of slices.

where N_i' is the effective normal force at the base and is given by

$$N_i' = N_i - u_i l_i \qquad (7.47)$$

where u_i is the pore water pressure. Since interslice forces are neglected, the normal force N_i at the base of the slice can be obtained by summing up forces in the direction perpendicular to the base. Therefore, N_i is given by:

$$N_i = W_i \cos \alpha_i \qquad (7.48)$$

Combining Equations 7.46–7.48 and substituting into Equation 7.45, we obtain an expression for the factor of safety F:

TABLE 7.10 Number of Unknowns and Equations in Ordinary Method of Slices

Unknowns		Equations	
Parameter	Number	Type	Number
Factor of safety	1	Summation of moments	1
Total unknowns = 1		Total equations = 1	

$$F = \frac{\sum\limits_{i=1}^{n} [c_i' l_i + (W_i \cos \alpha_i - u_i l_i) \tan \phi_i']}{\sum\limits_{i=1}^{n} W_i \sin \alpha_i} \tag{7.49}$$

The effective stress in Equation 7.49 has been derived by first resolving the weight W_i perpendicular to the base and then subtracting the force due to pore water pressure:

$$\sigma_i' = \frac{W_i \cos \alpha_i}{l_i} - u_i$$

This method has been found to produce unrealistically low or negative pressure at the slice base and should be avoided (Corps of Engineers 2003). A more reliable expression can be derived by first calculating the effective weight W_i'

$$W_i' = W_i - u_i b_i = W_i - u_i l_i \cos \alpha_i$$

and then determining the effective normal force N_i' by summing forces perpendicular to the base as follows:

$$N_i' = W_i' \cos \alpha_i \tag{7.50}$$

Substituting the expression for W_i' into Equation 7.50:

$$N_i' = W_i \cos \alpha_i - u_i l_i \cos^2 \alpha_i \tag{7.51}$$

Combining Equations 7.46 and 7.51 and substituting into Equation 7.45, an alternate expression for the factor of safety is obtained as follows:

$$F = \frac{\sum\limits_{i=1}^{n} [c_i' l_i + (W_i \cos \alpha_i - u_i l_i \cos \alpha_i^2) \tan \phi_i']}{\sum\limits_{i=1}^{n} W_i \sin \alpha_i} \tag{7.52}$$

Equation 7.52 is the recommended expression to use for the ordinary method of slices. If $\phi = 0$, then the factor of safety calculated by this method will be the same as the one calculated by the Swedish circle method presented in Section 7.6.2.2. Since the conditions of statics are not satisfied, the factor of safety calculated by this method is reported to be 10–60% below (conservative) the lower bound values obtained from other methods that completely satisfy static equilibrium (Lambe and Whitman 1979).

7.6.3.5 Simplified Bishop Method

Description. Developed by Bishop (1955), this method assumes that the interslice forces are horizontal (normal to the sides) and ignores the interslice shear forces. Typical slice geometry and the free body diagram are shown in Figure 7.19. The normal and shear forces at the base are obtained by summing up forces in the vertical direction ($\Sigma F_y = 0$) and employing the Mohr-Coulomb shear strength relationship along with the definition of the factor of safety given by Equation 7.12. The expression for the factor of safety is determined from moment equilibrium ($\Sigma M_y = 0$) about the center of the slip circle. The system of unknowns and number of equations are listed in Table 7.11.

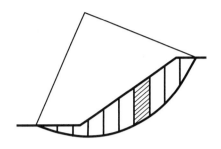

(a) Slope and typical slip surface

Mathematical Formulation. Referring to the free body of a slice in Figure 7.17, the summation of forces in the vertical direction gives:

$$N_i \cos \alpha_i + S_i \sin \alpha_i = W_i \quad (7.53)$$

Combining Equations 7.42 and 7.46, the shear force S_i at the base can be expressed in terms of the factor of safety F and the effective stress shear strength parameters as follows:

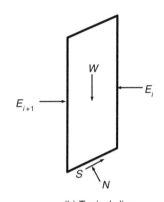

(b) Typical slice

FIGURE 7.19 (a) Typical slice geometry and (b) free body diagram for Bishop's simplified method.

$$S_i = \frac{1}{F} (c_i' l_i + N_i' \tan \phi_i') \quad (7.54)$$

Here, N_i' is the effective normal force at the slice base, and if u is the pore water pressure, then Equation 7.54 can be expressed as:

$$S_i = \frac{1}{F} [c_i' l_i + (N_i - u_i l_i) \tan \phi_i'] \quad (7.55)$$

TABLE 7.11 Number of Unknowns and Equations in Bishop's Simplified Method

Unknowns		Equations	
Parameter	Number	Type	Number
Factor of safety	1	Summation of moments	1
Normal force N	n	Vertical force equilibrium	N
Total unknowns = $n + 1$		Total equations = $n + 1$	

Combining Equations 7.53 and 7.55, we get:

$$N_i = \cos \alpha_i + \frac{1}{F} [c_i'l + (N_i - u_i l_i) \tan \phi_i'] \sin \alpha_i = W_i \qquad (7.56)$$

Simplifying Equation 7.56, we can find an expression for N_i as follows:

$$N_i = \frac{W_i - [c_i'l_i - u_i l_i \tan \phi_i'] \dfrac{\sin \alpha_i}{F}}{\cos \alpha_i + \dfrac{1}{F} \sin \alpha_i \tan \phi_i'} \qquad (7.57)$$

Combining Equations 7.45–7.47, the factor of safety in terms of moment equilibrium about the center of the circular slip surface takes the following form:

$$F = \frac{\sum c_i'l_i + (N_i - u_i l_i) \tan \phi_i'}{\sum W_i \sin \alpha_i} \qquad (7.58)$$

Substituting N_i from Equation 7.57 into Equation 7.58 and simplifying:

$$F = \frac{\sum \left[\dfrac{c_i'l_i \cos \alpha_i + (W_i - u_i l_i \cos \alpha_i) \tan \phi_i'}{\cos \alpha_i + (1/F) \sin \alpha_i \tan \phi'} \right]}{\sum W_i \sin \alpha_i} \qquad (7.59)$$

Since F appears on both sides of the equation, a trial-and-error procedure is needed. The convergence is reported to be rapid (Lambe and Whitman 1979). The numerator in Equation 7.59 is further simplified by defining a parameter m_{α_i} as follows:

$$m_{\alpha_i} = \cos \alpha_i + \frac{1}{F} \sin \alpha_i \tan \phi_i' \qquad (7.60)$$

Substituting in Equation 7.59, the factor of safety is given by:

$$F = \frac{\sum [c_i'l_i \cos \alpha_i + (W_i - u_i l_i \cos \alpha_i) \tan \phi_i'] (1/m_{\alpha_i})}{\sum W_i \sin \alpha_i} \qquad (7.61)$$

Variations of m_{α_i} with α_i for various values of $\tan \phi_i'/F$ are shown in Figure 7.20. Equation 7.61 provides the expression for the factor of safety by Bishop's simplified method and is recommended for general practice.

Although static equilibrium conditions are only partially satisfied by Bishop's simplified method, several investigators concluded that the factor of safety calculated by this method compares quite well with more rigorous methods which satisfy complete equilibrium (Fredlund

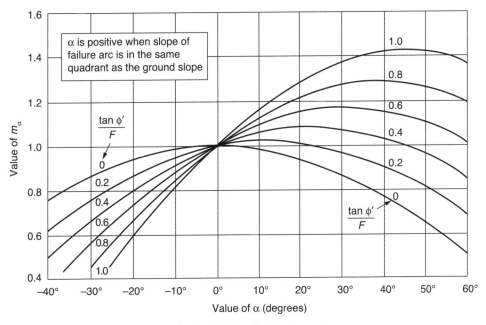

FIGURE 7.20 Variations in m_α with the slice base inclination angle α.

and Krahn 1977). Due to the fact that this method allows relatively rapid hand calculation with a sufficient degree of accuracy, it has been used worldwide as a popular and acceptable method for slope stability analysis.

Additional Known Forces. Bishop's simplified method can be used to include additional forces where the magnitudes are known and the orientations and locations are either known or assumed. These forces are in addition to the slice weight W_i, the interslice forces, and the shear force at the slice base. In this section, three additional types of forces are considered, as shown in Figure 7.21 and described below:

1. *Horizontal seismic force* kW_i—This force, where k is the seismic coefficient, acts through the center of gravity of the slice and has a moment arm y_{k_i} about the center of the slip circle.
2. *Reinforcement force* T_i—This is the force developed in the soil-reinforcing material used for mechanical slope stabilization. It intersects the failure surface, and although in most cases it is horizontally inclined, we will assume that it is oriented at an angle δ_i to the horizontal direction. The moment arms of the horizontal and vertical components of T_i about the center of the slip circle are y_{T_i} and x_{T_i}, respectively.
3. *External force* Q_i—This is the force due to water load acting normal to the top of the slice. The moment arms of the horizontal and vertical components of Q_i are y_{Q_i} and x_{Q_i}, respectively.

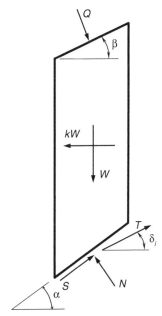

FIGURE 7.21 Additional known forces in Bishop's simplified method.

Summation of moments about the center of the slip circle was presented in Equations 7.41–7.43. Inclusion of the additional known forces will change the moment equilibrium equation as follows (resistive moments are positive):

$$r \sum \frac{l_i s_i}{F} - r \sum W_i \sin \alpha_i - \sum k W_i y_{s_i} + \sum T \cos \delta_i \cdot y_{T_i}$$

$$+ \sum T \sin \delta_i \cdot x_{T_i} + \sum Q_i \sin \beta_i \cdot y_{Q_i} \qquad (7.62)$$

$$- \sum Q_i \sin \beta_i \cdot x_{Q_i} = 0$$

If the net moment due to additional forces is expressed by M_{net}, then

$$r \sum \frac{l_i s_i}{F} - r \sum W_i \sin \alpha_i + M_{net} = 0 \qquad (7.63)$$

where

$$M_{net} = \sum T_i \cos \delta_i \cdot y_{T_i} + \sum T_i \sin \delta_i \cdot x_{T_i} - \sum k W_i y_{s_i}$$

$$+ \sum Q_i \sin \beta_i \cdot y_{Q_i} - \sum Q_i \cos \beta_i \cdot x_{Q_i} \qquad (7.64)$$

The factor of safety can be calculated from Equation 7.63 as follows:

$$F = \frac{r \sum l_i s_i}{r \sum W_i \sin \alpha_i - M_{net}} \qquad (7.65)$$

It follows from Equation 7.65 that there is an increase in the resistive moment and the factor of safety when M_{net} is positive. The opposite is true (the factor of safety decreases) if M_{net} is negative. Since the additional forces have known magnitudes and orientations, no additional assumptions are necessary, and the mathematical formulation will involve similar steps as outlined in Bishop's simplified method. Invoking the shear strength parameters in terms of effective stresses:

$$F = \frac{r \sum [c_i' l_i + (N_i - u_i l_i) \tan \phi_i']}{r \sum W_i \sin \alpha_i - M_{net}} \qquad (7.66)$$

Summation of forces in the vertical direction gives the following:

$$N_i \cos \alpha_i + S_i \sin \alpha_i + T_i \sin \delta_i - W_i - Q_i \cos \beta_i = 0 \qquad (7.67)$$

For simplicity, the vertical summation of the additional known forces is combined into a single term denoted by P_{net} as follows:

$$P_{net} = T_i \sin \delta_i - Q_i \cos \beta_i \qquad (7.68)$$

Substituting into Equation 7.67, we get:

$$N_i \cos \alpha_i + S_i \sin \alpha_i - W_i + P_{net} = 0 \qquad (7.69)$$

Combining Equations 7.55 and 7.69 and solving for N_i:

$$N_i = \frac{W_i - P_{net} - [c_i' l_i - u_i l_i \tan \phi_i'] \dfrac{\sin \alpha_i}{F}}{\cos \alpha_i + \dfrac{1}{F} \sin \alpha_i \tan \phi_i'} \qquad (7.70)$$

Substituting N_i into Equation 7.66, we get the modified expression for the factor of safety:

$$F = \frac{\sum \left[\dfrac{c_i' l_i \cos \alpha_i + (W_i - P_{net} - u_i l_i \cos \alpha_i) \tan \phi_i'}{\cos \alpha_i + (1/F) \sin \alpha_i \tan \phi'} \right]}{\sum W_i \sin \alpha_i - (M_{net}/r)} \qquad (7.71)$$

Equation 7.71 incorporates additional known forces due to slope reinforcement, seismic events, and external water load on top of the slope. Since the equilibrium of forces is only considered in the vertical direction, the contribution of the horizontal forces is included only indirectly through moment equilibrium condition combined into a single parameter M_{net}, as shown in Equation 7.64. Although the reinforcing element intersecting the failure surface generally tends to be horizontal, and only a horizontal reinforcement force is usually considered in practice, Equation 7.71 allows a provision for incorporating reinforcement forces that are inclined at any angle to the failure surface.

Example 1: Long-Term Stability
Figure 7.22a shows a slope with seepage conditions represented by a flow net, the slope geometry, soil properties, and a failure circle. Determine the factor of safety using the ordinary method of slices.

Solution. Long-term stability checks imply drained conditions, and an effective stress approach is used. The following steps are performed in this method, and the results are presented in Table 7.12:

Step 1. The region bounded by the slope surface and the failure surface is divided into a suitable number of vertical slices, as shown in Figure 7.22a.

Step 2. The weight of each slice W_i is determined from $W_i = \gamma b_i h_i$, where b_i is the width, h_i is the average height, and γ is the total unit weight.

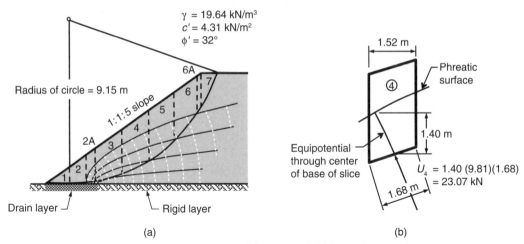

FIGURE 7.22 (a) Slope geometry, properties, and flow net and (b) determination of pore water pressure at the base of slice 4 (adapted from Lambe and Whitman 1979).

Step 3. The term $W \sin \alpha$ is computed for each slice and summed, where α is the inclination of the slice base.

Step 4. The pore water pressure u_i is determined along the failure arc by multiplying the pressure head at the middle of the slice base by the unit weight of water. The pore water force U_i at the base of the ith slice is given by $U_i = u_i l_i$, where l_i is the length of the slice base. Figure 7.22b shows the calculation of pore water force for slice 4.

Step 5. Equation 7.52 is used to calculate the factor of safety:

$$F = \frac{\sum_{i=1}^{n} [c_i' l_i + (W_i \cos \alpha_i - u_i l_i \cos \alpha_i^2) \tan \phi_i']}{\sum_{i=1}^{n} W_i \sin \alpha_i}$$

$$= \frac{4.31 \times 12.7 + 271.5 x \tan 32°}{179} = 1.25$$

Example 2: Long-Term Stability

Given the slope geometry and soil properties in Figure 7.22, determine the factor of safety using Bishop's simplified method.

Solution. Long-term stability checks imply drained conditions, and an effective stress approach is used. The following steps are performed in this method, and the results are entered in Table 7.13:

Step 1. Geometry parameters b_i, l_i, h_i, and α_i; the slice weights W_i; and the pore water pressures u_i at the slice base are determined as shown in Table 7.12. These parameters are not repeated in Table 7.13. Columns 2–5 are generated from these values.

TABLE 7.12 Calculation of the Factor of Safety by the Ordinary Method of Slices

Slice	b_i (m)	l_i (m)	h_i (m)	W_i (kN)	α_i (deg)	$W_i \sin \alpha_i$ (kN)	$W_i \cos \alpha_i$ (kN)	u_i (kN/m)	$u_i l_i \cos \alpha_i^2$ (kN)	$(W_i \cos \alpha_i - u_i l_i \cos \alpha_i^2)$ (kN)
1	1.37	1.34	0.49	13.2	−1.7	−0.4	13.2	0	0	13.2
2	0.98	0.98	1.28	24.6	2.8	1.2	24.6	0	0	24.6
2A	0.55	0.58	1.77	19.1	8.0	2.7	18.9	1.4	0.79	18.1
3	1.52	1.62	2.26	67.5	14.4	16.8	65.5	10.0	15.2	50.3
4	1.52	1.71	2.74	81.8	24.8	34.3	74.4	13.9	19.5	54.9
5	1.52	1.89	2.84	84.8	35.4	49.2	68.7	12.0	13.6	55.1
6	1.34	2.04	2.56	67.4	47.7	49.9	45.2	5.3	4.89	40.3
6A	0.18	0.37	2.04	7.2	55.1	5.9	4.1	0	0	4.1
7	0.98	2.23	1.16	22.3	60.5	19.4	10.9	0	0	10.9
Sum		**12.7**				**179**				**271.5**

TABLE 7.13 Determination of the Factor of Safety Using Bishop's Simplified Method

1	2	3	4	5	6			7		
	$W_i \sin \alpha_i$ (kN)	$c'_i l_i \cos \alpha_i$ (kN)	$u_i l_i \cos \alpha_i$ (kN)	$(W_i - u_i l_i \cos \alpha_i) \tan \phi$ (kN)	m_{α_i} Trial F			$\{(3) + (5)\}/m_{\alpha_i}$ Trial F		
Slice					$F = 1.25$	$F = 1.3$	$F = 1.35$	$F = 1.25$	$F = 1.3$	$F = 1.35$
1	−0.4	5.77	0	8.24	0.98	0.98	0.97	14.29	14.29	14.44
2	1.2	4.22	0	15.37	1.02	1.02	1.02	19.20	19.20	19.20
2A	2.7	2.47	0.8	11.43	1.05	1.06	1.05	13.23	13.11	13.23
3	16.8	6.76	15.7	32.36	1.09	1.09	1.08	35.88	35.88	36.22
4	34.3	6.69	21.6	37.61	1.11	1.11	1.1	39.90	39.90	40.27
5	49.2	6.64	18.5	41.42	1.10	1.09	1.08	43.69	44.09	44.5
6	49.9	5.92	7.3	37.55	1.04	1.03	1.02	41.79	42.20	42.61
6A	5.9	0.91	0	4.49	0.98	0.97	0.95	5.51	5.56	5.68
7	19.4	4.73	0	13.93	0.93	0.91	0.92	20.06	20.50	20.28
Sum	**179**							**233.60**	**234.78**	**236.46**

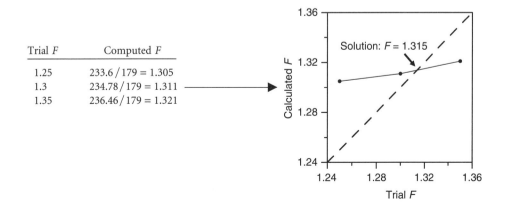

Trial F	Computed F
1.25	$233.6 / 179 = 1.305$
1.3	$234.78 / 179 = 1.311$
1.35	$236.46 / 179 = 1.321$

Solution: $F = 1.315$

Step 2. A trial factor of safety is assumed, and the parameter m_α is determined using Equation 7.60. Three trials are conducted using $F = 1.25$, $F = 1.3$, and $F = 1.35$; the corresponding m_α values are presented in column 6.

Step 3. The sum of columns 3 and 5 is divided by the m_α values and entered in column 7.

Step 4. The factor of safety is calculated using Equation 7.61 for each value of the trial factor of safety.

Step 5. The computed vs. trial factor of safety is then plotted, and the correct solution is graphically determined using Equation 7.61:

$$F = \frac{\sum [c_i' l_i \cos \alpha_i + (W_i - u_i l_i \cos \alpha_i) \tan \phi_i'] (1/m_{\alpha_i})}{\sum W_i \sin \alpha_i}$$

$$= \frac{(\text{Column } 7)}{(\text{Column } 2)}$$

Example 3: End-of-Construction (Short-Term) Stability of Embankments
Figure 7.23 shows an embankment on a clay foundation, associated material properties, and a failure circle extending into the foundation soil. Determine the factor of safety using Bishop's simplified method.

Solution. The following steps are used to determine the factor of safety. End-of-construction or short-term stability calls for undrained conditions, and the total stress approach is followed. Table 7.14 presents the results of the analysis.

Step 1. The failure zone is divided into 10 vertical slices. The failure arcs in slices 6 through 10 pass through the foundation soil. The shear strength properties in these slices along the failure arc will be given by the foundation layer properties, and the unit weights required for slice weight calculation will use the unit weights of both the embankment and the foundation soil. These slices are therefore denoted in two ways. For example, slice 6 is called 6E and 6F, to signify the embankment portion E and the foundation portion F, respectively.

Step 2. From the slope and slice geometry, the slice width b_i, average slice height h_i, and slice base inclination α_i are determined. Using the total unit weights of the embankment and the foundation soils, the slice weights W_i are obtained and entered in column 5.

Step 3. Quantities required in Equation 7.61 are determined and entered in columns 10 and 11. Note that the pore water pressure term is zero.

Step 4. The parameter m_α is determined using Equation 7.60 for three trial values of the factor of safety: $F = 1.0$, $F = 1.5$, and $F = 2.0$. These values are entered in column 12.

Step 5. The sum of columns 10 and 11 is divided by the m_α values and entered in column 13.

Step 6. The factor of safety is calculated using Equation 7.61 for each value of the trial factor of safety. Note that the pore water pressure term is zero.

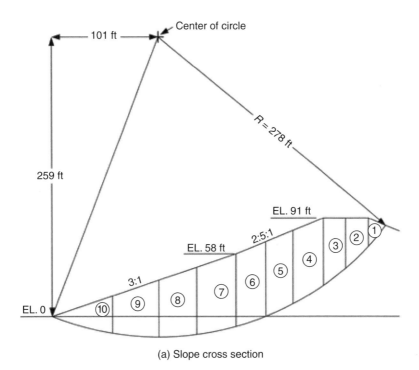

(a) Slope cross section

| | Total unit weight, γ (pcf) | c (psf) | φ (degrees) |
Material			
Embankment	135	1780	5
Foundation	127	1600	2

(b) Material properties

FIGURE 7.23 End-of-construction stability check for an embankment (adapted from Corps of Engineers 2003).

Step 7. The computed vs. trial factor of safety is then plotted, and the correct solution is graphically determined.

7.6.3.6 Force Equilibrium Methods

Both the ordinary method of slices and Bishop's simplified method satisfied moment equilibrium and either did not or only partially satisfied force equilibrium. Three force equilibrium methods that satisfy force equilibrium in both the horizontal and vertical directions but do not satisfy moment equilibrium were listed in Table 7.9:

1. *Corps of Engineers method* (1970, 2003)—The U.S. Army Corps of Engineers uses the modified Swedish method, in which the interslice forces are assumed to be parallel to the average embankment slope given by the inclination of the line joining the toe and the crest. This assumption sometimes produces a factor of safety that is highly unconservative compared to rigorous methods which satisfy complete equilibrium.

2. *Janbu's simplified method* (1954a, 1954b)—In this method, the interslice forces are assumed to be horizontal, and no interslice shear force is considered. It has been

TABLE 7.14 Determination of the Factor of Safety for End-of-Construction Stability

1	2	3	4	5	6	7	8	9	10	11	12			13		
											m_{α_i} Trial F			$[(10)+(11)]/m_{\alpha_i}$ Trial F		
Slice	b_i (ft)	l_i (ft)	h_i (ft)	W_i (kips)	α_i (deg)	$W_i \sin \alpha_i$ (kips)	c_i (ksf)	ϕ_i (deg)	$c_i l_i \cos \alpha_i$ (kips/ft)	$W_i \tan \phi$ (kips)	$F=1.0$	$F=1.5$	$F=2.0$	$F=1.0$	$F=1.5$	$F=2.0$
1	17	25.4	13	30	48	23	1.78	5	30.2	2.62	0.74	0.71	0.70	44.35	46.22	46.38
2	22	30.0	36	108	43	73	1.78	5	39.1	9.44	0.79	0.77	0.76	61.44	63.03	63.86
3	22	27.5	54	158	37	95	1.78	5	39.1	13.82	0.85	0.83	0.82	62.25	63.75	64.53
4	28	32.6	65	251	31	128	1.78	5	49.8	21.95	0.90	0.89	0.88	79.72	80.61	81.53
5	26	28.4	69	241	24	99	1.78	5	46.2	21.08	0.95	0.94	0.93	70.82	71.57	72.34
6E	26	—	64	241	—	75	—	—	—	—	—	—	—	—	—	—
6F	26	27.3	5	16	18	5	1.60	2	41.6	0.56	0.96	0.96	0.96	43.92	43.92	43.91
7E	36	—	52	259	—	49	—	—	—	—	—	—	—	—	—	—
7F	36	36.6	13	61	11	12	1.60	2	57.6	2.13	0.99	0.99	0.98	60.33	60.33	60.94
8E	36	—	40	193	—	14	—	—	—	—	—	—	—	—	—	—
8F	36	36.0	18	82	4	6	1.60	2	57.6	2.86	1.0	1.0	0.99	60.46	60.46	61.07
9E	44	—	26	156	—	−14	—	—	—	—	—	—	—	—	—	—
9F	44	44.1	17	97	−5	−8	1.60	2	70.4	3.38	0.99	0.99	0.99	74.52	74.52	74.52
10E	54	—	10	73	—	−19	—	—	—	—	—	—	—	—	—	—
10F	54	55.9	8	56	−15	−14	1.60	2	86.4	1.95	0.96	0.96	0.96	92.03	92.03	92.03
Sum						524								649.86	656.48	661.66

Trial F	Computed F
1.00	649.86/524 = 1.240
1.50	656.48/524 = 1.252
2.00	661.66/524 = 1.262

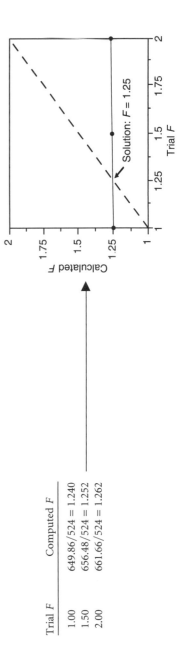

reported that this assumption may lead to significant underestimation of the factor of safety, and as a result, correction factors need to be applied to the computed factor of safety (Janbu 1973).

3. *Lowe and Karafiath's method* (1960)—In this method, the inclinations of the interslice forces are functions of the individual base inclinations α_i and therefore are not constant, as the above two methods are, but change with each slice. This procedure has been reported to produce a factor of safety with reasonable accuracy that lies within 10% of the values computed by the complete equilibrium procedures.

One advantage of the force equilibrium methods is that they can be used for more complex, noncircular slip surfaces, which often are needed to represent layered stratigraphy, such as weak interfaces between different soils or between soils and other materials such as geosynthetics. The general procedure used in a force equilibrium method is outlined below for the modified Swedish method. Figure 7.24 shows the slope configuration, the free body of a typical slice, and the assumed orientation of the interslice forces inclined at an angle θ parallel to the average embankment slope. Table 7.15 presents the system of unknowns and the number of available equations for this method.

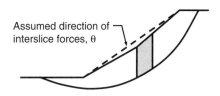

(a) Slope and typical slip surface

Force equilibrium methods require an iterative trial-and-error procedure for determination of the factor of safety using either a graphical or a numerical approach. A brief step-by-step description of the graphical solution is presented next.

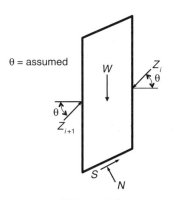

(b) Typical slice

FIGURE 7.24 (a) Typical slice geometry and (b) free body diagram for modified Swedish method.

Graphical Solution for Modified Swedish Method
The modified Swedish method involves repeated assumptions of trial values of the factor of safety, construction of force vector polygons for each slice for each trial factor of safety, and checking if force equilibrium is achieved after all force polygons are constructed for each trial. The correct factor of safety is obtained when the force vector polygon closes at the last slice or, in other words, the "error in closing" is minimized to zero. The procedure is summarized in Figure 7.25 and described below (after Corps of Engineers 2003).

TABLE 7.15 Number of Unknowns and Equations in the Modified Swedish Method

Unknowns		Equations	
Parameter	Number	Type	Number
Factor of safety	1	Horizontal force equilibrium	n
Normal force N	n	Vertical force equilibrium	n
Interslice force Z	$n-1$		
Total unknowns = $2n$		Total equations = $2n$	

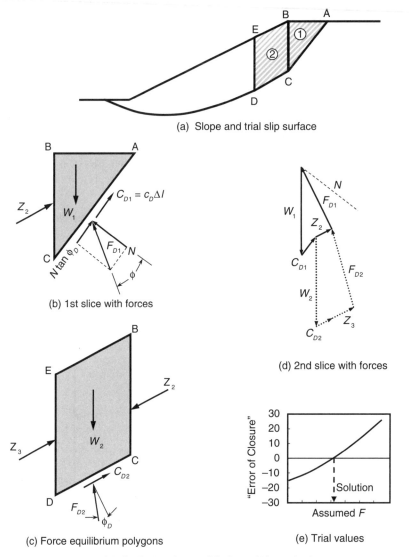

(a) Slope and trial slip surface

(b) 1st slice with forces

(c) Force equilibrium polygons

(d) 2nd slice with forces

(e) Trial values

FIGURE 7.25 Graphical solution for modified Swedish method.

Step 1. Trial Factor of Safety

Figure 7.24 shows the system of forces in a typical slice. This includes the slice weight W_i; the interslice forces Z_i and Z_{i+1} at the right and left boundaries, respectively; and the normal force N_i and the shear force S_i at the base. If F is the trial factor of safety, then

$$S_i = \frac{1}{F} (c_i l_i + N_i \tan \phi_i) \tag{7.72}$$

In terms of the developed shear strength parameters, $c_{d_i} = c_i / F$ and $\tan \phi_{d_i} = \tan \phi_i / F$:

$$S_i = c_{d_i} l_i + N_i \tan \phi_{d_i} \tag{7.73}$$

Step 2. Components of the Force Polygon

For the first slice ABC (Figure 7.25b), the components of the force polygon include (1) the cohesive force $C_{D1} = c_{D1}\Delta l$ acting parallel to the base; (2) the force F_{D1}, which is the resultant of the normal force N and the shear force $N \tan \phi_D$ inclined at an angle ϕ_D with the normal direction; (3) the interslice force Z_2 inclined at an angle θ with the horizontal; and (4) the slice weight W_1 along the vertical direction.

Step 3. Construction of the Force Polygon

As shown in Figure 7.25c, the following sequence is used to construct the force equilibrium polygon:

1. The weight vector W_1 is drawn first in the vertical direction.
2. The cohesive force vector C_{D1} is drawn starting at the tip of the weight vector and parallel to the base inclination.
3. A line representing the resultant force F_{D1} is drawn such that its tip meets the tail of the weight vector and makes an inclination ϕ_D with the normal direction.
4. A line representing the interslice force Z_2 is drawn from the tip of the force vector C_{D1} in the assumed inclination θ (parallel to the embankment slope). The intersection of the two lines drawn in steps 3 and 4 determines the tip of Z_2 and the tail of F_{D1}, thus defining the magnitude of both of these force vectors.
5. The process is continued with the second slice BCDE, which contains the forces shown in Figure 7.25d. The force polygon for the second slice (drawn with dashed lines) is constructed by starting the weight vector W_2 from the tip of C_{D1} and attaching the tip of F_{D2} to the tail of F_{D1}.
6. Progressing slice by slice in a similar fashion, the last slice (nth slice) is reached at the toe of the slope. Since there is no other force on the left of the last slice, the force polygon must close with the resultant force F_{Dn}. Since the assumed factor of safety is generally not the correct one, the force polygon does not close, and an additional (fictitious) force Z_{n+1}, called the "closure error," is required to close the force polygon.
7. Steps 1 through 6 are repeated with a different assumed factor of safety, and a plot of the "closure error" vs. factor of safety is developed, as shown in Figure 7.25e.
8. The intersection of a horizontal line through the zero "closure error" and the curve defines the correct factor of safety.

If the slope is submerged, and shear strength is expressed in terms of effective stresses, the system of forces will include additional known forces due to water pressure, as shown in Figure 7.26.

For an interior slice, the complete set of forces consists of (1) the slice weight W; (2) the forces due to water pressure on the right side U_R, on the left side U_L, and at the slice base U_B; (3) the interslice forces Z_i and Z_{i+1}; (4) the developed or mobilized shear force components $C'_D \Delta l$; and (5) the resultant F'_D of the effective normal force N' and the frictional component of the shear strength $N' \tan \phi'_{D_i}$. If the slice is submerged, an additional force P will act on the top of the slice. Since the weight and all the water forces in (1) and (2) are known, they can be expressed as a single resultant R by drawing a force polygon as shown in Figure 7.26b. This resultant is vertical if there is no seepage or flow through the slope. Figure 7.26c shows the combined force vector polygon for the ith slice. The procedure is continued for all slices, and the "closure error" or force imbalance is determined. The correct factor of safety is obtained by repeating the process with a different assumed factor of safety as described earlier.

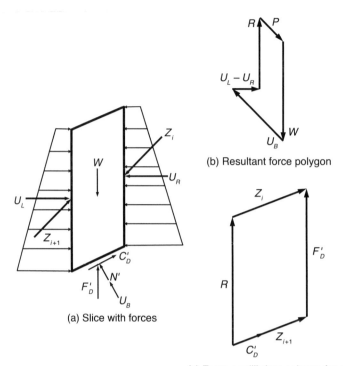

(b) Resultant force polygon

(a) Slice with forces

(c) Force equilibrium polygon for slice

FIGURE 7.26 Graphical solution for modified Swedish method including forces due to water pressure.

7.6.3.7 Complete Equilibrium Methods

In these methods, requirements for static equilibrium are completely satisfied; these include horizontal force equilibrium ($\sum F_x = 0$), vertical force equilibrium ($\sum F_y = 0$), and moment equilibrium ($\sum M = 0$). Three such methods and their assumptions were listed in Table 7.9: (1) Spencer's (1967, 1973) method, (2) the Morgenstern-Price (1965) method, and (3) Sarma's (1973) method. All of these methods require computers due to their complexity, and hand calculations are impractical. Only a brief description of Spencer's method is provided below.

Spencer's Method. In Spencer's (1967) method, it is assumed that the interslice forces are all parallel; that is, they all have an equal inclination θ with the horizontal. This angle is solved as an unknown in the calculations. The procedure is suitable for noncircular slip surfaces. The method involves an iterative trial-and-error procedure in which the factor of safety F and the angle θ are repeatedly assumed until complete equilibrium conditions are satisfied for all slices.

A noncircular slip surface and the free body diagram of a typical slice are shown in Figure 7.27. The system of forces

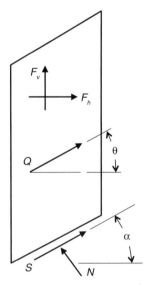

FIGURE 7.27 System of forces acting on a slice for Spencer's method.

acting on the slice includes: (1) the resultant of all known horizontal and vertical forces, denoted by F_h and F_v, respectively; (2) the resultant of the left and right interslice forces, denoted by Q_i, which is inclined at the same inclination θ as the interslice forces; and (3) the normal force N and the shear force S at the slice base. The summation of horizontal forces, summation of vertical forces, and the Mohr-Coulomb shear strength criterion at the slice base obtained from Equation 7.55 can be used to find an expression for the resultant interslice force Q as follows (Duncan and Wright 2005):

$$Q = \frac{-F_v \sin \alpha - F_h \cos \alpha - (c'l/F) + (F_v \cos \alpha - F_h \sin \alpha + ul)(\tan \phi'/F)}{\cos(\alpha - \theta) + [\sin(\alpha - \theta) \tan \phi'/F]} \quad (7.74)$$

Note that the subscript denoting the ith slice has been dropped from Equation 7.74 for clarity. Complete force equilibrium dictates that:

$$\sum Q_i = 0 \quad (7.75)$$

Similarly, complete moment equilibrium dictates that the sum of moments about any point will be zero. Accordingly, summation of the moment about the origin of the Cartesian coordinate system (Figure 7.27) produces:

$$\sum Q_i (x_i \sin \theta - y_i \cos \theta) = 0 \quad (7.76)$$

Substitution of Q_i from Equation 7.74 into Equations 7.75 and 7.76, the force and moment equilibrium equations, produces two equations with two unknowns, namely the factor of safety F and the interslice force inclination θ. These unknowns are solved by a trial-and-error procedure in which F and θ are repeatedly assumed until the errors (force and moment imbalance) are within acceptable limits.

7.6.4 Method of Slices: A Comparison Study

Fredlund and Krahn (1977) performed a systematic study to compare various commonly used slope stability analysis procedures based on the method of slices, but extended them to include composite failure surfaces and external loadings due to surcharge, submergence, and earthquake. A 40-ft-high 2:1 slope was analyzed, as shown in Figure 7.28. Six different cases representing various combinations of geometry, properties, and water conditions were considered for each of the methods. The University of Saskatchewan SLOPE computer program (Fredlund 1974) was used in this study to calculate the factor of safety.

The results are summarized in Table 7.16 for the following five methods which were discussed previously: (1) ordinary method of slices, (2) Bishop's simplified method, (3) Spencer's method, (4) Janbu's simplified method, and (5) the Morgenstern-Price method. Various methods can be compared by plotting the factor of safety against λ, which is a ratio of the interslice shear and normal forces. In Spencer's method, λ is equal to the tangent of the angle between the resultant interslice force and the horizontal. For Bishop's simplified method, which satisfies overall moment equilibrium, $\lambda = 0$. λ is also zero for Janbu's simplified method (uncorrected), where there is no interslice shear force. Fredlund and Krahn (1977) plotted the

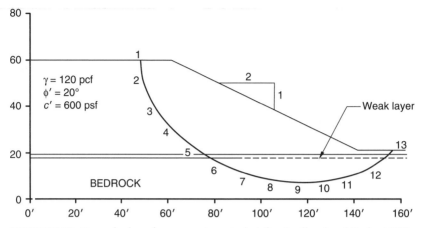

FIGURE 7.28 Example slope for comparison study (after Fredlund and Krahn 1977).

TABLE 7.16 Calculation of the Factor of Safety by Various Methods

Slope Properties	Ordinary Method	Bishop's Simplified Method	Spencer's Method (λ)	Janbu's Simplified Method	Morgenstern-Price Method (λ)	% Difference
Case 1						
2:1 slope, 40 ft high, $\phi = 20°$, $c = 600$ psf	1.928	2.080	2.073 (0.237)	2.041	2.076 (0.254)	0.19
Case 2						
Same as case 1 with a thin weak layer, $\phi = 10°$, $c = 0$	1.288	1.377	1.373 (0.185)	1.448	1.378 (0.159)	0.07
Case 3						
Same as case 1 except $r_u = 0.25$	1.607	1.766	1.761	1.735	1.765	0.05
Case 4						
Same as case 2 except $r_u = 0.25$ for both layers	1.029	1.124	1.118	1.191	1.124	0.0
Case 5						
Same as case 1 except a piezo-metric line	1.693	1.834	1.830	1.827	1.833	0.05
Case 6						
Same as case 2 except a piezo-metric line for both layers	1.171	1.248	1.245	1.333	1.250	0.16

After Fredlund and Krahn (1977).

Note: r_u is the pore pressure ratio, defined as the ratio of pore water pressure to total vertical stress at any depth.

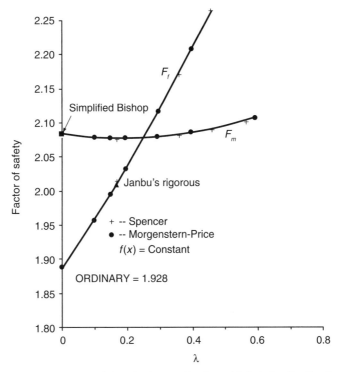

FIGURE 7.29 Variation in factor of safety with λ (after Fredlund and Krahn 1977).

variations in factor of safety with λ for case 1 conditions in Table 7.16, as shown in Figure 7.29. Two types of curves were obtained: (1) the F_f curve, representing the methods that satisfy overall force equilibrium, and (2) the F_m curve, representing the methods that satisfy overall moment equilibrium. The following conclusions can be drawn from Figure 7.29:

1. Factor of safety with respect to moment equilibrium is relatively insensitive to the assumptions related to the interslice forces. Accordingly, Bishop's simplified method, the Morgenstern-Price method, and Spencer's method all produce similar values for the factor of safety.
2. Factor of safety with respect to overall force equilibrium is very sensitive to the assumptions related to the interslice forces.
3. The intersection of these curves represents a unique combination of the factor of safety and λ that satisfies complete (both force and moment) static equilibrium conditions.

The comparison study summarized here provides some insight into the relative magnitudes of the factor of safety calculated by the most common methods of slices under various slope conditions and materials, as well as the sensitivity of the various analysis procedures to the assumptions related to the interslice forces. It also was found that the average difference between the factor of safety calculated by Bishop's simplified method and the Morgenstern-Price method for all six cases is on the order of only 0.1%, as shown in Table 7.16.

7.6.5 Solutions Using Slope Stability Charts

7.6.5.1 Background Information

First published by Taylor (1937, 1948), slope stability charts provide a quick and efficient way to determine an approximate value of the factor of safety as a preliminary estimate before embarking on a more detailed computer analysis for planning, design, and back-analysis purposes. Since the introduction of Taylor's pioneering charts, a number of well-known advancements were successively made in the chart solutions during the 1960s and continued through the 1990s. A summary of these methods is presented in Table 7.17.

Some of the benefits of using chart solutions are as follows:

1. Fast and efficient preliminary analysis of slope stability before more rigorous computer analysis is undertaken
2. Rapid verification of computer-generated results as a quality control step for subsequent more detailed computer analysis
3. Back calculation of the shear strengths of failed slopes by assuming a factor of safety of unity
4. The ability to make quick comparisons between various design alternatives

It is argued that the accuracy of the design charts is usually as good as the certainty with which shear strength parameters can be estimated for conducting slope stability analysis.

TABLE 7.17 Well-Known Slope Stability Charts

Reference	Parameters	Slope Inclination	Methodology
Taylor (1937, 1948)	c_u c, ϕ	0–90°	Both $\phi = 0$ and c-ϕ analyses
Bishop and Morgenstern (1960)	c, ϕ, r_u	11–26.5°	Bishop's method
Gibson and Morgenstern (1962)	c_u	0–90°	$\phi = 0$ analysis, c_u linearly increasing with depth
Spencer (1967)	c, ϕ, r_u	0–34°	Spencer's method
Janbu (1968)	c_u c, ϕ, r_u	0–90°	$\phi = 0$ analysis, Janbu's generalized procedure of slices
Hunter and Schuster (1968)	c_u	0–90°	$\phi = 0$ analysis, c_u linearly increasing with depth
Chen and Giger (1971)	c, ϕ	20–90°	Limit analysis
O'Connor and Mitchell (1977)	c, ϕ, r_u	11–26°	Extended Bishop and Morgenstern (1960)
Hoek and Bray (1977)	c, ϕ	0–90°	Friction circle and three-dimensional block analysis
Cousins (1978)	c, ϕ, r_u	0–45°	Extension of Taylor (1948)
Charles and Soares (1984)	ϕ	26–63°	Bishop's method, nonlinear Mohr-Coulomb failure envelope
Barnes (1991)	c, ϕ, r_u	11–63°	Extended Bishop and Morgenstern (1960)

Adapted from Abramson et al. (2002).

While it is recommended that engineers continue to make use of slope stability charts, it is also important to understand the limitations of the chart solutions. These charts have been developed for simple homogeneous soils, assuming circular slip surfaces, and two-dimensional limit equilibrium analysis. In order to incorporate nonhomogeneous soil conditions (variability in soil properties), it is necessary to approximate the real conditions in the slope with a fictitious equivalent slope that has homogeneous properties using various averaging techniques, as described below.

7.6.5.2 Equivalent Homogeneous Slope

An equivalent homogeneous slope can be obtained by averaging the shear strength parameters c and ϕ and the unit weight γ. Assume the cross section of a slope where three soil layers have heights of h_i, shear strength properties of c_i and ϕ_i, and unit weight of γ_i. The approximate location of the critical slip surface is determined from the charts (described later) and is considered known. The central angle of the arc subtended by the various layers or soil zones is given by δ_i. This is obtained by first drawing the slope cross section and the critical slip surface, and then actually measuring the δ_i angles with a protractor. The layer heights h_i and the subtended angle δ_i are used to determine weighted averages for the shear strength and unit weight for one "homogeneous" slope for use in the chart solutions.

Averaging Shear Strength. The average cohesion c_{avg} and the average friction angle ϕ_{avg} are obtained as follows:

$$\phi_{avg} = \frac{\sum \delta_i \phi_i}{\sum \delta_i} \tag{7.77}$$

$$c_{avg} = \frac{\sum \delta_i c_i}{\sum \delta_i} \tag{7.78}$$

Averaging Unit Weight. The average unit weight γ_{avg} is given by:

$$\gamma_{avg} = \frac{\sum \gamma_i h_i}{\sum h_i} \tag{7.79}$$

Averaging Undrained Shear Strength. If an embankment is resting on a weak clay ($\phi = 0$) foundation with an undrained shear strength c_u, a portion of the slip circle passes through the clay foundation. In that case, it is preferable to approximate the shear strength of the embankment soil by an equivalent undrained shear strength and then determine an average undrained shear strength for both the embankment and the foundation soil, as follows:

$$c_{u(avg)} = \frac{\sum \delta_i c_{u(i)}}{\sum \delta_i} \tag{7.80}$$

The usual practice to determine the equivalent undrained shear strength of the embankment soil is to calculate the average normal stress on the portion of the embankment that is enclosed within the slip surface and then estimate the shear strength from the failure envelope. This shear strength is assumed to be the undrained shear strength of the embankment soil and is used in Equation 7.80.

7.6.5.3 Chart Procedures

The methodology for using slope stability charts is presented for the following four types of slopes (adapted from Corps of Engineers 2003):

1. Slopes in soils with $\phi = 0$ and uniform strength
2. Slopes in soils with $\phi > 0$, $c > 0$ and uniform strength
3. Slopes in soils with $\phi = 0$ and strength linearly increasing with depth
4. Infinite slopes in soils with $\phi > 0$, $c = 0$ and $\phi > 0$, $c > 0$

Procedure 1. Soils with $\phi = 0$ and Uniform Strength
Slope stability charts for soils with $\phi = 0$ are presented in Figure 7.30 (Janbu 1968). Adjustment factors needed for surcharge loading, submergence and/or seepage, and tension cracks are provided in Figures 7.31–7.33, respectively. These charts can be used to determine the factor of safety for a variety of slip circles extending to any depth, and multiple possibilities should be examined to arrive at a conservative estimate. As shown in Figure 7.30, the critical circle may be a slope circle, a toe circle, or a deep circle, and the following criteria can be used to determine which possibilities exist:

1. If there is water outside the slope, a circle passing above the water may be critical.
2. If a weaker layer underlies a stronger layer, then the critical circle passes through the weaker layer.
3. If a stronger layer underlies a weaker layer, then the critical circle may be tangent to the base of either layer.

The following steps are performed for each potential critical circle:

Step 1. Calculate the depth factor $d = D/H$, where D is the depth of the lowest point of the slip circle from the toe of the slope and H is the height of the slope measured from the toe. If the slip circle is entirely above the toe, then the intersection of the circle and the slope is considered to be an adjusted toe, and all parameters (D, H, and H_w) are adjusted accordingly in the calculations.

Step 2. Determine the center of the critical circle given by the coordinates X_0, Y_0 from the bottom charts in Figure 7.30, and draw the circle to scale on the slope cross section.

Step 3. Obtain an average undrained shear strength $c_{u(\text{avg})}$ from Equation 7.80; this is simply denoted by c in Figure 7.30.

Step 4. Calculate the parameter P_d in Figure 7.30 as follows:

$$P_d = \frac{\gamma H + q - \gamma_w H_w}{\mu_q \mu_w \mu_t} \tag{7.81}$$

where γ = average unit weight of the soil, H = surcharge pressure on the soil, γ_w = unit weight of water, H_w = height of external water level above the toe (Figure 7.32), μ_q = surcharge adjustment factor (Figure 7.31), μ_w = submergence adjust-

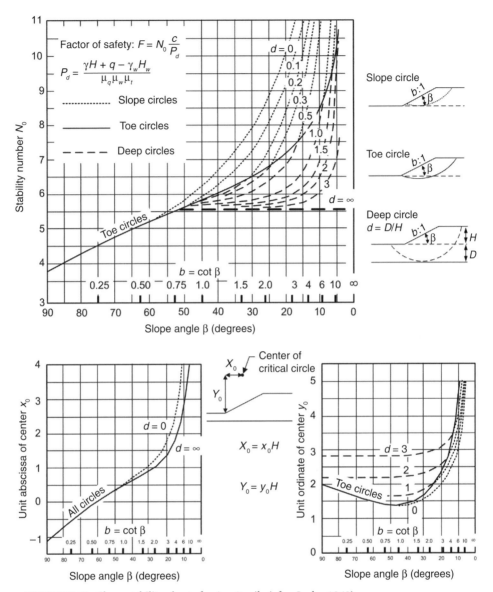

FIGURE 7.30 Slope stability charts for $\phi = 0$ soils (after Janbu 1968).

ment factor (Figure 7.32), and μ_t = tension crack adjustment factor (Figure 7.33). If there is no surcharge, no external water above the toe, and no tension cracks, then the factors μ_q, μ_w, and μ_t are all taken as unity.

Step 5. Obtain the stability number N_0 from the upper chart in Figure 7.30.

Step 6. Calculate the factor of safety as follows:

$$F = \frac{N_0 c}{P_d} \tag{7.82}$$

where c is the average shear strength given by $c_{u(\text{avg})}$ in Equation 7.80.

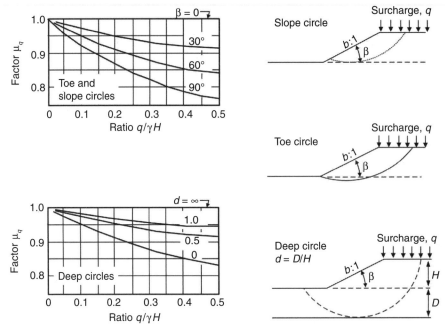

FIGURE 7.31   Surcharge adjustment factor for $\phi = 0$ and $\phi > 0$ soils (after Janbu 1968).

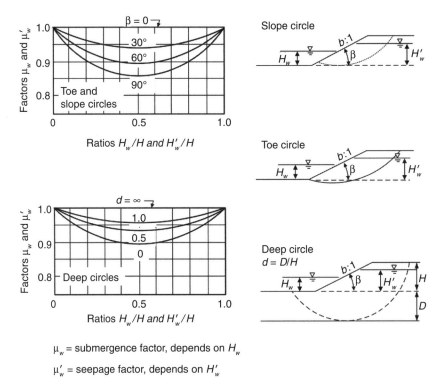

μ_w = submergence factor, depends on H_w

μ_w' = seepage factor, depends on H_w'

FIGURE 7.32   Submergence and seepage adjustment factor for $\phi = 0$ and $\phi > 0$ soils (after Janbu 1968).

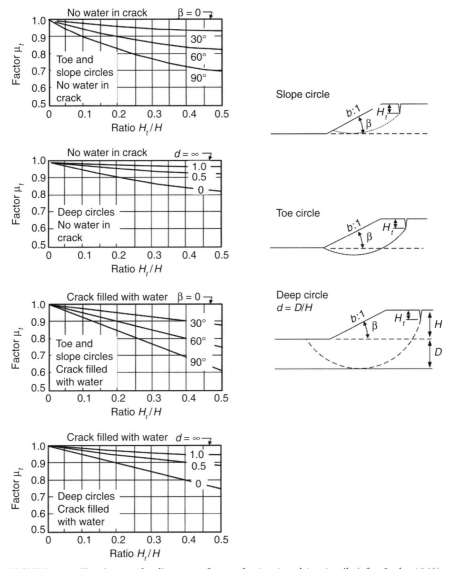

FIGURE 7.33 Tension crack adjustment factors for $\phi = 0$ and $\phi > 0$ soils (after Janbu 1968).

Procedure 2. Soils with $\phi > 0$, $c > 0$ and Uniform Strength

The slope stability charts for $\phi > 0$ soils are presented in Figure 7.34. These charts can be used for both total and effective stress analysis. The adjustment factors for surcharge pressure, seepage and submergence, and tension cracks are obtained from the charts presented in Figures 7.31–7.33 for $\phi = 0$ soils.

The stability numbers in Figure 7.34 are calculated using a toe circle, which is mostly the case for slopes in uniform soils with $\phi > 0$. In nonhomogeneous soils, or if water is present outside the slope, a circle other than the toe circle may be critical. In estimating the location of the critical circle, the following criteria may be used:

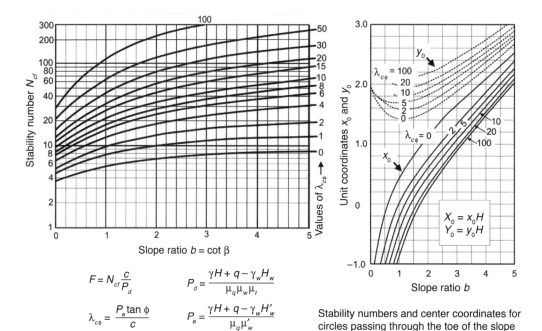

$$F = N_{cf}\frac{c}{P_d} \qquad P_d = \frac{\gamma H + q - \gamma_w H_w}{\mu_q \mu_w \mu_t}$$

$$\lambda_{c\phi} = \frac{P_e \tan \phi}{c} \qquad P_e = \frac{\gamma H + q - \gamma_w H'_w}{\mu_q \mu'_w}$$

Stability numbers and center coordinates for circles passing through the toe of the slope

FIGURE 7.34 Slope stability charts for $\phi > 0$ soils (after Janbu 1968).

1. If there is water outside the slope, a circle passing above the water may be critical.
2. If a weaker layer underlies a stronger layer, then the critical circle may be tangent to the base of the weaker layer.
3. If a stronger layer underlies a weaker layer, then the critical circle may be tangent to the base of either layer, and both possibilities should be examined.

The following steps are performed for each potential critical circle:

Step 1. Calculate the parameter P_d according to Equation 7.81. If the slip circle passes above the toe of the slope, then the point of intersection of the circle and the slope is taken as the toe of the slope for calculation of H and H_w. If there is no surcharge, no external water above the toe, and no tension cracks, then the factors μ_q, μ_w, and μ_t are all taken as unity.

Step 2. Calculate the parameter P_e as follows:

$$P_e = \frac{\gamma H + q - \gamma_w H'_w}{\mu_q \mu'_w} \qquad (7.83)$$

where H'_w is the height of water or the average level of the piezometric level within the slope and μ'_w is the seepage correction factor obtained from Figure 7.32. For a steady seepage condition, H'_w is related to the position of the phreatic surface beneath the crest of the slope, as shown in Figure 7.35.

Step 3. Calculate the dimensionless parameter $\lambda_{c\phi}$ as follows:

$$\lambda_{c\phi} = \frac{P_e \tan \phi}{e} \qquad (7.84)$$

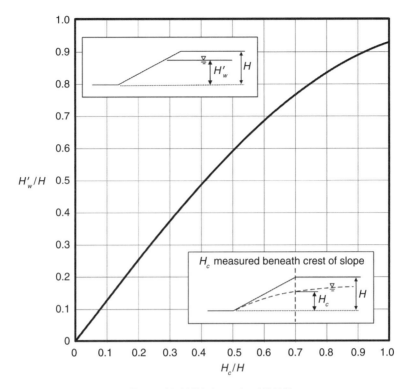

Enter with H_c/H, determine H'_w/H from curve

FIGURE 7.35 Steady seepage adjustment factor for $\phi > 0$ soils (after Duncan et al. 1987).

where ϕ and c represent the average values which can be determined from Equations 7.77 and 7.78, respectively. If $c = 0$, $\lambda_{c\phi}$ is infinity, and the charts for an infinite slope (procedure 4) should be used. Steps 3 and 4 (given below) are actually iterative, and judgment is used to estimate the initial values of c and $\tan \phi$ in Equation 7.84 to obtain $\lambda_{c\phi}$.

Step 4. Determine the coordinates of the center of the slip circle using the right chart in Figure 7.34. Plot the circle on the slope cross section, and calculate the average values of c and ϕ using Equations 7.77 and 7.78. Recalculate the value of $\lambda_{c\phi}$, and continue the iterative process until the value of $\lambda_{c\phi}$ is stabilized.

Step 5. Determine the value of the stability N_{cf} from the left chart in Figure 7.34 using the value of the slope angle β and $\lambda_{c\phi}$ obtained from step 4.

Step 6. Calculate the factor of safety as follows:

$$F = N_{cf} \frac{c}{P_d} \tag{7.85}$$

Procedure 3. Soils with $\phi = 0$ and Strength Linearly Increasing with Depth

Figure 7.36 shows the chart for soils in which the undrained shear strength ($\phi = 0$ condition) linearly increases with depth, reaching a value of C_b at the bottom of the slope. The following are the steps for determining the factor of safety:

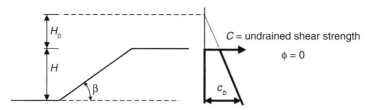

Steps:

1. Extrapolate strength profile upward to determine value of H_0 where strength profile intersects zero

2. Calculate $M = H_0/H$

3. Determine stability number from chart below

4. Determine c_b = strength at elevation of toe of slope

5. Calculate $F = N \dfrac{c_b}{\gamma(H + H_0)}$

Use $\gamma = \gamma_{buoyant}$ for submerged slope

Use $\gamma = \gamma_{total}$ for no water outside slope

Use average γ for partly submerged slope

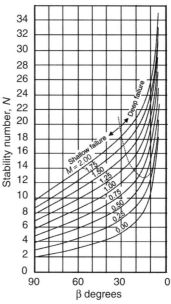

FIGURE 7.36 Slope stability charts for $\phi = 0$ soils with strength increasing as a function of depth (adapted from Hunter and Schuster 1968).

Step 1. Establish the linear variation of the undrained shear strength with slope height H, and extrapolate the line for determining the height H_0 that corresponds to "zero" strength.

Step 2. Calculate the parameter M as follows:

$$M = \frac{H_0}{H}$$

Step 3. Calculate the stability number N from the chart using M and the slope angle β.

Step 4. Calculate the factor of safety as follows:

$$F = N \frac{C_b}{\gamma(H + H_0)} \tag{7.86}$$

where γ is the unit weigh of the soil. For partially submerged soils, γ is taken as the weighted average unit weight as defined earlier. The total unit weight is used in the case of a dry slope above water. If the slope is submerged, the buoyant unit weight should be used.

Procedure 4. Infinite Slopes

Figure 7.37 presents the charts for the analysis of infinite slopes in two types of soil conditions: (1) shallow sliding in cohesionless soils and (2) sliding of a relatively thin layer of residual soil overlying a firmer soil. Both total and effective stress analyses can be performed. The steps for using the charts in effective stress analysis of infinite slopes are as follows:

Step 1. Calculate a pore pressure ratio r_u:

$$r_u = \frac{u}{\gamma H} \tag{7.87}$$

where u is the measured or estimated pore water pressure at depth H and γ is the total unit weight of the soil.

For seepage parallel to the slope, r_u can be calculated from

$$r_u = \frac{X}{T} \frac{\gamma_w}{\gamma} \cos^2 \beta \tag{7.88}$$

where X is the perpendicular distance between the surface of seepage and the sliding surface and T is the perpendicular distance between the slope surface and the sliding surface.

A more critical condition is the seepage emerging from the slope, as shown in Figure 7.37. For this condition, r_u can be calculated as follows:

$$r_u = \frac{\gamma_w}{\gamma} \frac{1}{1 + \tan \beta \tan \theta} \tag{7.89}$$

where θ is the angle of the seepage with the horizontal.

Step 2. Obtain the dimensionless parameters A and B using the bottom charts in Figure 7.37.

Step 3. Calculate the factor of safety as follows:

- In terms of effective stress:

$$F = A \frac{\tan \phi'}{\tan \beta} + B \frac{c'}{\gamma H} \tag{7.90}$$

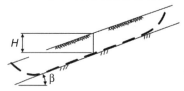

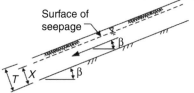

γ = total unit weight of soil

γ_w = unit weight of water

c' = cohesion intercept

ϕ' = friction angle

r_u = pore pressure ratio = $u/\gamma H$

u = pore pressure at depth H

Seepage parallel to slope

$$r_u = \frac{X}{T} \frac{\gamma_w}{\gamma} \cos^2 \beta$$

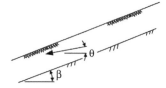

Steps:

1. Determine r_u from measured pore pressure or formulas at right
2. Determine A and B from charts below
3. Calculate $F = A \dfrac{\tan \phi'}{\tan \beta} + B \dfrac{c'}{\gamma H}$

Seepage emerging from slope

$$r_u = \frac{\gamma_w}{\gamma} \frac{1}{1 + \tan \beta \tan \theta}$$

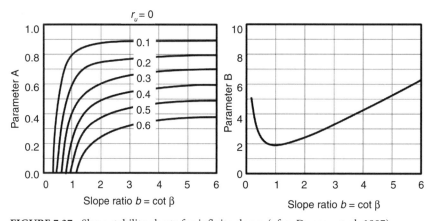

FIGURE 7.37 Slope stability charts for infinite slopes (after Duncan et al. 1987).

• In terms of total stress:

$$F = \frac{\tan \phi}{\tan \beta} + B \frac{c}{\gamma H} \qquad (7.91)$$

where c' and ϕ' are the shear strength parameters in terms of effective stress, c and ϕ are the shear strength parameters in terms of total stress, and H is the vertical distance between the slope surface and the sliding surface.

7.6.6 Important Practical Questions

Some important issues that often are raised while formulating a slope stability analysis strategy may include the following: (1) Should a two- or three-dimensional analysis be performed? (2)

Should a limit equilibrium or finite element analysis be performed? (3) Should peak or residual strength be used in the analysis? Answering these questions requires careful consideration of several factors, such as (1) the actual increase in the level of accuracy achieved by employing a more rigorous (and hence more expensive) analysis and the practical significance of the increased accuracy and (2) the appropriateness and correctness of the approach selected for each particular case. Of course, the correct approach, once selected, also must be applied correctly for a successful slope analysis. These topics are briefly discussed in the following sections with reference to practical case histories reported in the literature. The following discussion is intended to aid the user of this handbook in making informed decisions on a case-by-case basis.

7.6.6.1 Two-Dimensional vs. Three-Dimensional Analysis

Conventional slope stability analysis used in practice employs a two-dimensional (2-D) idealization of a slope failure, which is actually a three-dimensional (3-D) shape in the field. In the 2-D analysis, the underlying assumption is that the slope is infinitely wide, the end effects are negligible, and therefore a plain strain condition can be used for stability calculations. In a 3-D analysis, the familiar plane strain case used in the method of slices is extended from the x-y plane to the third dimension (z direction) by converting the slice into columns (Hovland 1977). During the last 40 years, numerous researchers have incorporated the 3-D effects in the analysis of slopes and landslides and compared the 3-D factor of safety F_3 with the corresponding 2-D factor of safety F_2. Duncan (1996) provided an excellent tabular summary of these efforts during a 23-year span (1969–1992), beginning with such early works as Anagnosti (1969), Baligh and Azzouz (1975), and Giger and Krizek (1975) and extending to works during the early 1990s such as those by Seed et al. (1990) and Leshchinsky and Huang (1992). Further research and development in the theory and practice of 3-D slope stability analysis was continued in recent years by, for example, Stark and Eid (1998), Soong et al. (1998), Koerner and Soong (1999), Stark et al. (2000), Huang et al. (2002), Loehr et al. (2004), Xie et al. (2006), and Cheng and Yip (2007). However, to date, the geotechnical community is still debating the applicability and usefulness of 3-D analysis as opposed to 2-D analysis in everyday practice, with many researchers and practitioners often advocating quite different views. Details of these discussions are available in the literature cited above and are beyond the scope of this chapter. Only some key points (both in favor and against) are summarized below:

1. A 2-D analysis is a conservative approach. As reported by Duncan (1996), the overwhelming majority of studies that involve both fundamental research as well as case histories indicate that $F_3 > F_2$.
2. A 3-D analysis may be particularly useful in situations where there is a complex slope geometry or topography, unusual distribution of shear strengths and/or pore water pressure within the potential sliding mass, and in the case of translational failure through underlying weak materials and/or geosynthetic interfaces such as those found in waste landfill slides (Stark and Eid 1998). However, it is often argued that a carefully selected sufficient number of 2-D sections can produce a minimum factor of safety comparable to or lower than the 3-D factor of safety, even in such complicated scenarios.
3. Comparison of 2-D and 3-D is only valid when minimum factors of safety are compared for both cases (Stark and Eid 1998). It should be noted that the maximum or highest slope typically selected for a 2-D analysis may not always produce the minimum

factor of safety, as was clearly demonstrated by Seed et al. (1990) in the case of a well-known landfill slope failure. In an actual failure (which is truly 3-D), the minimum factor of safety is known and is equal to unity. Bearing that in mind, 2-D analysis should be performed such that the minimum factor of safety is obtained in order to conduct a correct comparative evaluation with the corresponding 3-D factor of safety.

4. For back analysis of the shear strength of an actual failed slope, it is sometimes recommended that a 3-D analysis be performed to capture the end effects and then that shear strength be used in the remedial design. The 3-D analysis may result in an increase of as much as 30% in the back-calculated shear strength compared to a 2-D analysis (Stark and Eid 1998). Skempton (1985) suggested a 3-D correction factor given by $1/\{1 + KD/B\}$ (where K is the coefficient of earth pressure mobilized at failure and D and B are the average depth and width of the failure mass, respectively) to be applied to the 2-D back-calculated shear strength. This will typically increase the 2-D back-calculated shear strength by approximately 5%.

5. It is generally known that a 3-D analysis involves considerably higher complexity in its formulation and implementation compared to a 2-D analysis. It is sometimes argued that the impact of the large number of simplifying assumptions needed to obtain a statically determinate solution with the method of columns may actually be equivalent to the 3-D effects themselves (Duncan and Wright 2005). From a practical standpoint, the accurate determination of shear strength, pore pressure, and field drainage conditions is by itself a quite challenging task which engineers must accomplish with a sufficient degree of confidence in order to correctly perform a 2-D analysis. The additional challenges brought on by a 3-D analysis may make it more prone to error, which can outweigh the anticipated benefits or the improved "accuracy" expected from a more rigorous and expensive procedure.

6. Some of the well-known commercially available 3-D slope stability software includes 3D-PCSTABL (Thomaz 1986), CLARA 2.31 (Hungr 1988), and TSLOPE3 (Pyke 1991). A comparative evaluation of these programs was presented by Stark and Eid (1998). Recently, O. Hungr Geotechnical Research released a Windows version called CLARA-W (2001), which incorporates extensions and methodologies suggested by Lam and Fredlund (1993) and Hungr (1997).

7.6.6.2 Limit Equilibrium vs. Finite Element Analysis

As indicated in the previous section, 2-D limit equilibrium methods are the most widely used slope stability analysis techniques in everyday engineering practice. Limit equilibrium methods, even when a 3-D analysis is performed, do not have any provision for calculating stresses and deformations in the slope mass at discrete locations. Computation of stresses and deformation can be accomplished by the finite element technique. The core of the finite element method involves discretization of the slope domain and performing incremental analysis to realistically simulate field conditions such as nonlinear stress-strain behavior, changes in geometry, and sequential changes in excess pore water pressure (generation or dissipation) with progress and type of construction. It is, therefore, obvious that many complicated field conditions can be successfully modeled with the finite element method, thus increasing the accuracy of the slope analysis and improving the safety of the design. However, such a sophisticated analysis is quite expensive and time consuming and often requires specialized training and high-end computing resources, which may not be practical for every design firm or for every project.

A comprehensive summary of the use of the finite element method for the analysis of a large number of dams, embankments, and slopes during the period 1967–1990 was presented by Duncan (1996). In that paper, the author compiled more than 80 case studies in four different categories based on the following four types of constitutive (stress-strain) relationships: (1) linear-elastic material, (2) multilinear-elastic material, (3) hyperbolic-elastic material, and (4) elastoplastic and elastoviscoplastic material. Relative advantages and disadvantages of these models are discussed, along with the principal findings from each of those case studies. Users of this handbook are strongly encouraged to consult this reference if a finite element analysis is warranted for a particular project.

7.6.6.3. Peak vs. Residual Strength and the Concept of Progressive Failure

The development of progressive failure in clay slopes and whether the peak or residual shear strength should be used in stability analysis have been discussed and debated since the early 1960s, and a large body of literature exists on these topics (Skempton 1964, 1970; 1985; Bjerrum 1967; Mesri and Cepeda-Diaz 1986; LaRochelle 1989; Ramsamooj and Lin 1990; Stark and Eid 1994; Mesri and Shahien 2003; Stark et al. 2005). In his 1964 Rankine lecture, Skempton (1964) stated: "…as we shall see, from the analysis of actual slips in clays, the values of the shear strength parameters as determined by conventional tests do not necessarily bear any relation to the values which must have been operative in the clay at the time of failure. This conclusion, which has now been established beyond the slightest doubt, is obviously one of immense practical significance." Standard laboratory shear strength tests generally involve determination of the peak load-carrying capacity, and the test is often discontinued once the peak stress is overcome as indicated by a drop in capacity. However, analysis of numerous case histories of slope failures in overconsolidated clays demonstrates that the mobilized shear strength on the slip plane resembled a post-peak (reduced) shear strength close to a residual value. Accordingly, the factor of safety calculated using the laboratory peak value will be greater than unity even for failed slopes. Some key concepts regarding the fully softened and residual shear strengths for use in stability analysis are summarized here.

7.6.6.3.1 Mechanisms

Heavily overconsolidated stiff clays often are fissured and sometimes are called "brittle" soils, which exhibit significant reduction in strength when sheared slowly beyond the peak. At large displacements during a drained direct shear test of an undisturbed specimen, the strength drops from the peak value to a minimum constant value called the residual strength, as shown schematically in Figure 7.38. For stiff, fissured London clays, the peak value was reported to be at 0.1- to 0.25-in. displacements, while the residual values were obtained at a displacement of approximately 10 in. The post-peak reduction in drained strength occurs in two stages (Figure 7.39): (1) in the first stage, a "fully softened" or "critical state" is reached due to an increase in water content (called dilatancy) and (2) in the second stage, the residual strength is reached due to reorientation of the clay particles parallel to the direction of shearing. In the fully softened stage, particles are randomly oriented with predominantly edge-to-face interaction, while the residual stage represents predominantly face-to-face interaction of particles. The particle reorientation takes place only in clays that have platy clay minerals, and the above two-stage process takes place only in clays that have a clay fraction (percentage by weight finer than 0.002 mm) that exceeds 20–25% (Skempton 1985). The fully softened peak corresponds to the peak strength of a remolded normally consolidated soil, and at large displacements, the

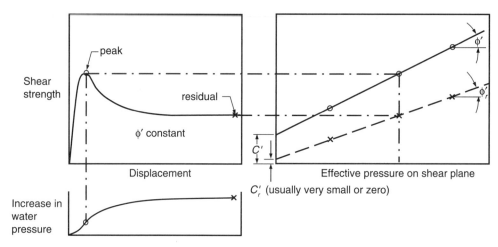

FIGURE 7.38 Peak and residual shear strengths in laboratory tests (after Skempton 1964).

normally consolidated peak strength reduces to the same residual strength as in the undisturbed specimen. These conditions are schematically shown in Figure 7.39.

In stiff overconsolidated clays and shales, the local shear stress may exceed the local peak shear strength, which has fallen to a residual value due to the presence of fissures and discontinuities that act as stress concentrators or due to water inflow or changes in pore water pressure. When the peak shear strength is exceeded at a point in the slope, a redistribution of stresses occurs in the vicinity of that point, thus causing the surrounding areas to be overstressed as well. As a result of this continued overstressing and stress redistribution, a phenom-

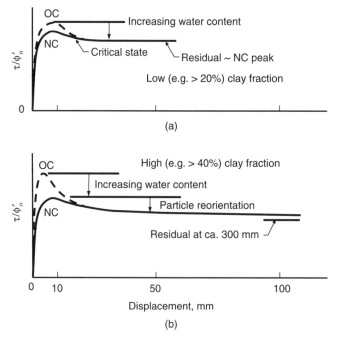

FIGURE 7.39 Mechanisms underlying residual strength development (after Skempton 1985). OC = overconsolidated and NC = normally consolidated.

enon of progressive failure is initiated within the slope. At some point, the entire system becomes overstressed, causing an abrupt failure or landslide. Higher overconsolidation and peak-to-residual shear strength ratios will increase the likelihood of progressive failures in slopes (Bjerrum 1967).

7.6.6.3.2 Some Practical Guidelines

1. The residual shear strength condition exists in cohesive slopes that contain a pre-existing shear surface, such as in old landslides, in bedding shears and folded strata, and in sheared joints or faults (Skempton 1985). Mesri and Shahien (2003) reported that part of the slip surface even in first-time slides and the entire failure surface of reactivated slides exist at the residual condition. The residual strength is a drained shear strength, and therefore an effective stress approach should be followed in stability analysis.

2. Both fully softened and residual shear strengths are fundamental soil properties and can be determined using a direct shear or a ring shear test (Stark and Eid 1993). The relationship between effective normal stress and fully softened or residual shear strength is characterized by curved failure envelopes, with no shear strength at zero effective normal stress. Therefore, a secant friction angle corresponding to the average effective normal stress acting on the slip surface is recommended for use in stability analysis (Stark et al. 2005).

3. The secant friction angles that correspond to fully softened and residual conditions depend on the soil index properties such as the plasticity index. Figure 7.40 shows the empirical correlation between the secant friction angle and plasticity index at an effective normal stress of 50 kPa. Also plotted in Figure 7.40 are the mobilized friction angles back calculated from case histories of slope failures in different soft to stiff clays, encompassing both first-time and reactivated slides (Mesri and Shahien 2003). It has been found that the mobilized residual friction angles represented by square symbols in the figure are well within the ranges of the empirical relationships for residual conditions. For first-time slides in homogeneous slopes or portions of first-time slopes not assumed to be in residual conditions, the back-calculated mobilized values are mostly within or near the ranges for fully softened conditions. Figure 7.40 demonstrates a sufficient degree of reliability of these empirical relationships for the prediction of fully softened and residual friction angles within a practical range of effective normal stresses for use in stability analysis.

7.7 Slope Stabilization Methods

Engineers should recognize that conventional slope stability analysis differs from the stability analysis of a failed slope or landslide. A routine slope stability analysis of embankments, dams, fills, and cuts is an *a priori* process, which means that the factor of safety is calculated as part of the design prior to construction. A factor of safety value of at least 1.5 generally is used in practice for such routine analysis. On the other hand, stability analysis of landslides is an after-the-fact phenomenon in which the factor of safety, defined as the ratio of the shear strength to the destabilizing shear stresses, is actually 1.0. Moreover, since the geometry of the slip surface is known, and the actual pore water pressure along the slip surface can be measured or reliably estimated, the effective stress can be determined. Knowing that the factor of safety at the onset of failure was unity, the mobilized shear strength acting along the slip surface can

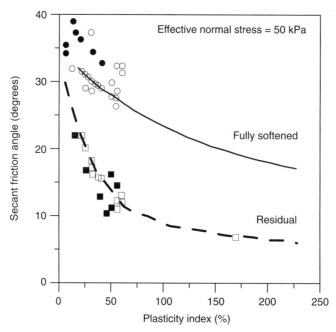

FIGURE 7.40 Empirical relationship between secant friction angle and plasticity index compared with back-calculated friction angle from actual slope failures (after Mesri and Shahien 2003). Back-calculated secant friction angle for (■) reactivated slides, (□) first-time slides with parts of observed failure surface assumed to be in residual condition, (●) entire slip surface of first-time slides in homogeneous slopes, and (○) portions of slip surfaces in first-time slides not in residual condition.

thus be determined in terms of effective stresses. This process is called back analysis and provides extremely valuable information for the design of a remedial measure for a slope.

The need for slope stabilization arises from primarily two types of scenarios: (1) potentially unstable slopes or potential landslides, as predicted from visual observations, monitoring, historical data, and engineering analysis, and (2) actual landslides, where the slope failure has already occurred. The first case is a preventive action, whereas the second case is a remedial or corrective measure. In both cases, stability can be achieved by either increasing the shear strength (resistive forces), decreasing the destabilizing stresses (driving forces), or both. According to Cornforth (2005), the definition of slope or landslide stabilization is providing or restoring "permanent stability under current and reasonably foreseeable future conditions." Although this definition implies a lot of individual judgment, and may be subject to case-by-case interpretations, the broad meaning is quite clear: to ensure safety and stability of a slope to the best of one's engineering abilities.

It should be noted that a relatively inexpensive yet prudent decision may be to simply relocate the planned project (if possible) such that the unstable location of the slope is completely avoided. Realignment of roadways is such an example. If avoidance is not an option, then a comprehensive slope stabilization project needs to be undertaken. Detailed descriptions of various ground improvement methods applicable to slope and landslide stabilization are available in the National Highway Institute reference manual (Elias et al. 2005). Excellent descriptions also are provided in other geotechnical literature (Cornforth 2005; Abramson et al. 2002; Sabatini et al. 1997; Rollins 1994). A summary is provided in Table 7.18

for quick reference, where these methods are presented under the following five major categories: (1) earthwork construction, (2) drainage control, (3) erosion control, (4) retaining walls, and (5) soil reinforcement. In addition to these five major methods, the shear strength and the bearing capacity of a slope also can be improved by various chemical and/or mechanical stabilization methods, such as soil-cement fill, grouting, deep soil mixing with cement, deep injection with lime, vibrocompaction, stone columns, and deep dynamic compaction. Information on these techniques is available in the literature (Elias et al. 2005).

TABLE 7.18 Major Categories of Slope Stabilization Methods

Category/Procedure	Conceptual Diagram

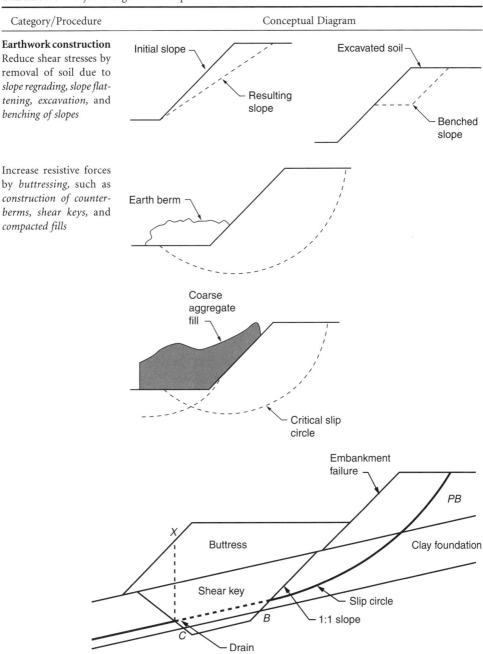

Earthwork construction Reduce shear stresses by removal of soil due to *slope regrading, slope flattening, excavation,* and *benching of slopes*

Increase resistive forces by *buttressing,* such as *construction of counterberms, shear keys,* and *compacted fills*

TABLE 7.18 Major Categories of Slope Stabilization Methods (continued)

Category/Procedure	Conceptual Diagram

Drainage control
Increase shear strength by dewatering the slope using *horizontal drains* and *trench drains* with collector pipes and *interceptor drains*

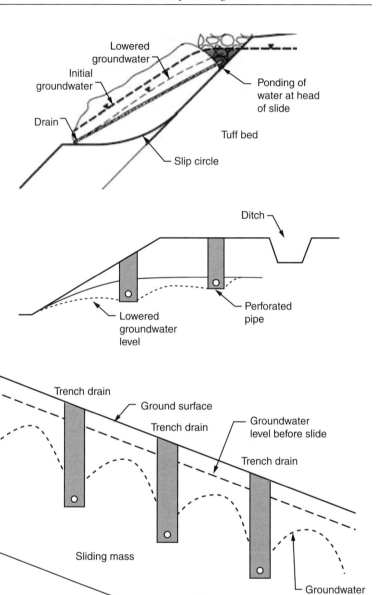

TABLE 7.18 Major Categories of Slope Stabilization Methods (continued)

Category/Procedure	Conceptual Diagram
Prevent the flow (seepage) of destabilizing water through the slope by constructing various *seepage barriers* such as *slurry trench cutoff walls, grout curtains,* and *soil-cement cutoff walls* using deep soil mixing	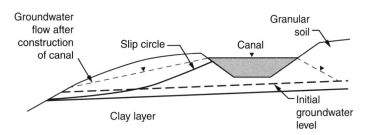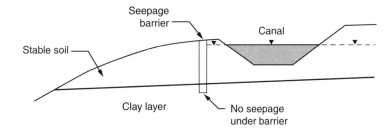
Erosion control Prevent loss of soil from the slope surface by constructing protective layers such as *geocomposite filters, riprap, gabion mattresses, chunam plaster, shotcrete,* and *vegetation*	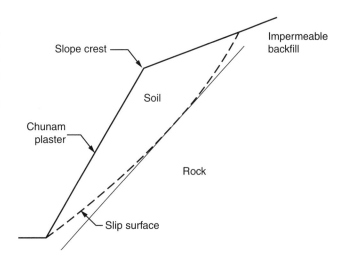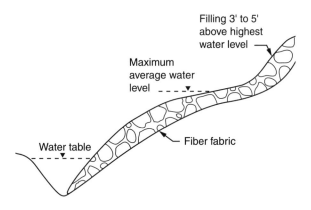

TABLE 7.18 Major Categories of Slope Stabilization Methods (continued)

Category/Procedure	Conceptual Diagram

Retaining walls
Provide temporary or permanent support to the slope by constructing structural walls, such as *gravity* and *cantilever* walls, *driven piles* and *drilled shaft* walls, and *tieback anchor* walls

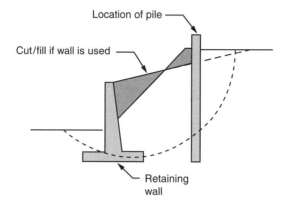

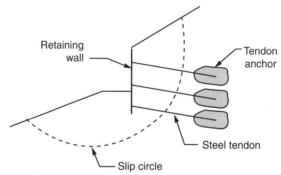

Soil reinforcement
Improve the shear resistance and tensile capacity of the slope mass by inclusion of synthetic reinforcing elements such as metallic strips, steel rods, geosynthetic grids, and fabrics; examples include *soil nailing, micropiles, reticulated micropiles,* and *mechanically stabilized earth walls*

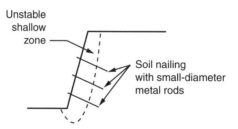

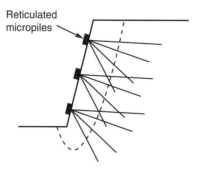

TABLE 7.18 Major Categories of Slope Stabilization Methods (continued)

Category/Procedure	Conceptual Diagram
Soil reinforcement (continued)	

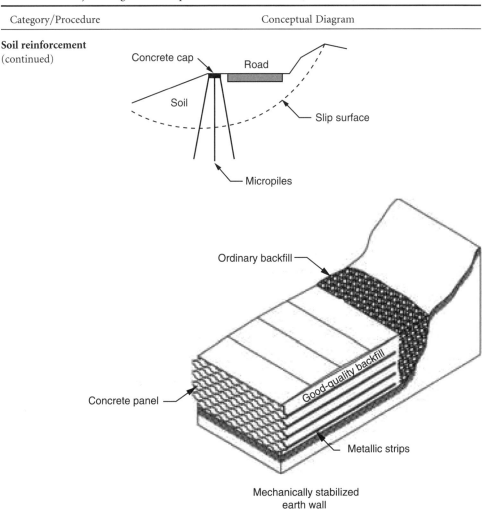

Mechanically stabilized
earth wall

References

Abramson, L.W., Lee, T.S., Sharma, S., and Boyce, G.M. (2002). *Slope Stability and Stabilization Methods*, 2nd edition, John Wiley and Sons, 712 pp.

Anagnosti, P. (1969). Three dimensional stability of fill dams. *Proceedings, 7th International Conference on Soil Mechanics and Foundation Engineering*, A.A. Balkema, Netherlands.

Baligh, M.M. and Azzouz, A.S. (1975). End effects on stability of cohesive slopes. *J. Geotech. Eng. Div. ASCE*, 101(11):1105–1117.

Barnes, G.E. (1991). A simplified version of the Bishop and Morgenstern slope stability charts. *Can. Geotech. J.*, 28(4):630–632.

Bishop, A.W. (1955). The use of slip circles in the stability analysis of earth slopes. *Geotechnique*, 5(1):7–17.

Bishop, A.W. and Bjerrum, L. (1960). The relevance of the triaxial test to the solution of stability problems. *Proceedings of the ASCE Research Conference on the Shear Strength of Cohesive Soils,* Boulder, CO, 437–501.

Bishop, A.W. and Morgenstern, N.R. (1960). Stability coefficients for earth slopes. *Geotechnique,* 10(1):129–150.

Bjerrum, L. (1967). Progressive failure in slopes of overconsolidated plastic clays and clay shales. *J. Soil Mech. Found. Eng.,* 93(5):3–49.

Charles, J.A. and Soares, M.M. (1984). Stability of compacted rockfill slopes. *Geotechnique,* 34(1): 61–70.

Chen, W.F. and Giger, M.W. (1971). Limit analysis of slopes. *J. Soil Mech. Found. Div. ASCE,* 97(SM-1):19–26.

Cheng, Y.M. and Yip, C.J. (2007). Three-dimensional asymmetrical slope stability analysis extension of Bishop's, Janbu's, and Morgenstern-Price's techniques. *J. Geotech. Geoenviron. Eng.,* 133(12):1544–1555.

CLARA-W (2001). Windows Version, O. Hungr Geotechnical Research, Vancouver, BC.

Cornforth, D.H. (2005). *Landslides in Practice: Investigation, Analysis, and Remedially Preventative Options in Soils,* John Wiley and Sons, 596 pp.

Corps of Engineers (1970, 2003). *Slope Stability,* Engineer Manual, EM-1110-2-1902, Department of the Army, Washington, D.C.

Cousins, B.F. (1978). Stability charts for simple earth slopes. *J. Geotech. Eng. Div. ASCE,* 104(GT-2):267–279.

Duncan, J.M. (1996). State of the art: limit equilibrium and finite element analysis of slopes. *J. Geotech. Eng.,* 122(7):577–596.

Duncan, J.M. and Wright, S.G. (2005). *Soil Strength and Slope Stability,* John Wiley and Sons, 297.

Duncan, J.M., Buchignani, A.L., and De Wet, M. (1987). *Engineering Manual for Slope Stability Studies,* Charles E. Via, Jr. Department of Civil Engineering, Virginia Polytechnic Institute and State University, Blacksburg.

Elias, V., Welsh, J., Warren, J., Lukas, R., Collin, J.G., and Berg, R.R. (2005). *Ground Improvement Methods,* Report No. FHWA-NHI-04-001, National Highway Institute, Federal Highway Administration, Washington, D.C., 1022.

Fellenius, W. (1922). *Statens Jarnjvagars Geoteknniska Commission,* Stockholm.

Fellenius, W. (1936). Calculation of the stability of earth dams. *Transactions of the 2nd Congress on Large Dams,* International Commission on Large Dams of the World Power Conference, Vol. 4, 445–462.

Fredlund, D.G. (1974). *Slope Stability Analysis, User's Manual CD-4,* Department of Civil Engineering, University of Saskatchewan, Saskatoon.

Fredlund, D.G. and Krahn, J. (1977). Comparison of slope stability methods of analysis. *Can. Geotech. J.,* 14(3):429–439.

Gibson, M. and Morgenstern, N.R. (1962). A note on the stability of cuttings in normally consolidated clays. *Geotechnique,* 12(3):212–216.

Giger, M.W. and Krizek, R.J. (1975). Stability analysis of vertical cut with variable corner angle. *Soils Found.,* 15(2):63–71.

Highway Research Board (1958). *Landslides and Engineering Practice,* Special Report 29, National Research Council, Washington, D.C., 232 pp.

Hoek, E. and Bray, J.W. (1977). *Rock Slope Engineering,* Institute of Mining and Metallurgical Engineering, London.

Holtz, R.D. and Kovacs, W.D. (1981). *An Introduction to Geotechnical Engineering,* Prentice Hall, Englewood Cliffs, NJ, 733 pp.

Hovland, J.H. (1977). Three-dimensional slope stability analysis method. *J. Geotech. Eng. Div. ASCE,* 103(GT9):970–986.

Huang, C.-C., Tsai, C.-C., and Chen, Y.-H. (2002). Generalized method for three-dimensional slope stability analysis. *J. Geotech. Geoenviron. Eng.*, 128(10):836–848.

Hungr, O. (1997). Slope stability analysis (keynote paper). *Proceedings 2nd Pan-American Symposium on Landslides,* Rio de Janeiro, International Society for Soil Mechanics and Geotechnical Engineering, 3:123–136.

Hungr, O. (1988). *CLARA: Slope Stability Analysis in Two or Three Dimensions,* O. Hungr Geotechnical Research, Vancouver, BC.

Hunter, J.H. and Schuster, R.L. (1968). Stability of simple cuttings in normally consolidated clays. *Geotechnique,* 18(3):372–378.

Janbu, N. (1954a). *Stability Analysis of Slopes with Dimensionless Parameters,* Thesis for the Doctor of Science in the field of Civil Engineering, Harvard University Soil Mechanics Series, No. 46.

Janbu, N. (1954b). Application of composite slip surface for stability analysis. *Proceedings of the European Conference on Stability of Earth Slopes,* Stockholm, Vol. 3, 43–49.

Janbu, N. (1968). *Slope Stability Computations, Soil Mechanics and Foundation Engineering Report,* The Technical University of Norway, Trondheim.

Janbu, N. (1973). Slope stability computations. *Embankment Dam Engineering—Casagrande Volume,* R.C. Hirschfeld and S.J. Poulos, Eds., John Wiley and Sons, New York, 47–86.

Koerner, R.M. and Soong, T-Y. (1999). Assessment of 10 landfill failures using 2-D and 3-D stability analysis procedures. *Proceedings of the Second Austrian Geotechnical Conference,* H. Brand, Ed., Vienna, Austria, 41 pp.

Lam, L. and Fredlund, D.G. (1993). A general limit equilibrium model for three-dimensional slope stability analysis. *Can. Geotech. J.,* 30:905–919.

Lambe, T.W. and Whitman, R.V. (1979). *Soil Mechanics, SI Version,* John Wiley and Sons, 533.

LaRochelle, P. (1989). Problems of stability: progress and hopes. *The Art and Science of Geotechnical Engineering,* E.T. Cording et al., Eds., Prentice Hall, Englewood Cliffs, NJ, 269–290.

Leshchinsky, D. and Huang, C. (1992). Generalized three dimensional slope stability analysis. *J. Geotech. Eng.,* 118(11):1748–1764.

Loehr, J.E., McCoy, B.F., and Wright, S.G. (2004). Quasi-three-dimensional slope stability analysis method for general sliding bodies. *J. Geotech. Geoenviron. Eng.,* 130(6):551–560.

Lowe, J. and Karafiath, L. (1960). Stability of earth dams upon drawdown. *Proceedings of the First Pan American Conference on Soil Mechanics and Foundation Engineering,* Mexico City, 537–552.

Mesri, G. and Cepeda-Diaz, A.F. (1986). Residual shear strength of clays and shales. *Geotechnique,* 36(2):269–274.

Mesri, G. and Shahien, M. (2003). Residual shear strength mobilized in first-time slope failures. *J. Geotech. Geoenviron. Eng.,* 129(1):12–31.

Morgenstern, N.R. and Price, V.E. (1965). The analysis of the stability of general slip surfaces. *Geotechnique,* 15(1):77–93.

Morgenstern, N.R. and Price, V.E. (1967). A numerical method for solving the equations of stability of general slip surfaces. *Comput. J.,* 9(4):388–393.

Morgenstern, N.R. and Sangrey, D.A. (1978). Methods of stability analysis. *Landslides—Analysis and Control,* R.L. Schuster and R.E.J. Krizek, Eds., Transportation Research Board Special Report 176, National Research Council, Washington, D.C., 155–171.

O'Connor, M.J. and Mitchell, R.J. (1977). An extension of the Bishop and Morgenstern slope stability charts. *Can. Geotech. J.,* 14(1):144–151.

Peck, R.B. (1967). Stability of natural slopes. *J. Soil Mech. Found. Div. ASCE,* 93(SM4):403–417.

Pyke, R. (1991). *TSLOPE3: User's Guide,* Taga Engineering Systems and Software, Lafayette, CA.

Ramsamooj, D.V. and Lin, G.S. (1990). Prediction of progressive failure in heavily overconsolidated slope. *J. Geotech. Eng.,* 116(9):1368–1380.

Rollins, K.M. (1994). *In-situ Deep Soil Improvement*, American Society of Civil Engineers, 160.

Sabatini, P.J., Elias, V., Schmertmann, G.R., and Bonaparte, R. (1997). *Earth Retaining Systems*, Geotechnical Engineering Circular No. 2, Report FHWA-SA-96-038, Federal Highway Administration, Washington, D.C., 161.

Sarma, S.K. (1973). Stability analysis of embankments and slopes. *Geotechnique*, 23(3):423–433.

Seed, R.B., Mitchell, J.K., and Seed, H.B. (1990). Kettlemen Hills waste landfill slope failure. 11. Stability analysis. *J. Geotech. Eng.*, 116(4):669–690.

Skempton, A.W. (1948). The $\phi = 0$ analysis of stability and its theoretical basis. *Proceedings of the 2nd International Conference on Soil Mechanics and Foundation Engineering*, Rotterdam, Vol. 1, 72–78.

Skempton, A.W. (1954). The pore pressure coefficients A and B. *Geotechnique*, 4(4):143–147.

Skempton, A.W. (1964). Long-term stability of clay slopes, *Geotechnique*, 14(2):77–102.

Skempton, A.W. (1970). First-time slides in overconsolidated clays. *Geotechnique*, 20(3):320–324.

Skempton, A.W. (1985). Residual strength of clays in landslides, folded strata and the laboratory. *Geotechnique*, 35(1):3–18.

Soong, T.Y., Hungr, O., and Koerner, R.M. (1998). Stability analyses of selected landfill failures by 2-D and 3-D methods. *Proceedings of the 12th GRI Conference, Lessons Learned from Geosynthetic Case Histories*, Geosynthetic Institute, Folsom, PA, 296–329.

Spencer, E. (1967). A method of analysis of the stability of embankments assuming parallel inter-slice forces. *Geotechnique*, 17(1):1–26.

Spencer, E. (1969). Circular and logarithmic spiral slip surfaces. *J. Soil Mech. Found. Div. ASCE*, 95(1):227–234.

Spencer, E. (1973). Thrust line criterion in embankment stability analysis. *Geotechnique*, 23:85–100.

Stark, T.D. and Eid, H.T. (1993). Modified Bromhead ring shear apparatus, *ASTM Geotech. Test. J.*, 16(1):100–107.

Stark, T.D. and Eid, H.T. (1994). Drained residual shear strength of cohesive soils. *J. Geotech. Eng.*, 120(5):856–871.

Stark, T.D. and Eid, H.T. (1998). Performance of three-dimensional slope stability methods in practice. *J. Geotech. Geoenviron. Eng.*, 124(11):1049–1060.

Stark, T.D., Eid, H.T., Evans, W.D., and Sherry, P.E. (2000). Municipal solid waste slope failure. II. Stability analyses. *J. Geotech. Geoenviron. Eng.*, 126(5):408–419.

Stark, T.D., Hangseok, C., and McCone, S. (2005). Drained shear strength parameters for analysis of landslides. *J. Geotech. Geoenviron. Eng.*, 131(5):575–588.

Taylor, D.W. (1937). Stability of earth slopes. *J. Boston Soc. Civ. Eng.*, 24:197–246.

Taylor, D.W. (1948). *Fundamentals of Soil Mechanics*, Wiley, New York, 700 pp.

Terzaghi, K.V. (1936a). Stability of slopes of natural clay. *Proceedings of the First International Conference on Soil Mechanics and Foundation Engineering*, Cambridge, MA, Vol. 1, 161–165.

Terzaghi, K. (1936b). The shearing resistance of saturated soils. *Proceedings of the First International Conference on Soil Mechanics*, Vol. 1, 54–56.

Thomaz, J.E. (1986). *A General Method for Three Dimensional Slope Stability Analysis*, Information Report JHRP-86-4, Purdue University, West Lafayette, IN.

Varnes, D.J. (1978). Slope movement types and processes. *Landslides—Analysis and Control*, R.L. Schuster and R.E.J. Krizek, Eds., Transportation Research Board Special Report 176, National Research Council, Washington, D.C., 11–33.

Whitman, R.V. and Bailey, W.A. (1967). Use of computers for slope stability analysis. *J. Soil Mech. Found. Div. ASCE*, 93(4):475–498.

Xie, M., Esaki, T., and Cai, M. (2006). GIS-based implementation of three-dimensional limit equilibrium approach of slope stability. *J. Geotech. Geoenviron. Eng.*, 132(5):656–660.

8

Expansive Clays

by

Thomas M. Petry
Missouri University of Science and Technology, Rolla, Missouri

8.1 Introduction

"Expansive" clay soils are found worldwide, on all continents, including all 50 states of the U.S. An expansive clay is one that exhibits significant and possibly damaging volume change potential when its moisture content changes, ranging from less than 2.5 cm (1.0 in.) to over 50 cm (20 in.). Depending on the levels of moisture content in the soil mass and the physico-chemical environment causing either gain of moisture or loss of moisture, these volume changes can be either increases or losses of volume. The clay itself and its physicochemical makeup are only part of how much volume change will occur. The other factors that affect

FIGURE 8.1 Typical distress caused by expansive clays.

determining volume change which occurs are discussed in detail below. This chapter, as part of a handbook for engineering practice, includes what is essential to know for predicting behavior, but not an exhaustive discussion of all the theoretical and scientific details about these clays and their behavior. Figures 8.1–8.3 are photographs of damage caused by expansive clay soils.

As will be discussed along the way, expansive (or contractive) clay behavior is affected by the type(s) of clay minerals present, the percent by weight of the soil that is clay-sized particles (as little as 10% to nearly 100%), the particular soil chemical properties of the clay and soil, and the level of moisture, or moisture content, in the soil. In addition, the denseness in terms of how the particles of clay are packed will be a factor, as well as how these particles are arranged relative to one another. The particular stress history of the soil mass is an important factor, as are the stresses that will exist in the soil mass during the lifetime of a project. It is well known that the amount of "negative" potential energy, or "suction," in the clay and soil is a significant part of the overall stress situation in the clay soil mass. All of these factors will be discussed as this chapter progresses. What is important to the practicing geotechnical engineer is how to determine the overall site conditions and to test enough samples in ways that provide properties useful in predicting behavior of the soil mass during the life of the project of interest.

Although most soil properties are obtained by testing using standardized tests, each test, using relatively undisturbed or remolded samples, must be done in such a way as to represent actual field conditions in order to obtain results that can reliably be used to predict field behavior. Each of these will be discussed below. After more than 37 years of experience with expansive clays, the author has come to find that there are as many ways to predict expansive

FIGURE 8.2 Differential movement between wall and column caused by expansive clay soil.

FIGURE 8.3 Structural damage caused by expansive clay soil.

clay behavior using test results as there are those who strive to do so and that expansive clay behavior is most reliably predicted by those who have experienced a particular clay soil and the particular location worldwide where it is found. Therefore, the discussion below is of a general nature and includes guides for engineering practice using the simplest and least expensive of tests, properly done to achieve the most reliable results. Anyone who wants to delve into the depths of the theoretical behavior of these clays and examine the many theories available is directed to *Fundamentals of Soil Behavior* by Mitchell and Soga (2005).

8.2 Basic Causes of the Problem

8.2.1 Clay Minerals

Clay minerals are known as hydrous aluminum silicates. These minerals generally are made from stacks of two types of sheets: silica tetrahedral sheets and alumina octahedral sheets. They are rightly named as sheets, because each is just a few angstroms thick and can be thousands of angstroms wide in each of its other dimensions. Each type of clay mineral family consists of stacks of these elementary sheets in differing arrangements. The clays that are expansive in nature consist of silica tetrahedral sheets that contain substitutions of aluminum ions for some of the silica ions and alumina octahedral sheets with substitutions of either iron or magnesium for some of the aluminum ions. As can be understood by considering each of these ions and their natural charges, those that are present affect the clay behavior differently. Silica has a +4 charge and aluminum has a +3 charge, while iron can have and magnesium has a +2 charge. The substitutions described above, therefore, cause the silica or alumina sheets to have a net negative charge for each substitution. The basic reason why clay minerals are expansive starts first with their inherent negative charges. It follows that when more substitutions are present, clay will have a higher potential to be problematic, since more moisture will be required, in addition to balance charges in the clay soil.

8.2.2 Associated Cations

There are many cations, or positively charged ions, present in the atmosphere and in the soil of differing types and concentrations. In clay soils, these cations provide sources of positive charges to assist in offsetting the negative charges in the clays mentioned above. When these cations are close enough to the clay mineral surfaces, they essentially become part of the overall charge system of the clay and are associated with the clay in the cation exchange complex (CEC). The remaining cations are part of the soil chemistry not closely associated with the clay and are part of those ions in the pore water system of the soil.

The most abundant cations found in soils are sodium, potassium, calcium, magnesium, and iron, followed by several others including silicon and aluminum. The particular cations associated with a clay and in the pore water of the soil is dependent on the chemical history of the soil. Cations can be moved into and out of the pore water of a soil by various forces, but the movement of water through soil voids is the most likely cause. Exchange of cations into and out of the CEC can happen when the concentrations of cations in the pore water are high enough relative to the type and concentration of cations in the CEC. In some cases, this is done to improve the behavior of clays by artificially raising concentrations of desirable cations in the pore water of a clay soil. An illustration of the lyotropic scale is shown below, where the cations on the right of the listing will more easily exchange for those on the left of them on the scale:

Na < Li < K < Rb < Cs < Mg < Ca < Ba < Cu < Al < Fe < Th

As would be expected, sodium is a cation which requires significant water associated with it to be satisfied. It is something experienced by anyone who takes in more sodium than normal and gains weight because of the water held in their body by the sodium present. It turns out that a clay with primarily sodium in its CEC and pore water requires the most water to satisfy its physicochemical needs. On the other hand, a clay soil that contains mostly calcium will have a very significantly lower need for water.

Another phenomenon associated with cations and clays is well known to those who have used a hydrometer test to determine the percent clay-sized particles there are in a soil. A sodium hexametaphosphate solution is used to "disperse" the clay particles so that they will act "individually." The key part of this solution is the sodium, which, when the overall concentration of cations is small, can cause this dispersion to take place. Also, when too much of the solution is placed in the test cylinder with the soil, the clay "flocculates" or forms flocks of particles that fairly quickly fall to the bottom of the cylinder. Unfortunately, there are clays in nature that "disperse" without any addition of the solution and cause many problems.

8.2.3 Water Layers

Because of the structure of a water molecule, the two hydrogen atoms are located near one end of it and the oxygen atom is found near the other end. This causes the water molecule to act as sort of a "bar magnet," with positive and negative ends. Many of the behaviors of water, including its overall molecular structure in fluid and solid states, are caused by water molecules being this way. This phenomenon holds a stream of water together somewhat and is partly responsible for the surface tension capability of water.

Along with the cations present near and away from clay mineral surfaces, there are even many more water molecules present, some associated with the clay mineral surfaces and broken bonds, some associated with cations, and others associated with each other. The water molecules tend to form "layers" of water inside of and around clay particles. Those layers most closely associated with the clay are very tightly held and are more difficult to move around than those found farther away from the clay surfaces. How tightly these water layers are associated with the clay becomes less and less the farther they are from the clay surfaces. There are a few water layers that remain with the clay even when heated to normal oven temperatures of 100°C. These are called the "adsorbed" water layers. Surrounding them are the layers associated either strongly or progressively nearly not at all with the clay, making up the rest of what is called the "highly viscous" water layers. Outside of these layers, the water in the soil is not considered associated with the clay at all.

8.2.4 Cation-Water Effects

The overall concept of cation and water association with clay minerals is as follows: in order to balance the charge imbalance caused by the substitutions of ions in the mineral layers, both cations and water molecules act together. The cations present act as part of the clay makeup, and the amount of water needed to complete the balance is determined by the particular clay mineral makeup and the types and concentrations of cations present in the soil. If there is insufficient water present to complete the balance of charges, the soil will have a net negative energy with the potential to bring available water to it, with the result being volume increase.

Neither clay mineralogy nor soil chemical makeup normally is determined during geotechnical engineering investigations, because of the expense and time required to do so. These are very significant factors which affect clay soil behavior and, instead, their effects on behavior are measured as described below.

8.2.5 Atterberg Limits and Indexes

Certain of the behavior limits first named by Atterberg and expanded upon by many are useful in indicating the potential of a clay to have expansive characteristics. The most often used is the plasticity index (PI). Chen (1988) provided a chart of expected expansive behavior relative to a clay soil's PI, as shown in Table 8.1. If a clay has a PI greater than 15, one must suspect some expansive behavior, and if it has a PI of 55 or more, one should expect the clay to have highly expansive behavior. However, there are some nonexpansive clays that have a high liquid limit (LL) because of the amount of clay in the soil, while there are some expansive clays that have a relatively low LL since there is so little clay present. If the PI is divided by the percent of clay-sized particles present by weight, the activity (A) is the result. In reality, this magnifies the PI according to the

TABLE 8.1 Classification of Expansive Soils Based on PI

Swelling Potential	PI
Low	0–15
Medium	10–35
High	20–55
Very high	35 and above

After Chen (1988).

percent clay by weight, giving, in a sense, the PI of the clay itself. As shown in Figure 8.4, those materials that have an activity of over 1 are suspected of being expansive.

Another useful property derived using Atterberg limits is the liquidity index (LI). The LI is found by subtracting the plastic limit (PL) from the soil's moisture content and then dividing this by the soil's PI. In reality, it describes where the soil's moisture content lies relative to the zone of moisture contents where the soil acts with plasticity. If this number is a negative number, for instance, the soil is drier than its PL, and if it is a positive number, the soil's moisture content is between its PL and LL. Seldom is the LI above 1 or a soil mois-

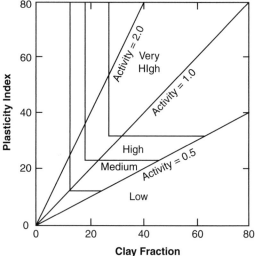

FIGURE 8.4 Potential expansive nature of clay.

ture content above its LL. It has been determined by experience that, generally, a clay soil, if the moisture is available, would have a natural LI of 0.15–0.2, resulting in a moisture content slightly higher than its PL. This is most often significantly less than a saturated moisture content for that soil mass.

The shrinkage limit (SL) is described as that moisture content where, as the soil is drying, volume change ceases, thereby becoming the lower limit of volume change. In theory, this is correct, but in terms of how it is generally measured, as the saturated water content at that point, it is more moist than the actual limit of shrinkage. The difficulty is in the way the water content is determined and in how shrinkage occurs in soils. Shrinkage includes the removal of water by desiccation and the resulting capillary forces at the fringes of the clod, reducing the volume of the clod. Since capillary forces cannot occur in a saturated soil, using the saturated moisture content to define the SL is incorrect and the value determined is larger than the real moisture content at the lower limit of volume change. The standardized SL found is useful, however, in that if a soil has a moisture content below this amount, the soil may be highly expansive. In fact, the author has measured a swell pressure of 20 kg/cm^2 (20 tsf) in such soils.

8.3 Grain-to-Grain Structures

Since clay minerals are sheet-like, it is not surprising that clay particles are generally flake-shaped. They can be made up of only one fundamental stack of a clay mineral or can have many of these layered one on another. Since they are sheet minerals, their orthogonal dimensions are very large, up to 2 μm in width, relative to their thickness. The arrangement of these flake-shaped particles in nature can vary from a "house-of-cards" random arrangement to a "deck-of-cards" parallel arrangement. Depending on the mode of their deposition as a clay mass and the stress history after deposition, these particles can exhibit many particle-to-particle structures. Also, stresses history can change random orientations into nearly parallel ones over geologic time. It is important to realize that these are descriptions of clay-grain-to-clay-grain structures and do not represent the complexity of real soil mass structures.

In reality, groups of clay particles or grains do become arranged as described, and then these groups interact with other clay particle groups, and silt, sand, and gravel particles, to form the overall soil grain structure in a soil mass. One needs to visualize the groups of clay grains interacting with the other grains that are deposited with them. This becomes even more complicated when these soil masses are remolded as a result of construction processes. The overall soil structure contains grains, groups of grains, and clods of varying sizes.

When sufficient stresses are impressed long enough on the clay soil mass, particles that are or become parallel in their relative positions can be compressed and held at these positions to form diagenetic bonds, which tend to "lock" particles together. Only with sufficient weathering cycles, including shrink and swell, can these bonds be released. Following this release, the clay may have significantly more potential to swell. Also, more parallel clay particle arrangements, when the bedding planes are at or near horizontal, present the most significant swell potential. As will be discussed later, the ease with which moisture can enter a drier clay is affected a great deal by soil mass grain-to-grain structure.

8.4 Clay Moisture Potentials

When clay soils lack sufficient moisture to balance the physicochemical charges present, they exhibit negative moisture potential. This negative potential is described as soil suction. There are two sources of this negative potential. The first and most important is called matric, or matrix, suction, and the source is the physicochemical need of the clay for moisture to balance charges. The second is osmotic suction and occurs when capillary potentials and cation differential potentials are present. Moisture will tend to move in a water system to even out cation concentrations, so when a higher concentration is located in one part of the soil mass than in another, moisture tends to move toward the larger cation concentration, to even out the differences. In addition, physical capillary spaces can have water in them, and the capillary tension moves the water. The sum of soil suction is called total suction of the soil mass. These potentials are measured in units of pF, which is the log base 10 of the equivalent height of a water column in centimeters that would cause the same amount of positive pressure. Therefore, a pF of 6 in a soil would represent a negative potential equivalent to 10^6 cm of water pressure, which is 10,000 m of water head potential!

Of course, there can be positive water potentials in clay soils as well. However, this would happen only when the soil is saturated, and the clay soil's negative potential would be essentially zero at that point. Therefore, the negative moisture potentials in clay soils can be far more important than positive ones. Clay soils have a pF of 2.0–2.5 at what is called field capacity and have a pF of about 3.4 at their PL and 5.5 at their SL.

8.5 Moisture and Water Movements

Moisture in soil masses can exist in the same forms know in the atmosphere: fluid, gaseous, and solid (frozen). Both fluid and gaseous or vapor water can move in the soil or be transferred when the moisture potential gradients are high enough. Moisture potential gradients occur when there is a difference in moisture potential between two locations in the soil mass some distance apart. The gradient is the potential difference divided by the distance between the points. As moisture is caused to move from one place to another in the soil, there are moisture potential losses. The gradient has to be sufficient to overcome the losses that occur in order for the moisture to move. Part of moisture energy potential is the vertical position differences between the two points in consideration. It is a fact that moisture does move in any direction where the gradient is high enough. Also, desiccation of the surface and near-surface clay soils does increase their negative potential, further causing moisture to move vertically toward the surface.

Part of the regular desiccation of near-surface soils is the evapotranspiration that happens through grass, plants, and trees. The roots of grass and plants can have significantly negative moisture potential and thereby cause moisture to leave the soil and move into the grass and plants and, eventually, to the atmosphere through leaves. The potential to take moisture out of soils is greatest in trees. Research has proven that a tree's roots can spread as far away from the tree as the tree is tall or more and desiccate the clay soils. In fact, this phenomenon is being used to take pollutants from the ground as well.

When moisture moves from one part of a soil mass to another, it does not have to be in fluid form. In fact, much of the moisture movement in clay soil masses is actually by transfer from one part of the clay to another slowly as vapor. This transfer of moisture continues as long as the moisture potential gradient making it happen is large enough. Of course, once a clay soil reaches the moisture level where its charge imbalance is satisfied, the moisture potential will be such that the transfer of moisture will stop. The soil may continue to become wetter, from some source of water, but the clay's potential to move moisture will have ceased to exist. Because of the many sources of moisture that can exist in a soil mass and be provided by weather, it is difficult to keep a clay soil from transferring moisture to it when it is relatively dry, even when it is under a structure. Figure 8.5 shows the drying effects of evapotranspiration and the depth of moisture changes in subgrades.

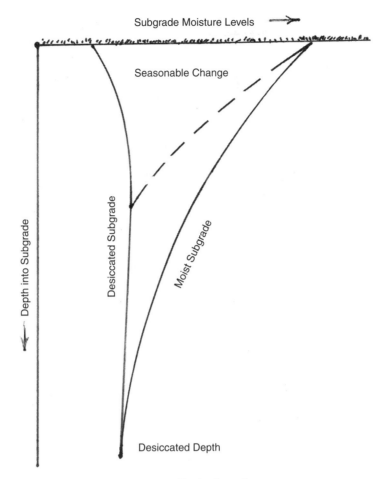

FIGURE 8.5 Water content profiles in the active zone.

8.6 Moisture and Soil Mass Structures

Another source of wetting is due to the nature of the deposition of the clay soil, bedding planes that occur, and changes to bedding plane orientation due to tectonic movements. Other paths which can bring moisture into a clay from the atmosphere can be present because a clay soil is deposited in layers along with soils with much higher permeability, such as silts and sands. This can happen in near-shoreline deposits and in deltaic deposits, for instance. The layered nature of alluvial soils can also contain clays among silts and sands or even gravels. These more permeable materials then provide easier paths for moisture to reach clays. This phenomenon is most problematic when the clay soils are relatively dry and dense because of their geologic history.

Clay soils, weathered from shales especially, tend to have bedding planes normal to the axis of applied stress history. They also have micro grain-to-grain parallel structures of particles. Although these soil masses have relatively low permeabilities perpendicular to their bedding planes (vertical), moisture does move into the clay and, because of particle alignment, most swell is perpendicular to their bedding planes. Some of these weathered shales have experienced enough tectonic movements to cause minor faults in them, thereby providing paths which will bring moisture deeper into the clay soils. In extreme cases, where the shales have been shifted by tectonic forces, they have had their bedding planes oriented to near vertical. This would cause the vertical permeability of these clay soils to be much greater than if the bedding planes were near horizontal. The near-vertical bedding planes also increase the likelihood of vertical cracks in the ground that provide easy avenues for moisture to deeply invade the clay soil mass. In the prediction of depth of moisture change in a clay soil mass, one must take into account the orientation of all cracks and fissures in the soil mass and not forget that all owners wish to beautify their properties with artificial watering that will hasten at-depth moisture increase in the clay soil.

8.7 Weathering Effects

In light of the discussion above, one would expect that expansive clay subgrades do not have behavior dictated only by clay-grain-to-clay-grain interactions because of their actual micro- and macrostructure derived from their deposition and geologic histories. They can definitely be overconsolidated because of their past stress history, especially when weathered from a shale. Weathering over time causes other changes to the macrostructure also. Nearly all climates have periods of drought or drying and periods of wetting or moistening. As these cycles progress, the clay mass is subjected to changes in moisture contents to some depth. During drying cycles, this means shrinkage of surface and near-surface clay and development of cracks. If the drying is severe enough, these cracks can penetrate many feet into the soil mass and become near horizontal, as well as near vertical. As moisture re-enters the clay soil mass, the clays take on moisture initially along the cracks into which water penetrates. Eventually, these crack channels of moisture movement are swelled closed, and further moisture transfers from areas of more moisture to those of less, until, if the moisture source continues, the clay soil mass comes to a sort of equilibrium moisture regime status. This happens when the moisture potential differences do not provide enough gradient to move the moisture further. A phenomenon similar to this also happens under structures, and this point may be called the equilibration of moisture for the situation that exists long enough.

Since the micro- and macrostructure of clay masses are not homogeneous in all directions to start with and are further disrupted by the natural shrink-swell that happens over time, these materials do not have the same properties in all directions and no longer are only parallel and perpendicular to their original bedding planes. This means that the overall effect results in a clay soil subgrade that has many interconnected cracks and fissures that never go away. Also, the clay soil now is made up of irregular-shaped blocks of soil mass. Figures 8.6 and 8.7 show how this phenomenon looks at the surface and when clay subgrades are cut open.

When loss of moisture at or near the surface of a clay soil subgrade occurs, the amount of water in the clay lessens and capillary forces cause compression forces of considerable size to shrink the soil mass. Over many repetitions of shrinkage, these forces result in significant increase in effective stresses in the soil mass. If these forces are great enough and are applied over long times in the history of the soil mass, they cause additional overconsolidation to occur. This means that a clay weathered from a shale, being overconsolidated, will become further overconsolidated by this phenomenon. Of course, this depends on the climate affecting the clay soil mass. The climates that result in the most cycles of shrink and swell from weather are the semiarid climates.

FIGURE 8.6 Typical surface cracking on expansive clay.

FIGURE 8.7 Typical cracked and fissured subgrade of expansive clay.

8.8 Swelling and Shrinking

8.8.1 Swelling

Swelling, or the tendency for volume increase upon taking on moisture, begins when the current moisture content of the clay is below that which provides the desired charge balance within the clay. As moisture is caused to move into the clay, water molecules are forced into the spaces between clay mineral sheets. These act in a fashion similar to forcing oil into the cylinder of a jack, pushing things apart. Soil scientists and engineers agree that a large portion of the swell that occurs is inside clay particles, between clay mineral sheets. This is why swelling can express such large pressures. The physicochemical forces can be extreme and can cause swelling that can break large reinforced concrete grade beams. The amount of swell also has been observed as extreme, an example of which is the ground surface rising over 0.91 m (30 in.).

The swelling and pressure potential is affected by several factors. The first is the clay mineral and soil chemical situation. Second is the amount of moisture needed to reach the balance of charges for the clay. This moisture content is likely to be a fairly low LI and the soil is not saturated. Third is the particular effective stress in the soil mass where the swelling is happening, where more stress reduces the actual swell that can happen, as seen in Figure 8.8. Fourth is the denseness of the structure in the soil. This has two parts. A clay soil with a higher dry unit weight can express more swell. The denseness of the clay soil structure also affects its permeability and the ease with which moisture can enter the clay. A clay subgrade with light

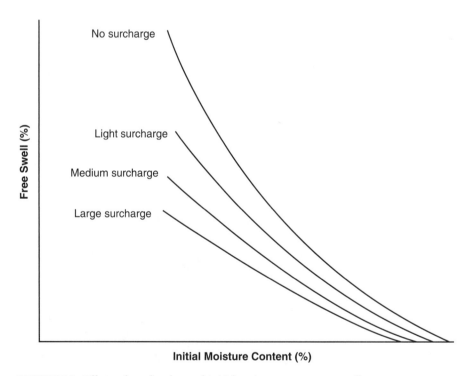

FIGURE 8.8 Effects of overburden and initial moisture content on swell.

loads, which also has been compacted dense and dry of optimum, represents a subgrade with the most swelling potential. Lastly, the phenomenon of clay swelling is not elastic in nature, such that each swell event will be different, even for the same soil and environmental factors discussed above.

8.8.2 Shrinkage

Shrinkage occurs when the moisture potential energy outside of the clay causes a sufficient gradient to move water out of the clay. The amount of shrinkage that occurs is affected by a few factors. First is the moisture content where shrinkage initiates. If this is above the real soil SL, then shrinkage can occur. If it is below the real SL, no shrinkage will occur. Also, the higher the initial moisture content, the more shrinkage potential a clay soil has. Second is the denseness of the clay soil structure. A denser structure will allow less shrinkage to occur. Third is the interconnected cracks and fissures in the clay soil mass, as these will dictate where and in what direction the shrinkage will occur. Lastly, the phenomenon of shrinkage in clay soils is not elastic, in that each shrinkage event will differ, even for the same soil and changes in environmental factors discussed above.

8.8.3 *In Situ* Situations

When predicting behavior of clay soil subgrades, swell is normally what is being considered. The amount the surface or a project element may be moved upward by swell is called the potential vertical rise. This swell is affected by the factors described above in all their possible combinations. Use of Atterberg limits alone, initial moisture content alone, or any of the other variable properties of the soil mass alone cannot accurately provide these predictions. The layers of the soil subgrade that have the potential to expand are those that are drier than their desired moisture content, which will have moisture available and have a low enough overburden pressure on them to allow swell. Surface layers are the most likely, then, to swell, and layers deeper in the soil mass are less likely to do so, because of both their moisture contents and overburden stresses. Somehow the geotechnical engineer must, with sampling and testing data, be able to determine the depth of clay soil that will expand and how much it will expand.

Although shrinkage in part of the clay soil supporting a project, coupled with swell in some other part of the supporting clay soil, will cause the most damaging effect, shrinkage generally is not determined or predicted. Part of the reason is because there is no standardized test to provide data to predict it. Therefore, the geotechnical engineer must somehow prevent shrinkage from affecting the project while predicting swell and recommending how to deal with it.

The prediction of time for swell to occur, or rate of expansion, has been investigated, and there are some reverse consolidation time procedures available. However, the author has not seen them applied in practice since the availability of moisture and time when it will be available within the project life are not predictable with certainty.

8.9 Shear Strength

The subject of shear strength of clay soils has undergone much discussion in the literature in the recent past. Some state their belief that these soils have cohesion and some state that cohesion does not actually exist. It would seem most useful to look at the way shear strength can be developed in clays and how it varies as *in situ* situations change. Shear strength must come from interaction of particles in the soil or clods or even blocks of material. In each case, it is what happens between the surfaces of particles that provides shear strength.

First, there is some form of frictional resistance between particles. This friction happens when sufficient effective stress exists between these particles. Effective stress in an expansive clay soil comes from applied loading to the clay by project elements, from the effective unit weight of soil materials supported by the particles, and from the negative moisture potential within the clay itself. Since the first two of these sources of stress are discussed at length elsewhere, only the third will be discussed here. Negative moisture potential, or suction, in clays is variable with the particular clay soil and its soil chemical makeup and with the moisture content of the soil. In fact, when these soils are fairly dry, they can express significant soil suction, but as they become more moist, this suction essentially approaches zero. As discussed above, this happens at a moisture content somewhat above the PL, at an LI of 0.15–0.2. Thus, above that moisture level, the clay must rely on the other sources of confinement to develop friction. In nature, an expansive clay soil can exist at very dry to even saturated conditions, so that the geotechnical engineer must predict the correct expected moisture conditions during the life of a project in order to recommend the shear strength to use for design for friction components of shear strength.

The second factor involved in shear strength for clays is cohesion, or interactive shear strength that is independent of the effective stress between particles. When a clay contains little or no soil suction yet shows enough strength to adhere together and resist the shear forces applied during the LL test, then cohesion must exist between the particles of the soil that are interacting. It is well known that this cohesion reduces as the clay contains more moisture as it progressively becomes wetter in the zone of plastic behavior, from the PL to the LL. The other factor that affects the cohesion developed during shearing is the rate at which shear occurs. During tests that utilize the quickest rates of shear, the cohesion is highest, and during very slow shearing, it does not exist. Some would say that this is due to soil suction dissipation during differing drainage conditions while shearing, but soil suction is not something that dissipates because it is caused by the physicochemical need of the clay, which does not change appreciably with constant moisture content. The reality is that the water molecules that associate with clays in what is called the double water layer are more difficult to deform relative to one another than water that is not associated with the clay. Shearing between these associated water molecules is what causes cohesion, a property that likely will increase as more tightly associated water molecules around the clay are sheared.

Dry unit weight differences affect shear strength because more densely packed particles, clods, etc. interact with friction and cohesion more, whereas more loosely packed materials interact less. The shrink and swell of expansive clay soils cause their overall denseness to reduce, thereby causing lower shear strengths. Swell, in itself, as more moisture comes into the clay and the soil structure becomes looser, causes significant loss of shear strength. Shrinkage in an expansive clay soil mass will cause cracks, fissures, clods, and blocks of material to form, all reducing the contacts that can provide shear strength. This is most damaging in slopes cut into expansive clays, where cycles of shrinking and swelling over time will reduce the soil mass strength to its lowest friction potential. Many slopes have had shallow slope failures along cracks and fissures as well as along bedding planes because a friction angle of 25°, for instance, becomes half of that over time. Figure 8.9 shows a bridge approach slope in an expansive clay failing and pushing on the bridge supports, and Figure 8.10 shows a second failure of an expansive clay slope where the first one covered several lanes of an urban interstate highway.

FIGURE 8.9 Slope movement under bridge and against supports.

FIGURE 8.10 Typical slope failure in expansive clay.

8.10 Variations of Properties

Some time ago, the author and associates performed research to find out just how variable properties of all types are within a project site (Petry et al. 1980). The site chosen was in a borrow area of the Eagle Ford clay shale that had weathered to a highly active clay soil. The site was 5 acres in size, and a four-application, four-replication set of plots were laid out on the site. Within each plot, holes located randomly were sampled to 1.8 m (6 ft). Relatively undisturbed and disturbed samples were taken from each of 80 holes and were subjected to testing to determine all their normal physical and selected chemical properties. The most interesting facts determined are that the chemical properties of the soil varied much more than its physical properties. However, all properties varied far more than expected, and in many cases the statistical variance of a property exceeded its statistical mean. What this indicates to geotechnical engineers is that they need to expect all properties within the soil mass of a project to vary significantly and that they need to sample randomly and use statistical analyses to predict properties used in recommendations for design.

8.11 Geotechnical Investigations

8.11.1 Philosophy

The purpose of a geotechnical investigation is to develop the necessary information to make recommendations for a project as to how to design, construct, and maintain the project relative to the geotechnical situations found. As economies have grown stressed and competition has increased among those who perform geotechnical investigations, pressure has grown to limit the site and testing part of the investigation yet provide truly reliable and effective recommendations. This situation also may become more stressful for geotechnical engineers as they are part of a design-build team that is required to provide timely responses to requests for recommendations. This can be extremely difficult when the materials one is investigating are expansive clay soils that contain the kinds of variability in behavior and properties as discussed above. In addition, the geotechnical engineer of record is legally expected to use the "standard of care" for the area where the project is located as the minimum plan followed.

8.11.2 How Many Borings

It is important for the geotechnical engineer to review all information—historical, geological, topographical, soil origins, and engineering—before a necessary visit to the site. Although a basic number of borings is dictated by the size of the project area and the importance of the project, additional borings may be warranted because of what is found during these surveys of information. The expected vertical and horizontal variability of site materials must be considered. When planning to use statistical analyses of the materials and properties found, the least significant statistical number of three data points must be considered. It is best to randomly locate borings across the site and to place borings where significant parts of the project will be located. The more important the project and the results of possible damage to it caused by the expansive clay subgrade, the more information needed. Lower numbers of borings usually dictate more interpretation and prediction of properties and usually result in more conservative recommendations.

8.11.3 The Site Soil Profile

The properties needed for an expansive clay soil subgrade include all the factors explained above that affect behavior. The depth to "seasonal" moisture change or the level into the subgrade where expected volume change will cease must be determined across the site. The depth to rock or nonactive layers must be found as well. The depth to the water table is essential to predicting moisture change and must be found if a water table exists or will exist during the life of a project. The presence of cracks, fissures, faults, slickensides, bedding planes, and other planes of possible weakness needs to be determined.

The clay soil in the zone expected to change volume normally is sampled using some kind of relatively thin-walled sampling device to obtain relatively undisturbed samples. The best knowledge possible of how these materials will act as intact materials is essential. Below that depth, sampling usually occurs every 1.5–3 m (5–10 ft) until the material that will be used to support project loads is explored at least 1.5 m (5 ft) more after it is located.

Properties that are determined and used to create a "profile" with depth include, at the least, the following: moisture content, dry unit weight, Atterberg limits, swelling behavior, and shear strength. In addition, the LI can be calculated and soil suction properties may be found. It is not necessarily standard practice to conduct each type of test on each sample taken from the field, although that would provide the best data set to use for predictions. Usually, economy of work to be performed vs. usefulness of the properties to be determined dictates which samples are tested for which properties. At least three results for all properties for each layer of material tested are needed to provide statistically significant information. This "characterization" of the site profile is the basis for predictions and recommendations to be made.

Testing that has to be done on relatively undisturbed samples includes determination of dry unit weights; studies of cracks, fissures, etc.; swell testing; and shear strength testing. It is extremely important that the samples used for swell and shear strength testing represent the actual expansive clay soil subgrade in their moisture content and dry unit weight. It is unfortunate that the stresses that were part of their environment *in situ* have been removed, but this can be overcome by proper testing techniques.

8.11.4 Other Site Information

Other information gathered during site visits and boring and sampling adds to the soil profile information when the geotechnical engineer considers the whole site and project and how they interact. An important feature of all sites is topography. Topography dictates how surface and subsurface drainage moves and how it will affect the project. Another important feature is outcroppings of materials and rock, which, when considered, are keys to layering, bedding plane directions, and the types of materials to expect under a site. The indicated behavior and possible problems observed for project structures around the site can be helpful as well. Surface cracking and observed fissures, etc. can provide information on possible moisture movement and directions of possible volume change.

Perhaps the most overlooked features on a site are the types and amounts of vegetation and trees naturally occurring on the site. Considering the climate that affects the site soils, this information can be used to estimate the depth to water and the active nature of a clay soil. It has been established by research and observation that the roots of a tree, especially trees that favor clayey soils, which grow quickly and have shallow root systems, can spread as far away from the tree as it is tall at any point of maturity. Figure 8.11 shows a tree whose root has grown out to where a source of moisture was located under the structure. Considering the physico-

FIGURE 8.11 Typical tree root, bush, and poor exposure of building exterior.

chemical energy levels of tree roots, they tend to dry out a "bowl" of the clay under and around them. When the tree is removed, this "bowl" of dry soil is left and requires special procedures during construction to improve and use for support of the intended project. It has been observed that bushes that are a significant size, vertically and horizontally over 1.2 m (4 ft), have a lesser but important influence on site subgrade clay soils. Noting existing types and sizes of site vegetation will be important to the future of a project, just as much as noting anything else on the site that will affect the project.

8.12 Swell Testing

The actual amount of swell that occurs during swell testing is dependent on all the factors explained above. However, the sample used, if taken from the soil mass and protected correctly, should represent the particular clay soil and soil chemistry situation found there. The moisture content and dry unit weight of the sample must be preserved as well to provide the best results. However, one factor has been changed during the sampling and preparation processes normally used. The *in situ* stress has been removed, changing the stress history of the clay soil and changing how it will swell during the test. Research reported by the author and associates (Petry et al. 1992) has shown the differences this can make for a highly overconsolidated and highly active clay soil. However, the procedures recommended by this research have not found their way into standardized testing, because of the additional time and expense associated with their use. The best that can be done, then, is to start with a sample that best represents the *in situ* situation, except for the sampled stress situation.

Some practicing geotechnical engineers conduct essentially a swell pressure test, inundating the sample with water in the process, followed by reducing the stress to the expected project levels. It is then assumed that the swell that occurs represents field behavior. This procedure does dramatically change the stress history of the clay before the swelling portion of the procedure and cause internal swelling that can significantly change the clay structure and will result in swell amounts not representative of actual field behavior. Others allow the sample to swell, while being inundated with water and having essentially no stress on it. They follow this procedure with compressing the clay sample to its original height and obtaining what is called a swelling pressure. The first part of this procedure opens up the structure of the clay such that it cannot be compressed to where it was during the compression part of the procedure, so that the swelling pressure cannot represent actual field behavior either.

Proper swell testing procedures begin with samples that represent the *in situ* situation well, followed by placement of expected project overburden stresses on them and inundation of the sample with water. If one wishes to know the swelling pressure potential of a clay soil, then the overburden pressure is increased as the clay begins to swell, so that no swell can occur. The highest stress that has to be applied to keep the clay soil from swelling is then the swell pressure potential for that sample. It has been determined that addition of moisture slowly to a sample in a swelling test normally will not prolong the test, even though it is much harder to do, and may well cause the clay to exhibit more swelling potential. This is believed to occur because of the way that swelling opens up the clay soil structure to allow more water to be taken into the clay. This procedural change has not been adopted for standardized testing because of the expense of doing it. Also, the difference in the swell results are not significant enough to warrant this type of procedure.

The overall philosophy of swell testing is, therefore, fairly simple. Proper swell testing is done using an intact sample from the clay soil mass, taken from the depth at which swell potential is needed, and the expected project overburden pressure is applied. Then the sample is inundated with water and allowed to swell or the pressure to stop swelling is determined.

One more comment on swell testing has to be made. Those expansive clay soils that have had considerable and long-term overburden stresses applied to them may well have diagenetic bonds between clay mineral stacks. Normal swell testing, not conducted over long periods of time, such as weeks, likely will not result in accurate swelling potentials. In cases such as this, when clearance below project structural elements is based on normal swell test results, the long-term swell behavior can be catastrophic for projects. It has been known to result in rise of the ground surface over 0.6 m (2 ft) more than expected.

Settlement is not usually a problem for expansive clay soils, especially when they are overconsolidated. However, if during swell testing with project loads applied the results are compression, not swell, a consolidation test must be performed on these materials and can be done as an extension of the swell test.

8.13 Shear Strength Testing

When considering shear strength testing of expansive clay soils, one must determine what the likely loading situation will be and how quickly the loads will be applied. Most foundation design situations are based on the unconfined compression test, and cohesion only is assumed to be the result. As it turns out, this usually is a conservative approach considering the real shear strength of the same clay soil. This is acceptable also because of the relatively fast application of loads to subgrade soils. Given the opportunity and funds to do so, a better test would be the consolidated undrained triaxial test with pore pressure measurement. This would provide a better understanding of the shear strength, both friction and cohesion, of the soil.

If the shear strength is being determined for the design of a slope, the testing must be done radically differently. Since expansive clay soils in a slope, over time with shrinkage and swelling, experience a significant loss of shear strength, the test method must provide the lowest shear strength for the soil. This is the residual shear strength as determined using the direct shear method. The sample must be saturated and consolidated under the chosen overburden stress, then sheared at a very slow rate, so as to not develop cohesion or pore pressure. Also, the amount of deflection is caused to be very large relative to the sample diameter; this is achieved by moving the shearing device back and forth. Then the sample parts are set back on top of each other and a very slow shear test is done. This procedure will provide a realistic friction angle, the residual friction angle, for the soil. A slope is then designed to have sufficient safety using the residual friction angle.

To obtain the correct shear strength for a clay soil, it is paramount that the sample tested be intact, relatively undisturbed, or remolded so as to, as closely as possible, represent the *in situ* clay soil. Then the test chosen must place this sample in the same situation of saturation, drainage, rate of shear, and confinement as the *in situ* conditions expected for the project. The project situations chosen for the test also must represent the worst-case scenarios. Then the results of the test can be relied upon to predict the behavior of the soil for the project.

8.14 How to Deal with Expansive Clays

8.14.1 Alternatives

How a geotechnical engineer deals with expansive clays depends a good deal upon when he or she becomes involved with a project. If involvement starts during the process of site selection, it can be possible to avoid having the project even be supported by these problematic clay soils. Most often, however, the geotechnical engineer becomes part of the project engineering team at a later stage, when the project and site location are already set. In this case, the geotechnical engineer has two types of recommendations to offer. The first is where the behavior of the expansive clay soil is predicted and the recommendations for design and construction are a result of using these clays as they exist. The second includes possible further investigation, but may result in an optimal situation where methods of ground improvement are used to make the expansive clay soil into a material much more economical to use. An available third option, although not always used correctly, includes replacement of the expansive clay soil with a material much less potentially damaging to the project.

8.14.2 Design for Use

Designing a project to use the expansive clay soil subgrade at the site is not always possible for all project elements, especially any sidewalks, driveways, and parking areas. Project structure supports must be founded on as stable and unchanging a material as possible. Structures need to be supported on drilled shafts, normally belled or underreamed and founded well below the zone of expected moisture change in the subgrade. These drilled shafts are made of concrete reinforced enough to resist the pullout forces of an expanding subgrade acting on the shaft and be able to safely support the downward loads applied. Grade beams or other structural elements supported on these drilled shafts need to be constructed with at least twice the clearance under them and over the clay surface as the amount of vertical rise predicted for the subgrade upon wetting. All of the supports for the project major structures must be built in

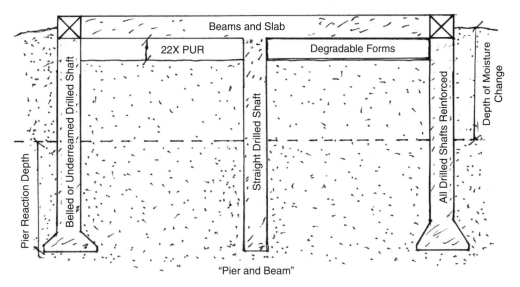

FIGURE 8.12 Typical pier and beam foundation for expansive clay.

this same way, so that the subgrade cannot swell and push up on the structure. Figure 8.12 illustrates this design concept.

There are existing structures that are supported by drilled shafts without this clearance being provided. In almost all cases, they have suffered catastrophic damage from differential swell. In addition, proper use of swelling testing to predict volume increase must include consideration of the need to release diagenetic bonds during the test process. Major damage has occurred when this was not done, for the sake of testing economics, and clay soil weathered from a highly overconsolidated shale has literally lifted the floor in areas between drilled shaft supports.

Another significant factor is proper selection of the depth of the subgrade that will become wetted over time. Those who do not remember that project owners want the areas adjacent to their structures to have grass and other vegetation, and will install watering systems to provide the best environment for foliage to grow, may regret it with time. This is most critical when the bedding planes of the expansive clay subgrade are near vertical in orientation. Predicting the depth of possible moisture change and, therefore, possible swelling requires knowledge of the standard of care and practice in the area where the project will be constructed.

The prediction of potential vertical rise is also best done using methods and properties which are utilized where the project is located. There are nearly as many differing methods, based on differing types of information, to predict potential vertical rise as there are locations where the data used to develop the method were measured. The best methods are those used by the geotechnical engineering community where the project is to be built.

8.14.3 Removal and Replacement

Removal of a depth of expansive clay soil in a subgrade, perhaps to the level where swelling is not expected, and replacement with less active soils have been used for many years. First, the geotechnical engineer must determine what depth of clay to replace to render the subgrade inactive enough to prevent damage to the project. Second, the soil that is to be compacted into the subgrade hole left by the clay removal must be selected. There are those who would place free-draining materials, like gravel or sand, in this hole. Unfortunately, this causes a "bathtub" effect, where drainage and leaking waters of any kind under the structure will collect and will cause deeper wetting than originally predicted. This nearly always results in significant distortion and damage to the structure. Similar effects can happen when "select fill" of clayey sand is backfilled into the hole.

The only way to overcome these effects is to provide collection and removal of any moisture that may move under the structure. This sort of system is used under basements in expansive clays with good success. In these cases, the water collection drain and removal system extends beyond the exterior basement walls so that water that moves into the soils around the basement can be removed as well.

The best soil to be compacted into the hole where the expansive clay was removed is either a moisture-stable clay, preferably one that is less active than the soil removed, or a clay soil that has been otherwise treated to dramatically reduce its activity. These materials are discussed below.

A key factor to remember when backfilling soils that will support part or all of a project is to make sure that when they are compacted in place, they have proper activity, shear strength, and compressibility. This can only be determined by preparing these soils to the gradations of clods and particles such as will exist when they are remolded into the subgrade and then compacting them to the specifications that will be used for the project. This may require larger than normal compacted samples for cutting down or testing, since field gradations are often much coarser than for normally prepared materials in the testing laboratory.

For large open structures, such as warehouses, another type of foundation is used. The structural elements are supported on drilled shafts and grade beams, such as discussed above. In between columns and walls, the floor is made of slabs-on-grade supported by moisture-stable or otherwise treated and improved clay soils. Figure 8.13 shows this type of system. It is important that the soil-supported slabs be separated functionally from the superstructure elements using construction joints that will allow the slabs to move independent of these elements.

Retaining walls that are used to support expansive clay materials must be built with proper drains installed and with backfills of free-draining and nonexpansive soils. The clay surface behind them must be cut back to form a 45° or lower slope, so that swell that occurs will not topple the wall. This concept is shown in Figure 8.14.

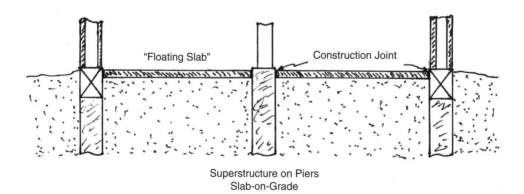

FIGURE 8.13 Floating slabs between superstructure support foundation.

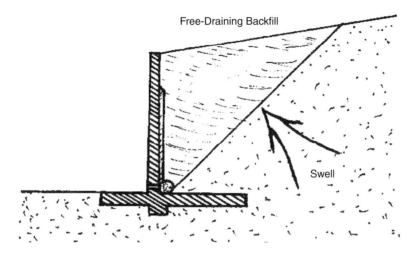

FIGURE 8.14 Proper construction of backfill and retaining wall.

8.14.4 Overpowering the Clay

There are opportunities that allow placing loads on expansive clay subgrades to overcome their natural tendencies for swell. This can occur when the actual contact pressures under shallow or deep foundation elements are sufficient to meet or exceed the swelling pressure of the supporting clay. When expansive clay soil subgrades are moderately active and the amount of overconsolidation is moderate to low, this is a very real possibility. In these cases, the compressibility of the founding clay soil must be known and the amount of settlement that will occur must be predicted to be within acceptable limits.

8.14.5 Improvement of Expansive Clays

Ground improvement itself is a broad subject, but it can be narrowed down some when expansive clay soils are concerned. The idea is to apply methods and/or agents to make the clay soil activity not a problem for projects. There are three main categories of ground improvement: mechanical reworking of the soil without addition of a chemical agent, reworking of the soil with the addition of a chemical agent, and reworking the soil with an agent which binds the clay soil particles together. These categories are covered below.

8.14.5.1 Compaction

The most frequently used method to improve clay soils is proper remolding so as to dramatically reduce their swelling potential. In reviewing compaction data for many clays, it has been noted that when applying standard Proctor compaction energy, the optimum water content is very close to the PL of the clay soil. This is not too surprising when one considers that these soils act with plasticity above this moisture content and would not be compactible when in the plastic state. If this is tied with the moisture content where expansive clays will not express swell, which is slightly above their PL, then compaction wet of optimum would significantly reduce swell potential. Actually, when dealing with expansive clays, engineering practice includes the recommendation that compaction water contents vary from the standard Proctor optimum to 4% over that amount.

Field compaction procedures are applied to produce dry unit weights of at least 90 or 95% of the maximum from the compaction test, be it standard or modified Proctor. The practice for expansive clay soils should be dry unit weights between 90 and 95% of the maximum from the standard Proctor test. Compaction of expansive clay soils to high dry unit weights and dry of optimum moisture will result in a subgrade with excessive swell potential.

Compaction in the field is dramatically different than in the laboratory. The largest difference is that in the field the clay soil is made up of clods and particles much larger than those used for most laboratory testing. If real-world behavior is to be predicted using laboratory tests, then the material used should be as coarsely graded as the field soils, which will require the use of much larger compacted specimens. It is very possible that the results can in this way much more accurately represent the soil mass upon which a project will be founded and may result in very different recommendations by the geotechnical engineer.

8.14.5.2 Proper Slope Angles

As discussed earlier, slopes cut into expansive clay soils are susceptible to significant loss of shear strength. Hence, in the worst case, semiarid climate slopes must have lower angles than normally expected. If the slopes are not laid back at slope angles dictated by the residual direct shear test friction angle of these soils, then they will experience near-surface to deep slope failures. This residual friction angle determined by a slow rate of strain and relatively large amounts of strain can be even less than that from other types of drained tests.

8.14.5.3 Moisture Content Control and Prewetting

Because changes in moisture content are the primary cause of both swelling and shrinking of an expansive clay subgrade, it stands to reason that not allowing moisture content to change and bringing the moisture levels in the subgrade to desired levels and making them as stable as possible would prevent many of the problems associated with use of these problematic clay soils. One application of this has already been discussed in relation to the proper moisture contents for compaction. The problem is not only how to develop the desired levels of moisture but also keeping them at the appropriate level over the life of the project.

Subgrade clay soils can be compacted at desired moisture levels, but this is not done for *in situ* natural subgrades that need to be brought to desired moisture levels. Ponding of water on the surface has been tried but is no more effective than irrigation watering of the surface for the same length of time. Once the surface layers of the clay are moist and swell closes avenues for moisture intrusion, transfer of the added moisture over time is the phenomenon that controls wetting of the subgrade. Wetting from the surface in most cases will take months to achieve. The most efficient method to add moisture at depth in expansive clay subgrades has been the injection of water that contains a surfactant, using probes that can be inserted to the desired depth for moisture addition. If the expansive clay subgrade contains interconnected cracks and fissures, this process provides moisture within the clay clods and leaves water in the cracks and fissures. Done properly, with probes inserted at 1.5-m (5-ft) horizontal intervals and using multiple injections, a clay soil subgrade can be brought to desired moisture levels within a few days. The injected clay will, of course, express swell during and after this process for about a week or so. The effectiveness of this process is checked one day after each injection pass by undisturbed continuous sampling of the injected soil mass and swell testing of the samples. When the swell noted is reduced to an acceptable amount, the process is complete.

Control of established desired moisture levels in the subgrade includes protection from and/or practical elimination of harmful drying or wetting effects. Protection can take the form of slabs-on-grade, the cover afforded by the structure, or use of vertical moisture movement barriers. Once a subgrade has the desired moisture profile, application of concrete flatwork, such as sidewalks and driveways or a slab-on-grade for the structure, the structure-soil situation will be much more stable than if this moisture addition had not been done.

If concrete flatwork is placed on an expansive clay subgrade that is not moist enough, eventually, in almost every case, the clay soil under these sidewalks, etc. will become moist, and significant damage due to differential movements will occur. However, the use of concrete flatwork extended out far enough away from a structure will prevent drying effects from reaching clays that support the project elements.

The most effective moisture movement barriers are placed vertically adjacent to the structure and extend at least half the depth to where moisture change can happen in the subgrade. These vertical barriers are best made of well-densified lean concrete at least 6 in. thick or geomembranes made of polyethylene or polypropylene at least 30 mil thick. In all cases, the barriers must be sealed to the slab or structure to prevent drying from intruding near to the ground surface. A concrete-type barrier is shown in Figure 8.15. Depending where they have been used with great success around the world, these barriers are placed just outside of the structure or take the form of relatively thin vertical concrete walls that help to support the structure along its perimeter. When a basement is part of the structure, its exterior walls form such a barrier, but the likelihood of moisture working its way under the basement must be counteracted by efficient drains and pumping of water collected well away from the structure.

If such protective measures are not included adjacent to sidewalks, driveways, etc., the fact that they have been constructed on moisture-improved clay will not prevent drying effects from causing damage eventually. The owner must decide whether to place barriers against moisture change or rebuild these areas periodically. It is an interesting fact that roadway pavements supported by expansive clays that have vertical edge geomembranes installed, with drains outside of them, require heavy maintenance or repair at intervals many times longer than those that have drains only.

Unwanted and potentially damaging wetting of clay foundation soils can occur when sources of concentrated wetting are not controlled. These sources include improper grading, which causes drainage to move toward a structure, either from surface or subsurface sources. Surface grading of at least 1% away from a structure will move surface water away. In such

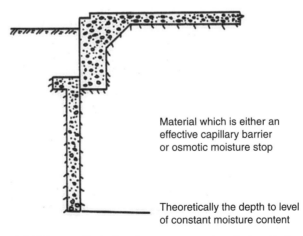

Material which is either an effective capillary barrier or osmotic moisture stop

Theoretically the depth to level of constant moisture content

FIGURE 8.15 Vertical moisture movement barrier installation.

cases, a system of swales to drains must be installed to ensure that, even during heavy rainfall, all water will move away from the structure. Subsurface water can move through porous layers, along bedding planes, or on top of much less pervious layers. Subsurface drains must be placed, using a geomembrane on the side toward the structure, to intercept, collect, and carry away water. In addition, it is not wise to place too steep a slope near the perimeter of a structure, so as to allow easier drying access to the clays under the structure.

Plans for watering grass around a structure should include watering uniformly all the way to the perimeter of the structure. As described earlier, trees and relatively large shrubs should not be placed too near to the perimeter of a structure or flatwork, due to their ability to differentially dry out the clays supporting these elements. It is also unwise to water these trees and shrubs in such a way as to differentially wet clays that support flatwork or structures. The greatest damaging forces from expansive clays come from differential drying or wetting effects, which must be eliminated.

Often overlooked sources of wetting are tap water leaking either inside or outside of a structure, wastewater leaking when pipes crack or break, and roof runoff water, especially when concentrated by downspouts. Just because a faucet leaks outside of a structure does not preclude the leak from dramatically affecting structure foundation clay soils. Leaks of pressurized water under a structure likely will lead to damaging differential movements of the structure. These leaks must be fixed immediately upon discovery. Wastewater leaks are much harder to determine, but almost always are under a structure and therefore cause significant damage. Wastewater systems, especially if suspected of leaking, must be checked periodically for leaks. The water that comes from rain falling on roofs can amount to relatively large volumes and often is concentrated around a structure where downspouts are located. All roof runoff must be taken away from a structure, or damage from concentrated wetting of the clay soils that support the structure can likely happen.

Differential drying effects must be eliminated, in so far as possible, for all projects, even when vertical moisture movement barriers are installed. These barriers include too steep a slope away from a structure near to its perimeter, planting trees at a distance closer to a structure than the height to which they will grow at maturity, and allowing shrubs to grow too large and too close to a structure. The differential drying caused by how a structure is oriented will likely occur on the south and especially southwest sides. It is also best to construct foundation elements, such as slabs, grade beams, and footings, as deep into the subgrade as feasible to help limit exposure of foundation clay soils to drying.

8.14.5.4 Addition of Agents

Agents are added to expansive clay soils by intimate mixing and injection. They are added to, first, overcome the volume change characteristics of the clay and, second, to provide added strength to the treated clay soils. Almost all of the agents added to improve shrink-swell tendencies are ionic in nature. They are mixed with layers of the clay to cause exchange of cations associated with the clay, which in turn will reduce the volume change potential of the clay. Those agents most successfully used and proven both in the laboratory and field are lime, or calcium hydroxide, and potassium in differing compounds. Portland cement, because it contains some lime, can have similar effects when mixed with finely pulverized clay soils. The calcium in lime exchanges for other, more active cations such as sodium and causes the clay to have significantly reduced plasticity and swell potential. Potassium-containing compounds result in the potassium ions becoming part of the clay mineral structure, reducing the swell potential of the clay without changing its PI. Certain polyquaternary amines will exchange into a clay and make it hydrophobic, rather than hydrophilic, as it once was. Another agent useful in combination with lime or Portland cement is class C fly ash, which contains calcium oxide. Lime kiln dust also has been applied. Other agents are claimed to reduce clay soil volume change potential, but the author has yet to observe their success.

Proper addition of sufficient quantities of lime or Portland cement, and combinations of lime or Portland cement with class C fly ash, will cause pozzolan cementation compounds to be formed in a clay soil, thereby providing additional shear strength in these materials. This is particularly desirable when the treated clay soils are used to support pavement systems. In this application, as well as those mentioned above for the layers into which agents are mixed, the improvements in clay soil behavior are effective only in the soils treated and not those beneath them in the subgrade. Therefore, the geotechnical engineer must assess how deep these treatments must be applied for the project being considered and the expansive clay subgrade involved. Treatment of one layer at the surface does not change how deeper layers will act over time.

It is not the purpose of this discussion to consider how chemical agents are tested for whether they will provide the desired results or how they should be applied in the field. The next chapter on soil improvement covers more of these details. It is important to say that proper testing and application of chemical agents are paramount to understanding how they can improve clay behavior and how successful their use will be in the field. Any agent that will be used must cause the clay soil to exhibit the properties desired, when applied at an economical rate, and the tests used must represent field conditions as closely as possible.

Injection of agents to improve behavior in expansive clay soil subgrades has been applied with limited success. The largest part of the success is due to the addition of moisture during the injection process. If this moisture is prevented from being lost over time, the subgrade is significantly improved in behavior. Lime slurries, lime–fly ash slurries, potassium-containing compounds, and other agents called ionic have been injected in expansive clay subgrades that contain cracks and fissures. Research conducted by the author using lime–fly ash slurry to form vertical moisture movement reduction curtains has been successful. Laboratory injection of potassium compounds has resulted in a clay soil with less swell potential and that has reduced properties to transfer moisture between layers of clay with significantly differing moisture potentials. Injections of other agents have not led to such success.

The application of agents to improve expansive clay soil subgrades has been successful when the agents are tested by methods that use soils prepared as in the field, applied as in the field, and compacted as in the field. These successes have occurred when proper field methods and equipment are used and proper specifications are followed. The geotechnical engineer can then depend on the results of these applications to make recommendations based on proper assessment of how the improved clay soil will act over time relative to the project.

References

Chen, F.H., *Foundations on Expansive Soils,* American Elsevier Science, New York, 1988.

Mitchell, J.K. and Soga, K. (2005). *Fundamentals of Soil Behavior,* 3rd edition, John Wiley & Sons, Hoboken, NJ.

Nelson, J.D. and Miller, D.J. (1992). *Expansive Soils, Problems and Practice in Foundation and Pavement Engineering,* John Wiley & Sons, New York.

Petry, T.M., Armstrong, J.C., and Tsai, C.T. (1980). Relationships and variations of properties of clay. *The Proceedings of the 4th International Conference of Expansive Soils,* ASCE, June 16–18, 18 pp.

Petry, T.M., Sheen, J.-S., and Armstrong, J.C. (1992). Effects of pre-test stress environments on swell. *Proceedings of the 7th International Conference on Expansive Soils,* Vol. 1, ASCE, ISSMFE, TRB, and National Science Foundation, August, 6 pp.

Ground Improvement

by
Thomas M. Petry
Missouri University of Science and Technology, Rolla, Missouri

9.1 Introduction

The subject of this chapter is the improvement of soils so that they can be utilized for project purposes. For many years, this has been called soil stabilization, but the word *stable* has legal significance as meaning permanently not moving or causing any kind of problem for a project. Soil stabilization must now be referred to in legal circumstances as ground improvement. As such, the subject of this chapter is soil stabilization or ground improvement, and the reader must realize that the actual permanence involved varies depending on the soil and how it is improved.

Soil stabilization (ground improvement) is any process of altering unsuitable *in situ* or "borrowed" soil to improve selected engineering characteristics, at a lower cost and with better quality control than can be obtained by replacement, bridging over, or bypassing the unsuitable material.

The common ways of dealing with unsatisfactory soils include bypassing the soil and/or site, removing and replacing the unsuitable soil, redesigning the project, and treating or reworking the unsuitable soil to improve the selected properties. The last of these is soil stabilization.

When considering how to characterize soil stabilization methods and processes, there are three major categories of ground improvement. The first is mechanical stabilization or improvement. This category includes ways to improve soil properties without the addition of agents. The second is chemical stabilization or improvement, which, as the name implies, involves the addition of chemical agents. The third category is physical stabilization or improvement. This group of methods includes adding agents or energy to bind the soil particles and clods and partially or fully fill the voids between them.

The scope of this chapter includes basic descriptions of and reasoning for soil stabilization methods, along with how to evaluate them. Each method is discussed in sufficient detail to allow application, and further details are not included.

During evaluation of alternative ground improvement methodologies, the geotechnical engineer must consider how well each will deliver the selected properties and the permanence of the improvement. First, the method must be compatible with the soil material. Second, it must result in the desired ground improvements. In addition, the results must be as permanent as required by the project. Next, it should be possible for a local contractor to perform the methodology and to do so reasonably safely. Finally, the methodology should be relatively economical to accomplish.

9.2 Geotechnical Investigations for Ground Improvement

Geotechnical investigations for ground improvement projects are initiated by owners, contractors, engineers, lawyers, or insurance companies. An owner who initiates a project is likely to have an architect/engineering firm contact the geotechnical engineering firm, and the project can be either new or remedial. A contractor who initiates a project may well contact the geotechnical engineer directly, whether the project is new (less likely) or remedial (more likely). An engineer who initiates a project will contact the geotechnical engineer directly, and the project is likely to be remedial. A lawyer who initiates a project most likely will contact the geotechnical engineer directly, and the project most likely will be remedial. An insurance company that initiates a project most likely will contact the geotechnical engineer directly, and the project will likely be remedial. Communication among the geotechnical engineer, the client, and the contractor who is to accomplish the methodology is important in any project, but it is paramount in a project that includes ground improvement. It is important to remember that just because a project is remedial and the client indicates that time is of the essence, it still takes the same amount of time to conduct the investigation. Shortcuts, especially in ground improvement projects, where curing of specimens often is involved, usually result in less than satisfactory results and eventual litigation.

Ground improvement geotechnical investigations are likely to be more extensive, take more time, and be more expensive than site investigations that do not include ground improvement. The increased time and expense should be offset by the improvements, rescuing the project, and the ultimate cost of using the unsuitable material. In addition, the "standard of care" for use of ground improvement methodologies likely will require special construction specifications.

Ground improvement geotechnical investigations for new projects may well involve more testing that can require curing times, and therefore they take more time and are more expensive. They also may require special testing techniques which are more sophisticated and, as a result, cost more. Also, testing for ground improvement may proceed only when the initial testing indicates the need for it.

Geotechnical investigations for remedial projects are larger in scope than new project investigations because of the need for exploration to determine the nature of the problems, substantiate responsibilities, and possibly prepare for expert witness testimony. These tasks are in addition to providing a solution for remediation of the project.

9.3 Mechanical Stabilization (Ground Improvement)

Mechanical stabilization (improvement) consists of any methodology, with or without devices placed in or on the soil mass, that improves selected engineering properties of the soil mass without the addition of agents or other particle-binding energy. In other words, no chemical or binding effects are included in the methodology. The following is a partial listing of the most prominent methodologies:

1. Blending of materials
2. Replacement of materials
3. Compaction and/or reworking
4. Preloading or preconsolidation
5. Change of slope geometry
6. Control of surface and subsurface water
7. Control of moisture contents and retention of moisture
8. Erosion control
9. Mechanically stabilized earth and earth reinforcement
10. Slope drainage
11. Control of frost effects and permafrost effects
12. Electroosmosis

9.3.1 Blending of Materials

Blending is the improvement of the gradation of soils to meet the criteria of filter design, base course specifications, or to provide a material which is either less permeable or stronger and less compressible. The process consists of mixing two or three naturally occurring soils and/or crushed stone to form the desired composite. It usually is not feasible to improve shrink-swell behavior of clays or to dilute the chemicals present or overcome chemically related problems in the soil by blending.

The aggregate or coarse fraction consists of those grains larger than an arbitrary limit, usually taken as larger than either a No. 40 or No. 200 U.S. Series sieve. In either case, this includes only gravel and sand. The binder or fines fraction includes those grains that are smaller than the arbitrarily set limit as stated above. These materials always include silts and clays if they are present and also sands if present.

The purpose of the aggregate fraction is to provide internal friction and relative incompressibility, and ideally it must be well graded and have angular particles. The function of the binder is to provide cohesion and imperviousness, and it should have some plasticity to develop high cohesion, but have little shrink-swell behavior. The best binders are those smaller than a No. 40 U.S. Series sieve, which are CL soils with a liquid limit less than 40 and a plasticity index of 5–15.

The relative amounts of aggregate and binder determine the physical properties of the compacted blended soils that result. Without binder, these soils usually have high internal friction and are relatively incompressible, because loading is carried by grain-to-grain contact. In such cases, cohesion is negligible and the soil permeability is relatively high.

When there is a small percentage of binder, some of the binder is trapped in the voids and compressed by compaction while only partially filling the voids. Compaction of the fines is variable. There is a sharp increase in cohesion and a sharp decrease in friction from binder between particles. There is a small increase in compressibility, yet there is still relatively high permeability. There is a real danger of the binder being eroded out by seepage. There is a sharp increase in capillary potential, which can cause frost problems. The strength at maximum dry unit weight is about three-fourths of the binder value and two-thirds of the aggregate value.

An optimum amount of binder is present when all the voids are filled with well-compacted binder material at compacted dry unit weight and there is still grain-to-grain contact of aggregate. At these higher binder percentages, friction decreases sharply to that of the binder and cohesion increases slowly to the binder value. This becomes more of a problem at high binder percentages. The resultant compressibility is not a problem until there is too much binder to fit in the voids between the aggregate particles.

Optimum binder percentages are determined as follows. The proportions of the mix are set so that the total binder (from all sources) is from 75 to 90% of that required to fill the voids at maximum dry unit weight. The binder required for maximum strength is about 20–27% and is less than that required for maximum dry unit weight.

The design of the mixture usually comes down to the following steps. The aggregate is compacted to maximum dry unit weight and the binder is compacted to maximum dry unit weight, and the amount of compacted binder needed to "fill" the aggregate voids is computed. Then the aggregate-binder mixes are compacted with increasing binder percentages until a maximum dry unit weight is obtained. This likely will require differing moisture contents to determine. Samples are always made and tested for the desired properties. Once the mixture gradation is determined, other ways of proportioning may be used to match the final gradation needed. Blending is often used to manufacture a filter material for drains, etc. In all cases, this requires blending of soils to match a gradation.

9.3.2 Compaction

Compaction is artificial densification of soil masses or soil layers for one or more of four reasons. The first is to build up the ground surface with what is called a fill. Second, and similar, is backfilling a trench or area behind a subsurface wall. Compaction also can be used to improve soil materials in place or to rework nonuniform soil materials so that they provide more uniform support.

Compaction is best done in layers thin enough to allow the compaction effort to reach all the soil of the layer as the energy for compaction is applied to the surface of the layer. Deep compaction is done by dropping very large and heavy weights from considerable heights onto the soil mass, by inserting compacting probes into the soil mass, or by applying large and deep fills on top of the soil mass to preconsolidate it.

In every case, a laboratory-established denseness standard is developed for the soil being used, and appropriate field compaction methods and equipment are used to densify the soil. Following this, the denseness is checked against the standard and either accepted or rejected. These tests are conducted for each layer compacted and for each 233–465 m² (2500–5000 ft²) of each layer.

To set a compaction standard for sands, the lab standard includes determination of the largest (e_l) and smallest (e_d) void ratios that can occur for the sand. These are found by carefully filling containers of a known volume with the loosest dry sand possible and finding the corresponding void ratio, followed by vibrating the cylinder and adding sand to find the densest void ratio. The vibration is accompanied by placing a relatively light overburden on the sand to aid in densifying it. Depending on the region of the world where this is done, differing standards are available that specify the testing process and equipment.

The corresponding field specifications are a desired range of relative density (D_R) using the following formula:

$$D_R = \left(\frac{e_l - e_{nat}}{e_l - e_d} \right) \times 100\%$$

where e_{nat} is the natural void ratio in the field. It is well known that sands with a relative density below 33% are considered loose, whereas sands with a relative density between 34 and 66% are medium dense. Those with a relative density above 66% are considered dense. Normally used field specifications are between 75 and 85% D_R.

It is important to note that sands without sufficient fines will not respond correctly to the impact compaction tests used for cohesive soils. The result will be a dry unit weight that will place the sand in the medium-dense range, well below what is needed for its use in any project.

Sands are best densified dry with vibration and some load to assist in the process or saturated and vibrated with some assisting normal load. Relatively thin layers are densified dry or saturated using a device that vibrates them and applies a relatively low normal load to aid in the process. The thin layers are checked for their relative density as described above.

Aggregate materials that are lacking in fines must be densified in a fashion similar to sands. They will not respond to compaction testing normally used for silts and clays and, like sand, will not provide proper compaction field standards with these tests. Aggregates normally do not have such wide ranges of void ratios when compacted, and proper use of a vibration test will result in adequate data to develop field specifications.

Silts, clays, and other materials with sufficient fines require manipulation at their optimum moisture level to be properly compacted. The laboratory standard used normally is developed by a drop hammer compaction test, originally developed by Proctor, a Los Angeles County engineer who wanted to determine the possible and desired denseness for field work. He first did full-scale field tests to see what was possible and then developed the standard Proctor test, which closely matched what he saw in the field.

The standard Proctor-type test uses a 943-cc (1/30-ft³) mold, 10.2 cm (4 in.) in diameter and 11.4 cm (4.5 in.) high. The drop hammer weighs 24.5 kN (5.5 lb) and drops 30.5 cm (1 ft) to the layer, which becomes about 3.8 cm (1.5 in.) thick. The hammer face is 5.1 cm (2 in.) so the layer thickness is correct to allow all the energy to affect the layer of soil. There are three layers then, and 25 drops of the hammer are applied to each one. The total energy applied is about 594 MN-m/m^3 (12,400 ft-lb/ft^3).

Eventually, construction equipment became large and efficient enough that more compaction was possible and desired, so the modified Proctor test was developed. The same mold was used as for the standard Proctor test, but five layers were used and the number of hammer blows per layer was kept at 25. The hammer weighs 45.6 kN (10 lb) and is dropped 45.7 cm (18 in.). The resultant compaction energy is about 2.7 GN-m/m^3 (56,300 ft-lb/ft^3).

The increase in compaction energy from standard to modified levels in the Proctor-type test normally provides about a 10% increase in maximum dry unit weight for a more than fourfold increase in effort.

It was found that the dry unit weight–water content curve did peak at a maximum dry unit weight, and that happened at an optimum water content. The modified effort also causes the optimum water content to be reduced by about 5% water content compared to the standard effort optimum. The right, or wet, side of the curve has been determined to be roughly parallel to the zero air voids curve, and the equation for dry unit weight (γ_d) is

$$\gamma_d = \frac{G_s \gamma_w}{1 + \omega G_s}$$

where G_s = specific gravity of the solids, γ_w = unit weight of water, and ω = water content, which is a plot of the dry unit weight when the water content would result in saturation.

The compaction curve is the shape it is because of soil behavior. When very dry, the soil is resistant to densifying, and although air can be driven out, the resistance to particle rearrangement keeps the soil from becoming more dense. When the water content of the soil is approaching the optimum, the soil is less resistant to particle manipulation and the air can still be driven out to allow densification. When the soil is just at the optimum water content, the reduction in resistance to densification is very good and the amount of air that can be driven out is at the maximum. Just past the optimum water content, the water available in the soil is starting to prevent the removal of air from the voids, so densification cannot be as great. Significantly past the optimum, the soil has too much water in the way of removing the air and the soil is too easily manipulated. Pore pressure builds up upon hammer blows and the soil displaces instead of densifying.

It has been the experience of many who work with clays that the optimum water content for standard Proctor testing is at about the plastic limit for a clay. This adds to the impossibility of compacting the soil above this moisture level since the soil is acting with plasticity.

A Proctor curve developed for a cohesionless soil with no fines, like a sand, will have two peaks, one where the water lubricates the process of densification until the amount of water allows for apparent cohesion to get in the way of densification and the second at the zero air voids, where the soil is saturated. The largest dry unit weight found by this method will not approach the 75% D_R level needed for proper compaction, so the test is invalid for any material that acts in this way. To determine whether the Proctor-type analysis is proper for a soil, the test must be run to see how the soil responds.

The specifications used for soils that respond to Proctor-type tests include a percent of the maximum dry unit weight to achieve in the field and a range of water content around the optimum which will assist in achieving that dry unit weight and cause the soil to have the required characteristics. Basically, it is unwise to require higher dry unit weights than necessary because of the expense, and it is wise to optimize water content for best results.

When a clay is compacted dry of optimum, it takes more effort to compact it, it costs less since less water is needed, the soil weights are less, and the working conditions are better. The soil is more permeable than if compacted wet of optimum. The swell potential of the clay is significantly more than if compacted wet of optimum, and there will be lower volume change when smaller loads are applied. However, there will be higher volume change when larger loads are applied. Finally, generally the soil will have a greater shear strength. The geotechnical engineer must decide which properties are desired and specify the correct moisture content range to develop these properties.

The Proctor-type curve for dry unit weight vs. water content is least "peaked" for a clay soil and the optimum water content is some distance from the zero air voids. If the soil is less plastic to nonplastic, the curve will become sharply "peaked," and this peak will be much closer to the zero air voids. This seems logical, perhaps, since the peak for a sand is at the zero air voids.

The range of water content that is acceptable and most efficient for compaction of a silt is, therefore, much smaller than it would be for a clay to achieve the same percentage of the maximum dry unit weight. In fact, it is common to specify that a silt have a water content within ±2% of the optimum, while a clay can be compacted within ±4% of the optimum. Typical specifications read: "the dry unit weight must be at least 90% (or 95%) of the maximum dry unit weight" and "the water content must be within $\pm x\%$ of optimum water content." To minimize swell potential, the dry unit weight may be given as between 90 and 95% of the maximum dry unit weight and the water content as at least the optimum water content to 4% above the optimum.

The field testing to determine if these specifications have been met may be done by a few different means. A cylinder of known volume can be driven into the compacted soil layer to determine the dry weight and water content of the contents. A sand cone device can be used to determine the volume of the hole out of which a sample is taken and then the sample is dried to find its dry weight and water content. Also, a nuclear densimeter and moisture gauge can be employed to find the dry unit weight and water content. This last method is by far the most accurate and most widely used procedure.

To achieve properly compacted materials, a set of specifications should include the following. First, the contractor must clean all the "A"-horizon topsoil from the site, followed by proof rolling the fill foundation material and densifying it, if needed. The lift size should not exceed 30 cm (1 ft), unless very large and heavy rollers are available. The dry unit weight and water content must be checked every lift and at the intervals stated earlier. The water content must be within the ranges discussed above, and the dry unit weight must be within the proper range as discussed above.

The type of roller can be specified or left up to the contractor, as indicated below. For marginally plastic or marginally cohesive soils, a steel-wheel roller normally is used. A pneumatic roller can be used for silts and soils of low plasticity. Vibratory rollers can be more efficient when used to compact silts and low-plasticity clays. A pad foot roller may be used for low-plasticity clays, but a sheepsfoot roller is most often used for clays. Generally, the more plastic a soil is, the more it must be compacted from the bottom up, not the top down; therefore, penetration is needed. Vibration does not work in compaction of clays.

The site and fill must be well drained at all times, especially at the end of the day. There are two important things to remember. First, nothing will be constructed unless contracted, specified, and checked. Second, when calculating movement of soil from borrow to truck to site, etc., all quantities should be calculated based on moving the weight of dry soil, since this does not change, as water content does.

9.3.3 Densification of Deep Layers

Relatively thick or deep layers of sands can be densified using a vibroflot, which saturates and vibrates at depth, or by pile driving into saturated sands. Results from these methods are determined using some sort of standardized penetration test in the field. This type of at-depth densification is done by specialty contractors and involves coordination by the geotechnical engineer with the contractor.

Thick layers of marginally cohesive soils, such as silts, can be compacted using impact compaction. Very large weights are lifted by cranes and dropped from significant heights onto the top of the soil layer, densifying pockets of material. The cranes move over the project site to provide densified materials. This very specialized compaction is done by specialist contractors and involves coordination by the geotechnical engineer with the contractor.

Clay subgrades can be densified while in place using preloading. A fill is built on top of the clay to provide overburden. This process of increasing the effective stress in deep soil layers of silt or clay soils will preconsolidate the soil mass before placement of a structure. Normally, the depth of the fill is determined by the amount of preloading desired, and the top layers of the overburden are removed down to the level at which the project is to be built when the preconsolidation is complete. Any time a large fill is constructed, it is prudent to utilize settlement plates to monitor the progress of compression and to install piezometers to monitor pore pressure during the process. To aid in the drainage of water as the subgrade compresses, vertical strip drains are used and connected to a drainage layer in the fill to carry water away. The geotechnical engineer would be responsible for monitoring movements and pore pressure during the process.

9.3.4 Improvement of Slope Stability

Slopes for all uses normally are constructed using the naturally occurring soils at the project site. In view of how many slopes fail each year, it appears that many of them are not designed. The geotechnical engineer of record is responsible for recommending proper slope designs for the soil and site conditions that will exist during the life of the project. The shear strength used for such designs must be appropriate for the material involved. Not taking into account erosion control (discussed later), slopes of gravels, sands, and even silts are designed using the lowest friction angle expected for these materials and with sufficient factors of safety. As discussed in Chapter 8, expansive clays require special testing and consideration of shear strength. A slope made of a clay that will expand and contract with the climate will eventually have a shear strength equal to its residual friction angle, which may well be one-half of the peak friction angles measured. Slope stability can be improved, therefore, by using the correct friction angles for any soil, as long as the slope is allowed to drain properly. Berms of free-draining materials can be placed on a slope at and near its bottom to increase the safety of the slope in most cases. Slopes that are experiencing saturation and loss of strength, and which start to slide, cannot be corrected in this manner.

9.3.5 Water Content Stabilization

Changes in water content of clays cause either swelling or shrinkage. One of the most effective ways to stabilize these clays is stabilization, in so far as possible, of their water content. No change in water content means no volume change. It is difficult to exclude water from the clays because of their large negative moisture potential, so it is best to establish the desired water content and provide for its stability. As described in Chapter 8, the target water content where clays are adequately satisfied is where their liquidity index is about 0.15–0.2.

Methods used for moisture addition include ponding of water on the site, sprinkling water on the ground surface at the site, or injection of water that contains a surfactant. Of these, injection is used where the clay subgrade is well fractured and has fissures caused by shrink-swell from climate events. Injection is done by contractors using specially developed devices that push probes into the ground and control water flow to them. The addition of water by this means can be done in less than a week, even if multiple injection passes are needed. The swell potential of the injected soil is tested after each addition of water, and the target swell percentage is usually less than 1%. Once the water content profile is brought to the needed level, the rest of the method consists of maintaining that water content profile for the life of the project.

Part of the process of maintaining water content involves removing concentrated wetting and drying effects. Concentrated wetting effects include poor drainage. Correct drainage provides positive drainage away from project structures at 1–3% slopes, and cutoff drains and trench drains are used to carry water away. Watering of the site must provide a uniform distribution of moisture, especially near structures, concrete flatwork, and pavements. Roof runoff can be a significant amount of water and must be controlled as well. Plumbing trenches usually are backfilled with materials that allow movement of water along them, from inside leaks and outside sources. These trenches should be backfilled with clay soils that are moisture stable, and all plumbing leaks, from either pressurized or wastewater sources, must be fixed immediately upon being noticed.

Concentrated drying of foundation clay soils can be caused by the following effects. The structure or pavement can be exposed by poor backfill of soil against it and by having grades too steep near to it as well. The portion of the geographic location that dries the most is the southwest corner. It is wise to place trees for shade somewhat away from this corner or to use embedment of the structure in that area. Trees and large bushes (over 1.2 m [4 ft] in size) placed too close to any project structure or pavement will lead to differential drying. It has been noted from experience that a tree's roots, especially trees that are fast growing and that spread horizontal roots close to the ground surface, can spread as far away from the trunk as the tree is tall at maturity.

The most effective devices to maintain water stability in a clay subgrade are moisture movement barriers. They can be horizontal pavement or slabs-on-grade, including sidewalks and driveways. To be effective, they must be sealed to the structure, encapsulating the soil beneath them. They need to extend far enough horizontally to prevent moisture loss beneath the structure. Vertical and horizontal high-density polyethylene or polypropylene sheets usually 30 mils thick have proven to be very effective. Well-densified lean concrete also has proven to work well. All of these moisture movement barriers must extend far enough away from or deep enough below the structures to which they are sealed to prevent significant moisture loss in foundation soils. These devices are shown in Chapter 8 on expansive clays.

9.3.6 Control of Moisture

Moisture control includes managing water entering and/or exiting a soil mass for the purpose of making the mass more stable. Types of moisture control include the following:

- Proper location of the structure
- Grading the surface for drainage
- Installation of subsurface drains
- Installation of moisture barriers
- Lowering the water table
- Electroosmosis
- Prewetting the soil mass
- Removal of drying effects
- Removal of differential wetting

The decision must be made to hold water away from or inside the soil mass. Moisture exclusion is used when the existing or expected moisture would cause the soil mass to become unstable or cause exceedingly high loads on the structure. It is used for fills, retaining wall backfills, and soft or loose soils. It also may be a part of erosion protection.

Moisture entrapment is used when moisture buildup is likely to occur, such as in soils of high activity, and it is necessary to control the buildup and exit of moisture to ensure stability of the soil mass and the structure. It is used for fills and foundation soils, as discussed above.

Drainage involves guiding surface and subsurface waters to where desired—not necessarily where they would naturally go. The general concept is that water follows the easiest path to locations of lower hydraulic energy. It flows downhill, but it needs to get there the way that causes the least problem for a project or structure.

Drainage processes and devices most naturally include slopes, as it only takes a 1% slope to move water, but the area must be maintained. A swale or shallow depression that leads water somewhere can be very effective. Ditches are deeper depressions used to move water where it will not be harmful. Geotextile filter fabrics can be used to control the movement of soil particles along with the water. Gutters and downspouts often are used to collect and direct roof runoff. Drop inlets can be located where water appears to pond, to move it away, as can drains that contain only aggregate or a pipe with aggregate. Geotextile filter fabrics commonly are used to make a "sock" around slotted pipe and to wrap other drain materials to separate soil particles and keep them from entering the drain. Interceptor drains are used to stop water, usually below the ground, from moving toward a structure. Strip and edge drains are being used along pavements to collect water, and drainage piping is used to move all the water to an outlet.

In some cases, bentonite clay is used to seal the soil, and sheets that contain bentonite are applied for that purpose as well. When deep cutoffs are needed, slurry walls or concrete walls are built to control subsurface waters.

9.3.7 Control of Frost and Permafrost (Ground Freezing)

The problems associated with freezing ground include heaving and the formation of cracks and thawing from the surface down and related instability. This is a particularly difficult problem to address for rural roads. The alternatives are to keep traffic off during thawing or to design to overcome a possible problem. The basic problem is the depth of frost penetration, the soils supporting the roadway, and the depth to the water table.

The heave that occurs is caused by ice expansion, which would be about 9% at a maximum. This turns out to be no higher than the minor shrink-swell of clay soils. The real problem is the ice lens formation and growth that occur when water is progressively brought to the freezing front and the soil splits open; lenses form in these splits when water is fed from the water table by capillaries in the soil.

Silts or heavily silty soils are the problem soils because of their capillary rise capacity. The worst of these are fine silts with particles <10 m, followed by those that are fairly uniform and contain 20% <20-m particles. If the soils are uniform, then this drops to 10% <20 m. When the soils are well graded, it takes only 3% <20 m. No problems happen when there is 1% <20 m. These frozen soils have sufficient strength to support loads since the strength of ice at 0°C is about 90 kPa (10 tsf) and at −10°C is 500 kPa (56 tsf).

Instability problems occur when the defrosting enters the roadway and subgrade from the top and the lenses of ice defrost, resulting in an oversaturated material that has little shear strength. The pavement, lacking proper support, and the base materials, not able to drain fast enough, fail under normal axle loads. Therefore, it is wise to not load pavement during the thaw period. Another prime way to overcome this problem is to separate the freezing and water. This can be done by lowering the water source or raising the pavement structure. A layer of coarse materials above the water source can interrupt the capillary action, and a layer of well-densified lean clay can be a capillary cutoff. Clays are reported by many to have large capillary rise capacities, but the truth is that they transfer moisture slowly by osmotic forces and do not have capillary capacities unless cracked or poorly compacted.

In areas where there is permafrost (permanently frozen ground), two types of problems normally can occur. Of course, there are many unique landforms in these areas, but this discussion focuses on fairly uniform permafrost situations. The permafrost may or may not extend all the way to the ground surface. Aerial photographs and the types of trees growing in a location can assist greatly in determining the depth to the permafrost.

If the permafrost extends to the surface, it is extremely important that projects placed on it do not disturb it by thermal pollution. The melted permafrost likely will not have the strength needed to support project structures. The solution is to somehow insulate the permafrost from heating effects or to install some system that will permanently keep the permafrost frozen.

In areas where the permafrost may exist at some depth into the ground and there is a zone of nonfrozen soil above it, a water table also can exist in this zone. When changes to topography are expected as part of project construction, the shape of the permafrost table below the topography can be such that water will pond in the nonfrozen subgrade and eventually lead to instability of the subgrade. The way to overcome this sort of problem is to shape the topography such that the underlying permafrost table will drain water away from the project. In addition to adhering to these simple rules, it is wise for the geotechnical engineer to seek expert assistance when dealing with permafrost.

Situations not normally thought to be problematic can occur when installing and operating a freezer for a warehouse, store, or restaurant. If there is insufficient insulation between the freezer floor and the ground supporting it, the ground will become frozen to some depth below the floor. This can result in frost heave and/or movement of moisture toward the area of the soil being affected. Moisture naturally moves from higher temperatures to lower temperatures, even in clays. Also, capillary forces can move water horizontally as well as vertically. When it is expected that ground freezing may occur, steps should be taken to either insulate the ground from the source of freezing or design the structure for it.

9.3.8 Erosion Control

Erosion is defined as the separation of particles from the soil mass. This is followed by transportation of particles, which will not happen if erosion is reduced or eliminated. Therefore, erosion, by this definition, is the reason for loss of materials. The causes of erosion are well known and many. They include the effects of the velocity of flow of water and/or wind as well as the velocity of impact of drops of water and particles already eroded. Freeze-thaw surface action can, by itself, loosen and remove particles from the soil mass. The seldom discussed surface reaction to water caused by air expulsion when water enters voids faster than a lean clay can expand to accept the water into its voids can cause what is known as "sheet" erosion of an exposed slope. The erosion due to soil grain structure that occurs in loess and other cohesionless soils is the reason for piping of silts and sands. Water flowing through their grain-on-grain structure will move these soil particles since they do not have the cohesion to hold them together. This can be most damaging in clays as well when the soil-water disequilibrium of dispersive clay causes clay particles to disperse into the water and be carried away, followed by the silt and sand particles left behind.

Dispersive clay erosion occurs when mainly sodium ions are present in the clay soil and relatively low concentrations of cations of all types are present. If these clays cannot swell the cracks and fissures present in them when they are relatively dry, there are many channels for the dispersion to take place, and the soil mass is in danger of being washed away in relatively pure water. Vegetation and shrink-swell behaviors cause these cracks to be present, and the porosity of the soil, when it is not dense enough, will enhance the erosion process. Many slopes and earth dams have failed because of this phenomenon. Figure 9.1 shows a slope eroded because of surface reaction to water and loss of clay from the dispersive clay phenomenon.

Remedies for erosion follow what makes sense for the situation. First, reduction of flow and/or impact can greatly reduce erosion. The types of remedies used are changes of slope, building of berms, and terracing of slopes. It is difficult to cut off water completely, but clay layers can be used to stop water. Covering surfaces with plastics, asphalt emulsions, and various types of paving blocks can be very helpful in overcoming erosion. The most frequently used method is vegetation of the surface to protect the soil and reduce velocity of flow and impacts. Hay is utilized to help protect a surface while grass is grown on it. Cross-hatched wires and degradable mats of different materials are being used for this purpose as well.

Since the dispersive clay phenomenon is physicochemically based, many of the above remedies will not work to reduce erosion of these problematic materials. Agents that are discussed below which can change the soil's tendency for dispersion need to be employed, along with well-densified materials. In addition, proper construction techniques must be employed to reduce porosity. Methods to redirect water away or downstream filters in the soil mass to trap the clay particles also will help.

FIGURE 9.1 Slope with dispersive clay and surface reaction erosion.

Stabilization of surface layers of soil exposed to water can be accomplished using several methods. In some cases, the whole soil mass must be stabilized as well. Lime and Portland cement (discussed below) are used to improve soil properties and reduce or eliminate erosion. Gabions, or cages made of fencing material, are used to enclose large rocks and make erosion-resistant surfaces for soils. Large blocks of concrete and even massive pods are used when the erosion forces are much larger than can effectively be overcome by surface treatments.

9.4 Chemical Modification and Stabilization

Chemical modification and/or stabilization involves major applications of chemicals to improve the behavior of soils, but it generally does not cause the soil particles to be bound together. Although commonly called chemical "stabilization," the more correct term is chemical modification. Chemical agents can be added by spraying on the surface of the layer to be treated, by intimate mixing of the agent with the soil layer to be treated, or by injection of chemical slurry into the soil subgrade. The effects realized are normally a mix of both physical and chemical or physicochemical. The idea is to effect an improvement of the physicochemical environment and/or surrounding particles. Because the most chemically active types of particles are clay, this kind of treatment is mostly used for clays.

Chemical modification is used to improve soil workability, making the soil easier to use as a construction material. It is used to reduce plasticity and shrink-swell potential. If clays are dispersive, it is used to flocculate their particles. When clays are difficult to compact, chemicals can be added to slightly disperse their particles and assist the process. Chemicals are used to treat surface soils to cause waterproofing and dust-proofing as well. In addition, when increased amounts of some chemicals used to modify soils are applied, they act to physically stabilize the soil. This is covered later in this chapter.

9.4.1 How Soils Are Affected by Chemicals

In some cases, chemicals affect the behavior of the sand and silt portions of soils, but this is mainly for waterproofing and dust-proofing only. Sand and silt fractions have relatively little capacity to hold onto any chemicals, and they are affected by chemicals in the pore water which physically change their behavior. As will be discussed, the clay fraction is very "chemically involved" and has the capacity to hold onto cations, etc.

Clay soils are a composite of one or more basic clay minerals, which are hydrous aluminum silicates with substitutions. There are two other substances in clay fractions. The first is very small fragments of pure rock minerals such as quartz, feldspar, calcite, pyrite, mica, etc. The second is other particles ("gunk") without crystalline structure, called allophane. Free silica and aluminum in allophane may affect clay chemical reactions. In addition, organic materials may be present that can dramatically affect chemical reactions.

Clay particles obtain initial charges in at least three ways. These include broken bonds in the clay, isomorphic substitutions and/or inner clay structure imperfections, and hydrogen bonding (replacement of O for OH or OH for O). Broken bonds around the edges of silica-aluminum units create unbalanced charges, which are balanced by counterions adsorbed or attracted near the clay particle surface. These are the predominate charge source in kaolinite and halloysite. In other clay minerals, such as illite, chlorite, and smectite, broken bonds are not a major charge source.

Charge imbalance from isomorphic substitutions occurs when lower valence ions replace higher valence ions in clay particle mineral structures and often cause structure imperfections. Hydrogen ions of exposed hydroxyl groups may be replaced by other cations or simply may be leached out. The result is clay particles made of charge-deficient (or negatively charged) clay minerals.

The clay-water system cannot exist with a net electrical charge. Particle negative charge is balanced by cations that exist in pore water or adsorbed on and in the particle. Cations are attracted to the particle and repulsed by each other. The charge gradient in pore water aligns dipolar water molecules, causing the double water layer. The thickness of and charge gradient in the double water layer depend on the following and must become balanced if possible. They depend on the total charge of the particle and the type and concentration of cations present. The pH of the soil-water system also has an effect.

The amount of water interlayer depends on the interlayer charge imbalance and the type and concentration of cations present there. The imbalance of charges there has to be satisfied and the osmotic pressure must be high. The electro- or physicochemical potential energy is also high, and the water moving into the interlayers of the clay causes >90% of swelling.

The counterions associated with clay particles/layers may be replaced by or exchanged with other ions in solutions, such as magnesium, calcium, sodium, potassium, carbonates, sulfates, and nitrates. Charges on clay particles may be measured by the number of exchangeable cations associated with the clay. This ion replacement behavior is called cation exchange capacity (CEC). CEC is measured on the total exchange complex internal and external to clay layers and particles. It is measured in milliequivalents of calcium per 100 g of soil. Examples of CEC for clays are listed below:

Kaolinite	3–15
Halloysite	5–10
Montmorillonite	80–150

Illite	10–40
Vermiculite	100–150
Chlorite	10–40

The CEC of different clay minerals varies and also may vary as a result of particle size, temperature, type and concentration of cations present, the pH of the soil, and the percentage of clay in the soil (CEC is given per 100 g soil). The CEC is determined by exchanging all cations in the exchange complex to calcium, followed by exchanging all the calcium out and measuring the amount of calcium exchanged out.

Other soil chemistry properties important to chemical stabilization include the soluble salts in pore water. These are given as milliequivalents of magnesium, calcium, potassium, sodium, and other ions per liter of saturation extract.

The sodium absorption ratio can be used to predict dispersion. It is the ratio of the sodium in the pore water extract compared to the square root of the sum of the concentrations of calcium and magnesium divided by 2, all in milliequivalents per liter of saturation extract. The exchangeable sodium percentage (ESP) also is useful in predicting dispersion of clays. It is the ratio of the sodium in the exchange complex to the CEC, all given in milliequivalents per 100 g of soil. The ESP can be amplified to represent the ESP of the clay by dividing the ESP of the whole soil by the percent clay in the soil. These indicators, in fact all chemical indicators, must be modified or analyzed knowing the total physicochemical situation or environment.

Almost every chemical (and/or base) imaginable has been added to soil in attempts to improve its engineering properties. Engineers could have saved a lot of effort if they had done proper research into soil chemistry. It turns out that very few chemicals are economically feasible. Those shown to be most useful include calcium, potassium, sodium, a combination of calcium and sodium, Portland cement, lime kiln dust, and fly ash plus calcium. Others that have shown promise are combinations of potassium and lignosulfonates and polyquaternary amines.

Chemical reactions, then, can only occur with clay, and really only three things can happen. First, there can be base exchange/ion crowding as the result of changing the type and concentration of cations in the pore water and associated with the clay. Second is dissolution of silica and alumina caused by very low or high pH conditions (<2 or >12) or by addition of NaCl. At a pH of about 10 or so, carbonates can cause precipitation of magnesium and calcium. Third is the formation of insoluble gels on crystals that contain the given mixture plus Si, Al, Ca, H, O, and other ions in the pore water as adsorbed to the clay before treatment. All but the last of these effects is called modification; the last is called stabilization.

There are other chemical and even biological agents that have been proposed for use in clays to improve their behavior. They are promoted by suppliers to engineers and even political groups that have authority over funds for construction. Unfortunately, not all of these agents are successful in improving clay behavior, even though claims are made as to how they act to improve the clay. The Committee on Chemical and Mechanical Stabilization of the Transportation Research Board recognized the problems involved in knowing how and if an agent would actually work in the field to improve clays. The committee published a guide on how to approach testing of chemical agents for use in soils which explains very thoroughly what information is needed from suppliers and how to go about testing chemical agents for soils (TRB 2005). In addition, a journal article by members of the committee covers similar information (Petry and Das 2001). It is recommended that anyone approached by a company that sells chemical agents, especially those not well known to work, should review either or both of these documents before testing and recommending the use of an agent.

9.4.2 Lime Treatment of Clays

Lime is a source of calcium ions, which happen to easily exchange into a clay for cations that make the clay more active. Lime is applied to clay soils most of the time by intimate mixing in layers of soil, but also by injection to significant depths into well-fractured and fissured clay subgrades. There are basically two types of lime used for most applications. When lime is produced by driving away carbon dioxide from limestone-type rock, the result is quicklime or CaO. Although usable in dry form, quicklime has to be hydrated either during the mixing process or as it is made into hydrated lime or $CaOH_2$ prior to mixing. The most useful form of lime now being added to clay soils is a slurry suspension of quicklime, often hydrated in the field to save the cost of transporting the heavier hydrate to the field. Lime tends to be a preferred choice, since it is relatively inexpensive compared to the results achieved. Modifying a clay soil normally takes less than 6% lime by dry weight of soil.

There is also danger involved in handling quicklime in the field; when it is exposed to water, it slakes down and hydrates in a process that produces extreme heat. In addition, a slurry of lime as used in the field treatment of clay soils has a pH of about 12.5. Therefore, safety measures must be employed in the field for both of these conditions. Also, lime in its dry forms must be protected from the atmosphere in so far as possible, since extended exposure will cause the carbon dioxide in the air to recombine with the lime and render it useless.

The relatively high pH of a lime suspension is an aid in determining how much lime to add to clay soils to modify or chemically "fix" them. In the laboratory, small samples of soil are treated with no lime and progressively higher percentages of lime by dry weight of soil. The lime and soil mixtures are wetted to form a slurry that can be tested for its pH. The pH of the lime-soil slurries is plotted to determine the percent lime that will provide a pH of about 12.5. When this point is reached, enough lime has been added to fully exchange out other cations for calcium ones and there is sufficient lime left over to cause the pH to be about 12.5. The higher pH of lime added to the clay soil also maintains a better pH environment for the exchange process and dissolving of clay particles, a part of the reaction used for lime stabilization that will be explained later. The National Lime Association can be contacted for further information about lime and treatment of soils with lime.

The exchange of calcium cations for others reduces the need for water to offset clay mineral negative charges and thereby dramatically reduces the double water layers of the clay particles. The improvement in the plasticity characteristics of the clay soil is dramatic. There have been cases of clay soils with a plasticity index of 70 where this property has been reduced to nearly 0 and certainly below 15. The other result is that the swell potential of the clay disappears. The clay soil then acts more like a silt material and tends to shed water rather than absorb it. Since improvements to the plasticity index can be measured, Atterberg limits testing of the treated soil is recommended to verify the results of lime treatment. Often, a series of these tests, done on the clay soil with no and increasingly greater lime percentages, is used to verify the percent lime to be added in the field. Improvements to swell behavior sometimes are tested to further verify the results of lime treatment.

Another, not well-recognized, effect of lime treatment is the crowding of the ions exchanged out of the soil in the pore waters of the soil. This cation crowding is what causes some of the waterproofing accomplished by lime treatment and may be the factor that is responsible for about half of the reduction in plasticity index. Since the clay particles have much smaller double water layers and have so many ions in the pore water surrounding them, the clay soil is very flocculated. This likely accounts for the silt-like texture of the soil. Some lime stabilization (discussed later in this chapter) also may occur as a result of the modification process with lime. Because of the changed nature of the treated clay soil, it is imperative that compaction tests be done to determine its treated maximum dry unit weight and optimum water content. Lime modification normally reduces the achievable denseness and results in about a 5% increase in optimum water content for compaction. If desired, testing for changed shear strength also can be done.

The presence of organic materials in the layer to be treated with lime can detrimentally affect the results of lime treatment. It has been determined that the presence of as little as 3% organics can dramatically and negatively affect lime-clay reactions and that 5% organics can essentially negate the improvements sought by lime treatment. This happens because the organic materials are physicochemical in nature, as is the clay, and they use up the calcium from the lime, so that it is impractical to add enough to achieve the desired results.

Lime slurry pressure injection has been employed for many years to improve clay subgrades. Probes are inserted into the subgrade, and the slurry is injected into the cracks and fissures present in the soil mass. Eventually, there is some modification of the clays between these lime-filled seams, but the changes are far from what is achieved by intimate mixing. Probably the greatest benefit derived from this method is the water that is added to the soil during the process and after. Soil mass structural improvements occur in some cases. Overall, it is mainly a moisture stabilization method for most clays.

9.4.3 Construction Processes for Lime Treatment

The objective of construction is to achieve in the field what has been conceived, tested, and designed for use. This means that laboratory programs should simulate field conditions, using the type of lime, application method, and water to be used in the field. The same degree of pulverization and compaction should be used, and the conditions of mellowing and curing should be the same as in the field. Deciding what depth of soil to treat is crucial to setting the construction sequencing. The geotechnical engineer must remember that the only material whose properties are changed is the material treated in this process. Deeper layers are not improved. Prior to treatment, the materials to be treated must be brought to the finished grade and be free of organic materials in so far as possible. Layers to be treated are commonly 0.3 m (1 ft) thick and can be as deep as 0.45 m (18 in.) or even deeper, if the equipment is large enough and powerful enough. When deeper layers need to be treated, the materials are moved to the soils which are too deep (or thick) for treatment of a layer and brought back over the treated and compacted layer to form a second treated and compacted layer.

Before lime is added in any form to the layer to be treated, pretreatment activities can be performed to improve the final product. One of these is to make sure that at least some pulverization is done to facilitate treatment; another is the addition of water to the soil, so that when the lime is added, the level of moisture will be as specified.

The type of lime to be added must be specified and may vary for differing conditions. Dry hydrate can be applied faster than slurry and can help to dry out soils that are naturally too wet. However, dry hydrate will easily be blown around, causing environmental problems, and may well require that more water be brought to the site. Dry quicklime can be more economical since less lime needs to be brought to the site, and it may provide faster reactions, drying out soils even faster than hydrate. However, quicklime will require more water to be brought to the site, does not hydrate easily and uniformly in the layer, and is more caustic than hydrate. The advantages of lime slurry, particularly that made at the site using quicklime, are that it is dust free and easily uniformly spread. It may require more manipulation, however, especially in wet soils. Most applications are using slurries made from quicklime at or near the site.

Following lime addition, preliminary mixing and watering are done to uniformly distribute the lime, pulverize the treated soil to pass 5 cm (2 in.) in size, and bring the water content to 5% above optimum for compaction of the treated soil. Mixing is done using pulvamixers to full depth in single or multiple passes. When the mixture appears to be uniform, it is lightly compacted with the mixer to allow for mellowing for 24–48 hours and possibly longer for CH clays. The water content during mellowing should be 5% above the optimum for compaction of the treated soil. In some cases, this is the only mixing done, so before compaction can proceed, the mixture must meet final pulverization specifications and be fully compacted.

The mellowing period allows modification to occur and helps further pulverization. It is recommended that a mellowing period be used. After this step, final mixing and pulverization proceed. During this step, the water content should be from 3 to 5% above the optimum. Final pulverization is to 100% passing 2 cm (1 in.) in size and at least 60% passing a No. 4 U.S. Series sieve. Compaction can follow and normally is started with either a sheepsfoot roller or pad foot roller and finished with a pneumatic roller or even a steel-wheel roller. Compaction specifications should call for dry unit weights at least 95% of the maximum from a standard Proctor-type test and water content at the optimum. The level of compaction energy used to develop the specifications may vary. Curing of each completed layer usually takes a week and the water content is maintained at the optimum.

Construction should be monitored for subgrade preparation, pretreatment, depth of cut and mixing, and pulverization. The lime used must meet quality standards, and lime quantities are checked per square meter (square yard) of layer. The uniformity of mixing can be checked using phenolphthalein. The water content is monitored prior to mellowing, during mixing and compaction. Compaction usually is checked by nuclear densimeters, which must be calibrated for the presence of more hydrogen molecules than normal. Finally, the curing conditions and time are verified.

The results of lime treatment are highly dependent on proper construction. Once in place for very long at all, it is very hard to determine the actual amount of lime added and is practically impossible after a long time. The quality of product achieved is based on specification and control during construction of the amount of lime added, water conditions throughout, mixing, pulverization, compaction, and cure.

9.5 Portland Cement Modification

Cement modification is the treatment of fine-grained soils with small amounts of Portland cement to improve their engineering properties, using usually ≤5% by dry weight of soil. For granular or silty fine-grained soils, it is used to improve compaction properties, develop a better "working table," and reduce or eliminate adverse plasticity. Usually the idea is to upgrade pit-run or dirty gravels to acceptable base materials. This also will increase bearing value as "strength" to acceptable values. Information about treatment of soils with Portland cement can be obtained from the Portland Cement Association.

For essentially plastic fine-grained soils, Portland cement modification is used to reduce plasticity (liquid limit drops, plastic limit increases, and shrinkage limit increases). There is an increase in effective particle size, from cementation of small particles together. Some base exchange occurs from cations liberated during the cement reaction, and there is an increased CEC of clay in the high-pH environment. Lime stabilization effects are claimed to occur, but this is questionable when the amount of calcium is not large and the pH is not as high as in lime treatment.

Reduction in volume change potential does occur to some degree, coupled with overall reduction in plasticity, when the degree of pulverization is great enough and the cement paste can surround clods and bind particles together, causing waterproofing. There is an increase in bearing value as "strength" and an increase in workability after treatment because of the base exchange and conglomeration of particles and clods. The material generally is upgraded to subbase quality.

For fine-grained basically granular or silty soils, the choice is probably between Portland cement and asphalt cement (discussed later). For plastic soils, the choice is probably between lime, Portland cement, and asphalt cement.

Comparing the effects of Portland cement vs. lime for modification reveals the following. Lime is better for reducing plasticity and volume change, with the potential difference small on lower plasticity index soil and larger when the soil plasticity index is larger. Portland cement is likely to produce higher "strength" and produce it faster. Neither works well for A-horizon soils because of the organics present. Portland cement requires more mixing but less water than lime and produces higher dry unit weights. Portland cement–treated material may well have to cure longer (3 days) to achieve sufficient strength to place another layer over it, whereas lime-treated material can be covered after 1 day.

For silty clays or low-plasticity clays, more "improvement" may be gained by Portland cement modification, whereas lime is probably a better modifier for high-plasticity clays. The proof is always in the mix design, using field pulverization standards and compaction specifications. A delay in compaction and remixing will work well for a lime-treated clay, while a Portland cement–treated material must be compacted before the cement sets up and cannot be disturbed after.

The mix design for Portland cement–modified soils is done using an Atterberg limits series of tests with differing percents of cement applied to the soil. This can be followed by a series of strength tests with samples prepared using field gradations, field application methods, and compaction standards, followed by field curing. Common strength tests used include CBR, unconfined compression, cohesiometer, and triaxial tests. The construction procedure for Portland cement modification is very close to that for lime treatment. The difference is that there can be only one mix and compaction cycle for Portland cement–treated soils and it must be done before cementation setup. The common field pulverization standard is 100% passing 5 cm (2 in.) in size and 55% passing a No. 4 U.S. Series sieve. Pulverization standards vary for differing locations and agencies, but finer pulverization is always better for chemically treated clay soils, especially during Portland cement treatment. Portland cement treatments and construction are discussed later in this chapter.

Sulfate-induced heave can be a problem in clay soils that are treated with either lime or Portland cement. This phenomenon occurs when the soluble sulfates in the soil are sufficient to cause the formation of ettringite and similar minerals that use up calcium ions, deplete the pH of the treated soil, and act to cause damaging three-dimensional heave. This is discussed at length in Section 9.8 on physical stabilization.

9.6 Fly Ash and Other Coal Combustion By-products

Fly ash is produced during combustion of coal in electrical power plants. It is taken out of the flu gases and may be used as a soil improvement material. There are basically two types of fly ash: class F and class C. They are produced by burning different coals. Both contain very fine particles of oxides that may have some benefit. Class C fly ash is the most useful type, since it contains significant calcium oxide or lime. Although it may be added to clay soils alone to try to modify them, it does not work well unless some lime is added with it. Commonly about 50% by dry weight of the lime that would be used to modify clays that react with both lime and fly ash can be replaced with class C fly ash, and nearly the same improvement results as if all lime was applied. In areas where much class C fly ash is produced, it would be prudent to test mixtures of lime and fly ash to see if they work as well as lime alone, since the cost of materials would be less.

Bottom ash is about 20% ash and the rest fly ash. Therefore, it may work in place of fly ash in some applications. This dark gray, granular, porous material is collected in water-filled hoppers at the bottom of furnaces and is a waste product, like fly ash. Its properties and likelihood of use are variable, depending on the source of the coal burned. Boiler slag is a molten bottom ash that is water quenched and fractures to form angular, coarse, hard, black, glassy material. It has approximately the same composition as fly ash, depending on the coal source as well. Both of these waste materials may provide beneficial effects when added to soils and in combination with agents that would cause binding materials to form.

Both lime kiln dust and cement kiln dust have been used in limited applications to modify clay soils. These waste products are combinations of lime or cement and fly ash. They and all waste products are quite variable and should be thoroughly tested to determine the benefits of adding them to soils.

9.7 Dust-Proofing and Waterproofing Agents

Various additives and agents have been used to reduce dust and/or stop water movement, usually for fine-grained soils. These procedures differ from conventional soil stabilization in that the soil usually will have the required engineering properties only if surface abrasion and/or moisture entrance/exit is prevented, keeping moisture out to reduce strength loss and holding particles in the soil mass.

The requirements for a good dust-proofing/waterproofing agent include rendering the soil immune to adverse effects of water, desiccation, traffic abrasion, and erosion. Also, the treatment must be effective in thin layers, less than 15 cm (6 in.). For economic reasons, it is preferable if less than 5 or 6% dust-proofing/waterproofing agent is needed. Finally, the treatment should retain the desired properties for at least one cycle of seasonal change without retreatment, including a heat and cold cycle. It also would be preferable if the dust-proofing/waterproofing agent improved soil engineering properties and/or could be applied as a penetration treatment, sprayed on the surface of the material.

Materials that have been used as dust-proofing and waterproofing agents fall into different categories. The first of these is asphaltic materials such as emulsified asphalt suspensions, road oils of asphalt cut back with solvents, MC and RC liquid asphalts, and bituminous surface treatments.

There are several chemicals that have been used, including those that are deliquescent, taking water from the air. Sodium chloride has been used for haul roads and other applications. Sodium silicate has been utilized mainly for dust control, as has ammonium chloride.

Rubber materials, polymers, and plastics of various types have been tried with varying success. Liquid latex has been sprayed on. Emulsified rubber (Petroset) has proven to be better at resistance to ultraviolet light as an erosion control agent. Neoprene liquid has been sprayed on as well. Calcium acrylate, a mixture of acrylic acid and lime, has been applied for these purposes. The waste product aniline furfural, which comes from distilling corncobs without air, is a resinous mixture that is highly toxic to touch and as a vapor. It has been applied with success in the field, however. Epoxies of all kinds have been applied, but they are resinous, expensive, and tend to deteriorate badly. Both vinyl and polyvinyl chloride have been tried, but are also expensive. Elastomeric polysulfides of all kinds have been tried as well. They have elastic properties and can be sprayed on in a thin coating, forming a sheet several molecules thick that does not deteriorate, like rubber, when weathered. Also, pectins, which are by-products of different fruit-canning processes, have been applied. Studies have been conducted to evaluate various chemicals as dust-proofing and waterproofing agents, but the best approach is to test a chemical before recommending its use for a particular project.

This is also true for other miscellaneous materials, such as discussed below. Enzymic products, such as Paczyme, which are biological agents, have shown some promise in some areas and on some soils. Wetting agents that decrease surface tension may assist in waterproofing. Normally these are surfactants or soaps. Both Portland cement and lime have been applied using lesser percentages than for modification. Bentonite has been applied to help control water, but cannot be used on a roadbed. Hydrophobic agents that tend to be absorbed on particles, rather than water, have been used, including silicates and stearates. Each particle must be coated in order for them to work well. Finally, many waste products, of course, have been applied. One class that has proven to be somewhat successful is chrome lignin and other lignosulfates, which come from the manufacture of paper and are highly acidic.

It is difficult to know just which agent to choose, especially among the more exotic ones. Therefore, it is wise to investigate the experience of others and utilize complete testing to determine how well an agent will work in the soil for the project under consideration. It is unwise to take the word of a company or person trying to sell an agent.

9.7.1 Sodium Chloride or Salt Applications

Salt (NaCl) stabilization historically has been used on soil-aggregate mixtures, typically GC soils to control dust, since it always increases surface tension in water. It can help compaction because the NaCl may reduce flocculation, which gets in the way of particle structure densification. It has been known to give increased "strength," believed to be from NaCl recrystallization in the voids of the soil, cementing particles together, which happens sometimes.

Sodium chloride works best on unsurfaced roads, where 1 or 2% salt is optimum for strength and lowers the freezing point, which helps in freeze-thaw situations. Obviously, NaCl (or any chemical) does not react much with SiO_2, so the reactions that occur must be with the clay fraction or the water in the soil pores.

This chapter has not addressed in detail how other lesser used chemical agents are tested for use or how well they may have done in the past. The author has tested potassium and related combinations such as potassium chloride, potassium hydroxide (with a pH in solution of 14), and potassium with ammonium lignosulfonate. Potassium appears to have a unique quality, because of its size and coordination number. It seems that it moves into the holes in the silica tetrahedral layer of the clays, making them less active and more like illite, a moderately active clay.

Testing on a particular polyquaternary amine, applied to compacted specimens by injection, has proven that it reduces swell potential significantly. This is believed to occur because the clay becomes less hydrophilic and more hydrophobic.

Claims have been made that the class of agents that are sulfonated oil products are ionic agents and exchange cations in the clay soil for hydrogen, thereby making the clay less active. The geotechnical engineer of record would have to test these agents to ensure that this improvement is achieved, since the author has not noted it in any testing done using this class of agents.

The remaining class of agents is enzymes. These biological agents are claimed to improve clay behavior as well. One would have to test them to be sure of this, within the guidelines mentioned earlier from the Committee on Chemical and Mechanical Stabilization of the Transportation Research Board. In all cases where chemical agents are concerned, the geotechnical engineer of record should personally ensure that an agent works to improve the soil of the project as desired and not only rely on the author's experience or anyone's claims.

9.8 Physical Stabilization

Physical stabilization is the improvement of selected properties of soils and soil aggregate mixtures by the addition of binding or cementing agents or methods. The most pronounced benefit is strength gain by cohesion or adhesion. The second most pronounced benefit is partial or total filling of voids and reduction of permeability. In addition, since most agents become relatively rigid, the modulus is higher and the compressibility becomes lower.

9.8.1 Ground Freezing

During ground freezing, although no agent is added, energy is added or used to freeze the soil and its water. Salt brine is circulated through pipes inserted into the ground to freeze it. The ice formed becomes the physical stabilizing agent. This procedure is most often used to allow the advancement of tunnels or excavations through soft and wet soils. It also is used to maintain frozen ground in permafrost areas. In all cases, the effects of "frost heave" must be considered and either accepted or eliminated.

9.8.2 Ground Furnacing

Furnacing is the application of intense heat to desiccate soil and produce limited fusion and vitrification of an *in situ* soil column. It starts by drilling a 4- to 5-in. hole to the desired depth. A burner that uses fuel oil or gas, along with compressed air, is introduced near the bottom to create a column of burning gases in the hole. The burner heats the soil at the bottom and sides of the hole to 2000°F or 1100°C. In porous soils, like loess, it is claimed that the burning gases penetrate the soil mass. Within a few feet, the soil is completely or partially made into a vitrified brick-like mass. Within 2–3 m (7–10 ft), the soil is stabilized. Beyond this, the soil is desiccated. The cost depends on soil moisture and fuel availability.

9.8.3 Chemical Cementing

Chemical stabilization is the bonding of soil particles with a cementing agent and is the result of a chemical reaction within the soil. This reaction may or may not include soil particles, although the bonding does involve intermolecular forces in the soil.

The first common type of chemical bonding agents is soluble silicates, usually sodium silicate solutions. In the presence of a weak acid or metallic salts, the silicate breaks down into sodium hydroxide and a colloidal silica gel or an insoluble silicate. The silica (SiO_2 and H_2O) is a viscous jelly-like mass that solidifies into silica with released water. Calcium silicate is an insoluble precipitate, reached after rapidly passing through the silica gel phase. The silica gel–soil mixture can be manipulated and rolled to form a membrane that becomes hard and impervious. The calcium and silicate precipitate fills the soil voids with an impervious binder in a flash reaction that does not permit manipulation. This form is injected into large voids to block the flow of water.

A delayed reaction is possible with the use of formamide, an organic reagent that slowly breaks down to form the acid that produces the colloidal silica gel. Gelling time can be controlled from minutes to several hours, during which the soluble silicate and the reagent remain in their initial condition. This process is used in injection stabilization, where the lower viscosity of the ungelled silicate permits greater penetration of low-permeability soils. The widest use of silicates is in stabilizing sands and rock to improve the strength and reduce water flow. Although they shrink upon drying and become brittle, they appear to be relatively permanent in a moist state (or before drying).

Organic monomers form a wide range of complex chemicals that are initially water soluble. A water solution of typically 10% can be readily mixed with the soil or, the most common practice, injected into the voids. A second chemical, termed the activator or catalyst, causes the molecules of the monomer to link together in a process called polymerization. The resulting polymer consists of a lattice of the linked organic molecules with water trapped between them. A similar reaction involves separate soluble ingredients which subsequently polymerize. In either case, the polymer is an elastic solid, by weight largely water, whose strength and rigidity are controlled by its chemistry and concentration. Some of the better known agents are discussed below.

The first to be extensively used in soil stabilization has the trade name AM-9. It is a mixture of acrylamide-methylene-bisacrylamide used in a 10% solution and has a viscosity of only 1.5 times that of water. It polymerizes into a rubbery gel similar to stiff gelatin. Rates of reaction can be controlled from a few minutes to 10 hours by choice and proportion of activators.

A second system, Terranier, is a water-soluble, low-molecular-weight phenolic flavonoid monomer derived from pine bark. There are two forms available: Terranier A, with a viscosity 20–30 times that of water, and Terranier C, with a viscosity 2 times that of water. The catalyst solutions of iron sulfate and formaldehyde produce polymerization, with the corresponding gel times from a few minutes to several hours. The dark-colored gel is rubbery to semirigid.

A third system has the trade name Cynaloc, which is a white viscous liquid. When diluted with an equal volume of water, as usually is done, it is 10–20 times more viscous than water.

The polymerization can be controlled from between a few minutes to an hour, and the resulting material is a relatively rigid solid resin.

A fourth system is called chrome-lignin, which utilizes the waste lignin black liquor from a sulfate paper manufacturer. Potassium or sodium dichromate reacts with the liquor to form an organic monomer, chrome-lignin, which slowly polymerizes into a brown gel. Typical concentrations are from 10 to 20% by weight. The rate of gel is controlled by temperature and concentration. Typical gel times are 5 minutes to 1 or 2 hours.

Solidified water polymers generally shrink greatly and lose their continuity upon drying, but maintain their strength if continuously wet. One exception is Cynaloc, which shrinks only slightly to a rigid strong solid that resembles a white plastic. The long-term stability of these organic polymers has yet to be established. Chrome-lignin has been proven to last more than a decade with no deterioration if in a wet environment. AM-9 has lasted nearly a decade in a wet environment.

Another kind of bonding agent is organic mixtures. Analine, a liquid coal tar derivative, and furfural, an organic liquid from refining corn products, can be mixed in a ratio of two parts to one and then react to form a deep-red viscous resin that hardens slowly by polymerization to a solid. One liquid is mixed with the soil, followed by the other and water, and then the soil must be immediately compacted. About 5% by weight of the resin has proved ample for rigidly stabilizing loose sands and similar soils so that they can be used as roads within a few hours of processing.

Finally, there are chemicals that react with soil to form cementing agents. One of these is phosphoric acid, which when added with a wetting agent acts on clay minerals to form aluminum phosphates. Acids react with carbonates and silicates to dissolve them and precipitate binders. Other acids have been tried as well, but they are not utilized much because of safety issues and economics.

Deep soil mixing is a special application of soil cements to form columns of physically stabilized materials. Soil cements as described above are used with specialized equipment to cut into the soil and mix the cement and soil. In some cases, this process is done with mechanical mixing blades, and in others, the cutting and mixing are done with jets of fluid that contain the soil-cementing materials. This kind of ground improvement is done solely by specialty contractors, and geotechnical engineers must interact with them to achieve the desired ground improvements.

9.8.4 Lime Stabilization

Lime stabilization involves the addition of more lime than is needed to modify the clay soil and enough lime to provide adequate strength gain. This is what makes it stabilization, beyond modification. The mechanisms are the same as for lime modification: cation exchange and ion crowding, flocculation and aggromulation, and carbonation. The larger amount of lime causes significant pozzolanic cementation, which controls stabilization.

Addition of lime over the optimum for modification (LMO) gives excess free calcium and hydroxyls. This provides a sustained pH that dissolves Al and Si from the clay. The Ca, OH, Al, and Si react to form gelatinous calcium aluminum hydrates (CAH), low-order calcium silicate hydrates (CSH I), and high-order calcium silicate hydrates (CSH II). Figure 9.2 shows clay that has been treated with lime and the effects of the high pH. Figure 9.3 is a photograph of pozzolan cementation as a result of lime treatment of a clay.

CAH provides very quick low-level strength and therefore does not contribute much. CSH I forms rapidly to provide rapid early strength gain with time, but not much long-term

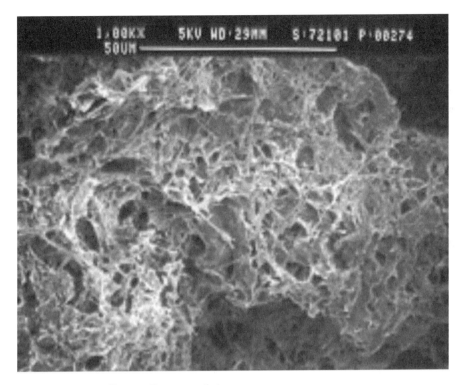

FIGURE 9.2   pH effects on lime-treated clay.

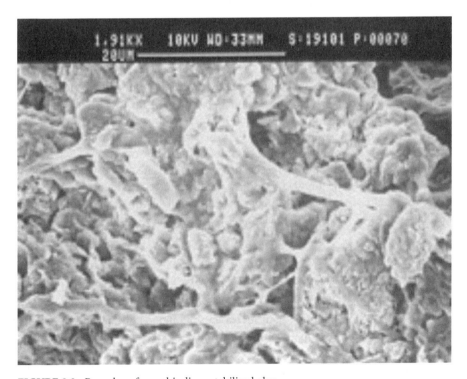

FIGURE 9.3   Pozzolans formed in lime-stabilized clay.

improvement in strength. CSH II forms slowly from CSH I and more Si to provide long-term and significant strength gains, forming a mineral called tobermonite. The CSH I to CSH II reaction is very pH dependent; thus there is an optimum amount of lime to provide the best lime stabilization, the LSO.

The LSO is defined as the lime content that provides maximum strength gain. The initial pH and change of pH are controlled by the lime content. CSH II forms at a slightly lower pH than CSH I. The formation of CSH I uses up Si, Ca, and OH, so the pH lowers. However, there needs to be enough Si for the formation of CSH II, so a sustained pH is needed. At the LSO, the best combination of CSH I and CSH II happens for strength. It has been noted that the LSO is normally at least as much lime as the LMO, but not more than twice as much.

The LSO depends on how curing is done and how strength is determined as well. Curing normally is done in a moist room for 7 days, 28 days, 6 months, or even 1 year. Specimens are tested as molded or soaked in water for differing periods of time. Normally, the geotechnical engineer follows the standard of care in the locale of the project. CAH forms within 24 hours. CSH I forms within 7–10 days. CSH II continues to increase from 10 days to 3 years or more, as long as the "ingredients" as well as the proper pH and water are present.

It is important not to expect the LSO to always be the same for a given clay soil. A minimum acceptable strength must be established, using standardized lab mix design criteria. It is best to correlate one's own field performance relative to laboratory performance experience with that of others and to continuously evaluate both lab and field performance, so as to develop engineering judgment on a regional basis. It is important to understand any changes that take place in the soil in the laboratory and the field and not be afraid to change one's mind.

A standard test that is inexpensive and accepted by the client and that gives reproducible results should be used. Also, sample preparation, lime addition, mixing, mellowing, curing, etc. should be standardized. A standard compaction procedure should be used for lime-treated soil, compacting all comparisons at the same standard, possibly with an average percent lime. Finally, a standard curing procedure (moist, soaked, or both) should be used, with time of curing standard (28 days). The procedures must be easy for technicians to follow and any possible errors should be on the safe side.

In summary, there is no unique lime content. For modification, a pH test and plasticity index series testing are used. For stabilization, the pH test gives the lowest percentage and the LSO may be as much as two times the LMO. It is necessary to run a mix design to determine the lime content, and the answer depends on the problem presented by the project and the soils present. Testing in the laboratory must be done using fresh lime of the type used in the field. Applications require an additional 1% by dry weight of CaO to provide the equivalent treatment of $CaOH_2$. Also, for field work, an additional 1% is added to account for the inconsistencies and errors that might happen there. Finally, construction of lime-stabilized clays is the same as for lime-modified clay soils.

9.8.5 Agents with Lime

Among the agents that have been tried in combination with lime stabilization, one that is beneficial is sodium chloride. Addition of this agent causes improved compaction characteristics, improved moisture retention, and increased solubility of silica. The improvement in compaction is a greater maximum dry unit weight and decreased optimum water content than for lime-treated soil alone. A decrease in treated swell has been measured, perhaps from further ion crowding. The largest improvement, however, is in the strengths that result from treatment. The optimum salt content, when added to lime treatment, is only 1 or 2% by dry weight of soil. The salt is added in the compaction water during construction.

Lime has been applied using the percentages for modification to reduce plasticity and shrink-swell of clays, followed by either Portland cement or asphalt to provide a gain in shear strength. Although these combination applications are more expensive, the benefits may well be worth the effort and cost. The addition of Portland cement after lime modification has been most popular where there is a need for lower activity from lime treatment and higher, earlier strength gain from the Portland cement. The construction procedures for the lime addition are the same as for lime alone, but the addition of the Portland cement, along with compaction and finishing, must be done as for soil cement, as will be described below.

The addition of fly ash along with lime has been shown to be beneficial, both in property changes and economics. As described above, calcareous fly ash contains significant calcium oxide. Replacement of 25–67% of lime with fly ash in lime stabilization produces the same modification and stabilization effects, and sometimes more quickly. The construction procedures when using fly ash are the same as for lime alone.

9.8.6 Soil Cement (Portland Cement)

Soil cement is the treatment of soil with Portland cement to increase stability and compressive strength. It is a mixture of pulverized soil and measured amounts of Portland cement and water, compacted to high dry unit weights. Its uses are many:

- A base course for roads, streets, airports, shoulders, and parking areas
- A subbase for rigid or flexible pavement and soil cement pavements
- Road widening and construction of storage areas
- Reconstruction of failing granular bases
- Patching
- Slope protection for earth dams and embankments
- Earth dam cores
- Reservoir linings
- Ditch linings
- Stabilized subgrades
- Foundation leveling

There are three types of mixtures of soil and Portland cement. The first, cement-modified soil, was discussed in Section 9.4 on chemical stabilization. The other two types are compacted soil cement and plastic soil cement. Almost any type of soil can be improved by soil cement treatment. The soil must be readily pulverized fine enough so that the cement paste can coat particles and bond them together. It must be easy to mix Portland cement with the soil. Finally, the soil cement must be able to be built under a wide range of weather conditions.

Sandy and gravelly soils with about 10–35% silt and clay combined have the most favorable characteristics and generally require the least amount of cement. Almost all granular materials work well, if they contain 55% or more passing a No. 4 U.S. Series sieve. Sandy soils deficient in fines, such as some beach, glacial, and windblown sands, can make good soil cement but require more Portland cement than those most suitable as discussed above. Also, traction is a problem except when the mixture is damp, and the surface is "tender" and prone to tracking.

Silty and clayey soils make satisfactory soil cement, but those that contain high clay contents are harder to pulverize. If they can be pulverized sufficiently, the soil cement is suitable. The more clayey, the higher the percent Portland cement needed to make suitable soil cement, and construction is more dependent on weather conditions.

There are four soil cement requisites for satisfactory characteristics and serviceability: (1) an adequate quantity of Portland cement can be incorporated with the soil, (2) the proper weight of water can be mixed uniformly with the soil cement mixture, (3) the moist soil cement mixture can be compacted to a proper dry unit weight before cement hydration, and (4) not more than 45% of the soil is larger than a No. 4 U.S. Series sieve or any of it larger than 5 cm (>2 in.).

Soil cement testing requirements vary depending on the soil being treated and the size of the project. When dealing with major projects, more time and more resources are available for testing. Small and emergency projects can be handled using simpler and shorter testing regimens. All projects involve sampling of the soil to be treated, soil identification tests, and preparation of the sample for testing. If the project is major and the soil is a sand, an abbreviated set of testing is done. The set includes determining the percent cement using a chart, compaction testing with cement in the soil, and then compression testing. For major projects, a complete series of detailed tests is done for soils other than sands. This series includes compaction tests with cement in the soil, preparing specimens for both wet-dry and freeze-thaw testing, followed by compression testing. Specimens for wet-dry and freeze-thaw tests are cured for 7 days in a moist environment and strength specimens are cured up to 28 days. As a result of these tests, the percent cement is known to provide both durability and strength.

For both major and small projects, an abbreviated process can be used if the soil being tested has been used by the geotechnical engineer before. The soil identification test results and soil series information are used to determine the percent cement used before. It is probably wise for the geotechnical engineer to make specimens and test them using the percent cement chosen. Soil cement mixtures for small and emergency projects are determined by how well the soil, treated with varying percents of cement, performs in two rapid tests. Specimens would need to be made, so compaction testing is done first using an average percent cement expected to be used. Specimens are compacted and cured. Then they undergo testing using the "pick" test and "click" test, to see if they are hardened sufficiently. All of these testing procedures and choices are published by the Portland Cement Association. It is important to realize that all specimens must be compacted such that the Portland cement will not set up before compaction is complete.

Strength tests generally are conducted as supplementary by some, but they should be performed to determine the strength parameters for design. Which test determines the percent cement to be added depends on how and where the material is to be used. The freeze-thaw test applies the most destructive forces on the treated soil. The wet-dry test is perhaps the least destructive, except for clays, where it may be the determining factor. Actually, the choice of test depends on what kind of forces would cause the most difficulty for the treated soil during the life of the project.

Strength specimens are broken at ages of 2, 7, and 28 days. They are kept at 100% humidity and then broken after a soak of 4 hours using compression at 138 kPa/sec (20 psi/sec). It is assumed by some that the strengths achieved normally are such that strength testing is of minor importance, only used to find the rate of hardening. Since the strength achieved is high enough for many uses, the percent Portland cement can be determined by durability tests. The strengths expected vary between 2 and 5.5 MPa (300–800 psi).

In glaciated areas in the northern U.S. and in eastern and southeastern coastal areas, some sandy soils require inordinately high percentages of Portland cement, because of organics or other deleterious materials. This may be corrected by adding normally reacting materials such as crushed rock or by adding small percentages of $CaCl_2$. The material can be added to even a 50-50 mixture, which is then tested by compression, followed by freeze-thaw and wet-dry tests. Usually only 0.6–1.0% $CaCl_2$ by dry weight is needed. Adding enough material to reach a compressive a strength of 1.7 MPa (250 psi) ensures that the mix will pass the other tests. There is no advantage to adding this agent to other than poorly reacting sands. $CaCl_2$ is added in the water or to the cement as a dry powder. It also can be added in dry form or in a solution before the cement is added.

Plastic soil cement, often worked similar to concrete, is a thorough mixture of soil and Portland cement with sufficient water to give the consistency of plastering mortar. The soils used are lighter textured, and those with 30% passing a No. 200 U.S. Series sieve are not used. Since the amount of water included is higher than for compacted soil cement, about 4% additional Portland cement is needed for hardening. The expected dry unit weight is about 15% less than the maximum for compaction. A test can be done to determine the dry unit weight to expect by rodding and dropping. Specimens are molded using these methods to compact them and tested after cure. When erosion resistance is needed, an additional 2% Portland cement is added above that determined from testing.

The procedure for construction of soil cement is similar to those described for lime treatment, with some very important differences. The process is fairly simple. It starts by shaping the roadway or subgrade to crown, which will help keep it dry, and grading. The roadway or compacted subgrade is scarified to smaller than 5 cm (<2 in.) in size. The materials are pulverized as necessary so that at least 55% passes a No. 4 U.S. Series sieve. The soil is prewetted, if needed, but kept dry of optimum for ease of working the soil. Then the soil is shaped for mixing and sometimes windrowed.

The soil is processed using high-speed rotary mixers. The Portland cement is uniformly spread and mixed with a pulvamixer. Water is applied as needed to bring the mixture to the optimum water content. Sometimes a pug mill is used to mix the soil and Portland cement away from where it will be compacted. Compaction initially is done using a sheepsfoot roller or pad foot roller, if the treated soil is a clay. Pneumatic rollers are used for less plastic soils, or steel-wheel rollers can be used to finish the layer. Light watering is applied to maintain the optimum water content during compaction and afterward to promote curing. Final grading and finishing are done soon after compaction, and transverse joints are cut vertically through the thickness of the layer at the end of the day. The finished layer is cured by spraying with water to prevent evaporation or is covered. Bituminous materials are sprayed on the surface of the layer to promote curing.

During the few days of curing, traffic loads on the layer are kept below construction loads and the bituminous surface is sanded to stop pickup of treated materials. Rainfall that occurs before the bituminous materials are added should be drained off. If it rains before the layer is completed, the layer must be finished as soon as possible. After the bituminous materials are sprayed on the finished layer, there will be no problems due to rainfall. The layer must be protected from freezing for 7 days and cannot be placed when the temperature is below 4°C (40°F).

It is well known that soil cement should not be placed over expansive clays. Even though this is not done, there can be problems from cracking of the soil cement subgrade that can cause reflective cracking in flexible pavements. Lime and fly ash are the best additives to help control this cracking. Sulfates and expansive cements are second best, and calcium chloride can provide some benefits as well.

Kaoline soil cement shrinks faster than montmorillonite soil cement. Longer curing increases shrinkage in sandy soils, but the reverse is true in clayey soils. An optimum proportion of cement gives the best results. Molding moisture has the greatest effect on shrinkage, and optimum to slightly above it should be used. Shrinkage can be reduced by improved compaction.

The most recent studies have shown that precracking the layer of soil cement, with microcracks that will heal later, can significantly reduce reflective cracking in flexible pavements over soil cement. This procedure was first proposed in Austria. Three to five passes of a vibratory roller 1–3 days after placement causes the microcracks. It is believed that the microcracks introduced in the stabilized base will minimize severe shrinkage cracks and that the cracked base will gain strength with time.

9.8.7 Sulfate-Induced Heave

Sulfate-induced heave occurs in layers of clay soils that have sufficient sulfates present and that are treated with either lime, fly ash, Portland cement, or similar agents. It has been the cause of millions of dollars of damage to roadways, taxiways, and runways. What appears to happen is the development of humps across and lengthwise along the roadways. In reality, as will be explained further below, minerals forming and hydrating in the treated soils express swell in all directions, thereby buckling the treated layer supporting the pavement. Figures 9.4–9.6 show the damage that can be caused by sulfate-induced heave.

FIGURE 9.4 Results of sulfate-induced heave on roadway.

FIGURE 9.5 Buckling of roadway by sulfate-induced heave.

FIGURE 9.6 Sulfate-induced heave in interstate highway subgrade.

The phenomenon was first reported by Hunter (1988) and Mitchell (1986) as affecting streets in Las Vegas in the late 1980s. It occurs when relatively high amounts of soluble sulfates are present in the clay to be treated. There has to be at least 10% clay in the soil. Water also has to be available to fully hydrate the minerals forming. Supposedly because of the possible presence of soluble sulfates, it can occur anywhere west of the Mississippi River and has many times.

Although there are many miles of unaffected subgrade in Texas, that is where most of the damage has occurred. It has been seen repeatedly along the eastern edge of the Eagle Ford shale formation in Dallas County, at Laughlin Air Force Base in Del Rio, in Arlington, and in the fast-growing community of Frisco, as well as along the I-45 interstate at Ennis. These are just a few of the actual accounts, since not all occurrences are reported and remedial actions are taken to overcome it. In addition to damage of pavements in Texas, sulfate-induced heave has occurred more than once in Las Vegas, several places in California, and was a considerable problem during construction of the new Denver International Airport.

During conventional lime, fly ash, or Portland cement treatment of these clays, the sulfates use the calcium and alumina available, because of these agents and their pH, to form new unwanted minerals, the main one being ettringite. This mineral formation also uses hydroxyls to make the stalk-like crystals that expand to two or three times the volume of their constituents upon formation. When formed and water is available, the crystals expand several times more in size and cause extreme swell, more than any expansive clay is capable of doing. As a result, the normally occurring CAH and CSH are not formed, because of the lack of calcium, alumina, and a pH to ensure solubility of silica. The resultant "treated" soil appears to be lime modified, but that is all, and it has little shear strength. Layers of the ettringite-laden clay soil expand in all directions, forcing pavements up and sideways, and buckling is one of the unsightly results.

Ettringite has been identified as the expansive mineral that causes the phenomenon. Pyrite, naturally available in the soils, is a contributing factor, in that it weathers to become sulfate compounds, such as gypsum. There is a 241% expansion when the mono- to trisulfoaluminate transfer occurs. The crystal expansion pressure is reported to be as much as 241 MPa (35,000 psi). After the mineral forms, it can hydrate and expand 227%. The three-dimensional expansion occurs when the stalk-like crystals act to push the particles of the soil apart.

The most difficult part of dealing with this phenomenon is finding where the soluble sulfates are in the layer of clay soil to be treated, because they appear in totally random locations, horizontally and vertically, and in highly varying amounts. Normally, the soil subgrade is sampled at horizontal intervals and somewhere on the roadbed to be treated. Unfortunately, this sampling interval can be set to every 15 m (50 ft) and the presence of potentially damaging sulfates easily can be missed. One method that has shown promise in Texas is the use of a magnetometer to locate salts concentrations as the magnetometer is moved over the ground, close to the surface. High readings indicate the presence of salts, which, in Texas at least, are mostly sulfate salts. These locations are marked and sampled for sulfate determination. It is important to note that this method has not yet been proven to work elsewhere.

The next difficulty is the actual determination of soluble sulfates in the samples taken. Normally, an amount of dry soil suspected of containing sulfates is added to a solution, so that the sulfates can become soluble and measurable. The author recommends using as pure water as possible, distilled and demineralized, and adding 10 times the weight of water to the weight of dry soil. This suspension is then mixed and time is allowed for the sulfates to become soluble. The liquid portion of the suspension is then tested to find the weight of sulfates present. There are some standards for this testing, but the best is probably that used by chemists for this purpose. The results are given in percent of the weight of soil that is sulfates or in parts per million. Less than 2000 ppm or 0.2% soluble sulfates is considered not likely to be damaging. When the sulfates are from 2000 to 5000 ppm or 0.2 to 0.5%, some moderate damage can occur. When the soluble sulfates are between 5000 and 10,000 ppm or 0.5 and 1.0%, moderate to somewhat severe damage will likely occur. When the soluble sulfates are over 10,000 ppm or 1.0%, severe damage most likely will happen. To determine whether significant ettringite formation and hydration will occur, with accompanying three-dimensional swell, testing is needed. This can be done in the field, of course, but a three-dimensional swell test has been developed by the author which predicts field damage that can occur. This test is described in Little and Petry (1992). Figures 9.7 shows this test in process, and Figure 9.8 shows the results of significant sulfate-induced heave in laboratory specimens.

There are three possible ways to deal with clay soils which are to be treated that have significant soluble sulfates. The first is to isolate where the problematic soils are along the subgrade and remove and replace them with clay soils without sulfates. A second method that can be used is to treat the soils with the stabilizing agent of choice and also with an additive that will promote the formation of pozzolans over ettringite. Two agents that have shown some promise and that must be used in a mix design to be verified for use are potassium hydroxide to hold the pH high and any source of amorphous silica that can boost pozzolan

FIGURE 9.7 Three-dimensional swell test for sulfate-induced heave.

FIGURE 9.8 Three-dimensional swell test results in sulfate-laden, lime-treated soil.

formation. The idea in each case is to create an environment that favors pozzolans over ettringite.

The third and most widely used field method to overcome sulfate-induced heave is a combination of prewetting to solubilize the sulfates before the agent of choice is added, followed by keeping the mixture at a water content of 5% above the optimum for compaction, if possible, to promote formation and hydration of any ettringite that can form. Ideally, this formation would occur during the mellowing period before final mixing, so that once formed and hydrated, the ettringite can become part of the treated subgrade and no more heave will occur. It has been determined that once formed and hydrated, this troublesome mineral does not reverse itself and can be a part of an acceptable treated layer.

The construction procedure for this method starts by prewetting the subgrade to 5% above the optimum at least 3 days before adding the agent and involves some mixing of soil and water to promote sulfate solubility. The agent is added and mixed, and the treated layer is kept at moisture levels above the optimum during mellowing. If sulfate-induced heave occurs, then a second amount of agent is added to make sure all the ettringite has formed and hydrated and to form pozzolans, if possible. After another mellowing period without any three-dimensional heave, the final mixing is done and any desired agent is then added. The layer is then compacted and kept moist, at the optimum water content, for the curing period. Using this method and double and even triple treatments, many sulfate-laden subgrades have been successfully treated.

9.8.8 Asphalt Stabilization

Soil asphalt stabilization is the improvement of soil properties by the addition of low percentages of bitumen, to provide strength gain by binding and improve waterproofing. Bitumens consist of nonaqueous systems of hydrocarbons completely soluble in carbon disulfide. Asphalt has primary components that are natural or refined petroleum bitumens or combinations of natural and refined bitumens. Tars are bitumen condensates produced by destructive distillation of organic materials such as coal, oil, lignite, peat, and wood and most often are used in asphalt. References are available from the Asphalt Institute.

Asphalt is too viscous to be incorporated with soils. Its fluidity has to be increased to mix. This can be done by heating, emulsifying in water, or cutting back with some volatile solvent and then heating. Tars are not emulsified but instead are cut back and heated to 27–104°C (80–220°F).

Nearly every inorganic soil can be treated with asphalt, with which it can be mixed. It works best, however, in certain soils. These include soils that are greater than 50% smaller than a No. 4 U.S. Series sieve and 35–100% smaller than a No. 40 U.S. Series sieve. Also, 10–20% of their particles should be smaller than a No. 200 U.S. Series sieve. This percentage may be as high as 50%. The liquid limit must be less than 40 and the plasticity index less than 10. The maximum diameter of grains must be less than one-third the lift thickness as well.

The expected stabilizing effects include binding soil particles together, which works well only in cohesionless soils to improve strength. Another is protecting soil from deleterious effects of water, or "waterproofing," which works best in water-sensitive, cohesive soils by plugging the voids.

Soil asphalt quality is affected by several factors. The first of these is the nature of the soil. Acid organic matter found in forest and river bottom soils is detrimental to soil asphalt. Neutral and basic organic material from arid and semiarid regions does not have a lot of influence, however. Fine-grained soils from arid regions that are high in pH dissolved salts do not respond well. Finally, plastic clays are difficult to treat because of mixing problems and the amount of asphalt required.

The next factor that affects quality is the amount and type of asphalt added. Within limits, the more asphalt used, the better the quality, and normally 1–5% up to 10% is used. For fine-grained soils, increasing the asphalt greatly improves only waterproofing and not strength. Too much asphalt makes fine-grained soils gooey and they cannot be compacted. Type of asphalt will not be discussed further here except to say that low-penetration asphalt or liquid asphalt usually is used and low-heat asphalts are the norm. Mixing also affects quality, and the more thorough the mixing, the better the stabilized soil.

Another factor that affects quality is the compaction conditions. The denseness of the soil asphalt is dependent on the volatiles content and amount and type of compaction. There is an optimum volatiles content for compaction. The final dry unit weight achieved can vary by 5%, with volatiles changing from, say, 5 to 11%. The type of compaction can vary the dry unit weight achieved by about 4% as well. As a general rule, the lower the volatiles, the higher the strength for cured specimens. For samples saturated after cure, the curve shows the most strength at or near the optimum dry unit weight and volatiles content. Water picked up probably causes this result. Plastic soils do not show this correlation. The percent volatiles for best dry unit weight and for strength are different.

Admixtures, such as phosphorous pentoxide or certain amines, may be needed to improve the rewet strength of the soil asphalt. Field performance results are inconclusive. However, when 2% was added to soils with about 5% asphalt, the strength gain was 50–200%.

The last factor that affects quality is the cure conditions. The longer the period of cure and the warmer the temperature of cure, the greater the volatiles lost. The longer the period of immersion, the greater the water pickup. The strength of any given soil asphalt mixture is inversely related to the volatiles content at the time of the test.

Designing soil asphalt includes several steps. The first is determining the thickness of the base to be treated, followed by the type, grade, and amount of asphalt. Next are the compaction and cure conditions in the field. The thickness is usually that which provides a CBR of about 80%, or 80% resistant to penetration testing as an ideal rock base material. Usually 15–20 cm (6–8 in.) of soil asphalt is commonly used in the U.S. If used for anything other than a base, a proper strength test is conducted to determine the thickness.

The type, grade, and amount of asphalt are the next decisions. The asphalt should be as heavy and warm as can be handled. Type, grade, and amount are selected on the basis of a laboratory test program, designed to determine effect on stability. A typical formula used to determine the amount of asphalt is

$$(\%) \; P \;=\; (0.015A \;+\; 0.02B \;+\; 0.03C \;+\; 0.09D)$$

where A = the percent soil larger than a No. 10 U.S. Series sieve, B = the percent soil smaller than a No. 10 and larger than a No. 40 U.S. Series sieve, C = the percent soil smaller than a No. 40 and larger than a No. 200 U.S. Series sieve, and D = the percent soil passing a No. 200 U.S. Series sieve.

Compaction conditions are found by testing as well. The molding volatiles content equals the volatiles in the asphalt plus water to be added, determined from laboratory test results. The stability of cured and immersed samples is determined at various percent volatiles, and the optimum provides the greatest strength. It is important to remember that shortcuts are not reliable and a full mix design is best.

The proper construction sequence and control are imperative to provide soil asphalt that will perform as expected. The construction sequence starts with pulverization of the soil to be treated and addition of water necessary for mixing. The bitumen is then added and mixed with equipment similar to that used for lime or Portland cement treatment. Next the soil is aerated to bring the volatile content to what is needed for compaction. The treated layer is then compacted and finished. Aeration follows during curing to bring the volatiles content to the optimum for strength. A surface cover is applied to hold the volatiles content where set.

In order to provide the expected quality of soil asphalt layer, close construction control is necessary during mixing, compacting, drying, and applying surface protection. The tests performed include determination of (1) water content before and during the process, (2) bitumen content after mixing, and (3) density after compaction.

The optimum volatiles content for compaction usually is much greater than that for stability, and good mixing may require an even higher content, especially in clayey soils.

Soil asphalt is used to bind soil particles together (coarse grained) and/or to waterproof the mixture (fine grained). The asphalt applied is either cut with some volatiles or emulsified. Soil asphalt is applied in several forms. It is applied in an intimate mix, where all particles are coated with asphalt, or as a waterproofing to provide stability to granular soils with some fines. In phase stabilization, groups of particles or clods are covered with asphalt. Membrane stabilization is used to cover or surround compacted soil with an asphalt layer.

Asphalt stabilization is a method of physical stabilization where no migration or chemical reactions occur and works best on chemically inert soils. It works well for situations where flexibility is needed with increased strength and waterproofing. It can be used with lime-modified or -stabilized soils. Finally, it can be used as a penetration treatment through which vegetation can grow.

References

Hunter, D. (1988). Lime-induced heave in sulfate-bearing clay soils. *J. Geotech. Eng.,* 114(2):150–167.

Little, D.N. and Petry, T.M. (1992). Recent developments in sulfate-induced heave in treated expansive clays. *Proceedings of the Second Interagency Symposium on Stabilization of Soils and Other Materials,* sponsored by U.S. Army Corps of Engineers, Bureau of Reclamation, Soil Conservation Service, Federal Highway Administration, Environmental Protection Agency, and NAVFAC, November 2–5, Metairie, LA, 14 pp.

Mitchell, J.K. (1986). Practical problems from surprising soil behavior. *J. Geotech. Eng. Div. ASCE,* 112(3):259–289.

Petry, T.M. and Armstrong, J.C. (1989). *Stabilization of Expansive Clay Soils,* Transportation Research Record 1219, Transportation Research Board, 10 pp.

Petry, T.M. and Berger, E. (2005). Sampling and testing when sulfates are suspected and clay is to be treated. *Proceedings of the 2nd International Symposium of Treatment and Recycling of Materials for Transport Infrastructures,* October 24–26, Paris.

Petry, T.M. and Berger, E.A. (2006). *Impact of Moisture Content on Strength Gain in Lime-Treated Soils,* included in the CD of papers presented at the 85th Annual Meeting of the Transportation Research Board, January.

Petry, T.M. and Das, B. (2001). *Evaluation of Chemical Modifiers/Stabilizers for Chemically Active Soils—Clays,* Transportation Research Record 1757, Transportation Research Board, 7 pp.

Petry, T.M. and Glazier, E.J. (2005). The effect of organic content on lime treatment of highly expansive clay. *Proceedings of the 2nd International Symposium of Treatment and Recycling of Materials for Transport Infrastructures,* October 24–26, Paris.

Petry, T.M. and Lee, T.W. (1989). *Comparison of Quicklime and Hydrated Lime Slurries for Stabilization of Highly Active Clay Soils,* Transportation Research Record 1190, Transportation Research Board, 7 pp.

Petry, T.M. and Little, D.M. (1993). *Update on Sulfate Induced Heave in Lime and Portland Cement Treated Clays; Determination of Potentially Problematic Sulfate Levels,* Transportation Research Record 1362, Transportation Research Board, 6 pp.

Petry, T.M. and Wohlgemuth, S.K. (1989). *The Effects of Pulverization on the Strength and Durability of Highly Active Clay Soils Stabilized with Lime and Portland Cement,* Transportation Research Record 1190, Transportation Research Board, 8 pp.

Petry, T.M. and Zhao, H. (2006). A study of injection of chemical agents in an expansive clay. *Proceedings of the Fourth International Conference on Unsaturated Soils,* Geotechnical Institute of ASCE, April 2–5, Carefree, AZ.

TRB (2005). *Evaluation of Chemical Stabilizers,* Committee on Chemical and Mechanical Stabilization, Subcommittee Thomas M. Petry (Chair) and Khaled Sobhan, Transportation Research Board Circular, Transportation Research Board, Washington, D.C.

10

Site Investigation and *In Situ* Tests

by
Sanjay Kumar Shukla
Edith Cowan University, Perth, Australia

Nagaratnam Sivakugan
James Cook University, Townsville, Australia

10.1 Introduction

Unlike other civil engineering materials, **soils** and **rocks** have significant variability associated with them. Their engineering properties can vary dramatically within a few meters in an area of proposed construction. A thorough and comprehensive **site investigation** (*aka site exploration* or *site characterization*) is therefore a prerequisite for design of all civil engineering structures and is one of the most important steps in a foundation design. Site investigation refers to the appraisal of the surface and subsurface conditions at a proposed construction site. Information on surface conditions is necessary for planning construction techniques. Information on subsurface conditions at a site is used to plan, design, and construct the foundations of structures and other underground works. A typical site investigation includes preliminary studies such as a desk study and **site reconnaissance**, geophysical surveys, drilling **boreholes**, *in situ* testing, **sampling** and laboratory testing of samples, and groundwater observations and measurements. Desk study involves collection of as much existing information as possible about the site through geological maps, aerial and satellite photographs, soil survey reports, site investigation reports of nearby sites, etc. Site reconnaissance consists of a walk-over survey to visually assess the local conditions such as site access, adjacent properties and structures, topography, drainage, etc.

The properties of soils are determined by either laboratory or *in situ* testing or a combination thereof. Both approaches have advantages and limitations in their applicability. The sampling, transportation, and specimen preparation usually subject the specimen to strains that alter the soil structure. For this reason, realistic determination of *in situ* properties by laboratory tests can be difficult. *In situ* testing is useful for measuring soil properties in their undisturbed condition without the need for sampling. *In situ* tests become more useful in soils which are sensitive to disturbance and in subsoil conditions where the soils vary laterally and/or vertically. The results of *in situ* testing also are used in construction, monitoring the performance of structures, and back analysis. The standard penetration test and static **cone** penetration test are the two most popular *in situ* tests that are widely used in deriving soil parameters for most routine geotechnical and foundation engineering designs. The penetration-type tests form the *logging methods* or *sounding methods* of subsurface exploration and usually are fast and economical. In such penetration tests, a penetration tool attached to a rod is made to penetrate overburden deposits by means of dynamic or static loading, and a continuous or semicontinuous record of the resistance to penetration is obtained. Other specialized *in situ* tests that form the *specific methods* of subsurface investigation are the vane shear test, pressuremeter test, dilatometer test, plate load test, borehole shear test, and K_0 stepped blade test. Specific methods often are slower and more expensive to perform than logging methods and normally are carried out to obtain specific soil parameters, such as undrained shear strength or deformation modulus. The logging and the specific methods often are complementary in their use (Canadian Geotechnical Society 2006). Many of these *in situ* tests are described in this chapter; however, more details can be found in the relevant standards.

All the findings are presented to the **client** in the form of a site investigation report, which consists of a site plan, several boring logs which summarize the soil and rock properties at each test pit and borehole, and the associated laboratory and *in situ* test data. The extent of a site investigation program for a given project depends on the type of project, the importance of the project, and the nature of the subsurface materials involved. The level of investigation should be appropriate to the proposed site use and to the consequences of failure to meet the performance requirements. For example, a large dam project usually would require a more thorough site investigation than would be required for a highway project. A further example is loose sands or soft clays, which usually require more investigation than is required for dense sands or hard clays. The site investigation project can cost about 0.1–1% of the total construction cost of a project. The lower percentage is for smaller projects and for projects with less critical subsurface conditions; the higher percentage is for large projects and for projects with critical subsurface conditions.

10.2 Objectives of Site Investigation

The purpose of a site investigation is to conduct a scientific examination of a site in order to collect as much information as possible, at minimal cost, about the existing topographical and geological features of the site (for example, the exposed overburden, the course of nearby streams/rivers, the rock outcrop, the hillock or valley, vegetation, etc.) and mainly the subsurface conditions underlying the site. Investigation of the subsurface conditions at the site for the proposed construction of an engineered system is essential before the foundation design is finalized. Subsurface investigation is needed basically to provide the following:

1. Sequence and extent of each soil and rock stratum underlying the site and likely to be affected by the proposed construction
2. Engineering geological characteristics of each stratum and geotechnical properties (mainly strength, compressibility, and permeability) of soil and rock which may affect design and construction procedures of the proposed engineered systems and their foundations
3. Location of the **groundwater table** (or *water table*) and possible harmful effects of soil, rock, and water on materials to be used for construction of structural elements of the foundation

The above information is used in determining the type of foundation and its dimensions, estimating the load-carrying capacity of the proposed foundation, and identifying and solving the construction, environmental, and other potential problems, thus enabling the foundation engineer to arrive at an optimum design with due consideration given to the subsurface material characterization.

10.3 Stages of Site Investigation

A site investigation generally is accomplished sequentially in four phases. Information obtained in each phase of investigation may disclose problems which require further investigation in the next phase. All four phases of investigation as described below are not essential for all projects.

10.3.1 Collection of Available Information

All the preliminary details of the proposed engineered system (e.g., an 11-story building), including its dimensions, location, loadings, functional requirements, intended construction method, starting date, estimated period of construction, and related local building code regulations, are collected. The information related to the behavior of existing structures, if any, adjacent to the site, as well as information available through local experience, also should be collected, along with other sources of information, including maps (geological/topo-graphical/agronomy), aerial and satellite photographs, hydrological data, soil manuals, records of trial pits and boreholes in the vicinity, and related publications.

10.3.2 Site Reconnaissance

Site reconnaissance is carried out in the form of a site inspection and study of the various available sources of information. A visit to the site is made to obtain information on local topography, such as evidence of erosion or landslides, excavation, recent fills, soil and rock characteristics in the existing open cuts, type and behavior of adjacent structures, water level in nearby streams/rivers and wells, flood marks, etc. Inquiries should be made regarding previous use of the site, such as underground workings in the form of coal mines, quarries, ballast pits, mineral workings, old brick fields, etc. Information about the removal of overburden by excavation, erosion, or landslides gives an idea of the amount of preconsolidation of the soil strata. Rock outcrops may give an indication of the presence of bedrock. Wells, at the site or in the vicinity, give useful indications of the groundwater conditions. Flood marks of rivers may indicate their highest water levels. Tidal fluctuations may be of importance.

The information obtained from site reconnaissance will assist in planning the preliminary and the detailed investigations described below. It also is useful in determining the method of investigation, field tests to be carried out, and the logistics of investigation.

10.3.3 Preliminary Site Investigation

This phase of investigation identifies the areas that need further investigation. It consists of obtaining information about the depth and thickness of each subsurface stratum, types of soil and rock in each stratum, and the location of the groundwater table. The investigation is carried out by making a limited number of test pits or boreholes. A few **undisturbed samples** are collected for laboratory testing to determine permeability, compressibility, and shear strength of the soil/rock. **Disturbed samples** are collected from various depths for visual classification and for determination of index properties. Standard penetration and cone penetration tests also are conducted to complement the soil parameters derived from the laboratory tests. Geophysical investigation of the site by the *electrical resistivity method* or *seismic refraction method* provides a simple and quick means of obtaining useful information about subsurface strata. Strength and settlement correlations with index properties are very useful at this phase of investigation.

10.3.4 Detailed Site Investigation

The objective of a detailed site investigation is to determine the geotechnical properties of strata which are shown by preliminary investigation to be critical. In the case of soils, for most projects, the geotechnical properties of interest are grain size distribution, specific gravity, consistency limits, *in situ* bulk unit weight, natural moisture content, permeability, shear strength parameters, and consolidation parameters. For rocks, the properties of importance are specific gravity, porosity, water absorption, and compressive strength. This phase of investigation includes a **drilling** program with boreholes in addition to those made in the preliminary investigation phase and more detailed soil and rock sampling for laboratory testing. A standard penetration test, plate load test, *in situ* vane shear test, field permeability test, or any other field test may be conducted as per the requirement of the specific problem. More advanced means of logging boreholes by radioactive methods fall under the detailed investigation. If the foundation soil near the ground surface is soft to medium stiff, it is a good practice to extend at least one borehole to competent rock, especially if the structure is heavy or its performance requires proper settlement control.

10.4 Methods of Subsurface Investigation

There are several methods of subsurface investigation (see Table 10.1); however, the commonly used methods are making test pits, trenches, and boreholes at the site of the proposed structure.

TABLE 10.1 Subsurface Investigation Methods

Method	Mode of Operation	Applicability
Geophysical methods		
Electrical resistivity method	Measurements of variations in the apparent resistivity as measured on the ground	Alluvial deposits, weathered and fissured rock, buried channels, and groundwater
Seismic refraction method	Measurements of velocities of compressional waves from the travel time curves of seismic waves	Alluvial deposits, weathered and fissured rock, buried channels, and groundwater
Field tests (logging methods or subsurface sounding methods)		
Standard penetration test	Variation in the engineering properties is correlated with the number of blows required for unit penetration of a standard penetrometer by a drive hammer at a desired elevation	Best suited for sands; not applicable to soft to firm clays
Static cone penetration test	A cone penetrometer is advanced by pushing, and the static force required for unit penetration is correlated to the engineering properties	Best suited for sand, silt, and clay; not applicable to gravels
Dynamic cone penetration test	A cone penetrometer is driven by a standard hammer, and the dynamic force required for unit penetration is correlated to the engineering properties	Best suited for sands; not applicable to clays
Test pits and trenches	Undisturbed samples can be collected and *in situ* tests can be performed	All types of soil and rock deposits
Drifts (or tunnels)	Undisturbed samples can be collected and *in situ* tests can be performed along with exploration of geological formations in hills	All types of soil and rock deposits
Shafts	Exploration at a great depth or to extend the exploration below riverbeds by means of tunnels	All types of soil and rock deposits
Boring/drilling	Holes are bored into the ground to obtain soil samples and rock cores for visual inspection and laboratory testing	All types of soil and rock deposits

TABLE 10.1 Subsurface Investigation Methods (continued)

Method	Mode of Operation	Applicability
Field tests (specific methods)		
In situ unit weight and natural moisture content	The unit weight and the moisture content are measured by suitable methods	For all types of soil and rock deposits
Plate load test	A steel plate is loaded at the desired elevation and the settlement is measured under each load until a desired settlement takes place or foundation soil failure occurs	Best suited for sand and clay
Vane shear test	A vane is advanced into the *in situ* soil at the desired elevation and the torque required to rotate the vane is measured	Best suited for clays; not applicable to sands and gravels
Borehole shear test	A rapid, *in situ* direct shear test performed on the walls of a borehole	Best suited for soils and weak rocks
Pressuremeter test	Commonly consists of horizontal expansion of a membrane mounted on a relatively long probe placed in a slightly oversized, prebored hole through injection of water	Best suited for soft rock, dense sand, gravel, and till; not applicable to soft sensitive clays, loose silts, and sands
Flat dilatometer test	A blade of a standard design is advanced into the ground using common field equipment; soon after penetration, the membrane attached to the blade is inflated using gas pressure, and pressure readings are taken	Best suited for sand and clay; not applicable to gravel
K_0 stepped blade test	A blade with four steps is penetrated into the soil in a borehole, and soil pressures are measured	Best suited for clays of soft to medium consistency
In situ California bearing ratio test	The resistance to penetration of a metal piston in a soil mass is measured	All types of soil deposits
Borehole logging	A soil/rock formation parameter (temperature/spontaneous electric current/natural radioactivity/resistance to electric current/velocity of sound propagation/reaction to gamma-ray bombardment/reaction to neutron bombardment) is continuously recorded along the depth in the borehole	All types of soil and rock deposits

10.4.1 Test Pits and Trenches

Test pits and trenches are excavations into the ground that permit visual inspection of the subsurface conditions of the soils and rocks in place. Where desired, good-quality undisturbed blocks or tube sampling and *in situ* tests can easily be carried out. Moreover, investigation by test pits and trenches is relatively inexpensive.

Pits and trenches may be excavated manually with hand tools such as a pickaxe and shovel or mechanically by power excavation equipment such as a backhoe (see Figure 10.1a). The depth should be according to the requirements of investigation and generally is limited to a few meters below the groundwater table. In dry ground, pits and trenches generally are economical

(a)   (b)

FIGURE 10.1 (a) Trench excavation with power excavation equipment (backhoe) and (b) manual excavation of a test pit with a spade.

in comparison to boreholes up to a depth of about 5 m, depending upon the location. Unsupported pits and trenches are rarely dug to a depth exceeding 3 m except in the case of hard soils. The top of the pit should be kept large enough so that its dimensions at the bottom are at least 1.2 m × 1.2 m, which is sufficient to provide necessary working space (see Figure 10.1b). The width of a trench should be at least 1.2 m.

For deep pits and trenches, the walls should be supported by a suitable sheeting and bracing system, and they must be ventilated to prevent accumulation of dead air. When water is encountered in a pit, a suitable dewatering system may be required for further progress.

Undisturbed samples from test pits should be obtained from each stratum if the nature of the deposit permits. For this purpose, a pillar of suitable dimensions (e.g., 40 cm × 40 cm) should be left undisturbed at the center of the pit to collect undisturbed samples of the required size from each stratum, showing a change of formation. Special care should be taken to preserve the natural moisture content of the samples.

It should be noted that trenches are similar to pits in all respects, except that they are continuous over a length and provide continuous exposure of the subsurface along a desired line or section. They are best suited for exploration on slopes.

10.4.2 Boreholes

A borehole may be defined as a small-diameter hole, usually vertical, drilled at a site primarily to obtain soil and rock samples. In addition, the hole is utilized for the *in situ* determination of such engineering properties as permeability and shear strength. Use of boreholes is the only direct practical method of subsurface exploration to greater depths. Two common problems with boreholes are caving of the walls and heaving of the bottom of the hole. The latter occurs to some extent in all holes, whether above or below the groundwater table, due to the stress release caused by removal of material from the hole. However, it is most serious in the case of holes below the groundwater, since water seeping into the bottom of a hole from the surrounding area can result in considerable disturbance to the soil to be sampled. This disturbance normally is minimized by maintaining the level of the **drilling fluid** in the hole at all times at or above the groundwater table. By this arrangement, any seepage will be from the hole to the surrounding area and will stabilize rather than disturb the base of the hole. Caving of the borehole wall, particularly the portion below the groundwater table, can take place in both soil and rock. The wall can be stabilized by lining with drive pipe or **casing** or by means of drilling fluids, grouting, or freezing. Lining a borehole with drive pipe or casing is the most effective method of supporting the walls of a borehole. Drilling fluid in its simplest form is merely water. More commonly, the term refers to mixtures of water and a thixotropic substance such as bentonite, generally 6% bentonite by weight of water (U.S. Army Corps of Engineers 1972). The primary advantages of using drilling fluid are its lower cost compared to casing and its tendency to minimize stress relief in the soil adjacent to the borehole wall. A major disadvantage is that it cannot be used for borings in which permeability and pressure tests are to be performed. Grout is often used to stabilize portions of boreholes which pass through deposits such as gravel, boulders, and highly fractured rock, which are extremely susceptible to caving. There are several methods of boring or drilling into ground, as described below.

10.4.2.1 Auger Boring

Often auger boring is the simplest and most economical method of subsurface investigation and soil sampling up to a depth of about 6 m in alluvial deposits, which can stand unsupported. The soil samples obtained from such borings are highly disturbed. This boring method is useful for identification of changes in the soil strata, determination of groundwater level, and advancement of a borehole for spoon and tube sampling. Several types of hand-operated and machine-operated augers are available (Figure 10.2), which are commonly used in routine applications, and range in size from 1 through 48 in. (25.4 through 1219 mm). Boreholes may be advanced by rotating the auger while at the same time applying a downward pressure on it to assist in obtaining penetration. The auger is withdrawn from the borehole, and the soil is collected for examination and tests. The empty auger is returned to the hole and the procedure is repeated. A steel pipe, called casing, may be required to prevent the borehole walls from sloughing or caving in when the hole is extended below the groundwater table. The casing is advanced by driving by means of a "monkey" suspended from a winch, but it is not driven to a depth greater than the top of the next sample to be collected. Hand-operated augers generally are used for advancing holes to depths of 3–5 m. However, boreholes up to about 50–60 m can easily be made by machine-operated augers.

10.4.2.2 Wash Boring

In this method, before advancing a borehole, a short casing, 2–3 m in length, is driven into the ground to prevent caving of surface soils. The casing is cleaned out by means of a chopping bit attached to the lower end of a drill rod, which is kept inside the casing. Water is pumped through the drill rod, and it exits at high velocity through holes in the bit. The water rises between the casing and drill rod, carrying suspended soil particles, and overflows at the top of the casing through a "T" connection into a container, from which the effluent is recirculated back through the drill rod (Figure 10.3). The hole is advanced by raising, rotating, and dropping the bit into the soil at the bottom of the hole. Drill rods, and if necessary casing, are added as the depth of the boring increases.

(a)

Auger

(b)

FIGURE 10.2 (a) Hand-operated auger and (b) machine-operated auger.

The wash boring method is quite rapid for advancing holes in soft to stiff cohesive soils and fine sand but is not suitable for gravel and boulders. The change of stratification can be inferred from the rate of progress and color of the wash water. Because heavier particles of different soil layers remain in suspension within the casing pipe and get mixed up, this method is not suitable for obtaining samples for classification; however, undisturbed samples can be

FIGURE 10.3 Wash boring method.

obtained by attaching a tube **sampler** to the end of the drill rod and driving it into the soil to the desired depth by hammering or jacking.

10.4.2.3 Percussion Drilling

In this method, a bit or a chisel attached to a drill rod is lifted, rotated slightly, and dropped repeatedly onto the bottom of the hole. Water is circulated using a pump to bring the debris (soil and rock cuttings) to the ground surface at certain time intervals. Casing is required to prevent caving of the borehole wall. Samples may be obtained at intervals using suitable tools, but they are not reliable, particularly in the case of soils, because of high disturbance by the action of this method of drilling. As the tools are meant for rapid drilling by pulverizing the soil and rock deposit, they are not suitable for careful investigation. However, this is the only method suitable for drilling boreholes in boulder and gravelly strata.

10.4.2.4 Rotary Drilling

In this method, a drill bit attached to the end of a hollow drill rod is rotated under pressure to advance the hole by cutting action. If the wall of the hole tends to cave in, drilling fluid is pumped continuously down the hollow drill rod and the mud suspension returns to the surface through the annular space between the rod and the wall of the hole, along with the formation of the mud cake on the wall of the hole. The mud cake thus formed provides sufficient strength in conjunction with the hydrostatic pressure of the mud suspension against the wall so that the cavity is maintained without any protective casing. The mud pressure also tends to seal off the water flow into the hole from any permeable water-bearing strata.

Rotary drilling is the most rapid method of advancing boreholes in rock masses unless they are highly fissured; however, it also can be used for all other soils. In this method, **cores** from rock as well as from concrete and asphalt pavements may be obtained by the use of coring tools (coring bit and core catcher). Coring tools should be designed so that continuous recovery of core in sound rock is achieved. It is important to ensure that boulders or layers of cemented soils are not mistaken for bedrock. This necessitates core drilling to a depth of at least 3 m in bedrock in areas where boulders are known to occur.

Open boreholes are a hazard and should be backfilled when they are no longer required. Backfilling generally is done with locally available soil; however, under certain circumstances, backfilling with grout is advisable, especially when it is essential to prevent the movement of water from one stratum to another and to prevent piping of material to the surface through the borehole. Such circumstances can arise when investigating the ground in landslide-prone areas, downstream of dams and proposed embankments, and at proposed locations of structures (Lowe and Zaccheo 1975).

10.4.3 Selection of Test Pits and Boreholes

Every meter a borehole is advanced costs money. Therefore, good care is required in selecting the right number of boreholes and limiting the depth to what is absolutely necessary. Determination of the number of test pits and boreholes and their depth for a project is governed by the subsurface material variability, type of project and loadings, performance requirements, foundation type selected, and budget availability. The minimum depth is related to the depth at which the increase in stress within the soil mass caused by the foundation loads is small and will not cause any significant settlement. The basis for determining the spacing of boreholes is less logical; spacing is based more on variability of site conditions, experience, and judgment. More test pits and boreholes and closer spacing generally are recommended for sites located in less developed areas where previous experience is sparse or nonexistent (Canadian Geotechnical Society 2006). The number of test pits and boreholes must be sufficient so that a geotechnical consultant can make an economical design recommendation with an adequate margin of safety.

In spite of these facts, there are no clear-cut criteria for determining the number of test pits or boreholes. For a compact building site covering an area of about 0.4 ha (= 4000 m^2), one borehole or trial pit in each corner and one in the center should be adequate. Additional boreholes or test pits may be required in very uneven sites, where fill areas have been made, or when the soil varies laterally. For buildings, a minimum of three boreholes or test pits, where the surface is level and the first two boreholes or test pits indicate regular stratification, may be adequate. A single borehole may be sufficient for a concentrated foundation such as a tower base in a fixed location with the hole made at that location. For very large areas, the geological nature of the terrain will help in deciding the number of test pits or boreholes. Cone penetration tests, if possible, may be performed at every 50-m interval by dividing the area into a grid pattern, or geophysical methods may be adopted to decide on the number of boreholes or test pits. A general rule of thumb for approximate spacing of boreholes is as follows:

Type of Project	Spacing (m)
Multistory building	10–30
One-story industrial plant	20–60
Highways	250–500
Dams and dikes	40–80
Residential subdivision	250-500

For residential subdivisions, often test pits are adequate. If boreholes are required, they can be spaced at 250- to 500-m intervals as suggested above.

Similar to the number of test pits or boreholes, there are no binding rules for the depth of exploration. However, exploration should be continued to a depth at which the loads of the engineering system can be carried by the stratum in question without undesirable settlement and shear failure. In any case, the depth to which seasonal variation or frost penetration affects the soil strata at a site should be regarded as the minimum depth of exploration at that site. Boreholes should be advanced to depths where the net increase in the vertical effective stress due to the proposed structure is about 10% of what is applied at the surface or where it is about 5% of the current effective overburden stress, using the smaller value unless bedrock is encountered (American Society of Civil Engineers 1972). In line with this guideline, the depth of exploration for a building with a width of 30.5 m would be approximately as follows (Sowers and Sowers 1970):

No. of Stories	Depth of Exploration (m)
1	3.5
2	6
3	10
4	16
5	24

For hospitals and office buildings, Sowers and Sowers (1970) suggested the following rule to determine the depth of exploration for light steel and narrow concrete buildings

$$D_e = 3S^{0.7} \tag{10.1a}$$

and for heavy steel or wide concrete buildings

$$D_e = 6S^{0.7} \tag{10.1b}$$

where D_e is the depth of exploration (in meters) and S is the number of stories.

As a general rule of thumb, the depth of investigation normally is 1.5 times the width of the footing/structure below the foundation level/bearing level. In certain cases, it may be necessary to take at least one borehole or test pit to twice the width of the footing below the foundation level. If a number of loaded areas are in close proximity, the effect of each is additive. In such cases, the whole area may be considered to be loaded and exploration should be carried out up to 1.5 times the least lateral plan dimension of the building. When deep excavation is anticipated, the depth of investigation should be at least 1.5 times the depth of excavation. For important (or high-rise) structures, it is common to extend at least one of the boreholes to the bedrock or to competent (hard) soil, particularly if there are intermediate strata of soft or compressible materials. The minimum depth of core drilling into the bedrock is about 3 m. If the bedrock is irregular or weathered, the core drilling may have to be deeper. In the case of road cut, the depth of investigation can be equal to the bottom width of the cut. For fill, the depth of investigation is whichever is the greater of 2 m below ground level or equal to the height of the fill. For highway and airport pavements, the minimum depth of investigation is generally 1.5 m below the proposed subgrade elevation. It should be noted that the depth of exploration at the start of the investigation work may be modified during the drilling operation as exploration proceeds, depending on the subsurface conditions encountered.

10.5 Sampling and Laboratory Testing

Soil and rock samples representing each subsurface stratum are obtained for visual identification and laboratory testing to determine engineering properties. There are two types of samples: disturbed samples and undisturbed samples. In disturbed soil samples, often the natural structure of the *in situ* soil is destroyed, although the natural moisture content can be preserved with suitable precautions. Such samples may be obtained in the course of excavation and boring. Disturbed samples of clayey soils may be unsuitable for shear strength measurements unless they are required for fill. Such samples also are not suitable for consolidation and hydraulic conductivity tests. Disturbed, but representative, samples generally are used for classification and tests to determine index properties. These samples may not be truly representative, especially when taken from below the groundwater table. To procure good-quality samples, where possible, the groundwater level may be lowered by means of pumping.

Undisturbed samples have natural structure and moisture, and they truly represent the *in situ* soil mass in terms of their properties. For most rocks, undisturbed samples are easily obtained, but for soils they can only be obtained by special methods. Soil samples obtained by auger boring and wash boring methods are highly disturbed. For cohesive soils of all types, it is possible with most strata to procure undisturbed samples as *chunk* or *tube samples,* which are very satisfactory for examination and laboratory testing purposes. *Chunk* or *block samples* are taken where clay is exposed in test pits, and tube samples may be obtained in test pits as well as in boreholes from the desired depths by pressing a well-designed *thin-walled tube sampler* into the *in situ* soil. Undisturbed sampling of sands, especially below the water table, is not always an easy task, but special methods can be adopted for this purpose. Wash samples obtained from percussion and rotary drilling methods in rock masses are highly disturbed, whereas rock samples obtained as cores or blocks are undisturbed.

To collect undisturbed samples, properly designed sampling tools are required, which differ for cohesive and cohesionless soils and for rocks. The fundamental requirement of a sampling tool is that on being forced into the ground, it should cause as little displacement, remolding, and disturbance as possible. The degree of disturbance is mainly controlled by the design features of the tool cutting shoe/edge and inside wall friction. A typical cutting shoe/edge with a sampling tube is shown in Figure 10.4. Clearance ratios and area ratio are defined in terms of the inside and outside diameters of the sampling tube (D_{it} and D_{ot}) and cutting shoe (D_i and D_o), respectively, as follows.

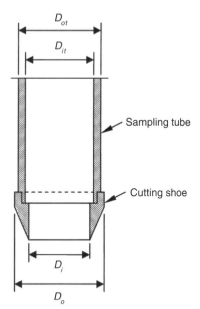

FIGURE 10.4 Cutting shoe attached to a sampling tube.

Inside clearance ratio:

$$C_i \ (\%) \ = \ \frac{D_{it} \ - \ D_i}{D_i} \times 100 \qquad (10.2a)$$

Outside clearance ratio:

$$C_o \ (\%) \ = \ \frac{D_o \ - \ D_{ot}}{D_{ot}} \ \times \ 100 \qquad (10.2b)$$

Area ratio:

$$A_R \ (\%) \ = \ \frac{D_o^2 \ - \ D_i^2}{D_i^2} \ \times \ 100 \qquad (10.2c)$$

The inside clearance ratio allows for elastic expansion of the soil as it enters the tube, reduces frictional drag on the sample from the wall of the tube, and helps retain the sample. Generally it should be between 1 and 3%. The outside clearance ratio facilitates the withdrawal of the sampler from the ground, and it should not be much greater than the inside clearance ratio. The area ratio is kept as low as possible, consistent with the strength requirements of the sampling tube. For a good-quality undisturbed sample, it must be less than 10%. The wall friction can be reduced by a smooth finish on the sample tube and oiling the tube properly, in addition to providing suitable inside clearance. To procure an undisturbed sample, it also is necessary for the valve attached to the sampling tool to have a large orifice to allow the air and water to escape quickly and easily when driving the sampler. The **recovery ratio**, defined as the ratio of the length of the sample within the sampling tube to its depth of penetration, expressed as a percentage, should be at least 96% for an undisturbed sample.

Soils are commonly sampled using *thin-walled (Shelby) open tube samplers*, *split-barrel samplers*, or *piston samplers*. The thin-walled open tube sampler is an ordinary seamless steel tube with an outside diameter of 50–150 mm and its lower edge chamfered to make penetration easy. The most common form of the thin-walled (Shelby) tube sampler has outside diameters of 50.8 mm (2 in.) and 76.2 mm (3 in.). The Shelby tube with a 50.8-mm (2-in.) outside diameter has an inside diameter of about 47.63 mm (1⅞ in.), with an area ratio of 13.75%. Depending on the requirements of undisturbed sampling, a thin-walled tube sampler with a separate cutting shoe also may be used. Attachment of the head to the tube is kept concentric and coaxial to ensure uniform application of force to the tube by the sampler insertion equipment. The tube also can be attached to a drill rod for obtaining samples from the bottom of a borehole.

A 35-mm-inner-diameter × 457- to 610-mm-long *split-barrel sampler*, also referred to as the *split-tube* or *split-spoon sampler*, is a modified form of the open tube sampler where the sampling tube is split into two halves held together by the cutting edge and the sampler head, as seen in Figure 10.5. The sampler head contains a venting area, which is required to avoid sample compression. This sampler makes removal of the sampler easier and provides penetration resistance, if used in a standard penetration test (see Section 10.7), which may be utilized to correlate *in situ* properties such as unit weight, shear strength, and load-bearing capacity of the foundation soil. The area ratio for the split-barrel sampler is about 110%, implying that the samples from this sampler are highly disturbed. Disturbed samples generally are used for visual identification, soil classification, and preliminary laboratory tests.

A piston sampler consists of a thin-walled sampling tube fitted with a piston. The sampler is attached to the lower end of a hollow drilling rod, through which passes an inner rod that operates the piston. To begin with, the sampler is lowered to the bottom of the borehole with

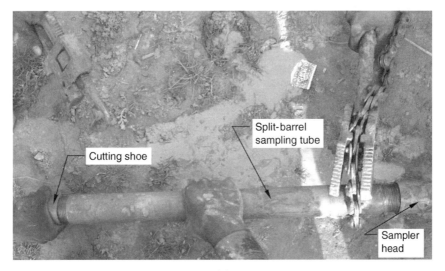

(a)

(b)

FIGURE 10.5 Split-barrel sampling tube: (a) separating the sampling tube from the cutting shoe and the drilling rod and (b) samples in the two halves of the sampling tube.

the piston locked in the lower position. The piston incorporates a seal which prevents water and debris from entering the tube. As the piston is held against the soil at the bottom of the hole, it is unlocked and the tube is driven down into the soil for the full length of travel of the piston. The piston is now locked at the top of the tube and the whole assembly is withdrawn to the surface, where the sampler head and the piston are removed before waxing and sealing the tube. The piston sampler generally is available in sizes ranging from 35 to 100 mm internal diameter, producing sample lengths of up to 600 mm. Piston samplers generally are required for sampling very soft silts and clays.

Undisturbed rock samples are obtained from open test pits in the form of blocks dressed to a size convenient for packing (e.g., 90 mm × 75 mm × 50 mm). Samples in the form of cylindrical cores are obtained by means of rotary drills with a coring bit. To obtain cores of the rock, a core barrel is attached to a drilling rod. A coring bit is attached to the bottom of the barrel. The cutting elements may be diamond, tungsten, carbide, and so on. Various types of core barrels are available (Das 2007); however, the NX type is commonly used in routine site investigation work, giving core samples of a diameter equal to 2⅛ in. (53.98 mm). Core drills are so designed that continuous recovery of core in sound rock is achieved. It is important to ensure that boulders or layers of cemented soils are not mistaken for bedrock. This necessitates core drilling to a depth of at least 3 m in bedrock in areas where boulders are known to occur.

The number of undisturbed samples required depends on the importance of the investigation, which is governed by the type of structure. In general, soil samples are obtained at every change in stratum and at intervals not exceeding 1.5 m within a continuous stratum. In important investigations such as the foundation for an earth dam, continuous core sampling may be necessary.

The procedures for preserving soil and rock samples immediately after they are obtained in the field and the accompanying procedures for transporting and handling the samples require proper care so that the desired inherent conditions can be maintained for some period of time. The procedures for preserving samples depend on the type of samples obtained, the type of tests and engineering properties required, the fragility and sensitivity of the soil, and the climatic conditions. Where disturbed samples are required for testing, or where it is desirable to keep them in good condition without loss of moisture for some period (e.g., 1–2 weeks) immediately after being taken from the test pit or the borehole, they should be placed in labeled airtight containers with a minimum of air space. For an undisturbed sample in a tube, both ends of the sample should be cut and removed to a depth of about 25 mm. Molten wax layers are then applied to each end to give a plug about 25 mm thick. If the sample is very porous, a layer of waxed paper should first be placed over the ends of the sample. Any space left between the end of the tube and the top of the wax layer should be tightly packed with sawdust or other suitable material, and a close-fitting lid or screw cap should be placed on each end of the tube. If the samples are transported, the labeled containers or tubes encased in cushioning material (sawdust, rubber, foam, etc.) should be carefully packed in wood, metal, or other type of suitable boxes/containers to prevent damage during transit. Samples are handled in the same orientation in which they were sampled, including during transportation, with appropriate markings on the boxes/containers. The samples should always be stored in cool rooms, preferably with a high humidity (e.g., 90%). More details about preserving and transporting soil samples can be found in ASTM D4220.

The drill core is the sample record for the subsurface geology at the borehole location, so it is preserved for some period of time, varying from as short as 3 months to several years, even 10 years. For large and critical structures, it may be necessary to retain the core for many years for re-examination and testing required at some later time. Some countries have regulations governing the disposition and storage of core samples. The extent and type of preservation required depend on the geologic characteristics and the intended testing of the rock samples. This is best done in core boxes, which are usually 1.5 m long and divided longitudinally by light battens to hold four to six rows of cores, as shown in Figure 10.6. The depth of the box and the width of the compartments should be such that there can be no movement of the cores when the box is closed and transported. If vibration or variations in

FIGURE 10.6 Rock cores in a core box.

temperature may subject samples to unacceptable conditions during transport, the samples are placed in suitable core boxes that provide cushioning or thermal insulation. The properties of soft rocks depend to some extent on their moisture content. Representative samples of such rocks should therefore be preserved by coating them completely with a thick layer of wax after removing the softened skin. Core photography in color is performed on all cores to permanently record the unaltered appearance of the rock. Based on the length of rock core recovered from each run, the following quantities may be calculated for a general evaluation of the rock quality encountered:

$$\text{Core recovery} = \left(\frac{\text{Length of the core recovered}}{\text{Total length of the core run}} \times 100 \right) \% \quad (10.3)$$

and

$$\text{RQD} = \left(\frac{\sum \text{Lengths of intact pieces of recovered core} \geq 100 \text{ mm}}{\text{Total length of the core run}} \times 100 \right) \% \quad (10.4)$$

where RQD is the **rock quality designation**. A core recovery of 100% indicates the presence of intact rock; for fractured rocks, the core recovery will be smaller than 100%. RQD is used to define the quality of the rock mass as given in Table 10.2.

It is important that a sample be accurately identified with the test pit and borehole and the depth below reference ground surface from which it was taken. A waterproof identification tag is placed inside the container, and an identification number is also marked outside the container and box.

TABLE 10.2 Relation between RQD and *In Situ* Rock Quality

RQD (%)	Rock Quality
<25	Very poor
25–50	Poor
50–75	Fair
75–90	Good
>90	Excellent

If groundwater is encountered in a borehole, the water level in the borehole is maintained at or above the groundwater table during the drilling and sampling operation to avoid any instability. The position of the **groundwater level**, or groundwater levels if there is perched groundwater or piezometric surfaces if there is artesian groundwater, is identified at the site. The variability of these positions over both short and long time periods is studied. If a test pit has been excavated or an open well exists near the site of investigation, measurement of the depth of the water table as well as collection of water samples does not present any difficulty. However, if water samples are to be collected from a borehole drilled at the site, some difficulty is expected due to the narrowness of the borehole, caving in of the sides, etc. In the authors' experience, however, the water table depth measurement in boreholes stabilized with casing or bentonite slurry is easily done by lowering a metal measuring tape/rope/cable with a weight attached to the lower end. The weight ensures plumbness and permits some feel for obstructions. The size of the weight should be such that its displacement of water causes an insignificant rise in the borehole water level; otherwise a correction is required for the displacement. An electrical measuring device, if available, also can be used conveniently without the need for any correction in the measured value. Boreholes can be observed with a camera without any difficulty, as shown in Figure 10.7. A borehole camera is very useful for photographing the stratification in drilled boreholes. Where casing is used, the depth to the groundwater level in a borehole after its completion is determined both before and after the casing is pulled. In sands, the water level is determined at least 30 min after the boring is completed; in silts, the level is taken after at least 24 h. In clays, accurate determination of the water level is not possible unless pervious seams are present. In spite of this fact, water level in clays is taken after at least 24 h. A stabilized borehole water level reflects the pressure of groundwater in the earth material. Under suitable conditions, the groundwater water level in the borehole and the groundwater table will be the same. For boreholes with casing or drilling mud, the water level in the borehole may not accurately reflect the groundwater table location. Interpretation and application of groundwater level in boreholes should therefore be done carefully.

FIGURE 10.7 Observation of a borehole and measurement of water table.

For laboratory tests on undisturbed samples, the samples are carefully taken out of the sampling tubes without causing any disturbance to the samples. If the tubes are oiled inside before use, it is quite possible for samples of a certain moisture range to be pushed out by means of suitably designed piston extruders. If the extruder is horizontal, there should be a support for the sample as it comes out from the tube so that it will not break. All extruding operations must be in one direction, that is, from cutting edge to the head of the sampling tube. For soft clay samples, pushing with an extruder piston may result in shortening or distortion of the sample. In such cases, the tube may be cut by means of a high-speed hacksaw in proper test lengths, which can directly be used for the desired tests. After the sample is extruded, it is kept in either a humidity chamber or a desiccator and removed only when actual testing is carried out, to avoid possible loss of moisture.

Samples of soils and rocks are tested in the laboratory to determine their engineering properties depending on the phase of the investigation. For example, during the site reconnaissance phase, visual classification of soils and rocks usually is sufficient, but for the detailed site exploration phase, several tests as given in Table 10.3 are conducted, keeping the design needs of the structure under consideration. For laboratory tests, the size and type of sample required are dependent upon the tests to be performed, the relative amount of coarse particles present, and the limitations of the test equipment to be used. For example, 1–15 kg of a fine-grained soil (or nongravelly soil) is sufficient for its laboratory test analysis, whereas a large quantity (e.g., 50–100 kg) may be required for the analysis of a coarse-grained soil (or gravelly soil).

TABLE 10.3 Laboratory Tests That Can Be Conducted on Samples for a Detailed Site Exploration

Materials		Tests/Properties
Soils	Physical tests	Visual classification
		Natural moisture content
		Unit weight
		Specific gravity
		Grain size analysis
		Consistency limits (liquid limit, plastic limit, shrinkage limit)
		Permeability test
		Consolidation test
		Shear strength (unconfined compression, triaxial compression, direct shear)
		Swelling index test
	Chemical tests	Soluble salt content: chlorides and sulfates
		Calcium carbonate content
		Organic matter content
Groundwater	Chemical analysis using pH determination	
	Bacteriological analysis	
Rocks	Visual examination	
	Petrographic examination	
	Unit weight	
	Specific gravity	
	Water absorption	
	Porosity	
	Unconfined compressive strength	
	Shear strength	
	Brazilian tensile strength	

10.6 Geophysical Methods

Geophysical methods can be used to determine the distributions of physical properties (e.g., elastic moduli, electrical resistivity, density, magnetic susceptibility, etc.) at depths below the ground surface that reflect the local subsurface characteristics of the materials (soil/rock/water). These methods may be used for investigation during the reconnaissance phase of a site investigation program since they provide a relatively rapid and cost-effective means of deriving areally distributed information about subsurface stratification. The geophysical investigation can optimize detailed investigation programs by maximizing the rate of ground coverage and minimizing the drilling and field testing requirements. Since geophysical investigations some-times may be prone to major ambiguities or uncertainties in interpretation, these investiga-tions often are verified by drilling or excavating test pits. In fact, geophysical investigation methods may be used to supplement borehole and outcrop data and to interpolate between boreholes.

A wide range of geophysical methods are available for subsurface investigation, for each of which there is an operative physical property to which a method is sensitive (Dobrin 1976; Kearey et al. 2002). The type of physical property to which a method responds clearly deter-mines its range of applications. Seismic refraction/reflection and ground-penetrating radar methods can be used to map soil horizons and depth profiles, water tables, and depth to bedrock in many situations. Electromagnetic induction, electrical resistivity, and induced polarization (or complex resistivity) methods may be used to map variations in water content, clay horizons, stratification, and depth to aquifer/bedrock. The magnetic method is very suitable for locating magnetite and intrusive bodies such as dikes in subsurface rocks. Other geophysical methods such as gravity and shallow ground temperature methods may be useful under certain specific conditions. Crosshole shear wave velocity measurements can provide soil and rock parameters for dynamic analyses.

Seismic and electrical resistivity methods are routinely used in conjunction with boring logs for subsurface investigation; these methods are therefore described in some detail in this section.

10.6.1 Seismic Methods

Seismic methods require generation of shock or seismic waves, which are parcels of elastic strain energy that propagate outward from a seismic source such as an earthquake, an explo-sion, or a mechanical impact. Sources suitable for seismic investigation usually generate short-lived wave trains, known as pulses, which typically contain a wide range of frequencies. Except in the immediate vicinity of the source, the strains associated with the passage of a seismic pulse are small and may be assumed to be elastic. Based on this assumption, the propagation velocities of seismic pulses are determined by the elastic moduli and densities of the materials through which they pass. There are two groups of seismic waves: *surface waves* and *body waves*. Surface waves in the form of *Rayleigh waves* and *Love waves* can propagate along the boundary of a solid. Body waves can propagate through the internal volume of an elastic solid and may be of two types: *compressional waves* (longitudinal, primary or P-waves), which propagate by compressional and dilational uniaxial strains in the direction of wave travel with particles oscillating about fixed points in the direction of wave propagation, and *shear waves* (trans-verse, secondary or S-waves), which propagate by a pure shear strain in a direction perpen-dicular to the direction of wave travel with individual particles oscillating about fixed points in a plane at right angles to the direction of wave propagation. The velocity v_p of a P-wave is given by

$$v_p = \sqrt{\frac{K + \frac{4}{3}G}{\rho}} \qquad (10.5)$$

where K is the bulk modulus of elasticity, G is the shear modulus of elasticity, and ρ is the density of the subsurface material. The velocity v_s of an S-wave is given by:

$$v_s = \sqrt{\frac{G}{\rho}} \qquad (10.6)$$

From Equations 10.5 and 10.6, the ratio v_p/v_s is obtained as

$$\frac{v_p}{v_s} = \sqrt{\frac{1 - \nu}{\frac{1}{2} - \nu}} \qquad (10.7)$$

where ν is Poisson's ratio of the subsurface material. Since Poisson's ratio for rocks typically is about 0.25, $v_p \approx 1.7 v_s$; that is, P-waves always travel faster than S-waves in the same medium.

Seismic methods generally use only P-waves, since this simplifies the investigation in two ways. First, seismic/shock detectors, which are insensitive to the horizontal motion of S-waves and hence record only the vertical ground motion, can be used. Second, the higher velocity of P-waves ensures that they always reach a detector before any related S-waves and hence are easier to recognize (Kearey et al. 2002).

Seismic methods make use of the variation in elastic properties of the strata which affect the velocity of shock/seismic waves traveling through them, thus providing dynamic elastic moduli determinations in addition to mapping of the subsurface horizons. The required shock waves are generated within the subsurface materials, at the ground surface or at a certain depth below it, by striking a plate on the soil/rock with a hammer or by detonating a small charge of explosives in the soil/rock. The radiating shock waves are picked up by the vibration detector (e.g., geophone), where the travel times are recorded. Either a number of geophones are arranged in a line or the shock-producing device is moved away from the geophone to produce shock waves at intervals. Figure 10.8 shows the travel paths of primary waves in a simple geological section involving two media (e.g., the soil underlain by bedrock) with respective primary wave velocities of v_1 and v_2 $(>v_1)$ separated at a depth z. From the seismic source S, the energy reaches the detector D at the ground surface by three types of ray path. The *direct ray* travels along a straight line through the top layer from the source to the detector at velocity v_1. The *reflected ray* is obliquely incident on the interface and is reflected back through the top layer to the detector, and its entire path is within the top layer at velocity v_1. The *refracted ray* travels obliquely down to the interface at velocity v_1, along a segment of the interface at the higher velocity v_2, and backs up through the upper layer at velocity v_1.

The travel time t_{dir} of a direct ray is given simply by

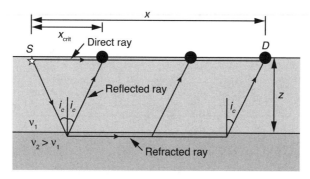

FIGURE 10.8 Seismic/shock ray paths from a near-surface source to a surface detector for a two-layer system.

$$t_{\mathrm{dir}} = \frac{x}{v_1} \qquad (10.8)$$

where x is the distance between the source S and the detector D.

The travel time of a reflected ray is given by:

$$t_{\mathrm{refl}} = \frac{\sqrt{x^2 + 4z^2}}{v_1} \qquad (10.9)$$

The travel time of a refracted ray is given by

$$t_{\mathrm{refr}} = \frac{z}{v_1 \cos i_c} + \frac{x - 2z \tan i_c}{v_2} + \frac{z}{v_1 \cos i_c} \qquad (10.10)$$

where i_c is the critical angle of incidence, expressed as:

$$i_c = \sin^{-1}\left(\frac{v_1}{v_2}\right) \qquad (10.11)$$

Substitution of Equation 10.11 into Equation 10.10 yields:

$$t_{\mathrm{refr}} = \frac{x}{v_2} + \frac{2z\sqrt{v_2^2 - v_1^2}}{v_1 v_2} \qquad (10.12)$$

Time-distance curves for direct, reflected, and refracted rays are illustrated in Figure 10.9. By suitable analysis of the time-distance curve for reflected or refracted rays, it is possible to compute the depth to the underlying layer, such as the bedrock. This provides two independent seismic methods, namely the *seismic reflection method* and the *seismic refraction method*, for locating the subsurface interfaces. The seismic refraction method is especially useful in determining depth to rock in locations where successively denser strata are encountered, that

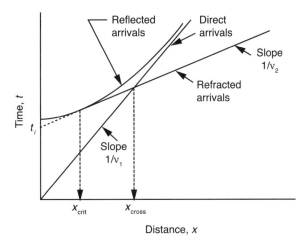

FIGURE 10.9 Time-distance curves for seismic/shock waves from a single horizontal discontinuity.

is, when the velocity of shock or seismic waves successively increases with depth. This method is therefore commonly used in site investigation work. From Figure 10.9, it is evident that the first arrival of seismic energy at a surface detector offset from a surface is always a direct ray or a refracted ray. The direct ray is overtaken by a refracted ray at the *crossover distance* x_{cross}. Beyond this crossover distance, the first arrival is always a refracted ray. Since critically refracted rays travel down to the interface at the critical angle, there is a certain distance, known as the *critical distance* x_{crit}, within which refracted energy will not be returned to the surface. At the critical distance, the travel times of reflected rays and refracted rays coincide because they follow effectively the same path. In the refraction method of site investigation, the detector should be placed at a sufficiently large distance to ensure that the crossover distance is well exceeded so that refracted rays are detected as first arrivals of seismic energy. In general, this approach means that the deeper a refractor, the greater the range over which recordings of refracted arrivals need to be taken.

In Figure 10.9, the intercept on the time axis of the time-distance plot for a refracted ray, known as the *intercept time* t_i, is given by:

$$t_i = 2z \frac{\sqrt{v_2^2 - v_1^2}}{v_1 v_2} \tag{10.13}$$

Since t_i can be determined graphically as shown in Figure 10.9 or numerically from the relation $t_i = t_{refr} - x/v_2$, Equation 10.13 can be used to determine the depth to bedrock as:

$$z = \frac{t_i}{2} \frac{v_1 v_2}{\sqrt{v_2^2 - v_1^2}} \tag{10.14}$$

The seismic reflection method may be useful in delineating geological units at depths. Normally recordings are restricted to small offset distances, well within the critical distance for reflecting interfaces of main interest. This method is not constrained by layers of low seismic velocity and is especially useful in areas of rapid stratigraphic changes.

10.6.2 Electrical Resistivity Method

The electrical resistivity method is useful in determining the depth to bedrock and anomalies in the stratigraphic profile, in evaluating stratified formations where a denser stratum overlies a less dense medium, and in locations of prospective sand-gravel or other sources of borrow material. This method is based on the determination of the subsurface distribution of electrical resistivity of earth materials from measurements on the ground surface. Resistivity parameters also are required for the design of grounding systems and cathodic protection for buried structures. The resistivity of a material is defined as the resistance (Ω) between the opposite faces of a unit cube of the material. If the resistance of a conducting cylinder with length L and cross-sectional area A is R, the resistivity ρ (Ω-m) is

$$\rho = R \frac{A}{L} \qquad (10.15)$$

The current I is related to the applied voltage V and the resistance R of the material by Ohm's law as:

$$I = \frac{V}{R} \qquad (10.16)$$

Each soil/rock has its own resistivity dependent upon water content, compaction, and composition. Certain minerals such as native metals and graphite conduct electricity via the passage of electrons. Most of the rock-forming minerals are, however, insulators, and electric current is carried through a rock mainly by the passage of ions in the pore water. Thus, most rocks conduct electricity by electrolyte rather than electronic processes. It follows that porosity is the major control of the resistivity of rocks, and the resistivity generally increases as porosity decreases. However, even crystalline rocks with negligible intergranular porosity are conductive along cracks and fissures. The range of resistivities among earth materials is enormous, extending from 10^{-5} to 10^{15} Ω-m. For example, the resistivity is low for saturated clays and high for loose dry gravel or solid rock (see Table 10.4). Since there is considerable overlap in resistivities between different earth materials, identification of a rock is not possible solely on

TABLE 10.4   Resistivity of Subsurface Earth Materials

Subsurface Earth Materials	Mean Resistivity (Ω-m)
Marble	10^{12}
Quartz	10^{10}
Rock salt	10^{6}–10^{7}
Granite	5000–10^{6}
Sandstone	35–4000
Moraines	8–4000
Limestone	120–400
Clays	1–120

the basis of resistivity data. Strictly speaking, Equation 10.15 refers to electronic conduction, but it still may be used to describe the *effective resistivity* of a rock, that is, the resistivity of the soil/rock and its pore water. Archie (1942) proposed an empirical formula for effective resistivity as

$$\rho = a\eta^{-b}S^{-c}\rho_w \tag{10.17}$$

where η is the porosity, S is the degree of saturation, ρ_w is the resistivity of water in the pores, and a, b, and c are empirical constants. ρ_w can vary considerably according to the quantities and conductivities of dissolved materials.

Normally one would expect a fairly uniform increase in resistivity with geologic age because of the greater compaction associated with increasing thickness of overburden. There is no consistent difference between the range of resistivities of igneous and sedimentary rocks, although statistically metamorphic rocks appear to have a higher resistivity than either of the other rocks (Dobrin 1976).

The test involves sending direct currents or low-frequency alternating currents into the ground and measuring the resulting potential differences at the surface. For this purpose, four metal spikes are driven into the ground at the surface along a straight line, generally at equal distances; one pair serves as current electrodes and the other pair as potential electrodes (Figure 10.10). The resistivity can be estimated using the following equation (Kearey et al. 2002):

$$\rho = \frac{2\pi V}{I\left[\left(\dfrac{1}{r_1} - \dfrac{1}{r_2}\right) - \left(\dfrac{1}{R_1} - \dfrac{1}{R_2}\right)\right]} \tag{10.18}$$

where V is the potential difference between electrodes P_1 and P_2; r_1 and r_2 are the distances from potential electrode P_1 to current electrodes C_1 and C_2, respectively; and R_1 and R_2 are the distances from potential electrode P_2 to current electrodes C_1 and C_2, respectively.

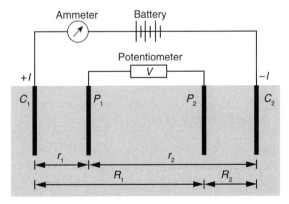

FIGURE 10.10 Generalized form of the electrode configuration used in the electrical resistivity method. C_1 and C_2 are current electrodes, and P_1 and P_2 are potential electrodes.

When the ground is uniform, the resistivity calculated from Equation 10.18 should be constant and independent of both electrode spacing and surface location. When subsurface inhomogeneities exist, however, the resistivity will vary with the relative positions of the electrodes. Any computed value is then known as the apparent resistivity ρ_a and will be a function of the form of the inhomogeneity. Equation 10.18 is thus the basic equation for calculating the apparent resistivity for any electrode configuration. The current electrode separation must be chosen so that the ground is energized to the required depth and should be at least equal to this depth. This places practical limits on the depths of penetration attainable by normal resistivity methods due to the difficulty in laying long lengths of cable and the generation of sufficient power. Depth of penetration of about 1 km is the limit for normal equipment.

There can be several configurations of electrodes, but the *Wenner configuration* is the simplest in that current and potential electrodes are maintained at an equal spacing a (see Figure 10.11). Substitution of this condition, that is, $r_1 = a$, $r_2 = 2a$, $R_1 = 2a$, and $R_2 = 2a$, in Equation 10.18 yields:

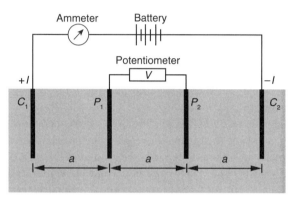

FIGURE 10.11 Wenner electrode configuration used in the electrical resistivity method. C_1 and C_2 are current electrodes, and P_1 and P_2 are potential electrodes.

$$\rho_a = 2\pi a \ \frac{V}{I} \qquad (10.19)$$

In the study of horizontal or near-horizontal overburden soil-bedrock interfaces, the spacing a is gradually increased about a fixed central point. Consequently, readings are taken as the current reaches progressively greater depths. This technique, known as *vertical electrical sounding*, also called *electrical drilling* or *expanding probe*, is used extensively to determine overburden thickness and also to define horizontal zones of porous media. To study the lateral variation of resistivity, the current and potential electrodes are maintained at a fixed separation and progressively moved along a profile. This technique, known as *constant separation traversing* (also called *electrical profiling*), is used to determine variations in bedrock depth and the presence of steep discontinuities.

10.7 Standard Penetration Test

The standard penetration test (SPT) originally was developed in 1927 in the U.S. for granular soils and is one of the oldest and most commonly used *in situ* tests (ASTM D1586; AS 1289.6.3.1). The test is useful for site exploration and foundation design and provides a qualitative guide to the *in situ* properties of soil and samples for classification purposes. The purpose of the test is to drive a split-barrel sampler to obtain a representative soil sample and a measure of the resistance of the soil to penetration of the sampler. A schematic diagram of

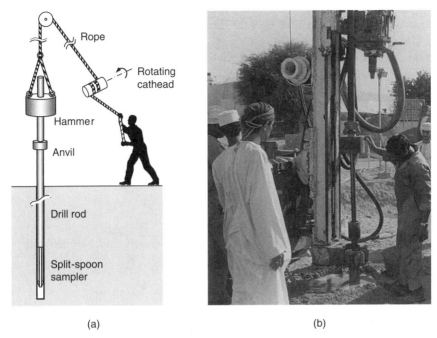

(a)   (b)

FIGURE 10.12 Standard penetration test setup: (a) schematic of rotating cathead arrangement and (b) photograph of automatic tripping arrangement.

an SPT setup, using an old-fashioned rotating cathead, is shown in Figure 10.12a. These days, an automatic tripping mechanism (Figure 10.12b) is used instead of the cathead and rope arrangement to raise and release the hammer. A 35-mm-internal-diameter × 50-mm-outer-diameter split-barrel sampler at the bottom of the borehole, connected to the anvil through drill rods, is driven into the ground by repeatedly dropping a 63.5-kg hammer over a distance of 760 mm onto the anvil. The number of blows required to achieve three subsequent 150-mm penetrations is recorded. The number of blows required to penetrate the final 300 mm is termed the *blow count, standard penetration resistance,* or **N-value** at that depth. For example, if successive blow counts of 5, 7, and 12 are observed for each 150-mm penetration, then standard penetration resistance is $N = 7 + 12 = 19$. The boring is advanced incrementally to permit intermittent or continuous sampling. Typically, test intervals and locations selected are 1.5 m or less in homogeneous strata, with test and sampling locations at every change of strata. The N-values obtained are plotted with depth, where the data points are connected by straight lines. For successful completion of the test, the drilling fluid level within the borehole should be maintained at or above the *in situ* groundwater water table at all times during drilling, removal of drill rods, and sampling.

If the soil stratum below the water table consists of very fine or silty sand, due to the buildup of excess pore water pressure during driving, which in turn reduces the effective stress, the blow count is overestimated. Here, the measured blow count $N_{measured}$ must be reduced to N using the following equation (Terzaghi and Peck 1948):

$$N = 15 + \frac{1}{2}(N_{measured} - 15) \qquad (10.20)$$

Due to the variability associated with the choice of SPT equipment and the test procedure worldwide, various correction factors are applied to the blow count (N). The two most important correction factors are the *hammer efficiency correction factor* (E_h) and the *overburden pressure correction factor* (C_N). The actual energy (E_a) delivered by the hammer to the split-barrel sampler can be significantly less than the theoretical value (E_{th}), which is the product of the hammer weight and the drop height. Kovacs and Salomone (1982) reported that the actual hammer efficiency $\eta_h \ (= 100 E_a / E_{th})$ is on the order of 30–90%. Most SPT correlations are based on a hammer efficiency of 60%, and therefore the current practice is to accept an efficiency of 60% as the standard (Terzaghi et al. 1996). Therefore, E_h is defined as:

$$E_h = \frac{\eta_h}{60} \tag{10.21}$$

The blow count $(N_1)_{60}$ corrected for overburden pressure and hammer efficiency is expressed as:

$$(N_1)_{60} = C_N E_h N \tag{10.22}$$

where C_N is the ratio of the measured blow count to what it would be at an overburden pressure of 1 t/ft² (1 kg/cm² or 100 kPa). Several expressions have been proposed for C_N, the most popular of which is (Liao and Whitman 1986)

$$C_N = 9.78 \sqrt{\frac{1}{\sigma'_{vo} \ (\text{kPa})}} \tag{10.23}$$

where σ'_{vo} is the effective overburden pressure at the point of measurement. It should be noted that the overburden correction generally is applied for granular soils only. N_{60} refers to the SPT value without overburden correction.

Two other correction factors are the *borehole diameter correction factor* (C_b) and the *drill rod length correction factor* (C_d), given in Tables 10.5 and 10.6. These are discussed in detail by Skempton (1986). When using samplers with liners, the blow count is overestimated and a multiplication factor of 0.8 is recommended in dense sands and clays and 0.9 in loose sands (Bowles 1988).

While the standard penetration test gives the blow count, laboratory tests on sands are carried out on the basis of relative density. The interrelationships among blow count, relative density, friction angle, and Young's modulus are discussed below.

TABLE 10.5 Borehole Diameter Correction Factor (Skempton 1986)

Borehole Diameter (mm)	Correction Factor, C_b
60–120	1.00
150	1.05
200	1.15

TABLE 10.6 Drill Rod Length Correction Factor (Skempton 1986)

Rod Length (m)	Correction Factor, C_d
0–4	0.70
4–6	0.85
6–10	0.95
>10	1.00

Using Meyerhof's (1957) approximation,

$$\frac{N_{60}}{D_r^2} = a + b\sigma'_{vo}$$
(10.24)

where D_r is the relative density, and a and b are site-dependent parameters.
Skempton (1986) suggested that for sands with relative density $D_r > 35\%$:

$$\frac{(N_1)_{60}}{D_r^2} \approx 60$$
(10.25)

Here, $(N_1)_{60}$ should be multiplied by 0.92 for coarse sands and 1.08 for fine sands. Kulhawy and Mayne (1990) suggested the following:

$$\frac{(N_1)_{60}}{D_r^2} \approx 70$$
(10.26)

Peck et al. (1974) suggested a relationship between N_{60} and friction angle ϕ for granular soils, shown in Figure 10.13, which is widely used in granular soils for estimating the friction angle from the blow count. Wolff (1989) expressed this relation as:

$$\phi \ (\text{deg}) = 27.1 + 0.3N_{60} - 0.00054N_{60}^2$$
(10.27)

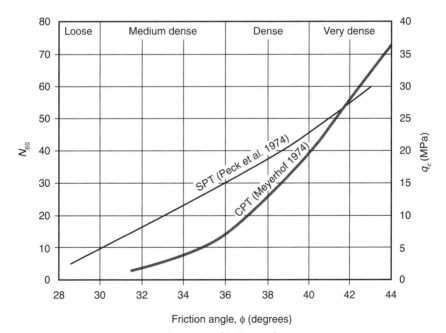

FIGURE 10.13 Penetration resistance vs. friction angle.

Hatanaka and Uchida (1996) provided a simple correlation between ϕ and $(N_1)_{60}$ for granular soils as:

$$\phi = \sqrt{20(N_1)_{60}} + 20 \tag{10.28}$$

Schmertmann (1975) proposed an $(N)_{60}$-ϕ-σ'_{vo} relation graphically for granular soils, which can be expressed as (Kulhawy and Mayne 1990):

$$\phi = \tan^{-1}\left[\frac{N_{60}}{12.2 + 20.3\left(\dfrac{\sigma'_{vo}}{p_a}\right)} \right]^{0.34} \tag{10.29}$$

where p_a is the atmospheric pressure (= 101.3 kPa). The friction angles estimated from Equation 10.27 are quite conservative compared to those derived from Equation 10.28 or 10.29. The differences can be quite large for large values of D_r.

Young's modulus (E) is an essential parameter for computing deformations, including settlement of foundations. Leonards (1986) suggested that for normally consolidated sands

$$E \ (\text{kg/cm}^2) \approx 2q_c \ (\text{kg/cm}^2) \approx 8N_{60} \tag{10.30}$$

where q_c is the static cone penetration resistance.

Kulhawy and Mayne (1990) suggested that

$$\frac{E}{p_a} = \alpha N_{60} \tag{10.31}$$

where atmospheric pressure p_a is in the same units as E; $\alpha = 5$ for sands with fines, 10 for clean normally consolidated sands, and 15 for clean overconsolidated sands.

In spite of its simplicity and the large historical database, the SPT has numerous sources of uncertainty and error, making it less reproducible. Lately, static cone penetration tests, using piezocones, have become increasingly popular because they offer better rationale, reproducibility, and give continuous measurements. An SPT is not very reliable in cohesive soils, due to the pore pressure developments during driving that may affect the effective stresses temporarily, and therefore any correlations in clays should be used with caution. A rough estimate of undrained shear strength (c_u) can be obtained from (Hara et al. 1971; Kulhawy and Mayne 1990):

$$\left(\frac{c_u}{p_a}\right) = 0.29N_{60}^{0.72} \tag{10.32}$$

On the basis of the regression analysis of 110 data points, Mayne and Kemper (1988) obtained the following relationship:

	*Very loose	Loose	Medium dense	Dense	Very dense
#D_r (%) 0	15	35	65	85	100
*N_{60}	4	10	30	50	
##$(N_1)_{60}$	3	8	25	42	
**ϕ' (degrees)	28	30	36	41	
##$(N_1)_{60}/D_r^2$		65	59	58	

FIGURE 10.14 Classification of granular soils based on relative density: # = Gibbs and Holtz (1957), * = Terzaghi and Peck (1948), ## = Skempton (1986), and ** = Peck et al. (1974).

$$OCR = 0.193 \left(\frac{N_{60}}{\sigma'_v} \right)^{0.689} \tag{10.33}$$

where OCR is the overconsolidation ratio of a natural clay deposit and σ'_v is the effective vertical stress.

Based on relative density, granular soils can be classified as shown in Figure 10.14. Also given in the figure are the N_{60}, $(N_1)_{60}$, ϕ', and $(N_1)_{60}/D_r^2$ values. Skempton (1986) suggested that the N values given by Terzaghi are based on SPT rigs with an energy rating of 45%, and hence the N_{60} values reported in Figure 10.14 need to be multiplied by 0.75.

10.8 Static Cone Penetration Test

The static cone penetration test, or simply cone penetration test (CPT), also known as the Dutch cone penetration test, originally was developed in the Netherlands in 1920 and can be used successfully to determine the penetration resistance as the *end-bearing resistance* and *side-friction resistance* during the steady penetration of a solid cone into the soil at a rate of 20 mm/s (ASTM D3441; AS 1289.6.5.1). The test is applicable to most soils, except gravelly soils, soil fills containing stones and brick bats, and soils with standard penetration resistance N greater than 50. Boreholes are not necessary to perform this test. The test is a valuable sounding method of recording variation in the *in situ* penetration resistance of soils in cases where the *in situ* unit weight is disturbed by boring operations, thus making the SPT values unreliable, especially under water. Experience indicates that a complete static CPT up to depths of 15–20 m can be completed in a day with manual operation of the equipment, making it one of the most inexpensive and fastest methods available for subsoil investigation. The major advantages of the CPT over the SPT are its continuous profile and the higher accuracy and repeatability it provides; subsequently, if a good CPT-SPT correlation exists, very comprehensive equivalent SPT values can be obtained (Canadian Geotechnical Society 2006).

Comparing a CPT probe with an SPT probe, the split-barrel sampler is replaced by a probe that consists of a solid cone with a 60° apex angle and a base diameter of 35.7 mm, resulting in a projected area of 10 cm^2. The cone is attached to a drill rod with a friction sleeve that has a surface area of 150 cm^2, which is advanced into the soil at a constant rate of 10–20 mm/s.

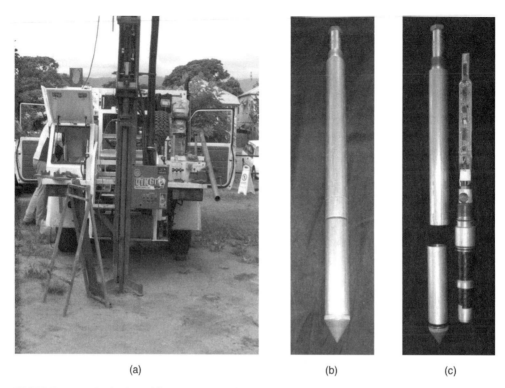

(a)   (b)   (c)

FIGURE 10.15 Static CPT: (a) mini test rig, (b) external view of piezocone, and (c) internal view of piezocone.

A mini test rig is shown in Figure 10.15a. A variety of cone penetrometers, including mechanical, electrical, and electronic, are available; however, the mechanical penetrometer is the most common. It operates incrementally, using a telescopic penetrometer tip, resulting in no movement of the push rod during measurement of the resistance components. Design constraints for mechanical penetrometers preclude a complete separation of the end-bearing and side-friction components. Electrical and electronic cone penetrometers measure cone resistance by a force transducer fitted to the cone, and the friction force on the friction sleeve is measured separately from the cone resistance. A cone that consists of a piezometer built in for pore pressure measurements is called a *piezocone,* shown in Figure 10.15b and c. Here, the three measurements that are taken continuously as the cone is pushed into the soil are *cone resistance* (q_c), *friction resistance* (f_s), and *pore water pressure* (u). Resistance as well as pore water pressure may be determined continuously, as shown in Figure 10.16, or at desired depth intervals, usually at equal increments of depth. In the latter case, resistance and pore water pressure values are plotted at depths corresponding to the depths of measurement and can be connected with straight lines as an approximation for a continuous graph. When determining the penetration of a soil under a pavement or for design of pavement depth, 25- to 30-mm intervals are appropriate. Intervals of 150–200 mm are more appropriate for other applications. The **friction ratio** f_R, defined as

$$f_R = \frac{f_s}{q_c} \times 100\% \qquad (10.34)$$

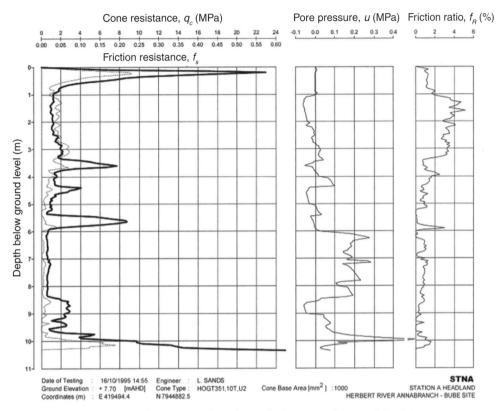

FIGURE 10.16 Static CPT data measured continuously (courtesy of Leonard Sands).

is a useful parameter in identifying the soil. Values for f_R are in the range of 0–10%, with granular soils at the lower end and cohesive soils at the upper end of the range. Using the pair of values for q_c and f_R, the soil type can be identified from Figure 10.17.

The undrained shear strength (c_u) of clays can be estimated from (Schmertmann 1975)

$$c_u = \frac{q_c - \sigma_{vo}}{N_k} \tag{10.35}$$

where σ_{vo} is the total overburden pressure and N_k is known as the cone factor, which varies in the range of 14–25 and can be obtained through calibration. The lower end of the range applies to normally consolidated clays and the upper end to overconsolidated clays. It depends on the penetrometer and the type of clay and increases slightly with plasticity index. Based on test data from Aas et al. (1984), N_k can be estimated by (Bowles 1988)

$$N_k = 13 + 0.11\text{PI} \pm 2 \tag{10.36}$$

where PI is the plasticity index of the soil.

Mayne and Kemper (1988) suggested an N_k of 15 for an electric cone and 20 for a mechanical cone. They proposed equations for estimating the effective preconsolidation pressure (σ_p') and the OCR of a clay (in MPa) as

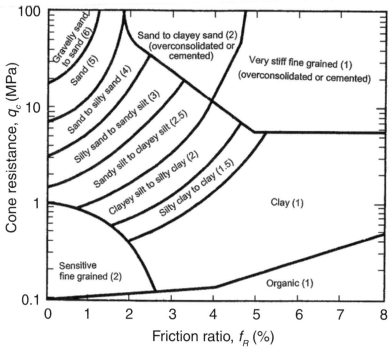

FIGURE 10.17   Soil classification from a piezocone (adapted from Robertson et al. 1986). $(q_c/p_a)/N_{60}$ values shown in parentheses.

$$\sigma'_p = 0.243 q_c^{0.96} \tag{10.37}$$

and

$$\text{OCR} = 0.37 \left(\frac{q_c - \sigma_{vo}}{\sigma'_{vo}} \right)^{1.01} \tag{10.38}$$

where σ_{vo} and σ'_{vo} are total and effective vertical stress, respectively.

Classification of clays based on undrained shear strength and the corresponding consistency terms are given in Table 10.7. Also given in the table are the approximate borderline values of $(N_1)_{60}$ and q_c/p_a and a field identification guide.

Variation of ϕ with q_c for granular soils, as proposed by Meyerhof (1974), is shown in Figure 10.13. The dependence of q_c on overburden stress is not incorporated here and therefore this must be used with caution.

Kulhawy and Mayne (1990) showed that the q_c-σ'_{vo}-ϕ relationship in sands, proposed by Robertson and Campanella (1983), can be approximated by:

$$\phi = \tan^{-1} \left[0.1 + 0.38 \log \left(\frac{q_c}{\sigma'_{vo}} \right) \right] \tag{10.39}$$

TABLE 10.7 Consistency Terms for Clays with $(N_1)_{60}$ and q_c Values

Consistency[a]	c_u (kPa)[a]	$(N_1)_{60}$[a]	q_c/p_a[b]	Field Identification Guide[c]
Very soft	<12	0–2	<5	Exudes between fingers when squeezed in hand; can easily be penetrated several centimeters by fist
Soft	12–25	2–4		Can be molded by light finger pressure; can easily be penetrated several centimeters by thumb
Firm	25–50	4–8	5–15	Can be molded by strong finger pressure; can be penetrated several centimeters by thumb with moderate effort
Stiff	50–100	8–15	15–30	Cannot be molded by fingers; can be indented by thumb but penetrated only with great effort
Very stiff	100–200	15–30	30–60	Readily indented by thumbnail
Hard	>200	>30	>60	Can be indented by thumbnail with difficulty

[a] Terzaghi and Peck (1948).
[b] McCarthy (2007).
[c] AS 1726 (1993), Canadian Geotechnical Society (1992).

Schmertmann (1970) proposed that the modulus of elasticity $E = 2q_c$ for sands, and later Schmertmann et al. (1978) suggested $E = 2.5q_c$ for axisymmetric loading in sands and $E = 3.5q_c$ for plane strain loading.

Geotechnical engineers do not always have the luxury of availability of both SPT and CPT data. When only one type of data is available, it is useful to have some means of converting it to the other. Ratios of q_c/N_{60} for different soils, as given by Sanglerat (1972) and Schmertmann (1970, 1978) are shown in Table 10.8. Robertson et al. (1983) presented the variation of $(q_c/p_a)N_{60}$ with a mean grain size of D_{50}, and the upper and lower bounds are shown in Figure 10.18. The soil data were limited to D_{50} less than 1 mm. Also shown in the figure are the upper and lower bounds proposed by Burland and Burbidge (1985) and the average values suggested in the *Canadian Foundation Engineering Manual* (Canadian Geotechnical Society 2006) and by Kulhawy and Mayne (1990) and Anagnostopoulos et al. (2003).

TABLE 10.8 Ratio of q_c/N_{60}

Soil	q_c (kg/cm²)/N_{60}
Silts, sandy silts, slightly cohesive silt-sand mix	2[a] (2–4)[b]
Clean fine to medium sands and slightly silty sands	3–4[a] (3–5)[b]
Coarse sands and sands with little gravel	5–6[a] (4–5)[b]
Sandy gravel and gravel	8–10[a] (6–8)[b]

After Sanglerat (1972) and Schmertmann (1970, 1978).
[a] Values proposed by Sanglerat (1972) and reported in Peck et al. (1974).
[b] Values suggested by Schmertmann (1970, 1978) and reported by Holtz (1991) in parentheses.

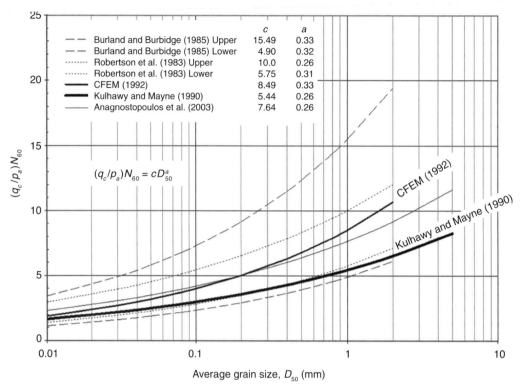

FIGURE 10.18 $(q_c/p_a)/N_{60}$ vs. D_{50} variation in granular soils.

All the curves in Figure 10.18 take the following form:

$$\left(\frac{q_c}{p_a}\right)/N_{60} \approx cD_{50}^a \tag{10.40}$$

where a and c are constant parameters.

Kulhawy and Mayne (1990) approximated the dependence of the q_c/N_{60} ratio on D_{50} (in mm) as:

$$\left(\frac{q_c}{p_a}\right)/N_{60} \approx 5.44D_{50}^{0.26} \tag{10.41}$$

Based on an extensive database of 337 points, with test data for D_{50} as high as 8 mm, Anagnostopoulos et al. (2003) noted that for Greek soils:

$$\left(\frac{q_c}{p_a}\right)/N_{60} \approx 7.64D_{50}^{0.26} \tag{10.42}$$

Kulhawy and Mayne (1990) also suggested that q_c/N_{60} can be related to the fines content in a granular soil as:

$$\left(\frac{q_c}{p_a}\right) / N_{60} \approx 4.25 - \frac{\% \text{ fines}}{41.3} \tag{10.43}$$

10.9 Dynamic Cone Penetration Test

The dynamic CPT is used to determine the resistance of different soil strata to dynamic penetration of a cone and thereby obtain an indication of their relative strengths or densities or both. This test is subject to all of the limitations of the SPT and should not be used for quantitative evaluation of soil density and other parameters. The main advantage of this test is that it is fast and economical, and a continuous resistance vs. depth profile is obtained that can provide a visual relationship of soil type or density variations. This test is thus commonly used for conducting a reconnaissance survey of wide areas in a short time, which enables selective *in situ* testing or sampling for a typical soil profile. It can provide useful data for local conditions where reliable correlations have been established. In the test procedure, a metallic cone with or without thread is attached to a drilling rod, and a hammer of known weight is allowed to fall freely from a standard height to drive the cone into the ground. The number of hammer blows required for a specific penetration of the cone is measured. The cone without thread is left in the ground after completion of the test. There is wide variation in the dimensions of the cone and the test procedures adopted in different countries. Moreover, limited literature is available on the dynamic CPT.

For shallow pavement and floor slab applications, ASTM D6951 recommends a procedure for determining the penetration resistance of a soil using a steel cone with a 60° included angle and 20-mm base diameter driven with an 8-kg hammer, dropped over a height of 575 mm. During the lifting and release of the hammer, the total penetration for a given number of blows or the penetration per blow is manually recorded. If the cone does not advance more than 2 mm in five blows, the test is terminated, and the test rig is moved to the next location. The depth of probing varies with application. For typical roadwork, an investigation to a depth of 900 mm generally is adequate. AS 1289.6.3.2 describes the procedure for determining the penetration resistance of a soil (in millimeters per blow) using a steel cone with a 30° included angle and 20-mm base diameter driven with a 9-kg hammer, dropped over a height of 510 mm. The dynamic CPT data generally can be used to determine *in situ* **California bearing ratio** (CBR) from the correlations available locally for *in situ* soils or for *in situ* compacted soils. For example, the U.S. Army Corps of Engineers recommends the use of penetration (in millimeters) per blow, denoted by DCP, for estimating CBR as

$$\text{CBR} = \frac{292}{\text{DCP}^{1.12}} \tag{10.44a}$$

for all soils except for CH soils and CL soils below a CBR of 10. For these soils, the following equations are recommended by the U.S. Army Corps of Engineers (De Beer et al. 1989). For CH soils

$$\text{CBR} = \frac{1}{0.002871 \times \text{DCP}} \tag{10.44b}$$

and for CL soils with a CBR < 10

$$\text{CBR} = \frac{1}{(0.017019 \times \text{DCP})^2} \tag{10.44c}$$

If distinct layering exists within the soil, a change in slope on a graph of cumulative penetration blows vs. depth is observed for each layer. Although the exact interface is difficult to define because of a transition zone between the layers, the layer separation can be defined by the intersection of the lines representing the average slope of adjacent layers. Once the layer thicknesses have been defined, the average penetration rate per layer can be calculated and then the correlations (Equations 10.44a–10.44c) can be used for estimating CBR values of individual layers.

IS 4968 (Parts I and II) recommends a cone with a 60° included angle and 50- or 62.5-mm base diameter driven with a 65-kg mass, dropping 750 mm, as shown in Figure 10.19, and suggests measuring the number of blows for every 100-mm penetration of the cone for foundation applications. The dynamic cone penetration resistance (N_{cd}) is expressed as the number of blows per 300-mm penetration and reported as a continuous record of blows for every 300-mm penetration either in tabular form or as a graph of N_{cd} vs. depth, as shown in Figure 10.20. Correlations between the dynamic cone penetration values and penetration resistances obtained by other penetration methods such as SPT or static CPT may be developed for a given site by conducting the latter tests adjacent to (about 3–5 m away) the location of the dynamic CPT. For a 62.5-mm cone driven dry up to a depth of 9 m (without bentonite slurry), the following relationships for medium to fine sands have been developed by the Central Building Research Institute, Roorkee, India. Up to a depth of 4 m,

$$N_{cd} = 1.5(N_1)_{60} \tag{10.45a}$$

and for depths of 4–9 m,

$$N_{cd} = 1.75(N_1)_{60} \tag{10.45b}$$

When a 62.5-mm cone is driven into the ground by circulating bentonite slurry, then the following relationship is suggested:

$$N_{cd} = (N_1)_{60} \tag{10.45c}$$

Equations 10.44 and 10.45 should be used with caution. In fact, selection of the appropriate correlation for a specific application is a matter of professional judgment.

10.10 Plate Load Test

The plate load test involves loading a square or circular plate, usually at the elevation of the proposed footing or pavement and under the same loading conditions expected in service. This test is conducted to estimate the ultimate bearing capacity of the soils and pavement components (unbound base and subbase layers), in either the compacted condition or the natural state, provided the soil strata and pavement components are reasonably uniform. The test will directly give the ultimate bearing capacity if it is conducted on a full-size footing;

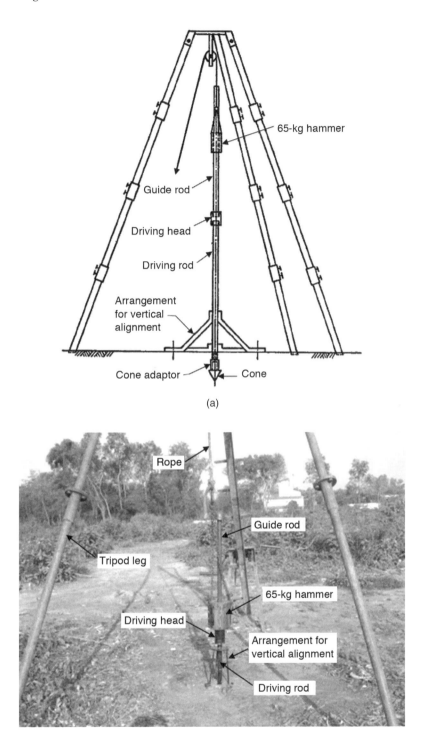

(a)

(b)

FIGURE 10.19 Dynamic CPT: (a) typical assembly of equipment and (b) test arrangement at the site of the proposed Bharat Heavy Electricals Limited (Jagadishpur, Sultanpur District, India) centralized stamping unit and fabrication plant (after Shukla 2006a).

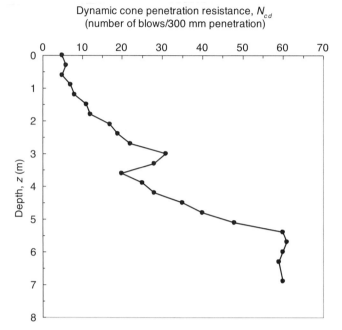

FIGURE 10.20 Presentation of dynamic cone penetration data obtained at the site of the proposed Bharat Heavy Electricals Limited
(Jagadishpur, Sultanpur District, India) centralized stamping unit
and fabrication plant (after Shukla 2006a). Soil strata up to 4.5-m
depth from the ground surface consist of silty clays of low plasticity
(CL) underlain by clayey silts with none to low plasticity (ML) up to
7-m depth.

however, this is not usually done since a very large load is required. The data also can be used
to estimate the settlement of the footing or pavement and to determine the **modulus of
subgrade reaction** (*aka subgrade modulus*) of the soils and pavement components. Figure
10.21 presents the essential features of the plate load test.

To conduct the test for foundation applications, a pit is excavated to the depth at which
the test is to be performed. The pit should be at least five times as wide as the test plate. It
should have a carefully leveled and cleaned bottom at the base level, so that loads are transmitted over the entire contact area of the undisturbed soil. An effort should be made to
maintain the moisture condition as per the field requirements. If the test has to be conducted
under the estimated worst conditions, the soil may be prewetted to the desired extent to a
depth under the bearing plate of not less than twice its diameter. If the water table is within
1 m below the footing, the plate load test should be performed at the water table level. If the
water table is above the footing level, it should be lowered to the footing level before the plate
is placed.

Except in the case of roadwork and circular footings, square plates normally are adopted.
For clayey and silty soils and for loose to medium-dense sandy soils with an SPT value less than
15, a 450-mm square plate or concrete block should be used. In the case of dense sandy or

(a)

(b)

FIGURE 10.21 Plate load test setup: (a) reaction loading platform supported on walls and (b) equipment at the site of the proposed raft foundation of an aqueduct-cum-bridge along the alignment of the 70-km-long Bansagar feeder channel, Sidhi District, Madhya Pradesh, India (after Shukla 2006b).

gravelly soils with an SPT value in the range of 15–30, three plates 300–750 mm in size should be used depending on practical considerations of type of load application and maximum grain size. The width of the plate should be at least four times the maximum size of the soil particles present at the test location. The thickness of the plate should not be less than 25 mm if the plate is made of mild steel. If concrete blocks are used, they must have a depth of not less than two-thirds their width. When there are three or more test locations, the distance between test locations should not be less than five times the width or diameter of the largest plate used in the tests.

The load can be applied by gravity loading, a reaction loading platform supported on walls, or a reaction loading beam/truss attached to anchor piles. The loading platform should be supported at points as far away from the test area as practicable, generally at a minimum distance of 3.5 times the width or diameter of the test plate or 2.5 m, whichever is greater, with a height of 1 m or more above the bottom of the pit to provide sufficient working space. For reaction loading, a hydraulic jack of sufficient capacity, but not less than 50 t (500 kN), is used, along with a pressure gauge, electronic load cell, or proving ring to measure the force exerted by the jack.

A minimum seating pressure of, for example, 7 kPa can be applied and removed before starting the load test. The load is applied on the plate in cumulative equal increments up to 100 kPa or one-tenth of the estimated ultimate bearing capacity of the soil without impact, fluctuation, or eccentricity. Settlements are recorded from dial gauges accurate to at least 0.02 mm as soon as possible before and after the application of each load increment and at equal time intervals while the load is constant, which will provide not less than six settlement measurements between load applications. At least two dial gauges resting at diametrically opposite ends of the plate should be fixed with the help of a reference beam placed over firm ground as far as practicable, but not less than 2.5 m from the center of the loaded area. Each load increment should be kept for not less than 1 h or until the rate of settlement is appreciably reduced to a value of 0.02 mm/min. The test should be continued until a settlement of 25 mm under normal circumstances or 50 mm in special cases (such as dense gravel or gravel and sand mixture) is obtained or until the capacity of the testing apparatus is reached. After the load is released, the elastic rebound of the soil should be recorded for a period of time at least equal to the duration of the load increment. At the end of the test, it is necessary to excavate soil below the test plate to a depth equal to twice the dimension of the plate so as to examine and record the subsoil profile.

Figure 10.22 shows typical pressure vs. settlement curves. Curve A generally is obtained for dense cohesionless soils or stiff cohesive soils. Curve B is obtained when loose to medium-dense cohesionless soils or soft to medium-stiff cohesive soils are loaded. For very loose cohesionless soils or very soft clayey soils, results like curve C are obtained. Since failure is well defined with a distinct peak in curve A, the ultimate bearing capacity of the soil for the plate (q_{up}) is obtained without any approximation, as shown in the figure. In the case of curve B, however, where the yield point is not de-

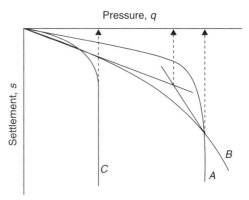

FIGURE 10.22 Typical pressure-settlement curves obtained from plate load tests.

fined, the ultimate bearing capacity for the test plate can be determined by intersection of tangents, as shown in the figure. It also can be determined by the intersection of the two straight lines obtained in the plot of pressure vs. settlement, both to logarithmic scales as recommended in IS 1888.

Since the width (or diameter) B_p of the test plate generally is smaller than the width B of the full-size footings, extrapolation of the test results is required to obtain the ultimate bearing capacity (q_u) of the soil for the full-size footings. Any such extrapolation is not considered a standard approach; however, the ultimate bearing capacity for the full-size footings can be approximated (Bowles 1988; Das 1998) for cohesionless soils as

$$q_u = q_{up} \left(\frac{B}{B_p} \right) \tag{10.46}$$

and for cohesive soils as

$$q_u = q_{up} \tag{10.47}$$

In general, Equation 10.46 should be used if B/B_p is less than about 3. The settlement of the full-size footing also can be approximated (Bowles 1988; Das 1998) from the settlement of the test plate at the same applied pressure for cohesionless soils as

$$s = s_p \left[\frac{2B}{B_p (B + B_p)} \right]^2 \tag{10.48}$$

and for cohesive soils as

$$s = s_p \frac{B}{B_p} \tag{10.49}$$

where s is the settlement of the full-size footing (mm), s_p is the settlement of the test plate (mm), B is the width of the footing (m), and B_p is the width of the test plate (m). It has been found that Equation 10.48 is valid for medium and dense cohesionless soils. Use of this equation for loose cohesionless soils may lead to underestimation of settlement. Equations 10.48 and 10.49 can be used to determine the bearing pressure for soils that correspond to an allowable settlement of the footing (e.g., 50 mm); the bearing pressure thus obtained is called the *safe bearing pressure*.

The plate load test has a few limitations. The test results reflect only the character of the soil located within a depth less than twice the width of the test bearing plate. Since foundations generally are larger than the test plates, the settlement and shear resistance will depend on the properties of a much thicker stratum. Moreover, this method does not give the ultimate settlements, particularly in the case of cohesive soils, because of the short duration. The settlement measured is mainly the immediate settlement. Consolidation settlement, which constitutes a major part of the total settlement, cannot be predicted through this test. Hence the plate load test is not of much relevance in cohesive soils, and therefore Equations 10.47 and 10.49 generally are of limited use in foundation design.

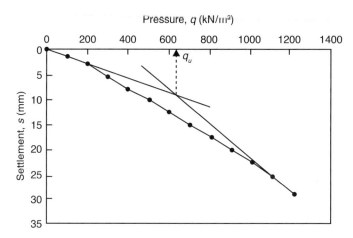

FIGURE 10.23 Plate load test results obtained at the site of the proposed raft foundation of an aqueduct-cum-bridge along the alignment of the 70-km-long Bansagar feeder channel, Sidhi District, Madhya Pradesh, India (after Shukla 2006b).

Another limitation is related to the effect of size of the footing. For clayey soils, the bearing capacity (from shear consideration) of a larger foundation is almost the same as that of the smaller test plate, whereas in dense sandy soils, the bearing capacity increases with the size of the footing. Presently no available methods provide absolute values for the bearing capacity of soil *in situ* against which this test method can be compared. Reproducibility of test results also is not possible, mainly because of the variability of the soil and the resulting disturbance of the soil under the loaded test plate. In spite of these major limitations, plate load tests are occasionally used.

The first author conducted a plate load test on May 6–7, 2006 at the foundation site of the raft foundation (Pier No. P-7) of the proposed aqueduct-cum-bridge at km 46.615 of the Bansagar feeder channel, Sidhi District, Madhya Pradesh, India (Figure 10.21). The test was conducted on a square test plate with a 700-mm side and a 7-kPa seating pressure and 50-kN load increment under saturated condition of foundation soil. The plate load test results are shown in Figure 10.23. Using a double tangent method, the ultimate bearing capacity (q_{uP}) of the foundation soil for the test plate is 640 kN/m². The soil strata at the foundation level in the test pit consist of a sand-gravel-cobble matrix with some percentage of silt and clay-sized particles. The *in situ* total unit weight of the foundation material was 22.6 kN/m³.

The plate load test data can be used to determine the modulus of subgrade reaction of *in situ* soil, which is required in the evaluation and design of structures such as airport and highway pavements and raft foundations. The modulus of subgrade reaction (k_s) is defined as the ratio of applied pressure on the horizontal surface of the foundation soil to the corresponding settlement of the surface. It is determined as the slope of the secant drawn between the point corresponding to zero settlement and a specific point corresponding to s_0 (generally 1.25 mm) settlement of a load-settlement curve obtained from a plate load test using a larger (preferably 750-mm) test plate. Figure 10.24 explains the method for determining k_s. The modulus of subgrade reaction is not a unique property of the soil. It is affected by several factors, such as the size, shape, embedded depth, and material characteristics of the footing.

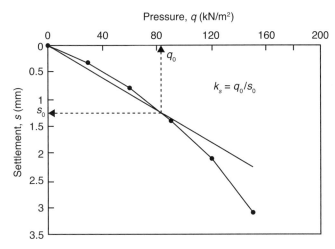

FIGURE 10.24 Determination of modulus of subgrade reaction from plate load test results.

Hence, the modulus of subgrade reaction obtained from a plate load test in the field has to be appropriately corrected before being used for evaluation and design of structures. Extrapolating results of a plate load test to the actual structure is a real problem. Noting that k_s is inversely proportional to the footing size in clays, Terzaghi (1955) proposed the following formula when the contact pressures are less than one-half of the ultimate bearing capacity:

$$\frac{k_s}{k_{sp}} = \frac{B_p}{B} \tag{10.50}$$

where k_{sp} is the plate load value of the subgrade modulus, using a plate of dimension B_p, and k_s is the value to be used under the actual footing of width B. In granular soils, k_s decreases with increasing B, but is not inversely proportional, so it was proposed to use the following:

$$k_s = k_{sp} \left(\frac{B + 1}{2B} \right)^2 \tag{10.51}$$

The work of Bond (1961) indicates that Equation 10.51 may not give reliable values of k_s for medium-dense to dense sands (relative density greater than 40%). There is a possibility that this equation can largely underestimate the value of k_s over the usual range of B of 5–10 ft as reported by Bowles (1975).

To obtain the modulus of subgrade reaction for a rectangular plate/footing of dimensions B and $L = mB$ using the subgrade modulus (k_{sp}) of a square plate, Terzaghi (1955) proposed the following:

$$k_s = k_{sp} \left(\frac{m + 0.5}{1.5} \right) \tag{10.52}$$

In the past, attempts have been made to propose the value of the modulus of subgrade reaction using laboratory test data. Using the stress-strain modulus E_s of the soil from laboratory triaxial tests, Vesic (1961a, 1961b) proposed the following:

$$k_s' = 0.65 \sqrt[12]{\frac{E_s B^4}{EI}} \frac{E_s}{1 - v^2} \qquad (10.53)$$

where $k_s' = k_s B$ and E, I, and v are, respectively, the modulus of elasticity, the moment of inertia, and Poisson's ratio of the footing material. Since the twelfth root of any value multiplied by 0.65 will be close to 1 for all practical purposes, Equation 10.53 reduces to (Bowles 1988):

$$k_s = \frac{E_s}{B(1 - v^2)} \qquad (10.54)$$

The value of k_s also can be obtained from the CBR value of the soil subgrade using (Nascimento and Simoes 1957)

$$k_s = 10\text{CBR} \qquad (10.55)$$

for the CBR value at 0.1-in. (2.54-mm) penetration of the plunger in a CBR test.

It is observed that for a given load intensity applied by a civil engineering structure to a saturated foundation soil, the settlement of the structure increases with time. Consequently, the modulus of subgrade reaction, calculated as the contact pressure (or applied pressure) per unit soil settlement, decreases. In view of this field observation, Shukla (2008) presented a time-dependent expression for the modulus of subgrade reaction of the saturated cohesive foundation soil that takes into consideration the primary consolidation aspect. This expression involves parameters that can be determined by conducting one-dimensional laboratory consolidation testing on undisturbed soil specimens.

10.11 Field Vane Shear Test

The field vane shear test, shown schematically in Figure 10.25, is commonly used for determining the *in situ* undrained shear strength of saturated clays, especially of soft to medium consistency. The test is most suitable for sensitive saturated cohesive soil deposits which are highly susceptible to sampling disturbance. The vane used in the test consists of two rectangular blades that are perpendicular to each other, as seen in Figure 10.25. The blades are as thin as possible, consistent with the strength requirements, so that the vane causes as little remolding and disturbance as possible to the soil when inserted into it. The area ratio (A_r) of the vane as calculated from the following equation is kept as low as possible and generally should not exceed 12% for typical vanes:

$$A_r = \frac{8t(D - d) + \pi d^2}{\pi D^2} \times 100\% \qquad (10.56)$$

where D is the overall diameter of the vane, d is the diameter of the central rod, and t is the thickness of the vane blades. The *Canadian Foundation Engineering Manual* (Canadian Geotechnical Society 2006) recommends that for acceptable results, the blade thickness should not exceed 5% of the vane diameter D.

The vane is pushed without a hole from the ground surface or inserted from the bottom of a borehole to the desired depth where the test is to be carried out. It is rotated at the rate of 0.1° per second by applying a torque at the surface through a torque applicator/meter that also measures the torque. This rotation will initiate shearing of the clay along a cylindrical surface surrounding the vane. The undrained shear strength of the undisturbed clay can be determined from the maximum applied torque (T) by

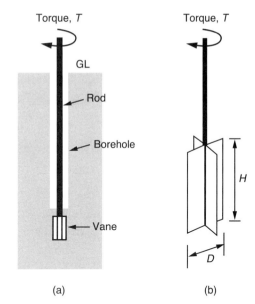

FIGURE 10.25 Principle of the vane shear test: (a) vane inside the borehole and (b) geometrical details of vane.

$$c_u = \frac{2T}{\pi d^2 (H + D/3)} \tag{10.57}$$

where H and D are, respectively, the height and breadth of the rectangular blades (i.e., height and diameter of the cylindrical surface sheared), which are typically of the 2:1 ratio, with D in the range of 38–100 mm for field vanes. Miniature vanes are used in laboratories to determine undrained shear strength of clay samples still in sampling tubes. The test can be continued by rotating the vane rapidly after shearing the clay, to determine the remolded shear strength. The test can be carried out at depths even greater than 50 m.

For better accuracy of results, it is necessary to measure or calibrate the friction from both soil and bearings in the torque rods and deduct it from the maximum applied torque. Back analysis of several failed embankments, foundations, and excavations in clays has shown that the vane shear test overestimates the undrained shear strength for design. A reduction factor (λ) has been proposed to correct the shear strength measured by vane shear test, and the correct shear strength is given by

$$c_{u(\text{corrected})} = \lambda c_{u(\text{FVST})} \tag{10.58}$$

where the factor λ can be estimated as follows (Bjerrum 1972):

$$\lambda = 1.7 - 0.54 \log[\text{PI } (\%)] \tag{10.59}$$

Morris and Williams (1994) suggested that for PI > 5:

$$\lambda = 1.18 \exp(-0.08\text{PI}) + 0.57 \tag{10.60}$$

Since a vertical vane blade usually shears a cylindrical mass of soil with its edges perpendicular and ends parallel to the horizontal bedding plane, the measured c_u may not be the same as that required in designs. Moreover, the maximum torque is used, which means that the shear strength is computed from one reading only. The vane shear test generally is not recommended alone for subsoil exploration. It is, however, a useful test for soils that are sensitive to sampling disturbance.

The measured field vane shear strength can be correlated with the preconsolidation pressure and the OCR of the clay deposit. Mayne and Mitchell (1988) suggested the following empirical relationship for estimating the preconsolidation pressure of a natural clay deposit:

$$\sigma'_c = 7.04 [c_{u(\text{FVST})}]^{0.83} \qquad (10.61)$$

where σ'_c is the preconsolidation pressure (kPa) and $c_{u(\text{FVST})}$ is the field vane shear strength (kPa). The OCR also can be correlated to $c_{u(\text{FVST})}$ as

$$\text{OCR} = \beta \, \frac{c_{u(\text{field})}}{\sigma'_v} \qquad (10.62)$$

where σ'_v is the effective overburden pressure and β is a factor defined by Hansbo (1957) as

$$\beta = \frac{222}{w\,(\%)} \qquad (10.63a)$$

by Larsson (1980) as

$$\beta = \frac{1}{0.08 + 0.0055(\text{PI})} \qquad (10.63b)$$

and by Mayne and Mitchell (1988) as

$$\beta = 22[\text{PI}\,(\%)]^{-0.48} \qquad (10.63c)$$

10.12 Borehole Shear Test

The borehole shear test is a rapid *in situ* direct shear test performed on the walls of a 76-mm-diameter borehole in soils and weak rocks. The borehole is usually vertical but may be inclined or horizontal. It is preferably made by pushing a thin-walled (Shelby) sampling tube, but it also can be augered or drilled with care to avoid any significant change in natural soil conditions. The purpose of the test is to obtain independent measurements of soil friction and cohesion using a shear head, which is an expandable probe consisting of a double-acting fluid cylinder, diametrically opposed sharply grooved shear plates, and a pulling yoke for attaching pull rods (Figure 10.26). The length of the shear plate is 64 mm for standard plates and 20 mm for smaller high-pressure plates.

In order to accomplish the test objective, the shear head is lowered into a borehole to the test depth (Figure 10.27). A constant normal force is then applied to the plates using compressed gas (usually carbon dioxide or nitrogen), causing the plates to contact the walls of the borehole. Before initiating shear failure, the soil is allowed to consolidate (i.e., to dissipate excess pore water pressures) for at least a minimum of 5 min. In saturated clays, a consolidation time of up to 15 min may be necessary to dissipate the excess pore water pressures. To initiate soil shear failure, the shear head is pulled axially along the borehole at a constant rate, and the pulling force is monitored until it reaches a peak value indicative of shear failure. Care should be taken so that the normal pressure is the same during consolidation and shear. The cycle of consolidation and shear phases is repeated at least four times by increasing the normal force to provide a set of test results for interpreting the shear strength parameters of the soil. Two types of tests may be conducted: (1) stage tests in which the normal stress is increased incrementally after each peak shear force has been recorded without changing the position of the shear head and (2) fresh shearing in which the shear head is retracted and

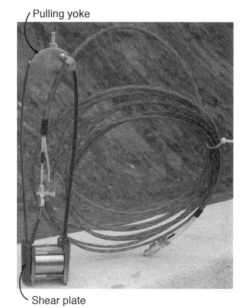

FIGURE 10.26 Shear head for use in borehole shear test (courtesy of Professor David White, Iowa State University).

removed after the peak shear force is recorded so that the shear plates may be cleaned and a new shearing surface tested. The time required for each test sequence varies with soil type and type of test and can range from 20 min to 2 h, usually averaging about 1 h (Lutenegger 1987; Lutenegger and Timian 1987). The normal stress (σ_n) and shear stress (τ_f) are calculated from the respective forces and the plate areas as

FIGURE 10.27 Borehole shear test at the site (courtesy of Professor David White, Iowa State University).

$$\sigma_n = \frac{F_n}{A} \tag{10.64}$$

and

$$\tau_f = \frac{F_s}{2A} \tag{10.65}$$

where F_n is the normal force applied on each shear plate, A is the area of a shear plate, and F_s is the maximum pulling force acting on two shear plates.

The normal stress and shear stress readings are plotted to give the Mohr-Coulomb failure envelope of the soil, as in a direct shear test. A least squares linear regression analysis is performed on these data to give the friction angle and cohesion. It should be noted that in a properly conducted test, drainage is essentially complete, and the test yields effective stress shear strength parameters in a fraction of the time required for comparable laboratory tests.

10.13 Pressuremeter Test

A pressuremeter test provides a stress-stain response of the *in situ* soil in terms of an arbitrary modulus of deformation called the *pressuremeter modulus* (E_p) for use in foundation analysis and design. The test is applicable to a wide variety of soil types, weathered rock, and low- to moderate-strength intact rock. There are two types of pressuremeter tests: the *prebored pressuremeter test* and the *self-boring pressuremeter test*. A carefully prepared borehole is required for a prebored pressuremeter test, whereas with a self-boring pressuremeter test, the borehole is advanced by a mechanical or jetting device that sits inside the hollow core of the probe.

A prebored pressuremeter test (Figure 10.28), originally developed by Menard (1956), is an *in situ* stress-strain test performed on the wall of a borehole using an inflatable cylindrical probe that is expanded by applying water pressure from a reservoir to apply a uniform radial stress to the borehole wall. To obtain realistic test results, disturbance to the borehole wall during drilling must be minimized. The test is suitable for both cohesive and cohesionless soils, but it is not used for high-pressure testing in rocks.

The test involves inserting the probe into a borehole and lowering it to the desired test depth.

FIGURE 10.28 Pressuremeter test apparatus arrangement at the test site.

The borehole diameter should be close to that of the probe to ensure adequate volume change capability. Figure 10.29 illustrates the basic components of this device. The probe consists of three parts: top and bottom guard cells and a measuring cell in the middle. The flexible wall of the probe consists of a rubber membrane fitted with an outer flexible sheath or cover which takes the shape of the borehole as pressure is applied. Typical diameter D of the pressuremeter probe varies from approximately 32 to 74 mm. The probe that is most commonly used has a diameter of 58 mm and a length of 420 mm. ASTM D4719 recommends that the diameter of the borehole lie in the range $1.03D$ to $1.2D$. When the probe is inflated by applying gas (usually nitrogen) pressure in guard cells and

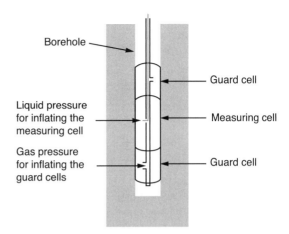

FIGURE 10.29 Basic components of pressuremeter test apparatus in a borehole.

water pressure in the measuring cell, it presses against the unlined wall of the borehole and causes volumetric deformation. The water pressure is applied under equal pressure increments or equal volume increments. The increase in volume of the measuring cell is determined from the movement of the gas-water interface in the control cylinder; readings normally are taken at 15, 30, 60, and 120 s after a pressure increment has been applied (Craig 1997). The test is terminated when yielding of the borehole wall becomes relatively large. The radial stress on the borehole wall is equal to the water pressure applied. Minimum spacing (center to center of the probe) between consecutive tests should not be less than 1.5 times the length of the inflatable part of the probe; however, the spacing commonly varies from 1 to 3 m. The testing can take place continuously and no withdrawal is required between tests.

The instrument should be calibrated before each use to compensate for pressure losses and volume losses associated with rigidity of the probe walls and expansion of the probe components, respectively. Plotting the corrected volume (V) of the measuring cell and pressure (p) results in a smooth curve called a *pressuremeter test curve*, shown in Figure 10.30. A pseudoelastic linear section occurs on this curve between pressures p_i and p_f. The pressure p_i is necessary to achieve initial contact between the cell and the borehole wall and to recompress soil disturbed or softened as a result of boring. The pressure p_f corresponds to the onset of plastic strain in the soil. Eventually a *limit pressure p_l* is approached at which continuous expansion of the borehole wall can be expected.

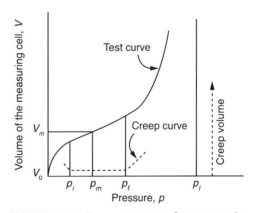

FIGURE 10.30 Pressuremeter test data presentation.

A *creep curve*, obtained by plotting the volume change between the 30- and 120-s readings against the corresponding pressure, helps in fixing the values of p_i and p_f where significant breaks occur in the shape. The datum or reference pressure for the interpretation of pressuremeter results is p_m, which is equal to the *in situ* total horizontal stress in the soil before boring. V_m is the corrected volume of the measuring cell that corresponds to pressure p_m. V_0 is the volume of the uninflated measuring cell at ground surface. At any stage during a test, the corrected volume V that corresponds to pressure p is referred to as the current volume of the measuring cell. The pressuremeter modulus E_p is calculated from the slope of the pseudoelastic portion of the curve using the expression (ASTM D4719)

$$E_p = 2(1 + \nu)V_m \left(\frac{dp}{dV} \right)_{p=p_m}$$ (10.66)

where $\left(\dfrac{dp}{dV} \right)_{p=p_m}$ is the reciprocal of the rate of change of the corrected volume V of the measuring cell with respect to applied pressure p at $p = p_m$ and ν is Poisson's ratio of the soil.

For practical purposes, the pressure where the probe volume reaches twice the original soil cavity volume is the limit pressure p_l. Usually this pressure is not obtained by direct measurements during the test due to limitations in the probe expansion or the excessively high pressure required to reach this state. However, if the test is conducted to read sufficient plastic deformation, the limit pressure can be determined by extrapolation in a $1/V$ vs. p plot. ASTM D4719 gives more details on testing procedures. The ratio E_p/p_l of the pressuremeter modulus E_p to the limit pressure p_l tends to be a constant characteristic of any given soil; typical values are shown in Table 10.9.

If a pressuremeter test is compared with a borehole shear test, it can be noted that the lateral shear stress in a borehole shear test is applied around only part of the circumference of a borehole, allowing drainage, and it is held constant for some time to allow the soil to consolidate, unlike stress application in a pressuremeter test.

The results obtained from pressuremeter tests have been correlated with various soil parameters in the past. Kulhawy and Mayne (1990) suggested a correlation between the preconsolidation pressure σ'_c and the limit pressure p_l as:

$$\sigma'_c = 0.45 p_l$$ (10.67)

Baguelin et al. (1978) proposed that

$$c_u = \frac{(p_l - p_i)}{N_p}$$ (10.68a)

TABLE 10.9 Typical Prebored Pressuremeter Test Values

Type of Soil	Limit Pressure, p_l (kPa)	E_p/p_l
Soft clay	50–300	10
Firm clay	300–800	10
Stiff clay	600–2500	15
Loose silty sand	100–500	5
Silt	200–1500	8
Sand and gravel	1200–5000	7
Till	1000–5000	8
Old fill	400–1000	12
Recent fill	50–300	12

After Canadian Geotechnical Society (2006).

where c_u is the undrained shear strength of a clay and

$$N_p = 1 + \ln\left(\frac{E_p}{3c_u}\right) \tag{10.68b}$$

Typical values of N_p vary between 5 and 12, with an average of about 8.5.

Ohya et al. (1982) correlated E_p with standard penetration resistance N_{60} for clay as

$$E_p \ (\text{kN/m}^2) = 1930(N_{60}^{0.63}) \tag{10.69}$$

and for sand as

$$E_p \ (\text{kN/m}^2) = 908(N_{60}^{0.66}) \tag{10.70}$$

10.14 Flat Dilatometer Test

The flat dilatometer test (DMT), developed in Italy in 1980, is currently used to obtain information about a soil's *in situ* stratigraphy, stress, strength, compressibility, and pore water pressure for use in predicting settlements of shallow foundations, compaction control, detecting slip surfaces in clay slopes, predicting the behavior of laterally loaded piles, evaluating sand liquefiability, estimating consolidation/flow coefficients, and selecting soil parameters for finite element analyses. The test procedure and the original correlations were described by Marchetti (1980). ASTM D6635 and Marchetti et al. (2001) provide the details of the test apparatus, testing procedure, and empirical correlations based on field experience.

The flat dilatometer is a stainless steel blade that has a thin, flat, expandable, circular steel membrane mounted flush on one face with a retaining ring (Figure 10.31). The nominal dimensions of the blade are 95 mm wide and 15 mm thick. The blade has a cutting edge to penetrate the soil. The apex angle of the edge is 24–32°. The lower tapered section of the tip is 50 mm long. The blade can safely withstand up to 250 kN of pushing thrust. The circular steel membrane is 60 mm in diameter, with a thickness of 0.20 mm (0.25-mm-thick membranes are sometimes used in soils that may cut the membrane). When at rest, the membrane is flush with the surrounding flat surface of the blade. The blade is connected, by an electropneumatic tube running through the insertion rods, to a control unit (Figure 10.32) on the surface. The

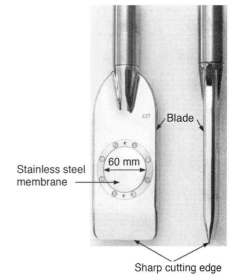

FIGURE 10.31 The flat dilatometer: front and side views (adapted from Marchetti et al. 2001).

control unit is equipped with pressure gauges, an audiovisual signal, a calibration syringe, a valve for regulating gas (generally dry nitrogen) flow from a gas pressure tank, and vent valves.

A suitable load cell, just above the blade or at the top of the rods, is required to measure the thrust applied during blade penetration. As a result of an applied internal gas pressure, the membrane expands into the soil in an approximately spherical shape along an axis perpendicular to the plane of the blade. The blade is advanced into the ground using common field equipment (e.g., push rigs normally used for the CPT or drill rigs). The blade can also be driven using an SPT hammer and rods, but quasi-static push at a rate of 10–30 mm/s is by far preferable. Pushing

FIGURE 10.32 Control unit of the DMT (after Marchetti et al. 2001).

the blade with a 20-t penetrometer truck is most effective up to 100 m of profile per day (ASTM D6635; Marchetti 1980; Marchetti et al. 2001; Totani et al. 2001). A general layout of the DMT is shown in Figure 10.33.

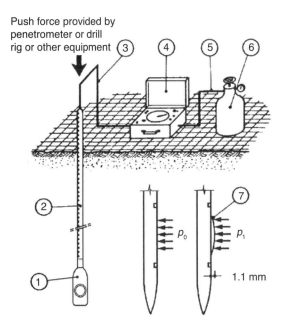

FIGURE 10.33 General layout of the DMT: 1 = dilatometer blade, 2 = push rods (e.g., CPT), 3 = pneumatic-electric cable, 4 = control box, 5 = pneumatic cable, 6 = gas tank, and 7 = expansion of the membrane (after Marchetti et al. 2001).

The DMT starts by inserting the dilatometer, with the membrane facing the horizontal direction, into the ground to a desired depth. Soon after penetration, the operator inflates the membrane using gas pressure and takes two readings within a minute: the A pressure, required to just begin to move to the membrane ("liftoff"), and the B pressure, required to move the center of the membrane 1.1 mm against the soil. A third reading C ("closing pressure") also can optionally be taken by slowly deflating the membrane soon after B is reached. The probe is then advanced to the next depth, which typically is 0.2 m further down, and another set of readings is taken. The test is progressed to the desired depth. The membrane expansion is not a load-controlled test (apply the load and observe settlement) but rather is a displacement-controlled test (fix the displacement and measure the required pressure). Pressure readings A, B, and C are corrected by the membrane stiffness pressures ΔA and ΔB determined by calibration (with a calibration syringe before and after each DMT sounding) to take into account the membrane stiffness and are converted into corrected A pressure p_0, corrected B pressure p_1, and corrected C pressure p_2, respectively (Totani et al. 2001). For corrected A pressure:

$$p_0 = 1.05(A - Z_M + \Delta A) - 0.05(B - Z_M - \Delta B) \tag{10.71}$$

For corrected B pressure:

$$p_1 = B - Z_M - \Delta B \tag{10.72}$$

For corrected C pressure:

$$p_2 = C - Z_M + \Delta A \tag{10.73}$$

Z_M is the gauge pressure deviation from zero when vented to atmospheric pressure. If ΔA and ΔB are measured with the same gauge used for current readings A and B, Z_M is taken as zero. The test data are reduced to obtain the following three DMT indices:

1. Material index, I_D:

$$I_D = \frac{p_1 - p_0}{p_0 - u_0} \tag{10.74}$$

where u_0 is the preinsertion pore water pressure acting at the center of the membrane, often assumed as hydrostatic below the water table. The I_D value depends on the rigidity, pore water pressure generation, and permeability of the soil, thus providing an indication of soil type.

2. Horizontal stress index, K_D:

$$K_D = \frac{p_0 - u_0}{\sigma'_{v0}} \tag{10.75}$$

where σ'_{v0} is the preinsertion effective overburden pressure at the center of the membrane, generally calculated from unit weights estimated using the DMT results. K_D is related to the lateral stress ratio K_0 and gives information about the stress history of the soil mass and thus is used for predicting the OCR.

3. Dilatometer modulus, E_D:

$$E_D = 34.7(p_1 - p_0) \tag{10.76}$$

The dilatometer modulus E_D is not Young's modulus E, and it should be used only after combining it with K_D. It is the basis for evaluating the *in situ* drained modulus and compressibility behavior of the soil mass.

DMT test data have been used to develop a series of empirical correlations, some of which are described below (ASTM D6635; Marchetti et al. 2001; Totani et al. 2001):

1. Soil type: for clay $I_D < 0.6$ and for sand $I_D > 1.8$.
2. Coefficient of earth pressure at rest, K_0:

$$K_0 = \left(\frac{K_D}{1.5} \right)^{0.47} - 0.6 \qquad (10.77)$$

for clay with $I_D < 1.2$ and for sand and silt with $I_D \geq 1.2$.
3. DMT tip bearing, q_D:

$$q_c \approx \frac{q_D}{(1.1 + 0.1)q_c} \qquad (10.78)$$

where q_c is the cone resistance determined from a static cone penetrometer or piezocone. It should be noted that the DMT tip bearing is the axial thrust at the end of the dilatometer blade divided by the projected cross-sectional area of the blade normal to the penetration. This tip bearing may be used to evaluate stratigraphy.
4. OCR:

$$OCR = (0.5K_D)^{1.56} \qquad (10.79)$$

for clay with $I_D < 1.2$ and for sand and silt with $I_D \geq 1.2$.
5. Undrained shear strength, c_u:

$$c_u = 0.22\sigma'_{v0}(0.5K_D)^{1.25} \qquad (10.80)$$

for clay with $I_D \leq 0.6$.
6. Drained plane strain angle of internal friction, ϕ_{ps}:

$$\phi_{ps} = 28° + 14.6° \log K_D - 2.1°(\log K_D)^2 \qquad (10.81)$$

for $I_D > 1.8$.
7. Vertical drained constrained modulus of soil deformation, M:

$$M = R_M E_D \qquad (10.82a)$$

For $I_D \leq 0.6$:

$$R_M = 0.14 + 2.36 \log K_D \qquad (10.82b)$$

For $0.6 < I_D < 3.0$:

$$R_M = R_{M0} + (2.5 - R_{M0}) \log K_D \qquad (10.82c)$$

with

$$R_{M0} = 0.14 + 0.15(I_D - 0.6) \qquad (10.82d)$$

For $I_D \geq 3.0$:

$$R_M = 0.5 + 2 \log K_D \qquad (10.82e)$$

For $K_D \geq 10$:

$$R_M = 0.32 + 2.18 \log K_D \qquad (10.82f)$$

R_M must be limited to 0.85.

8. Horizontal coefficient of permeability, k_h:

$$k_h = \frac{c_h \gamma_w}{M_h} \qquad (10.83)$$

where $M_h \approx K_0 M$ and γ_w is the unit weight of water.

9. Equilibrium pore pressure in freely draining soils, u_0:

$$u_0 \approx p_2 \approx C - Z_M + \Delta A \qquad (10.84)$$

An example of DMT results is shown in Figure 10.34. Comparative studies have indicated that DMT results (in particular K_D) are noticeably reactive to factors that are scarcely felt by other *in situ* tests (CPT, SPT, etc.), especially in sands, such as stress state/history, aging, cementation, and structure. An additional feature of the DMT is that it provides two independent parameters, whereas most of the penetration tests provide just one "primary" parameter, as penetration resistance for interpretation. This test method applies best to sands, silts, clays, and organic soils that can be penetrated with the dilatometer blade, preferably using the

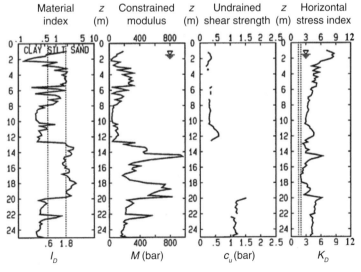

FIGURE 10.34 Graphical presentation of DMT results (1 bar = 100 kPa) (adapted from Marchetti et al. 2001).

pseudostatic push. Test results for soils that contain primarily gravel-sized particles and larger may not be useful at the present understanding of the test. When properly performed at suitable sites, this test provides a rapid means of characterizing subsurface conditions. Since this method tests the soil *in situ* and soil samples are not obtained, soil samples from parallel borings may be obtained for correlation purposes, but prior information or experience may preclude the need for borings.

10.15 K_0 Stepped Blade Test

The K_0 stepped blade test was presented by Handy et al. (1982) as a technique for rapidly and accurately determining the *in situ* lateral stress in soils within a range of 10 kPa or less. The K_0 *stepped blade* has four steps spaced 100 mm apart, incrementally increasing in thickness from 3 to 7.5 mm (Figure 10.35). Each step carries a pneumatic pressure cell that is flush with the surface and has a membrane cover that comes in contact with the soil when penetrated into it. The cell has two chambers that are inflated with slightly different gas pressures so that when the soil pressure is reached, the membrane is lifted and allows a cross-leak that is measured by a differential pressure gauge, at which time the gas pressure is also read. As the gas pressure equals the soil pressure, no additional calibration is needed.

The test is conducted in a borehole at the desired depth. The first blade step is pushed into undisturbed soil in the bottom of the hole, and the soil pressure is measured on that step. The second blade step then is pushed, and pressures are measured on step 2 and then step 1. The third step is then pushed, and pressures on steps 3, 2, and 1 are measured in that order. The fourth step is pushed, and the process is repeated. A fifth step of the same thickness as step 4 but without a pressure cell is pushed, and all cells again are read to provide additional data. The sequence of reading is intended to allow approximately the same drainage, that is, consolidation time for each step penetration. If consolidation is prevented by poor drainage, all steps should measure a limit pressure, which is a measure of soil strength, but does not differentiate between internal friction and cohesion. The total number of measurements at five different

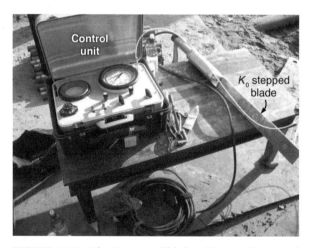

FIGURE 10.35 The K_0 stepped blade with control unit and other accessories (courtesy of Professor David White, Iowa State University).

penetration depths is $4 + 4 + 3 + 2 + 1 = 14$ and usually requires about 5 to 10 min. The instrument is then removed from the hole, the hole is advanced to the next test depth, and the test sequence is repeated.

When the blade is pushed stepwise into a soil mass, it introduces different levels of disturbance and allows corresponding soil pressure measurements, which are extrapolated to obtain a hypothetical pressure on a zero-thickness blade (Figure 10.36). A linear relationship is needed for an extrapolation to be most effective, and such a relationship is found between blade step thickness and the logarithm of pressure. A justification for this observation may be found in the linear void ratio–logarithm of pressure relationships commonly found from consolidation tests (Handy 2008). The implication is that when such a relationship exists, soil next to the blade must be consolidating such that its modulus is increasing linearly with pressure. Thus, a consolidating soil response (not elastic or shearing response) is required for a successful extrapolation to obtain *in situ* lateral stress; therefore, the test data are accepted only if there is an increase in pressure with each increase in step thickness. The K_0 stepped blade measures total stress; therefore, to know the effective stress, the level of the groundwater table should be determined. The role of pore water pressures generated during testing also needs to be measured in order to use the stepped blade as a practical tool for geotechnical investigations.

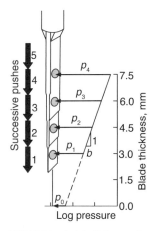

FIGURE 10.36 Schematic diagram showing the extrapolation principle of the K_0 stepped blade (adapted from Handy 2008).

Based on a series of controlled laboratory tests on compacted soil, Handy et al. (1982) suggested that the initial stress condition could be described by a simple expression as

$$p_0 = a p_t e^{-bt} \tag{10.85}$$

where p_0 is the *in situ* stress, p_t is the pressure on a blade of thickness t, and a and b are regression coefficients. The value of a is tentatively assumed to be 1.0. The value of b is most likely an indicator of the generation of excess pore water pressure in the soil and not drained soil compressibility, as originally thought. Equation 10.85 indicates that a plot of blade thickness vs. logarithm of measured pressure would be linear with slope b and intercept log p_0, as shown in Figure 10.36. Handy (2008) suggested that to obtain effective stress, one should subtract the static pore water pressure calculated from the elevation in relation to a groundwater table. Effective stresses are used to calculate K_0 as

$$K_0 = \frac{\sigma'_h}{\gamma'} \tag{10.86}$$

where σ'_h is the horizontal effective stress at depth z below the ground level, and γ' is the submerged unit weight of soil at that depth. The K_0 values obtained from this test in soft to medium-consistency clays compare reasonably well with the flat dilatometer and laboratory odometer correlations of K_0; however in overconsolidated clays, the K_0 stepped blade results are much higher than those given by other techniques (Lutenegger and Timian 1986).

10.16 *In Situ* California Bearing Ratio Test

The California bearing ratio (CBR) test measures the penetration resistance of an approximately 50-mm-diameter and at least 100-mm-long metal piston pushed into a soil mass at a rate of penetration of about 1 mm/min. The test value is reported as the ratio of the penetration load of the soil to that of a standard material, expressed in percent. This ratio is popularly known as the CBR and is usually calculated for 2.5- and 5-mm penetration of the piston. The standard material generally is crushed rock which has a penetration load of 13.43 and 20.14 kN, corresponding to the 2.5- and 5-mm penetration values, respectively. The CBR at 2.5-mm penetration is generally greater than that at 5-mm penetration, in which case the former is taken as the CBR for the design purpose. When the CBR at 5-mm penetration is greater than that at 2.5-mm penetration, the test should be repeated. If the check test gives a similar result, the CBR for the design purpose is then taken as the CBR determined at 5-mm penetration. In brief, it can be stated that the CBR of a soil mass is the larger of the two CBR values determined at 2.5- and 5-mm penetrations.

The CBR test is an ad hoc penetration test, and the test values generally are used for evaluation of the strength of subgrade, subbase, and base course materials, including recycled materials, for use in road and airfield pavements. The test gives empirical strength values which may not be directly related to fundamental properties governing the strength of soil. The CBR values obtained from the test are used in conjunction with empirical curves based on experience for the design of flexible pavements. The test may be performed in the laboratory on undisturbed samples or samples recompacted to field density. However, for more realistic values for design purposes, the test is performed in field conditions of the soil.

The test area for the *in situ* CBR test is leveled by removing from the surface loose and dried materials which are not representative of the soil to be tested. A loaded truck equipped with a metal reaction beam fitted at the rear end, as shown in Figure 10.37a, is located so that the center of the beam rests directly over the soil surface to be tested. The reaction beam may be

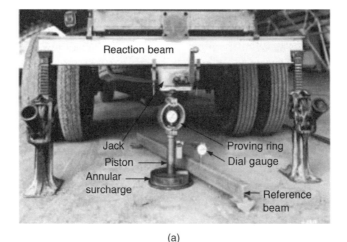

(a)

FIGURE 10.37 Setup for CBR *in situ* test: (a) loaded truck with test arrangement (adapted from ASTM D4429) and (b) schematic diagram (adapted from AS 1289.6.1.3).

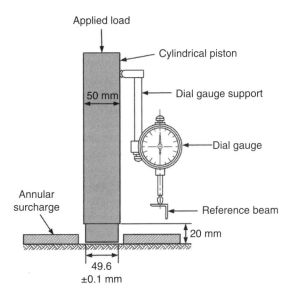

FIGURE 10.37(b)

placed approximately 0.5 m above the ground to provide a reaction of about 50 kN. A mechanical screw jack is used to apply the load, and a proving ring is used to measure the applied load. The piston and the dial gauge are positioned suitably, as shown in Figure 10.37. An annular metal surcharge plate weighing approximately 4.5 kg is placed beneath the penetration piston so that when the piston is lowered, it will pass the central hole in the surcharge plate. The load is applied by the mechanical screw jack, and the deflection of the proving ring usually is recorded at each 0.5-mm penetration increment, to a final penetration of 5 mm. At the completion of the test, a sample from the point of penetration is obtained to determine its water content. Unit weight determination also should be made at a location about 100–150 mm away from the point of penetration. Typical load vs. penetration curves are shown in Figure 10.38. In some instances, the stress-penetration curve may be concave upward initially because of surface irregularities or other causes, in which case the zero point should be adjusted and all the measurements should then be carried out with respect to the corrected zero point, as shown in Figure 10.38. A rough estimate of the *in situ* CBR value can be made by penetrating the soil mass with the thumb or indenting the soil with the thumbnail, as described in Table 10.10. Based on the *in situ* CBR value obtained from the test, the soil subgrade can be described as very poor to good (see Table 10.11).

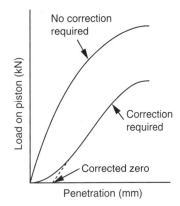

FIGURE 10.38 Typical load-penetration curves.

In the past, attempts have been made to correlate CBR with basic soil properties. The suggested CBR correlations can be used in design computations in the absence of more reliable data. Wilson and Williams (1950) developed a relation between CBR and load-bearing capacity as

TABLE 10.10 Estimating CBR Value by Identification Procedure

Approximate CBR Value	Identification Procedure
<2	Easily penetrated with thumb
2–3	Moderate effort to penetrate with thumb
3–6	Indented by thumb
6–16	Indented by thumbnail
>16	Difficult to indent with thumbnail

$$\text{CBR} = \frac{q_u \ (\text{kPa})}{68.95} \approx \frac{q_u \ (\text{kPa})}{70} \quad (10.87)$$

where q_u is the ultimate bearing capacity. This relation is based on the fact that the stress applied to the piston to give a standard penetration of 2.5 mm often is close to the ultimate bearing capacity. For soils of low stiffness, the ultimate bearing capacity might not reach even a 5-mm penetration because the shear strength of the soil is mobilized only in the

TABLE 10.11 Description of the Soil Subgrade Quality

CBR Value	Description
<3	Very poor
3–4.5	Poor
4.5–6.5	Fair
6.5–10	Medium
>10	Good

zone local to the edge of the piston, as reported by Hight and Stevens (1982). Therefore, the CBR values predicted from Equation 10.87 using the ultimate bearing capacity may be overestimated.

Nascimento and Simoes (1957) related CBR to the modulus of subgrade reaction by analyzing test results. It was concluded that the modulus of subgrade reaction (in kN/m^3) is $\frac{1}{8} \times 10^4$ to $\frac{1}{4} \times 10^4 \times \text{CBR}$ for soft materials and $\frac{1}{8} \times 10^4$ to $\frac{1}{3} \times 10^4 \times \text{CBR}$ for hard materials.

Based on experimental results over a wide range of saturated clays, Black (1962) suggested the following relation between saturated and unsaturated CBR values:

$$\text{Unsaturated CBR} = (\text{Degree of saturation})^{2.3} \times \text{Saturated CBR} \quad (10.88)$$

Heukelon and Klomp (1962) summarized the work on widespread field vibration testing of soil and granular materials. They showed that the general trend is represented by the relationship

$$E_r \ (\text{MPa}) = 10.34 \times \text{CBR} \quad (10.89)$$

where E_r is the resilient Young's modulus and CBR is expressed as a percentage. The scatter of results is such that individual results could differ from this by a factor of two.

Nielson et al. (1969) suggested a correlation between Young's modulus of elasticity of soil (E) and CBR (in percent) by using the theory of elasticity. From their study, they proposed the relation

$$E \ (\text{kPa}) = \frac{0.75\pi a (1 - v)^2}{(1 - 2v)} \times 689.5 \text{CBR}$$

$$\approx \frac{0.75\pi a (1 - v)^2}{(1 - 2v)} \times 700 \text{CBR} \quad (10.90)$$

where a is the radius of the plunger and ν is Poisson's ratio of the soil. Substituting $\nu = 0.25$ and $a = 0.975$ in., Equation 10.90 gives:

$$E \text{ (kPa)} = 1792.7\text{CBR} \approx 1800\text{CBR} \qquad (10.91)$$

From a laboratory study on fine sand, silty sand, sand-gravel with fines, coarse sand, and a well-graded soil mixture, the following relation was proposed (Nielson et al. 1969):

$$E \text{ (kPa)} = 2151.24\text{CBR} \approx 2150\text{CBR} \qquad (10.92)$$

It was pointed out that Equation 10.92 does not hold well for granular soils with large amounts of clay binders, which produce extremely high CBR values.

Kate (1980) made an attempt to compare the results of CBR obtained in the field and laboratory by considering 20 sites distributed in an area of about 1500 km^2 with mostly silty clays. It was shown that the laboratory CBR values were always higher than the field CBR, thus overestimating the strength of the soil, possibly due to the confinement of soil in the mold. A relation between the laboratory and field CBR values was suggested as:

$$\text{CBR}_{(lab)} = 1.35 \times \text{CBR}_{(field)} \qquad (10.93)$$

Hight and Stevens (1982) analyzed a CBR test on saturated clays using finite element techniques and showed that the CBR did not correlate consistently with either the strength or the static stiffness of clays. In stiff intact clays, the CBR reflects only undrained strength, whereas when the stiffness of clay is low, the CBR depends on both stiffness and undrained strength. They pointed out that the correlation between stiffness and CBR should be treated with caution, and it was recommended that the CBR value for saturated or nearly saturated clays of low permeability should be examined in conjunction with the full load-penetration curve obtained from the test.

Lister and Powell (1987) suggested the following correlations between the resilient Young's modulus of elasticity and the CBR:

$$\begin{aligned} E_r \text{ (MPa)} &= 10 \times \text{CBR} && \text{for CBR} \leq 5 \\ &= 17.6 \times \text{CBR}^{0.64} && \text{for CBR} > 5 \end{aligned} \qquad (10.94)$$

The Indian Roads Congress recommends the correlations in Equation 10.94 for the design of flexible pavements (IRC: 37).

10.17 Test Methods for *In Situ* Unit Weight Determination

The term *in situ unit weight* refers to the volumetric weight, usually expressed as kilonewtons per cubic meter (kN/m^3), of a soil in its natural or compacted condition. The *in situ* unit weight of soil is required for stability analysis of geotechnical structures, such as foundations, retaining walls, embankments, slopes, etc. Generally, for coarse-grained soils, the greater the unit weight, the higher the shear strength and stiffness, and hence there is less of a tendency for settlement or instability. For compacted earth fills, it is standard practice to determine the *in situ* unit weight of the soil after it is compacted to establish whether the compaction effort

FIGURE 10.39 Determination of *in situ* unit weight by sand cone method.

has been adequate as part of a construction quality control program. *In situ* unit weight of soil is also required for estimating the cut-and-fill quantities in earthworks.

Of the direct test methods employed for *in situ* unit weight determination, the *sand replacement method* (*aka sand cone method*, ASTM D1556) and *rubber balloon method* (ASTM D2167) have a long history of use, especially on soils that do not deform under the pressures imposed during the test. These methods are limited to soils in unsaturated condition, and they may require special care in soils that consist of unbonded granular materials in which a small hole with stable sides cannot be obtained. In these methods, a test hole is made (Figure 10.39), and the total weight (W) of the soil removed from the hole and its moisture content (w) are determined. In the sand replacement method, the volume (V) of the hole is determined by filling it with a standard sand of known density and measuring the mass of sand required to fill the hole. In the rubber balloon method, the volume of the hole is determined by expanding a rubber balloon with water and measuring the volume of water required to fill the hole. In the sand replacement method, the sand size should be large enough to avoid entering the voids of the soil. If the soil weight, water content, and volume of the hole are known, the *in situ* total unit weight and *in situ* dry unit weight can be calculated as

$$\gamma_t = \frac{W}{V} \tag{10.95}$$

and

$$\gamma_d = \frac{\gamma_t}{1 + \frac{w\,(\%)}{100}} = \frac{W}{V\left[1 + \frac{w\,(\%)}{100}\right]} \tag{10.96}$$

Another direct method, called the *core cutter method* (*aka drive cylinder method*, ASTM D2937), involves driving a steel cylinder with a hardened cutting edge into the ground using a protective dolly and a specially designed steel rammer (Figure 10.40). The cutter is then dug out and the soil trimmed off flush at each end. Since the volume of the cutter is known and the contained weight of the soil can be found by weighing, the total unit weight can easily be determined. At the same time, small specimens of soil are taken from either end, from which the water content is determined. In place of a steel cylinder, a thin-walled steel (Shelby) tube can be driven using a rammer or pushed into the soil by use of a drilling machine. The sample is then extruded in the laboratory to determine the unit weight and water content as well as strength. This method is suitable for natural and compacted cohesive soils which do not contain significant amounts of particles coarser than 4.75 mm. This method may not be applicable for soft, highly plastic saturated or other soils which are easily deformed, compressed during sampling, or which may not be retained in the core cutter. Because of sample disturbance and compaction during driving, any method that involves driving is less accurate than methods which do not depend on driving.

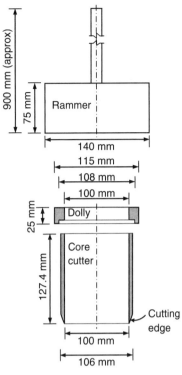

FIGURE 10.40 Core cutter, dolly, and rammer used in the core cutter method of determining *in situ* unit weight of soil (adapted from IS 2720, Part XXIX).

For cohesive and stabilized soils and where an irregular-shaped intact lump of soil has been obtained, the unit weight can be determined by the *water immersion method*, in which three weights are taken: the weight of the lump sample in air W, the weight of the lump sample coated with paraffin wax in air $W_{soil+wax}$, and the submerged weight of the lump sample coated with paraffin wax in water $W_{(soil+wax)sub}$. The unit weight is calculated as

$$\gamma_t = \frac{W}{\dfrac{W_{soil+wax} - W_{(soil+wax)sub}}{\gamma_w} - \dfrac{W_{wax}}{\gamma_{wax}}} \tag{10.97}$$

where γ_w is the unit weight of water, W_{wax} is the weight of paraffin wax in air, and γ_{wax} is the unit weight of paraffin wax.

The unit weight also can be determined by the *water displacement method*, which is similar to the water immersion method except that the volume of the waxed sample is found by lowering it into a container of water with a siphon outlet and measuring the volume of water displaced.

If the *in situ* soil consists of clay-sized to boulder-sized particles, the methods described above cannot be used to precisely determine the unit weight of soil. Figure 10.41 shows such

FIGURE 10.41 Soil consisting of clay-sized to boulder-sized particles at the raft foundation level for 500-m-long aqueduct-cum-bridge at chainage 46.615 km of the Bansagar feeder channel, Sidhi District, Madhya Pradesh, India (after Shukla 2006b).

a soil deposit found at the raft foundation level of a 500-m-long aqueduct-cum-bridge at chainage 46.615 km of the Bansagar feeder channel, Sidhi District, Madhya Pradesh, India (Shukla 2006b). The major steps of a simple practical method (which can be called the *water replacement method*), adopted by Shukla, also are shown in this figure. A test pit, approximately 1 m × 1 m in plan, is hand excavated in the soil to be tested, and all the soil particles from the pit are saved in a container to precisely determine the weight of the excavated soil. The volume of the pit is measured by covering it with a flexible plastic sheet and then filling it with water. The volume of the water required to fill the pit gives the volume of the *in situ* soil excavated from the pit. Equation 10.95 is used to calculate the *in situ* total unit weight of the soil. ASTM D4914 and ASTM D5030 describe the sand and water replacement methods, respectively, for determining the *in situ* unit weight of soil that contains large particles. These methods also can be found in AS 1289.5.3.5 and AS 1289.5.3.2, respectively.

A rapid, nondestructive, indirect method for *in situ* unit weight determination involves the use of a nuclear density meter (Figure 10.42). Both the total unit weight and water content of the *in situ* soil can be measured using controlled radiation techniques. This apparatus generally consists of a small shielded radiation source and a detector. The apparatus operates either at the ground surface (ASTM D6938) or in drilled holes (ASTM D5195). The meter allows gamma rays (photons) to be emitted into the soil. These photons collide with electrons in the soil, where some are scattered and some are absorbed. The quantity of photons that reach the detector relates to the unit weight. To determine water content, a neutron-emitting material

FIGURE 10.42 Nuclear density meter.

and a detector are used. The nuclear density meter requires daily calibration, which can be done by testing in concrete of known density or other materials such as limestone, granite, or aluminum. Although operation of a nuclear density meter requires trained and certified personnel, a significant advantage of the nuclear method compared to the direct methods of determining *in situ* unit weight is the rapid speed with which results are obtained. The nondestructive nature of the test method also allows repetitive measurements to be made at a single test location for statistical analysis and monitoring changes over time. Since this method utilizes radioactive materials, its use may be hazardous to the health of the operators unless proper precautions are taken.

10.18 Site Investigation Work and Report

Site investigation work is performed by an **investigator** on the basis of a client's written proposal, which includes a brief description of the job and location, structural loads and other requirements provided by the structural engineer and/or architect, and the scope of work. The investigator submits a geotechnical proposal that includes a tentative boring plan, testing program, and cost estimate. The geotechnical proposal should be carefully written because, if accepted by the client, it will be part of a legally binding contract. Any unusual conditions that may be anticipated based on the investigator's experience should be noted (Handy and Spangler 2007).

Borings represent a considerable part of the cost of a site investigation. The total cost of an investigation project may be calculated simply on the basis of the number of boreholes and their depths. A boring plan is therefore required and should include locations, depths, and tests that are anticipated to be performed in the field and in the laboratory. A plot of the relative cost per test vs. relative test accuracy, as shown in Figure 10.43, can be useful in estimating the total cost of an investigation project.

After a contract is negotiated, the investigation should be performed in a timely manner. If unusual or adverse site conditions are encountered, this information should be conveyed immediately to the client, the structural engineer, and/or the architect, so that they can make an appropriate change in the project plan or location.

Data from site investigations are used for many different purposes during the design period, during construction, and often after completion of a project. Investigation records are therefore kept in a systematic manner for each project as its *site investigation report*. The report should be clear, complete, and accurate.

A geotechnical site investigation consists of several test pits and boreholes. The detailed information obtained from each borehole is presented in graphical form in a *boring log*. The

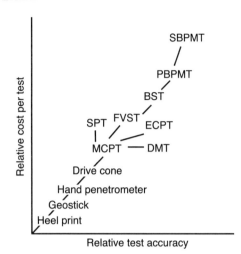

FIGURE 10.43 Cost vs. accuracy of common *in situ* test methods (adapted from Orchant et al. 1986). BST = borehole shear test, DMT = flat dilatometer test, ECPT = electric cone penetration test, FVST = field vane shear test, MCPT = mechanical cone penetration test, PBPMT = prebored pressuremeter test, SBPMT = self-boring pressuremeter test, and SPT = standard penetration test.

data collected from a test pit or exposed cut also can be presented in a similar way. The boring log generally should include the following details (Das 2007): name and address of the drilling company; driller's name; job description and number; number, type, and location of borings; date of boring; subsurface stratification; elevation of the water table and date observed; use of casing and mud losses; standard penetration resistance and depth of the SPT; number, type, and depth of soil samples collected; and in the case of rock coring, type of core barrel used and the actual length of coring, core recovery, and RQD for each run. Figure 10.44 shows a typical boring log. The log also may include certain laboratory and *in situ* test data such as unit weight, water content, shear strength parameters (angle of internal friction and cohesion), etc.

A site investigation report generally includes the following (Das 2007; AS 1726): the objectives of the investigation; a description of the proposed work; a description of the site location, including any structures nearby, drainage conditions, the nature of vegetation on the site and surrounding it, and any other features unique to the site; a description of the geological setting of the site; a description of the methods of investigation and testing used; the location and reduced levels of all boreholes and test sites; details of the number of borings, depths of borings, types of borings, etc.; a description of the water table conditions; recommendations regarding the foundation, including the type of foundation recommended, the allowable bearing pressure, and any special construction procedure that may be needed (alternative foundation design procedures also should be discussed in this portion of the report); conclusions and limitations of the investigation; and a disclaimer (i.e., a legal clause) related to the investigator's role and the contents of the report submitted to the client, as necessary. The following graphical presentations should be attached to the report: a site location map, a plan

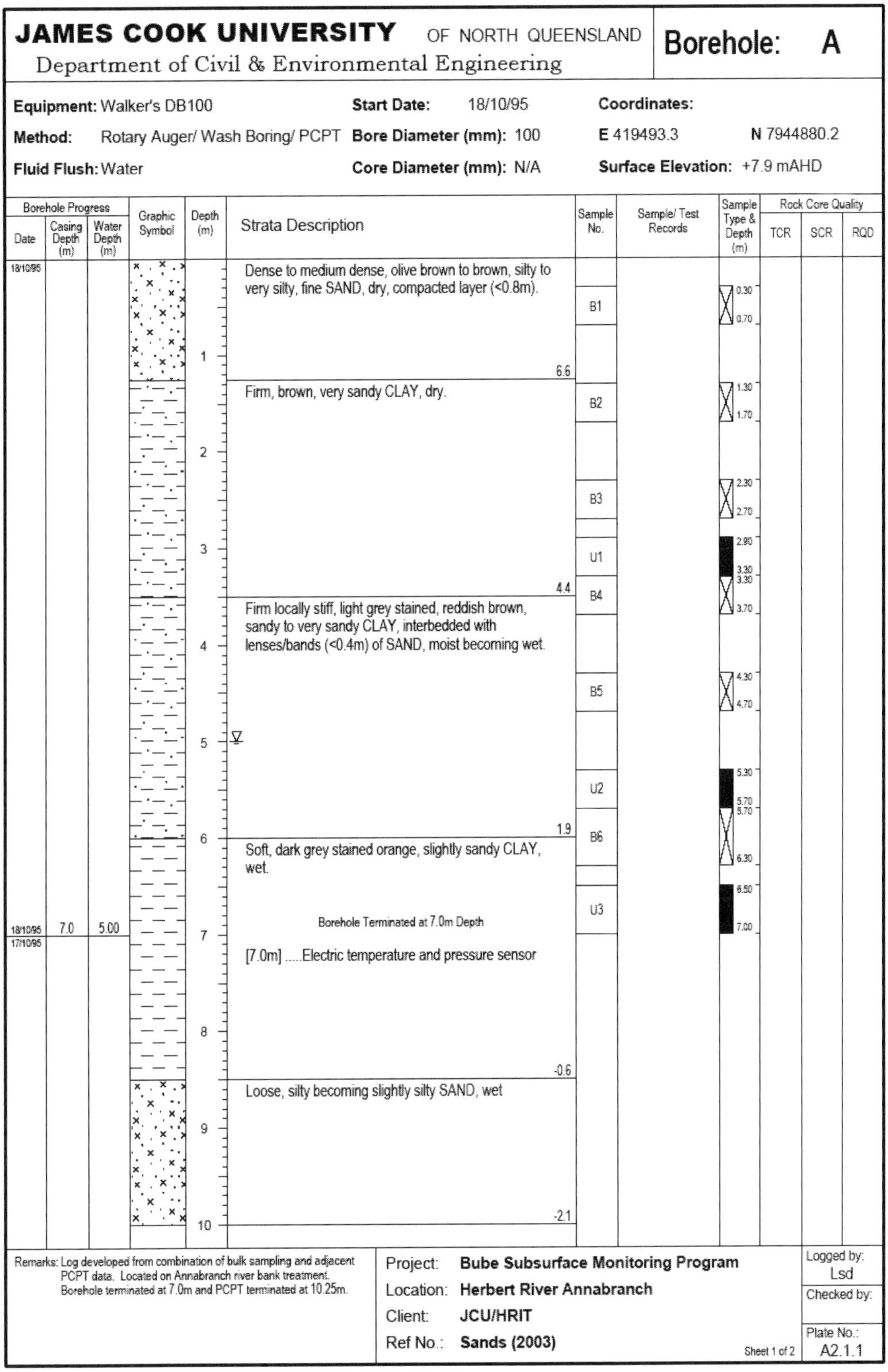

(a)

FIGURE 10.44 (a) Boring log and (b) symbols commonly used in boring log (courtesy of Leonard Sands).

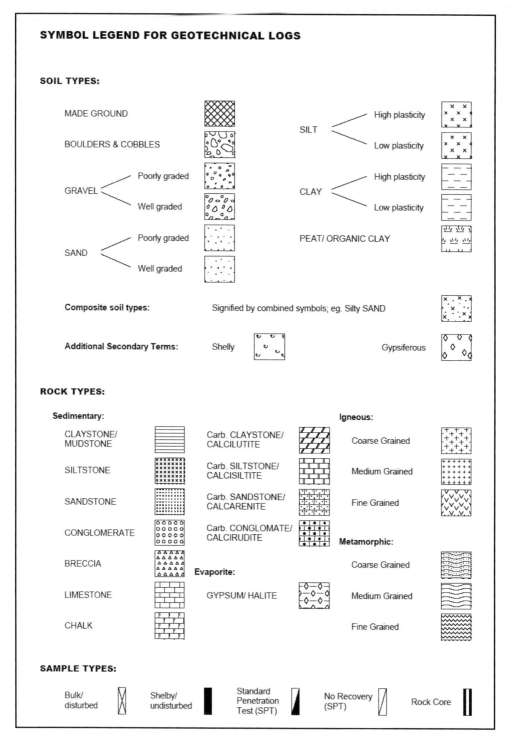

FIGURE 10.44(b)

view of the location of borings with respect to the proposed structure and those nearby, boring logs, laboratory test results, and other special graphical results.

The results of the investigation are interpreted in terms of the actual findings, and recommendations for design parameters should be made by experienced geotechnical engineers who are familiar with the purpose, conditions, and requirements of the site investigation study. Interpolation between investigated locations should be made on the basis of available geologic knowledge of the area. The use of geophysical techniques is a valuable aid in such interpolation. Geophysical data should be identified separately from laboratory test data or *in situ* test data. The site investigation report should contain recommendations related to necessary precautions in the design, construction, or renovation of the structure under consideration. It should be noted that recommendations or conclusions are derived from consideration of relevant available facts as well as interpretation, analysis, and individual judgment. Since the process involves interpretation and exercising judgment, opinions of professionals may differ, although substantial agreement is expected. It is important that the final report be thoroughly checked, preferably by an associate of the investigator, to minimize the possibility of misinterpretations, misstatements, or errors. The report can include an executive summary following the title page, which should include the title, date, and the client's and investigator's names and addresses.

10.19 Soil Variability

Subsurface conditions are positively defined only at the individual test pit, borehole, or open cut examined. Conditions between observation points may be significantly different from those encountered in the investigation. Unlike other civil engineering materials (e.g., concrete, steel, etc.), soils are nonhomogeneous, three-phase particulate materials, and the theories often are oversimplified. With limited site investigation data available, it becomes even more difficult to arrive at realistic deterministic solutions with confidence. A probabilistic approach to geotechnical problems is becoming increasingly popular, with risk analysis and reliability studies quite common in traditional geotechnical and mining engineering applications.

With very limited geotechnical data coming from the laboratory and *in situ* tests for a project, it is not possible to obtain realistic estimates of the standard deviation of the soil parameters. Typical values of the *coefficient of variation* (COV) reported in the literature can be used as the basis for estimating the standard deviation of the soil parameters. COV is calculated as:

$$\text{COV} = \frac{\text{standard deviation}}{\text{mean}} \times 100\% \qquad (10.98)$$

Harr (1987), Lee et al. (1983), and Baecher and Christian (2003) collected test data from various sources and presented the COV values. These are summarized in Table 10.12.

10.20 Field Instrumentation

In view of the limitations of site investigation and the simplified design steps, the use of instruments to monitor the performance of earth and earth-supported structures is increasingly becoming the way of life in large geotechnical projects. Geotechnical instrumentation is carried out for one or more of the following reasons:

TABLE 10.12 Suggested COV Values

Parameter	COV (%)	Reference
e or n	20–30	Lee et al. (1983), Baecher and Christian (2003)
D_r (sand)	10–40	Baecher and Christian (2003)
G_s	2–3	Harr (1987), Baecher and Christian (2003)
ρ or γ	3–10	Lee et al. (1983), Harr (1987), Baecher and Christian (2003)
Liquid limit	10–20	Lee et al. (1983), Baecher and Christian (2003)
Plastic limit	10–20	Lee et al. (1983)
PI	30–70[a]	Lee et al. (1983)
$w_{optimum}$	20–40[b]	Lee et al. (1983)
$\rho_{d,maximum}$	5	Lee et al. (1983)
k	200–300	Lee et al. (1983), Baecher and Christian (2003)
c_v	25–50	Baecher and Christian (2003), Lee et al. (1983)
C_c and C_r	20–40	Baecher and Christian (2003)
CBR	25	Lee et al. (1983)
ϕ-sand	10	Harr (1987), Lee et al. (1983)
ϕ-clay	10–50	Lee et al. (1983), Baecher and Christian (2003)
ϕ-mine tailings	5–20	Baecher and Christian (2003)
c_u or q_u	40	Harr (1987), Lee et al. (1983)
N (from SPT)	25–40	Harr (1987), Lee et al. (1983), Baecher and Christian (2003)
q_c (from CPT)	20–50[a]	Baecher and Christian (2003)
c_u (from vane shear test)	10–40	Baecher and Christian (2003)

After Lee et al. (1983) and Baecher and Christian (2003).

[a] Lower values for clays and higher values for sandy/gravelly clays.

[b] Lower values for clays and higher values for granular soils.

1. *Design verification*—To verify new or unconventional designs, particularly when simplified theories and assumptions are involved (e.g., pile load test)
2. *Construction control*—To monitor performance during construction so as to be able to alter or modify the design and procedures (e.g., deep excavations near buildings)
3. *Safety*—To warn of any impending failure (e.g., early warning systems for landslides)
4. *Legal protection*—To document strong evidence against any possible lawsuit (e.g., noise and vibrations due to pile driving)
5. *Verification of long-term performance*—To monitor in-service performance (e.g., drainage behind retaining walls)
6. *Advancing state-of-the-art*—To verify new developments in R&D and new design methodologies or construction techniques (e.g., new geosynthetic products)
7. *Quality control*—To verify compliance by the contractor (e.g., settlement of a compacted fill)

Whereas site investigation offers one-off measurements prior to construction, geotechnical instrumentation is used during or after construction, to monitor the ongoing performance of bridge abutments, retaining walls, foundations, and embankments. This includes monitoring deformation, pore water pressure, load, pressure, strain, and temperature. Brief descriptions of some of the common instruments used in geotechnical engineering are given below.

Piezometers are used for measuring water level and pore water pressure. They range from the simple and inexpensive Casagrande's open standpipe piezometer to more complex pneumatic, vibrating-wire, or hydraulic piezometers. *Settlement cells or plates* can be placed within embankments or foundations to monitor ongoing settlement. *Vertical inclinometers* are quite useful for monitoring lateral deformation near embankments on soft soils, landslides, and deflection of piles under lateral loading. *Horizontal inclinometers* can be used to determine the settlement profile beneath an embankment cross section. *Load cells* are used to monitor the load on tiebacks, rock anchors, soil nails, and piles. *Extensometers* measure deformation along the axis and can be used for measuring deformation in any direction, such as settlement, heave, or lateral displacement. *Strain gauges* can be mounted onto steel or concrete structures such as piles, bridges, and tunnel linings to monitor strain while in service. *Pressure cells* are useful for measuring horizontal or vertical earth pressures within the soil beneath foundations and embankments.

In using geotechnical instrumentation, it is necessary to understand properly what the following terms mean: accuracy, precision, resolution, and sensitivity. *Accuracy* is how close the mean value of a measurement is to the *true* value. *Precision* is how close a set of measurements is to the mean value (not necessarily the true value). It is a measure of repeatability or reproducibility. Precision should not be confused with accuracy. Precise measurements need not be accurate and vice versa. *Resolution* is the smallest change that can be detected by a readout device, whether digital or analog (e.g., 0.01 g on a digital balance). *Sensitivity* refers to the response of a device to a unit input. Dunnicliff (1993) discusses geotechnical instrumentation in great detail.

Defining Terms

Borehole: A deep hole of circular cross section made in the ground (soil/rock) to ascertain the nature of the subsurface materials.

California bearing ratio (CBR): A ratio of the penetration load of a soil that corresponds to the specified penetration of a standard cylindrical rod driven into the soil to the corresponding penetration load of a standard material, expressed in percent.

Casing: A steel pipe used to support a borehole wall and to stop fluid loss.

Client: The individual or organization commissioning a site investigation.

Cone: The cone-shaped point of a penetrometer tip, upon which end-bearing resistance develops.

Core: Cylindrical pieces of rock recovered by means of rotating a hollow steel tube (core barrel) equipped with a coring bit.

Disturbed sample: A sample that does not contain the natural structure of the *in situ* soil.

Drilling: The process of making a hole in the ground.

Drilling fluid: A solution of water and a thixotropic substance such as bentonite.

Friction ratio: The ratio of friction resistance to cone resistance, expressed in percent, in the cone penetration test.

Groundwater level: The level of the water table surrounding a borehole or well. The groundwater level can be represented as an elevation or as a depth below the ground surface.

Groundwater table (*aka water table*): The top surface of free water in the ground at which the water pressure equals atmospheric pressure. Each material below the groundwater table is saturated with water.

In situ: In its original place.

Investigator: The individual or organization responsible for performing the site investigation work and preparing and submitting the site investigation report.

Modulus of subgrade reaction (*aka subgrade modulus*): Ratio of pressure applied to the surface of a soil mass to the corresponding settlement of the surface. Determined as the slope of the secant drawn between the point corresponding to zero settlement and the point of 1.25-mm settlement on a load-settlement curve obtained from a plate load test on soil using a 750-mm-diameter or smaller loading plate with corrections for size of the plate used.

***N*-value**: The blow count representation of the penetration resistance of a soil, reported in number of blows per 300 mm. Equal to the sum of the number of blows required to drive the sampler over a depth interval of 150–450 mm.

Recovery ratio: The ratio of the length of the sample in a sampling tube to its depth of penetration.

Rock: A natural aggregate of mineral grains connected by strong and permanent cohesive forces; occurs in large masses and fragments.

Rock quality designation (RQD): A measure of the degree of fractures/joints in rock masses, defined as the ratio of the accumulated lengths (greater than or equal to 100 mm) of sound rock to the total core length drilled.

Sampler: A cylindrical steel barrel or tube used for collecting soil samples.

Sampling: The process of obtaining soil and rock samples.

Site investigation (*aka site exploration* or *site characterization*): Appraisal of the surface and subsurface conditions at a proposed construction site by analysis of information gathered by such methods as desk study, reconnaissance, geological and geophysical surveys, drilling boreholes, *in situ* testing, sampling, visual inspection, laboratory testing of samples of the subsurface materials, and groundwater observations and measurements.

Site reconnaissance: Walk-over survey to visually assess local conditions such as site access, adjacent properties and structures, topography, drainage, etc.

Soil: A natural aggregate of mineral grains that can be separated by such gentle means as agitation in water.

Undisturbed sample: A sample that contains the natural structure of the *in situ* soil.

References

Aas, G., Lacasse, S., Lunne, T., and Madshus, C. (1984). In situ testing: new developments. *NGM-84*, Linkoping, Vol. 2, 705–716.

American Society of Civil Engineers (1972). Subsurface investigation for design and construction of foundations of buildings. *J. Soil Mech. Found. Div. ASCE*, 98(SM5):481–490.

Anagnostopoulos, A., Koukis, G., and Sabatakakis, N. (2003). Empirical correlations of soil parameters based on cone penetration tests (CPT) for Greek soils. *Geotech. Geol. Eng.*, 21(4):377–387.

Archie, G.E. (1942). The electrical resistivity log as an aid in determining some reservoir characteristics. *Trans. Am. Inst. Mining Met. Eng.*, 146:54–52.

AS 1289.5.3.2 (2004). *Determination of the Field Density of a Soil—Sand Replacement Method Using a Sand Pouring Can, with or without a Volume Displacer*, Standards Australia, Strathfield, Australia.

AS 1289.5.3.5 (1997). *Determination of the Field Dry Density of a Soil—Water Replacement Method*, Standards Australia, Strathfield, Australia.

AS 1289.6.1.3 (1998). *Determination of the California Bearing Ratio of a Soil—Standard Field-in-Place Method*, Standards Australia, Strathfield, Australia.

AS 1289.6.3.1 (2004). *Determination of Penetration Resistance of a Soil—Standard Penetration Test*, Standards Australia, Strathfield, Australia.

AS 1289.6.3.2 (1997). *Determination of the Penetration Resistance of a Soil—9 kg Dynamic Cone Penetrometer Test*, Standards Australia, Strathfield, Australia.

AS 1289.6.5.1 (1999). *Determination of the Static Cone Penetration Resistance of a Soil—Field Test Using a Mechanical and Electrical Cone or Friction-Cone Penetrometer*, Standards Australia, Strathfield, Australia.

AS 1726 (1993). *Geotechnical Site Investigations*, Standards Australia, Strathfield, Australia.

ASTM D1556 (2000). *Standard Test Method for Density of Soil in Place by the Sand-Cone Method,* ASTM International, West Conshohocken, PA.

ASTM D1586 (1999). *Standard Test Method for Penetration Test and Split-Barrel Sampling of Soils,* ASTM International, West Conshohocken, PA.

ASTM D2167 (1994, reapproved 2001). *Standard Test Method for Density and Unit Weight of Soil in Place by the Rubber-Balloon Method,* ASTM International, West Conshohocken, PA.

ASTM D2937 (2004). *Standard Test Method for Density of Soil in Place by the Drive-Cylinder Method,* ASTM International, West Conshohocken, PA.

ASTM D3441 (2005). *Standard Test Method for Mechanical Cone Penetration Tests on Soil,* ASTM International, West Conshohocken, PA.

ASTM D4220 (1995, reapproved 2000). *Standard Practices for Preserving and Transporting Soil Samples,* ASTM International, West Conshohocken, PA.

ASTM D4429 (2004). *Standard Test Method for CBR (California Bearing Ratio) of Soils in Place,* ASTM International, West Conshohocken, PA.

ASTM D4719 (2000). *Standard Test Method for Prebored Pressuremeter Testing in Soils,* ASTM International, West Conshohocken, PA.

ASTM D4914 (1999). *Standard Test Methods for Density of Soil and Rock in Place by the Sand Replacement Method in a Test Pit,* ASTM International, West Conshohocken, PA.

ASTM D5030 (1989). *Standard Test Method for Density of Soil and Rock in Place by the Water Replacement Method in a Test Pit,* ASTM International, West Conshohocken, PA.

ASTM D5195 (2002). *Standard Test Method for Density of Soil and Rock In-Place at Depths Below the Surface by Nuclear Methods,* ASTM International, West Conshohocken, PA.

ASTM D6635 (2001). *Standard Test Method for Performing the Flat Plate Dilatometer,* ASTM International, West Conshohocken, PA.

ASTM D6938 (2008). *Standard Test Method for In-Place Density and Water Content of Soil and Soil-Aggregate by Nuclear Methods (Shallow Depth),* ASTM International, West Conshohocken, PA.

ASTM D6951 (2003). *Standard Test Method for Use of the Dynamic Cone Penetrometer in Shallow Pavement Applications,* ASTM International, West Conshohocken, PA.

Baecher, G.B. and Christian, J.T. (2003). *Reliability and Statistics in Geotechnical Engineering,* John Wiley & Sons, 605 pp.

Baguelin, F., Jezequel, J.F., and Shields, D.H. (1978). *The Pressuremeter and Foundation Engineering,* Trans Tech Publications, Clausthal, Germany, 618 pp.

Bjerrum, L. (1972). Embankments on soft ground. *Proceedings of ASCE Specialty Conference on Performance of Earth and Earth Supported Structures,* Vol. 2, Purdue University, West Lafayette, IN, 1–54

Black, W.P.M. (1962). A method of estimating the California bearing ratio of cohesive soils from plasticity data. *Geotechnique,* 12(4):271–282.

Bond, D. (1961). The influence of size on foundation settlement. *Geotechnique,* 11(2):121–143.

Bowles, J.E. (1975). Combined and special footings. *Foundation Engineering Handbook,* H.F. Winterkorn and H.-Y. Fang, Eds., Van Nostrand Reinhold, New York, chap. 16.

Bowles, J.E. (1988). *Foundation Analysis and Design,* 4th edition, McGraw-Hill, 1004 pp.

Burland, J.B. and Burbidge, M.C. (1985). Settlements of foundations on sand and gravel. *Proc. Inst. Civ. Eng.,* 78(1):1325–1381.

Canadian Geotechnical Society (1992). *Canadian Foundation Engineering Manual,* 3rd edition.

Canadian Geotechnical Society (2006). *Canadian Foundation Engineering Manual,* 4th edition, 488 pp.

Craig, R.F. (1997). *Soil Mechanics,* E & FN Spon, London.

Das, B.M. (1998). *Principles of Geotechnical Engineering,* 4th edition, PWS Publishing, Boston.

Das, B.M. (2007). *Principles of Foundation Engineering,* 6th edition, Thomson, Pacific Grove, CA.

De Beer, M., Kleyn, E.G., and Savage, P.F. (1989). Advances in pavement evaluation and overlay design with the aid of the dynamic cone penetrometer (DCP). *2nd International Symposium on Pavement Evaluation and Overlay Design,* Rio de Janeiro, Brazil.

Dobrin, M.B. (1976). *Introduction to Geophysical Prospecting,* McGraw-Hill, New York.

Dunnicliff, J. (1993). *Geotechnical Instrumentation for Monitoring Field Performance,* John Wiley & Sons, 577 pp.

Gibbs, H.J. and Holtz, W.G. (1957). Research on determining the density of sands by spoon penetration testing. *Proceedings of 4th International Conference on Soil Mechanics and Foundation Engineering,* Vol. 1, Butterworths, London, 35–39.

Handy, R.L. (2008). Evolution of geotechnical soil testing. I. Field tests. *Int. J. Geotech. Eng.,* 2(2):11–28.

Handy, R.L. and Spangler, M.G. (2007). *Geotechnical Engineering,* 5th edition, McGraw-Hill, New York.

Handy, R.L., Remmes, B., Moldt, S., Lutenegger, A.J., and Trott, G. (1982). In situ stress determination by Iowa stepped blade. *J. Geotech. Eng. Div. ASCE,* 108(GT11):1405–1422.

Hansbo, S. (1957). *A New Approach to the Determination of the Shear Strength of Clay by the Fall Cone Test,* Report No. 114, Swedish Geotechnical Institute.

Hara, A., Ohata, T., and Niwa, M. (1971). Shear modulus and shear strength of cohesive soils. *Soils Found.,* 14(3):1–12.

Harr, M.E. (1987). *Reliability-Based Design in Civil Engineering,* McGraw-Hill, 290 pp.

Hatanaka, M. and Uchida, A. (1996). Empirical correlation between penetration resistance and effective friction of sandy soil. *Soils Found.,* 36(4):1–9.

Heukelon, W. and Klomp, A.J.G. (1962). Dynamic testing as a means of controlling pavements during and after construction. *Proceedings of the First International Conference on Structural Design of Asphalt Pavement,* University of Michigan, Ann Arbor, 667–679.

Hight, D.D. and Stevens, M.G.H. (1982). An analysis of the California bearing ratio test in saturated clays. *Geotechnique,* 32(4):315–322.

Holtz, R.D. (1991). Stress distribution and settlement of shallow foundations. *Foundation Engineering Handbook,* 2nd edition, H.-Y. Fang, Ed., Van Nostrand Reinhold, New York, 166–222.

IRC: 37 (2001). *Guidelines for the Design of Flexible Pavements,* Indian Roads Congress, New Delhi.

IS 1888 (1982, reaffirmed 1997). *Method of Load Test on Soils,* Bureau of Indian Standards, New Delhi.

IS 2720, Part XXIX (1975, reaffirmed 1995). *Determination of Dry Density of Soils In-Place by the Sand Replacement Method,* Bureau of Indian Standards, New Delhi.

IS 4968, Part I (1976). *Dynamic Method Using 50 mm Cone without Bentonite Slurry,* Bureau of Indian Standards, New Delhi.

IS 4968, Part II (1976, reaffirmed 1997). *Dynamic Method Using Cone and Bentonite Slurry,* Bureau of Indian Standards, New Delhi.

Kate, J.M. (1980). Comparative studies of field and laboratory CBR results. *Indian Geotech. J.,* 10(3):203–214.

Kearey, P., Brooks, M., and Hill, I. (2002). *An Introduction to Geophysical Exploration,* Blackwell Science, London.

Kovacs, W.D. and Salomone, L.A. (1982). SPT hammer energy measurements. *J. Geotech. Eng. Div. ASCE,* 108(4):599–620.

Kulhawy, F.H. and Mayne, P.W. (1990). *Manual on Estimating Soil Properties for Foundation Design,* Final Report EL-6800, Electric Power Research Institute, Palo Alto, CA.

Larsson, R. (1980). Undrained shear strength in stability calculation of embankments and foundations on clay. *Can. Geotech. J.,* 17:591–602.

Lee, I.K., White, W., and Ingles, O.G. (1983). Soil variability. *Geotechnical Engineering,* Pitman, chap. 2.

Leonards, G.A. (1986). *Advanced Foundation Engineering—CE683,* Lecture Notes, Purdue University, West Lafayette, IN.

Liao, S.S. and Whitman, R.V. (1986). Overburden correction factors for SPT in sand. *J. Geotech. Eng. Div. ASCE,* 112(3):373–377.

Lister, N.W. and Powell, W.D. (1987). Design practice for bituminous pavements in the United Kingdom. *Sixth International Conference on the Structural Design of Asphalt Pavements,* University of Michigan, Ann Arbor.

Lowe III, J. and Zaccheo, P.F. (1975). Subsurface explorations and sampling. *Foundation Engineering Handbook,* H.F. Winterkorn and H.-Y. Fang, Eds., Van Nostrand Reinhold, New York, chap. 1.

Lutenegger, A.J. (1987). Suggested method for performing the borehole shear test. *Geotech. Testing J.,* 10(1):19–25.

Lutenegger, A.J. and Timian, D.A. (1986). In situ tests with K_0 stepped blade test. *Proceedings of In Situ '86,* ASCE, Geotechnical Special Publication No. 6, 730–751.

Lutenegger, A.J. and Timian, D.A. (1987). Reproducibility of borehole shear test results in marine clay. *Geotech. Testing J.,* 10(1):13–18.

Marchetti, S. (1980). In situ test by flat dilatometer. *J. Geotech. Eng. Div. ASCE,* 106(GT3):299–321.

Marchetti, S., Monaco, P., Totani, G., and Calabrese, M. (2001). *The Flat Dilatometer Test (DMT) in Soil Investigations,* A Report by the ISSMGE Committee TC16, International Conference on In Situ Measurement of Soil Properties, Bali, Indonesia, 41 pp.

Mayne, P.W. and Kemper, J.B. (1988). Profiling OCR in stiff clays by CPT and SPT. *Geotech. Testing J.,* 11(2):139–147.

Mayne, P.W. and Mitchell, J.K. (1988). Profiling of overconsolidation ratio in clays by field vane. *Can. Geotech. J.,* 25(1):150–158.

McCarthy, D.F. (2007). *Essentials of Soil Mechanics and Foundations,* 7th edition, Pearson Prentice Hall, 850 pp.

Menard, L. (1956). *An Apparatus for Measuring the Strength of Soils in Place,* Master's thesis, University of Illinois, Urbana.

Meyerhof, G.G. (1957). Discussion on research on determining the density of sands by spoon penetration testing. *Proceedings of 4th International Conference on Soil Mechanics and Foundation Engineering,* London, Vol. 3, 110.

Meyerhof, G.G. (1974). Penetration testing outside Europe. General report. *Proceedings of European Symposium on Penetration Testing,* Stockholm, Vol. 2.1, 40–48.

Morris, P.M. and Williams, D.T. (1994). Effective stress vane shear strength correction factor correlations. *Can. Geotech. J.,* 31(3):335–342.

Nascimento, U. and Simoes, A. (1957). Relation between CBR and modulus of strength. *Proceedings of the 4th International Conference on Soil Mechanics and Foundation Engineering,* Vol. 2, No. 4, 166–168.

Nielson, F.D., Bhandhausavee, C., and Yeb, K.-S. (1969). Determination of modulus of soil reaction from standard soil tests. *Highw. Res. Rec.,* 384:1–12.

Ohya, S., Imai, T., and Matsubara, M. (1982). Relationships between N value by SPT and LLT pressuremeter results. *Proceedings of the Second European Symposium on Penetration Testing,* Amsterdam, Vol. 1, 125–130.

Orchant, C.J., Trautmann, C.H., and Kulhawy, F.H. (1986). In situ testing to characterize electric transmission line routes. *Proceedings of In Situ '86 ASCE Special Conference on Use of In Situ Tests in Geotechnical Engineering,* Virginia Tech, Blacksburg, ASCE Geotechnical Special Publication No. 6, 869–886.

Peck, R.B., Hanson, W.E., and Thornburn, T.H. (1974). *Foundation Engineering,* John Wiley & Sons, New York.

Robertson, P.K. and Campanella, R.G. (1983). Interpretation of cone penetration tests. I. Sand. *Can. Geotech. J.*, 20(4):718–733.

Robertson, P.K., Campanella, R.G., and Wightman, A. (1983). SPT-CPT correlations. *J. Geotech. Eng.*, 109(11):1449–1459.

Robertson, P.K., Campanella, R.G., Gillespie, D., and Greig, J. (1986). Use of piezometer cone data. *Use of In Situ Tests in Geotechnical Engineering*, Geotechnical Special Publication No. 6, ASCE, 1263–1280.

Sanglerat, G. (1972). *The Penetrometer and Soil Exploration*, Elsevier, Amsterdam, 464 pp.

Schmertmann, J.H. (1970). Static cone to compute static settlement over sand. *J. Soil Mech. Found. Div. ASCE*, 96(SM3):1011–1043.

Schmertmann, J.H. (1975). Measurement of in situ shear strength. *Proceedings of Specialty Conference on In Situ Measurement of Soil Properties*, Raleigh, NC, Vol. 2, 57–138.

Schmertmann, J.H. (1978). *Guidelines for Cone Penetration Test Performance and Design*, Report FHWA-TS-78-209, Federal Highway Administration, Washington, D.C., 145 pp.

Schmertmann, J.H., Hartman, J.P., and Brown, P.R. (1978). Improved strain influence factors diagram. *J. Geotech. Eng. Div. ASCE*, 104(GT8):1131–1135.

Shukla, S.K. (2006a). *Field Testing Related to Subsoil Investigation at the Site for Centralized Stamping Unit and Fabrication Plant*, Report No. CE/06-07/November 15, 2006, Department of Civil Engineering, Institute of Technology, Banaras Hindu University, Varanasi, India.

Shukla, S.K. (2006b). *Allowable Load-Bearing Pressure for the Raft Foundation of Aqueduct-cum-Bridge along the Alignment of Bansagar Feeder Channel*, Report No. CE/06-07/June 7, 2006, Department of Civil Engineering, Institute of Technology, Banaras Hindu University, Varanasi, India.

Shukla, S.K. (2008). A time-dependent expression for modulus of subgrade reaction of saturated cohesive foundation soils. *Int. J. Geotech. Eng.*, 2(4):435–439.

Skempton, A.W. (1986). Standard penetration test procedures and the effects in sands of overburden pressure, relative density, particle size, ageing and overconsolidation. *Geotechnique*, 36(3):425–447.

Sowers, G.B. and Sowers, G.F. (1970). *Introductory Soil Mechanics and Foundations*, 3rd edition, Macmillan, New York.

Terzaghi, K. (1955). Evaluation of coefficient of subgrade reaction. *Geotechnique*, 5(4):297–326.

Terzaghi, K. and Peck, R.B. (1948). *Soil Mechanics in Engineering Practice*, John Wiley and Sons, New York.

Terzaghi, K., Peck, R.B., and Mesri, G. (1996). *Soil Mechanics in Engineering Practice*, 3rd edition, John Wiley & Sons, New York.

Totani, G., Marchetti, S., Monaco, P., and Calabrese, M. (2001). Use of the flat dilatometer test (DMT) in geotechnical design. *IN SITU 2001, International Conference on In Situ Measurement of Soil Properties*, Bali, Indonesia, 1–6.

U.S. Army Corps of Engineers (1972). *Soil Sampling*, Engineer Manual, EM 1110-2-1907, U.S. Government Printing Office, Washington, D.C.

Vesic, A.B. (1961a). Beams on elastic subgrade and Winkler's hypothesis. *Proceedings of the Fifth International Conference on Soil Mechanics and Foundation Engineering*, Paris, 845–850.

Vesic, A.B. (1961b). Bending of beams resting on isotropic elastic solid. *J. Eng. Mech. Div. ASCE*, 87(EM2):35–53.

Wilson, G. and Williams, G.M.J. (1950). Pavement bearing capacity computed by theory of layered systems. *Proc. Am. Soc. Civ. Eng.*, 116:750–769.

Wolff, T.F. (1989). Pile capacity prediction using parameter functions. *Predicted and Observed Axial Behaviour of Piles, Results of a Pile Prediction Symposium*, Evanston, IL, ASCE Geotechnical Special Publication, 96–106.

11

Vibration of Foundations

by
Braja M. Das
California State University, Sacramento, California

Nagaratnam Sivakugan
James Cook University, Townsville, Australia

11.1 Introduction

Foundations that support vibrating equipment experience rigid body displacements. The cyclic displacement of a foundation can have six possible modes (Figure 11.1):

1. Translation in the vertical direction
2. Translation in the longitudinal direction
3. Translation in the lateral direction
4. Rotation about the vertical axis (yawing)
5. Rotation about the longitudinal axis (rocking)
6. Rotation about the lateral axis (pitching)

In this chapter, the fundamentals of the vibration of foundations in various modes supported on an elastic medium will be developed. The elastic medium that supports the foundation is

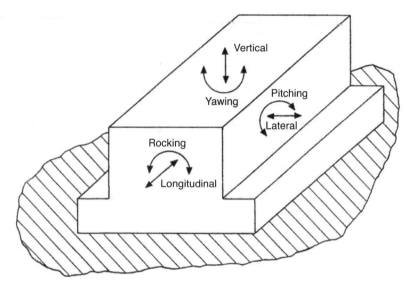

FIGURE 11.1 Six modes of vibration for a foundation.

considered to be homogeneous and isotropic. In general, the behavior of soils departs considerably from that of an elastic material. Only at low strain levels may soils be considered a reasonable approximation of an elastic material.

11.2 Vibration Theory: General

In this section, we will discuss the elements of vibration theory, an understanding of which is essential to the design of foundations subjected to cyclic loading.

11.2.1 Free Vibration of a Spring-Mass System

Figure 11.2 shows a foundation resting on a spring. Let the spring represent the elastic properties of the soil. The load W represents the weight of the foundation plus that which comes from the machinery supported by the foundation. Due to the load W, a static deflection z_s will develop. By definition,

$$k = \frac{W}{z_s} \tag{11.1}$$

where k = spring constant for the elastic support.

If the foundation is disturbed from its static equilibrium position, the system will oscillate. The equation of motion of the foundation when it has been disturbed through a distance z can be written from Newton's second law of motion as

$$\left(\frac{W}{g}\right)\ddot{z} + kz = 0$$

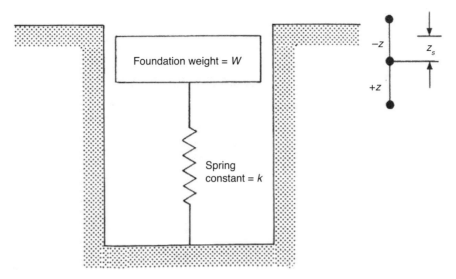

FIGURE 11.2 Free vibration of a spring-mass system.

or

$$\ddot{z} + \left(\frac{k}{m} \right) z = 0 \tag{11.2}$$

where

g = the acceleration due to gravity
$\ddot{z}$ = d^2z/dt^2
t = time
m = mass = W/g

The preceding equation can be solved to obtain the *frequency of vibration* (that is, the number of cycles per unit time) as

$$f = f_n = \frac{\omega_n}{2\pi} = \frac{1}{2\pi} \sqrt{\frac{k}{m}} \tag{11.3}$$

where

f = frequency of oscillation (cps)
f_n = undamped natural frequency (cps)
ω_n = undamped natural circular frequency (rad/s) = $\sqrt{k/m}$

Under idealized situations, the vibration can continue forever.

Example 1

A mass is supported by a spring. The static deflection of a spring z_s due to the mass is 0.4 mm. Determine the natural frequency of vibration.

Solution

$$k = \frac{W}{z_s}$$

However, $W = mg$ and $g = 9.81 \ \text{m/s}^2$, so

$$k = \frac{mg}{z_s}$$

$$f_n = \frac{1}{2\pi} \sqrt{\frac{k}{m}} = \frac{1}{2\pi} \sqrt{\left(\frac{mg}{z_s}\right)\frac{1}{m}} = \frac{1}{2\pi} \sqrt{\frac{g}{z_s}}$$

$$= \frac{1}{2\pi} \sqrt{\frac{9.81}{\left(\dfrac{0.4}{1000}\ m\right)}} = 24.9 \ \text{cps}$$

11.2.2 Free Vibration with Viscous Damping

In the case of *undamped free vibration* as explained above, vibration would continue once the system had been set in motion. However, in practical cases, all vibrations undergo a gradual decrease in amplitude with time. This characteristic of vibration is referred to as *damping*. Figure 11.3 shows a foundation supported by a spring and a dashpot. The dashpot represents the *damping characteristic* of the soil. The dashpot coefficient is equal to c. For free vibration of the foundation, the differential equation of motion can be given by:

$$m\ddot{z} + c\dot{z} + kz = 0 \tag{11.4}$$

The preceding equation can be solved to show three possible cases of vibration that are functions of a quantity called the damping ratio D. The damping ratio is defined as

$$D = \frac{c}{c_c} \tag{11.5}$$

where the critical damping coefficient c_c is

$$c_c = 2\sqrt{km} \tag{11.6}$$

- If $D > 1$, it is an *overdamped* case. For this case, the system will not oscillate at all. The variation of displacement z with time will be as shown in Figure 11.4a.
- If $D = 1$, it is a case of *critical* damping (Figure 11.4b). For this case, the sign of z changes only once.
- If $D < 1$, it is an *underdamped* condition. Figure 11.4c shows the nature of vibration with time for this case. For this condition, the *damped natural frequency* of vibration f can be given as

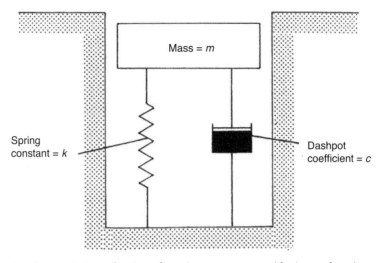

FIGURE 11.3 Free vibration of a spring-mass system with viscous damping.

$$f = \frac{\omega_d}{2\pi} \tag{11.7}$$

where the damped natural circular frequency ω_d (rad/s) is

$$\omega_d = \omega_n \sqrt{1 - D^2} \tag{11.8}$$

Combining Equations 11.7, 11.8, and 11.3,

$$f = f_m = \frac{\omega_n \sqrt{1 - D^2}}{2\pi} = f_n \sqrt{1 - D^2} \tag{11.9}$$

Example 2

For a machine foundation, $W = 70$ kN, $k = 12{,}500$ kN/m, and $c = 250$ kN-s/m. Determine:

  a. Whether the system is overdamped, underdamped, or critically damped
  b. The damped natural frequency

Solution
Part a

$$c_c = 2\sqrt{km} = 2\sqrt{k\left(\frac{W}{g}\right)} = 2\sqrt{(12{,}500)\left(\frac{70}{9.81}\right)} = 597.3 \text{ kN-s/m}$$

$$D = \frac{c}{c_c} = \frac{250}{597.3} = 0.419 < 1$$

Therefore the system is underdamped.

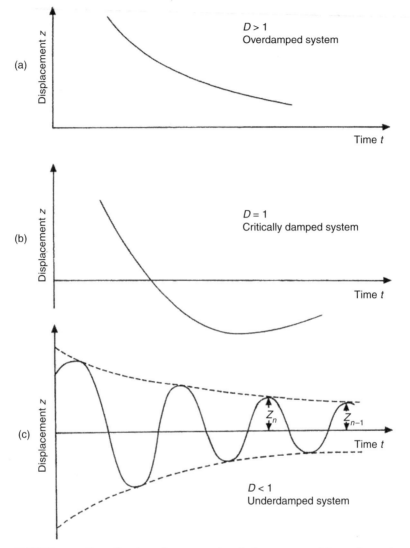

FIGURE 11.4 Free vibration of a mass-spring-dashpot system: (a) overdamped case, (b) critically damped case, and (c) underdamped case.

Part b
From Equation 11.9:

$$f_m = f_n \sqrt{1 - D^2} = \frac{1}{2\pi} \left(\sqrt{\frac{k}{m}} \right) \left(\sqrt{1 - D^2} \right)$$

$$= \frac{1}{2\pi} \left[\sqrt{\frac{12{,}500}{\left(\dfrac{70}{9.81} \right)}} \right] \left[\sqrt{1 - (0.419)^2} \right] = 6.05 \text{ cps}$$

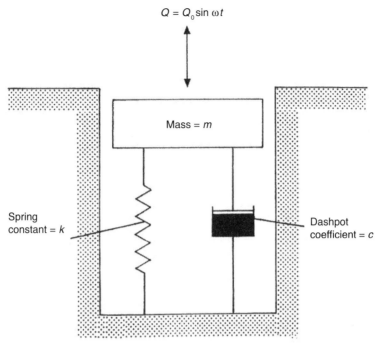

$$Q = Q_0 \sin \omega t$$

Mass = m

Spring
constant = k

Dashpot
coefficient = c

FIGURE 11.5 Steady-state forced vibration with damping.

11.2.3 Steady-State Forced Vibration with Damping

Figure 11.5 shows a foundation resting on a soil that can be approximated to be an equivalent spring and dashpot. This foundation is being subjected to a sinusoidally varying force $Q = Q_0 \sin \omega t$. The differential equation of motion for this system can be given by

$$m\ddot{z} + kz + c\dot{z} = Q_0 \sin \omega t \qquad (11.10)$$

where ω = circular frequency of vibration (rad/s).

Equation 11.10 can be solved to obtain the amplitude (i.e., maximum displacement) of vibration Z of the foundation as

$$Z = \frac{\left(\dfrac{Q_0}{k}\right)}{\sqrt{\left[1 - \left(\dfrac{\omega^2}{\omega_n^2}\right)\right]^2 + 4D^2\left(\dfrac{\omega^2}{\omega_n^2}\right)}} \qquad (11.11)$$

where $\omega_n = \sqrt{k/m}$ is the undamped natural frequency and D is the damping ratio.

Equation 11.11 is plotted in a nondimensional form as $Z/(Q_0/k)$ vs. ω/ω_n in Figure 11.6. Note that the maximum value of $Z/(Q_0/k)$ (and hence Z) occurs as

$$\omega = \omega_n \sqrt{1 - 2D^2} \qquad (11.12)$$

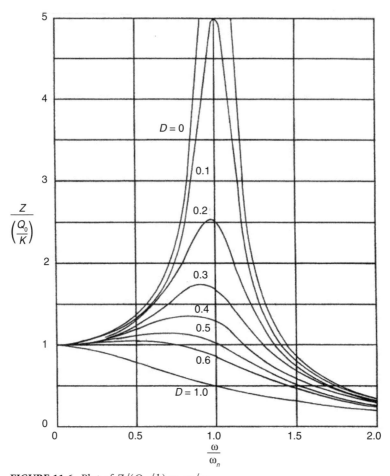

FIGURE 11.6 Plot of $Z/(Q_0/k)$ vs. ω/ω_n.

or

$$f_m = f_n \sqrt{1 - 2D^2} \qquad (11.13)$$

where f_m is the frequency at *maximum amplitude* (the *resonant frequency for vibration with damping*) and f_n is the natural frequency $= (1/2\pi)\sqrt{k/m}$. Hence, the *amplitude of vibration at resonance* can be obtained by substituting Equation 11.12 into Equation 11.11, or

$$
\begin{aligned}
Z_{\text{res}} &= \frac{Q_0}{k}\; \frac{1}{\sqrt{[1 - (1 - 2D^2)]^2 + 4D^2(1 - 2D^2)}} \\[2mm]
&= \frac{Q_0}{k}\; \frac{1}{2D\sqrt{1 - D^2}}
\end{aligned}
\qquad (11.14)
$$

11.2.4 Rotating Mass Type Excitation

In many cases of foundation equipment, vertical vibration of foundations is produced by counterrotating masses, as shown in Figure 11.7a. Since horizontal forces on the foundation at any instance cancel, the net vibrating force on the foundation can be determined to be equal to $2m_e e \omega t$ (where m_e = mass of each counterrotating element, e = eccentricity, and ω = angular frequency of the masses). In such cases, the equation of motion with viscous damping (Equation 11.10) can be modified to the form

$$m\ddot{z} + kz + c\dot{z} = Q_0 \sin \omega t \qquad (11.15)$$

$$Q_0 = 2m_e e\omega^2 = U\omega^2 \qquad (11.16)$$

$$U = 2m_e e \qquad (11.17)$$

In Equation 11.15, m is the mass of the foundation *including* $2m_e$. Equation 11.15 can be solved to find the amplitude of motion as:

$$Z = \frac{\left(\dfrac{U}{m}\right)\left(\dfrac{\omega^2}{\omega_n^2}\right)}{\sqrt{\left[1 - \left(\dfrac{\omega^2}{\omega_n^2}\right)\right]^2 + 4D^2\left(\dfrac{\omega^2}{\omega_n^2}\right)}} \qquad (11.18)$$

Figure 11.7b shows a nondimensional plot of $Z/(U/m)$ vs. ω/ω_n for various values of damping ratio. For this type of excitation, the angular resonant frequency can be obtained as

$$\omega = \frac{\omega_n}{\sqrt{1 - 2D^2}} \qquad (11.19)$$

or the damped resonant frequency f_m as

$$f_m = \frac{f_n}{\sqrt{1 - 2D^2}} \qquad (11.20)$$

The amplitude at damped resonant frequency (similar to Equation 11.14) can be given as:

$$Z_{res} = \frac{\left(\dfrac{U}{m}\right)}{2D\sqrt{1 - 2D^2}} \qquad (11.21)$$

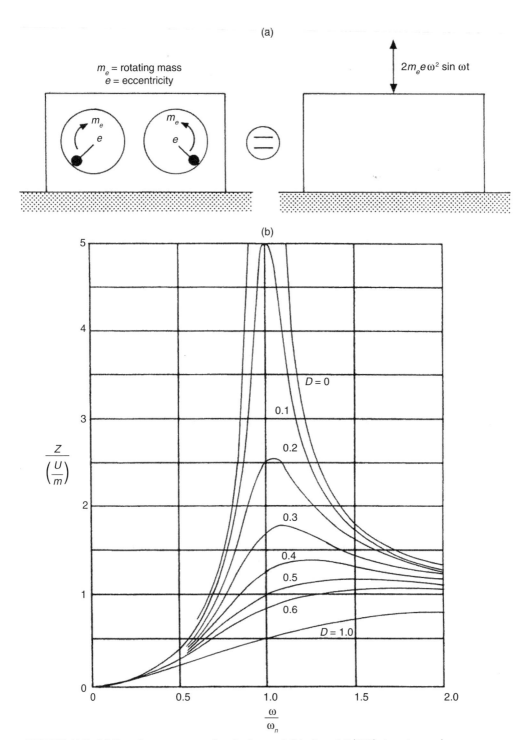

FIGURE 11.7 (a) Rotating mass type of excitation and (b) plot of $Z/(U/m)$ against ω/ω_n.

Example 3

Refer to Figure 11.5. The weight of the machine and foundation = 200 kN, spring constant k = 18×10^4 kN/m, damping ratio $D = 0.3$, Q (kN) = $Q_0 \sin \omega t$, $Q_0 = 60$ kN, and $\omega = 130$ rad/s. Determine:

a. The amplitude of motion Z
b. The resonant frequency of vibration with damping

Solution
Part a
From Equation 11.11:

$$Z = \frac{\left(\dfrac{Q_0}{k}\right)}{\sqrt{\left[1 - \left(\dfrac{\omega^2}{\omega_n^2}\right)\right]^2 + 4D^2\left(\dfrac{\omega^2}{\omega_n^2}\right)}}$$

From Equation 11.3:

$$\omega_n = \sqrt{\frac{k}{m}} = \sqrt{\frac{(18 \times 10^4 \text{ kN/m})}{\left(\dfrac{200 \text{ kN}}{9.81}\right)}} = 93.96 \text{ rad/s}$$

Hence:

$$Z = \frac{\left(\dfrac{60}{18 \times 10^4}\right)}{\sqrt{\left[1 - \left(\dfrac{130}{93.96}\right)\right]^2 + (4)(0.3)^2\left(\dfrac{130}{93.96}\right)^2}} = 0.00027 \text{ m} = 0.27 \text{ mm}$$

Part b
From Equation 11.13:

$$f_m = f_n \sqrt{1 - 2D^2}$$

$$f_n = \frac{\omega_n}{2\pi} = \frac{93.96}{(2)(\pi)} = 14.95 \text{ cps}$$

Thus:

$$f_m = (14.95)\sqrt{1 - (2)(0.3)^2} = 13.54 \text{ cps}$$

11.3 Shear Modulus and Poisson's Ratio

To solve practical problems in foundation vibration, relationships for the spring constant k and dashpot coefficient c are necessary. Those relationships presently available are functions of the shear modulus G and Poisson's ratio μ of various soils. In this section, we will discuss some of the available relationships for the shear modulus of sand and clayey soils.

11.3.1 Shear Modulus of Sand

At *low-strain amplitudes* ($\leq 10^{-4}\%$), the shear modulus G of sand was correlated by Hardin and Black (1968) as

$$G = \frac{6908(2.17 - e)^2}{1 + e} (\bar{\sigma}_0')^{0.5} \tag{11.22}$$

for round-grained soil and

$$G = \frac{3230(2.97 - e)^2}{1 + e} (\bar{\sigma}_0')^{0.5} \tag{11.23}$$

for angular-grained soil where

G = shear modulus (kN/m^2)
e = void ratio
$\bar{\sigma}_0'$ = average effective confining pressure (kN/m^2)

In the field,

$$\bar{\sigma}_0' \approx \frac{\sigma_v' + 2\sigma_v'(1 - \sin\phi')}{3} \tag{11.24}$$

where

σ_v' = vertical effective stress at a certain point in a soil mass
ϕ' = drained friction angle

Example 4

For a dry angular-grained sand deposit, the dry unit weight $\gamma = 17.5 \text{ kN/m}^3$, angle of friction $\phi' = 34°$, and specific gravity of soil solids $G_s = 2.67$. Estimate the shear modulus of the soil at a depth of 7 m from the ground surface.

Solution

$$\gamma_d = \frac{G_s \gamma_w}{1 + e}$$

$$\gamma_w = \text{unit weight of water} = 9.81 \text{ kN/m}^3$$

$$e = \frac{G_s \gamma_w}{\gamma_d} - 1 = \frac{(2.67)(9.81)}{17.5} - 1 \approx 0.497$$

At a depth of 7 m:

$$\sigma_v' = (17.5)(7) = 122.5 \text{ kN/m}^2$$

$$\overline{\sigma}_0' = \frac{\sigma_v' + 2\sigma_v'(1 - \sin\phi')}{3} = \frac{122.5 + (2)(122.5)(1 - \sin 30)}{3}$$

$$= 81.7 \text{ kN/m}^2$$

From Equation 11.23:

$$G = \frac{3230(2.97 - e)^2}{1 + e}(\overline{\sigma}_0')^{0.5} = \frac{3230(2.97 - 0.497)^2}{1 + 0.497}(81.7)^{0.5}$$

$$\approx 199,273 \text{ kN/m}^2$$

11.3.2 Shear Modulus of Clay

The shear modulus G, at low-strain amplitudes, of clay soils was proposed by Hardin and Drnevich (1972) in the form

$$G \text{ (kN/m}^2) = \frac{3230(2.97 - e)^2}{1 + e}(\text{OCR})^K [\overline{\sigma}_0' \text{ (kN/m}^2)]^{0.5} \quad (11.25)$$

where OCR is the overconsolidation ratio and K is a constant which is a function of the plasticity index (PI). The term $\overline{\sigma}_0'$ was defined by Equation 11.24. The recommended variations of K with PI are shown in Table 11.1.

11.4 Analog Solution for Vertical Vibration of Foundations

11.4.1 Constant Force Excitation

Lysmer and Richart (1966) provided an analog solution for vertical vibration of a rigid circular foundation. According to this solution, it was

TABLE 11.1 Recommended Variations of K with PI

PI (%)	K
0	0
20	0.18
40	0.30
60	0.41
80	0.48
≥100	0.50

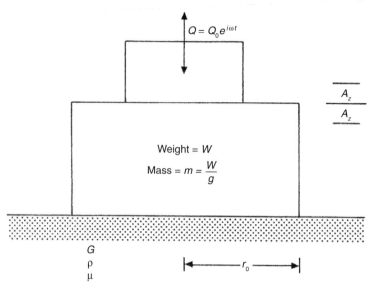

FIGURE 11.8 Vertical vibration of a foundation.

proposed that satisfactory results could be obtained within the range of practical interest by expressing the rigid circular foundation vibration (Figure 11.8) in the form

$$m\ddot{z} + c_z\dot{z} + k_z z = Q_0 e^{i\omega t} \tag{11.26}$$

where

$$k_z = \text{static spring constant for a rigid circular foundation} = \frac{4Gr_0}{1-\mu} \tag{11.27}$$

$$c_z = \text{dashpot coefficient} = \frac{3.4r_0^2}{1-\mu}\sqrt{G\rho} \tag{11.28}$$

m = mass of the foundation and the machine the foundation is supporting
r_0 = radius of the foundation
μ = Poisson's ratio of the soil
G = shear modulus of the soil
ρ = density of the soil

If a foundation is rectangular with a length L and width B, then the equivalent radius of a circular foundation can be given as:

$$r_0 \approx \sqrt{\frac{BL}{\pi}} \tag{11.29}$$

The resonant frequency f_m (frequency at maximum displacement) for *constant force excitation* can be obtained by solving Equations 11.26–11.28 (similar to solving Equation 11.10), or

$$f_m = \left(\frac{1}{2\pi} \right) \left(\sqrt{\frac{G}{\rho}} \right) \left(\frac{1}{r_0} \right) \sqrt{\frac{B_z - 0.36}{B_z}} \qquad \text{for } B_z \geq 0.3 \qquad (11.30)$$

where the mass ratio B_z is

$$B_z = \left(\frac{1 - \mu}{4} \right) \left(\frac{m}{\rho r_0^3} \right) \qquad (11.31)$$

The amplitude of vibration A_z at resonance for *constant force type excitation* can be determined from Equation 11.14 as

$$A_{z(\text{resonance})} = \left(\frac{Q_0}{k_z} \right) \left(\frac{1}{2D_z \sqrt{1 - D_z^2}} \right) \qquad (11.32)$$

where

$$k_z = \frac{4Gr_0}{1 - \mu}$$

The damping ratio D_z is

$$D_z = \frac{0.425}{\sqrt{B_z}} \qquad (11.33)$$

Substituting the above relationships for k_z and D_z into Equation 11.32 yields:

$$A_{z(\text{resonance})} = \frac{Q_0(1 - \mu)}{4Gr_0} \frac{B_z}{0.85 \sqrt{B_z - 0.18}} \qquad (11.34)$$

The amplitude of vibration at frequencies other than resonance can be obtained using Equation 11.11 as:

$$A_z = \frac{\left(\dfrac{Q_0}{k_z} \right)}{\sqrt{\left[1 - \left(\dfrac{\omega^2}{\omega_n^2} \right) \right]^2 + 4D_z^2 \left(\dfrac{\omega^2}{\omega_n^2} \right)}} \qquad (11.35)$$

The relationships for k_z and D_z are given by Equations 11.27 and 11.33 and

$$\omega_n = \sqrt{\frac{k_z}{m}} \qquad (11.36)$$

11.4.2 Rotating Mass Excitation

If a structure is subjected to vertical vibration due to rotating mass excitation, as shown in Figure 11.9 (similar to that shown in Figure 11.7a), the corresponding relationships will be as follows.

Resonant frequency:

$$f_m = \left(\frac{1}{2\pi}\right)\left(\sqrt{\frac{G}{\rho}}\right)\left(\frac{1}{r_0}\right)\sqrt{\frac{0.9}{B_z - 0.45}} \qquad (11.37)$$

Amplitude of vibration at resonance A_z:

$$A_{z\,(resonance)} = \frac{m_1 e}{m}\ \frac{B_z}{0.85\sqrt{B_z - 0.18}} \qquad (11.38)$$

where

m_1 = total rotating mass causing excitation
m = mass of the foundation and supporting machine

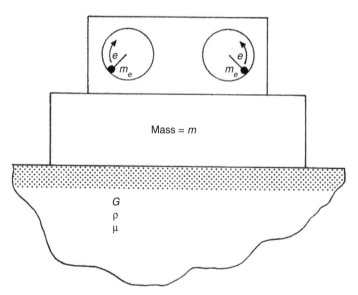

FIGURE 11.9 Foundation vibration (vertical) by a frequency-dependent exciting force.

Amplitude of vibration at frequencies other than resonance:

$$A_z = \frac{\left(\dfrac{m_1 e}{m}\right)\left(\dfrac{\omega}{\omega_n}\right)^2}{\sqrt{\left[1 - \left(\dfrac{\omega^2}{\omega_n^2}\right)\right]^2 + 4D_z^2\left(\dfrac{\omega^2}{\omega_n^2}\right)}} \tag{11.39}$$

Note that B_z, D_z, and ω_n are defined by Equations 11.31, 11.33, and 11.36, respectively.

Example 5

A foundation 6 m long and 2 m wide is subjected to a constant-force-type vertical vibration. The total weight of the machinery and foundation block $W = 670$ kN, unit weight of the soil $\gamma = 18$ kN/m³, $\mu = 0.4$, $G = 21,000$ kN/m², amplitude of the vibrating force $Q_0 = 7$ kN, and operating frequency $f = 180$ cpm. Determine:

   a. The resonant frequency
   b. The amplitude of vibration at resonance

Solution
Part a
This is a rectangular foundation, so the equivalent radius (Equation 11.29) is

$$r_0 = \sqrt{\frac{BL}{\pi}} = \sqrt{\frac{(2)(6)}{\pi}} = 1.95 \text{ m}$$

The mass ratio (Equation 11.31) is

$$B_z = \left(\frac{1-\mu}{4}\right)\left(\frac{m}{\rho r_0^3}\right) = \left(\frac{1-\mu}{4}\right)\left(\frac{W}{\gamma r_0^3}\right)$$

$$= \left(\frac{1-0.4}{4}\right)\left[\frac{670}{(18)(1.95)^3}\right] = 0.753$$

From Equation 11.30, the resonant frequency is

$$f_m = \left(\frac{1}{2\pi}\right)\left(\sqrt{\frac{G}{\rho}}\right)\left(\frac{1}{r_0}\right)\sqrt{\frac{B_z - 0.36}{B_z}}$$

$$= \left(\frac{1}{2\pi}\right)\left[\sqrt{\frac{21,000}{\left(\dfrac{18}{9.81}\right)}}\right]\left(\frac{1}{1.95}\right)\sqrt{\frac{0.753 - 0.36}{0.753}} = 6.3 \text{ cps} \approx 378 \text{ cpm}$$

Part b

From Equation 11.34:

$$A_{z(\text{resonance})} = \frac{Q_0(1-\mu)}{4Gr_0} \frac{B_z}{0.85\sqrt{B_z-0.18}}$$

$$= \left[\frac{(7)(1-0.4)}{(4)(21{,}000)(1.95)}\right]\left[\frac{0.753}{0.85\sqrt{0.753-0.18}}\right]$$

$$= 0.00003 \text{ m} = 0.03 \text{ mm}$$

11.5 Rocking Vibration of Foundations

11.5.1 Constant Force Excitation

Hall (1967) developed a mass-spring-dashpot model for rocking vibration of rigid circular foundations (Figure 11.10). According to this model,

$$I_0\ddot{\theta} + c_\theta\dot{\theta} + k_\theta\theta = M_y e^{i\omega t} \tag{11.40}$$

where

M_y = amplitude of the exciting moment
θ = rotation of the vertical axis of the foundation at any time t
I_0 = mass moment of inertia about the y-axis (i.e., axis perpendicular to the cross section passing through O) or

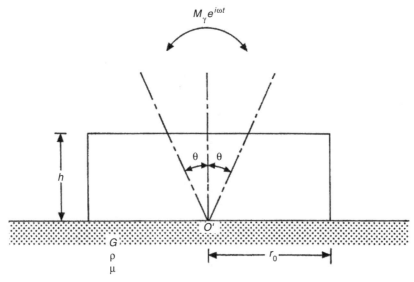

FIGURE 11.10 Rocking vibration of a foundation.

$$I_0 = \frac{W_0}{g} \left(\frac{r_0^2}{4} + \frac{h^2}{3} \right) \tag{11.41}$$

W_0 = weight of the foundation and machine
g = acceleration due to gravity
h = height of the foundation

$$k_\theta = \text{static spring constant} = \frac{8Gr_0^3}{3(1 - \mu)} \tag{11.42}$$

$$c_\theta = \text{dashpot coefficient} = \frac{0.8r_0^4 \sqrt{G}}{(1 - \mu)(1 + B_\theta)} \tag{11.43}$$

$$B_\theta = \text{inertia ratio} = \frac{3(1 - \mu)}{8} \frac{I_0}{\rho r_0^5} \tag{11.44}$$

Based on the solution of Equation 11.40, the resonant frequency f_m, the amplitude of vibration at resonant frequency $\theta_{\text{resonance}}$, and the amplitude of vibration at frequencies other than resonance θ are given by the following relationships:

$$f_m = \left(\frac{1}{2\pi} \sqrt{\frac{k_\theta}{I_0}} \right) \left(\sqrt{1 - 2D_\theta^2} \right) \tag{11.45}$$

$$D_\theta = \frac{0.15}{\sqrt{B_\theta} \, (1 + B_\theta)} \tag{11.46}$$

where D_θ is the damping ratio

$$\theta_{\text{resonance}} = \frac{M_y}{k_\theta} \frac{1}{2D_\theta \sqrt{1 - D_\theta^2}} \tag{11.47}$$

$$\theta = \frac{\left(\dfrac{M_y}{k_\theta} \right)}{\sqrt{\left[1 - \left(\dfrac{\omega^2}{\omega_n^2} \right) \right]^2 + 4D_\theta^2 \left(\dfrac{\omega^2}{\omega_n^2} \right)}} \tag{11.48}$$

$$\omega_n = \sqrt{\frac{k_\theta}{I_0}} \tag{11.49}$$

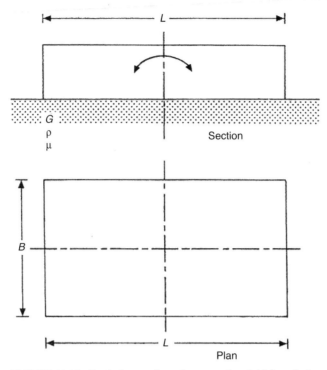

FIGURE 11.11 Equivalent radius of rectangular rigid foundation rocking motion.

In the case of rectangular foundations, the preceding relationships can be used by determining the equivalent radius as

$$r_0 = \sqrt[4]{\frac{BL^3}{3\pi}} \tag{11.50}$$

The definitions of B and L are shown in Figure 11.11.

11.5.2 Rotating Mass Excitation

Referring to Figure 11.12, for rocking vibration with rotating mass excitation, the relationships for f_m, $\theta_{\text{resonance}}$, and θ are as follows:

$$f_m = \left(\frac{1}{2\pi} \sqrt{\frac{k_\theta}{I_0}} \right) \left(\frac{1}{\sqrt{1 - 2D_\theta^2}} \right) \tag{11.51}$$

$$\theta_{\text{resonance}} = \frac{m_1 e z'}{I_0} \frac{1}{2D_\theta \sqrt{1 - D_\theta^2}} \tag{11.52}$$

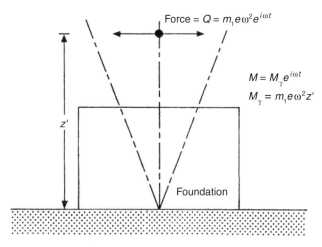

FIGURE 11.12 Rocking vibration due to rotating mass excitation.

$$\theta = \frac{\left(\dfrac{m_1 e z'}{I_0}\right)\left(\dfrac{\omega^2}{\omega_n^2}\right)}{\sqrt{\left[1 - \left(\dfrac{\omega^2}{\omega_n^2}\right)\right]^2 + 4D_\theta^2\left(\dfrac{\omega^2}{\omega_n^2}\right)}} \qquad (11.53)$$

The relationships for D_θ and ω_n are given in Equations 11.46 and 11.49, respectively.

Example 6

A horizontal piston-type compressor is shown in Figure 11.13. The operating frequency is 600 cpm. The amplitude of the horizontal unbalanced force of the compressor is 30 kN, and it creates a rocking motion of the foundation about point O (see Figure 11.13b). The mass moment of inertia of the compressor assembly about the axis $b'Ob'$ is 16×10^5 kg-m^2 (see Figure 11.13c). Determine:

   a. The resonant frequency
   b. The amplitude of rocking at resonance

Solution

The moment of inertia of the foundation block and the compressor assembly about $b'Ob'$ is

$$I_0 = \left(\frac{W_{\text{foundation block}}}{3g}\right)\left[\left(\frac{L}{2}\right)^2 + h^2\right] + 16 \times 10^5 \text{ kg-m}^2$$

Assume the unit weight of concrete is 23.58 kN/m^3.

$$W_{\text{foundation block}} = (8 \times 6 \times 3)(23.58) = 3395.52 \text{ kN} = 3395.52 \times 10^3 \text{ N}$$

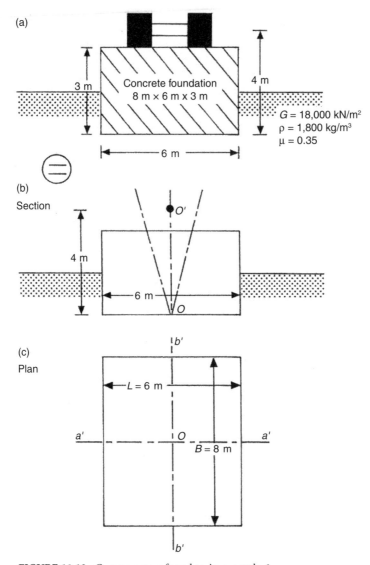

FIGURE 11.13 Compressor referred to in example 6.

$$I_0 = \frac{3395.52 \times 10^3}{(3)(9.81)} (3^2 + 3^2) + 16 \times 10^5 = 36.768 \times 10^5 \text{ kg-m}^2$$

From Equation 11.50, the equivalent radius of the foundation is

$$r_0 = \sqrt[4]{\frac{BL^3}{3\pi}} = \sqrt[4]{\frac{8 \times 6^3}{3\pi}} = 3.67 \text{ m}$$

Part a. Resonant Frequency

$$k_\theta = \frac{8Gr_0^3}{3(1 - \mu)} = \frac{(8)(18,000)(3.67)^3}{(3)(1 - 0.35)} = 3,650,279 \text{ kN-m/rad}$$

$$B_\theta = \frac{3(1 - \mu)}{8} \frac{I_0}{\rho r_0^5} = \frac{3(1 - 0.35)}{8} \frac{36.768 \times 10^5}{1800(3.67)^5} = 0.748$$

$$D_\theta = \frac{0.15}{\sqrt{B_\theta}\ (1 + B_\theta)} = \frac{0.15}{\sqrt{0.748}\ (1 + 0.748)} = 0.099$$

From Equation 11.51:

$$f_n = \left(\frac{1}{2\pi} \sqrt{\frac{k_\theta}{I_0}} \right) \left(\frac{1}{\sqrt{1 - 2D_\theta^2}} \right)$$

$$= \left(\frac{1}{2\pi} \sqrt{\frac{3,650,279 \times 10^3 \text{ N-m/rad}}{36.768 \times 10^5}} \right) \left[\frac{1}{\sqrt{1 - 2(0.099)^2}} \right]$$

$$= 5.05 \text{ cps} = 303 \text{ cpm}$$

Part b. Amplitude of Vibration at Resonance

$$M_{y(\text{operating frequency})} = \text{unbalanced force} \times 4 = 30 \times 4 = 120 \text{ kN-m}$$

$$M_{y(\text{at resonance})} = 120 \left(\frac{f_m}{f_{\text{operating}}} \right) = 120 \left(\frac{303}{600} \right)^2 = 30.6 \text{ kN-m}$$

$$(m_1 e \omega^2) z' = M_y$$

$$\omega_{\text{resonance}} = \frac{(2\pi)(303)}{60} = 31.73 \text{ rad/s}$$

$$m_1 e z' = \frac{M_y}{\omega^2} = \frac{30.6 \times 10^3 \text{ N-m}}{(31.73)^2} = 0.0304 \times 10^3$$

From Equation 11.52:

$$\theta_{resonance} = \frac{m_1 ez'}{I_0} \frac{1}{2D_\theta \sqrt{1 - D_\theta^2}}$$

$$= \left(\frac{0.0304 \times 10^3}{36.768 \times 10^5} \right) \left[\frac{1}{(2)(0.099) \sqrt{1 - (0.099)^2}} \right]$$

$$= 4.2 \times 10^{-5} \text{ rad}$$

11.6 Sliding Vibration of Foundations

Hall (1967) developed the mass-spring-dashpot analog for sliding vibration of a rigid circular foundation (Figure 11.14; radius = r_0). According to this analog, the equation of motion for the foundation can be given in the form

$$m\ddot{x} + c_x \dot{x} + k_x x = Q_0 e^{i\omega t} \tag{11.54}$$

where

m = mass of the foundation

$$k_x = \text{static spring constant for sliding} = \frac{32(1 - \mu)Gr_0}{7 - 8\mu} \tag{11.55}$$

$$c_x = \text{dashpot coefficient for sliding} = \frac{18.4(1 - \mu)}{7 - 8\mu} r_0^2 \sqrt{\rho G} \tag{11.56}$$

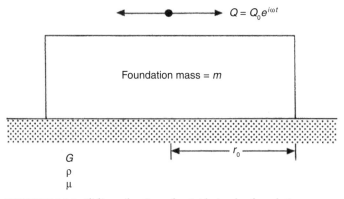

FIGURE 11.14 Sliding vibration of a rigid circular foundation.

For sliding vibration, the damping ratio in sliding D_x is

$$D_x = \frac{0.288}{\sqrt{B_x}} \qquad (11.57)$$

where the dimensionless mass ratio is

$$B_x = \frac{7 - 8\mu}{32(1 - \mu)} \frac{m}{\rho r_0^3} \qquad (11.58)$$

For rectangular foundations, the preceding relationships can be used by obtaining the equivalent radius r_0, or

$$r_0 = \sqrt{\frac{BL}{\pi}}$$

where B and L are the length and width of the foundation, respectively.

The resonant frequency f_m may be given as

$$f_m = \left(\frac{1}{2\pi} \sqrt{\frac{32(1 - \mu)Gr_0}{(7 - 8\mu)m}} \right) \sqrt{1 - 2D_x^2} \qquad (11.59)$$

for constant force excitation (i.e., Q_0 = constant) and

$$f_m = \left(\frac{1}{2\pi} \sqrt{\frac{32(1 - \mu)Gr_0}{(7 - 8\mu)m}} \right) \frac{1}{\sqrt{1 - 2D_x^2}} \qquad (11.60)$$

for rotating mass type of excitation.

Similarly, the amplitude at resonance is

$$A_{x(\text{resonance})} = \frac{Q_0}{k_x} \frac{1}{2D_x \sqrt{1 - D_x^2}} \qquad (11.61)$$

where $A_{x(\text{resonance})}$ = amplitude of vibration at resonance (for constant force excitation) and

$$A_{x(\text{resonance})} = \frac{m_1 e}{m} \frac{1}{2D_x \sqrt{1 - D_x^2}} \qquad (11.62)$$

where

m_1 = total rotating mass causing excitation

e = eccentricity of each rotating mass (for rotating mass excitation)

The amplitudes of vibration at frequencies other than resonance are

$$A_x = \frac{\left(\dfrac{Q_0}{k_x}\right)}{\sqrt{\left[1 - \left(\dfrac{\omega^2}{\omega_n^2}\right)\right]^2 + 4D_x^2\left(\dfrac{\omega^2}{\omega_n^2}\right)}} \tag{11.63}$$

for constant force excitation and

$$A_x = \frac{\left(\dfrac{m_1 e}{m}\right)\left(\dfrac{\omega}{\omega_n}\right)^2}{\sqrt{\left[1 - \left(\dfrac{\omega^2}{\omega_n^2}\right)\right]^2 + 4D_x^2\left(\dfrac{\omega^2}{\omega_n^2}\right)}} \tag{11.64}$$

where

$$\omega_n = \sqrt{\frac{k_x}{m}} \tag{11.65}$$

11.7 Torsional Vibration of Foundations

Similar to vertical, rocking, and sliding modes of vibration, the equation for the torsional vibration of a *rigid circular foundation* (Figure 11.15) can be written as

$$J_{zz}\ddot{\alpha} + c_\alpha\dot{\alpha} + k_\alpha\alpha = T_0 e^{i\omega t} \tag{11.66}$$

where

J_{zz} = mass moment of inertia of the foundation about the axis z–z

c_α = dashpot coefficient for torsional vibration

k_α = static spring constant for torsional vibration = $\dfrac{16}{3}Gr_0^3$ $\qquad$ (11.67)

α = rotation of the foundation at any time due to the application of a torque $T = T_0 e^{i\omega t}$

The damping ratio D_α for this mode of vibration was determined as (Richart et al. 1970)

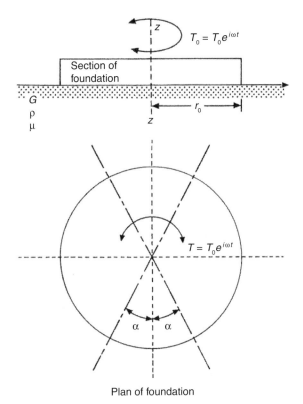

Plan of foundation

FIGURE 11.15   Torsional vibration of a rigid circular foundation.

$$D_\alpha = \frac{0.5}{1 + 2B_\alpha} \tag{11.68}$$

B_α is the dimensionless mass ratio for torsion at vibration:

$$B_\alpha = \frac{J_{zz}}{\rho r_0^5} \tag{11.69}$$

The resonant frequencies for torsional vibration are

$$f_m = \left(\frac{1}{2\pi} \sqrt{\frac{k_\alpha}{J_{zz}}} \right) \sqrt{1 - 2D_\alpha^2} \tag{11.70}$$

for constant force excitation and

$$f_m = \left(\frac{1}{2\pi} \sqrt{\frac{k_\alpha}{J_{zz}}} \right) \left(\frac{1}{\sqrt{1 - 2D_\alpha^2}} \right) \tag{11.71}$$

for rotating mass excitation (see Figure 11.15).

For constant force excitation, the amplitude of vibration at resonance is

$$\alpha_{resonance} = \frac{T_0}{k_\alpha} \frac{1}{2D_\alpha \sqrt{1 - D_\alpha^2}} \qquad (11.72)$$

and for rotating mass type of excitation is

$$\alpha_{resonance} = \frac{m_1 e \left(\dfrac{x}{2}\right)}{J_{zz}} \frac{1}{2D_\alpha \sqrt{1 - D_\alpha^2}} \qquad (11.73)$$

where

m_1 = total rotating mass causing excitation
e = eccentricity of each rotating mass (for rotating mass excitation)

For the definition of x in Equation 11.73, see Figure 11.16.

For a rectangular foundation with dimensions $B \times L$, the equivalent radius may be given by:

$$r_0 = \sqrt[4]{\frac{BL(B^2 + L^2)}{6\pi}} \qquad (11.74)$$

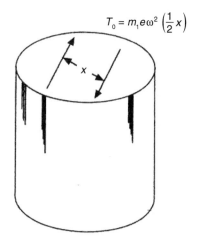

$$T_0 = m_1 e \omega^2 \left(\frac{1}{2}x\right)$$

FIGURE 11.16 Rotating mass excitation for torsional vibration.

Example 7

For a radar antenna foundation (shown in Figure 11.17) subjected to torsional vibration, $T_0 = 24.4 \times 10^4$ N-m, mass moment of inertia of the tower about the axis z–$z = 13.56 \times 10^6$ kg-m^2, and the unit weight of concrete used in the foundation = 23.68 kN/m^3. Calculate:

a. The resonant frequency for the torsional mode of vibration
b. The angular deflection at resonance

Solution
Part a

$$J_{zz} = J_{zz(tower)} + J_{zz(foundation)}$$

$$= 13.56 \times 10^6 + \frac{1}{2} \left[\pi r_0^2 h \left(\frac{23.58 \times 1000}{9.81} \right) \right] r_0^2$$

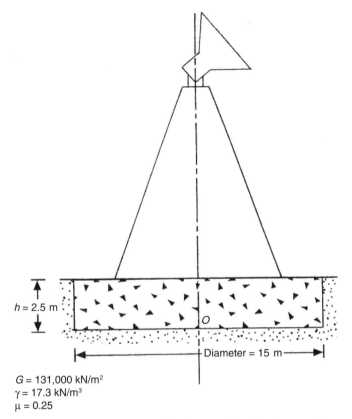

$h = 2.5$ m

Diameter = 15 m

$G = 131{,}000$ kN/m²
$\gamma = 17.3$ kN/m³
$\mu = 0.25$

FIGURE 11.17 Radar antenna foundation referred to in example 7.

$$= 13.56 \times 10^6 + \frac{1}{2}\left[(\pi)(7.5)^2(2.5)\left(\frac{23.58 \times 1000 \text{ N}}{9.81}\right)\right](7.5)^2$$

$$= 13.56 \times 10^6 + 29.87 \times 10^6 = 43.43 \times 10^6 \text{ kg-m}^2$$

$$B_\alpha = \frac{J_{zz}}{\rho r_0^5} = \frac{43.43 \times 10^6}{\left(\frac{17.3 \times 1000}{9.81}\right)(7.5)^3} = 1.038$$

$$D_\alpha = \frac{0.5}{1 + 2B_\alpha} = \frac{0.5}{1 + (2)(1.038)} = 0.163$$

$$f_m = \left(\frac{1}{2\pi}\sqrt{\frac{k_\alpha}{J_{zz}}}\right)\sqrt{1 - 2D_\alpha^2}$$

$$k_\alpha = \frac{16}{3} Gr_0^3 = \left(\frac{16}{3}\right)(131{,}000 \times 1000 \text{ N/m}^2)(7.5)^3 = 294{,}750 \times 10^6$$

$$f_m = \left(\frac{1}{2\pi} \sqrt{\frac{294{,}750 \times 10^6}{43.43 \times 10^6}}\right) \sqrt{1 - (2)(0.163)^2} = 12.76 \text{ cps}$$

Part b

$$\alpha_{\text{resonance}} = \frac{T_0}{k_\alpha} \frac{1}{2D_\alpha \sqrt{1 - D_\alpha^2}}$$

$$= \left(\frac{24.4 \times 10^4 \text{ N-m}}{294{,}750 \times 10^6}\right) \frac{1}{(2)(0.163) \sqrt{1 - (0.163)^2}}$$

$$= 0.257 \times 10^{-5} \text{ rad}$$

References

Hall, J.R., Jr. (1967). Coupled rocking and sliding oscillations of rigid circular footings. *Proceedings, International Symposium on Wave Propagation and Dynamic Properties of Earth Materials*, Albuquerque, NM, 139–148.

Hardin, B.O. and Black, W.L. (1968). Vibration modulus of normally consolidated clays. *J. Soil Mech. Found. Div. ASCE*, 94(SM2):353–369.

Hardin, B.O. and Drnevich, V.P. (1972). Shear modulus and damping in soils: design equations and curves. *J. Soil Mech. Found. Div. ASCE*, 98(SM7):667–692.

Lysmer, J. and Richart, F.E., Jr. (1966). Dynamic response of footings to vertical loading. *J. Soil Mech. Found. Div. ASCE*, 91(SM1):66–92.

Richart, F.E., Jr., Hall, J.R., and Woods, R.D. (1970). *Vibration of Soils and Foundations*, Prentice Hall, Englewood Cliffs, NJ.

12

Geosynthetics

by
Ahmet H. Aydilek
University of Maryland, College Park, Maryland

Tuncer B. Edil
University of Wisconsin-Madison, Madison, Wisconsin

12.1 Geosynthetic Structures and Manufacturing Types

Geosynthetics are polymeric man-made materials used to facilitate infrastructure and envi-ronmental projects. The utilization of geosynthetics in the construction industry has been growing continuously. It is a billion-dollar industry and more than 500 different geosynthetic products exist.

Currently, there are eight different types of geosynthetics on the market: geotextiles, geogrids, geomembranes, geocomposites, geonets, geosynthetic clay liners, geopipes, and geocells (Figure 12.1). The majority (~95%) of the geosynthetic products are manufactured from synthetic polymers. Polymers are chemically linked large molecules of carbon atoms with hydrogen or other atoms attached. Six different types of polymers are used in manufacturing geosynthetics: polyethylene, polypropylene, polyvinyl chloride, polyamide, polyester, and polystyrene.

Geotextiles are the most commonly used type of geosynthetic material. They are permeable fabrics which have the ability to separate, filter, reinforce, protect, or drain soils. The two types of geotextiles are woven and nonwoven. Geotextiles typically are grouped by their fiber type and manufacturing process. Woven geotextiles are fabricated from monofilament, multifila-ment, slit-film, and fibrillated yarn fibers, whereas the fiber types for nonwoven geotextiles are continuous filament and staple fiber (Figure 12.2). Woven geotextiles are manufactured using plain, twill, or satin weaving techniques. There are many other weaving techniques, such as basket, hopsack twill, triaxial, and leno weaves (Figure 12.3), but they are rarely used. Three major processes are used to bond the loose web of the nonwoven geotextile fibers: needle

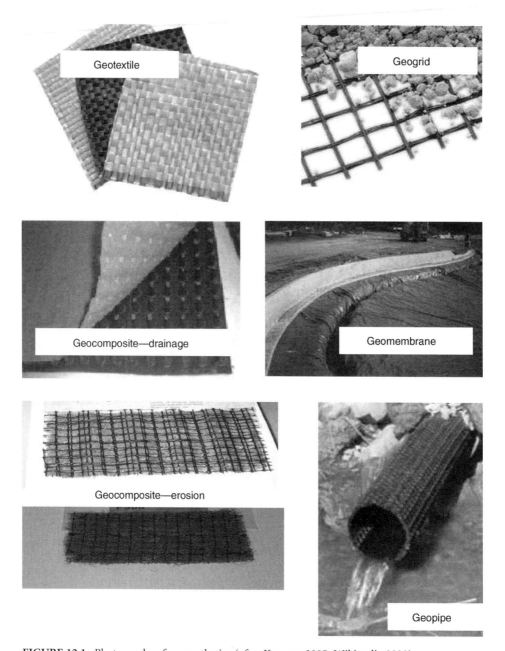

FIGURE 12.1   Photographs of geosynthetics (after Koerner 2005; Wikipedia 2008).

punching, thermal bonding (also called heat bonding or melt bonding), and chemical bonding. A fourth process, spun bonding, is used as a one-step complete manufacturing process from either the chemical or polymer stage to the finished geotextile on a roll.

Geomembranes are relatively impermeable membranes used in hydraulic barrier applications. Most of the geomembranes are manufactured from polyethylene and polyvinyl chloride using three different methods: extrusion, calendaring, and spread coating. All polyethylene geomembranes (high-density polyethylene, very-low-density polyethylene, etc.) are manufactured by two variations of the extrusion method: the flat die and circular die techniques. In the

Geonet

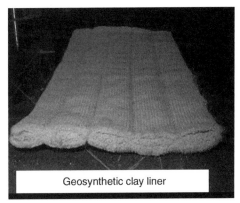

Geosynthetic clay liner

Geocell

FIGURE 12.1 *continued*

flat die technique, also called the cast sheeting technique, the polymer is forced into two horizontal lips and the thickness can be controlled from 0.7 to 3 mm. The thickness control is not precise in the circular die technique, which also is called the blown film technique. After the extrusion process, a high-friction surface can be obtained by texturing the geomembrane through a process of co-extrusion, lamination, or impingement. Polyvinyl chloride, scrim-reinforced, and some polyethylene (chlorosulphonated polyethylene) geomembranes are manufactured by the calendering method. After the polymer is mixed with additives, it is passed between two rotating rollers to form the final sheet. In the spread coating technique, molten polymer is spread as a coating onto a nonwoven or woven geotextile and the geomembrane is formed.

Geonets are grid-like materials and are used for their in-plane drainage capability. Geonets generally are used with one or two geotextiles on their upper and/or lower surfaces to prevent soil intrusion into the apertures, which would tend to block the in-plane drainage function of the material. *Geogrids* are polymeric materials that have an open grid-like appearance and consist of connected parallel sets of intersecting ribs with apertures that are large enough to interlock with the surrounding soil matrix. They are used for reinforcing soils and are manufactured as two types: uniaxial and biaxial geogrids. *Geopipes* are simply perforated or solid-wall polymeric pipes used for drainage of liquids. The extrusion technique is used in manufacturing geonets, geogrids, and geopipes.

Geosynthetic clay liners are prefabricated hydraulic barriers with bentonite clay incorporated between the geotextiles and/or geomembranes. They are used for liquid or solid waste

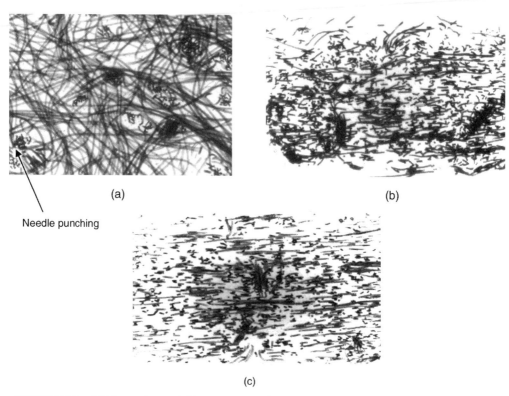

FIGURE 12.2   (a) Planar view and (b) cross-sectional view of staple fiber and (c) cross-sectional view of continuous filament nonwoven geotextile.

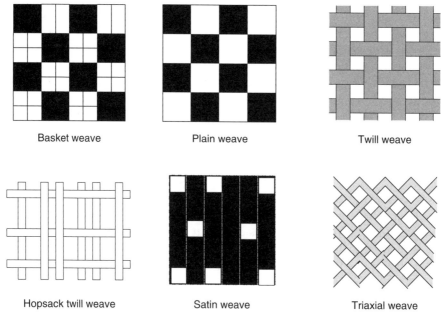

FIGURE 12.3 Types of woven geotextile manufacturing (after Smith 1993).

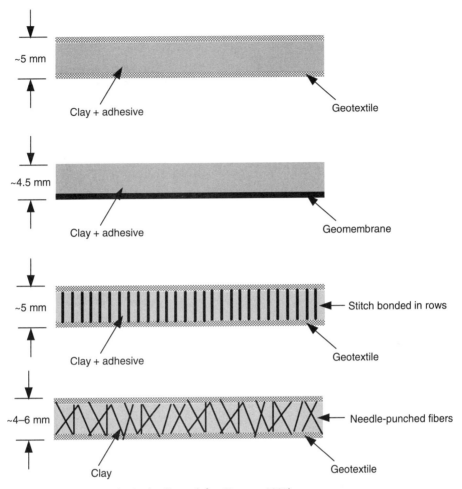

~5 mm

Clay + adhesive

Geotextile

~4.5 mm

Clay + adhesive

Geomembrane

~5 mm

Stitch bonded in rows

Clay + adhesive

Geotextile

~4–6 mm

Needle-punched fibers

Clay

Geotextile

FIGURE 12.4 Geosynthetic clay liners (after Koerner 2005).

containment. Currently four types of geosynthetic clay liners are available in North America, as shown in Figure 12.4. The top two shown in the figure are unreinforced geosynthetic clay liners and the bottom two are reinforced geosynthetic clay liners. The upper geotextiles in Figure 12.4 are usually woven and the lower ones are nonwoven.

Geocomposites consist of various combinations of geotextiles, geogrids, geonets, geomembranes, and other materials. They are used in drainage applications such as vertical (wick) drains, highway edge drains, and sheet drains; in erosion control systems; in containment systems as a moisture barrier; and in reinforcement applications (e.g., fibers and meshes). For instance, a prefabricated subsurface geocomposite drainage product consists of a geotextile filter material supported by a core, net, mesh, or spacer, and it collects liquids and/or gases and drains them off.

Geocells are relatively thick, three-dimensional networks constructed from strips of polymeric material. They are used in reinforcing walls and subbases in highway construction. In highway applications, they are placed on subsoil, filled with sand or gravel, and compacted. The surface is then sprayed with emulsified asphalt.

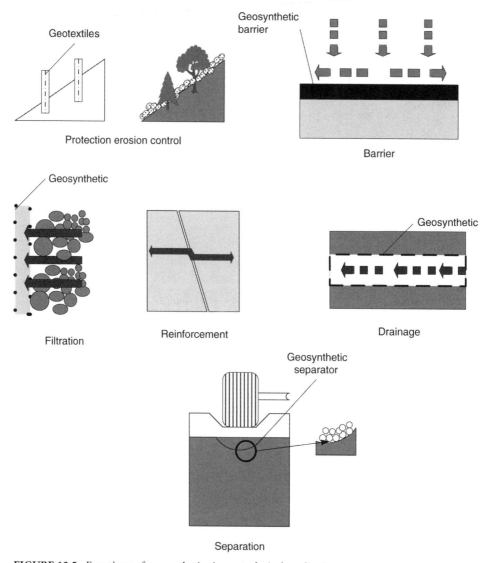

FIGURE 12.5 Functions of geosynthetics in geotechnical applications.

12.2 Functions of Geosynthetics

Geosynthetics perform six main functions: (1) filtration, (2) drainage, (3) reinforcement, (4) separation, (5) protection/erosion control, and (6) hydraulic barrier (Figure 12.5). These functions are described in detail in the following sections.

12.2.1 Filtration

Geotextiles are used primarily for filtration applications. Typical applications include dams, retaining structures (seepage control), and leachate collection systems (Figure 12.6). In a filtration application, the geotextile acts similar to a sand filter by allowing water to move

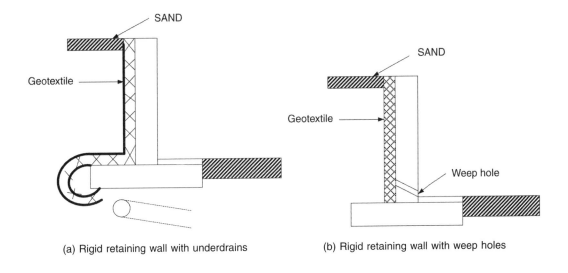

(a) Rigid retaining wall with underdrains   (b) Rigid retaining wall with weep holes

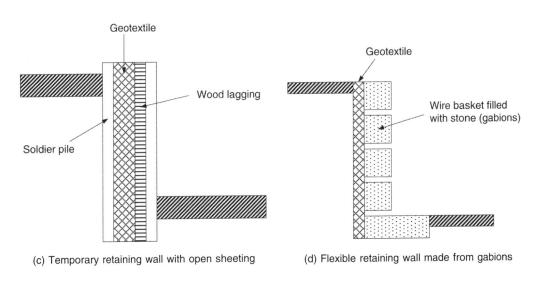

(c) Temporary retaining wall with open sheeting   (d) Flexible retaining wall made from gabions

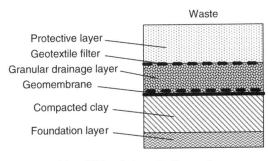

(e) Landfill leachate collection system

FIGURE 12.6 Geotextile filters in geotechnical and geoenvironmental engineering applications.

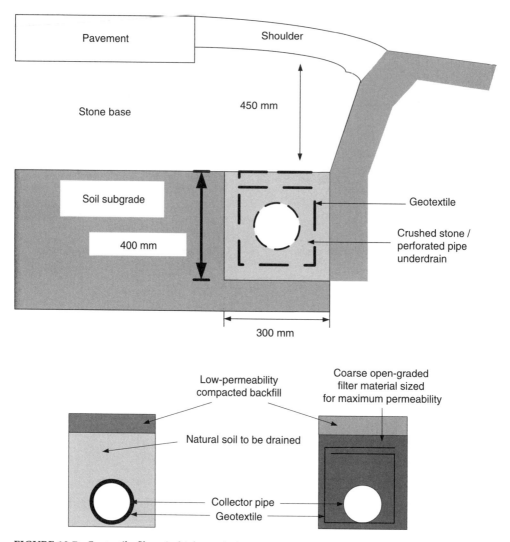

FIGURE 12.7   Geotextile filters in highway drainage systems.

through the soil while retaining upstream soil particles. Geotextiles are used to prevent soil erosion by keeping soils from migrating into drainage aggregate or pipes while maintaining flow through the system (Figure 12.7) as well as below riprap and other armor materials in coastal and stream bank protection systems (Figure 12.8).

12.2.1.1 Hydraulic Properties of Geotextiles

Hydraulic properties of geotextiles play an important role in designing for filtration applications. Three basic filter criteria are used for proper selection of a geotextile: (1) a retention requirement to prevent the migration of the soil particles through the geotextile, (2) a hydraulic conductivity requirement to ensure free flow of liquids through the geotextile, and (3) an anticlogging requirement to ensure that the geotextile will adequately meet the hydraulic conductivity and retention criteria throughout the life of the structure. Accordingly, filtration refers to adequate fluid flow with limited soil loss across the plane of the geotextile.

FIGURE 12.8 Geotextile filters in stream bank protection.

Pore structure parameters and permittivity are the main factors that affect filtration properties of geotextiles. Pore structure parameters include porosity, percent open area, and apparent opening size. Porosity is applicable to nonwoven geotextiles and can be calculated as

$$n = 1 - \frac{m}{\rho t} \tag{12.1}$$

where n is the porosity, m is the mass per unit area of the geotextile, ρ is the polymer density, and t is the geotextile thickness. Percent open area (POA) is applicable to woven geotextiles and is determined through counting the open areas in the geotextile:

$$POA = \frac{\text{Area of openings}}{\text{Total area of the geotextile sample}} \tag{12.2}$$

In order to determine apparent opening size (AOS), a series of beads with different but uniform diameters are sieved through the geotextile (ASTM D4751). AOS corresponds to the bead diameter when 5% of the beads of this diameter pass through the geotextile (i.e., O_{95}). This opening size also is termed the "largest opening in the geotextile." The dry sieving test (ASTM D4751) is sometimes used to determine pore sizes smaller than the AOS. However, this method is not very accurate for small pore sizes (e.g., smaller than O_{90}) due to electrostatic effects during testing. A better method to determine pore size is the bubble point test (ASTM D6767). In this method, the nonwetting fluid is extracted through the geotextile by applying a differential pressure. Then, the pressure is related to the pore size using

$$O = \frac{4T \cos \theta}{P} \tag{12.3}$$

where P is the absolute pressure being applied, O is the diameter of a pore that can be extruded by pressure P, T is the surface tension of a liquid against the sidewalls of a pore, and θ is the contact angle between the liquid and the pore wall. The bubble point test determines the constriction pore size (i.e., the smallest opening in a pore), albeit indirectly by approximating it from the measured minimum constriction area. The accuracy of the procedure described in ASTM D6767 has been verified for a wide range of pore sizes (Fischer 1994; Aydilek et al. 2007). Aydilek and Edil (2004) and Aydilek et al. (2007) showed that image analysis, an alternative technique, also can provide a direct and accurate measurement of pore sizes of geotextiles, particularly woven ones.

Permittivity of a geotextile is defined as its hydraulic conductivity divided by its thickness:

$$\psi = \frac{k}{t} \tag{12.4}$$

where k is the permeability (hydraulic conductivity), t is the thickness, and ψ is the permittivity of a geotextile. Since geotextiles have various thicknesses and compressibilities under applied loads, use of permittivity rather than permeability is considered to be more meaningful.

The retention performance of a geotextile is directly related to its pore structure. The other factors affecting the performance are type of flow and soil gradation. Piping of soil through geotextiles occurs if the large pore openings of the geotextile (e.g., O_{90}, O_{95}) are larger than the largest particles of the soil. This process usually is called internal erosion, and it changes the internal stability of the soil. Lafleur et al. (1989) suggested a piping rate of less than 0.25 g/cm^2 for granular filters, and this rate also is widely accepted for geotextile filters (Kutay and Aydilek 2005). The retention performance is particularly important in filtering contaminated soils and sludges and silt fence applications.

When the largest pore openings in the geotextile are much smaller than the smallest particles of the soil, then the fines in the soil close to the geotextile will be unable to pass through the geotextile. This will prevent the formation of an effective filter zone and may lead to blinding, blocking, or clogging of the geotextile (Figure 12.9). Blocking is encountered in woven geotextiles. Clogging is the intrusion of the soil particles inside the geotextile fibers and occurs in nonwoven geotextiles. Blinding refers to a soil buildup above the soil-geotextile interface that does not allow the passage of water flow.

Among various test methods to assess clogging performance, a widely used test is the gradient ratio (GR) test (ASTM D5101) (Figure 12.10). The method allows the determination of permeabilities and heads in the soil and soil-geotextile interface, as well as collection of the fines piped through the geotextile. GR is defined as the ratio of the hydraulic gradient at the soil-geotextile interface to the hydraulic gradient across the soil:

$$GR = \frac{i_{\text{soil-geotextile interface}}}{i_{\text{soil}}} \tag{12.5}$$

A GR greater than 1 signifies clogging according to the ASTM standard; however, a ratio up to 3 usually is acceptable (U.S. Army Corps of Engineers 1977; Haliburton and Wood 1982; Koerner 2005). ASTM D5101 requires 24-h testing before application of the next hydraulic gradient; however, recent studies have suggested that long-term testing is necessary to achieve stabilization (Fischer et al. 1999; Kutay and Aydilek 2005). In general, applied hydraulic gradient does not have a significant effect on filtration capacity of geotextiles.

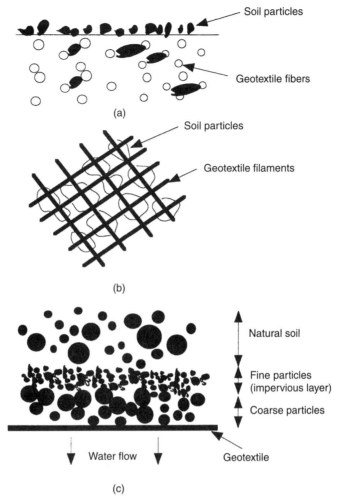

FIGURE 12.9 (a) Clogging, (b) blocking, and (c) blinding of a geotextile.

Biological clogging is a result of bacterial growth both on and in the fabric and is an important process in landfills, where enough nutrient and heat are supplied for bacterial growth. Geotextiles in solid waste landfill structures are exposed to a particular surrounding which affects their filtration performance. Especially in leachate filtration, access of bacteria to nutrients can impede the flow and cause the formation of ocher, bacterial adhesion, and biofilms, which ultimately may clog the filters. Test method ASTM D1987 (Figure 12.11) generally is used to determine the compatibility of soil/geotextile systems against biological clogging. The test is performed by recirculating leachate through the apparatus at a constant flow rate and measuring hydraulic conductivity intermittently. Test duration typically is set when stabilization of the flow rates is reached.

Koerner and Koerner (1990) conducted biological clogging tests using the leachates collected from six different landfills in the U.S. They tested 100 geotextile specimens and summarized the results as shown in Table 12.1. Based on their results, Koerner and Koerner (1990)

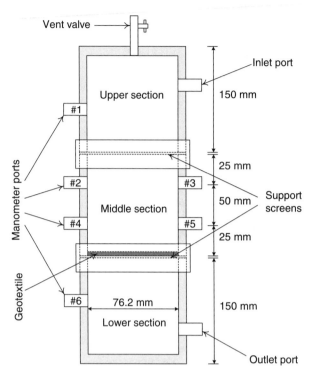

FIGURE 12.10   GR test apparatus.

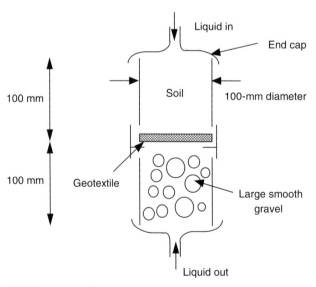

FIGURE 12.11   Biological clogging test device (after Koerner 2005).

TABLE 12.1 Type and Percent Clogging of Geotextiles Due to Biological Activity in Column Tests

Type of Clogging	% Flow Reduction	No. of Specimens
None	0–25	7
Minor (slow)	25–50	4
Moderate	50–75	38
Major (rapid)	75–95	36
Complete	95–100	15

After Koerner and Koerner (1990).

noted that geotextiles exhibit different clogging behavior as indicated by their flow rates, as shown in Figure 12.12. Backflushing of the clogged pipe and addition of biocide, a chemical substance capable of killing different forms of living organisms, are the two commonly used techniques in practice to increase the flow rates in landfill leachate collection systems.

12.2.1.2 Filtration Design

Geotextiles are used as filters in various applications, such as retaining walls, leachate collection systems, shoreline structures, and slopes, and they are expected to retain the majority of the soil particles, not to clog, and to have a high hydraulic conductivity such that it provides enough flow throughout the life of the structure. Luettich et al. (1992) also recommended that the survivability and durability of the geotextile during construction and throughout the life of the structure be taken into consideration and defined an eight-step filter design procedure:

Step 1. Define the application filter requirements:
- Identify the drainage media adjacent to the geotextile (e.g., voids, sharp contact points, etc.).
- Define the retention vs. hydraulic conductivity trade-off (i.e., retention will be important in the presence of a drainage material with little void volume, such as geonet, whereas for a gravel trench it is better to favor hydraulic conduction).

Step 2. Define the boundary conditions:
- Evaluate the confining stress (i.e., effect of high confining stresses on the retention performance of the geotextile).
- Define the flow conditions (i.e., steady state vs. dynamic).

Step 3. Determine the soil retention requirements:
- Define the soil grain size distribution.
- Define the soil Atterberg limits, density, and dispersion potential.
- Define the geotextile AOS (O_{95}).

Step 4. Determine the geotextile hydraulic conductivity requirements:
- Define the soil hydraulic conductivity (ASTM D5084).
- Define the hydraulic gradient for the application. Typical hydraulic gradients are given in Table 12.2.
- Determine the minimum allowable geotextile permittivity Ψ (ASTM D4491)

$$\Psi_{req} = \frac{k}{t} = \frac{q}{\Delta h/A} \qquad (12.6)$$

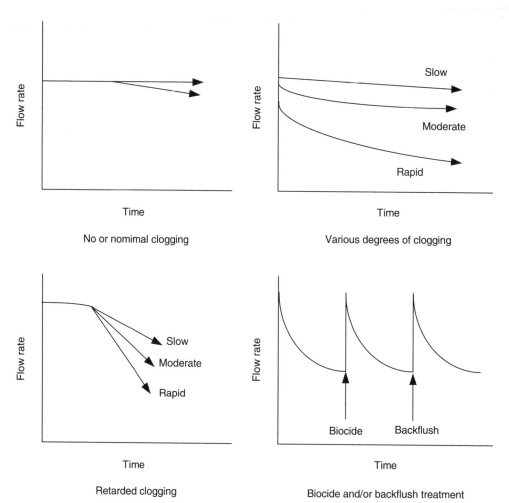

FIGURE 12.12 Typical biological clogging schemes for geotextiles (after Koerner 2005).

TABLE 12.2 Typical Hydraulic Gradients in Geotechnical Engineering Applications

Application	Typical Hydraulic Gradient
Standard dewatering trench	1.0
Vertical wall drain	1.5
Highway edge drain	1.0
Landfill leachate collection system	1.5
Dam toe drains	2.0
Dam clay cores	3.0–>10
Shoreline protection	10
Liquid impoundment with clay liners	>10

where q is the flow rate, Δh is the hydraulic head loss, and A is the cross-sectional area of the geotextile.

- Check against available allowable permittivity using

$$\Psi_{allowable} = \frac{k_g}{t} \left(\frac{1}{FS_{SCB} \times FS_{IN} \times FS_{CR} \times FS_{CC} \times FS_{BC}} \right) \quad (12.7)$$

where FS_{SCB}, FS_{IN}, FS_{CR}, FS_{CC}, and FS_{BC} are the partial factory factors for soil clogging-blinding, intrusion of adjacent materials into the geotextile, creep reduction, chemical clogging, and biological clogging, respectively. Theoretically, all factors of safety should be greater than 1.0; however, typically much greater values are recommended (Koerner 2005).

- Calculate the factor of safety:

$$FS = \frac{\Psi_{allowable}}{\Psi_{required}} \quad (12.8)$$

Step 5. Determine the anticlogging requirements:
- Use the AOS that satisfies the retention criteria. The criterion developed by Carroll (1983) is commonly used: $O_{95} < 2.5D_{85}$.
- For nonwoven geotextiles, use porosity $(n) > 30\%$.
- For woven geotextiles, use POA > 4%.

Step 6. Determine the survivability requirements/potential damage to the geotextile due to the adjacent materials and the construction technique (i.e., specify minimum index strength properties in regard to severity of the project).

Step 7. Determine the durability requirements, such as degradation of the geotextile due to exposure to sunlight and chemicals.

Step 8. Other design considerations:
- Intrusion of geotextile into the drainage layer
- Abrasion of the geotextile due to dynamic action
- Intimate contact of the soil and geotextile
- Biological and biochemical clogging factors

12.2.2 Drainage

Drainage refers to in-plane flow of water. Geonets, geocomposites, and nonwoven geotextiles are used for drainage applications. Typical applications include highway edge drains, landfill cover, and leachate collection systems (Figures 12.13 and 12.14).

12.2.2.1 Hydraulic Properties of Geosynthetics for Drainage Applications

The most essential property that affects the drainage performance of geosynthetics is transmissivity. Transmissivity is the amount of water flow within the plane of a geotextile under a certain hydraulic gradient, calculated as

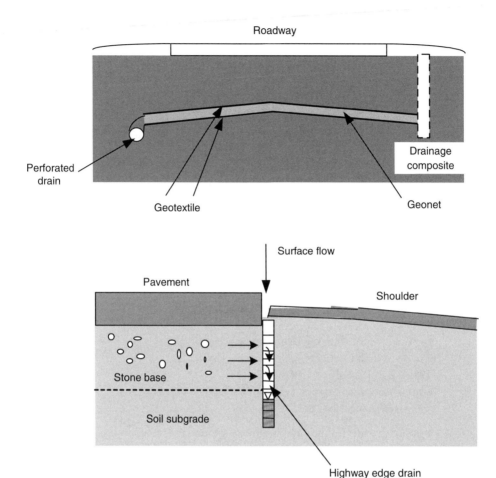

FIGURE 12.13 Geosynthetics in highway edge drains (after Koerner 2005).

$$\theta = kt = \frac{q}{iW} \qquad (12.9)$$

where θ is the transmissivity, q is the flow rate (amount of flow per unit area), W is the geotextile width, and the other terms are as defined previously.

Applied normal stress and hydraulic gradient have significant effects on the transmissivity values of geonets and needle-punched geotextiles. Another factor that affects drainage capacity is the creep of the geonet. Creep is highly dependent on polymer density of the geonet, temperature, and magnitude of the applied stress.

Geonets are used primarily in drainage applications. They are always used with other geosynthetics, mostly a geotextile or geomembrane, to prevent soil intrusion into their apertures. This intrusion can decrease the drainage capacity tremendously. Laboratory drainage tests should be run with composites (e.g., geonet + geomembrane or geonet + geotextile) to predict field performance (Figure 12.15).

Geocomposite drains are widely used in drainage applications. Main types include wick drains (prefabricated vertical drains), sheet drains, and highway edge drains. Wick drains (a

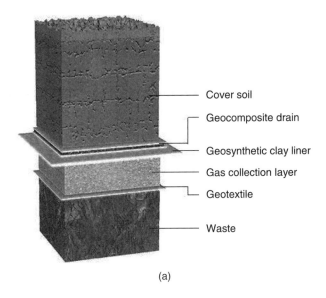

Cover soil
Geocomposite drain
Geosynthetic clay liner
Gas collection layer
Geotextile

Waste

(a)

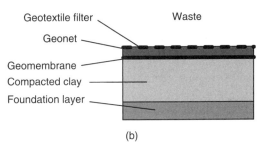

Geotextile filter   Waste
Geonet
Geomembrane
Compacted clay
Foundation layer

(b)

FIGURE 12.14 (a) Geosynthetics in a landfill cover system and (b) geonet or geocomposite drainage layer in a landfill leachate collection system.

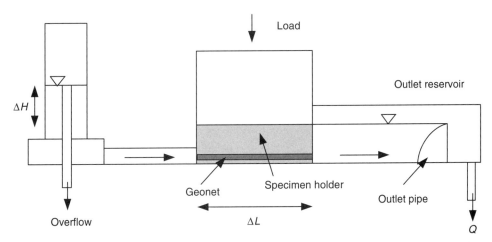

Load

Outlet reservoir

ΔH

Overflow   Geonet   Specimen holder   Outlet pipe

ΔL

Q

FIGURE 12.15 Transmissivity device for testing drainage capacity of geonets and geocomposites (after Koerner 2005).

plastic fluted core surrounded by a geotextile filter) have nearly replaced the conventional sand drains to accelerate the consolidation of soft clays. Geocomposites are clean, easy to place, and hard to clog. In some cases, the upper or lower geotextile may experience soil smear, and kinking of the drain element may cause a decrease in flow. Sheet drains perform the equivalent duty as geonets. They can be used in retaining walls, as drainage inceptors, and beneath floor slabs. Highway edge drains generally are used to drain the highway stone bases. The efficiency of the edge drain is dependent on pavement type, thickness of the stone base, system gradient, applied normal stress, and precipitation.

12.2.2.2 Drainage Design

Holtz et al. (1997) provided guidelines for drainage design with geosynthetics as follows:

Step 1. Evaluate the site conditions and critical nature of the application.

Step 2. Obtain soil samples from the site:
- Perform grain size distribution analysis.
- Perform field or laboratory hydraulic conductivity tests.

Step 3. Calculate anticipated flow into and through the drainage system:
- Use Darcy's law.
- Specific drainage systems include flow into trenches (Mansur and Kaufman 1962), horizontal blanket drains, and slope drains (Cedergren 1989).

Step 4. Determine geotextile requirements:
- Retention criterion: $\text{AOS} \leq BD_{85}$

For <50% passing No. 200 sieve		*For >50% passing No. 200 sieve*	
$B = 1$	For $C_u < 2$ or > 8	$B = 1$	For wovens
$B = 0.5 C_u$	For $2 < C_u < 4$	$B = 1.8$ and	For nonwovens
		$\text{AOS} < 0.3$ mm	
$B = 8/C_u$	For $4 < C_u < 8$		

- Clogging criterion:

Less critical/less severe	$O_{95} > 3D_{15}$
	Porosity $> 50\%$ or POA $> 4\%$
Critical/severe	Perform filtration test

- Hydraulic conductivity/permittivity criterion:

Severity of the project	Less critical/less severe	$k_{\text{geotextile}} > k_{\text{soil}}$
	Critical/severe	$k_{\text{geotextile}} > 10 \, k_{\text{soil}}$
Permittivity requirements	For <15% passing No. 200 sieve	$\Psi > 0.5$ s^{-1}
	For 15–50% passing No. 200 sieve	$\Psi > 0.2$ s^{-1}
	For >50% passing No. 200 sieve	$\Psi > 0.1$ s^{-1}

- Transmissivity requirements:
  Calculate the required transmissivity of the geosynthetic per Equation 12.9.
  Check against available transmissivity using

$$\theta_{\text{allowable}} = k_p t \left(\frac{1}{\text{FS}_{\text{SCB}} \times \text{FS}_{\text{IN}} \times \text{FS}_{\text{CR}} \times \text{FS}_{\text{CC}} \times \text{FS}_{\text{BC}}} \right) \quad (12.10)$$

where k_p is the in-plane hydraulic conductivity and t is the thickness of the geosynthetic. See Section 12.2.1 for the definition of partial FS values.

- Calculate the flow factor of safety:

$$FS = \frac{\theta_{allowable}}{\theta_{required}} \qquad (12.11)$$

Step 5. Collect samples of aggregate and geosynthetic before acceptance.

Step 6. Monitor installation during and after construction.

Step 7. Observe drainage system during and after storm events.

12.2.3 Reinforcement

Geotextiles and geogrids are used primarily for reinforcement. Typical applications are retaining structures such as walls (Figure 12.16), slopes, and embankments on soft ground (Figures 12.17–12.19). The geosynthetic acts as a tensile reinforcement element within a soil mass or in combination with the soil to produce a composite that has improved strength and deformation properties over the unreinforced soil.

12.2.3.1 Mechanical Properties of Geosynthetics for Reinforcement Applications

The most important property of geotextiles and geogrids for reinforcement applications is their tensile strength. Tensile strength tests are performed on geotextiles for two different purposes: quality control and determination of the load-strain characteristics. The grab strength test (ASTM D4632) is the most commonly used test for quality control. The test is conducted on a 100-mm-wide specimen with a 25-mm grip width. Load-strain characteristics typically are determined through wide-width tensile strength tests (Figure 12.20) conducted on 100-mm-long × 200-mm-wide specimens at a strain rate of 10 mm/min (ASTM D4595 for geotextiles and ASTM D6637 for geogrids). Typical load-strain curves for a geogrid and woven geotextile are given in Figure 12.21, along with example curves for two soils.

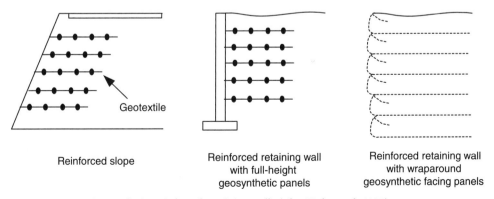

Reinforced slope

Reinforced retaining wall with full-height geosynthetic panels

Reinforced retaining wall with wraparound geosynthetic facing panels

FIGURE 12.16 Geosynthetic-reinforced retaining walls (after Holtz et al. 1997).

FIGURE 12.17 Embankment on soft ground.

In the field, geogrids are attached with wires, whereas geotextile rolls are joined together by some form of seaming. A common technique is "overlapping." A minimum overlap is 0.3 m, and greater overlap is required for specific applications. Another technique is "sewing" the geotextile rolls. The load transfer from one roll to the other roll is evaluated through a seam strength test. A 100-mm-long × 250-mm-wide specimen is tested (Figure 12.22), and seam efficiency is calculated through:

$$\text{Seam efficiency} = \frac{\text{Seam strength}}{\text{Wide-width tensile strength}} \times 100 \qquad (12.12)$$

The tensile strength properties of geogrids are different than those of geotextiles due to their different structure. Ribs and junctions (nodes) are the two main components of geogrids, and therefore they should be tested for strength separately. Furthermore, both directions should be tested in the case of biaxial geogrid specimens.

Creep is the deformation of a geotextile or a geogrid under a constant load and is determined through a creep test (ASTM D5262). Various factors affect the creep behavior of a geosynthetic, including temperature, humidity, testing duration, manufacturing method, and percentage of tensile strength applied (i.e., 20, 40, or 60%). Polymer type of the geosynthetic is another factor that affects creep. For instance, polyethylene and polypropylene are more prone to creep behavior than polyester. Geogrids manufactured from polyethylene or polypropylene typically are used as permanent reinforcement materials and therefore are more likely to exhibit creep behavior.

In order to analyze the field performance of geosynthetics in reinforcement applications, tests should be performed under specific design conditions and should refer to the particular

FIGURE 12.18   Road before (top) and after (bottom) reinforcement with geosynthetics.

soil of interest. These tests typically include interface shear tests and pullout tests. Geosynthetics often are used in structural fills as internal reinforcement (mechanically stabilized earth) or as a base reinforcement (embankments over soft foundations). Typical design methods for these applications require soil-geosynthetic interface strength properties. The most popular test setup is the soil-geosynthetic interface shear test (ASTM D5321 for large-scale interface

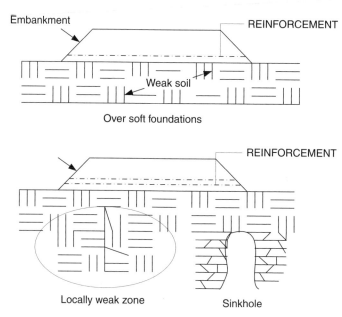

FIGURE 12.19 Typical cases when geosynthetic reinforcement is required for embankment construction.

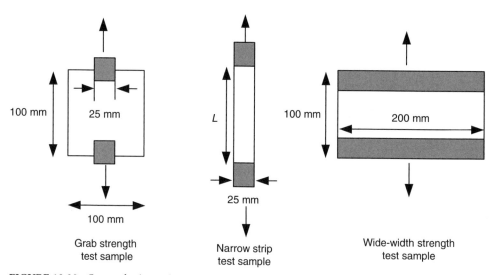

FIGURE 12.20 Geosynthetic tension test specimens (after Koerner 2005).

direct shear test) (Figure 12.23). The test consists of displacing soil subjected to a normal stress across a geosynthetic and measuring the resistance. Typical box dimensions are 300 mm × 300 mm. Due to its size, the test setup allows foundation soil heterogeneity to be taken into consideration.

Table 12.3 shows the soil-geotextile friction angles obtained for different cohesionless soils tested with various geotextiles. The soil-geotextile interface friction angle is likely to increase with angularity of sand particles. Sharma and Lewis (1994) compared the interface shear

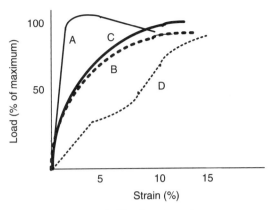

FIGURE 12.21 Typical load-strain performance curves
of geogrid, geotextile, and soils: A = graded granular fill,
B = clay, C = geogrid, and D = woven geotextile (after
Wrigley 1989).

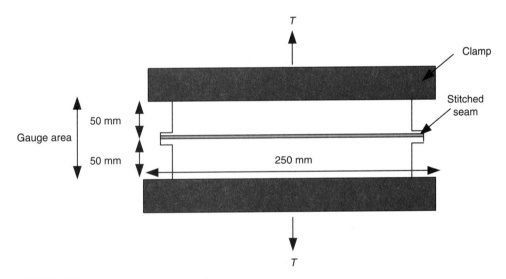

FIGURE 12.22 Seam strength test specimen.

properties of various geotextiles with two types of soil. Fine-grained cohesive clays provided
lower angles than sands in large-scale interface shear tests (Table 12.4).

An alternative test method occasionally used to determine the soil-geosynthetic interface
properties is the ring shear (or torsional ring shear) test. A circular specimen is subjected to
a normal stress and sheared. The test originally was developed for soil specimens (Skempton
1964) and modified for testing of soil-geosynthetic composites (Stark and Poeppel 1994).
Although the test method uses small interface surfaces, it provides continuous shear deformation and could be useful if unlimited deformation measurements are needed. Lower residual
shear strengths (when compared with those measured using a large-scale shear box) generally
are recorded due to allowed large continuous displacement.

The field performance of geotextiles in soil backfills often is determined through pullout
tests (Figure 12.24). In this test method, usually a geosynthetic is sandwiched between two soil

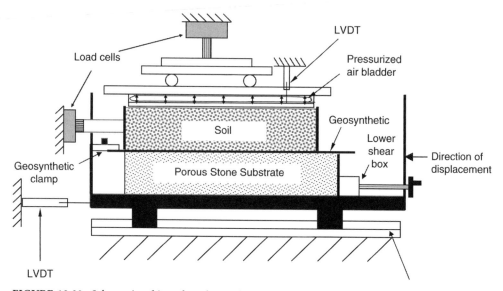

FIGURE 12.23 Schematic of interface direct shear test setup. (LVDT = linear variable differential transducer.)

TABLE 12.3 Effect of Soil Angularity on Soil-Geotextile Interface Friction Angle

Geotextile Type	Concrete Sand $\phi_{sand} = 30°$	Rounded Sand $\phi_{sand} = 28°$	Silty Sand $\phi_{sand} = 26°$
Woven, monofilament	26°	—	—
Woven, slit film	24°	24°	23°
Nonwoven, heat bonded	26°	—	—
Nonwoven, needle punched	30°	26°	25°

Compiled by Koerner (2005).

TABLE 12.4 Effect of Soil Type on Soil-Geotextile Interface Friction Angle

Geosynthetic Type	Sand	Clay
Woven geotextile	23–42°	16–26°
Nonwoven, needle-punched geotextile	25–44°	15–28°
Nonwoven, heat-bonded geotextile	22–40°	17–33°

layers under a normal stress and pulled out. The pullout test method evaluates the anchorage behavior of geosynthetics in reinforcement applications (e.g., geosynthetic-reinforced retaining walls).

The resistance of a geosynthetic to pullout has two main components: friction (all geosynthetics) and rib bearing (geogrids only). Friction develops between the upper and lower surfaces of the geosynthetic and the surrounding soil. Rib bearing is the passive resistance put forth against the transverse members of a geogrid by the soil. The soil provides this resistance by "strike-through," which means that the soil particles protrude through the apertures in the geogrid and cause bearing on the geogrid (Koerner 2005).

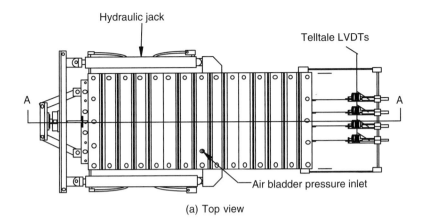

(a) Top view

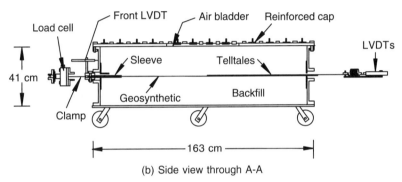

(b) Side view through A-A

FIGURE 12.24 Schematic of the pullout box. (LVDT = linear variable differential transducer.)

Because geosynthetics are extensible, progressive failure often occurs along the interface. The geosynthetic begins to move at the clamped end, but the magnitude of the displacement diminishes with distance from the clamp. Therefore, the interface friction angle is difficult to determine since the soil-geosynthetic interaction area constantly changes during the test. The progressive failure varies with normal stress. At low stresses, the geosynthetic fails progressively until the entire length of it displaces. At high normal stresses, the geosynthetic becomes anchored at a given distance from the front of the box, and only a portion of the interface experiences displacement and shearing resistance is not mobilized along the entire surface. The pullout data are evaluated using a parameter called the interaction coefficient, C_i:

$$C_i = \frac{\text{Maximum pullout resistance}}{2\,\text{Cross-sectional area of the specimen} \times \text{Shear strength of soil}}$$

$$= \frac{P}{2WL(c + \sigma_n \tan \varphi)}$$

(12.13)

where W and L are the specimen width and length, respectively, and c is cohesion. Typical conditions for a range of C_i values are

$C_i < 0.5$ Poor bond between the soil and geosynthetic, breakage of the geosynthetic
$C_i = 0.5\text{--}1.0$ Good bond between the soil and geosynthetic
$C_i > 1$ Interlocking between the soil grains and geosynthetic (geogrid)

Smoothness of the geosynthetic usually yields to a coefficient of less than 0.5. C_i is greater than 1 when an additional passive resistance generated by the interlocking between the soil grains and the geogrid is present.

12.2.3.2 Reinforcement Design

Geosynthetics typically are designed to reinforce embankments on soft ground, slopes, and walls. Holtz et al. (1997) provided guidelines for the design of geosynthetic-reinforced embankments on soft ground as follows:

Step 1. Define embankment dimensions and loading conditions.
Step 2. Determine engineering properties of the foundation soil and embankment fill material.
Step 3. Select factors of safety for design.
Step 4. Check against bearing capacity, rotational failure, and lateral stability (Figure 12.25). Classical bearing capacity theory is used for checking against bearing capacity failure. Factors of safety against rotational and sliding failures are given in Figures 12.26 and 12.27. If the factor of safety is lower than the minimum required for design, the strength of the geotextile T can be increased to provide an adequate factor of safety.
Step 5. Select a geosynthetic based on design strength, modulus, seam strength, soil-geosynthetic interface friction angle, and survivability requirements.
Step 6. Follow the construction sequence procedures recommended by Holtz et al. (1997).

Construction of embankments on soft clays and organic soils poses major challenges. The foremost problems faced in the construction of embankments on soft clays are bearing capacity failure, large lateral displacement, and excessive total and differential settlements. Several methods have been used to prevent these problems. Piled or column-supported embankment is, in many cases, faster than other methods and can be cost effective.

The conventional piled embankment system utilizes only arching of the granular soil in the embankment to transfer most of the loads to the piles instead of the subsoil (Figure 12.28). For this reason, the pile spacing has to be small and inclined piles are required at the outer edges of the embankment to counter the lateral thrust of the embankment. Recent development in geosynthetics has allowed the geosynthetics together with specified granular materials to be used as reinforcement at the base of the embankment to bridge the gaps in between the piles/columns that will transmit embankment loads onto the columns more efficiently. This layer is called the load transfer platform (LTP). LTPs also may be composed of reinforced concrete slab or, conventionally, thick unreinforced granular fill.

Currently there are two commonly used concepts for the design of a geosynthetic-reinforced LTP (GRLTP) for column-supported embankments (Jenner et al. 1998): (1) the tension membrane approach (catenary LTP) and (2) the improved arching approach (beam LTP) (see Figure 12.29). In the catenary LTP, only a single or at the most two layers of high-strength geogrids are used as reinforcement (BS 8006 1995). The reinforcement in essence behaves as a structural element, and any benefits resulting from the creation of a composite-reinforced soil mass are ignored. In the beam LTP, three or more layers of relatively lower strength

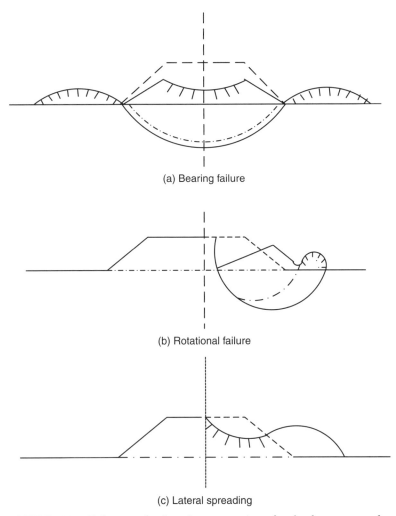

(a) Bearing failure

(b) Rotational failure

(c) Lateral spreading

FIGURE 12.25 Failure mechanisms in construction of embankments on soft ground (after Holtz et al. 1997).

geogrids are employed as reinforcement within a select aggregate. The beam LTP is claimed to be stiffer than the catenary LTP and therefore theoretically is able to take more load with less deflection (sagging) at the base of the LTP and consequently less deformation at the surface of the embankment than the catenary LTP. Han and Gabr (2002) and Pham et al. (2004) indicated that there are many factors that may affect the fill-LTP-column-subsoil interaction. The major contentious issues include the appropriate arching theory/method to use, the manner load by which is transferred from the geosynthetic or LTP to the columns, and the proportion of the load supported by the subsoil.

Abdullah and Edil (2007) constructed a heavily instrumented test embankment to evaluate the performance of different GRLTP designs for column-supported embankments. All the LTPs supported on columns performed satisfactorily in terms of total and differential settlements at the base of the embankment. The design concepts for the catenary and beam GRLTPs are somewhat different, but both design methods worked satisfactorily. The beam method was less expensive in this case because of the high cost of the high-strength geogrid in the catenary

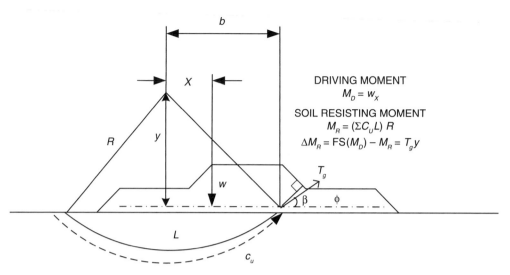

FIGURE 12.26 Rotational failure model for geosynthetic-reinforced embankments constructed on soft ground (after Holtz et al. 1997).

LTP, and it appeared to provide the closest to the field arching ratio (Figure 12.30) compared to other arching ratio methods.

Geotextiles and geogrids are increasingly being used in reinforcing highways, in particular the soft subgrades and bases. Subgrade soils have a major impact on the design, construction, structural response, and performance of pavements. Problems arise during construction with unstable subgrade when placing and compacting subbase and base materials and in providing adequate support for subsequent paving operations. Once the pavement is commissioned, pavement structural responses are influenced by the subgrade. The effects of a soft subgrade can be manifested as potholes, upheaval, rutting, and shoving (Yoder and Witczak 1975). Cracking and rutting are the principal types of distress encountered in flexible pavements (Huang 1993). These phenomena are particularly common in areas where soft and compressible subgrade is encountered.

One alternative is to use cellular confinement systems, often known as geocells, that increase the strength and stiffness of soils. The main contribution is the increase in bearing capacity of the shallow foundation. Geocells are three-dimensional mats consisting of polymeric webs that are collapsed into a sheet for shipment. When expanded at the job site, they can be filled with aggregates or other geomaterials to provide a stiff structural element. Bathurst and Jarrett (1988) proposed that the improvement is brought about by three mechanisms: (1) hoop strength in the cell wall that develops from the stiffness of the cell, (2) passive resistance by adjacent filled cells, and (3) friction developed between the soil and cell wall. Hoop stress and passive resistance are responsible for the improvement in lateral spreading, whereas friction increases the shear strength. The geocell-induced increase in shear strength (2τ) ultimately improves the bearing capacity of the foundation, as detailed in the following equation (Koerner 2005):

$$p = 2\tau + cN_c\zeta_c + qN_q\zeta_q + 0.5\gamma BN_\gamma\zeta_\gamma \qquad (12.14)$$

where p is the maximum bearing capacity stress; c is the soil cohesion; q is the surcharge load; B is the width of the applied pressure system (e.g., geocell); γ is the unit weight of the soil in

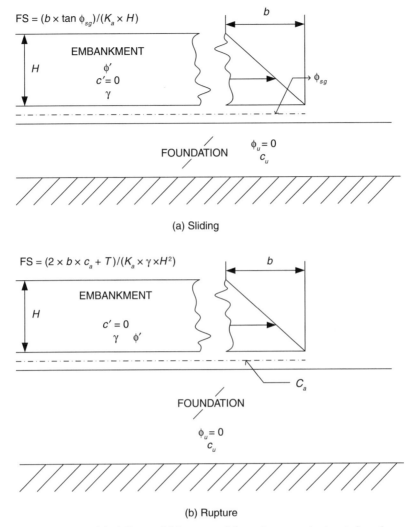

FIGURE 12.27 (a) Sliding and (b) rupture failures in geosynthetic-reinforced embankments constructed on soft ground (after Holtz et al. 1997).

the failure zone; N_c, N_q, and N_γ are the bearing capacity factors; and ζ_c, ζ_q, and ζ_γ are the shape factors to account for differences from the plane strain assumption.

Several approaches to geosynthetic-reinforced retaining wall design have been proposed, and the most commonly used one is the tieback wedge method. In this method, the classical Rankine earth pressure theory is combined with tensile-resistant tiebacks, and the reinforcement provided by the geosynthetic is assumed to extend beyond the Rankine failure plane. Holtz et al. (1997) listed the basic design steps for geosynthetic-reinforced retaining walls as follows, and further details can be found in the reference:

Step 1. Define design limits, wall dimensions, and loading conditions.
Step 2. Determine engineering properties of the foundation soil, reinforced soil, and backfill soil.
Step 3. Select factors of safety for design.

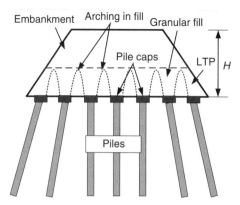

FIGURE 12.28 Conventional piled embankment system.

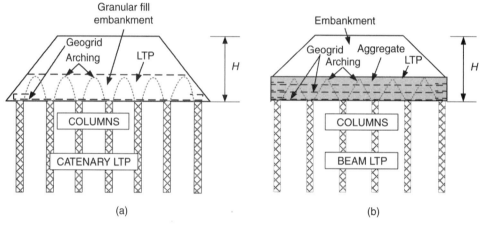

FIGURE 12.29 (a) Catenary and (b) beam-type reinforced LTP for column-supported embankments.

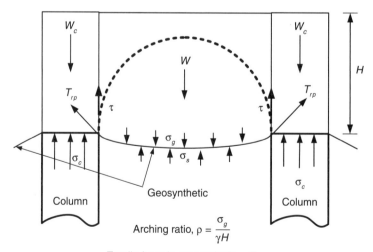

FIGURE 12.30 Forces and stresses acting during GRLTP design.

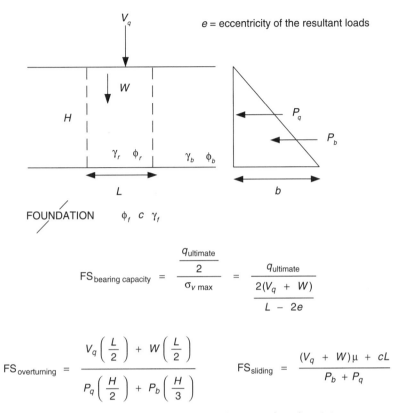

$$FS_{bearing\ capacity} = \dfrac{\dfrac{q_{ultimate}}{2}}{\sigma_{v\ max}} = \dfrac{q_{ultimate}}{\dfrac{2(V_q + W)}{L - 2e}}$$

$$FS_{overturning} = \dfrac{V_q\left(\dfrac{L}{2}\right) + W\left(\dfrac{L}{2}\right)}{P_q\left(\dfrac{H}{2}\right) + P_b\left(\dfrac{H}{3}\right)} \qquad FS_{sliding} = \dfrac{(V_q + W)\mu + cL}{P_b + P_q}$$

FIGURE 12.31 Distribution of forces in geosynthetic-reinforced retaining structures.

Step 4. Check bearing capacity of the foundation and sliding and overturning stability of the wall. Classical bearing capacity theory is used for checking against bearing capacity failures. Factors of safety against bearing capacity, rotational, and sliding failures are given in Figure 12.31.

Geotextiles and geogrids also are used in reinforcing unstable natural and man-made slopes (Figure 12.32). For most slopes, rotational (global) stability is of concern. This can be checked via well-known techniques, such as modified Bishop's analysis and the Janbu method. A number of slope stability analysis computer programs use these solution techniques (e.g., Stabl, XStabl). The stability of the unreinforced slope is first analyzed and the analysis is repeated by introducing geosynthetic reinforcement until an adequate factor of safety is established. Design considerations also include vertical spacing and anchorage depth of the geosynthetic. Similar to embankments, a sliding analysis should be conducted on reinforced slopes to evaluate if the reinforced mass is wide enough to resist sliding along the reinforcement. Further details on slope stability analysis are given in Chapter 6 of this handbook, and design details for reinforced slopes can be found in Holtz et al. (1997).

12.2.4 Separation

Geotextiles are used in separation applications. Roadways (unpaved and paved) and railroads are the primary application areas. The geosynthetic increases the stability and improves the

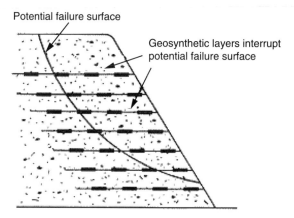

FIGURE 12.32 Geosynthetic-reinforced slope.

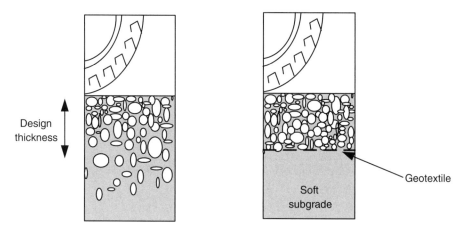

FIGURE 12.33 Geosynthetic separators in roadway applications (after Rankilor 1981; Holtz et al. 1997).

performance of the weak subgrade soil by separating the aggregate from the subgrade soil and keeping it clean (Figure 12.33).

12.2.4.1 Mechanical Properties of Geosynthetics for Separation Applications

Grab strength, puncture resistance, tensile strength, and seam strength of a geotextile are the most important properties for separation design. Puncture resistance is determined by the ASTM D4833 test (Standard Test Method for Index Puncture Resistance of Geomembranes and Related Products), in which an 8-mm-diameter steel rod penetrates into a geotextile that is clamped in an empty cylinder with a diameter of 45 mm. In addition to a decrease in strength, puncturing can cause significant strains in the geotextile. Figure 12.34 shows the laboratory-measured strain variations in a woven geotextile as a result of puncturing. The punctured area enlarges due to stretching over time, and lateral and axial strains develop in the specimen.

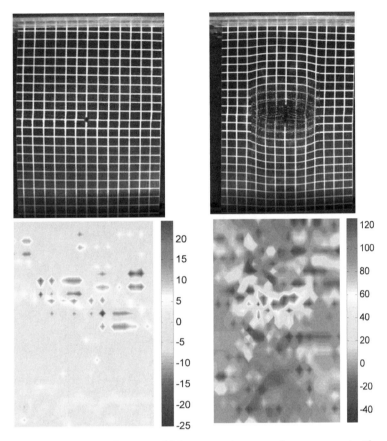

FIGURE 12.34 Development of failure around puncture in a woven geotextile (all axial strains are given as percentage values).

12.2.4.2 Separation Design

Geotextiles placed between the interface of the base course aggregate and subgrade soils function as separators and prevent mixing of these two layers. The separation function assists in preventing the formation of localized bearing failures that in particular occur in soft, weak subgrade soils (Figure 12.35a). Due to soft and weak subgrade soils, aggregate can move laterally and ruts may develop. The geotextile can provide a lateral restraint and prevent or minimize that movement (Figure 12.35b).

The separation function also is important in the case of initial construction and aggregate installation activities. The movements of aggregates usually occur when the geotextile is locked by the aggregate, and the grab tensile test is performed to evaluate the field conditions (Figure 12.36).

The required geotextile tensile strength for locking aggregate can be calculated using

$$T_{\text{req}} = P(d_v)^2 f(\varepsilon) \qquad (12.15)$$

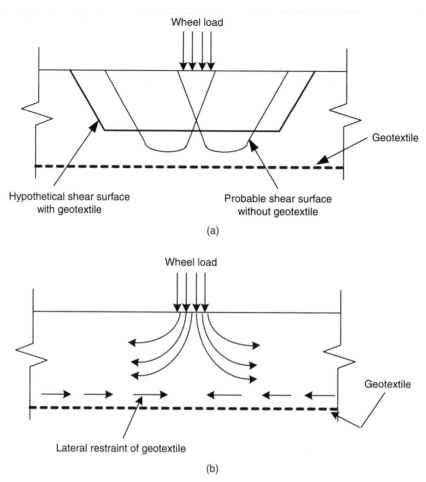

FIGURE 12.35 (a) Bearing capacity increase and (b) lateral restraint mechanism provided by geotextile in a roadway application (after Haliburton et al. 1981).

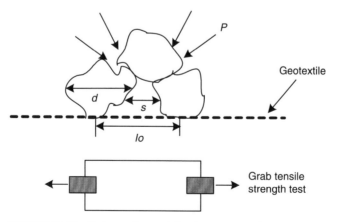

FIGURE 12.36 Grab tensile test (after Koerner 2005).

where T_{req} is the required grab tensile force, P is the applied pressure, d_v is the maximum void diameter $(d_v = 0.33d_{average\ stone\ diameter})$, and $f(\varepsilon)$ is the strain function of the geotextile. The strain function is equal to

$$f(\varepsilon) \ = \ \frac{1}{4} \left[\frac{2y}{b} \ + \ \frac{b}{2y} \right] \tag{12.16}$$

where b is the width of openings in the aggregate and y is the deformation of the openings.

For proper separation, the geotextile should have sufficient puncture and impact resistance to various objects. The puncture resistance (F_{req}) is calculated using the following relationship (Koerner 2005):

$$F_{req} \ = \ Pd_a^2 \left(\frac{h_h}{d_a} \right) (0.31d_a) \left(1 \ - \ \frac{A_p}{A_c} \right) \tag{12.17}$$

where P is the pressure exerted on the geotextile, d_a is the average diameter of the puncturing aggregate or sharp object, h_h is the protrusion height, and the ratio of A_p/A_c ranges from 0.3 to 0.8 depending on the type of aggregate.

12.2.5 Erosion Control

Geocomposites are used for erosion control. Typical applications include slope and stream bank protection and control of erosion in natural and man-made slopes. Geosynthetic functions during erosion control applications include provision of a medium for plant growth, filtration (beneath armor), and protection (from raindrop impact and rill erosion). The erosion control products protect the soils from erosion permanently or until vegetation can establish itself.

Two types of erosion control products exist: (1) temporary erosion control and revegetation materials (TERMs) and (2) permanent erosion control and revegetation materials (PERMs). TERMs are rolled out after preparation of the surface, and, except for the polymer, the natural products are completely biodegradable. Geofibers are mixed with soil to increase stability. Erosion control meshes and nets and erosion control blankets are biaxially oriented nets and often called rolled-sheet products. They are attached to the seeded ground by staplers, glue, or threading techniques. Fiber roving systems are formed of continuous yarns or fiber and are continuously fed over the surface before placement of the emulsified asphalt or soil stabilizer.

Two types of PERMs exist: soft (biotechnical-related) ones and hard (hard armor-related) ones. Among various soft PERMs, turf reinforcement mats are rolled out, filled with soil, and seeded. They provide strength to the surface soils, and plants provide additional strength. Erosion control and revegetation mats have similar duties, except they already contain the seeds. Geocellular confinement systems (geocells) also are increasingly being used for erosion control and in separation applications. They may be stacked to form walls, including seawalls, are very durable, and are easy to ship and store. Hard PERMs are made of stone or concrete and include gabions (wire boxes filled with rocks) and riprap (angular rocks). Typically they are underlain by a geotextile filter and are very effective in erosion control. Figure 12.37 shows

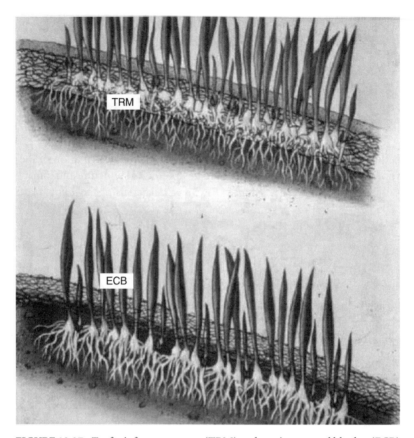

FIGURE 12.37   Turf reinforcement mat (TRM) and erosion control blanket (ECB) for erosion control (after Koerner 2004).

two types of erosion control products: a turf reinforcement mat and an erosion control blanket (Koerner 2004).

12.2.5.1 Erosion Control Design

Erosion of soils is affected by three distinct processes: detachment of soil particles and their transportation as well as deposition via environmental causes. There are various factors that affect each process, and a simplified approach by treating the eroded area as a slope or a ditch has been considered. Accordingly, geosynthetic rolled erosion control materials (RECMs) are designed for slope erosion and ditch erosion.

For slope erosion design, the Universal Soil Loss Equation is followed and the vegetative factor (C) is reduced significantly when RECMs are used:

$$E = RK(LS)CP \qquad (12.18)$$

where E is soil loss (t/km^2), R is a rainfall factor, K is an erodibility factor, LS is a gradient factor, and P is a conservation factor. All factors have dimensionless units. Table 12.5 provides a list of C factors for various RECMs. For ditch erosion, one of two approaches may be undertaken: the velocity approach or the shear strength approach. The velocity approach utilizes Manning's equation to estimate a required design velocity (V) based on slope of the

TABLE 12.5 Erosion Control Technical Council (ECTC) Recommendations (2002)

Category	Composition[a]	Time (months)	H-to-V (max.)	C Factor (for Universal Soil Loss Equation)	Allowable Shear Stress (Pa)
TERM	ECB	≤3	3:1	0.15	72
	ECB double	≤3	2:1	0.20	84
	ECB	≤12	3:1	0.15	72
	ECB double	≤12	2:1	0.20	84
	ECB	≤24	1.5:1	0.25	96
	ECB double	≤36	1:1	0.25	108
	TRM-A	NA	1:1	NA	288
	TRM-B	NA	0.5:1	NA	480

[a] ECB = erosion control blanket; TRM = turf reinforcement mat.

channel and hydraulic radius, which is equal to the cross-sectional area of flow divided by a wetted parameter:

$$V = \frac{R^{2/3} S^{1/2}}{n} \qquad (12.19)$$

where n is Manning's roughness coefficient, which typically ranges between 0.02 and 0.023 for unlined bare soils. A detailed summary of n values can be found in the *Hydraulic Reference Manual* (U.S. Army Corps of Engineers 2002). Calculated velocities based on Equation 12.19 are compared with the long-term velocities provided by Theisen (1992) to calculate factors of safety (Figure 12.38).

The shear approach calculates a required shear strength ($\tau_{required}$) based on depth of flow (d) and slope channel (S):

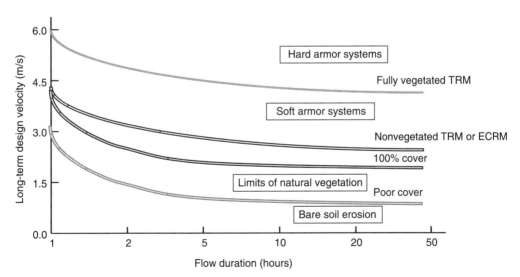

FIGURE 12.38 Design velocities for various erosion control systems. (TRM = turf reinforcement mat, ECRM = erosion control and revegetation mat.)

$$\tau_{\text{required}} = \gamma_w d S \qquad (12.20)$$

where γ_w is the unit weight of water. The strength determined from Equation 12.20 is compared to the allowable shear strength provided in Table 12.5 to calculate factors of safety for design.

12.2.6 Hydraulic Barrier

Geomembranes and geosynthetic clay liners (GCLs) are primarily used as hydraulic barriers in geotechnical and geoenvironmental construction. Geomembranes are used in dams, tunnels, canals, reservoirs, and landfills as liners and covers or for waterproofing the underground structures. They also are used as seepage barriers in highways and for control of contaminant migration during remediation applications. GCLs typically are used in landfill liners and cover systems. Mechanical and hydraulic properties of these geosynthetics are summarized below, and a detailed explanation of the design procedure for these materials is given in Chapter 13 of this handbook.

12.2.6.1 Mechanical Properties of Geosynthetic Clay Liners

Peel and puncture resistance tests are conducted to determine the mechanical properties of GCLs for hydraulic applications. An important property for design is internal shear strength, which often is determined through a direct shear test. The test typically is performed in the dry (as-received) state and hydrated state by subjecting the sample to normal effective stresses. Table 12.6 lists the effect of various hydrating liquids on shear properties of GCLs (U.S. EPA 1993).

Fox et al. (1998) and Eid et al. (1999) investigated the internal shear strength of unreinforced and reinforced GCLs under a range of different normal stresses and shear rates and reported modestly nonlinear peak shear strength (τ_p) and residual shear strength (τ_r) failure envelopes, which were caused by the ability of high normal stresses to enhance the orientation of the bentonite particles and the reinforced fibers in a direction parallel to shear. Fox et al. (1998) also determined that the contribution of needle-punched reinforcement to peak shear strength increased with normal stress; however, the contribution of stitch bonding was independent of normal stress. Three mechanisms may have caused this complex behavior: (1) high shear rates tear the reinforcing fibers more rapidly and increase the shear resistance; (2) as the shear rate increases, the effective normal stress decreases due to the induced excess pore water pressure

TABLE 12.6 Effect of Hydrating Liquid on GCL Shear Parameters

| | Testing State | | | | | |
| | Dry | | Free Swell | | Constrained Swell | |
Hydrating Fluid	c (kPa)	ϕ	c (kPa)	ϕ	c (kPa)	ϕ
Distilled water	6.9–68	26–42	3–9	0–23	3–7	16–37
Tap water	6.9–68	26–42	3–10	0–26	3–7	18–43
Mild leachate	6.9–68	26–42	3–14	4–20	5–8.3	18–43
Harsh leachate	6.9–68	26–42	3–12	0–32	4–7.6	13–45
Diesel fluid	6.9–68	26–42	5–6	29–46	4–6	24–51

in the GCL (i.e., inside the bentonite layer); and (3) the undrained frictional resistance of bentonite increases with increasing normal stress and contributes to the peak shear strength.

12.2.6.2 Mechanical Properties of Geomembranes

Typical geomembrane mechanical properties of interest in hydraulic barrier applications are tensile strength and rupture resistance. Tensile strength tests can be performed for quality control (e.g., grab tensile test), to determine load-strain characteristics during application (e.g., wide-width tensile test), or to evaluate seam integrity (e.g., peel test, wide-width tensile test, nondestructive methods). In order to evaluate seam efficiency, field-specific conditions can be employed, such as testing of dry, wet, dirty, or clean samples. Figure 12.39 shows the effect of polymer type on geomembrane stress-strain behavior during wide-width tensile tests.

Soil-geomembrane or geosynthetic-geomembrane interface shear strength can be important in designing landfill liners and covers or when geomembranes are used to line reservoirs or canals with steep slopes. In such applications, textured geomembranes typically are preferred over smooth ones for additional increase in frictional strength (Table 12.7). Previous studies also showed that the critical surface typically exists between the smooth geomembrane-geotextile and smooth geomembrane-clay interfaces (Table 12.8).

12.2.6.3 Hydraulic Properties of Geomembranes

Hydraulic transport of liquids through geomembranes occurs via either molecular diffusion or transport through the little gaps between the polymer materials. Three different types of tests are used to determine the hydraulic properties of geomembranes: (1) water vapor transmission test (ASTM E96), (2) solvent vapor transmission test, and (3) diffusion test. Diffusion tests can be batch-scale or large-scale column tests. The latter are likely to better represent field conditions.

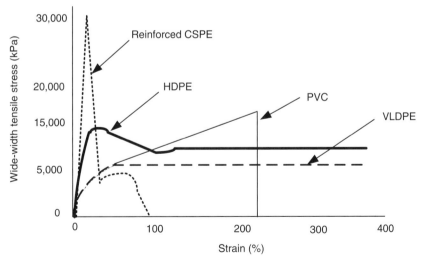

FIGURE 12.39 Stress-strain characteristics of different geomembranes subjected to wide-width tension tests (after Koerner 2005). (CSPE = chlorosulphonated polyethylene, HDPE = high-density polyethylene, PVC = polyvinyl chloride, and VLDPE = very-low-density polyethylene.)

TABLE 12.7 Soil-Geosynthetic Interface Friction Angles

Geomembrane Type	Sand	Clay
Polyvinyl chloride	21–33°	6–39°
High-density polyethylene	17–28°	5–29°
Textured high-density polyethylene	30–45°	7–35°
Very-low-density polyethylene	21–28°	—

After Sharma and Lewis (1994).

TABLE 12.8 Typical Values of Interface Shear Strength
(ISSMGE 1997)

Interface[a]	δ (°)
HDPE geomembrane–sand	15–28
HDPE geomembrane–clay	5–29
Textured HDPE geomembrane–sand	30–45
Textured HDPE geomembrane–compacted clay	7–35
Textured HDPE geomembrane–gravel	20–25
Geotextile-sand	22–44
Geotextile-clay	15–33
GCL-sand	20–25
GCL-clay	14–16
HDPE geomembrane–geonet	6–10
Textured HDPE geomembrane–geonet	10–25
HDPE geomembrane–geotextile	8–18
Textured HDPE geomembrane–geotextile	14–52
HDPE geomembrane–GCL	8–16
Textured HDPE geomembrane–GCL	15–25
Geotextile-geonet	10–27

[a] HDPE = high-density polyethylene.

Geomembranes are commonly used in landfill lining systems and are a vital part of the composite liners. Leakage through a composite liner is limited to (1) the imperfections (i.e., number of holes in the geomembrane), (2) the quality of the contact (i.e., poor, good, excellent) between the geomembrane and the underlying compacted clay liner or GCL (wrinkles are of concern), and (3) the radius of the wetted area through the underlying clay liner or GCL (Figure 12.40). Furthermore, seam integrity is likely to play an important role in advective and diffusive transport through liners (Giroud 1997; Foose et al. 2001).

12.2.6.4 Hydraulic Properties of Geosynthetic Clay Liners

GCLs are prefabricated bentonite clay layers incorporated between geotextiles and/or geomembranes and used as a barrier for liquid or solid waste containment. Three hydraulic properties play a role in the barrier performance of GCLs: (1) hydration, (2) free swell, and (3) hydraulic conductivity.

GCLs include a considerable amount of bentonite, and this low-permeability material is affected when exposed to different environments. The hydration capability of the bentonite

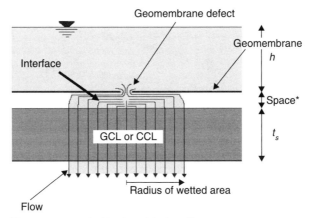

FIGURE 12.40 Transport through composite liners consisting of geomembrane. (CCL = compacted clay liner.)

decreases when the applied effective stress is increased and/or the environment contains ions and organic contaminants. Lin and Benson (2000) showed that free swell capacity of GCLs changes with permeant chemistry. Free swell is defined as the swelling of bentonite under zero normal stress. The same study also showed that wet-dry cycling had little effect on the swelling performance of the GCL specimens permeated with deionized water (DI) and tap water, but the swelling decreased and hydraulic conductivity increased when the specimens were permeated with $CaCl_2$ (Figure 12.41).

Jo et al. (2000) showed that swell capacity of nonprehydrated GCLs decreased with increasing salt concentration. At the same concentration, swell was largest with monovalent cations (i.e., Na^+, K^+, and Li^+) and smallest with trivalent cations (i.e., Al^{+3} and La^{+3}). A study conducted by Kolstad et al. (2004) also showed that relative abundance of monovalent and divalent cations in the same solution (i.e., multispecies solution) affects the swell amount in GCLs (Figure 12.42). Relative abundance of monovalent and divalent cations (RMD) was defined as

$$\text{RMD} = \frac{M_M}{\sqrt{M_D}} \qquad (12.21)$$

where M_M and M_D are the total molarity of monovalent and divalent cations in the solution, respectively.

Hydraulic conductivity is the most essential property in designing GCLs for barrier applications, and it is affected by the applied effective stress and hydraulic gradient, permeant chemistry, and seam integrity in the field. Jo et al. (2000) reported that GCLs permeated with trivalent and divalent solutions had higher hydraulic conductivity than GCLs permeated with monovalent solutions or DI water, and an increase in pH of the solution was accompanied by a decrease in hydraulic conductivity. Kolstad et al. (2004) showed that hydraulic conductivity of GCLs is likely to increase with increasing ionic strength and decreasing RMD.

Jo et al. (2000) related the hydraulic conductivity ratio (K_c / K_{DI}) to the bentonite swell ratio (H_b / H_{bs}) as measured in the free swell test. The hydraulic conductivity ratio was defined

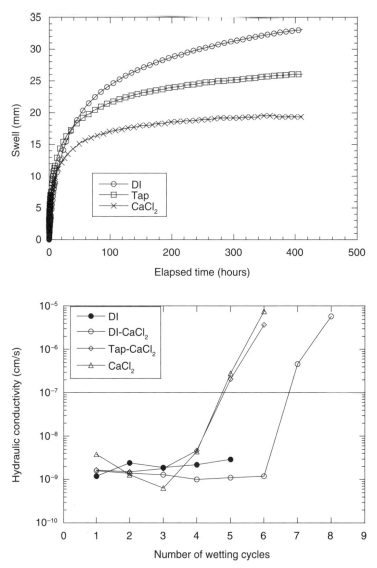

FIGURE 12.41 Effect of hydrating liquid on (top) swell and (bottom) hydraulic conductivity of GCLs.

as the ratio of hydraulic conductivity of a GCL subjected to an inorganic salt solution (K_c) to the hydraulic conductivity of a GCL tested with DI water (K_{DI}), and the bentonite swell ratio was defined as the ratio of the height of swollen bentonite in the GCL (H_b) to the height of bentonite solids initially in the GCL (H_{bs}). Hydraulic conductivity was practically constant when $H_b/H_{bs} \geq 4$ but increased rapidly for $H_b/H_{bs} < 4$. The results indicated that there is a strong relationship between hydraulic conductivity of a GCL and the amount of swell. A decrease in hydraulic conductivity was observed with increasing amount of swell for all GCL specimens (Figure 12.43).

It also has been shown that prehydration with tap or distilled water does not prevent the hydraulic conductivity of GCLs permeated with inorganic salt solutions from increas-

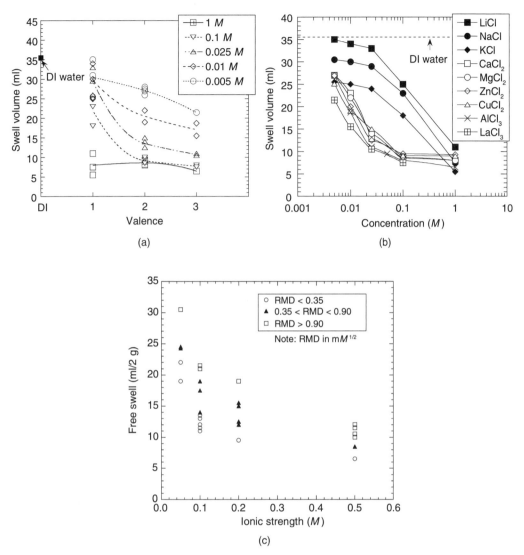

FIGURE 12.42 Effect of various factors on swell of GCLs permeated with (a and b) monovalent and (c) divalent cations.

ing substantially above the baseline hydraulic conductivities obtained with DI or tap water (~10^{-9} cm/s) (Vasko et al. 2001). A series of laboratory and field investigations indicated that the hydraulic conductivity of neither reinforced nor unreinforced GCLs is adversely affected by freezing and thawing (Kraus et al. 1997).

12.3 Durability and Aging of Geosynthetics

Geosynthetics are widely used in various geotechnical and geoenvironmental engineering projects, including waste containment facilities, with a design life ranging from temporary (1–5 years) to permanent (100+ years). Therefore, those materials should have resistance to

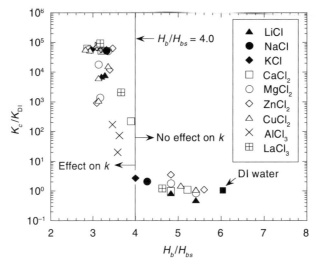

FIGURE 12.43 Relationship between hydraulic conductivity ratio and bentonite swell ratio for GCLs.

ultraviolet (UV) light and chemical, biological, and physical attacks (Wrigley 1989; Haxo and Haxo 1989).

12.3.1 Factors That Affect Durability

Various stresses act on geosynthetics during construction and service. Those stresses affect the long-term durability of the geosynthetic (Haxo and Haxo 1989):

- *Mechanical stresses*—Overburden, hydraulic head, physical action of precipitation, shear and tensile stresses on slopes, movement of support structures due to settlement, and puncturing and tear during construction cause tearing, cracking, or creep of the geosynthetic.
- *Chemical stresses*—Exposure to constituents of a waste liquid, UV light, oxygen, and ozone can cause breakdown of the polymer structure.
- *Biological stresses*—The result of biodegradation by microorganisms, biological stresses can cause the clogging of the filter/drainage media.

Geosynthetics in waste containment facilities may encounter certain conditions during construction which affect their long-term durability. Typical conditions include temperature, humidity, wind variation, exposure to UV light, stretching and puncturing of the geomembrane during placement, heat applied during seaming, and excessive traffic.

12.3.1.1 Factors That Affect the Durability of Geomembranes

Geomembranes are widely used in waste containment applications and various geoenvironmental engineering projects. Geomembranes must be durable and maintain their physical and mechanical properties over the design life of the specific facility.

The intrinsic durability of a geomembrane depends upon the polymer, the auxiliary compounding ingredients, and the manufacturing method (Haxo and Nelson 1984). How-

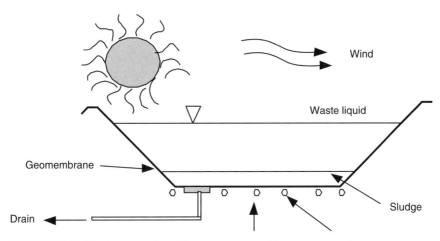

FIGURE 12.44 Environmental conditions encountered by an uncovered geomembrane in a liquid impoundment system (after Haxo and Haxo 1989).

ever, durability can vary greatly with respect to different degradation mechanisms with different exposures.

The principal agents aggressive to polymeric materials are heat, oxygen, moisture, atmospheric pollutants, chemicals, low temperatures, stress and strain, enzymes, and bacteria. In most exposures, two or more of these agents act together. Exposure may occur in three different types of environments (Figure 12.44):

- *Exposure to weathering*—Presence of oxygen, ozone, UV light, high humidity, low temperature, and fluctuating temperatures (particularly important during construction and surface impoundments)
- *Exposure to air-waste liquid interface*—Presence of an oily layer on the geomembrane, wind, and waves
- *Exposure to waste liquids*—Presence of water and organics (in turn, organics can partition to the geomembrane), strong acids or bases, and high waste temperature (Gulec et al. 2004)

The degrading agents at the bottom of a landfill are quite different than those that play a role in weathering. Due to an oxygen-free environment and the absence of UV light, oxidative and UV degradation are not significant in landfill liner applications.

Chemical degradation is an important factor that affects the durability of geomembrane liners due to possible moderate acidity and the dissolved organic and inorganic constituents of the leachate. Humid to wet conditions at the bottom of a waste disposal facility can result in swelling of a liner and leaching of the compounding ingredients.

12.3.1.2 Mechanisms of Polymer Degradation

Koerner et al. (1990) summarized the mechanisms of degradation that may be encountered in different exposures:

1. *UV degradation*—The light with the most sensitive wavelengths enters the molecular structure of the polymer, and free radicals are formed, which cause bond scissions in the structure of the polymer. Carbon black or chemical-based light stabilizers are effective

against UV degradation. A soil backfill or other types of covers may eliminate the UV degradation.

2. *Radiation degradation*—The basic mechanical properties of a polymer change above a certain radiation dose. When radioactive materials in the waste matrix are absent, this degradation is not encountered.

3. *Degradation by swelling*—Geomembrane durability is affected by swelling that occurs due to liquid adsorption. Although swelling does not result in any scission, or failure in the system, it may cause changes in physical and mechanical properties of the geomembrane. However, the amount of swelling is negligible in the most commonly used high-density polyethylene geomembranes.

4. *Degradation by extraction*—Polymers which have been compounded with the use of plasticizers and fillers exhibit this type of degradation by long-term extraction of one or more components of the compound.

5. *Degradation by delamination*—This may occur in geomembranes manufactured by calendering or spread coating. It is observed when liquid enters into the edge of the geomembrane and is drawn into the interface by capillary tension. Delamination causes the separation of the individual layers and destroys the composite action.

6. *Oxidation degradation*—Oxygen may create large-scale degradation when a free radical is created in the polymeric structure. The oxygen combines with the free radical to form a hydroperoxy radical and eventually causes chain scissions in the polymer. Antioxidation additives are added to the compound against oxidative degradation.

7. *Biological degradation*—Polymer degradation due to various biological lives is impossible due to the high molecular weights of the common resins used in geomembranes. Biological degradation usually is possible for plasticizers or additives within the resin. However, the amount of plasticizers and additives is very low in high-density polyethylene, and therefore biological degradation is not observed.

8. *Chemical degradation*—Chemical degradation in geomembranes has been studied with many different chemicals. However, complex waste streams such as leachate or mine acid drain are not usually addressed and must be considered on a site-specific basis. Depending on the type of leachate, the possible reactions are as follows: swelling, changes in physical and mechanical properties, and no reaction.

Most of the degradation mechanisms discussed for geomembranes also may be observed in geotextiles due to the same polymeric structure of both materials. However, biological degradation, which is not an important mechanism for geomembranes, may be a predominant mechanism for geotextiles with cellulosic fibers.

12.3.1.3 Test Methods to Assess the Compatibility of Geomembranes with Waste Liquids

- *U.S. EPA Method 9090* (1986): Compatibility Test for Wastes and Membrane Liners— Geomembrane is immersed in a chemical environment for a minimum period of 120 days at room temperature (23°C) and at 50°C. Geomembrane samples are periodically taken from the immersion tanks, and the change in their physical properties is quantified. These properties are tear resistance, puncture resistance, tensile strength, hardness, elongation at break, modulus of elasticity, volatile content, and specific gravity.

- *ASTM D5747*: Practice for Tests to Evaluate the Chemical Resistance of Geomembranes to Liquids

- *One-sided exposure* (Mitchell 1985)—This technique simulates conditions in a waste disposal facility by exposing the geomembrane to leachate on one side only and by simulating stresses due to the waste and leachate.

12.3.1.4 Types of Tests to Quantify Geomembrane Performance after Immersion

- *U.S. EPA Method 9090* (1986): Physical and Mechanical Property Tests—Thickness, mass, tear resistance, puncture resistance, tensile strength, and hardness tests
- *Transport property test*: Water vapor transmission, radioactive tracer transmission, water adsorption, water vapor adsorption, and benzene adsorption tests to evaluate the chemical changes after exposure
- Special analytical techniques

12.3.2 Lifetime Prediction of Geosynthetics

Since it is not feasible to run long-term tests, accelerated laboratory tests at high stress and high temperature using aggressive liquids usually are preferred to assess the durability of geosynthetics within a reasonably short period of time.

Pipe-Industry-Related Techniques. Stress-limit testing, the rate process method, and the Hoechst multiparameter approach are the techniques used in the pipe industry to predict the lifetime of pipes. A similar rate process method also is applied to geomembranes, in which they are subjected to stresses under high temperatures (Koerner 2005).

Basic Autoxidation Scheme. The basic autoxidation scheme has been adapted to develop a kinetic model for evaluating the mechanical degradation of polyolefin (polyethylene and polypropylene) geosynthetic products (Salman et al. 1998). Both geosynthetic materials are mechanically tested by wide-width strip test after accelerated testing.

Elevated Temperatures and Arrhenius Modeling. Mitchell and Spanner (1985) combined the compressive stress, elevated temperature, chemical exposure, and long testing time into one experimental device (Figure 12.45). Mechanical tests were performed to determine tensile strength and elongation, yield strength and elongation, and stress cracking behavior, and Arrhenius modeling was performed on the chemical data. Arrhenius modeling is the most common method used to extrapolate laboratory-based results into the future. This method enables the use of the results of chemical compatibility studies for the lifetime prediction of geosynthetic materials using

$$K = Ae^{-E/RT} \tag{12.22}$$

where K is the oxidative induction depletion rate, A is the pre-exponential factor (independent of temperature), E is the activation energy for the particular reaction, R is the gas constant (8.314 J/mol-K), and T is the absolute temperature (K = °C + 273). By plotting ln K against $1/T$, a straight line should be obtained with a slope of $-E_{act}/R$ that shows the particular property change (Figure 12.46). This plot is called an Arrhenius plot (Koerner et al. 1990).

Arrhenius modeling is used for interpretation of chemical degradation tests. The fraction (or property) retained is plotted vs. the incubation period for different temperatures (T_1, T_2, T_3). With the curves obtained, a designated value of the property is selected (e.g., 50%); from

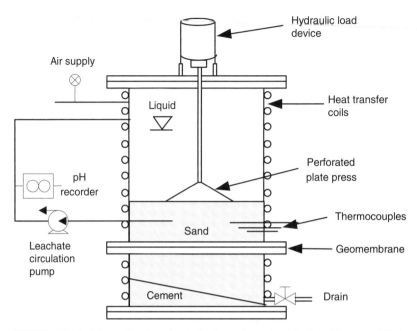

FIGURE 12.45 Schematic of accelerated aging column (Mitchell and Spanner 1985).

the resulting half-life values, a unique set of degradation times corresponding to each incubation period is obtained, and an Arrhenius graph is plotted. This plot is extrapolated to the desired temperatures expected at the site.

The critical assumption made in Arrhenius modeling is that the material behavior within the high-temperature incubation range (as indicated by the activation energy E_{act}) is constant and can be extrapolated to the lower temperature behavior of the material. However, instead of a single activation energy, often there is a distribution of activation energies, and the modeling results may not be satisfactory in the case of highly oriented geosynthetics such as

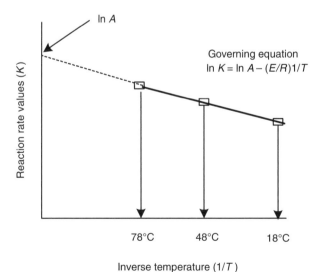

FIGURE 12.46 Arrhenius plot (after Koerner et al. 1990).

oriented geogrids, many geotextile fibers, and some geocomposites (Koerner et al. 1992). Application of this method has shown reasonable life spans for most geosynthetics under common applications. Some applications require indefinitely long lives, such as geomembranes used as liners and covers in mine tailings or radioactive waste disposal facilities (Gulec et al. 2004).

References

Abdullah, C.H. and Edil, T.B. (2007). Behaviour of geogrid-reinforced load transfer platforms for embankment on rammed aggregate piers. *Geosynth. Int.,* 14(3):141–153.

Aydilek, A.H. and Edil, T.B. (2004). Evaluation of woven geotextile pore structure parameters using image analysis. *Geotech. Test. J.,* 27(1):99–110.

Aydilek, A.H., D'Hondt, D., and Holtz, R.D. (2007). Comparative evaluation of geotextile pore sizes using bubble point test and image analysis. *Geotech. Test. J.,* 30(3):173–181.

Bathurst, R. and Jarrett, P. (1988). Large-scale model tests of geocomposite mattresses over peat subgrades. *Transp. Res. Rec.,* 1188:28–36.

BS 8006 (1995). *Code of Practice for Strengthened/Reinforced Soils and Other Fills,* British Standard Institution, London.

Carroll, R.G. (1983). *Geotextile Filter Criteria,* Engineering Fabrics in Transportation Construction, Transportation Research Record, 53 pp.

Cedergren, H.R. (1989). *Seepage, Drainage, and Flow Nets,* 3rd edition, John Wiley and Sons, New York, 465 pp.

Eid, H.T., Stark, T.D., and Doerfler, C.K. (1999). Effect of shear displacement rate on internal shear strength of a reinforced geosynthetic clay liner. *Geosynth. Int.,* 6(3):219–239.

Fischer, G.R. (1994). *The Influence of Fabric Pore Structure on the Behavior of Geotextile Filters,* Ph.D. dissertation, University of Washington, 502 pp.

Fischer, G.R., Mare, A.D., and Holtz, R.D. (1999). Influence of procedural variables on the gradient ratio test. *Geotech. Test. J.,* 22(1):22-31.

Foose, G., Benson, C.H., and Edil, T.B. (2001). Predicting leakage rates in composite liner systems. *J. Geotech. Geoenviron. Eng.,* 127(6):510–520.

Fox, P.J., Rowland, M.G., and Scheithe, J.R. (1998). Internal shear strength of three geosynthetic clay liners. *J. Geotech. Geoenviron. Eng.,* 124(10):933–944.

Giroud, J.P. (1997). Equations for calculating the rate of liquid migration through composite liners due to geomembrane defects. *Geosynth. Int.,* 4(3–4):335–348.

Gulec, S.B., Edil, T.B., and Benson, C.H. (2004). Effect of acidic mine drainage (AMD) on the polymer properties of an HDPE geomembrane. *Geosynth. Int.,* 11(2):60–72.

Haliburton, T.A. and Wood, P.D. (1982). Evaluation of the U.S. Army Corps of Engineers gradient ratio test for geotextile performance. *Proceedings of the Second International Conference on Geotextiles,* Las Vegas, Vol. 1, 97–101.

Haliburton, T.A., Laumaster, J.D., and McGuffer, V.C. (1981). *Use of Engineering Fabrics in Transportation Related Applications,* FHWA DTFH61-80-C00094.

Han, J. and Gabr, M.A. (2002). Numerical analysis of geosynthetics-reinforced and pile-supported earth platforms over soft soil. *J. Geotech. Geoenviron. Eng.,* 130(2):129–138.

Haxo, H.E. and Haxo, P.D. (1989). Environmental conditions encountered by geosynthetics in waste containment applications. *Proceedings of Durability and Aging of Geosynthetics Conference,* Drexel University, R.M. Koerner, Ed., Elsevier, 28–47.

Haxo, H.E. and Nelson, N.A. (1984). Factors in the durability of polymeric membrane liners. *Proceedings of International Conference on Geomembranes,* Denver, CO, Vol. II, June 20–24.

Holtz, R.D., Christopher, B.R., and Berg, R.R. (1997). *Geosynthetic Engineering,* Bi-Tech Publishers.

Huang, Y. (1993). *Pavement Analysis and Design,* Prentice Hall, Englewood Cliffs, NJ.

ISSMGE (1997). Report by International Society for Soil Mechanics and Geotechnical Engineering Technical Committee TC5 on Environmental Geotechnics, Bochum, Germany.

Jenner, C.G., Austin, R.A., and Buckland, D. (1998). Embankment support over piles using geogrids. *Sixth International Conference on Geosynthetics,* 763–766.

Jo, H.Y., Katsumi, T., Benson, C.H., and Edil, T.B. (2000). Hydraulic conductivity and swelling of non-prehydrated GCLs permeated with salt solutions. *J. Geotech. Geoenviron. Eng.,* 130(12): 1236–1249.

Koerner, R.M. (2004). *Geosynthetics in Erosion Control,* Geosynthetic Research Institute Lecture Notes, Drexel University, Philadelphia.

Koerner, R.M. (2005). *Designing with Geosynthetics,* 5th edition, Prentice Hall.

Koerner, G.R. and Koerner, R.M. (1990). Biological activity and potential remediation involving geotextile landfill leachate filters. *Geosynthetic Testing for Waste Containment Facilities,* R.M. Koerner, Ed., ASTM STP 1081, 313–334.

Koerner, R.M., Halse, Y.H., and Lord, A.E. (1990). Long-term durability and aging of geomembranes. *Proceedings of Waste Containment Systems: Construction, Regulation, and Performance Symposium,* San Francisco, R. Bonaparte, Ed., November 6–7.

Koerner, R.M., Lord, A.E., and Hsuan, Y.H. (1992). Arrhenius modeling to predict geosynthetic degradation. *Geotext. Geomembr.,* 11:151–183.

Kolstad, D.C., Benson, C.H., and Edil, T.B. (2004). Hydraulic conductivity and swell of nonprehydrated GCLs permeated with multi-species inorganic solutions. *J. Geotech. Geoenviron. Eng.,* 130(12):1236–1249.

Kraus, J.F., Benson, C.H., Erickson, A.E., and Chamberlain, E.J. (1997). Freeze-thaw cycling and hydraulic conductivity of bentonitic barriers. *J. Geotech. Geoenviron. Eng.,* 123(3):229–238.

Kutay, M.E. and Aydilek, A.H. (2005). Filtration performance of two-layer geotextile systems. *Geotech. Test. J.,* 28(1):79–91.

Lafleur, J., Mlynarek, J., and Rollin, A.L. (1989). Filtration of broadly graded cohesionless soils. *J. Geotech. Eng.,* 115(12):1747–1768.

Lin, L.C. and Benson, C.H. (2000). Effect of wet-dry cycling on swelling and hydraulic conductivity of GCLs. *J. Geotech. Geoenviron. Eng.,* 126(1):40–49.

Luettich, S.M., Giroud, J.P., and Bachus, R.C. (1992). Geotextile filter design guide. *Geotext. Geomembr.,* 11(4–6):355–370.

Mansur, C.I. and Kaufman, R.I. (1962). Dewatering. *Foundation Engineering,* G.A. Leonards, Ed., McGraw-Hill, 241–350.

Mitchell, D.H. (1985). Geomembrane compatibility tests using uranium acid leachate. *Geotext. Geomembr.,* 2:111–127.

Mitchell, D.H. and Spanner, G.I. (1985). *Field Performance Assessment of Synthetic Liners for Uranium Tailings Ponds,* Status Report, NUREG/CR-4023, PNL-5005, Nuclear Regulatory Commission, Battelle, PA (as cited by Koerner et al. 1990).

Pham, H.T.V., Suleiman, M.T., and White, D.J. (2004). *Numerical Analysis of Geosynthetic–Rammed Aggregate Pier Supported Embankments,* Vol. 1, Geotechnical Special Publication No. 126, ASCE, 657–664.

Rankilor, P.R. (1981). *Membranes in Ground Engineering,* John Wiley and Sons, Chichester, England, 377 pp.

Salman, A., Goulias, D., and Elias, V. (1998). Durability of geosynthetics based on accelerated thermo-oxidation testing. *J. Test. Eval.,* 26(5):472–480.

Sharma, H.D. and Lewis, S.P. (1994). *Waste Containment Systems, Waste Stabilization and Landfills,* John Wiley and Sons.

Skempton, A.W. (1964). Long-term stability of clay slopes. *Geotechnique,* 14(2):77–102.

Smith, J.L. (1993). *The Pore Size Distribution of Geotextiles,* M.S. thesis, Syracuse University, Syracuse, NY.

Stark, T.D. and Poeppel, A.R. (1994). Landfill liner interface shear strengths from torsional shear tests. *J. Geotech. Geoenviron. Eng.,* 120(3):597–616.

Theisen, M.S. (1992). The role of geosynthetics in erosion and sediment control: an overview. *Geotext. Geomembr.,* 11(4–6)199–214.

U.S. Army Corps of Engineers (1977). *Civil Works Construction Guide,* No. CW-02215, Office of Chief of Engineers, Washington, D.C.

U.S. Army Corps of Engineers (2002). *Hydraulic Reference Manual,* Version 3.1, HECS-RAS River Analysis, U.S. Army Corps of Engineers Hydrologic Engineering Center.

U.S. EPA (1993). *Report of Workshop on Geosynthetic Clay Liners,* U.S. Environmental Protection Agency, Washington, D.C., August.

Vasco, S.M., Jo, H.Y., Benson, C.H., Edil, T.B., and Katsumi, T. (2001). Hydraulic conductivity of partially prehydrated geosynthetic clay liners permeated with aqueous calcium chloride solutions. *Proceedings of Geosynthetics 2001,* Industrial Fabrics Association International, Portland, OR, 685–699.

Wikipedia (2008). www.wikipedia.org.

Wrigley, N.E. (1989). The performance of polymeric materials: plastics in the ground geomembrane. *Chem. Ind.,* pp. 414–420.

Yoder, E.J. and Witczak, M.W. (1975). *Principles of Pavement Design,* 2nd edition, John Wiley and Sons, New York.

13

Geoenvironmental Engineering

by
Nazli Yesiller
California Polytechnic State University, San Luis Obispo, California

Charles D. Shackelford
Colorado State University, Fort Collins, Colorado

13.1 Introduction

Several hundred million tons of wastes are generated on an annual basis in the U.S. and other parts of the world (OECD 2008). Wastes commonly are categorized based on differences in source, composition, physical and chemical properties, and potential level of risk as municipal solid waste, hazardous waste, agricultural waste, mining waste, medical waste, incinerator ash, coal power-plant ash, and radioactive waste. These various wastes contain contaminants that pose risk to human health and the environment. Thus, properly designed and constructed containment systems are required for safe disposal of these wastes. In addition, containment systems are used for storage and conveyance of liquids that range from water to various chemicals. Safe and economical storage of these liquids requires proper containment.

This chapter has been written to provide fundamental principles pertaining to a variety of design, construction, and analysis schemes in containment applications for geoenvironmental engineering. The specific topics presented in the chapter include clay mineralogy, natural and synthetic containment materials, waste containment systems including performance issues, contaminant transport, measurement of material properties, and vertical barriers. Whereas the focus of this chapter on containment applications is in line with the themes covered in this handbook, the various topics included in the chapter, such as clay mineralogy, contaminant transport, material properties, and vertical barriers, also are directly applicable to remediation of contaminated sites, which constitutes the second main branch within geoenvironmental engineering (Shackelford 2002).

13.1.1 Clay Mineralogy

An understanding of clay mineralogy is required in geoenvironmental engineering due to the ubiquity and significance of clay minerals in natural soils and engineered systems as well as the high potential for interaction between clay minerals and water or various chemicals. The presence of clay minerals and their specific properties render clay soils appropriate for barrier applications. Clay minerals also can interact extensively with chemicals, which can adversely affect the performance of containment barriers comprised of clay soils. The significance of clay mineralogy on soil behavior is described in detail in Lambe (1953, 1958), Grim (1959, 1968), and Yong and Warkentin (1975) and summarized by Holtz and Kovacs (1981).

13.1.1.1 Introduction

The characteristics of clay soils are distinctly different compared to other soil types (e.g., sands, gravels). These characteristics of clay soils result primarily from the unique and dominant behavior of the clay minerals that comprise the particles (solid phase) of clay soils. The clay minerals that comprise the particles of the clay soils are electrochemically active due primarily to the existence of typically net negative charges on the surfaces of the clay mineral particles. Also, individual soil particles comprised of clay minerals typically are small (<2–5 μm), which can result in very large surface areas per unit mass of dry clay soil or specific surface (as much as 800 m^2/g). The electrochemical activity and large surface area associated with clay mineral particles that comprise clay soils make these soils susceptible to interactions with liquids, including water, which can affect the properties of the soils, such as hydraulic conductivity (permeability), strength, and compressibility. Variations in water (or other liquid) content may have significant effects on the behavior of clay soils. In addition, clay soils typically are plastic materials, in that once deformed under load, the original shape is not recovered upon unloading (i.e., the deformed shape is more or less retained).

Clay minerals are formed due to the mechanical and chemical weathering of igneous and metamorphic rocks. The most common clay minerals (e.g., kaolinite, illite, and montmorillonite) are composed of hydrous aluminosilicates with additional metallic ions (e.g., Fe^{3+}, Fe^{2+}, Mg^{2+}, Ca^{2+}, Na^+, K^+) and have platy shapes. Tubular and stringy shapes also have been observed for less common clay minerals such as halloysite and attapulgite (palygorskite).

Net negative charges on the surfaces of clay mineral particles result primarily from two phenomena that occur at the molecular level: isomorphous substitution and edge dissociation. Isomorphous substitution is the replacement of a higher valence element within the crystalline structure (e.g., Al^{3+}) with a lower valence element (e.g., Mg^{2+}) at the time of crystalline formation (i.e., geologic time). Since this negative charge is internal to the crystalline structure and, therefore, is not accessible after crystalline formation, the negative charge must be balanced by an equivalent positive charge external to the clay particle surface in the form of freely exchangeable cations that are held electrostatically to the surfaces of the clay mineral particles. These cations are exchangeable in that they can be exchanged for other cations within the adjacent pore liquid with an equivalent amount of charge. The capacity of a clay mineral for such exchangeable cations is represented by the cation exchange capacity (CEC). The CEC typically is reported in either milliequivalents of exchangeable cation charge per 100 g of dry soil (i.e., meq/100 g) or centimoles of charge per kilogram of dry soil (i.e., $cmol_c/kg$), where 1 meq/100 g = 1 $cmol_c/kg$.

Edge dissociation is the dissociation of exposed hydroxyl groups at solid interfaces (e.g., $OH^- \rightarrow O^{2-} + H^+$) releasing the proton ($H^+$) into pore liquid. The process is pH dependent. The degree of dissociation is a function of the pH of the solution adjacent to the clay mineral particle, where dissociation increases with increasing pH. In this case, a net positive surface charge is dominant at relatively low pH when an excess supply of protons is present, whereas a net negative charge is dominant at relatively high pH, with the pH corresponding to zero net charge typically referred to as the "zero point of charge."

13.1.1.2 Mineral Types

Clay minerals can be categorized based on crystalline structure as determined by the type of unit, type of sheet, and arrangement and bonding of layers of the sheets present in the mineral. The two basic units in clay minerals are the silicon tetrahedron (one Si^{4+} surrounded by four O^{2-}) and the aluminum octahedron (one Al^{3+} surrounded by six OH^-). These units bond chemically to form sheets of tetrahedral units or octahedral units, and the sheets also chemically bond to form layers of sheets. The nature of the layers of sheets and the manner in which these layers are held together (i.e., interlayer bonding) determine the fundamental crystalline structure for the specific clay mineral, as well as the overall physical and chemical properties of the clay mineral.

In terms of the structure of clay minerals, one tetrahedral sheet chemically bonded to one octahedral sheet is referred to as a 1:1 clay mineral structure, whereas one octahedral sheet sandwiched between and chemically bonded to two tetrahedral sheets is referred to as a 2:1 clay mineral structure. In terms of interlayer bonding, the layers typically are held together by readily exchangeable hydrated cations (e.g., Na^+, K^+, Ca^{2+}, Mg^{2+}) or intermolecular interactions (e.g., dispersion forces, hydrogen bonding, van der Waals bonding).

Illite and chlorite, which has a structure and properties similar to those of illite, are the most abundant clay minerals in nature. However, kaolinite, illite, and montmorillonite are the three most commonly encountered clay minerals in engineering practice. Montmorillonite also is often referred to as smectite, as the term smectite predates the term montmorillonite (Grim 1968). The mineral structure and physical characteristics of kaolinite, illite, and montmorillonite are summarized in Table 13.1. As shown in Figure 13.1, specific minerals occupy specific locations on the plasticity chart.

Isomorphic substitution dominates the charge deficiency in montmorillonite, illite, and chlorite, whereas edge dissociation is prevalent in kaolinite as well as in other constituents within soils, such as metal oxides and metal oxyhydroxides. Interlayer bonding in montmorillonite is dominated by exchangeable, hydrated cations and is relatively weak, primarily because the isomorphic substitution occurs within the aluminum octahedral sheets, which are located relatively far from the interlayer regions within the crystalline structure that are accessible to exchangeable cations. This weak interlayer bonding is the reason clay soils that

TABLE 13.1 Typical Characteristics and Properties of the Common Clay Minerals Encountered in Engineering Practice

Mineral	Structure	Typical Particle Sizes (thickness × diameter in nm)	Extent of Isomorphic Substitution	Range of CEC (meq/100 g or cmol$_c$/kg)	Range of Specific Surface (m^2/g)	Water Sorption Capacity	Sensitivity to Pore Fluid Chemistry
Kaolinite	1:1	50–2,000 × 300–4,000	Low	3–10	10–20	Low	Low
Illite	2:1	30 × 10,000	Moderate	10–40	65–100	Moderate	Moderate
Montmorillonite	2:1	3 × 100–1,000	High	80–150	100–800	High	High

After Yong and Warkentin (1975).

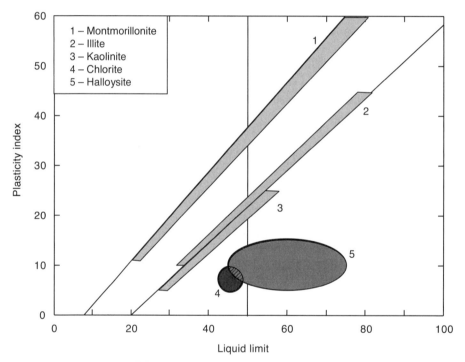

FIGURE 13.1 Location of clay minerals on plasticity chart (redrawn after Holtz and Kovacs 1981).

are dominated by montmorillonite, such as sodium bentonites, swell extensively in the presence of water. In contrast, isomorphic substitution in illite is located predominately in the silica tetrahedral sheets, which are relatively close to the interlayer regions within the crystalline structure, resulting in relatively strong interlayer bonding. This stronger interlayer bonding also reduces the accessibility of exchangeable cations from this region. Thus, even though illite has essentially the same crystalline structure as montmorillonite and, in some cases, greater surface charge deficiency than montmorillonite due to isomorphic substitution, water sorption capacity (i.e., swelling) and the CEC of illite are minimal compared to those of montmorillonite.

Finally, interlayer bonding in kaolinite generally is attributed to van der Waals bonding, which is also relatively weak. However, the significantly larger particles sizes associated with kaolinite and associated smaller surface areas (Table 13.1) render clays dominated by kaolinite less reactive than those dominated by montmorillonite, such that free swelling in kaolinite is also minimal relative to that for montmorillonite.

13.1.1.3 Diffuse (Electrostatic) Double Layer

The combination of the exchangeable, hydrated cations and bound water on the accessible surfaces of clay soil particles (i.e., interlayer space within individual clay mineral particles and space between individual clay particles) held in place by the electrical charge deficiency in the clay particles is referred to collectively as the electrostatic double layer or the diffuse double layer (DDL). The thickness of the DDL (t_{DDL}) is correlated to the dielectric constant of the liquid (ε) present in the pores (voids) of the clay soil, the valence of the cations (v) in the pore liquid, and the concentration of ions in the pore liquid (n_o), defined as the actual number of ions in the pore liquid (i.e., molar concentration of pore fluid multiplied by Avogadro's number, 6.02×10^{23} ions per mole), through the following relationship (Mitchell and Soga 2005):

$$t_{DDL} \propto \sqrt{\frac{\varepsilon}{n_o v^2}} \tag{13.1}$$

Thus, the thickness of the DDL increases with increasing ε and decreasing n_o and v in accordance with Equation 13.1. The DDL thickness also increases with decreasing temperature and increasing pH and anion adsorption (Lambe 1958; Mitchell and Soga 2005).

The importance of the DDL in the behavior of a clay mineral increases with decreasing particle size (increasing surface area). Thus, given the relative differences in particle sizes associated with each of the three primary clay minerals of interest (Table 13.1), the relative importance of the DDL in clay mineral behavior decreases in the order montmorillonite > illite > kaolinite. In fact, the effect of the DDL is likely to be significant only in the case of clay soils that contain appreciable amounts of montmorillonite, such as bentonite.

The presence of the DDL and the extent to which the DDL of adjacent particles occupies the interparticle void space affects the hydraulic conductivity of clays. A descriptive schematic of two clay soils with high and low t_{DDL} and resulting pathways for flow are presented in Figure 13.2.

13.2 Containment Materials

Containment systems are constructed using natural materials (i.e., soils) and/or manufactured synthetic (polymer) materials known as geosynthetics. Both soils and geosynthetics are used for a variety of functions, including as low-permeability barrier layers against transport of liquids and gases in containment systems, as drainage/filter media for conveyance and collection of liquids and gases in containment systems, and as layers to protect specific components of a containment system or to separate the containment systems from contained materials. The common barrier materials include low-permeability natural soils, including compacted clays and sand-bentonite mixtures, and the manufactured geosynthetics known as geosynthetic clay liners and geomembrane liners. The common drainage/filter materials

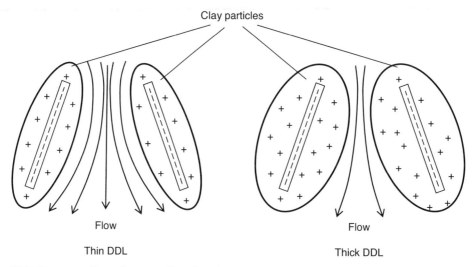

FIGURE 13.2 Effect of t_{DDL} on flow (based on Daniel 1994).

include high-permeability soils such as clean sands and gravels and geosynthetics such as geonets and geonet-geotextile composites (geocomposites). These materials have high liquid and gas conductivities. Basic information on soil drainage and filter materials is provided in Cedergren (1989) and on geosynthetic drainage and filter materials is provided in Chapter 12.

13.2.1 Compacted Soils

Compacted soils are the most traditional type of barrier material. Fine-grained soils such as clays and silts and amended soils such as sand-bentonite mixtures typically are densified by compaction in the field to construct compacted soil barriers with suitably low hydraulic conductivity, k. The k of compacted soil barriers is influenced significantly by both composition and compaction characteristics of the soil. Commonly, specific criteria for k are included in regulations for a compacted soil barrier based on permeation with a liquid (e.g., water and/or containment liquid). Typically, k must be less than or equal to 1×10^{-9} m/s (or 1×10^{-7} cm/s as commonly reported), although the limits on k will depend on a variety of factors, including the type of waste (e.g., municipal solid waste vs. tailings), the specific function of the barrier component (e.g., bottom liner vs. cover), and the specific regulations governing the safe disposal of the specific waste (e.g., federal vs. state). Although barriers made of compacted soils may contain soils other than clays, such as silts and sands, such barriers often are referred to collectively as compacted clay liners because of the inference of the low k associated with clays.

13.2.1.1 Composition

Low- and high-plasticity clays and silts (CL, CH, ML, MH)* are commonly used as compacted soil barriers. In addition, soils with high clay content such as clayey sands (SC) may be used (i.e., provided k is sufficiently low). The compacted soil barrier must contain a suitable fraction of clay-size particles because small clay particles reduce the pore sizes and

---

*All classifications are provided in accordance with the Unified Soil Classification System (USCS), ASTM D2487.

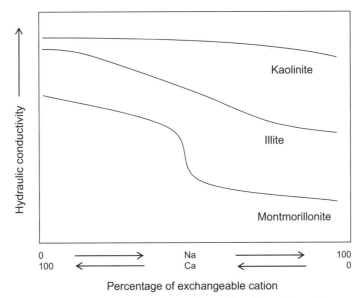

FIGURE 13.3 Effects of clay mineralogy and exchangeable cations on hydraulic conductivity (modified from Yong and Warkentin 1975).

interconnectivity of the pores that exist within the soil and, therefore, reduce the overall k of the soil. High clay mineral content is required in compacted soil barriers, as the presence of clay minerals allows for the development of a soil structure with high resistance to fluid migration. The type of clay mineral and the type of exchangeable cation predominant in the clay mineral can affect the k of soils comprised of the clay mineral, as indicated in Figure 13.3 (Yong and Warkentin 1975). High-swelling clay minerals (e.g., montmorillonite) tend to form "tight" soil structures in the presence of water (i.e., upon hydration) and correspondingly low k. However, high-swelling clay minerals also are more sensitive to pore-fluid chemistry, such that exposure to "strong" chemicals can result in shrinkage (i.e., reduction in t_{DDL}) and a higher k. Clays that contain monovalent cations (e.g., Na^+) also tend to form "tight" soil structures in the presence of water (i.e., upon hydration) due to swelling and correspondingly low k (Figure 13.3).

Compacted soil barriers also can be constructed using natural soils that do not contain a sufficient amount of fines but are amended with high-swelling bentonite. Bentonite is a natural soil that is dominated in composition by the montmorillonite clay mineral. The presence of a small amount (e.g., 5–10% by dry weight) of high-swelling sodium bentonite (i.e., bentonite which contains montmorillonite clay mineral with sodium [Na^+] as the dominant exchangeable cation) in an otherwise highly permeable material, such as clean sand, can significantly reduce the k to water to values that are at or below the regulatory maximum value. However, the high sensitivity of the montmorillonite clay mineral to pore-fluid chemistry also makes the sand-bentonite mixtures susceptible to chemical attack upon exposure to liquids that are chemically "strong," including some waste leachates.

The limiting index properties of soil that are likely to yield $k \leq 1 \times 10^{-9}$ m/s are provided by Benson et al. (1994b) on the basis of analysis of a database that includes compacted soil barriers at 67 North American landfills. The resulting criteria are summarized in Table 13.2. The effect of gravel content on k of compacted clayey soils has been evaluated on the basis of laboratory studies (Shelley and Daniel 1993). However, the upper limit on the gravel content

TABLE 13.2 Limiting Index Properties of Soils Likely to Achieve a
Geometric Mean Hydraulic Conductivity of $\leq 1 \times 10^{-9}$ m/s

Property	Maximum or Minimum Value (%)
Liquid limit	≥ 20
Plasticity index	≥ 7
Fines content (<0.075 mm)	≥ 30
Clay content (<2 μm)	≥ 15
Gravel content (>4.25 mm)	≤ 25

of 25% presented in Table 13.2 is based more on the likely difficulty of compacting soils with higher gravel contents in the field than on the ability of such soils to achieve low k. A practical upper limit on the plasticity index of 30 is recommended by Daniel and Koerner (2007) for barrier soils primarily on the basis that soils with a plasticity index greater than 30 likely will have low strength when wetted to high water saturation and, therefore, be difficult to compact.

The criteria listed in Table 13.2 allow for identification of potential soils for constructing compacted soil barriers with high likelihood of achieving suitably low k in the field. However, the criteria are meant only as guidelines to aid in the initial selection of soils considered for use as a compacted soil barrier. As such, there is no guarantee that low k will be achieved for soils that meet the criteria, nor is there any certainty that soils with characteristics outside of the provided limits will not achieve a suitably low k. Thus, once a potentially suitable soil is selected on the basis of the criteria noted in Table 13.2, the soil should be tested to determine k to verify the suitability.

13.2.1.2 Compaction and Hydraulic Conductivity

The k of compacted soil barriers is significantly influenced by compaction (e.g., Mitchell et al. 1965; Daniel and Benson 1990). The primary factors that affect k of compacted clays include: (1) type of compaction (e.g., kneading vs. static compaction), (2) energy of compaction (E), (3) the dry density (ρ_d, mass of solids per unit total volume of soil) of the compacted soil, and (4) the compaction or molding (gravimetric) water content (w). In general, lower k is achieved with kneading compaction, and k decreases with increasing E, ρ_d, and w. These factors are not necessarily mutually exclusive, since higher energy of compaction also typically results in higher dry density. However, the effect of dry density on k is minor relative to that of compaction water content on k. For example, k typically varies with dry density by less than an order of magnitude, whereas k of compacted clays can vary by two to four orders of magnitude or more as a function of compaction water content. In particular, a significant decrease in k of compacted clays typically occurs as w increases above the optimum water content, w_{opt} (i.e., $w > w_{opt}$).

The effect of compaction on the k of compacted clays has been explained on the basis of both microstructural behavior and macrostructural behavior (Figure 13.4). In terms of microstructural behavior (particle-scale), Lambe (1958) explained the behavior on the basis of two major particle arrangements or soil structures in fine-grained soils, viz. flocculated and dispersed. Soils have a flocculated structure with relatively large pores in clays compacted dry of w_{opt} and a dispersed structure with smaller pores in clays compacted wet of w_{opt}. These variations in the microstructure are used to explain the variations in k with w, where the larger void spaces between particles in clays compacted on the dry side of w_{opt} result in higher k and

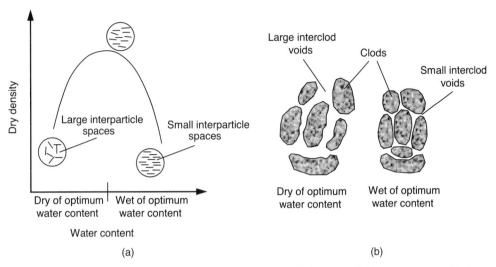

FIGURE 13.4 Compacted soil structures: (a) microstructural behavior and (b) macrostructural behavior (modified from Lambe 1958; Olsen 1962).

the smaller void spaces between particles in clays compacted wet of w_{opt} result in lower k (Figure 13.4a).

In terms of macrostructural behavior, Olsen (1962) proposed a clod theory to describe the effect of compaction on the k of fine-grained soils. In the clod theory, fine-grained soils are composed of particle agglomerations termed "clods." At lower water contents, the clods are relatively dry (i.e., hard) with high shear strength and, as a result, are difficult to compact, whereas at higher water contents, the wetter clods are relatively soft and more easily compacted. Thus, compacted fine-grained soils have large interclod voids dry of w_{opt} and small interclod voids wet of w_{opt}. These variations in the macrostructure have been used to explain the variations in k, where the larger interclod voids dry of w_{opt} resulting in higher k and the smaller interclod voids wet of w_{opt} resulting in lower k (Figure 13.4b).

An example of the effect of compaction water content and compaction energy on the macrostructure of compacted clay is presented in Figure 13.5. Additional depictions of compacted soil macrostructure are provided in Benson and Daniel (1990). Individual clods and interclod voids as well as boundaries between lifts are visible for the specimens compacted using lower energy and water content, whereas a uniform soil structure with no clods or interlift boundaries is observed for the specimens compacted at higher energy, in particular for the specimens compacted wet of w_{opt}. Thus, lower k is obtained for soil compacted with higher energy and higher water content.

13.2.1.3 Compaction Criteria

Two different approaches to specifying compaction criteria for compacted soil barriers have been used, with the primary objective of achieving a suitably low hydraulic conductivity (e.g., $k \leq 1 \times 10^{-9}$ m/s): a "traditional" approach and a "modern" approach. As shown in Figure 13.6a, the traditional approach is based on achieving a minimum percent compaction and a minimum water content based on a specified compaction energy, such as standard compaction energy (ASTM D698: Standard Test Methods for Laboratory Compaction Characteristics

FIGURE 13.5 Macrostructure of variably compacted soil specimens.

of Soil Using Standard Effort [12,400 ft-lbf/ft^3, 600 kN-m/m^3]) or modified compaction energy (ASTM D1557: Standard Test Methods for Laboratory Compaction Characteristics of Soil Using Modified Effort [56,000 ft-lbf/ft^3, 2700 kN-m/m^3]). This approach is based on that used for typical geotechnical applications pertaining to the dry density requirements with the primary objective of achieving high shear strength and low compressibility. However, the traditional approach does not take into account the likelihood of differences in compaction

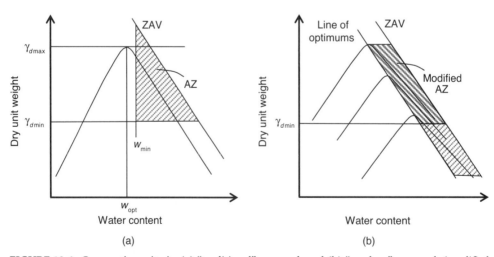

FIGURE 13.6 Compaction criteria: (a) "traditional" approach and (b) "modern" approach (modified from Daniel and Benson 1990).

energy between the laboratory and the field or the general variability in field compaction energy, both of which are factors that can significantly affect the k of a compacted soil barrier (see Daniel and Benson 1990).

In contrast, the "modern" approach (Figure 13.6b) takes into account the possible differences in compaction energy between the laboratory and the field, resulting in a zone of acceptable combinations of dry unit weight, γ_d ($= \rho_d \cdot g$), and compaction (molding) water content, w, referred to as an "acceptable zone" (AZ), that will provide a suitably low k regardless of compaction energy (Daniel and Benson 1990; Daniel and Wu 1993). In this approach, soils are compacted using a range of values for w and three different compactive efforts (energies): high effort, corresponding to ASTM D1557 (modified Proctor); medium effort, corresponding to ASTM D698 (standard Proctor); and low or reduced effort, which is the same as that specified in ASTM D698 except only 15 drops of the compaction hammer are used per loose lift of soil instead of the 25 specified in ASTM D698. After compaction, the compacted specimens are extruded from the compaction molds and placed in flexible-wall permeameters (described in Section 13.5.1) for measurement of k. The combinations of γ_d (or ρ_d) and w that provide suitably low k values (e.g., $k \leq 1 \times 10^{-9}$ m/s) then are used to develop the AZ. The AZ tends to fall between the line of optimums (i.e., a constant degree of saturation line that passes through the apexes on a series of compaction curves) resulting from the three compaction curves and the zero air voids (ZAV) curve (see Figure 13.6b).

The AZ developed solely on the basis of k testing then can be modified to include criteria for shear strength and shrinkage using similar analysis, where shear strength and volumetric strain tests are conducted to define the new boundaries of the AZ (e.g., Daniel and Wu 1993). Shear strength criteria can be established by determining the stress applied to the barrier system under the load of a waste mass. Shrinkage criteria can be established by determining strains associated with the onset and progression of cracking in compacted soils due solely to drying or to cyclic wetting and drying. The boundaries of the AZ are determined such that the combinations of w and γ_d (ρ_d) result in a compacted soil sufficiently wet to achieve low k and sufficiently dry to achieve high shear strength, low compressibility, and high shrinkage resistance. As an example, the AZ in Figure 13.6b, which initially included the entire region between the line of optimums and the ZAV, has been modified to include a minimum dry density requirement for strength considerations in the compacted soil for a bottom barrier system.

The significance of the AZ approach was demonstrated in the field using data from 85 full-scale compacted clay barriers (Benson et al. 1999). Measured k values based on field tests were always less than 1×10^{-9} m/s when the percentage of field-determined values for w and γ_d (ρ_d) on or above the line of optimums relative to total number of field-determined values of w and γ_d (ρ_d) for a given soil, or P_o, was greater than 90 (i.e., $P_o > 90\%$).

The recommended procedure for achieving low hydraulic conductivity ($k \leq 1 \times 10^{-9}$ m/s) of compacted clay liners in the field is as follows (Daniel and Benson 1990; Benson et al. 1999):

- Assess the effectiveness of potential soil(s) for barrier construction. Initial qualitative assessment can be made using the criteria provided in Table 13.2. Conduct laboratory compaction and hydraulic conductivity tests to generate an AZ for compaction similar to that presented in Figure 13.6b. Modify the AZ for shear strength and shrinkage criteria (or any additional criteria) as necessary by conducting additional tests or using existing information.
- Develop "modern" compaction specification criteria for construction in the field. Use the AZ defined on the basis of laboratory test results as the area bound by ZAV and the

line of optimums in the compaction specifications. A numerical value can be assigned to the line of optimums using the degree of saturation corresponding to optimum water content as follows:

$$S_{opt} = \frac{S_{opt1} + S_{opt2} + S_{opt3}}{3} \qquad (13.2)$$

where S_{opt} is the degree of saturation along the line of optimums and S_{opt1}, S_{opt2}, and S_{opt3} are the degrees of saturations corresponding to the optimum water contents based on the three compactive efforts used in the development of the AZ. Individual S_{opt} values can be calculated using the following equation:

$$S_{opt} = \frac{w_{opt}}{\left[\dfrac{\gamma_w}{\gamma_{dmax}} - \dfrac{1}{G_s} \right]} \times 100 \qquad (13.3)$$

where S_{opt} is the degree of water saturation at the optimum water content, w_{opt} is the optimum water content, γ_w is the unit weight of water, γ_{dmax} is the maximum dry unit weight, and G_s is the specific gravity of solids. Equation 13.3 also can be used to determine the degree of saturation for any given combination of water content and dry unit weight. The degree of saturation of the field soil should be $\geq S_{opt}$.

- Modify AZ with regard to shear strength or shrinkage criteria by specifying minimum and/or maximum w as well as minimum γ_d as necessary.
- Include criteria in compaction specifications related to obtaining uniform water content and maximum clod sizes of soils to be compacted. If field soils need to be wetted or dried for construction of a compacted barrier, sufficient time must be allowed for hydrating or dehydrating the soils. For wetting applications, field analysis indicated that k decreases with increasing hydration time (Benson et al. 1997). Initially dry clays should be hydrated for ≥ 24 h for CL soils and ≥ 48 h for CH soils in the field (Benson et al. 1997). Less hydration time may be used for initially moist soils. Hydraulic conductivity also decreases with decreasing clod size. Proper processing of soils for construction of compacted barriers includes working and discing the soil to achieve small clod sizes and uniform moisture distribution over sufficiently long hydration durations.
- Use of a kneading-type compactor (e.g., pad foot, sheepsfoot, and tamping foot) is recommended for achieving good interlift bonding between compacted soil layers. In addition, the length of the foot on the compaction equipment should be greater than or equal to the thickness of the loose layer of soil prior to compaction to ensure penetration completely through a compacted lift and good interlift bonding. In most applications, this requirement will limit loose lift thickness to less than about 200–250 mm.
- Moderately heavy to heavy compactors (weight ≥ 195 MN) should be used to achieve low k in the field. In addition, to ensure that the applied compaction energy fully penetrates the compacted soil, liners should be constructed in layers or lifts, with the compacted lift thickness no greater than about 150 mm. Thinner compacted layers may be required for lower energy compaction and/or compaction equipment with relatively short compaction feet.

- Hydraulic conductivity decreases with increasing barrier thickness, primarily due to a decrease in the probability of the presence of interconnected defects (e.g., desiccation cracks and poor interlift zones) throughout the thickness of a compacted clay liner (Benson and Daniel 1994). For nonhazardous containment applications (e.g., municipal solid waste landfills), a minimum barrier thickness ranging from 0.6 to 0.9 m typically is required, whereas a minimum barrier thickness of 0.9 m typically is required for hazardous waste containment (e.g., hazardous waste landfills).

13.2.2 Geomembranes

Geomembranes are thin (0.5–3.0 mm) polymeric sheets used as barriers against migration of fluids in containment systems. Geomembranes are used in containment facilities as single barriers or as part of composite barriers (described in Section 13.3). The use of geomembranes is required by regulation for various containment applications. The most commonly used geomembranes are high-density polyethylene (HDPE), linear low-density polyethylene (LLDPE), and polyvinyl chloride (PVC). Less common geomembranes include flexible polypropylene (fPP), ethylene propylene diene monomer (EPDM), and prefabricated bituminous geomembranes (i.e., asphalt-impregnated fabric/textile sheets [PBGM]). Reinforced geomembranes with improved mechanical properties, such as reinforced flexible polypropylene (fPP-R), reinforced ethylene propylene diene monomer (EPDM-R), reinforced chlorosulfonated polyethylene (CSPE-R), and reinforced ethylene interpolymer alloy (EIA-R), also have been used (Koerner 2005). The surfaces of geomembrane sheets may be smooth or textured to provide increased interface friction between the geomembrane and surrounding materials.

Geomembranes typically are manufactured into rolls and shipped to a site for installation. Typical roll widths range from approximately 6 to 7 m, although widths as narrow as 2 to 3 m are available. Typical lengths of manufactured geomembrane rolls range from approximately 70 to 80 m to over 500 m. The rolls are joined in the field by thermal or chemical seaming processes to cover large areas (Koerner 2005). Typically, thermal extrusion and fusion seams are used for polyolefin (polyethylene and polypropylene) geomembranes and PBGM, whereas chemical fusion and adhesive seams are used for PVC, CSPE-R, EIA, and EPDM. For example, a dual hot wedge thermal fusion seam commonly is used for HDPE and LLDPE geomembranes in waste containment applications (Figure 13.7a); thermal extrusion seams are used when transitioning from a textured geomembrane to a smooth geomembrane and around repairs or for limited-access areas in containment facilities (Figure 13.7b); and a solvent, such as methyl ethyl ketone, is used for fusion seaming of PVC (Figure 13.7c). Integrity of seams is paramount to the performance of geomembrane liners as barriers against the transmission of fluids. Flexible geomembranes, such as PVC and polypropylene, also are available in panels. Less field seaming is required for panels than sheets from rolls, which may be applicable for areas where *in situ* seaming is difficult. Also, factory seams are considered to be more uniform than field seams, as they are made in a more controlled and clean environment (Koerner 2005).

HDPE geomembranes typically are used as liners in bottom barrier systems due to their high resistance to chemical environments and good mechanical properties. In cover systems, differential settlement of underlying wastes may occur and a cover system may be required to conform to the shape of the variably deformed wastes. Greater flexibility allows for deformation of the geomembrane without development of excessive stress concentrations or rupture,

(a)

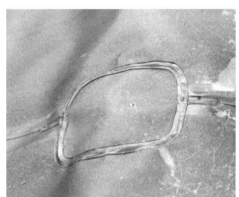

(b)

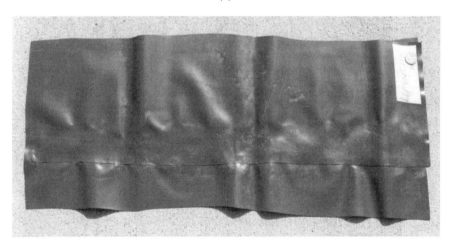

(c)

FIGURE 13.7 Geomembrane seams: (a) dual hot wedge thermal fusion seam, (b) thermal extrusion seam, and (c) sample from a chemically seamed PVC liner.

both of which could jeopardize the integrity of the cover system. LLDPE, PVC, and fPP geomembranes are preferred in cover applications due to their greater flexibility in comparison to other geomembranes. Although these geomembranes, in particular PVC, have relatively low chemical resistance, resistance to chemicals generally is not a primary concern for cover systems or in other applications where nonaggressive liquids are involved in the containment application (e.g., water conveyance canals). The thickness of geomembranes used as liners in bottom barrier systems typically ranges from 1.5 to 2.5 mm, whereas geomembrane thicknesses for cover applications typically range from 1.0 to 1.5 mm.

Geomembranes are highly resistant to transmission of water. Intact geomembranes (geomembranes without defects, such as holes or leaking seams) are essentially impervious to liquid-phase migration. In addition, water vapor transmission (WVT) rates are very low for geomembranes. For example, laboratory WVT rates as low as 0.006 g/m^2-d have been reported for a 2.4-mm-thick HDPE geomembrane (Koerner 2005). However, solvent vapor transmission (SVT) rates may be significantly higher than WVT rates for geomembranes. For example, an SVT rate of 15.8 g/m^2-d has been reported by Koerner (2005) for transmission of chloroform through 2.6-mm-thick HDPE. Both WVT and SVT rates as well as water and solvent permeabilities are lower for HDPE geomembranes in comparison to geomembranes comprised of other polymers.

Physical, mechanical, and endurance properties of geomembranes are determined for their use in containment systems. Additional examples of geomembrane tests as well as tests for determining integrity of seams are provided in Chapter 12. Timely covering of geomembranes subsequent to installation is critical for ensuring long-term performance. Degradation by oxidation and UV radiation is minimized by timely covering. In addition, thermal stresses in geomembranes are minimized by rapid placement of overlying layers. Exposure to high temperature differentials may generate large strains in geomembranes with high coefficients of thermal expansion and contraction.

13.2.3 Geosynthetic Clay Liners

Traditional or conventional geosynthetic clay liners (GCLs) are thin (~5- to 15-mm-thick), prefabricated (factory-manufactured) hydraulic barriers that consist primarily of a processed clay, typically sodium bentonite, or other low-permeability material that is either encased or "sandwiched" between two geotextiles or attached to a single polymer membrane (i.e., geomembrane) and held together by needle-punching, stitching, and/or gluing with an adhesive. The former type of traditional GCL often is referred to as a geotextile-encased GCL, whereas the latter type of traditional GCL often is referred to as a geomembrane-backed GCL. The pattern of stitching in stitch-bonded GCLs tends to be more uniform and systematic than that in needle-punched GCLs, which generally is more random. The hydraulic resistance of these conventional GCLs that do not include a geomembrane component is attributed to the bentonite component of the GCL, which swells in the presence of water to form a tight sealing layer.

GCLs that are stitch-bonded or needle-punched also are referred to as reinforced GCLs, whereas GCLs that are held together by mixing an adhesive (glue) with the bentonite to affix the bentonite to the adjacent geotextiles or a geomembrane are referred to as unreinforced GCLs. The presence of the stitched or needle-punched fibers in reinforced GCLs provides greater internal resistance to shear relative to unreinforced GCLs that rely essentially on the shear strength of the bentonite alone, which is relatively low in a saturated condition (e.g.,

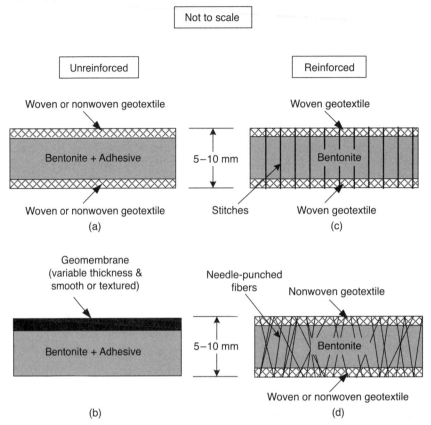

FIGURE 13.8 Schematic cross sections of conventional GCLs: (a) unreinforced, geotextile encased; (b) unreinforced, geomembrane backed; and (c, d) reinforced, geotextile encased (redrawn after Shackelford 2008).

Gilbert and Byrne 1996; Gilbert et al. 1997; Eid et al. 1999; Fox and Stark 2004). As a result, unreinforced GCLs usually are restricted to relatively flat slopes, such as the base of bottom barrier (liner) systems, whereas reinforced GCLs also can be used as liners or liner system components on the side slopes of waste containment systems. Needle-punching typically yields a stronger, more rigid GCL than stitch-bonding, although thermal fusing of fibers in stitch-bonding has been used to increase the internal resistance of stitch-bonded GCLs (Fox and Stark 2004). Schematic cross sections of conventional GCLs are shown in Figure 13.8.

A more recent type of GCL, often referred to as a geomembrane-backed, geotextile-encased GCL, essentially represents a combination of the two more traditional types of GCLs. Similar to the geotextile-backed GCL, this more recent type of GCL includes two hydraulically resistant materials, bentonite and a polymer sheet (e.g., polyethylene geofilm), and may be either unreinforced or reinforced. In the case of unreinforced GCLs of this type (Figure 13.9a), the polymer sheet is laminated (glued) to one of the two geotextiles of a conventional unreinforced, geotextile-encased GCL. In the case of reinforced GCLs (Figure 13.9b and c), the polymer is laminated (glued) to one of the two geotextiles of a conventional reinforced, geotextile-encased GCL.

GCLs typically are manufactured into rolls and shipped to site for installation in the form of panels or sheets, the dimensions of which are based primarily on the widths and lengths of

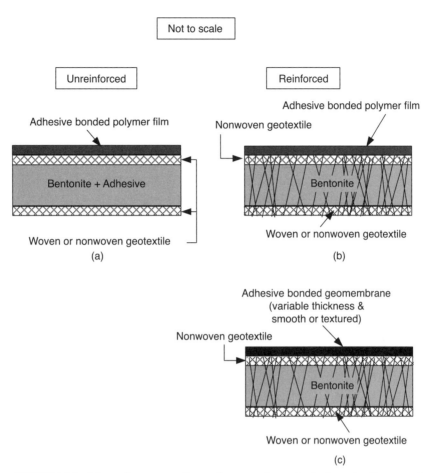

FIGURE 13.9 Schematic cross sections of geomembrane-backed, geotextile-encased GCLs: (a) unreinforced and (b, c) reinforced (redrawn after Shackelford 2008).

the rolls, which can vary. Typical roll or panel widths range from approximately 4.2 to 5.3 m, although widths as narrow as 2.4 m can be manufactured for some GCL products (U.S. EPA 2001). Typical lengths of manufactured GCL rolls range from approximately 30.5 to 61.0 m, although shorter panels can be used simply by cutting the rolls (U.S. EPA 2001).

Proper material installation and covering procedures are essential to meet the design intent for effective environmental containment and long-term performance (Richardson et al. 2002). Standard guidance regarding material handling, subgrade preparation, panel deployment, alignment, and overlapping and seaming is provided in relevant ASTM standards. In addition, guidelines are available from GCL manufacturers regarding GCL handling and installation.

Although exact installation procedures and recommendations may vary from manufacturer to manufacturer, installation generally consists of rolling out GCL panels on a prepared subgrade, with adjacent panels overlapped a minimum 150 mm (Estornell and Daniel 1992; Koerner 2005). For GCLs with nonwoven, needle-punched geotextiles on both the upper and lower surfaces, a bead of granular bentonite (typically ~0.4 kg/m) must be applied to the overlap of the adjacent panels to maintain the integrity of the sealing system. Subsequent to overlap treatment, a minimum of 300 mm cover soil (or geomembrane in some cases) usually is placed. The covering material is always placed during the same shift (same day) as the GCL is deployed to minimize the chance of unconfined hydration and possible damage to the GCL.

The two primary motivations driving the increasingly preferential use of GCLs in waste containment applications relative to alternative barriers or barrier components, such as compacted clay liners (CCLs) and geomembrane liners (GMLs), are (1) a savings in cost and (2) establishment of technical equivalency relative to CCLs (Koerner and Daniel 1995). The savings in cost results essentially from the ease of installation of GCLs relative to both CCLs and GMLs as well as from the maximization of disposal space due to the lower thickness of GCLs relative to CCLs. For example, the ability to seal containment facilities by simply overlapping adjacent GCL panels and placing dry bentonite between the panels favors the installation of GCLs relative to GMLs, where such adjacent panels must be welded thermally or chemically together to ensure an intact, continuous seam.

In terms of technical equivalency, there are a number of technical advantages that make GCLs preferable relative to CCLs and/or GMLs (Bouazza 2002). The primary technical justification probably has been the extremely low hydraulic conductivity, k, of GCLs when permeated with deionized water, which typically is less than approximately 3.0×10^{-11} m/s (Daniel et al. 1997). However, the potential for significant increases in k (one to several orders of magnitude) upon permeation with chemical solutions other than water is a concern (e.g., Shackelford et al. 2000; NRC 2007).

Another technical aspect that favors the use of GCLs is the greater self-healing capability of the bentonite in GCLs relative to CCLs constructed with typically lower plasticity natural clay soils and a generally greater ability to withstand relatively large differential settlements compared with CCLs. Small defects such as puncture holes up to 75 mm in diameter can be overcome upon hydration with water (U.S. EPA 2001). This self-healing capability generally leads to greater resistance of GCLs to increases in k resulting from climatological distress due to repeated freezing/thawing and/or wetting/drying cycles. However, there is concern about the possible reduction in swelling potential of the bentonite in traditional GCLs resulting from multivalent-for-monovalent cation exchange (e.g., Ca^{2+} for Na^+), which can lead to significant increases in k upon rehydration of the bentonite (Meer and Benson 2007; Benson et al. 2007; NRC 2007). Increases in k may result in release of contaminants through bottom barrier systems. Flexibility and self-healing capability of GCLs in comparison to CCLs favor use in cover systems placed over wastes with potential for large differential settlements such as municipal solid wastes.

13.3 Containment Systems

Containment systems are used to completely isolate the contained materials from the surrounding environment as well as to facilitate collection and removal of any by-products or effluents associated with the contained materials. The by-products commonly associated with waste containment include leachate and gas. Leachate is the contaminated liquid generated by decomposition of wastes and by infiltration of precipitation (rain, snowmelt) through a waste mass, which accumulates at the base of a containment facility. Leakage of leachate from a containment facility may cause contamination of the surrounding soils and groundwater. Gas is generated by the decomposition of organic fraction of municipal solid waste (MSW) or other organic wastes and is mainly composed of methane and carbon dioxide. Both methane and carbon dioxide are greenhouse gases, with the global warming potential of methane being 21 times greater than that of carbon dioxide. MSW landfills are one of the largest anthropogenic sources of methane in the atmosphere (U.S. EPA 2008). Leakage of gas from a containment facility contributes to air pollution. Also, methane is highly flammable and can be explosive in the presence of air. In addition, several components of landfill gas may be toxic and harmful to human health and the environment at elevated levels.

Another significant by-product of biological decomposition of organic components of MSW or other wastes is heat. In addition, chemical reactions that occur in wastes can result in significant heat production (e.g., heat production in ash landfills). Temperature controls organic waste decomposition and affects engineering properties of both wastes and containment materials. Elevated temperatures accelerate degradation of geosynthetic components of barrier systems and contribute to desiccation of earthen barrier materials (Rowe 2005). Leachate is generated in containment facilities for all types of wastes, whereas gas rich in methane and carbon dioxide and heat are generated only in containment facilities for wastes with high organics content and wastes undergoing significant exothermic reactions. Specific provisions are included in containment systems for leachate and gas management (including collection, removal, treatment, and beneficial use). However, provisions for management of heat or general temperature control for optimum performance typically are not included in containment systems.

13.3.1 Types and Configurations

Engineered containment systems consist of bottom (basal) liner systems and cover systems that completely encapsulate contained materials. Provisions are made for removal of leachate and gas as required (Figure 13.10). Bottom liner systems are placed beneath contained materials, whereas cover systems are placed over the contained materials. Covers may not be required or used for nonwaste containment applications, such as water conveyance canals, where only a bottom liner is needed. The sole use of covers for containment (i.e., without a liner system) may be considered for nonengineered contaminated sites. Side slopes below grade typically are constructed at horizontal:vertical inclinations ranging from 3:1 to 2:1, whereas the side slopes for cover systems typically are shallower (from 4:1 to 3:1).

Containment systems include alternating layers of materials with variable functions. Low-permeability barriers constructed using CCLs, GMLs, and/or GCLs resist movement of the contained materials and the by-products to the surrounding environment and infiltration of

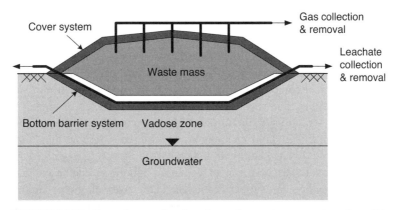

FIGURE 13.10 General scenario of engineered containment system for solid waste disposal.

TABLE 13.3 Individual and Composite Barriers Used in Containment Applications

Individual Barriers	Composite Barriers (Top/Intermediate/Bottom)
CCL	GML/CCL (common)
GCL	GML/GCL (common)
GML	GML/GCL/CCL (uncommon)
	GML/CCL/GML (uncommon)
	GML/GCL/GML (uncommon)

water or air into the contained materials. Blanket drainage and filter materials are used to collect and remove leachate and gas from contained wastes. Soil layers are placed between containment systems and contained materials to protect the containment systems.

The barrier systems used in containment applications can be categorized based on the number and arrangement of layers in a given barrier (individual or composite) and the total number of barriers in a containment system (single or double). The common types of individual or composite barrier systems are summarized in Table 13.3. Single barriers may be comprised of an individual barrier or a composite barrier. In general, composite barriers consist of a GML overlying and in intimate contact with either a CCL or a GCL. Double barriers consist of two single barriers (individual or composite) separated by a leak detection system, such as a layer of clean coarse-grained soil (sand or gravel) or a geosynthetic drainage layer. When the two barriers in a double barrier are both composite barriers, the barrier system is referred to more specifically as a double composite barrier system.

An important aspect of composite barriers is the requirement for the individual components of the composite barrier to be in intimate contact with each other (Daniel 1993). Composite barriers provide greater resistance to flow when such intimate contact is achieved between individual components of the composite barrier. For geomembranes, area for flow is low (holes or defects); however, there is no restriction to flow. For CCLs and GCLs, area for flow is high (entire surface area of the barrier); however, flow is restricted due to the low k of the individual barriers. The use of a GML overlying either a CCL or a GCL as a composite barrier results in a barrier with significantly reduced area for flow (due to the presence of the GML) and high resistance to liquid-phase flow (due to the presence of the CCL and/or GCL). If intimate contact is not established between the individual components of the composite barrier, lateral flow of liquid occurs between the barriers, which negates composite action. Composite barriers combine the advantages and eliminate the disadvantages of individual barriers.

In terms of composite barriers, uneven surfaces on a CCL can prevent good contact between the barrier layers. For this reason, the top surface of the CCL constructed using pad/tamping foot type of compactors must be smoothed using rubber tire or smooth drum compactors. Protrusions such as rocks, cobbles, large gravel particles, or organic matter such as tree stumps that can create gaps and/or penetrate an overlying GML need to be eliminated from the top surface of CCLs. Hand-picking may be required to remove these materials from the top surface of the CCL. The placement of a geotextile above a CCL with rocks or organic matter to protect the overlying GML against formation of holes should be avoided in the case of a composite barrier, because the permeable geotextile layer allows for lateral transfer of fluids between the CCL and GML. Thus, placing a geotextile between the CCL and overlying GML eliminates composite action in the barrier system. Similarly, for GML-GCL composites, lateral transfer of fluids may occur when a nonwoven, needle-punched geotextile in a geotextile-encased GCL is placed beneath an overlying GML. The nonwoven, needle-punched geotextile may create a relatively permeable layer between the overlying geomembrane and the underlying bentonite component of the GCL.

Double barrier systems provide redundancy relative to single barrier systems, resulting in more resistance to flow of liquids and better isolation of contained materials. Double barrier systems typically are used for hazardous waste containment applications. Additional barriers also may be used to provide further containment and isolation. The barriers are separated by filter/drainage layers in multiple barrier systems.

13.3.2 Bottom Barrier Systems

Bottom barrier systems (also commonly termed bottom or basal liner systems) are constructed prior to the placement of contained materials. These systems are placed over relatively flat subgrade surfaces at the bottom of the containment system as well as over inclined surfaces on sloping subgrade along sidewalls.

A typical single bottom barrier system consists of, from top to bottom: a protective soil layer, a blanket filter/drainage layer, and a barrier system (individual or composite). The filter/drainage layer, typically referred to as a leachate collection and removal system (LCRS), is used to collect and remove leachate in waste containment applications. This layer is not used in containment of liquids or for applications that involve nonsolid wastes. A typical double bottom barrier system consists of, from top to bottom: a protective soil layer, first blanket filter/drainage layer, first barrier system (individual or composite), second blanket filter/drainage layer, and second barrier system (individual or composite). The first filter/drainage layer is the LCRS and provides the same function as in a single bottom liner system. The second filter/drainage layer is used to collect and remove the liquids (e.g., leachate) that flow through the overlying barrier system and is referred to as the leak detection system (LDS) or sometimes as the secondary collection system.

The first (top) and second (bottom) barrier systems are called the primary and secondary barrier systems, respectively. The flow that occurs through the primary barrier system is detected, collected, and removed via the LDS. The amount of flow in the LDS can be used to assess the effectiveness of the primary barrier system. If the primary barrier system is a composite barrier that includes a CCL (e.g., GML/CCL), water may be expelled from the CCL into the LDS due to consolidation of the CCL under the load from the overlying wastes. The volume of this water must be estimated to make an accurate assessment of the effectiveness of the primary barrier system without the inclusion of such consolidation water. Minimum bottom liner system configurations required by regulations are provided in Figure 13.11 for both MSW and hazardous waste containment applications. These containment systems include landfills for solid wastes and surface impoundments for liquid wastes (without LCRS).

The LCRSs typically are constructed using granular materials with thicknesses between 300 and 600 mm. When geosynthetics are used for LCRS construction, these materials typically are overlain by granular protection layers with thicknesses ranging from 300 to 600 mm. In landfill applications, the maximum hydraulic head acting on bottom barrier systems typically is limited to 300 mm by regulatory requirements, primarily to minimize the hydraulic gradient driving liquid flow through the barrier. This restriction in hydraulic head is achieved by timely operation of pumps installed in sump collection areas at the base of landfills. The granular

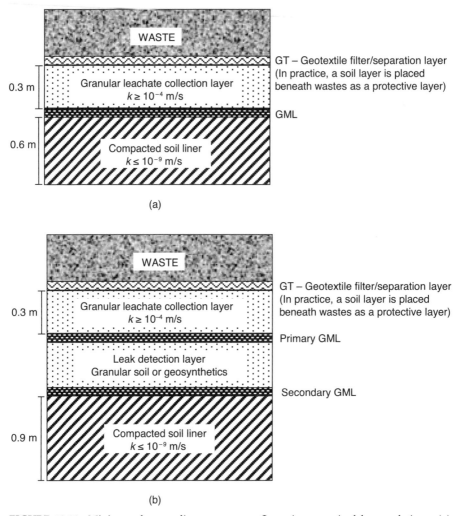

FIGURE 13.11 Minimum bottom liner system configurations required by regulations: (a) MSW and (b) hazardous waste.

LCRS materials may be replaced by geosynthetic alternatives along side slopes due to ease of placement above primary barriers along inclined surfaces. LDSs typically are constructed using geotextile-geonet composites due to ease of placement above secondary barriers.

GCLs typically are used as individual barriers or as part of composite barriers (GML/GCL) in primary (top) barriers due to ease of placement over LDSs. The use of a thinner GCL in place of a CCL in a primary composite barrier also provides for more disposal space. CCLs typically are preferred in secondary barriers due to a greater degree of redundancy in protection offered by thicker CCLs relative to thinner GCLs. Smooth geomembranes typically are used along the base of containment systems, whereas textured geomembranes are used along side slopes for improved interface friction between the geomembrane and overlying and/or underlying materials. When geosynthetic filter/drainage layers are used, geotextile-geonet-geotextile composites typically are specified along side slopes over textured geomembranes. In this case, the bottom geotextile in the three-layer geocomposite only serves to provide greater friction and,

TABLE 13.4 Relative Advantages and Disadvantages of CCLs, GCLs, and GMLs in Bottom Barrier Systems

Liner	Advantages	Disadvantages
CCL	• Greater resistance to punctures (thick) • Greater total attenuation capacity for chemical species (despite lower unit attenuation capacity than GCLs) • Greater experience with use of CCLs • Greater certainty for long-term performance	• Lower disposal capacity (thick) • Lower resistance to environmental distress (e.g., wet/dry and freeze/thaw cycles) • Difficult to repair • More cumbersome construction and QA/QC procedures
GCL	• Greater disposal capacity (thin) • More rapid and simpler construction than CCLs • More reproducible material properties (manufactured) • Greater resistance to environmental distress (e.g., wet/dry and freeze/thaw cycles) • Easier to repair • Higher unit attenuation capacity than CCLs • More straightforward QA/QC procedures	• Greater potential for puncture (thin) • Potential problems with integrity of panel seams • Greater potential incompatibility (increase in k) when subjected to chemical solutions • Greater potential for reduced swelling capacity and increases in hydraulic conductivity resulting from multivalent-for-monovalent cation exchange • Lower attenuation capacity than CCLs • Uncertain long-term integrity and performance
GML	• Greater disposal capacity (thin) • Construction can be easier and cheaper than compacted soils (based on site-specific conditions) • Reproducible material properties (manufactured) • High resistance to environmental distress (e.g., wet/dry and freeze/thaw cycles) • Virtually impervious to liquid flux	• Potential problems with puncture (thin) • Potential problems with integrity of panel seams • Rapid transport by vapor diffusion (in particular, volatile organic compounds) • Uncertain long-term integrity and performance

therefore, sliding resistance between the geocomposite and the underlying textured geomembrane. The interface friction between a geotextile and a textured geomembrane is significantly higher than that between a geonet and a textured geomembrane. Interface friction angles for various possible interfaces in bottom barrier systems are provided in Chapter 12. The relative advantages and disadvantages of CCLs, GCLs, and GMLs used in bottom barrier systems are summarized in Table 13.4.

13.3.3 Cover Systems

Cover systems are constructed over the contained materials when the containment systems are completely full. Earthen cover systems are used over solid contained materials, such as MSW. The cover systems are placed over relatively flat top surfaces and also over inclined side surfaces. The overall objective is to minimize the amount of infiltrating water that percolates through the cover, as such percolation ends up migrating through the underlying disposed waste, thereby leaching potentially toxic substances in the form of leachate that eventually can contaminate groundwater if not contained properly. Cover systems are designed to resist infiltration while promoting high surface runoff and maintaining stability (against wind/water erosion and slope failure). Cover systems represent the final defense against potential contamination from containment facilities and are required to perform for extended durations.

13.3.3.1 Conventional Covers

A typical conventional cover system consists of, from top to bottom: a vegetative soil layer, a protective soil layer, a blanket filter/drainage layer, and a barrier system (individual or composite). These layers may be underlain by a second blanket filter/drainage layer. A protective/foundation soil layer is used below the second filter/drainage layer (if used) or the barrier layer. A schematic drawing that illustrates the components of a conventional cover system is provided in Figure 13.12. The vegetative soil layer supports plant growth and prevents erosion along the top surface of the cover system. The protective soil layer provides a biotic barrier between the ground surface and the barrier system components against intrusion of plants from the vegetative layer as well as animals from the ground surface. In cold climates, this layer also serves as a frost protection layer, where the components of the barrier system are placed below the local frost depth. The filter/drainage layer allows for diverting and collecting the water entering the cover system due to precipitation or from surface runoff from surrounding areas. The barrier system prevents infiltration of water into the waste mass and also isolates the contained materials from the surrounding environment. Composite barriers with geomembrane components typically are required by regulation in the case where the bottom liner system includes a GML for MSW landfills as well as for all cases involving

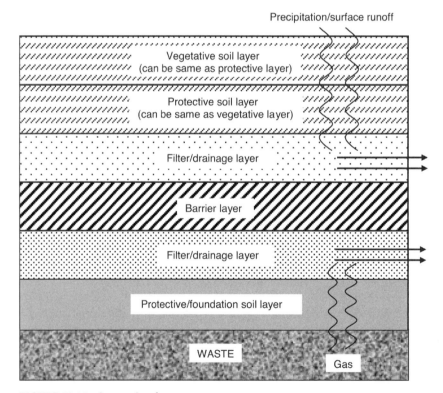

FIGURE 13.12 Conventional cover system.

hazardous and low-level radioactive waste landfills. The second filter/drainage layer placed beneath the barrier system is used to facilitate collection and removal of the gas generated by wastes in MSW landfills. The protective/foundation soil layer separates the cover from the wastes and provides a firmer base over the wastes for the construction of the overlying layers. Use of granular soil with high permeability for this layer facilitates migration of the landfill gas to the overlying gas collection layer.

The primary design consideration for conventional covers is to provide high resistance to infiltration of water. Low k is required in the barrier component of the cover system, similar to bottom liner systems. Materials and barrier configurations similar to those in bottom liner systems are used. Cover systems are subjected to low applied effective stresses, low hydraulic gradients, and potentially high changes in degree of saturation of earthen components. The cover systems also are subjected to high seasonal and diurnal temperature differentials and high thermal gradients, such that thermally driven moisture flow may occur in cover systems (Yesiller et al. 2008). Cover systems also are required to provide high resistance to gas emissions for facilities such as MSW landfills, where potentially harmful gases are generated. In general, gas conductivity is considered to follow similar trends and is affected by similar factors as hydraulic conductivity. Overall, conditions for maintaining a relatively constant and low k need to be investigated for cover system design.

Promoting surface runoff, preventing wind and water erosion, and maintaining slope stability are also important for cover systems. Cover vegetation is selected to minimize erosion and to promote evapotranspiration. Slopes are designed to cover wastes completely and to promote surface runoff. Berms and geosynthetic reinforcement materials may be used to improve stability of slopes. Single and composite barriers have similar stability along slopes for relatively shallow slopes ($<9°$), whereas single barriers perform better than composite barriers for steeper slopes ($>18°$). In particular, composite barriers with a GML/CCL/GML configuration have low resistance to sliding at high slope angles (Daniel and Koerner 1993). Interface friction angles are provided in Chapter 12 for various possible interfaces that may be present in cover systems.

The advantages and disadvantages of CCLs, GCLs, and GMLs in cover systems are summarized in Table 13.5. In terms of landfill gas management, lower cover gas emissions and higher landfill gas recovery rates were demonstrated for covers with CCLs and GMLs in comparison to a cover with GCL (Spokas et al. 2006). The GCL was not fully hydrated in the cover system, which resulted in higher gas conductivity compared to the cover systems that included a CCL and a GML.

13.3.3.2 Alternative Covers

Alternative covers are final covers that are designed and operated on the basis of the hydrologic water balance, as illustrated conceptually in Figure 13.13. The term "alternative cover" is used to distinguish covers based on water-balance principles from the more traditional or conventional covers designed on the basis of achieving a low saturated hydraulic conductivity to impede percolation of infiltrating water and increase surface runoff, such as those typically prescribed by regulations for closure of solid waste disposal facilities (e.g., landfills).

With respect to the water balance illustrated in Figure 13.13, amount of percolation (P_r) generated is given by the following expression:

$$P_r = P - (SRO + Q_i + ET + \Delta SWS) \tag{13.4}$$

TABLE 13.5 Relative Advantages and Disadvantages of CCLs, GCLs, and GMLs in Cover Systems

Material	Advantages	Disadvantages
CCL	• Greater resistance to punctures (thick) • Greater resistance than GCLs to adverse impact resulting from multivalent-for-monovalent cation exchange • Greater experience with use of CCLs • Greater certainty for long-term performance	• Lower resistance to environmental distress (desiccation, wet/dry cycles, and freeze/thaw cycles) • Lower resistance to differential settlement (from underlying wastes) • Lower disposal capacity (thick) • More difficult to repair • More cumbersome construction and QA/QC procedures
GCL	• Greater disposal capacity (thin relative to CCLs) • More rapid and simpler construction than CCLs • More reproducible material properties than CCLs (manufactured) • Greater resistance than CCLs to environmental distress (e.g., wet/dry and freeze/thaw cycles) • Greater resistance than CCLs to differential settlement (from underlying waste) • Easier to repair • More straightforward QA/QC procedures	• Greater potential for puncture (thin) • Potential problems with integrity of panel seams • Greater potential for reduced swelling capacity and increases in hydraulic conductivity resulting from multivalent-for-monovalent cation exchange • Uncertain long-term integrity and performance • If not fully hydrated (in arid areas or areas with cyclic wet/dry), relatively low resistance to gas migration and liquid percolation
GML	• Greater disposal capacity (thin relative to CCLs) • More rapid and simpler construction than CCLs • More reproducible material properties than CCLs (manufactured) • Greater resistance than CCLs to environmental distress (e.g., wet/dry and freeze/thaw cycles) • Greater resistance than CCLs to differential settlement • Virtually impervious to water flux	• Potential problems with puncture (thin) • Potential problems with integrity of panel seams • Rapid transport by vapor diffusion (in particular, volatile organic compounds) • Uncertain long-term integrity and performance

Modified after Daniel and Koerner (1993).

where P is precipitation, SRO is surface runoff, Q_i is intralayer flow, ET is evapotranspiration (evaporation plus vegetative transpiration), and SWS is the soil-water storage. Thus, in accordance with Equation 13.4, the amount of percolation generated for a given amount of precipitation can be minimized by maximizing SRO, Q_i, ET, and the change in SWS.

In the case of alternative covers, the primary design consideration is to ensure that there is sufficient water storage capacity, such that infiltrating water can be stored with little or no drainage during periods of elevated precipitation and limited evaporation and transpiration (e.g., winter), followed by subsequent release of the stored water during drier periods with greater evaporation and transpiration (e.g., summer). This concept is illustrated schematically in Figure 13.14. These types of covers are particularly well suited for regions with arid or semiarid climates, where potential evapotranspiration (PET) far exceeds precipitation (P) (e.g., $PET > 2P$), such as some western regions of North America (Shackelford 2005).

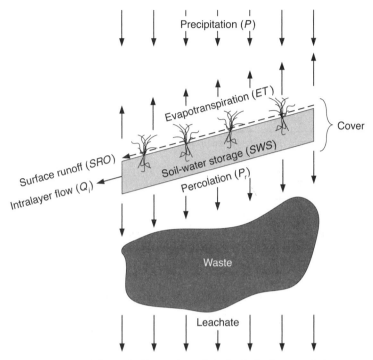

FIGURE 13.13 Schematic illustration of the hydrologic water balance applied to a cover for waste containment applications.

The primary motivations for the increasing interest in alternative covers are the typically lower cost and expected greater durability of alternative covers relative to traditional covers. Soils that have low potential for desiccation cracking and frost damage, such as silty sands, silts, silty clayey sands, clayey silty sands, and similar materials, typically are used for alternative covers. In general, this relatively wide range of soil types allows alternative covers to be constructed using locally available materials, resulting in greater congruity and harmony with nature.

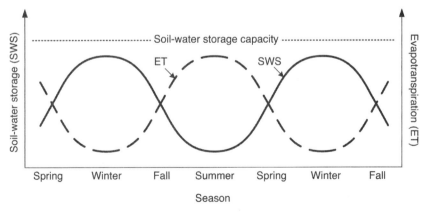

FIGURE 13.14 Concept of soil-water storage capacity for alternative covers (redrawn after Shackelford 2005).

TABLE 13.6 Equivalent Percolation Rates for Prescriptive Final Covers

Type of Traditional Cover	Maximum Annual Percolation (mm/yr)	
	Semiarid and Drier ($P/PET \leq 0.5$)	Humid ($P/PET > 0.5$)
Compacted clay (or lesser)	10	30
Composite	3	3

From Benson (1999) and Manassero et al. (2000).

The key criterion for replacement of a traditional cover with an alternative cover is the ability to show equal or better performance of the alternative cover relative to the traditional cover (Albright et al. 2004). In this regard, equivalent percolation rates have been recommended based on climatological conditions and the type of traditional cover to be replaced (Table 13.6). These recommended performance criteria are subject to change as new information becomes available regarding the performance of alternative covers in comparison to traditional covers.

Given the importance of soil-water storage, design of alternative covers generally consists of determining the thickness of cover needed to store infiltrating water and evaluating if the stored water can be removed during the growing season via evaporation and transpiration. The thickness varies with location as a function of meteorological conditions and vegetation. For example, relatively thin covers may be used in the desert southwestern U.S. (e.g., Arizona, Nevada, New Mexico), whereas much thicker covers probably are required in cool deserts where significant snowfall occurs (e.g., Montana, Wyoming, North Dakota) or in wetter climates. During preliminary design, hand calculations employing approximate methods are used to determine cover thickness (Stormont and Morris 1998; Khire et al. 2000; Benson and Chen 2003; Parent and Cabral 2005). The design is then checked and refined using numerical models (e.g., Khire et al. 2000; Zornberg et al. 2003; Ogorzalek et al. 2008; Bohnhoff et al. 2009).

Although emphasis in the design of alternative covers is placed on maximizing soil-water storage, vegetation is also important to the performance of the covers, since vegetation facilitates removal of stored water via transpiration. Consequently, the soils that are used for alternative covers also should be suitable for vegetation, at least for the portion of the cover where vegetation will be established. Vegetation should be established as soon as possible after construction is complete to minimize the potential for erosion and nurtured to maturity, which may require several years (e.g., 3–5 yr). Finally, since the soil is intended to be a medium for vegetation growth, only modest compaction should be used. Thus, unlike low-permeability compacted barriers, such as CCLs, the finer layers typically are compacted only to approximately 85% and at most 90% of maximum dry unit weight based on standard Proctor compaction (Benson et al. 1999; Zornberg et al. 2003). Such modest compaction also results in lower dry densities and correspondingly greater water storage capacities and ensures that the pore structure of the soil is not prone to large changes caused by shrinking and swelling or frost action. In the case where a fine-grained soil is used as the finer layer, compaction at a water content dry of the optimum water content may help reduce the potential for desiccation while simultaneously increasing the water storage capacity. The use of geosynthetic reinforcements to stabilize steep alternative covers is reported by Zornberg et al. (2001).

Alternative covers have been referred to by a variety of terms, including "water-balance covers," "evapotranspirative covers," "alternative earthen final covers," "monocovers," "store-

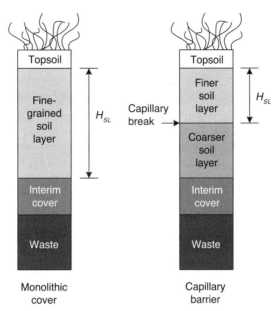

FIGURE 13.15 Schematic cross sections of the two primary types of alternative covers (modified after Shackelford 2005).

and-release covers," "soil-plant covers," and "phytocovers." However, despite this variety in terminology for alternative covers, the most common alternative covers can be classified as either monolithic or capillary barrier covers (Shackelford 2005), as illustrated in Figure 13.15. The primary difference between these two covers is that capillary barriers include a capillary break resulting from a finer textured soil overlying a coarser textured soil, whereas monolithic covers do not. The capillary break occurs under unsaturated soil conditions at the interface between the two adjacent layers. At low degrees of saturation, the unsaturated hydraulic conductivity of the coarser layer is lower than that of the finer soil. As a result, percolation into the coarser layer (and overall through the cover) is impeded. The water storage capacity of the finer soil is increased in a capillary barrier compared to a monolithic cover. Thus, the thickness of the storage layer, H_{SL}, of a capillary barrier cover conceptually can be less than that of a monolithic cover, all other factors being equal (see Figure 13.15).

In general, the effectiveness of a capillary barrier cover increases with an increase in the contrast in soil properties (e.g., unsaturated hydraulic conductivity) between the finer and coarser layers and when the water storage capacity is maximized. In contrast, the thickness of the underlying coarser layer in a capillary barrier cover is not nearly as important as that of the finer layer, since the coarser layer provides little storage capacity. The coarser layer only needs to be sufficiently thick to provide a good working platform for placement of the finer layer and, in some cases, adequate lateral drainage capacity. A layer 300 mm thick is generally adequate (Khire et al. 2000). However, since complete saturation of the finer, water storage layer during migration of the infiltration will destroy the capillary barrier effect, such that the cover will fail, provision must be made for adequate lateral drainage of infiltrating water (i.e., above the finer soil) to minimize the potential for saturation of the finer layer, particularly when the finer layer is relatively thin. This consideration may require the placement of a drainage layer (coarse-grained soil and/or geosynthetic) between the topsoil and the finer soil layer.

13.3.4 Example Landfill Design

Photos of various components of an MSW landfill with a single composite bottom barrier system and a conventional single composite cover system are presented in Figure 13.16. The bottom barrier system consisted of a GML/GCL barrier (Figure 13.16a) overlain by a geocomposite LCRS. A smooth 1.5-mm-thick HDPE geomembrane was used along the base, whereas a textured 1.5-mm-thick HDPE geomembrane was used on the side slopes. The LCRS consisted of a geocomposite with a geotextile filter layer and a geonet drainage layer along the base and a geotextile-geonet-geotextile composite along the side slopes (for improved friction against the underlying textured geomembrane) (Figure 13.16b). A protective sand layer with

(a)

(b)

FIGURE 13.16   Example MSW landfill.

(c)

(d)

FIGURE 13.16 (continued)

a thickness of 300 to 450 mm was placed over the LCRS (Figure 13.16c) prior to waste placement. The cover system consisted of, from top to bottom: a 150-mm-thick top/vegetative soil layer, a 750- to 900-mm-thick protective soil layer, a geonet-geotextile filter/drainage composite, a GML/CCL composite barrier layer, and a 300- to 600-mm-thick protective/foundation soil layer. Installation of the geocomposite over the geomembrane is shown in Figure 13.16d. Pipes used for the gas extraction system are visible in the figure.

13.3.5 Performance

13.3.5.1 Issues Affecting Performance

Development of secondary features such as cracks in a CCL due to environmental distress (air desiccation, wetting/drying cycles, freeze/thaw cycles) or mechanical distress (tensile strains resulting from differential settlement) can cause significant increases in k. Cover systems are affected by both environmental and mechanical distress, whereas environmental distress only is an issue for bottom barrier systems that are left exposed (without waste cover) for extended durations. Field data on air desiccation and freeze/thaw damage indicate that significant cracking and resulting increases in k may occur in earthen barriers (Montgomery and Parsons 1990; Corser and Cranston 1991; Benson and Othman 1993; Albrecht 1996; Benson and Khire 1995, 1997; Melchior 1997). For bottom barrier systems, placement of the first lift of wastes with a thickness of 3 to 5 m can insulate the underlying barrier and prevent temperature fluctuations that may result in environmental distress in the barriers (Hanson et al. 2005). In addition, heat generated in MSWs may cause desiccation of earthen components and affect long-term durability of geosynthetic components in bottom barriers (NRC 2007). The potential for desiccation depends on the thermal gradient acting on the barrier, texture of the underlying soils (with regard to moisture-suction relationships), and the location of the groundwater table.

13.3.5.2 Chemical Compatibility

The long-term performance of containment systems depends on maintaining the integrity of the systems in the presence of the contained chemicals. Thus, the compatibility between the barrier materials and the liquids being contained is an important consideration in terms of the long-term performance of the containment system.

The nonstandard liquids (i.e., liquids other than water) to which barriers can be exposed are categorized broadly as aqueous liquids, nonaqueous liquids, or mixed liquids (Shackelford 1994a). Aqueous liquids include inorganic solutions (acid, base, and salt solutions) and hydrophilic organic compounds (acids, bases, and neutral polar compounds). Examples of inorganic chemicals and maximum observed concentrations in various nonhazardous waste liquids are summarized in Table 13.7. Ranges of pH for the waste liquids also are presented in Table 13.7. The nonaqueous liquids mainly include hydrophobic organic compounds that are either lighter than water, referred to as LNAPLs (light nonaqueous-phase liquids), or denser than water, referred to as DNAPLs (dense nonaqueous-phase liquids). Mixed liquids are simply combinations of aqueous and nonaqueous liquids (i.e., separate aqueous and nonaqueous phases).

For earthen barrier components (CCLs and GCLs), compatibility typically is assessed by comparing the k of the barriers based on permeation with the liquids to be contained relative to k based on permeation with water. No changes or changes that are not sufficiently significant to increase k beyond the desired design or regulated value (e.g., $<1 \times 10^{-9}$ m/s) indicate

TABLE 13.7 Maximum Concentrations for Inorganic Chemicals and pH for Waste Liquids

Chemical Constituent or Parameter	Maximum Concentration (mg/L) Except for pH					
	Municipal Waste	Construction and Demolition Waste	Papermill Sludge	Municipal Incinerator Ash	Coal Burner Fly Ash	Iron Foundry Waste
Calcium (Ca)	2500	578	2400	3200	60	
Iron (Fe)	4000	172	950	121	0.24	<0.03
Magnesium (Mg)	780	192	6000	41	4.3	
Potassium (K)	3200	618	140	4300	29	
Sodium (Na)	6010	1290	4500	7300	50	
Sulfate (SO_4^{2-})	1850	<40	550	4900		5.1
pH	3.7–8.9	6.5–7.3	5.4–9.0	8.47–9.94	7.83–9.05	12.3

Data extracted from Bagchi (1994).

that the barrier component is essentially compatible with the contained liquids. For GMLs and other geosynthetics, compatibility typically is assessed by comparing the physical and mechanical properties of the as-manufactured geosynthetics that represent their condition at the time of installation with the same properties after extensive exposure to the contained chemicals. Various test methods for compatibility testing for geosynthetics are described in ASTM standards and U.S. EPA test methods. Little or no change in the properties typically indicates that the geosynthetic barrier component is compatible with the contained liquids.

Chemical incompatibility in terms of the k of earthen barriers has been assessed primarily in the form of research studies performed in the laboratory, although some compatibility testing has been performed in conjunction with the permitting of waste containment facilities. There are at least two reasons for the focus on research studies. First, aside from hazardous waste containment facilities, few environmental regulations specifically require compatibility testing as part of the permitting process. Second, when such compatibility testing has been required as part of the permitting process, the results typically are not published, but rather are contained in relatively obscure and/or restricted consulting and testing reports, such that the information is not generally readily available. Also, field studies on chemical compatibility are not readily available for at least three reasons (NRC 2007): (1) less control inherent in field tests, especially when dealing with potentially harmful liquids; (2) lack of evidence of chemical incompatibility in waste containment facilities that are performing satisfactorily; and (3) the extensive time scales associated with chemical compatibility processes (i.e., years or even decades may be required before there is any evidence of chemical incompatibility) (Shackelford 2005). Nonetheless, these restrictions do not preclude the importance of considering the potential for chemical incompatibility, especially in terms of the long-term performance of containment facilities.

In general, the results of the available information on the chemical incompatibility in terms of the k of earthen barriers indicate that qualitative predictions on barrier performance can be made based primarily on three considerations (NRC 2007): (1) the type and properties of the liquid to be contained, (2) the type and properties of the soil used as the barrier material, and (3) the physical conditions imposed on the barrier soil. In terms of the effects of nonstandard liquids on the k of earthen barriers, the potential for incompatibility increases with increasing concentration of chemical species (solutes) and/or increasing charge of ionic constituents in inorganic solutions and with decreasing pH. The pH can affect the k of earthen barriers both directly and indirectly (Shackelford 1994a). The direct effect results when solutions with low pH (<2) dissolve clay soils, resulting in the development of relatively large pores that cause increases in k (Bowders and Daniel 1987). The indirect effect results because the concentration of ionic species in solution tends to increase with decreasing pH.

Finally, high concentrations of hydrophilic organic compounds and pure-phase organic liquids (i.e., LNAPLs and DNAPLs) can result in a decrease in the dielectric constant of the liquid relative to that for water, resulting in shrinkage of the soil (i.e., decrease in t_{DDL} via Equation 13.1), which can cause cracking with concomitant increases in k (e.g., Mitchell and Madsen 1987; Shackelford 1994a). However, the migration of pure-phase nonpolar hydrophobic compounds (e.g., LNAPLs and DNAPLs) through a barrier soil initially at a high degree of water saturation (e.g., CCL) likely is restricted, because the magnitude of the pressure required to displace the water (wetting fluid) in the pore space of the soil with the nonpolar liquid (nonwetting fluid), known as the entry pressure, can be exceedingly high, such that the migration of these chemicals under typical field gradients is unlikely (Foreman and Daniel 1986; Broderick and Daniel 1990; NRC 2007).

In terms of the type and properties of the soil used as the barrier material, the potential for incompatibility generally increases as the activity, defined as the plasticity index divided by the percent of clay-size particles (<2 μm), of the soil increases (Shackelford 1994a; NRC 2007). Higher activity correlates with greater content of high-plasticity clay minerals, such as sodium montmorillonite, as well as with greater content of the smaller clay-sized particles, resulting in greater specific surface. The available evidence suggests that most natural clay soils that have been used as CCLs have relatively low activity, such that the potential for chemical incompatibility likely is small (e.g., Benson et al. 1999; NRC 2007). In contrast, the potential for chemical incompatibility in containment barriers that contain significant bentonite contents, such as compacted sand-bentonite and soil-bentonite liners, soil-bentonite backfills for vertical cutoff walls, and GCLs, may be particularly high. For example, as shown in Figure 13.17, the k of specimens of GCLs permeated with inorganic salt solutions has been shown to increase by several orders of magnitude in comparison to k with water. The factors affecting the k of GCLs are presented in more detail by Shackelford et al. (2000) and NRC (2007).

In terms of the physical conditions imposed on the barrier soil, three factors are particularly important (NRC 2007): (1) the stresses acting on the barrier soil, (2) the hydraulic gradient of the liquid across the barrier, and (3) the initial level or degree of hydration of the barrier soil. An increase in effective stress (due to stresses applied to soil or gradient in the liquid permeating through the soil) tends to reduce the susceptibility of the soil to chemical incompatibility. As a result, consolidation of barrier soils (e.g., from the weight of the overlying waste or due to high gradients) not only tends to decrease the k of the barrier soil but also to reduce the potential for chemical incompatibility.

A higher initial degree of water saturation of the barrier soil prior to permeation with a chemical solution, commonly referred to as prehydration, may provide increased resistance to chemical incompatibility. However, this effect is likely to be important only in the case of clay soils with a large swelling potential, such as sodium bentonite, and tends to be significant in such soils only at relatively high concentrations of chemical species in solution (Lee and Shackelford 2005b; NRC 2007). On a field scale, prehydration is becoming an increasing consideration in the use of GCLs.

The majority of data available for geomembranes includes determination of resistance of geomembranes to individual chemicals using laboratory tests. Polyethylene, in particular

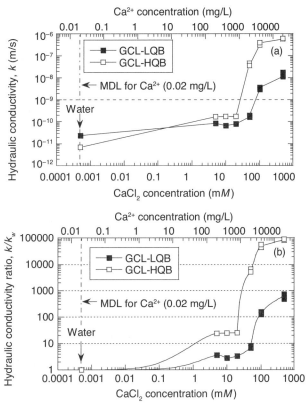

FIGURE 13.17 Results of hydraulic conductivity (k) tests for specimens of two GCLs with either lower quality bentonite (LQB) or higher quality bentonite (HQB): (a) k values (MDL = method detection limit) and (b) ratio of k based on any solution to k based on water (data from Lee and Shackelford 2005a).

HDPE, has high resistance to a variety of chemicals including organic compounds (hydrocarbons, oxygenated and chlorinated solvents, crude petroleum solvents, alcohols), organic and inorganic acids, heavy metals, and salts based on the results of such tests (Koerner 2005). Resistance of other types of geomembranes (PVC, polypropylene, CSPE-R, EPDM, EPDM-R) has generally been lower in comparison to HDPE, especially against organic compounds. Results of field exhumation studies also provide assessment of compatibility of geomembranes with contained liquids. Analyses of geomembranes exhumed from leachate lagoons and metal sludge impoundments after variable service periods (7–31 yr) did not indicate significant changes in the properties of the materials (NRC 2007). Another important consideration in assessment of compatibility is temperature as the resistance of geomembranes to various chemicals decreases as temperature increases (Koerner 2005).

13.3.5.3 Field Performance of Barrier Systems

Field-scale percolation rates for traditional covers were summarized by Benson (1999) based on data reported by Montgomery and Parsons (1990), Melchior (1997), and Khire et al. (1997). Field-scale percolation rates were 1–50 mm/yr (1–4% of precipitation) for semiarid to humid climates for intact compacted clay covers. Increased percolation rates of 30–150 mm/yr

TABLE 13.8 Average Monthly Liquid Seepage Rates from Primary (Top) Barriers in Double Barrier Systems

Type of Primary Barrier[a]	Liquid Seepage Rate (L ha^{-1} d^{-1})		
	Initial Stage	Active Waste Placement	Post-Closure Period
GML	307	187	127
GML/CCL	114	142	64.4
GML/GCL	133	22.5	0.3

Modified from NRC (2007).

[a] All barriers were overlain by sand layers for liquid collection.

(10–20% of precipitation) have been reported for cracked CCLs for similar climatic conditions. Low-plasticity clays compacted at water contents close to the optimum water content are recommended for construction of single barrier covers comprised of a CCL to limit potential volume change due to environmental and mechanical factors (Benson 1999). The percolation rates were 2–3 mm/yr for a GML/CCL barrier in a humid climate.

A comparison of seepage through the primary barrier component of double barrier systems is provided by Bonaparte et al. (2002) based on an extensive investigation of U.S. landfills. The data also are summarized in NRC (2007). The data (Table 13.8) indicated that composite barriers (GML/CCL and GML/GCL) performed better (i.e., less leakage) than single barriers (GML) in the primary system. The differences were particularly significant for post-closure periods at the landfills investigated. Less variation was present between the barrier systems for initial (shortly after waste placement) and active (during waste placement with regular daily cover and intermittent interim cover layers) waste placement periods.

Service lifetime of geomembranes, as well as other geosynthetics, is dependent on temperature. Lifetime estimates for HDPE geomembranes vary from several hundreds of years to a few tens of years as the temperature increases from 20 to 60°C (Rowe 2005). The temperatures of bottom barrier systems vary from 25 to 40°C in different climatic regions (Hanson et al. 2008). The temperatures in cover systems vary seasonally with air temperature fluctuations and range from 4 to 31°C (at the location of barrier components 1 m below the surface) in temperate climatic zones (Yesiller et al. 2008).

13.4 Contaminant Transport

Seepage of fluids (liquids and gases) occurs through earthen barriers in containment systems (i.e., CCLs and GCLs), and leakage occurs through defects in GMLs. Mass transport of fluids occurs through porous media via two primary processes: advection and diffusion. Advection is the chemical mass transport process that occurs in response to a hydraulic gradient in accordance with Darcy's law, and diffusion is the chemical mass transport process that occurs in response to a gradient in chemical concentration in accordance with Fick's law. Mass transport of contaminants also occurs through geomembranes in containment systems.

13.4.1 Seepage

The rate of steady-state fluid flow through a porous medium (e.g., soil) is determined using Darcy's law, which can be written as follows:

$$q = kiA \tag{13.5}$$

where q is the flow rate (L^3T^{-1}), k is the hydraulic conductivity (LT^{-1}), i is the hydraulic gradient (LL^{-1}), and A is the total cross-sectional area (solids plus voids) perpendicular to the direction of flow (L^2). The hydraulic gradient for one-dimensional steady-state flow may be defined as follows:

$$i = -\frac{\Delta h}{L} \tag{13.6}$$

where Δh is the change in total head (L) and L is the length of the porous medium across which the head change occurs parallel to the direction of the head change (L). The negative sign in Equation 13.6 is a mathematical sign convention that is required to make the hydraulic gradient positive ($i > 0$) when flow occurs in the direction of decreasing total head. The cumulative or total amount of flow, Q (L^3), can be calculated as follows:

$$Q = qt \tag{13.7}$$

where t is the total elapsed time for flow (T). Darcy's law is valid for laminar flow conditions.

13.4.2 Advection

Mass flux of chemical constituents dissolved in the aqueous phase (i.e., solutes) due to advection is given by the following expression:

$$J_a = vC \tag{13.8}$$

where J_a is the advective chemical mass flux ($ML^{-2}T^{-1}$), C is the concentration of a constituent in the fluid (ML^{-3}), and v is the discharge velocity or Darcy velocity (LT^{-1}) of the fluid. The discharge velocity, v, represents the volumetric flow rate of fluid, q (L^3T^{-1}), per unit total cross-sectional area of porous medium perpendicular to the direction of flow, A (L^2), in accordance with Darcy's law (Equation 13.5), as follows:

$$v = \frac{q}{A} = ki \tag{13.9}$$

Based on Equation 13.5, fluid flow is assumed to occur through the total cross-sectional area of the porous medium. However, in reality, fluid flow occurs only through the free (interconnected) pore spaces (voids) within the porous medium. Therefore, in order to determine the actual velocity of fluid flow through the porous medium, where velocity refers to change in macroscopic distance between two points per unit change in time, the relative percentage of pore volume to the total volume of the porous medium must be taken into consideration with respect to Equation 13.9, as follows:

$$v_s = \frac{v}{n} = \frac{ki}{n} \tag{13.10}$$

where v_s is the seepage or average linear velocity (LT^{-1}) and n is the total or bulk porosity of the porous medium. For coarse-grained soils, all of the pores typically are interconnected, whereas occluded (nonconductive) pore spaces are more likely in fine-grained soils. In the case where occluded pores are present, the percentage of pore space that is actually interconnected and, therefore, available to conduct fluid flow may be less than the total pore space within the medium, such that the following alternative definition for seepage velocity must be used:

$$v_s = \frac{v}{n_e} = \frac{ki}{n_e} \qquad (13.11)$$

where n_e is the effective porosity that represents the fraction of the total porosity that is available to conduct flow (i.e., $n_e \leq n$).

13.4.3 Diffusion

Diffusion describes the transport of fluid-soluble chemicals (i.e., solutes) via a chemical concentration gradient, from high chemical concentration to low chemical concentration. Chemical mass flux via diffusion is independent of fluid-phase migration due to advection. That is, chemical mass flux due to diffusion may occur in the same or opposite direction as advection and also may occur in the absence of advection (Shackelford 1989). When diffusion occurs in the same direction as advection, diffusion causes the chemical constituents within the fluid phase to disperse relative to the advective front. For this reason, diffusion is known as a dispersion process.

Diffusion through porous media is described by Fick's first law, which can be written for porous media as follows:

$$J_d = D^* i_c \qquad (13.12)$$

where J_d is the diffusive mass flux $(ML^{-2}T^{-1})$, D^* is the effective diffusion coefficient (L^2T^{-1}) with respect to a specific chemical constituent and porous medium, and i_c is the chemical concentration gradient (ML^{-4}) defined for one-dimensional steady-state diffusion as follows:

$$i_c = -\frac{\Delta C}{L} \qquad (13.13)$$

where ΔC is the change in solute concentration and L is the length of the porous medium over which the concentration change occurs parallel to the direction of concentration change (L). The negative sign in Equation 13.13 is a mathematical convention that is required to make the chemical gradient positive $(i_c > 0)$ when flux occurs in the direction of decreasing solute concentration.

Typical values of D^* for solutes diffusing in water-saturated compacted clays range from approximately 1×10^{-10} to 1×10^{-9} m^2/s (Shackelford and Daniel 1991). Values of D^* for aqueous-phase solutes diffusing through barriers comprised of significant amounts of high-swelling bentonite, such as GCLs, can be as much as one to four orders of magnitude lower than 1×10^{-10} m^2/s, whereas values of D^* for chemical constituents diffusing via the gas phase through water-unsaturated soils with continuous gas phase can be from one to four orders of magnitude higher than 1×10^{-9} m^2/s.

13.4.4 Retardation

Retardation refers to the reduction in the rate of chemical mass transport during migration through porous media due to adsorption of chemical constituents (solutes) from the liquid phase (e.g., pore water) to the solid phase (e.g., solid soil particles). Typical solutes that adsorb to clays include cations, such as the inorganic metal species (e.g., Na^+, K^+, Ca^{2+}, Mg^{2+}, Cd^{2+}, Pb^{2+}, and Zn^{2+}), and hydrophobic organic compounds, such as the BTEX (benzene, toluene, ethylbenzene, and xylene) compounds, although the mechanisms for the sorption of these two classifications of chemicals are not the same.

In terms of advective transport, a reactive transport velocity, v_r, for chemical constituents that are subject to retardation via adsorption may be defined as follows:

$$v_r = \frac{v_s}{R_d} \qquad (13.14)$$

where R_d is the retardation factor, which is equal to unity ($R_d = 1$, $v_r = v_s$) for nonreactive (nonadsorbing) solutes (i.e., conservative species) and is greater than unity ($R_d > 1$, $v_r < v_s$) for reactive (adsorbing) solutes. In accordance with Equation 13.14, the time associated with the transport of the reactive chemical constituent over a specific distance, L, will be increased relative to that based on the seepage velocity (i.e., $t = v_s \cdot L$) by an amount equal to $t \cdot R_d$.

The retardation factor also may be defined as follows:

$$R_d = 1 + \frac{\rho_d K_d}{n} \qquad (13.15)$$

where ρ_d is the dry density (ML^{-3}), and K_d is the distribution coefficient of the solute in the porous medium (L^{3}M^{-1}). The distribution coefficient, K_d, represents the ratio of the solute mass in the solid phase (i.e., sorbed mass) per unit mass of dry soil (MM^{-1}) relative to concentration of the same solute in the liquid phase (ML^{-3}). Use of Equation 13.15 implies linear, instantaneous, and reversible sorption.

Hydrophobic (nonpolar) organic compounds, such as the BTEX compounds, are largely immiscible in water (H_2O), but typically have aqueous solubilities that far exceed environmentally acceptable standards. The sorption of such compounds has been shown to correlate with the fraction of organic carbon in the solid phase of the soil, f_{oc}, where f_{oc} is defined as the mass of organic carbon in the soil, m_{oc}, normalized with respect to the mass of dry soil, m_s, or $f_{oc} = m_{oc}/m_s$. In this case, K_d in Equation 13.15 can be approximated as the product of the organic-carbon partition coefficient, K_{oc}, and f_{oc} (i.e., $K_d = K_{oc} \cdot f_{oc}$). K_{oc} represents the ratio of the sorbed mass per unit mass of organic carbon (MM^{-1}) relative to concentration of the same solute in the liquid phase (ML^{-3}). K_{oc} is correlated with the octanol-water partition coefficient, K_{ow}, such that $K_{oc} \approx K_{ow}$. Values of K_{ow} commonly are tabulated for a wide variety of organic compounds in standard chemistry textbooks and handbooks. In addition to the approximation in the correlation between K_{oc} and K_{ow}, the use of $K_{oc} \cdot f_{oc}$ for K_d in Equation 13.15 involves all the same assumptions inherent in the use of K_d.

13.4.5 Transient Solute Transport

Time-dependent or transient transport of solutes (e.g., contaminants) in a porous medium under the simultaneous effects of advection and diffusion can be estimated using the governing advective-diffusive-reactive partial differential equation. For the simplified case of one-dimensional transport of a reactive (sorbing) solute under steady-state flow conditions, the advective-diffusive-reactive equation may be written as follows:

$$R_d \; \frac{\partial C}{\partial t} \; = \; D^* \; \frac{\partial^2 C}{\partial x^2} \; - \; v_s \; \frac{\partial C}{\partial x} \tag{13.16}$$

Substitution of $v_s = 0$ in Equation 13.16 results in Fick's second law, which describes transient diffusion of a reactive solute through a porous medium, or:

$$R_d \; \frac{\partial C}{\partial t} \; = \; D^* \; \frac{\partial^2 C}{\partial x^2} \tag{13.17}$$

Several analytical solutions (e.g., van Genuchten 1981) and semianalytical or numerical solutions (e.g., Rowe and Booker 1985a, 1985b, 1986) to Equation 13.16 based on varying initial and boundary conditions exist for use in assessing solute transport through soil. In particular, the following analytical solution to Equation 13.16 probably is the simplest and most commonly used model for evaluating transport through CCLs (e.g., Shackelford 1990, 1992, 1993):

$$C(x, t) \; = \; \frac{C_o}{2} \left[\mathrm{erfc} \left(\frac{R_d x - v_s t}{2 \sqrt{R_d D^* t}} \right) + \exp \left(\frac{v_s x}{D^*} \right) \mathrm{erfc} \left(\frac{R_d x + v_s t}{2 \sqrt{R_d D^* t}} \right) \right] \tag{13.18}$$

where $C(x,t)$ is the concentration at distance x from the source at time t; C_o is the initial concentration of the solute in the influent, which is assumed to remain constant in time; and erfc is a common mathematical function known as the *complementary error function*. The analytical model given by Equation 13.18 also assumes that the porous medium is infinitely long and initially free of the solute (i.e., at $t = 0$ prior to the onset of solute transport) and that the seepage velocity is constant (i.e., in accordance with Equation 13.16). For the case of purely diffusive transport (i.e., $v_s = 0$), Equation 13.18 reduces to the analytical solution to Fick's second law given by Equation 13.17, or:

$$C(x, t) \; = \; C_o \; \mathrm{erfc} \left(\frac{R_d x}{2 \sqrt{R_d D^* t}} \right) \tag{13.19}$$

Examples of advective-diffusive transport from a compacted clay barrier with a thickness of 910 mm and a hydraulic conductivity of 1×10^{-9} m/s are provided in Figure 13.18. The temporal variation in the relative concentration at the base of the clay barrier, $C(L,t)/C_o$, is

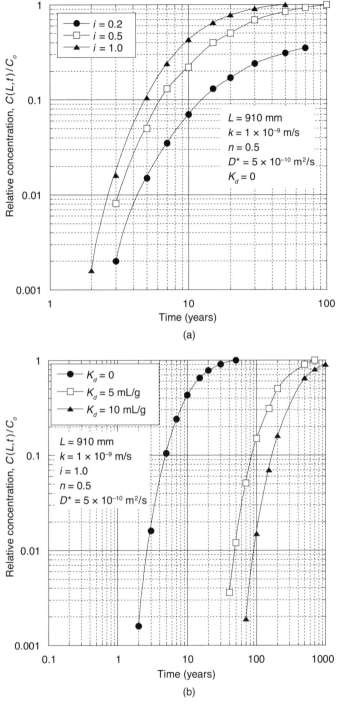

FIGURE 13.18 Transit times under advection and diffusion: (a) nonreactive solutes and (b) reactive solutes (replotted after Gray 1995).

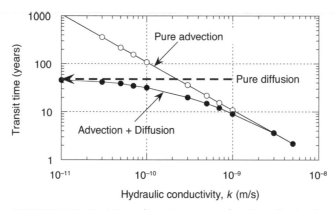

FIGURE 13.19 Variation of transit times as a function of hydraulic
conductivity (replotted after Shackelford 1988).

presented for a D^* value of 5.0×10^{-10} m^2/s and an n of 0.5. The times corresponding to solute
breakthrough through the compacted clay barrier at $C(L,t)/C_o = 0.01$ are relatively short (<10
yr) for all gradients ($i = 0.2$, 0.5, and 1.0) for no solute adsorption (i.e., $K_d = 0$). In case of
variable solute adsorption ($K_d = 0$, 5, and 10 mL/g), these breakthrough times increase
significantly with increasing solute adsorption.

The relative significance of diffusion increases as the k of the porous medium decreases,
such that diffusion becomes increasingly more predominant as the hydraulic conductivity of
the porous medium, k, decreases, all other factors being constant (Figure 13.19). In terms of
the results shown in Figure 13.19, diffusion reduces the transit time relative to that based on
pure advection at a k of 1×10^{-9} m/s, and diffusion becomes the prevailing transport
mechanism for values of k lower than about 2.0×10^{-10} m/s. Thus, purely advective transit
times are unrealistically high (unconservative) for porous containment barriers at such low k.
Other factors affecting the assessment of contaminant migration through engineered earthen
containment barriers are described in Shackelford and Rowe (1998).

13.4.6 Contaminant Mass Transport through Geomembranes

Liquid-phase mass transport of contaminants through geomembranes occurs by two primary
mechanisms: (1) advective transport through holes/defects in geomembranes and (2) diffusive transport through intact geomembranes. In advective mass transport, rates of leakage are
determined using equations developed based on analytical and empirical approaches (NRC
2007). The rates of leakage vary as a function of the number and sizes of holes in the
geomembrane, properties of the underlying material, and the applied hydraulic head.
Semiempirical approaches for calculating leakage through geomembranes are provided in
Giroud et al. (1998). Analytical approaches for estimating advective chemical mass flux
through geomembranes with defects are described in Rowe (2005). Diffusion of organic
compounds occurs through geomembranes via both the aqueous and gas phases, whereas
diffusion of inorganic chemical species through geomembranes is not expected (Katsumi et al.
2001; Edil 2003; NRC 2007).

Wrinkles that develop in geomembranes due to placement errors during construction or
due to thermal expansion under elevated temperatures can increase both advective and
diffusive transport. If wrinkles coincide with holes in a geomembrane, the potential for

advective contaminant migration increases. Stress cracking and thinning of sections at points of high tensile stress also allow for higher amounts of diffusive transport. More detail on contaminant mass transport through geomembranes can be found in the aforementioned references.

13.4.7 Contaminant Mass Transport through Composite Liners

Several studies have evaluated contaminant mass transport through composite liners consisting of a GML overlying either a CCL (GML/CCL) or a GCL (GML/GCL) (Rowe 1998, 2005; Katsumi et al. 2001; Foose 2002; Foose et al. 2002). These studies have involved both simplified analytical approaches (e.g., Katsumi et al. 2001; Foose 2002) and more sophisticated numerical modeling approaches (e.g., Foose et al. 2002). A detailed description of these assessments is beyond the scope of this chapter, but a brief summary of the overall conclusions follows.

Similar to contaminant mass transport through geomembranes, contaminant mass transport through geomembrane-based composite liners can occur via two pathways, based on the nature of the chemical solutions (e.g., leachate) being contained (e.g., Foose et al. 2002). In the case of leachates that contain primarily inorganic contaminants such as the heavy metals (e.g., Cd, Cu, Pb, Zn, etc.), the primary pathway for contaminant migration is through defects (e.g., holes, defective seams) in the overlying GML and subsequently through the underlying soil liner (i.e., CCL or GCL) via advection and/or diffusion, as illustrated in Figure 13.20a. This pathway results from the fact that intact geomembranes are impervious to liquid-phase migration, and inorganic chemicals cannot diffuse through intact geomembranes comprised of polymers. This pathway also may be dominant in the case of leachates that contain soluble organic compounds with low vapor pressures.

In the case of volatile organic compounds (VOCs) with high vapor pressures, such as the methane and the BTEX compounds, the primary pathway for contaminant transport through GML-based composite liners/barriers is via diffusion through the intact components, as illustrated in Figure 13.20b. In this case, the VOC within the leachate first must partition from the leachate into the geomembrane, diffuse through the geomembrane via a concentration gradient, partition out of the geomembrane into the pore water of the underlying soil, and finally diffuse through the underlying soil liner (i.e., CCL or GCL), as shown in Figure 13.20c.

The results of the aforementioned studies indicate that contaminant mass transport through composite liners is likely to be only a fraction of that through single liners comprised of the individual components (i.e., GML, CCL, or GCL), which is a conclusion similar to that based solely on leakage rates through composite liners (e.g., Foose et al. 2001). However, for inorganic contaminants, the mass flux through the composite liner is a function of the quality of the contact between the overlying GML and the underlying CCL or GCL, with the contaminant mass flux through the composite liner increasing as the quality of the contact decreases. Also, the mass flux of VOCs through composite liners tends to decrease as the thickness of the underlying soil liner increases, primarily due to the lower concentration gradient across thicker composite liners and the greater attenuation (sorption) capacity of the soil. For this reason, the mass flux of VOCs through GML/CCL composite liners tends to be lower than that through GML/GCL composite liners, a conclusion which is opposite to that based solely on leakage rates (Foose et al. 2001, 2002). Thus, a low leakage rate through a composite liner is a necessary but not sufficient condition to ensure low contaminant mass flux, as mass flux due to diffusion can be considerably greater than that due to advection in the case where diffusion dominates the transport process.

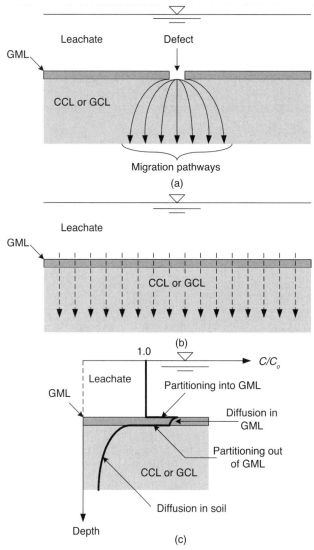

FIGURE 13.20 Contaminant transport pathways through composite liners: (a) advection and/or diffusion through defects in GML, (b) VOC diffusion through intact GML and underlying CCL or GCL, and (c) concentration profiles for pathway illustrated in (b) (redrawn after Katsumi et al. 2001; Foose et al. 2002).

13.5 Measurement of Material Properties

13.5.1 Hydraulic Conductivity

The rate of flow of water through low-permeability barrier materials (compacted soils and GCLs) is commonly regulated by limiting the maximum values for both the k of the barrier materials and the applied hydraulic gradient through the barrier systems via Equation 13.9. As a result, the most commonly determined hydraulic property for containment systems is the k

of the barrier materials. Both laboratory and field tests are used to determine k of compacted soils, whereas typically only laboratory tests are used for GCLs.

The primary factors that affect laboratory determination of k include degree of water saturation, applied effective stress, and hydraulic gradient:

- *Degree of water saturation* (S)—Hydraulic conductivity increases with increasing degree of water saturation for soils. At the completion of construction, the value of S for CCLs typically is high (80% $\leq S \leq$ 90%) due to the need to achieve a suitably low k (see Section 13.2.1.3), yet S is not 100%. However, S typically increases during the service life of these barriers, particularly in the bottom barrier systems, as they come in contact with the liquids from the overlying materials and also the variably saturated underlying subsurface soils. Determination of k for fully saturated conditions results in the highest expected value of k and, therefore, is conservative (high).

- *Applied effective stress* (σ')—Hydraulic conductivity decreases as σ' increases. In general, σ' initially is low for bottom barrier systems prior to the onset of waste filling of a containment facility but increases as the overburden stress increases due to waste filling. In contrast, σ' for cover barrier systems is relatively low throughout the life of the cover system. Values of σ' representative of field conditions should be used in laboratory testing, when using methods that allow for control of σ'. Use of values of σ' higher than those expected in the field could result in compression/consolidation of the specimens and, therefore, unconservative (low) measured k values.

- *Hydraulic gradient* (i)—The hydraulic gradient can have a contrasting effect on k. For example, application of excessively high i can result in excessively high σ' in a specimen in the case of permeation using setups with stress control (e.g., flexible-wall cells), resulting in unconservative (low) k, whereas application of excessively high i in setups where σ' cannot be controlled (e.g., rigid-wall cells) can hydraulically fracture the soil specimen, resulting in conservative (high) k values. The actual values of i acting on barriers for waste containment facilities such as landfills typically are relatively low ($i \leq$ 2), whereas higher i values are relevant for liquid containment facilities. Therefore, measurement of k using values of i that are similar to field values provides the most representative values of k. However, testing durations increase as i decreases and, therefore, values of i higher than expected in the field typically are used in laboratory testing. As a result, an upper limit on i of approximately 30 commonly is used for testing low-permeability soils (i.e., $k = 1 \times 10^{-9}$ m/s). As noted by Shackelford et al. (2000), the measurement of k is affected to a greater extent by the magnitude of σ' than by the magnitude of i. For this reason, much higher values of i (e.g., $50 < i < 550$) than those used for soils have been used to measure k of GCLs without adversely affecting the measured k value, primarily due to the relatively low thickness of the GCLs (~5–15 mm) compared to the thickness of laboratory compacted soil specimens, which typically is on the order of 116 mm (Shackelford et al. 2000).

Laboratory hydraulic conductivity tests are conducted using two general categories of permeability cells or permeameters: rigid-wall cells and flexible-wall cells. The type of the cells refers to the containment condition applied to the test specimen being permeated (Daniel 1994). Schematic representations of the two categories of permeameters are provided in Figure 13.21. Rigid-wall testing includes determination of k on a specimen housed in a solid, rigid mold. Compaction molds, segments of soil-sampling tubes such as the Shelby tube, consolidation rings, or other types of specimen holders with rigid walls have been used. A top cap and

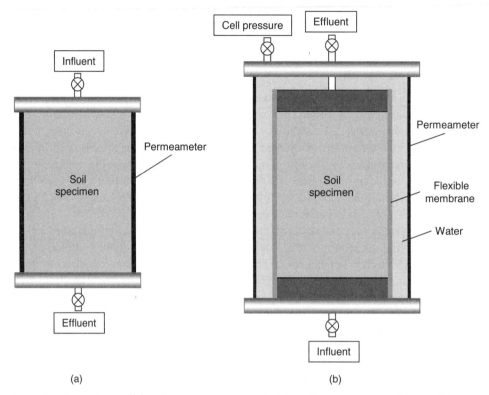

FIGURE 13.21 Types of laboratory permeameters: (a) rigid-wall permeameter and (b) flexible-wall permeameter.

a bottom platen, which allow for permeation of liquids through the specimen during a test, are used to hydraulically seal the solid mold. Flexible-wall testing includes determination of k on a specimen housed in a flexible impermeable membrane placed within a solid test chamber. A top cap and a bottom platen are used to hydraulically seal the specimen and the test chamber, which allow for permeation of liquids through the specimen during a test. The test chamber is filled with water that surrounds the specimen and allows for application of a confining pressure to the specimen via the flexible membrane and, therefore, control of σ'. For both tests, inflow and outflow pressures can be adjusted such that desired hydraulic gradients can be applied to a specimen during a test.

Rigid-wall testing is relatively simple (Figure 13.21a) and has less potential for specimen disturbance, as extraction of a test specimen is not required if the test is conducted using a compaction mold or a Shelby tube mold. While an unsaturated specimen can be inundated with water in rigid-wall testing to increase the degree of saturation, provisions do not exist for verification of complete saturation prior to a test. The complete state of stress cannot be controlled in a rigid-wall test.

Flexible-wall testing is more complicated than rigid-wall testing, but also allows for more control over the testing conditions (Figure 13.21b). For example, the degree of water saturation can be determined and controlled via a procedure known as back pressure saturation (e.g., see Daniel 1994). Also, effective stresses applied to a specimen can be controlled. Sidewall leakage may result in the measurement of high k that does not accurately represent the rate of flow from a specimen. While sidewall leakage between the test specimen and the test mold may occur in a rigid-wall test, the presence of the flexible membrane around a test specimen in a flexible-wall test minimizes or prevents the potential for such short-circuiting of the permeant liquid. Double-ring top caps and bottom platens can be used in rigid-wall test setups to separate the flow through the inner portion of the specimen from that through the outer portion of the permeameter to detect, but not control, the existence of sidewall leakage. These setups reduce the errors in interpretation of test results associated with sidewall leakage.

Liquids other than water that may come in contact with a barrier material also can be used in hydraulic conductivity testing (e.g., Shackelford 1994a; Shackelford et al. 2000). Hydraulic conductivity tests in which nonstandard liquids are used as the permeant liquids typically are referred to as compatibility tests, since the primary objective is to determine if the test specimen is compatible with the permeant liquid such that no significant increase in k is observed (i.e., relative to k based on permeation with water). Both rigid- and flexible-wall test systems also can be used for compatibility testing. Special provisions may be required for testing liquids that pose risks to health and environment or that may damage equipment components. Bladder accumulators that minimize contact of test liquids with operators and equipment can be used for these cases (Figure 13.22).

In compatibility testing, continuing a test until chemical equilibrium between the effluent and influent has been obtained is critical to ensure that all possible reactions between the soil and permeant liquid have occurred and the true equilibrium value for k has been obtained (e.g., Bowders 1988; Shackelford 1994a; Shackelford et al. 1999, 2000). In some cases, the requirement to obtain complete chemical equilibrium may result in impractical test durations (e.g., months or years). In such cases, an assessment of the potential impact resulting from premature termination of the test must be undertaken (e.g., Jo et al. 2005). This and other considerations required when performing compatibility testing are covered in more detail by Bowders et al. (1985), Shackelford (1994a), and Shackelford et al. (2000). In addition, similar

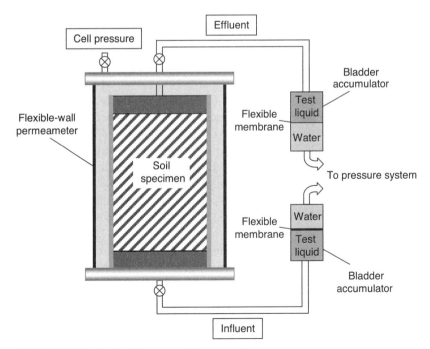

FIGURE 13.22 Hydraulic compatibility test setup.

to determination of k with water, representative results are obtained when compatibility tests are conducted under conditions that are similar to expected field conditions (e.g., hydraulic gradient and applied effective stress).

Field hydraulic conductivity tests are used to obtain representative values for k and to verify the suitability of soils and construction procedures to achieve low k in the field. Field tests are conducted on surface and near-surface soils (Daniel 1989; U.S. EPA 1993; Trautwein and Boutwell 1994). Tests are conducted on bottom barriers or covers subsequent to completion of construction or alternatively on test pads (smaller scale prototype barriers) specifically constructed for use in determination of properties of compacted barriers.

Field hydraulic conductivity tests can be broadly categorized into two groups based on the amount of soil tested: small-scale tests and large-scale tests. Small-scale tests include borehole and probe tests. In borehole tests, a casing is advanced into a small borehole (approximately 100–200 mm in diameter) and water is allowed to enter the surrounding compacted soils under falling or constant-head conditions. The common borehole tests are the Boutwell permeameter and constant-head permeameter tests. In probe tests, probes that provide an indication of hydraulic conductivity (typically through electrical or mechanical means) are pushed or driven into the compacted soils to obtain a profile of conductivity with depth.

Large-scale tests include infiltrometer and underdrain tests. The primary types of infiltrometer tests are single-ring, double-ring, sealed single-ring, and sealed double-ring infiltrometer tests (Figure 13.23a). As the name suggests, in infiltrometer testing, water is ponded over the soil and allowed to infiltrate into the soil through the infiltrometer casing. The sealed double-ring infiltrometer (SDRI) provides the most representative infiltration measurements in the field due to: (1) the use of a sealed inner ring for the measurements that prevents evaporative losses; (2) the use of two rings, which allows for development of vertical flow beneath the inner ring, thus preventing lateral spreading; and (3) the use of a plastic bag filled with water as the inflow reservoir placed under a constant level of water in the annular space between the rings, thus preventing pressure (and gradient) differences during a test. Hydraulic conductivities determined from moderate- to long-term (~50 d to 9–10 mo) SDRI tests provide representative field k values when the wetting front penetrates the compacted soil underneath to a significant degree (greater than one-third to one-half the thickness of the test soil) (Benson et al. 1997).

Underdrain (also termed lysimeter) tests include determination of amount of flow through a barrier by collecting the percolation emanating from the bottom of the barrier using a reservoir (underdrain, lysimeter) placed beneath the barrier (Figure 13.23b). The tests require placement of the underdrain beneath the barrier prior to construction of the barrier and, therefore, are relatively difficult to implement in comparison to infiltrometer tests. When an underdrain is used to determine k of an overlying barrier (e.g., CCL), some consideration must be given to the potential for underestimation of k due to the potential for a capillary barrier effect of that interface between the overlying barrier and the underdrain (e.g., Chiu and Shackelford 2000).

Borehole and probe tests can be used on flat areas or slopes, require short testing times, and generally are low cost. However, relatively small amounts of soil are tested using these tests such that representative measurements of k may not be obtained. Infiltrometer and underdrain tests can be used to test large amounts of soil and thus provide representative measurements of k on a field scale. However, these tests typically are more complicated to conduct, time-

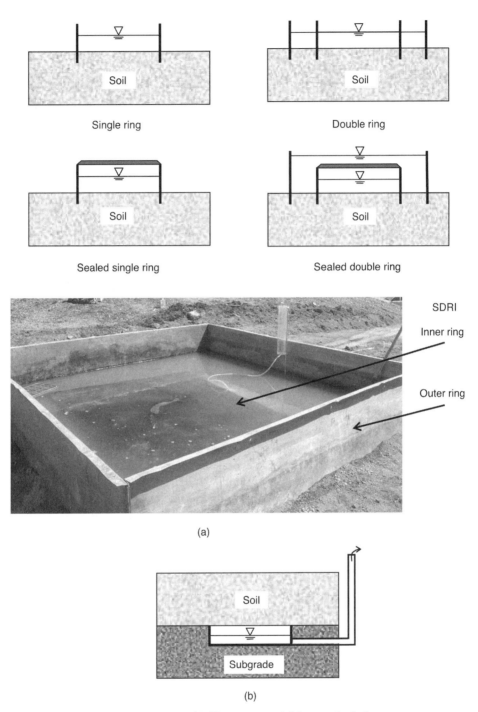

(a)

(b)

FIGURE 13.23 Field tests: (a) types of infiltrometers and (b) an underdrain.

consuming, and costly in comparison to borehole and probe tests. Significant differences in k between laboratory and field tests were observed in comparative studies (Benson et al. 1994a, 1997). Macro features that typically cannot be fully represented in laboratory tests control k in the field. Highest differences between laboratory and field conductivities (i.e., $k_{laboratory} \ll k_{field}$) are expected when inadequate soil preparation, processing, compaction, and quality control procedures are used, even though the water content and dry unit weight of field compacted soils may be similar for "good" and "poor" construction practices (Benson et al. 1997; Daniel and Koerner 2007).

13.5.2 Diffusion Testing

In order to evaluate contaminant transport through engineered barriers, both the effective diffusion coefficient, D^*, and the retardation factor, R_d, must be determined for site-specific soils and chemical constituents (e.g., see Equations 13.18 and 13.19). In terms of determining D^*, several different methods can be used (Shackelford 1991). In particular, the single-reservoir, decreasing source concentration method has been used extensively to evaluate compacted clay soils considered for construction of CCLs (e.g., Hong et al. 2009).

As illustrated schematically in Figure 13.24, the single-reservoir, decreasing source concentration test consists of placing a chemical solution (e.g., actual or simulated leachate) within a reservoir attached to a column of soil so that chemical species (solutes) will diffuse from the reservoir into the soil (Figure 13.24a). Samples of chemical solution in the reservoir are obtained as a function of time to evaluate the decrease in solute concentration vs. time in the reservoir (Figure 13.24b). At the end of the test ($t = t_f$), the reservoir is removed, and the soil specimen is extracted from the column and sliced into incremental segments. Each segment then is squeezed (consolidated) to extract samples of the pore water from which the solute concentration can be measured, providing the solute concentration distribution within the soil at the end of the test (Figure 13.24c). An analytical or numerical solution to Fick's second law for diffusion (Equation 13.17) based on appropriate initial and boundary conditions then is regressed (fitted) against both the reservoir concentration data (Figure 13.24b) and the soil

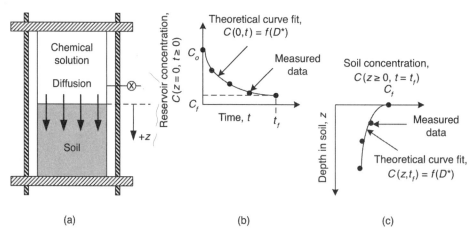

FIGURE 13.24 Schematic of setup and typical results for single-reservoir, decreasing source concentration diffusion test: (a) testing cell, (b) reservoir concentrations vs. time during test, and (c) soil concentrations vs. depth at end of test (redrawn after Shackelford 1991).

concentration data (Figure 13.24c) to determine the best-fit D^* values. The two values of D^* based on the two sets of concentration data then can be compared. Although these two D^* values conceptually should be the same, experience has indicated that differences in the two D^* values can occur. In general, the D^* value based on the soil concentration profile (Figure 13.24c) is considered to be the more reliable value (e.g., Hong et al. 2009). A number of factors associated with this and other methods of measuring D^* are discussed by Shackelford (1991).

13.5.3 Measurement of Retardation Factor

The other primary parameter required for evaluating contaminant transport through engineered barriers via either analytical modeling (e.g., Equations 13.18 and 13.19) or numerical modeling is the retardation factor, R_d. The retardation factor generally can be determined in two ways: indirectly through Equation 13.15 by determining the distribution coefficient, K_d, from the results of batch equilibrium sorption tests (BESTs) (e.g., Roy et al. 1992), and directly via column testing (e.g., Shackelford 1994b).

The general procedure for performing a BEST consists of: (1) placing a known volume (V_l) of a chemical solution with known initial concentrations (C_o) of individual solutes of interest in contact with a known dry mass of the soil of interest (m_s), (2) mixing the resulting suspension for a specified duration (typically 24 or 48 h), (3) extracting a sample of the resulting mixed suspension and centrifuging the sample to separate the solid and liquid phases (supernatant), and (4) measuring the final equilibrium concentration (C_f) of the same solute from a sample of the supernatant of the centrifuged sample. The concentration of the solute adsorbed during the procedure normalized with respect to the dry mass of the soil (C_s) then is determined by difference in accordance with the following equation:

$$C_s = \frac{(C_o - C_f)V_l}{m_s} \tag{13.20}$$

This procedure is repeated several more times using the same values for V_l and m_s but chemical solutions with different initial solute concentrations (i.e., different values for C_o) to develop a graph of C_s (y-axis) vs. C_f (x-axis), known as a *sorption isotherm* (test is conducted at constant temperature). Such a graph typically will be linear at relatively low concentrations and eventually become nonlinear as the values of C_o and, therefore, C_f increase (e.g., see Shackelford 1993). The slope of the linear portion of this graph is defined as K_d. Once K_d is determined from the results of the BEST, the value of R_d can be calculated from Equation 13.15 using the dry density (ρ_d) and the porosity (n) of the *in situ* soil (e.g., CCL).

Typically, the ratio of m_s to the mass of the chemical solution, m_l ($= V_l \cdot \rho_l$, where ρ_l is the density of the chemical solution, usually assumed to be that of water, or 1 g/mL), known as the *soil:solution ratio,* used in the test is such that m_l must be greater than m_s in order to facilitate a sufficient amount of supernatant to sample. The limiting (lowest) soil:solution ratio recommended by Roy et al. (1992) is 1:4, and values as high as 1:100 or 1:200 are common. This range of soil:solution ratios is much greater than that typically encountered in the field, where the gravimetric water content, w ($= m_l/m_s$), is less than unity (<100%). This difference in soil:solution ratios represents a limitation in the application of K_d values determined from a BEST. Other limitations in the application of R_d values based indirectly on K_d values determined from BESTs are discussed by Cherry et al. (1984).

Although there are some variations in column testing procedures, a traditional approach consists of (Shackelford 1994b): (1) establishing steady-state solvent (usually water) flow through the column of the porous medium (e.g., compacted soil), (2) continuously introducing into the influent a chemical solution with known concentrations (C_o) of chemical species, and (3) measuring the concentrations (C_e) of the same chemical species appearing in the effluent as a function of time. The results of this approach are plotted as "effluent breakthrough curves," or plots of relative effluent concentration, C_e/C_o vs. time. The measured effluent breakthrough curves then are evaluated using an appropriate transport model (typically an analytical model) to determine the value of R_d. An alternative to this procedure based on solute mass instead of solute concentration is described by Shackelford (1995). Because the soil in the columns can be placed at more realistic soil-solution ratios, and column tests are conducted under more realistic conditions involving solvent flow, values of R_d determined from column tests generally are considered to be more reliable than those based on BESTs. More detail regarding column testing procedures can be found in Shackelford (1994b).

13.6 Vertical Barriers

In the context of geoenvironmental engineering, vertical barriers are installed in the subsurface to contain contaminated soils and groundwater as well as nonengineered (i.e., unlined) waste disposal sites. Also, vertical barriers can be used to divert groundwater flow around contaminated sites. Vertical barriers may provide short-term containment until soil or groundwater is cleaned up for cases with highly toxic contaminants or relatively small-scale contamination. These barriers may be used for long-term containment for relatively low levels of contamination or for large-scale contaminated areas such as old dumps or mine tailings storage areas.

13.6.1 Vertical Cutoff Walls

Vertical cutoff walls are the most common type of vertical containment system. Cutoff walls are constructed by first excavating a narrow trench in the subsurface, which is filled with slurry to maintain trench stability. Then, the slurry is displaced and replaced with a permanent backfill. Trench widths between 0.6 to 1.5 m are used, with typical widths of 0.9 m, and trench depths can be well in excess of 50 m. The types of vertical cutoff walls are soil-bentonite (SB), cement-bentonite (CB), plastic concrete (PC), and composite (SBC) (Sharma and Reddy 2004). CB walls are constructed by filling the trench with a slurry, which is allowed to harden in place, in contrast to SB walls, where the slurry is replaced with the backfill. The composite vertical cutoff walls include a geosynthetic barrier within SB walls for additional resistance to fluid transport. SB cutoff walls are the most common type of vertical barrier used in the U.S. The amount and type of typical slurry and backfill components are provided in Table 13.9.

The slurry placed in the trench in the first step of construction allows for maintaining the stability of the trench. The slurry exerts outward hydrostatic pressure higher than that of surrounding groundwater since the density of the slurry is higher than the density of water, which prevents inward movement of trench walls. In addition, the slurry forms a thin, low *k* layer over the walls of the trench. This thin layer, termed filter cake, provides resistance to fluid transport both during the initial stage of construction and also when the trench is backfilled. The backfill materials are placed in the trench subsequent to full formation of the filter cake. Density of the slurry is monitored to obtain required hydrostatic pressure and viscosity of the slurry is monitored to ensure proper filter cake formation.

TABLE 13.9 Slurry and Backfill Materials and Compositions

Component	Bentonite Slurry (%)[a]	SB Backfill (%)	CB Slurry[b] (%)	CB Backfill[c] (%)
Water	93–96	25–35	63–81 (76)	55–70
Bentonite	4–7	2–4	4–7 (6)	6
Soil	—	61–73 (10–40 fines)		
Cement	—	—	15–30 (18)	18

Adapted from U.S. EPA (1984), Evans (1993), and Sharma and Reddy (2004).
[a] All percentages are provided on a weight basis.
[b] Typical values.
[c] Backfill contains 30–40% solids.

SB walls have lower k and strength than CB walls. Hydraulic conductivities typically vary between 1×10^{-7} and 1×10^{-10} m/s for SB walls, whereas hydraulic conductivities in the range 1×10^{-5} and 1×10^{-8} m/s are typical for CB walls (Evans 1993). In general, SB walls have better chemical compatibility than CB walls and resist a wider range of chemicals (Sharma and Reddy 2004). Soils excavated during construction can be reused in backfill mix for SB walls, reducing the need for disposal. However, the quality of locally available soils may not be appropriate for backfill construction. Densities of slurry and backfill are monitored to ensure that the backfill is denser (≥ 240 kg/m^3) than the slurry for proper replacement during construction. Hydraulic conductivity (with water and other liquids) of backfills is determined to ensure resistance to fluid transport.

Measurement of the k of SB cutoff walls can be performed in the laboratory or *in situ* (NRC 2007). Laboratory tests are performed on remolded specimens using the American Petroleum Institute filter press or flexible-wall permeameters. Laboratory tests also can be performed on undisturbed samples obtained by sampling the wall after construction using flexible-wall permeameters, although this sampling is difficult due to the soft nature of the materials (Britton et al. 2004). *In situ* tests include slug tests, piezocone soundings with pore pressure dissipation measurements, and large-scale pumping and injection (Britton et al. 2004). Britton et al. (2004) reported that laboratory methods involving the filter press and falling head procedures performed on remolded and undisturbed samples, respectively, yielded the lowest values for hydraulic conductivity, followed by field methods (piezocone and piezometer), whereas large-scale *in situ* measurement of k resulted in higher values than either the laboratory or field measurements. Testing of SB walls using slug tests (instantaneous change in water level, followed by monitoring until the water level returns to static conditions) with a push-in piezometer tip is recommended as an efficient method that limits disturbance while testing a relatively large volume of material (Britton et al. 2005; Choi and Daniel 2006a, 2006b).

Vertical cutoff walls are constructed in various configurations (U.S. EPA 1984). The vertical configurations include hanging walls, whereby the base of the wall is located within a relatively high k formation (e.g., aquifer) above a low k formation (e.g., clay, bedrock), and keyed-in walls, whereby the base of the wall extends into a low k stratum (e.g., clay, bedrock) beneath the contained materials. The horizontal configurations include circumferential walls, whereby the wall completely surrounds the contaminated area; upgradient walls, whereby the wall is located up gradient of the contaminated zone such that fresh (uncontaminated) groundwater is redirected around the wall before coming into contact with the contaminated zone; and downgradient walls, whereby the wall is located down gradient of the contaminated

TABLE 13.10 Summary Comparison of Vertical Cutoff Wall Configurations

Vertical Configuration	Horizontal Configuration		
	Circumferential	Upgradient	Downgradient
Keyed-in	• Most common • Expensive • Most complete containment • Vastly reduced leachate generation	• Not common • Used to divert groundwater around site in steep-gradient situations • Can reduce leachate generation • Compatibility not critical	• Used to capture miscible or sinking contaminants for treatment • Inflow not restricted, may raise groundwater table • Compatibility very important
Hanging	• Used for floating contaminants moving in more than one direction (such as on a groundwater divide)	• Rare • May temporarily lower groundwater table behind wall • Can stagnate leachate but not halt flow	• Used to capture floating contaminants for treatment • Inflow not restricted, may raise groundwater table • Compatibility very important

From U.S. EPA (1984).

zone such that groundwater is intercepted after flowing through the contaminated zone. In the case of downgradient walls, the contaminated groundwater must be collected and treated. A summary comparison of the configurations is provided in Table 13.10.

Short-term performance considerations for SB and CB vertical cutoff walls include defective material and "windows" due to caving or trapped low-quality material at joints between panels. Medium- and long-term performance considerations include chemical incompatibility, desiccation above the water table, and cracking and deterioration. Failure of vertical cutoff walls can be attributed to two primary mechanisms: (1) changes in the material properties after the wall has been constructed and (2) defects in the constructed wall, including entrapped sediment, improperly mixed backfill, inadequate excavation of the key, and formation spalling (Evans 1991). Property changes within cutoff walls result from the same mechanisms that affect CCLs: desiccation, freeze/thaw, chemical incompatibility, and excessive deformations. The possible effects of chemical incompatibility described with respect to earthen barriers and GCLs apply equally well to SB cutoff walls. However, issues such as the type of soil and stress conditions may be more critical in the case of SB walls because of the use of high-activity sodium bentonite in the slurry and the lack of significant waste overburden, respectively.

The possible scenarios for chemical transport through vertical cutoff walls are depicted in Figure 13.25. Pure diffusion (Figure 13.25a) is an ideal case, diffusion with advection (Figure 13.25b) occurs when unmitigated buildup of contaminated water retained by the wall is allowed to occur, and diffusion against advection (Figure 13.25c) occurs when the contaminated water inside the containment area is drawn down (e.g., by pumping or drainage) to induce inward flow of water.

13.6.2 Other Vertical Barriers

In addition to vertical cutoff walls, a variety of other types of vertical barriers may be considered for use in containment applications. These barriers include such components as sheet-pile walls, geomembrane panels, and permeation and jet grouting using a wide variety of

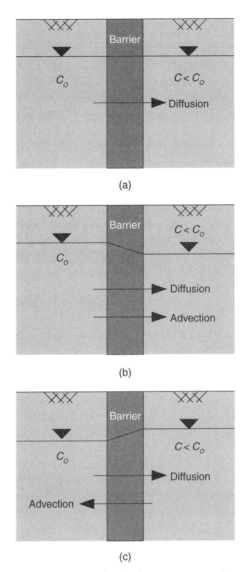

FIGURE 13.25 Chemical transport scenarios for vertical barriers: (a) pure diffusion, (b) diffusion with advection, and (c) diffusion against advection (redrawn after Shackelford 1993; NRC 2007).

chemical-based grouts and materials, as well as extraction and injection trenches (Rumer and Mitchell 1995; NRC 2007). However, very little data on field performance exist for these types of barriers.

Acknowledgment

The authors are indebted to Professor John J. Bowders of the University of Missouri for his careful review of and comments on a draft version of this chapter.

References

Albrecht, B. (1996). *Effect of Desiccation on Compacted Clay*, MS Thesis, University of Wisconsin, Madison.

Albright, W.H., Benson, C.H., Gee, G.W., Roesler, A.C., Abichou, T., Apiwantragoon, P., Lyles, B.F., and Rock, S.A. (2004). Field water balance of landfill final covers. *J. Env. Qual.*, 33:2317–2332.

Bagchi, A. (1994). *Design, Construction, and Monitoring of Sanitary Landfills*, John Wiley & Sons, Hoboken, NJ.

Benson, C.H. (1999). Final covers for waste containment systems: a North American perspective. *Proc. XVII Conference of Geotechnics of Torino, Control and Management of Subsoil Pollutants*, Italian Geotechnical Society, Torino, Italy, 1–32.

Benson, C.H. and Chen, C. (2003). Selecting the thickness of monolithic earthen covers for waste containment. *Soil and Rock America 2003*, Verlag Gluck auf GMBH, Essen, Germany, 1397–1404.

Benson, C.H. and Daniel, D.E. (1990). Influence of clods on hydraulic conductivity of compacted clay. *J. Geotech. Eng.*, 116(8):1231–1248.

Benson, C.H. and Daniel, D.E. (1994). Minimum thickness of compacted soil liners. II. Analysis and case histories. *J. Geotech. Eng.*, 120(1):153–172.

Benson, C.H. and Khire, M.V. (1995). Earthen covers for semi-arid and arid climates. *Landfill Closures, GSP No. 53*, J. Dunn and U. Singh, Eds., ASCE, 201–217.

Benson, C.H. and Khire, M.V. (1997). Earthen materials in surface barriers. *Barrier Technologies for Environmental Management*, National Academy Press, National Research Council, Washington, D.C., D79–D89.

Benson, C.H. and Othman, M. (1993). Hydraulic conductivity of compacted clay frozen and thawed in place. *J. Geotech. Eng.*, 119(3):276–294.

Benson, C.H., Hardianto, F., and Motan, E.S. (1994a). Representative specimen size for hydraulic conductivity assessment of compacted soil liners. *Hydraulic Conductivity and Waste Contaminant Transport in Soil*, ASTM STP 1142, D.E. Daniel and S.J. Trautwein, Eds., ASTM, West Conshohocken, PA, 3–29.

Benson, C.H., Zhai, H., and Wang, X. (1994b). Estimating hydraulic conductivity of compacted clay liners. *J. Geotech. Eng.*, 120(2):366–387.

Benson, C.H., Gunter, J.A., Boutwell, G.P., Trautwein, S.J., and Berzanskis, P.H. (1997). Comparison of four methods to assess hydraulic conductivity. *J. Geotech. Geoenviron. Eng.*, 123(10):929–937.

Benson, C.H., Daniel, D.E., and Boutwell, G.P. (1999). Field performance of compacted clay liners. *J. Geotech. Geoenviron. Eng.*, 125(5):390–403.

Benson, C.H., Thorstad, P.A., Jo, H.-Y., and Rock, S.A. (2007). Hydraulic performance of geosynthetic clay liners in a landfill final cover. *J. Geotech. Geoenviron. Eng.*, 133(7):814–827.

Bohnhoff, G.L., Ogorzalek, A.S., Benson, C.H., Shackelford, C.D., and Apiwantragoon, P. (2009). Field data and water-balance predictions for a monolithic cover in a semiarid climate. *J. Geotech. Geoenviron. Eng.*, 135(3):333–348.

Bonaparte, R., Daniel, D.E., and Koerner, R.M. (2002). *Assessment and Recommendations for Improving the Performance of Waste Containment Systems*, EPA Report EPA/600/R-02/099, Environmental Protection Agency, Office of Research and Development, National Risk Management Research Laboratory, Cincinnati, OH.

Bouazza, A. (2002). Geosynthetic clay liners. *Geotext. Geomembr.*, 20(1):3–17.

Bowders, J.J. (1988). Discussion of "termination criteria for clay permeability testing." *J. Geotech. Eng.*, 114(8):947–949.

Bowders, J.J. and Daniel, D.E. (1987). Hydraulic conductivity of compacted clay to dilute organic chemicals. *J. Geotech. Eng.,* 113(12):1432–1448.

Bowders, J.J., Daniel, D.E., Broderick, G.P., and Liljestrand, H.M. (1985). Methods for testing the compatibility of clay liners with landfill leachate. *Hazardous and Industrial Solid Waste Testing: Fourth Symposium, ASTM STP 886,* J.K. Petros, W.J. Lacy, and R.A. Conway, Eds., ASTM, West Conshohocken, PA, 233–250.

Britton, J.P., Filz, G.M., and Herring, W.E. (2004). Measuring the hydraulic conductivity of soil-bentonite backfill. *J. Geotech. Geoenviron. Eng.,* 130(12):1250–1258.

Britton, J.P., Filz, G.M., and Herring, W.E. (2005). Slug tests in soil-bentonite cutoff walls using a push-in piezometer tip. *Waste Containment and Remediation, GSP 142,* A. Alshawabkeh et al., Eds., ASCE, Reston, VA (CD version only).

Broderick, G.P. and Daniel, D.E. (1990). Stabilizing compacted clay against chemical attack. *J. Geotech. Eng.,* 116(10):1549–1567.

Cedergren, H.R. (1989). *Seepage, Drainage, and Flow Nets,* 3rd edition, John Wiley & Sons, Hoboken, NJ.

Cherry, J.A., Gillham, R.W., and Barker, J.F. (1984). Contaminants in groundwater: chemical processes. *Groundwater Contamination,* National Academies Press, Washington, D.C., 46–64.

Chiu, T.-F. and Shackelford, C.D. (2000). Laboratory evaluation of sand underdrains. *J. Geotech. Geoenviron. Eng.,* 126(11):990–1001.

Choi, H. and Daniel, D.E. (2006a). Slug test analysis in vertical cutoff walls. I. Analysis methods. *J. Geotech. Geoenviron. Eng.,* 132(4):429–438.

Choi, H. and Daniel, D.E. (2006b). Slug test analysis in vertical cutoff walls. II. Applications. *J. Geotech. Geoenviron. Eng.,* 132(4):439–447.

Corser, P. and Cranston, M. (1991). Observations on the performance of composite clay liners and covers. *Proc., Geosynthetic Design and Performance,* Vancouver Geotechnical Society, Vancouver, British Columbia, 16.

Daniel, D.E. (1989). In situ hydraulic conductivity tests for compacted clay. *J. Geotech. Eng.,* 115(9):1205–1226.

Daniel, D.E. (1993). Landfills and impoundments. *Geotechnical Practice for Waste Disposal,* D.E. Daniel, Ed., Chapman and Hall, London, 97–112.

Daniel, D.E. (1994). State-of-the-art: laboratory hydraulic conductivity test for saturated soils. *Hydraulic Conductivity and Waste Contaminant Transport in Soil,* ASTM STP 1142, D.E. Daniel and S.J. Trautwein, Eds., ASTM, West Conshohocken, PA, 30–78.

Daniel, D.E. and Benson, C.H. (1990). Water content-density criteria for compacted soil liners. *J. Geotech. Eng.,* 116(12):1811–1830.

Daniel, D.E. and Koerner, R.M. (1993). Cover systems. *Geotechnical Practice for Waste Disposal,* D.E. Daniel, Ed., Chapman and Hall, London, 455–496.

Daniel, D.E. and Koerner, R.M. (2007). *Waste Containment Facilities—Guidance for Construction Quality Assurance and Construction Quality Control of Liner and Cover Systems,* 2nd edition, ASCE, Reston, VA.

Daniel, D.E. and Wu, Y.-K. (1993). Compacted clay liners and covers for arid sites. *J. Geotech. Eng.,* 119(2):223–237.

Daniel, D.E., Bowders, J.J., and Gilbert, R.B. (1997). Laboratory hydraulic conductivity testing of GCLs in flexible-wall permeameters. *Testing and Acceptance Criteria for Geosynthetic Clay Liners,* ASTM STP 1308, L.W. Well, Ed., ASTM, West Conshohocken, PA, 208–226.

Edil, T.B. (2003). A review of aqueous-phase VOC transport in modern landfill liners. *Waste Manage.,* 23(7):561–571.

Eid, H.T., Stark, T.D., and Doerfler, C.K. (1999). Effect of shear displacement rate on internal shear strength of a reinforced geosynthetic clay liner. *Geosynth. Int.,* 6(3):219–239.

Estornell, P. and Daniel, D.E. (1992). Hydraulic conductivity of three geosynthetic clay liners. *J. Geotech. Eng.*, 118(10):1592–1606.

Evans, J.C. (1991). Geotechnics of hazardous waste control systems. *Foundation Engineering Handbook*, H.-Y. Fang, Ed., Van Nostrand Reinhold, New York.

Evans, J.C. (1993). Vertical cutoff walls. *Geotechnical Practice for Waste Disposal*, D.E. Daniel, Ed., Chapman and Hall, London, 430–454.

Foose, G.J. (2002). Transit-time design for diffusion through composite liners. *J. Geotech. Geoenviron. Eng.*, 128(7):510–520.

Foose, G.J., Benson, C.H., and Edil, T.B. (2001). Predicting leakage through composite landfill liners. *J. Geotech. Geoenviron. Eng.*, 127(6):510–520.

Foose, G.J., Benson, C.H., and Edil, T.B. (2002). Comparison of solute transport in three composite liners. *J. Geotech. Geoenviron. Eng.*, 128(5):391–403.

Foreman, D.E. and Daniel, D.E. (1986). Permeation of compacted clay with organic chemicals. *J. Geotech. Eng.*, 112(6):669–690.

Fox, P.J. and Stark, T.D. (2004). State-of-the-art: GCL shear strength and its measurement. *Geosynth. Int.*, 11(3):141–175.

Gilbert, R.B. and Byrne, R.J. (1996). Shear strength of reinforced geosynthetic clay liners. *J. Geotech. Eng.*, 122(4):259–266.

Gilbert, R.B., Scranton, H.B., and Daniel, D.E. (1997). Shear strength testing for geosynthetic clay liners. *Testing and Acceptance Criteria for Geosynthetic Clay Liners*, ASTM STP 1308, L.W. Well, Ed., ASTM, West Conshohocken, PA, 121–138.

Giroud, J.P., Soderman, K.L., Khire, M.V., and Badu-Tweneboah, K. (1998). New developments in landfill liner leakage evaluation. *Proceedings of the Sixth International Conference on Geosynthetics*, IFAI, Roseville, MN, 261–268.

Gray, D.H. (1995). Containment strategies for landfilled wastes. *Geoenvironment 2000*, Vol. 1, Y.B. Acar and D.E. Daniel, Eds., ASCE, Reston, VA, 484–498.

Grim, R.E. (1959). Physico-chemical properties of soils: clay minerals. *J. Soil Mech. Found. Div.*, 85(SM2):1–17.

Grim, R.E. (1968). *Clay Mineralogy*, 2nd edition, McGraw-Hill, New York.

Hanson, J.L., Yesiller, N., and Swarbrick, G.E. (2005). Thermal analysis of GCLs at a municipal solid waste landfill. *Waste Containment and Remediation*, GSP 142, A. Alshawabkeh et al., Eds., ASCE, Reston, VA, 1–15 (CD version only).

Hanson, J.L., Yesiller, N., and Oettle, N.K. (2008). Spatial variability of waste temperatures in MSW landfills. *Proceedings, 2008 Global Waste Management Symposium*, Penton Media, 1–11 (CD version only).

Holtz, R.D. and Kovacs, W.D. (1981). *An Introduction to Geotechnical Engineering*, Prentice Hall, Upper Saddle River, NJ.

Hong, C.S.-J., Davis, M.M., and Shackelford, C.D. (2009). Non-reactive solute diffusion in unconfined and confined specimens of a compacted clay soil. *Waste Manage.*, 29(1):404–417.

Jo, H.-Y., Benson, C.H., Shackelford, C.D., Lee, J.-M., and Edil, T.B. (2005). Long-term hydraulic conductivity of a geosynthetic clay liner (GCL) permeated with inorganic salt solutions. *J. Geotech. Geoenviron. Eng.*, 131(4):405–417.

Katsumi, T., Benson, C.H., Foose, G.J., and Kamon, M. (2001). Performance based design of landfill liners. *Eng. Geol.*, 60(1–4):139–148.

Khire, M.V., Benson, C.H., and Bosscher, P.J. (1997). Water balance modeling of earthen final covers. *J. Geotech. Geoenviron. Eng.*, 123(8):744–754.

Khire, M.V., Benson, C.H., and Bosscher, P.J. (2000). Capillary barriers: design variables and water balance. *J. Geotech. Geoenviron. Eng.*, 126(8):695–708.

Koerner, R.M. (2005). *Designing with Geosynthetics*, 5th edition, Prentice Hall, Upper Saddle River, NJ.

Koerner, R.M. and Daniel, D.E. (1995). A suggested methodology for assessing the technical equivalency of GCLs to CCLs. *Geosynthetic Clay Liners*, R.M. Koerner, E. Gartung, and H. Zanzinger, Eds., Balkema, Rotterdam, 73–98.

Lambe, T.W. (1953). The structure of inorganic soil. *Proceedings*, ASCE, Vol. 79, Separate No. 315, 49 pp.

Lambe, T.W. (1958). The structure of compacted clay. *J. Soil Mech. Found. Div.*, 84(SM2):1654-1–1654-34.

Lee, J.-M. and Shackelford, C.D. (2005a). Impact of bentonite quality on hydraulic conductivity of geosynthetic clay liners. *J. Geotech. Geoenviron. Eng.*, 131(1):64–77.

Lee, J.-M. and Shackelford, C.D. (2005b). Concentration dependency of the prehydration effect for a geosynthetic clay liner. *Soils Found.*, 45(4):27–41.

Manassero, M., Benson, C.H., and Bouazza, A. (2000). Solid waste containment systems. *International Conference on Geotechnical and Geoenvironmental Engineering (GeoEng2000)*, Vol. 1, Melbourne, Australia, November 19–24, Technomic Publ. Co., Lancaster, PA, 520–642.

Meer, S.R. and Benson, C.H. (2007). Hydraulic conductivity of geosynthetic clay liners exhumed from landfill final covers. *J. Geotech. Geoenviron. Eng.*, 133(5):550–563.

Melchior, S. (1997). In situ studies on the performance of landfill caps. *Proceedings, International Containment Technology Conf.*, Institute for International Cooperative Environmental Research, Tallahassee, FL, 365–373.

Mitchell, J.K. and Madsen, F.T. (1987). Chemical effects on clay hydraulic conductivity. *Geotechnical Practice for Waste Disposal '87*, GSP 13, R.D. Woods, Ed., ASCE, Reston, VA, 87–116.

Mitchell, J.K and Soga, K. (2005). *Fundamentals of Soil Behavior*, 3rd edition, John Wiley & Sons, Hoboken, NJ.

Mitchell, J.K., Hooper, D.R., and Campanella, R.G. (1965). Permeability of compacted clay. *J. Soil Mech. Found. Div.*, 91(SM4):41–65.

Montgomery, R. and Parsons, L. (1990). The Omega Hills cover test plot study: fourth year data summary. *Proceedings 22nd Mid-Atlantic Industrial Waste Conference*, Drexel University, July 24–27, Technomic Publ. Co., Lancaster, PA.

NRC (2007). *Assessment of the Performance of Engineered Waste Containment Barriers*, National Research Council, The National Academies Press, Washington, D.C.

OECD (2008). Environmental Data Compendium 2006/2007 Waste, http://www.oecd.org/dataoecd/60/59/38106368.pdf.

Ogorzalek, A.S., Bohnhoff, G.L., Shackelford, C.D., Benson, C.H., and Apiwantragoon, P. (2008). Comparison of field data and water-balance predictions for a capillary barrier cover. *J. Geotech. Geoenviron. Eng.*, 134(4):470–486.

Olsen, H.W. (1962). Hydraulic flow through saturated clays. *Clays Clay Miner.*, 9(2):131–161.

Parent, S.-E. and Cabral, A. (2005). Material selection for the design of inclined covers with capillary barrier effect. *Waste Containment and Remediation*, GSP 142, A. Alshawabkeh et al., Eds., ASCE, Reston, VA (CD version only).

Richardson, G.N., Thiel, R., and Erickson, R. (2002). GCL design series. 3. GCL installation and durability. *Geosynthetics Fabric Report*, Vol. 20, No. 7, IFAI, Minneapolis, http://www.ifai.com/Geo/GFR_OnLine-ArticleHome.cfm.

Rowe, R.K. (1998). Geosynthetics and the minimization of contaminant migration through barrier systems beneath solid waste. *Proceedings, Sixth International Conference on Geosynthetics*, Vol. 1, IFAI, Minneapolis, 27–102.

Rowe, R.K. (2005). Long-term performance of contaminant barrier systems, 45th Rankine Lecture. *Geotechnique*, 55(9):631–678.

Rowe, R.K. and Booker, J.R. (1985a). 1-D pollutant migration in soils of finite depth. *J. Geotech. Eng.*, 111(4): 479–499.

Rowe, R.K. and Booker, J.R. (1985b). 2-D pollutant migration in soils of finite depth. *Can. Geotech. J.*, 22(4):429–436.

Rowe, R.K. and Booker, J.R. (1986). A finite layer technique for calculating three dimensional pollutant migration in soil. *Geotechnique*, 36(2):205–214.

Roy, W.R., Krapac, I.G., Chou, S.F.J., and Griffin, R.A. (1992). *Batch-Type Procedures for Estimating Soil Adsorption of Chemicals*, EPA/530/SW-87/006-F, U.S. Environmental Protection Agency, Office of Solid Waste and Emergency Response, Washington, D.C.

Rumer, R.R. and Mitchell, J.K., Eds. (1995). *Assessment of Barrier Containment Technologies: A Comprehensive Treatment for Environmental Remediation Applications*, National Technical Information Service, Springfield, VA.

Shackelford, C.D. (1988). Diffusion as a transport process in fine-grained barrier materials. *Geotech. News*, 6(2):24–27.

Shackelford, C.D. (1989). Diffusion of contaminants through waste containment barriers. *Transportation Research Record No. 1219*, Geotechnical Engineering 1989, Transportation Research Board, Washington, D.C., 169–182.

Shackelford, C.D. (1990). Transit-time design of earthen barriers. *Eng. Geol.*, 29(1):79–94.

Shackelford, C.D. (1991). Laboratory diffusion testing for waste disposal—a review. *J. Contam. Hydrol.*, 7(3):177–217.

Shackelford, C.D. (1992). Performance-based design of soil liners. *Engineering Proceedings of the Mediterranean Conference on Environmental Geotechnology*, Cesme, Turkey, M.A. Usmen and Y.B. Acar, Eds, A.A. Balkema, Rotterdam, Netherlands, 145–153.

Shackelford, C.D. (1993). Contaminant transport. *Geotechnical Practice for Waste Disposal*, D.E. Daniel, Ed., Chapman and Hall, London, 33–65.

Shackelford, C.D. (1994a). Waste-soil interactions that alter hydraulic conductivity. *Hydraulic Conductivity and Waste Contaminant Transport in Soil*, ASTM STP 1142, D.E. Daniel and S.J. Trautwein, Eds., ASTM, West Conshohocken, PA, 111–168.

Shackelford, C.D. (1994b). Critical concepts for column testing. *J. Geotech. Eng.*, 120(10):1804–1828.

Shackelford, C.D. (1995). Cumulative mass approach for column testing. *J. Geotech. Eng.*, 121(10):696–703.

Shackelford, C.D. (2002). Geoenvironmental engineering. *Encyclopedia of Physical Science and Technology*, 3rd edition, Vol. 6, Academic Press, San Diego, 601–621.

Shackelford, C.D. (2005). Environmental issues in geotechnical engineering. *16th International Conference on Soil Mechanics and Geotechnical Engineering*, Vol. 1, Osaka, Japan, September 12–16, Millpress, Rotterdam, Netherlands, 95–122.

Shackelford, C.D. (2008). Selected issues affecting the use and performance of GCLs in waste containment applications. *Geosynthetics and Environment, Proceedings of the Geotechnical Engineering Conference of Torino (XXI Edition)*, Politecnico di Torino, Torino, Italy, November 27–28, 2007, M. Manassero, and A. Dominijanni, Eds., Pàtron Editore, Bologna (CD version only).

Shackelford, C.D. and Daniel, D.E. (1991). Diffusion in saturated soil. II. Results for compacted clay. *J. Geotech. Eng.*, 117(3):485–506.

Shackelford, C.D. and Rowe, R.K. (1998). Contaminant transport modeling. *3rd International Congress on Environmental Geotechnics*, Vol. 3, Lisboa, Portugal, September 7–11, P. Seco e Pinto, Ed., Balkema, Rotterdam, 939–956.

Shackelford, C.D., Malusis, M.A., Majeski, M.J., and Stern, R.T. (1999). Electrical conductivity breakthrough curves. *J. Geotech. Geoenviron. Eng.*, 125(4):260–270.

Shackelford, C.D., Benson, C.H., Katsumi, T., Edil, T.B., and Lin, L. (2000). Evaluating the hydraulic conductivity of GCLs permeated with non-standard liquids. *Geotext. Geomembr.*, 18(2–4):133–161.

Sharma, H.D. and Reddy, K.R. (2004). *Geoenvironmental Engineering,* John Wiley & Sons, Hoboken, NJ.

Shelley, T.L. and Daniel, D.E. (1993). Effect of gravel on hydraulic conductivity of compacted soil liners. *J. Geotech. Eng.,* 119(1):54–68.

Spokas, K., Bogner, J., Chanton, J.P., Morcet, M., Aran, C., Graff, C., Moreau-Le Golvan, Y., and Hebe, I. (2006). Methane mass balance at three landfill sites: what is the efficiency of capture by gas collection systems? *Waste Manage.,* 26(5):516–525.

Stormont, J.C. and Morris, C.E. (1998). Method to estimate water storage capacity of capillary barriers. *J. Geotech. Geoenviron. Eng.,* 124(4):297–302.

Trautwein, S.J. and Boutwell, G.P. (1994). In-situ hydraulic conductivity tests for compacted soil liners and caps. *Hydraulic Conductivity and Waste Contaminant Transport in Soil,* ASTM STP 1142, D.E. Daniel and S.J. Trautwein, Eds., ASTM, West Conshohocken, PA, 184–226.

U.S. EPA (1984). *Slurry Trench Construction for Pollution Migration Control,* EPA 540/2-84-001, U.S. Environmental Protection Agency, Cincinnati, OH.

U.S. EPA (1993). *Quality Assurance and Quality Control for Waste Containment Facilities,* EPA 600/R/-93/182, U.S. Environmental Protection Agency, Washington, D.C.

U.S. EPA (2001). *Geosynthetic Clay Liners Used in Municipal Solid Waste Landfills,* EPA530-F-97-002, Fact Sheet (Revised Issue), U.S. Environmental Protection Agency, Washington, D.C., 8 p. (http://www.epa.gov/garbage/landfill/geosyn.pdf).

U.S. EPA (2008). *Inventory of U.S. Greenhouse Gas Emissions and Sinks: 1990–2006,* 430-R-08-0025.

van Genuchten, M.Th. (1981). Analytical solutions for chemical transport with simultaneous adsorption, zero-order production, and first-order decay. *J. Hydrol.,* 49(3–4):213–233.

Yesiller, N., Hanson, J.L., Oettle, N.K., and Liu, W.-L. (2008). Thermal analysis of cover systems in municipal solid waste landfills. *J. Geotech. Geoenviron. Eng.,* 134(11):1655–1664.

Yong, R.N. and Warkentin, B.P. (1975). *Soil Properties and Behavior,* Elsevier, New York.

Zornberg, J.G., Somasundaram, S., and LaFountain, L. (2001). Design of geosynthetic-reinforced veneer slopes. *Proceedings, International Symposium on Earth Reinforcement* (IS Kyushu 2001), Vol. 1, Fukuoka, Japan, November 14–16, A.A. Balkema, 305–310.

Zornberg, J.G., LaFountain, L., and Caldwell, J. (2003). Analysis and design of evapotranspirative cover for hazardous waste landfill. *J. Geotech. Geoenviron. Eng.,* 129(6):427–436.

14

Railway Track Bed Foundation Design

by
Gurmel S. Ghataora
The University of Birmingham, Birmingham, U.K.

Michael Burrow
The University of Birmingham, Birmingham, U.K.

14.1 Introduction

The need to move goods and raw materials cheaply, over long distances and often through difficult ground, led to the development of railways. Soon afterward, their role as a means of transporting large groups of people was realized. Examples of using railways to open up large parts of a country to development include the construction of the railway line that connected the east and west coasts of the United States. The expansion and later the defense of the British Empire were made easier by the railways, because men and materials could be moved across

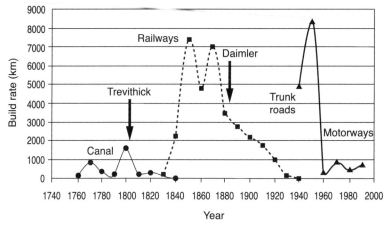

FIGURE 14.1 Build rate of railways in the U.K. (after Lowson 1998).

vast stretches of land. Later, in Europe, Asia, and the U.S., railways were used extensively during the two world wars. In the 1950s in the developed world, the improvements in the design and construction of cars and the associated improvements in road infrastructure led to the automobile becoming a more popular mode of transport. In addition to offering point-to-point travel, it gave greater freedom. Nevertheless, in terms of transporting large quantities of materials and people over long distances, railways are perhaps still the most efficient mode of transport.

The development of railways in relation to that of other modes of transport in the U.K. is shown in Figure 14.1. The trend shown is likely to be similar to that in many other parts of the developed world. More recently in many parts of the world, new high-speed lines have been constructed.

During the first 100 or so years during which the majority of the track was constructed, most of the attention was given to the rolling stock and parts of the track that lay above the ballast. Less attention was given to the track support system, which includes the ballast, subballast, and subgrade. With the development of the science of soil mechanics and the need to run higher speed trains with greater axle loads, much more attention has been given to the track support system. However, it is worth noting that perhaps the majority of trains run on track that was built half a century or more ago. In instances where older track is subjected to either faster trains or heavier axle loads or both, the track support system may require a great deal of maintenance in order to maintain acceptable line and level.

In the following sections, brief definitions of the components of the track support system are given, followed by structural design procedures, problems associated with existing track, methods of remediation, and site investigation. Although various form of slab track systems and joined sleeper systems such as the ladder system (e.g., Walui et al. 1997) are being developed, this chapter deals only with conventional ballasted track, since the former systems are not used widely.

This chapter is organized such that site investigation is the last section. The basis for this is that it is necessary to know something of the behavior of materials, potential problems, design methods, remedies, and how the various properties required can be measured before planning an effective site investigation.

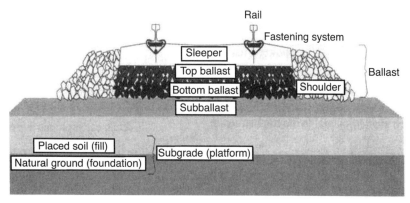

FIGURE 14.2   A typical section of ballasted track (Selig and Waters 1994).

14.2 Definitions

The railway track structure combines a number of components (Figure 14.2) in a structural system that is intended to withstand the combined effects of traffic loading and climate such that the subgrade is adequately protected and railway vehicle operating costs, safety, and comfort of passengers are kept within acceptable limits. Although there is no internationally accepted convention for describing the various components, the track support system typically is comprised of the rail, a fastening system, rail pads, sleepers, ballast, subballast, and subgrade. A typical layout adapted from Selig and Waters (1994) is shown in Figure 14.2, and that used by Network Rail in the U.K. is shown in Figure 14.3. The latter is similar to those given by the International Union of Railways.

The overall functional requirements of the track bed (Figure 14.3) are to impart long-term stability (in terms of track geometry) and to protect the subgrade in a cost-effective manner. In order to comply with these requirements, it needs to meet a range of structural requirements. The most significant of these are stiffness and strength. For example, Hunt (1993) demonstrated that track stiffness can affect the running cost of trains. Furthermore, research has shown that there is a theoretical optimal track stiffness to which a railway line should be designed, constructed, and maintained. Below the optimum, excessive track displacements occur; above it, unacceptable track deterioration may take place. Railway track that is too stiff can cause load concentrations, as the train load is distributed over fewer supports; this in turn can lead to increased ballast attrition and create variations in track stiffness and therefore differential settlement (Brandl 2001b; Selig and Waters 1994). Differential settlement can result in increased train-induced dynamic forces, which in turn worsen track geometry, thus accelerating the deterioration of the entire track structure. Track that is not stiff enough, however, may lead to excessive rates of settlement and various types of subgrade-related failure (see below).

The contribution of various layers to the load-bearing capacity of the track support system is shown in Figure 14.4 and discussed further in the following sections. From Figure 14.4, it can be seen that the subgrade has the most significant influence on the overall performance of the track, contributing approximately 40% of the load-bearing capacity.

The performance of various layers that make up the track bed, however, is affected by a number of factors. The most significant of these are listed Table 14.1.

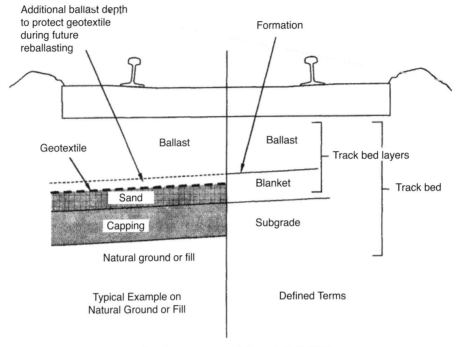

FIGURE 14.3 Definitions of track components (Network Rail 2005).

14.3 Track Bed

The load applied to the sleeper ultimately is carried by the subgrade. Good track design ensures that each of the track support layers (track bed–ballast, subballast, subgrade, and any other layers) can carry the required load such that track line and level are maintained commensurate with the planned maintenance regimen.

The function and behavior of each of the support layers are described briefly in the following sections.

14.3.1 Ballast

The rail and sleeper "ladder" frame is supported by ballast, which helps to transmit the load to the subgrade soil. Ballast provides flexible support in both the vertical and horizontal directions. The particulate nature of construction of this layer enables track to be realigned relatively easily. Ballast should be free draining and should be stable under dynamic loading (i.e., have adequate interparticle friction), and it normally is comprised of hard durable rock that complies with the following requirements.

14.3.1.1 Particle Size Distribution of Ballast

Ballast normally is comprised of particles ranging in size from 1.18 to 63 mm, with the majority of particles in the 28- to 50-mm size range. A comparison of particle sizes for British (Network Rail), German (Deutsche Bahn AG), Indian (Indian Railways), and Australian (Australian Rail Track Corporation) systems is given in Table 14.2, and the American Railway Engineering and Maintenance Right-of-Way Association (AREMA 2007) recommendations for particle size

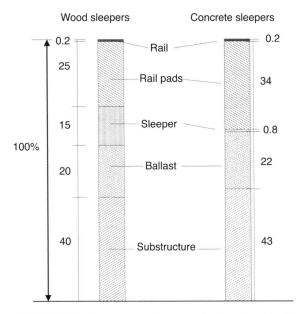

FIGURE 14.4 Average percentage contribution of each load-bearing permanent way element to overall behavior of the track (after Lichtberger 2005).

TABLE 14.1 Factors That Affect Behavior

Layer	Significant Factors That Affect Behavior
Ballast	Presence of fines, permeability
Subballast	Composition and permeability
Subgrade	Classification, compaction, and water content

After Lowson (1998).

TABLE 14.2 Comparison of Particle Size Distribution of Ballast in Europe, India, and Australia

Network Rail[a]		Deutsche Bahn AG[b]		Indian Railways[c]		Australian Rail Track Corporation[d]	
Size (mm)	Cumulative % Passing	Size (mm)	Cumulative % Passing	Size (mm)	Cumulative % Passing	Size (mm)	Cumulative % Passing
63	100	63	100	65	95–100	63.0	100
50	97–100	50	65–100	40	40–60	53.0	85–100
37.7	35–65	40	30–65	20	0–2	37.5	20–65
28	0–20	31.5	0–25			26.5	0–20
14	0–2	25				19.0	0–5
1.18	0–0.8					13.2	0–2
						4.75	0–1
						0.075	0–1

[a] Network Rail (2000) Track Ballast and Stoneblower Aggregate, Network Rail Standard NR/SP/TRK006.
[b] Lichtberger (2005).
[c] Indian Railways (2004).
[d] Australian Rail Track Corporation, Ballast Specification (ARTC 2007).

TABLE 14.3 Recommendations for Particle Size Distribution of Ballast in the United States (AREMA 2007)

Size No.[a]	Nominal Size Square Opening	Percent Passing									
		3"	2½"	2"	1½"	1"	¾"	½"	⅜"	No. 4	No. 8
24	2½"–¾"	100	90–100		25–60		0–10	0–5	—	—	—
25	2½"–⅜"	100	80–100	60–85	50–70	25–50	—	5–20	0–10	0–3	—
3	2"–1"	—	100	95–100	35–70	0–15	—	0–5	—	—	—
4A	2"–¾"	—	100	90–100	60–90	10–35	0–10	—	0–3	—	—
4	1½"–¾"	—	—	100	90–100	20–55	0–15	—	0–5	—	—
5	1"–⅜"	—	—	—	100	90–100	40–75	15–35	0–15	0–5	—
57	1"–No. 4	—	—	—	100	95–100	—	25–60	—	0–10	0–5

[a] Gradation Numbers 24, 25, 3, 4A, and 4 are main line ballast materials. Gradation Numbers 5 and 57 are yard ballast materials.

distribution of ballast are given in Table 14.3. Unlike the U.K., where only one range of sizes is acceptable, finer ballast particles are permissible in the U.S. for certain types of track (see Table 14.3).

14.3.1.2 Shape, Strength, and Durability of Ballast

In order for ballast to fulfill functional and structural requirements, it has to comply with a range of physical properties. While most countries have their own standards, the engineering properties required for ballast usually are similar. A comparison of Network Rail and Australian Railways requirements for ballast is given in Table 14.4, and U.S. and Deutsche Bahn AG requirements are given in Tables 14.5 and 14.6, respectively.

14.3.2 Subballast

The term "subballast" sometimes is synonymous with "blanket" layer. Subballast is placed between the ballast and subgrade, and it invariably is comprised of granular material with the following specific functions:

1. To prevent the ballast from punching into the subgrade. This is done by ensuring that the subballast layer is of finer gradation than the ballast.

TABLE 14.4 Properties of Ballast for the U.K. (Network Rail 2005) and Australian Railways (ARTC 2007)

		Maximum %	
		Network Rail	Australian Rail Track
Shape	Flakiness index	40	30
	Elongation index	40	30
Strength	Aggregate crushing value	22	25
Durability	Wet attrition value	4	6

TABLE 14.5 Recommended Limiting Values for Ballast Material in the U.S. (AREMA 2007)

Property	Granite	Traprock	Quartzite	Limestone	Dolomitic Limestone	Blast Furnace Slag	Steel Furnace Slag	ASTM Test
Percent material passing No. 200 sieve	1.0%	1.0%	1.0%	1.0%	1.0%	1.0%	1.0%	C 117
Bulk specific gravity[a]	2.60	2.60	2.60	2.60	2.65	2.30	2.90	C 127
Absorption percent	1.0	1.0	1.0	2.0	2.0	5.0	2.0	C 127
Clay lumps and friable particles	0.5%	0.5%	0.5%	0.5%	0.5%	0.5%	0.5%	C 142
Degradation	35%	25%	30%	30%	30%	40%	30%	[b]
Soundness (sodium sulfate) 5 cycles	5.0%	5.0%	5.0%	5.0%	5.0%	5.0%	5.0%	C 88
Flat and/or elongated particles	5.0%	5.0%	5.0%	5.0%	5.0%	5.0%	5.0%	D 4791

[a] The limit for bulk specific gravity is a minimum value. Limits for the remainder of the tests are maximum values.

[b] Materials having gradations containing particles retained on a 1-in. sieve shall be tested by ASTM C 535. Materials having gradations with 100% passing a 1-in. sieve shall be tested by ASTM C 131. Use grading most representative of ballast material gradation.

TABLE 14.6 Deutsche Bahn AG Requirements for Ballast (Lichtberger 2005)

Ballast Material	Los Angeles Test	Aggregate Impact Value	Impact Resistance	Deval Test
Bassalt	8.7–9.5	10	10.2–11.7	10.3–13.8
Porphyr	10.3	10	11.9	11.1
Sandstone	12.5	11	14	9.8
Limestone	13.7–23	15–23	16.3–21.3	5.9

2. To prevent plastic failure of the subgrade by being thick enough so that stresses from the ballast layer are reduced to levels that can be sustained by the underlying subgrade soils.
3. To prevent migration of fines into the ballast layer. In order to fulfill this function, the subballast layer may be designed as a filter layer. AREMA (2007) provides guidance on the design of the subballast layer as a filter layer based on the U.S. Bureau of Reclamation recommendations (see Table 14.7). In addition, it is recommended that the maximum particle size of the subballast should not exceed the largest ballast particle and no more than 5% of the former should be smaller than 60 micron.

TABLE 14.7 Requirements for Filter Material

Character of Filter Material	Ratio R_{50}	Ratio R_{15}
Uniform grain size distribution ($U = 3-4$)	5–10	
Well-graded to poorly graded (nonuniform) subrounded grains	12–58	12–40
Well-graded to poorly graded (nonuniform) angular particles	9–30	6–18

$R_{50} = \dfrac{D_{50} \text{ of the filter material}}{D_{50} \text{ of material to be protected}}$ $R_{15} = \dfrac{D_{15} \text{ of the filter material}}{D_{15} \text{ of material to be protected}}$

Note: Grain size curves (semilogarithmic plot) of subballast and the subgrade should be approximately parallel in the finer range of sizes.

Extracted from AREMA (2007).

Note: This table was prepared especially for earth dam design, and since its use here is for a different purpose, the values may be slightly exceeded. In the event soil in the subgrade may be subjected to piping, position and maximum percentage value of *D* for the subballast to be less than 5X D_{85} of the subgrade soil. The subgrade in this case should be well graded.

Although there are a number of formulae for the design of filters, Terzaghi's (1922) criteria, where the D_{15} size of the filter should lie between $4 \times D_{15}$ of the soil and $4 \times D_{85}$ of the soil, seems to be the most widely used.

The subballast layer thus is a subgrade protection layer, and it needs to be constructed to meet certain criteria. It must prevent seasonal variation of moisture in the subgrade, protecting it from shrinkage and swelling. To this end, the subballast layer should be comprised of material of a suitable particle size, it should be placed to adequate thickness, and it must be compacted to a suitable density. In addition, some subsoils may be susceptible to weakening due to frost action. These soils may be identified in terms of their coefficient of uniformity ($Cu = d_{60}/d_{10}$), as shown in Figure 14.5 (Lichtberger 2005). Alternatively, guidance provided by the U.S. Army Corps of Engineers (1984) relating frost susceptibility to soil particle size (shown in Figure 14.6) may be used.

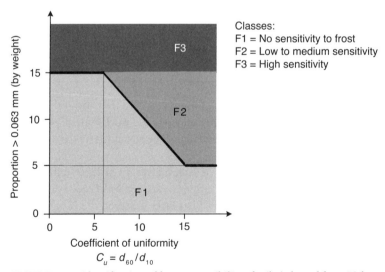

FIGURE 14.5 Identification of frost susceptibility of soils (adapted from Lichtberger 2005).

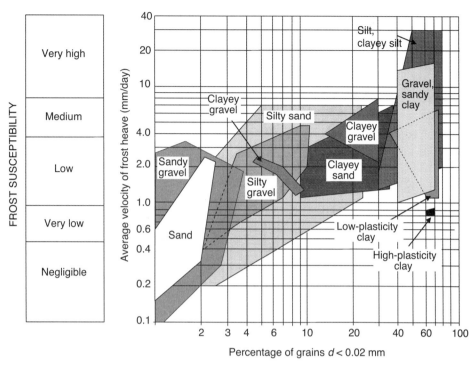

FIGURE 14.6 Relationship between size fraction below 0.02 mm and frost susceptibility of soils (after U.S. Corps of Engineers 1984).

The International Union of Railways (UIC 1994) recommends the use of Casagrande's frost susceptibility criteria. Its guideline states that the critical percentage (by weight) of particles with a diameter less than 0.02 mm is 10 and 3% for uniformly graded ($Cu < 5$) and well-graded ($Cu > 15$) soils, respectively. In addition, the guideline states that frost susceptibility of soils may be estimated from the sub-2-mm fraction of the soil, as shown in Figure 14.7.

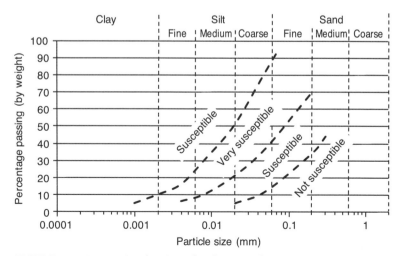

FIGURE 14.7 International Union of Railways guidance on estimation of frost susceptibility from particle size distribution (after UIC 1994).

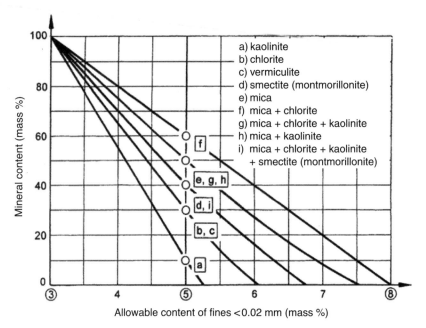

FIGURE 14.8 Mineral-composition-based criteria for nonfrost-susceptible soils: permissible sub-0.02-mm content in unbound layers in roads pavement (after Brandl 2001b).

Further, the recommendation states that frost susceptibility depends on geological conditions, mineralogy, and chemistry of soils, as well as the shape of finer particles. Brandl (2001b) has shown that some minerals are more frost susceptible than others. In general terms, he suggests that minerals such as carbonates, quartz, and feldspar exhibit neutral behavior. Minerals that show reduced frost susceptibility are essentially laminated silicates and include clays in the following groups: kaolinite, chlorite, vermiculite, and smectite. In addition, weathering results for mica and iron hydroxides are included. He provides a design chart (shown in Figure 14.8) that gives allowable mineral content for material with a diameter less than 0.02 mm in the unbound layers for use in road pavements. These guidelines can usefully be applied to railways.

Prior to placing and compacting the subballast layer to the desired density, the subgrade first should be compacted to the required density, and its surface must be planed and inclined at a suitable grade to ensure that water does not pond. An example from Indian Railways is given in Table 14.8. In addition, it is suggested that if the load is increased, the blanket thickness should be increased. Lichtberger (2005) suggests that a minimum layer thickness of 200 mm must be used where the elastic modulus of the subgrade is below 50 MN/m². In instances where the value of the modulus of earth formation drops below 10 MN/m², the subgrade may be covered with an additional protective layer, which can be dimensioned depending on train velocity (shown in Figure 14.9).

14.3.3 Subgrade

Ultimately, all the loads (static and dynamic) placed on the track by trains are carried by the subgrade. In a properly designed track foundation, the key functions of the overlying layers are to protect the track from inundation with water, the effects of weather such as frost, and excessive stresses, strains, and deformations. A general description of the subgrade and its

TABLE 14.8 Application and Thickness of Subballast for Axle Loads of Up to 22.5 t

| No Subballast Required | Thickness of Subballast Layer | | |
	0.45 m	0.6 m	1.0 m
• Rocky beds except those that are very susceptible to weathering (e.g., rocks consisting of shales and other soft rocks, which become muddy after coming into contact with water) • Well-graded gravel (GW) • Well-graded sand (SW) • Soils conforming to specifications of blanket material	• Poorly graded gravel (GP) with a coefficient of uniformity more than 2 • Poorly grade sand (SP) with a coefficient of uniformity more than 2 • Silty gravel (GM) • Silty gravel–clayey gravel (GM-GC)	• Clayey gravel (GC) • Silty sand (SM) • Clayey sand (SC) • Clayey silty sand (SM-SC) • Thickness to be increased to 1 m if plasticity index exceeds 7	• Silt with low plasticity (ML) • Silty clay with low plasticity (ML-CL) • Clay with low plasticity (CL) • Silt with medium plasticity (MI) • Clay with medium plasticity (CI) • Rocks that are very susceptible to weathering

After Indian Railways (2003).

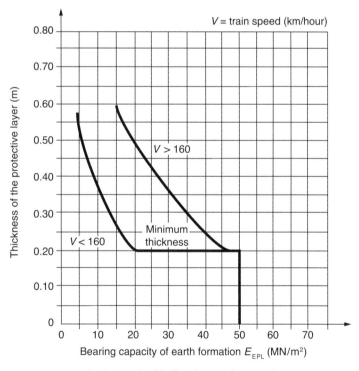

FIGURE 14.9 Thickness of subballast layer (after Lichtberger 2005).

impact on track is presented here; a more thorough description of the mechanics of the subgrade may be found in Selig and Waters (1994).

The impact of subgrade soils on the general performance of a track has been recognized by a number of researchers (Li and Selig 1998a; Selig and Waters 1994; Lichtberger 2005). The key aim of track design is to ensure that the stiffness of the subgrade layer is consistent and lies within an acceptable range of values. Variable and low stiffness results in increased maintenance to ensure adequate track geometry. It also may result in reduction in line speed. If the subgrade is excessively stiff, then measures may have to be taken to reduce it, such as introducing a layer below the subballast.

14.4 Failure of Rail Sleeper Support System

14.4.1 Ballast Failure

The contamination of the ballast by a variety of materials causes it to lose its functional and structural integrity. The contamination may result from the attrition of ballast under the action of repeated loading, the migration of fines from the subgrade, spillage of products carried by the trains, and it may be wind blown. Typical sources of contamination are listed in Table 14.9, from which it can be seen that most of the fines arise from the degradation of the ballast itself.

It is worth noting that Network Rail (2005), in its code of practice on formation treatments, states that ballast degradation, where ballast breaks down due to the mechanical action of both traffic and maintenance, is the foremost cause of track problems in the U.K. The second most common cause of failure relates to the migration of fines from the subgrade soil into the ballast.

The degree of ballast degradation can be measured using a fouling index (F_1). Tung (1989) proposes the following relationship between F_1 and the percentage of various materials passing two different sieve sizes:

$$F_1 = P_4 + P_{200}$$

where P_4 = percentage passing a 4.75-mm sieve and P_{200} = percentage passing a 0.075-mm sieve. Fouling categories based on the fouling index are shown in Table 14.10.

The angular nature of ballast gives it high interlock, resulting in an internal friction angle that may be as high as 65°. However, any contamination can result in a reduction of the angle of internal friction, leading to reduced shear strength and giving a lower bearing capacity. Furthermore, ballast contamination results in the reduction of the angle of spread of load. If there is a gap in the pressure footprint at the subgrade level, then plastic flow of material from the area of subject to higher pressures can occur. For example, for a 600-mm sleeper spacing at a ballast depth of 300 mm, if the load spread is 45°, the pressure at the ballast/subgrade layer may be considered to be nearly uniform. If the angle of load spread is reduced to 30°,

TABLE 14.9 Typical Sources of Ballast Contamination

	Selig and Waters (1994)	Sharpe (2005)
Ballast	76%	0.2 kg/sleeper/MGT[a]
Underlying granular layer	7%	
Surface	3%	4 kg/m²/yr (coal fines)[b]
Sleeper	1%	
Tamping		4 kg/tamp/sleeper[a]
Airborne		0.2–10 kg/sleeper/yr[c]

[a] Depending on ballast type. MGT = million gross tonnes.
[b] For example, coal spillage near power station.
[c] Depending on the area.

then there will be a 154-mm-wide strip at the subgrade/ballast surface between the sleepers that will not be subjected to any pressure from a passing train. Thus, it is possible that, under repeated load from a passing train, plastic flow of soil in the central, unloaded area can occur, especially under wet conditions. For a lower load spread of 30°, either about 520-mm-deep ballast may have to be used or the sleepers may have to be positioned at reduced spacing in order to achieve a near uniform pressure distribution.

TABLE 14.10 Ballast Categories and Fouling Index

Category	Fouling Index (F_1)
Clean	<1
Moderately clean	1–<10
Moderately fouled	10–20
Fouled	20–<40
Highly fouled	≥40

After Tung (1989).

14.4.2 Subgrade Failure

In general terms, the failure of the track bed may be defined as its inability to maintain line and level. The causes of failures can be related to subgrade type, groundwater condition, depth of construction, loading, and speed, among other factors. A summary of various types of subgrade failures and their causes is given in Table 14.11 (Selig and Waters 1994). The first four types

TABLE 14.11 Major Subgrade Problems

Type	Causes	Features
Progressive shear failure	• Repeated overstressing of subgrade • Fine-grained subgrade soils • High water content	• Squeezing near subgrade surface • Heaves in crib and/or shoulder • Depression under ties
Excessive plastic deformation (ballast pocket)	• Repeated loading • Soft or loose soils	• Differential subgrade settlement • Ballast pockets
Attrition with mud pumping	• Repeated loading of subgrade by ballast • High ballast:subgrade contact stress • Clay-rich rocks or soils • High water contact at subgrade surface	• Muddy ballast • Inadequate subballast • Poor ballast drainage
Liquefaction	• Repeated loading • Saturated silt and fine sand	• Large displacement • More severe with vibration • Can happen in subballast
Massive shear failure (slope stability)	• Weight of train, track, and subgrade • Inadequate soil strength	• High embankment and cut slope • Caused by increased water content
Consolidation settlement	• Embankment weight • Saturated fine-grained soils	• Increased static soil stress as in newly constructed embankment
Frost action (heave and softening)	• Periodic freezing • Frost-susceptible soils	• Occurs in winter/spring period • Rough track surface
Swelling/shrinkage	• Highly plastic soils • Changing moisture content	• Rough track surface
Slope erosion	• Running surface and subsurface water • Wind	• Soil washed or blown away
Soil collapse	• Water inundation of loose soil deposits	• Ground settlement

After Selig and Waters (1994).

of failure primarily are due to repeated traffic loading, the next two types are due to self-weight, and the remaining problems are due to environmental factors. Modes of failure associated with repeated dynamic loading, which is considered to be a major source of problems for poorly designed track, are summarized in Table 14.11.

Fine-grained cohesive soils with high moisture contents are particularly problematic when they are subjected to repeated dynamic loading (Li and Selig 1995). Associated track problems manifest themselves in the form of the migration of fines from the subgrade soil into the overlying ballast (as described above), the progressive shear failure of soil initiated under the heavily loaded parts of the sleeper, and heave at the track side. Plastic deformation of the soil under a sleeper leads to the formation of an uneven subgrade surface, resulting in the formation of pockets that may act as reservoirs for water. Plastic deformation and the formation of uneven subgrade surface are shown in Figures 14.10 and 14.11, respectively.

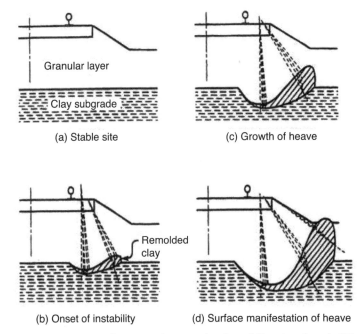

FIGURE 14.10 Development of progressive shear failure in subgrade (Li and Selig 1995).

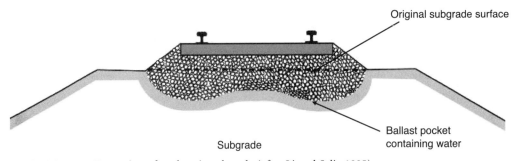

FIGURE 14.11 Formation of pockets in subgrade (after Li and Selig 1995).

FIGURE 14.12 Migration of fines from subgrade into ballast.

In terms of the migration of fines, it is commonly believed that the phenomenon occurs due to excessive repeated loading of soft subgrade soils. This is not always the case, as can be seen in Figure 14.12, which shows a section of track at a railway station that has started to show signs of fines migration within 18 months of renewal.

As it also is generally accepted that plastic failure is associated with softened subgrade, it is therefore essential to ensure that adequate drainage is maintained throughout the life of the track. For example, Ghataora et al. (2006) have shown the adverse effect of poor drainage on the strength of subgrade materials (see Figure 14.13).

In terms of permanent settlement, Freeme and Servas (1985) have shown that the inundation of fine soils leads to increased deformation compared to granular soils (Figure 14.14).

It is worth noting that certain soil types are more prone to specific types of problems. AREMA (2007) compiled a comprehensive list of soil groups and applications, including

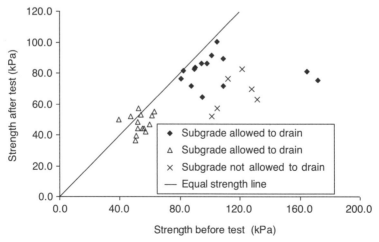

FIGURE 14.13 Effect of drainage on the strength of subgrade (Ghataora et al. 2006).

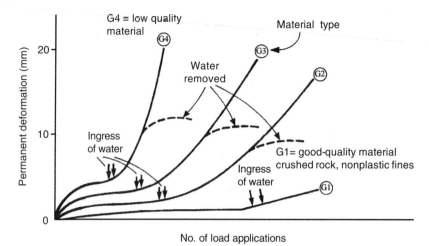

FIGURE 14.14   Effect of change in water content on permanent deformation of granular materials (adapted from Freeme and Servas 1985).

TABLE 14.12   Failure Modes Associated with Soil Groups

Failure Mode	Problem Soil Groups	Soil Groups That Pose Slight to No Problem
Pumping	Clay and organic soils are considered to be the worst *Soil groups*: ML, CL, MH, CH, OH, and PT	Essentially granular soils ranging in size from silt to gravel *Soil groups*: GW, GP, GM, GC, SW, SP, SM, and SC
Frost heave	Essentially silts, clays, and organic soils *Soil groups*: ML, CL, MH, CH, OH, and PT	Essentially granular soils ranging in size from silt to gravel *Soil groups*: GW, GP, GM, GC, SW, SP, SM, and SC

Adapted from Li and Selig (1995).

identification of problems. Table 14.12, an extract from the AREMA list, shows soils that are prone to pumping and frost action.

14.5 Track Bed Remediation

Ideally, the railway track system should be designed so that its various components do not fail under the action of repeated loads. However, this ideal situation is difficult to achieve in practice, as many railway lines are used well beyond their intended design life and also are subjected to loads and speeds for which they were not originally designed. In such cases, the track bed may exhibit the various signs of degradation as described above. While the ballast material lends itself to maintenance, the subgrade is more problematic. Where degradation of the latter occurs, various remediation techniques may be used, as summarized in Table 14.13.

However, if soils are found to be have inadequate properties for supporting the railway track, it may be necessary to stabilize them using a variety of remedial methods. Stabilization of track can be divided into three categories: drainage, mechanical stabilization, and chemical

TABLE 14.13 Track Bed Problems and Possible Remediation Techniques

Problem	Possible Remedy
Frost susceptible	• Add adequate thickness of cover layer • Replace with frost-resistant material
Exhibits excessive settlement	• Densification • Stabilization • Drainage • Lime/cement piles • Concrete piles
Susceptible to pumping	• Use subballast layer • Use sand blanket with geotextile at the ballast/sand interface • Replace upper layers of subgrade with suitable material • Soil stabilization (with lime and/or cement or other compounds) • Use geocomposites (only a few are designed to completely replace the sand layer)
Resilient modulus	• Compaction together with suitable drainage • Soil stabilization

Adapted from AREMA (2007).

stabilization. These measures are described in the following sections for track (stability of embankments or other earthworks is not included).

14.5.1 Drainage

The ability of soil to support a load, in terms of bearing capacity and limiting settlement, is reduced with increases in its moisture content. It is necessary, therefore, to ensure that any new drainage system is designed adequately and that older track not only is maintained but is reviewed periodically to take into account changes in land use and climate (Hay 1982). For example, Freeme and Servas (1985) showed the effect of changes in water content in terms of increase in permanent deformation of granular materials in road pavement (see Figure 14.14). Further, Hornych et al. (1998) showed that increase in moisture content results in a decrease in resilient modulus and an increase in plastic strain (see Figure 14.15).

Cedergren (1987) investigated the effect of saturation of a road pavement on its useful life. His findings, shown in Figure 14.16, suggest that if the pavement is saturated for only about

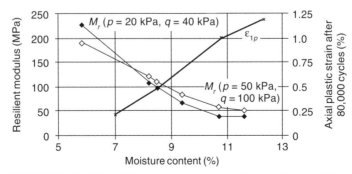

FIGURE 14.15 Effect of increase in moisture on both resilient modulus and plastic strain (Hornych et al. 1998).

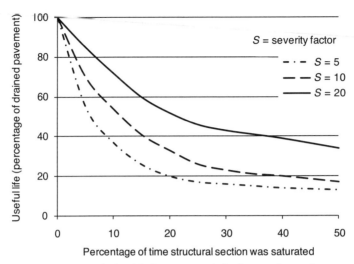

FIGURE 14.16 Relationship between period of saturation and pavement life (after Cedergren 1987). (Severity factor is the anticipated damage during the wet period relative to the dry period.)

10% of the time (5 weeks per annum), then there is an approximate 50% reduction in pavement life. These findings are equally applicable to railway tracks, and thus the importance of maintaining low moisture in the track support layers is clear.

The track bed may be comprised of both granular (ballast and subballast) and fine-grained materials (e.g., clayey subgrade). The two types of materials show very different responses to increases in moisture, but fine-grained materials are affected most significantly. In general, increases in the water content of the track support layers can result in the following problems:

- Loss of strength, particularly of fine-grained soils
- Softening of fine-grained subgrade soils (particularly clays) can result in plastic failure and reduction in resilient modulus, both of which lead to increased deformation and hence loss of track geometry and increased generation of fines due to ballast attrition
- Fine-grained soil can migrate into the overlying ballast (mud pumping)
- An increase in volume of soils prone to volume change

Water enters the railway track system from the following sources:

- Precipitation (rainfall)
- Surface flow (water entering the track system from the sides)
- Rising groundwater
- Capillary water

Water from precipitation, surface flow, and groundwater is influenced by gravity, particularly in granular soils, and may be removed by a suitable trench drain. Capillary water is influenced by pore size, and Cedergren (1989) suggests that drains should be installed to keep the free water surface approximately 1.6 m below the top of the subgrade.

Most of the drainage systems designed for railways are intended for surface water and gravitational water in soils. The effect of capillary water normally is taken into account in design implicitly by using the soaked strength of subgrade materials.

In order to design a suitable drainage system for a railway, first it is necessary to estimate the amount of water entering the system by the four processes listed above. Hay (1982) and ARTC (2006) describe the use of a rational method that takes into account the drainage area, the intensity of rainfall, and a runoff factor. For the design of surface runoff, AREMA (2007) lists 28 factors that need to be considered and in addition to the above includes a Soil Conservation Service Curve Number. Having determined the total flow, then using Manning's formula it is possible to determine the suitable drain required to drain the surface water in an open channel or a pipe, as follows:

$$Q = \left(\frac{1}{n}\right) \times A \times R^{0.67} \times S^{0.5}$$

where Q = flow (m³/s), n = roughness coefficient, A = cross-sectional area (m²), R = wetted perimeter, and S = slope of drain.

For aggregate fill, the cross-sectional area may be estimated using Darcy's equation:

$$Q = k \times i \times A$$

where Q = flow (m³/s), k = permeability of the aggregate (for an aggregate 20–60 mm in size, k may range from 0.1 to 1 m/s), i = hydraulic gradient, and A = cross-sectional area (m²).

Where possible, drains should be at a gradient of between 1:200 to 1:100 such that they are self-cleaning.

It is worth noting that ARTC (2007) class 1 track is designed for a 25-ton axle load (maximum), and drains should be designed for a 1-in-50-year storm return. For lower classification track, drains are designed for an average storm return period of as low as 5 years.

The function of the subsurface drains is to lower the water table under the track to an acceptable level. Often, these drains are positioned next to the track, in an area known as the cess, and are comprised of slotted pipes bedded in granular material in a trench. The granular material often is wrapped in a geotextile to prevent the fines from the surrounding soil from clogging up the pipe surround. However, the geotextile needs to be designed with care since filter cake can form on its outer surface and prevent its proper function.

While it is important to size the drains in terms of capacity and plan their layout, many railway organizations have standards that describe drainage systems for railways. As an example, the Indian Railways Geotechnical Engineering Directorate (Indian Railways 2003) specifies the use of trench backfill of a specific particle size depending on the nature of the surrounding material (see Table 14.14). An example of a drain specified by Network Rail (2005) for use where a sand blanket is installed is shown in Figure 14.17.

It should be noted that for subsurface drains to be effective, it is essential to ensure their continuity between the undertrack drainage layers and subsurface drains. Where the side drain is located in the cess area, it is essential to ensure that the ballast shoulders are periodically cleaned to allow water to flow away from underneath the track.

14.5.2 Mechanical Stabilization of Subgrade Soils

14.5.2.1 Compaction

The bearing capacity of some soils may be improved through compaction, by packing together particles of soil, reducing void space, and increasing the solid content per unit volume. In

TABLE 14.14 Trench Backfill Dependency on Material in Which Trench Is Made

Sieve Size (mm)	Backfill Grading for Trench Surround Material		
	Fine Silt/Clay	Coarse Silt to Medium Clay	Gravely Clay
53	—	—	100
45	—	—	97–100
26.5	—	100	—
22.4	—	95–100	50–100
11.2	100	48–100	20–60
5.6	92–100	28–54	4–32
2.8	83–100	20–45	0–10
1.4	59–96	—	0–5
0.71	35–40	6–18	—
0.355	14–40	2–9	—
0.18	3–5	—	—
0.09	0–5	0–4	0–3

Adapted from Geotechnical Engineering Directorate, 2003, Ministry of Railways, India.

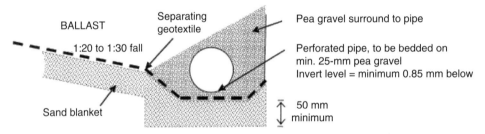

FIGURE 14.17 Integral drain where sand blanket is installed (Network Rail 2005).

general, compaction results in improvements in strength, volume stability, and reduction permeability. The compaction of soils is described in detail by Hausmann (1990) and with particular reference to roads and railway by Brandl (2001b, 2001c).

It is conventional wisdom to ensure that the subgrade is compacted to at least 95% of its maximum dry unit weight in the upper layers, perhaps the top 1 m. Below this, compaction equivalent to 90% of maximum dry unit weight may be adequate. Careful consideration needs to be given to the behavior of the compacted soils under repeat loading for a variety of moisture contents.

14.5.2.2 Use of Geotextiles, Geogrids, and Geocomposites

Geotextiles, geogrids, and composites have successfully been incorporated into railway track over the last 30 years to fulfill a range of roles that include separation, filtration, reinforcement, and drainage. Many examples of such applications are available, and only materials that affect the performance of the track are discussed here: those placed at the ballast subgrade interface. Their principal role is to prevent the migration of fines from susceptible subgrades (particularly those made from clay) to the overlying ballast layer. Some of these materials also may affect the stiffness of track. However, there is little published information on this aspect, and there is no satisfactory method of designing these separators. The choice of material is invariably based on laboratory and field trials. In the U.K., Network Rail has a range of standard solutions for existing track, and it categorizes the use of geosynthetics in conjunction

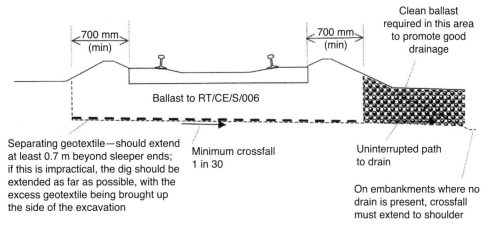

FIGURE 14.18 Example of standard application of geotextile for remediating track (Network Rail 2005).

with their standard solution. An example of one such solution is given in Figure 14.18 (Network Rail 2005). As these solutions are intended for use in the U.K., some care is required in using them elsewhere.

14.5.3 Chemical Stabilization

Chemical stabilization of railway track subgrade may be undertaken to improve the engineering properties of soils. The most commonly used techniques involve the use of either lime or cement or a combination of both.

Lime can be used to improve workability (it makes soft soils firmer and less moisture sensitive) and can result in increases in strength and volume stability. Quicklime normally is used for soil stabilization, since the hydration of lime results in reduction of moisture content of the soils and the heat of hydration during slaking helps to accelerate the cementitious reaction. Lime stabilization is considered to be suitable for stabilizing clayey soils with a plasticity index greater than 10% and clay content greater than 10%. For soils with plasticity less than 10%, cement stabilization may be used. Often, both lime and cement are used. In such cases, the application of lime is followed by the use of cement, where lime is used as a modifier to improve the workability of the soils.

The presence of sulfates in soils has a deleterious effect on lime-stabilized soils. Soluble sulfates below 0.3% do not present a risk, and concentrations higher than 0.8% are considered unacceptable in the U.K. (National Lime Association 2001). Guidelines on acceptable sulfate content vary in other countries.

14.6 Comparison of Design Methods*

The structure of a conventional railway track, described in Section 14.3, should be designed to withstand the damaging effects of railway traffic and climate, so that the subgrade is adequately protected and that vehicle operating costs, safety, and passenger comfort are kept

*This section is based on an article published by the authors in the *Journal of Rail and Rapid Transit* (Burrow et al. 2007b) and is reproduced in part here by kind permission of the journal.

within acceptable limits during the design life (Burrow et al. 2004; McElvaney and Snaith 2002).

The cumulative effect of repeated traffic loads deteriorates the track substructure over time. However, while the ballast lends itself to periodic maintenance to adjust track line and level, subgrade-related problems are less easily rectified. Consequently, a primary objective of design is to protect the subgrade from the types of failure described in Section 14.4.2 and summarized in Table 14.11. Of these, track problems related to subgrade attrition, progressive shear failure, and an excessive rate of settlement through the accumulation of plastic strain are associated with the uppermost part of the subgrade, where cyclic shear stresses are likely to be at their highest. Attrition may be prevented using an appropriately thick sand blanket layer, and progressive shear failure occurs at stress levels below that, causing massive shear. Therefore, foundation design procedures should explicitly prevent progressive shear failure and excessive plastic deformation. Several approaches may be adopted to help prevent these failure modes, including using nonballasted track forms, introducing an asphalt layer, increasing the flexural rigidity (EI) of the rail, and using techniques such as soil stabilization to permit higher stresses (Stirling et al. 2003). Usually, however, the use of track bed layers of appropriate thickness is likely to be effective and economical (Li and Selig 1998a).

To this end, there are a number of design procedures, including standards issued by infrastructure operators and research published in the literature. As the structural properties of the ballast and subballast layers are similar, such procedures usually recommend a single thickness for the track bed layers, and the proportion of ballast and subballast is not specified. As ballast is more expensive than subballast material, it is assumed that a minimum thickness of ballast, usually between 0.2 and 0.3 m, will be used to facilitate maintenance operations which are carried out periodically to readjust the line and level of the track.

A comparison of six design procedures under several theoretical operating conditions is presented below. Those considered are (1) from the U.S., a method proposed by Li et al. (1996); (2) from Europe, the International Union of Railways Standard UIC 719 R (UIC 1994); (3) from the U.K., a method developed by British Rail Research (Heath et al. 1972) and (4) the current Network Rail code of practice (Network Rail 2005); (5) from India, the Indian Ministry of Railways guidelines (Indian Railways 2004); and (6) from Japan, the West Japan Railway Company standards for high-speed and commuter lines (WJRC 2002a, 2002b).

14.6.1 Design Procedures

14.6.1.1 Li et al. Method

The method proposed by Li et al. (1996) aims to prevent both progressive shear failure and excessive plastic deformation. This is achieved by limiting the stresses in the subgrade such that plastic strain is of an acceptable level. Subgrade stresses are determined using an analytical model of the track system, whereas the allowable stresses are determined from an equation that relates plastic strain to the number of loading cycles. For design purposes, the track bed is considered to be a single homogeneous granular layer.

A three-dimensional, multilayer elastic model known as GEOTRACK (Selig and Waters 1994) was built to determine the subgrade stress distribution under various traffic loadings. The model simplifies the track substructure as a single granular layer overlying a homogeneous subgrade. To account for the increase in track loading that results from track and vehicle irregularities, Li et al. suggest that dynamic loads should be used. Where this information is unavailable, they prescribe the use of the following empirical equation, suggested by the American Railway Engineering Association (AREA), to modify static wheel loads:

$$K = 1 + \frac{0.0052 \times V}{D} \tag{14.1}$$

where K is the ratio of dynamic to static wheel loads, V is the train speed (km/h), and D is the wheel diameter (m).

To determine allowable plastic strains and deformations under repeated loading, cyclic load triaxial tests were conducted on various fine-grained soils (Li and Selig 1996). From these tests, it was found that the subgrade cumulative plastic strain (ε_p) could be related to soil deviator stress (σ_d) and the number of repeated stress applications (N) as follows:

$$\varepsilon_p\,(\%) = a \left(\frac{\sigma_d}{\sigma_s} \right)^m N^b \tag{14.2}$$

where σ_s is the compressive strength of the soil and a, b, and m are parameters dependent on the soil type. Integrating over the depth of the deformable part of the subgrade, the total cumulative deformation can be determined as:

$$\rho = \int_0^T \varepsilon_p dt \tag{14.3}$$

where T is the subgrade layer depth in meters.

For design purposes, Li et al. suggest that ε_p and ρ should be limited to 2% and 25 mm, respectively. These values are used for the comparisons described below.

Equations 14.2 and 14.3, together with GEOTRACK, were used to produce two sets of design charts. The charts in the first set give a minimum thickness of the track bed layers to prevent progressive shear failure and are functions of track bed layer and subgrade-resilient moduli (defined as the repeated deviator stress divided by the recoverable [resilient] axial strain), soil type, and traffic loading. The charts in the second set, which additionally are a function of subgrade depth, give thickness of the track bed layers to prevent excessive plastic deformation.

14.6.1.2 International Union of Railways Method

The International Union of Railways (UIC) Code UIC 719 R (UIC 1994) is a set of recommendations for the design and maintenance of the track substructure. Specifications are given for a single thickness of the ballast and subballast (i.e., track bed layers) and for the prepared subgrade (Figure 14.19). UIC 719 R specifies that the substructure may contain some or all of the following layers: ballast, a granular subballast, a geotextile, and a prepared subgrade (Figure 14.19).

The combined thickness of the granular layer (i.e., track bed layers) is determined from the type of soil forming the subgrade, traffic characteristics, track configuration, and quality and thickness of the prepared subgrade. No information is given on how the individual thicknesses

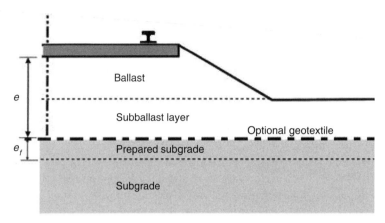

FIGURE 14.19 Calculation of the minimum thickness of the track bed (after UIC 1994).

of the ballast and subballast should be determined. The prepared subgrade is the layer below the subballast which has been treated to improve its engineering properties. Its inclusion in the design is optional, unless the subgrade requires improvement (see below). A geotextile also may be used.

The type of soil forming the subgrade is classified according to a simple system based primarily on the percentage of fines in the soil. There are four quality classes of soil: QS0 for soil that is deemed to be unsuitable without improvement; QS1 for "poor" soils that are considered acceptable in their natural condition subject to adequate drainage and mainte-nance, although improvement should be considered; QS2 for soils of "average" quality; and QS3 for soils that are considered to be "good." Poorer quality soils require thicker track bed layers.

To characterize the traffic using a line, the specifications suggested in UIC 714 (UIC 1989) are used. UIC 714 classifies a particular line as a function of the tonnage hauled, tonnage of tractive units, line speed, traffic mix (i.e., freight and/or passenger), and wear effects of vehicles. According to the classification determined using UIC 714, lines that carry faster and heavier traffic are required to have thicker track bed layers.

14.6.1.3 British Rail Research Method

British Rail Research developed a method that sought to protect against subgrade failure by excessive plastic deformation (Heath et al. 1972). To this end, a series of design charts were produced to relate the required thickness of the track bed layers to a measure of the strength of the subgrade, known as the threshold stress. The charts were developed by combining traffic-induced subgrade stresses predicted from a linear elastic model of the track system with soil threshold stresses determined by laboratory testing.

A single-layer elastic model of the track (i.e., the track bed layers and subgrade are treated as homogeneous) was developed to predict the stress distribution in the subgrade for various assumed sleeper spacings and contact pressures. Measurements of stresses at a site on the U.K.'s East Coast Main Line were used to verify the model.

In order to determine a suitable material parameter for use in design, a series of cyclic triaxial compression tests were performed on London Clay. The results of the tests indicated the existence of a threshold stress, above which repeated load applications cause large permanent deformations that increase exponentially with the number of loading cycles. Below this threshold stress, the plastic strain associated with each load cycle reduces until a stable condition is reached, where the permanent deformations are small.

14.6.1.4 Network Rail Code of Practice

Recommendations for the thickness of the track bed layers in the U.K. network are incorporated in the Network Rail Code of Practice NR/SP/TRK/9039: Formation Treatments (Network Rail 2005). The code recognizes that the condition of the railway substructure affects track geometry and maintenance requirements. Based on this premise, and where track geometry has been adequate in the past without the need for excessive maintenance, the code suggests that the subgrade possesses adequate strength and stiffness. Where this has not been the case, the required thickness of the track bed layers can be determined from a chart given in the code.

The chart relates the required thickness of the track bed layers to undrained subgrade modulus (or Young's modulus) for three different values of the dynamic sleeper support stiffness (30, 60, and 100 kN/mm per sleeper end). The values of the dynamic sleeper support stiffness relate to minimum requirements for existing main lines both with and without geogrid reinforcement and new track, respectively.

No technical details of how the chart was derived are given, although the document states that it was "derived using a combination of empirical data and multilayer elastic theory."

14.6.1.5 Indian Railways Method

The Indian Railways (2004) method is a set of guidelines provided by the Indian Ministry of Railways. The guidelines specify that the substructure should consist of a ballast layer, together with a subballast layer (known as a blanket layer).

While no recommendation is given in the Indian Railways 2004 publication, the thickness of the ballast layer in Indian railways is between 0.15 and 0.25 m in the majority of lines and up to 0.3–0.35 m in newer heavily trafficked lines (http://www.irfca.org/).

The Indian guidelines describe the following functions of the subballast:

1. Reduction of traffic-induced stresses to a tolerable limit on the top of subgrade, thereby preventing subgrade failures under adverse critical conditions of rainfall, drainage, track maintenance, and traffic loadings.
2. Prevention of the penetration of ballast into the subgrade and also prevention of upward migration of fine particles from the subgrade into the ballast under adverse critical conditions during service.
3. Facilitate drainage of surface water and reduce moisture variation in the subgrade, thereby reducing track maintenance problems.
4. Prevention of mud pumping by separating the ballast and subgrade soil. Thus, accumulation of negative pore water pressure in the soil mass, which is responsible for mud pumping, is avoided.
5. The appropriate thickness of the blanket layer is specified for axle loads of up to 22.5 t according to the predominant soil type in the uppermost 1 m of the underlying subgrade. Table 14.15 summarizes the required thickness of the blanket layer.

14.6.1.6 West Japan Railway Method

West Japan Railway Company (WJRC) has issued construction and maintenance standards for Shinkansen and commuter lines (WJRC 2002a, 2002b). The Shinkansen lines are of standard

TABLE 14.15 Application and Thickness of Subballast for Axle Loads of Up to 22.5 t

	Thickness of Subballast Layer		
No Subballast Required	0.45 m	0.6 m	1.0 m
• Rocky beds except those that are very susceptible to weathering (e.g., rocks consisting of shales and other soft rocks, which become muddy after coming into contact with water) • Well-graded gravel (GW) • Well-graded sand (SW) • Soils conforming to specifications of blanket material	• Poorly graded gravel (GP) with a coefficient of uniformity more than 2 • Poorly graded sand (SP) with a coefficient of uniformity more than 2 • Silty gravel (GM) • Silty gravel–clayey gravel (GM-GC)	• Clayey gravel (GC) • Silty sand (SM) • Clayey sand (SC) • Clayey silty sand (SM-SC) • Thickness to increase to 1 m if plasticity index exceeds 7	• Silt with low plasticity (ML) • Silty clay with low plasticity (ML-CL) • Clay with low plasticity (MI) • Rocks that are very susceptible to weathering

After Indian Railways (2004).

TABLE 14.16 Required Depth of Track Bed Layers for the West Japan Railway Company

Line	Annual Tonnage (MGT[a]/yr)	Required Track Bed Layer Depth (mm)
Shinkansen	NA	300
Commuter lines	$10 \leq$ MGT	250
	$10 >$ MGT	200

[a] MGT = million gross tonnes.

gauge (i.e., 1435 mm) and are dedicated to high-speed passenger trains operating at average speeds of 200 km/h. The commuter lines, on the other hand, use a narrow gauge (1067 mm) and may carry mixed traffic. For both types of line, the required depth of the track bed layers is given in Table 14.16. The substructure is assumed to have a bearing capacity (σ_b) of 288 kPa, and where it is less than this value, ground improvement is required. (Note that a bearing capacity of 288 kPa equates to a compressive strength [σ_s] of approximately 112 kPa, assuming a cohesion model plastic solution to a simple strip footing where $\sigma_b = 2.57\sigma_s$).

14.6.2 Comparison of Design Procedures

A comparison of the design methods was made by determining the combined thickness of the track bed layers specified by each method under a variety of conditions relating to:

- Subgrade
- Axle load
- Speed
- Cumulative tonnage

A summary of the factors accounted for in these comparisons is given in Table 14.17, and the results are presented in Figures 14.20–14.23, respectively. For Indian Railways, it was assumed that a 300-mm layer of ballast is used in addition to the specified blanket layer thickness.

TABLE 14.17 Factors Accounted for in the Design Procedures Reviewed

	Li et al.	UIC 719 R	British Rail	Network Rail Code 039	Indian Railways	WJRC
Static axle load	From GEOTRACK model used to formulate their design charts	Yes	From an elastic model—charts only go up to an axle load of 24 t	No—but 25.4-t axle load limit on U.K. network	No	No
Sleeper type, length, and spacing	Via GEOTRACK	Yes	No difference in stresses found for sleeper spacings of 630–790 mm	No	No	No
Rail section	Via GEOTRACK	No	No	No	No	No
Speed	By using a dynamic axle load (can use the AREA equation)	Yes	No—field results showed response was quasi-static up to 100 km/h, but could be incorporated by using a dynamic axle load	Via minimum requirements for the dynamic sleeper support stiffness; also, 125 mph is fastest speed on U.K. network		Crude variation—Shinkansen has greater depth than commuter lines
Annual tonnage	Yes	Yes	No	No	No	For commuter lines only
Cumulative tonnage	From annual tonnage multiplied by the design life	No	No	No	No	No
Subgrade condition	Charts are provided for different subgrade types in terms of the resilient modulus and soil strength	Yes (using soil quality determined primarily from the number of fines in the soil)	Using a threshold stress for the material in question	Undrained subgrade modulus or undrained shear strength soil	Yes (using soil classification)	Bearing capacity of subgrade assumed to be 288 kPa; otherwise ground improvement must be carried out

14.6.2.1 Subgrade

For the study, the subgrade was assumed to be a clay soil with a high percentage of fines and high plasticity and is typical of problematic soils in the U.K. The condition of the soil was represented by its resilient modulus, and it was assumed that it could vary from 15 to 100 MPa depending on seasonal variations and water content. As some of the procedures use measures of soil condition other than the resilient modulus, it was necessary to have a means of determining the resilient modulus from these measures in order to be able to compare the procedures. For the clay considered herein, an empirically founded relationship between the resilient modulus E_s and the ultimate compressive strength (Li et al.'s method) σ_s of the form $E_s \approx 250 \times \sigma_s$ was used (see Selig and Waters 1994). The threshold strength (British Rail method) was related to the resilient modulus by assuming that it was equal to half of the compressive strength (Selig and Waters 1994). European Standard UIC 719 R and Indian Railways do not give guidelines that relate the measure of soil quality used in the standard to any engineering measures of soil performance, such as the resilient modulus or strength. Consequently, it was assumed that the quality of the subgrade was class QS1 and that the subgrade had not been prepared. Young's modulus, necessary for the Network Rail code, was assumed to be equal to the resilient modulus, which is a more common means of expressing the modulus of materials subjected to repeated applications of stress. For the WJRC standard, the subgrade was assumed to have been improved to the minimum required bearing capacity of 288 kPa (see above).

In terms of traffic loading, two different scenarios were considered. Both assumed a mixture of 50% freight and 50% passenger traffic, the characteristics of which were representative of a Class 66 locomotive pulling fully laden wagons traveling at 125 km/h with axle loads of 250 kN and a high-speed locomotive-hauled passenger train with an axle load of 170 kN (Fox et al. 2004). However, the passenger train was assumed to travel at 200 km/h for one scenario and at 300 km/h for the other. These were considered to be representative of conditions on a main line in the U.K. and a high-speed line such as the Channel Tunnel Rail Link (CTRL), respectively. While Li et al.'s method and UIC 719 R account for mixed traffic, the other procedures do not. Consequently, for these procedures, a train with a 250-kN axle load traveling at 200 km/h and another traveling at 300 km/h were used to represent the traffic. A design life of 60 yr was used with a design loading of 900 MGT (million gross tonnes, i.e., 15 MGT/yr for 60 yr), as this is similar to the CTRL (Lord et al. 1999). The results of the study are shown in Figure 14.20. For the case of a passenger train traveling at 300 km/h, only the results using Li et al.'s procedure are shown, as the thicknesses of track bed layers determined using the other four methods are the same for a passenger train traveling at 200 km/h compared to 300 km/h.

14.6.2.2 Axle Load

The axle load study was carried out for a design on a clay subgrade with a resilient modulus of $E_s = 40$ MPa, and the results of the study are shown in Figure 14.21. The relationships described above for the subgrade condition study were used to determine the other required measures of soil strength. To simulate freight traffic, a train speed of 125 km/h was chosen, with axle loads varying from 140 to 350 kN; the latter figure is just above the current 343-kN (35-t) limit in the U.S. The current axle load limit is 250 kN (25.4 t) in the U.K. and 221 kN (22.5 t) in India, respectively. Therefore, for the British Rail, Network Rail, and Indian Railways comparisons, the load was limited to these respective values. The design life was chosen to be 60 yr, with a design loading of 900 MGT.

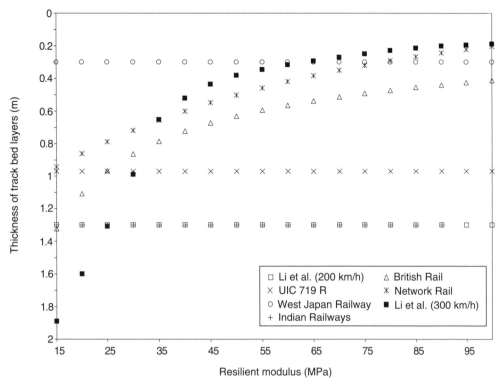

FIGURE 14.20 Variation in design thickness with subgrade condition.

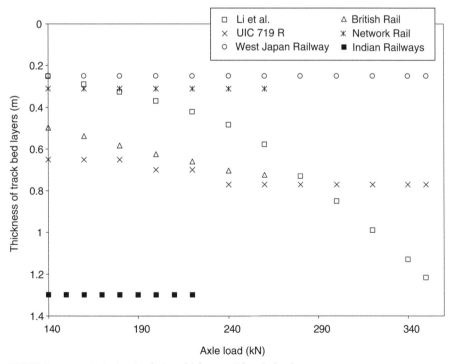

FIGURE 14.21 Variation in design thickness with axle load.

For the Network Rail procedure, the desired dynamic sleeper end stiffness was assumed to be 100 kN/mm per sleeper end, as this corresponds to a line speed greater than 160 km/h (Network Rail 2005). The WJRC standards for commuter traffic were used in the study, as the chosen line speed was 125 km/h (cf. Table 14.17).

14.6.2.3 Speed

The study of design thickness with speed used a high-speed locomotive with a 170-kN axle load (this is similar to the Eurostar high-speed trains operating on the CTRL). The required thickness of the track bed layers was determined for speeds between 80 and 350 km/h (see Figure 14.22).

For WJRC, design thicknesses appropriate to Shinkansen trains were used for speeds above 200 km/h and design thicknesses for commuter lines were used for all other speeds. The subgrade conditions and design life were the same as those for the axle load comparison described above.

14.6.2.4 Cumulative Tonnage

The cumulative tonnage study used a Class 66 locomotive pulling fully laden wagons traveling at 125 km/h with axle loads of 250 kN and the same subgrade conditions used for the axle load comparison. The cumulative tonnage was varied from 30 to 900 MGT, with an assumed annual tonnage of 15 MGT/yr (Figure 14.23).

From the comparisons shown in Figures 14.20–14.23, two general observations may be made:

1. For each comparison, there is a large variation in the specified thickness of the track bed layers among the procedures.
2. The design thickness specified by each procedure is a function of at least one of the four variables considered (subgrade resilient modulus, axle load, speed, and cumulative

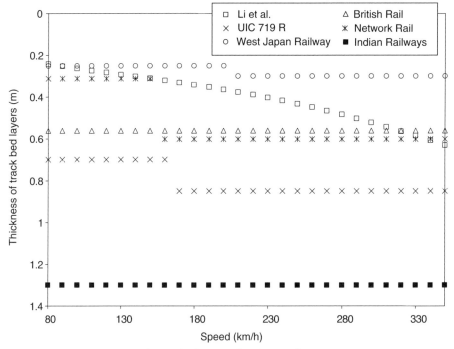

FIGURE 14.22 Variation in design thickness with train speed.

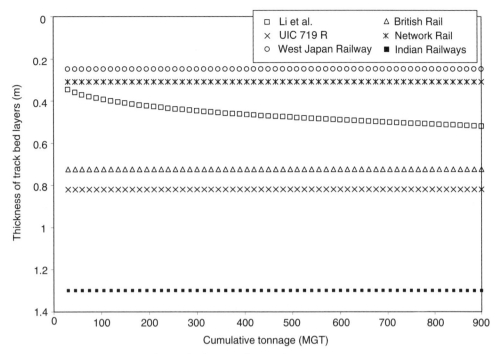

FIGURE 14.23 Variation in design thickness with cumulative tonnage.

tonnage). However, only Li et al.'s method gives a variation in required thickness with all of the variables.

The reasons for these differences, although complex, may be explained in part with reference to the approach to design adopted. Each of the procedures has varying amounts of empirical and analytical elements. In an analytical approach to design, two main processes are combined (Ullidtz 2002). In the first process, stresses, strains, and deflections induced by traffic loading in the subgrade are determined using an analytical model of the track system. In the second process, critical, or allowable, stresses, strains, and deflections are determined, often from experimentation on the subgrade soils. Induced stresses, strains, or deflections are compared with the allowable to formulate the design. This approach is summarized in Figure 14.24.

14.6.3 Characterization of Traffic

14.6.3.1 Axle Loads

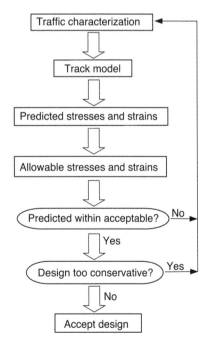

FIGURE 14.24 Analytical design (McElvaney and Snaith 2002).

Traffic characterization requires the magnitude, frequency, configuration, and duration of all loads to be modeled. In terms of the magnitude of the wheel loads, both static and dynamic components should be considered (see above). Research suggests that the dynamic component is a function of speed, vehicle mass, and sources of irregularity in the wheel, running surface, or vertical track geometry. Dynamic effects have been shown to significantly increase track loading, especially where vehicles that have out-of-round wheels are operating at high speeds (Esveld 2001; Shenton 1984). Since the stresses transmitted to the subgrade, and therefore the strains and deflections in the subgrade, are a function of the load, it may be expected that the thickness of the track bed layers specified by any procedure is a function of the axle load. For the static component, it is evident from Figure 14.21 that this is the case for all of the procedures considered except Indian Railways, WJRC, and Network Rail. Further, it can be seen from Figure 14.21 that even for the procedures that consider the static load as a design parameter, the thicknesses of the track bed layers recommended are not in close agreement. The reasons for this involve the analytical models used and how these simulate the distribution of the train load through the substructure, as discussed in Section 14.4.2.1.

As for the dynamic component, because it is a function of speed, the comparison of thickness of the track bed layers with design speed (Figure 14.22) may be used to help determine whether dynamic loading has been considered.

As described above, Li et al.'s method considers dynamic effects (Equation 14.1). This is reflected in Figure 14.22, which shows that the thickness of the track bed layers for this method is a continuous function of speed. However, the formula described by Equation 14.1 is based on empiricism and is believed to overestimate the dynamic increment (and therefore the required thickness of the track bed layers) at high speeds (Lord et al. 1999).

The recommended thicknesses of the track bed layers determined using UIC 719 R, the Network Rail standard, and the WJRC standards are irregular functions of speed. For UIC 719 R, the increase in traffic loading with speed is taken into account using UIC 714, and for the Network Rail standard, higher speed lines are required to have a greater sleeper support stiffness and consequently a greater thickness of the track bed layers. As mentioned earlier, the WJRC standard for commuter lines is not a function of speed; however, a thicker track bed layer is specified for the faster Shinkansen lines. Whether the requirement to increase the thickness with speed for both the Network Rail and WJRC procedures is due to setting higher standards for ride quality for high-speed lines or whether it is because the procedures recognize that damage to the subgrade increases with speed, therefore necessitating thicker track bed layers, is unclear.

In the British Rail method, design thickness is not a function of speed. However, it was recognized that dynamic loads should be considered, and it was reported that work was being undertaken to this end, although the results of this work were unavailable for incorporation in the procedure when it was published.

14.6.3.2 Traffic Mix

The traffic using a particular line may be a mixture of trains with different axle loads traveling at different speeds, and for design purposes, it is important to consider the effect of this spectrum of loads on the system. However, only the Li et al. and UIC 719 R methods account for the variation in load that may occur; the designs of the other procedures are based on a single-axle load.

To account for the variation in traffic, Li et al. convert the estimated spectra of wheel loads over the design life to the number of repetitions of a single design load that causes an equivalent amount of damage (Li and Selig 1998b). UIC 719 R adopts the approach developed by the UIC which enables daily traffic to be represented in terms of a theoretical traffic load.

14.6.4 Analytical Model and Layer Characterization

14.6.4.1 Analytical Model

In an analytical design procedure, it is assumed that the railway substructure can be modeled as a system of elastic layers characterized by two properties: the elastic modulus and Poisson's ratio. The elastic modulus usually is taken to be the resilient modulus, and it can be determined from laboratory tests on the materials, either directly or from an analysis of the response measured *in situ* (Brown 2003). Poisson's ratio usually is estimated.

Three of the procedures state that an elastic model was used to formulate their designs. Li et al.'s model consisted of separate track bed and subgrade layers with stress-state-dependent resilient moduli. The model used to develop the British Rail procedure represented the substructure by a single layer with a single value of resilient modulus, while the Network Rail code used multilayer elastic theory. The effect of these models on thickness of the track bed layers can be seen with reference to Figure 14.20. This figure shows that while all three procedures give similar functions of design thickness with resilient modulus, the thicknesses recommended are not the same. The thickness of the track bed layers given by British Rail is greater than the other two models, partly because the British Rail single-layer model neglects both the effect of the much higher stiffness of the track bed layers and also the changes in resilient modulus that occur with loading. Given the limited information available about Network Rail's model, it is difficult to determine whether the differences in design thickness in relation to other procedures given by the Network Rail code are due to the model used or to other factors.

It is not known if models were used to formulate the specifications given by the UIC, Indian Railways, and WJRC; however, as mentioned previously, it is believed that these standards are largely based on empiricism.

14.6.4.2 Layer Characterization

Since weaker subgrades can withstand lower stress levels, it would be expected that the thickness of the track bed layers recommended by a procedure is a function of the engineering properties of the subgrade (Figure 14.20). This is the case for the Li et al., British Rail, and Network Rail procedures; however, it is not the case for the UIC 719 R, Indian Railways, and WJRC standards. As mentioned above, the UIC 719 R and Indian Railways specifications are a function of the soil type. As described previously, the WJRC recommendations are based on the requirement that the subgrade bearing capacity be greater than 288 kPa. However, this may result in an overly conservative design when the bearing capacity is greater than this value.

14.6.5 Design Method and Material Performance

In an analytical approach to design, it is necessary to determine a measure of material behavior that can be compared to the subgrade stresses, strains, or deflections predicted by the analytical model to formulate the design. For the procedures described here, Li et al. use plastic strain and cumulative deformation (Equations 14.2 and 14.3), UIC 719 R and Indian Railways use a measure of type, the British Rail method uses threshold stress, the Network Rail code uses Young's modulus, and WJRC uses a nominal subgrade bearing capacity.

The contrasting measures of material performance used by Li et al. and the British Rail procedure are of particular interest. Li et al.'s procedure incorporates a model of material fatigue under cyclic loading to determine allowable stresses and strains. Such models relate subgrade deformation to the expected number of applications of load, using relationships of the form of Equation 14.2, and are widely used in the design of roads (McElvaney and Snaith 2002). The British Rail procedure, on the other hand, uses the threshold stress concept, which suggests that provided the stresses in the subgrade are always less than the threshold stress, the subgrade may in theory undergo an infinite number of load cycles before failure. Evidently, using the former approach, it is necessary to specify a design life, whereas for the latter it is not. This is demonstrated by Figure 14.23, which shows that the design thickness recommended by Li et al.'s method only is a function of cumulative traffic (and therefore design life). The concept of threshold stress is not well understood. However, its use in both railway and highway engineering is currently gaining credence as more research is undertaken to better understand the concept (Brown 2003).

14.6.6 Case Studies

In the U.K., lines exist with a thickness of track bed layers from less than 300 mm to 1000 mm or more. For example the thickness of the track bed layers (ballast + subballast + prepared subgrade) on ballasted sections of the CTRL, whose track bed layer design was based on UIC 719 together with French TGV best practice, is approximately between 0.85 and 1 m (O'Riordan and Phear 2001). The lower thickness is for sections of the track where the subgrade is Folkestone sand and the upper thickness is for sections on heavily overconsolidated clay subgrades. In addition, for the sections constructed on heavily overconsolidated clay subgrades, 0.65 m of the clay below the base of the prepared subgrade was dug out and replaced with Folkestone sand sandwiched between geotextiles. Since opening in September 2003, there have been no reported problems with the substructure, and it has been suggested that the interval between planned tamping and realignment maintenance of every 3 yr be increased (Schofield and Franklin 2005). This would therefore indicate that the thickness of the track bed layers on the CTRL is adequate.

   In terms of traffic loading and subgrade conditions, the CTRL sections constructed on overconsolidated clays correspond approximately to a track designed for trains traveling at 300 km/h on a subgrade, with a compressive strength of 100 kPa (O'Riordan and Phear 2001). Using these values and assuming a resilient modulus value of 25 MPa (see Section 14.3.1), the thicknesses of the track bed layers determined from all of the procedures are given in Table 14.18. For the sake of completeness, the thickness of track bed layers given by the Indian Railways code is given, although it is recognized that the code was not developed with high-speed rail in mind (although it is similar to the thickness given by Li et al.'s method). From this table, it can be seen that the track bed layer thickness recommended by the British Rail procedure is similar to the UIC 719 R recommendation, although the former was not produced with high-speed lines in mind. The Li et al. and Indian Railways procedures, on the

TABLE 14.18 CTRL Track Bed Layer Thickness

	Depth of Track Bed Layers (m)				
Li et al.	UIC 719 R	British Rail	Network Rail	Indian Railways	WJRC
1.31	1.00[a]	0.97	0.79	1.30	0.30[a]

[a] Including prepared subgrade.
[b] Subgrade improvement required, as bearing capacity of soil is likely to be less than the minimum of 288 kPa.

other hand, give values approximately 30% greater than the UIC 719 R value, which for the former may be attributed in part to the use of the AREA equation (Equation 14.1), which it is believed may overestimate dynamic wheel loading at very high speeds (Lord et al. 1999). However, as mentioned above, the overconsolidated clay subgrade on the CTRL was replaced with sand to a depth of 0.65 m, which is likely to have somewhat larger compressive strength and resilient modulus values than the clay. Therefore, it may be argued that a truer thickness of the track bed layers for the CTRL built on overconsolidated clay is somewhere between 1 and 1.65 m (i.e., 1 + 0.65 m), depending on the engineering properties of the Folkestone sand. The thicknesses given by the Li et al. and Indian Railways procedures are within these limits. Network Rail's procedure gives a value approximately 20% smaller than UIC 719 R, which may suggest that the code may not be suitable for designing high-speed lines. The WJRC procedure gives a very low thickness compared to UIC 719 R, but the subgrade would require improvement to achieve the required 288-kPa bearing capacity (see Section 14.2.5).

A further example is a mixed traffic line near Leominster in Herefordshire, U.K. (Brough et al. 2003). The track bed layer thickness varies along the site from approximately 900 to 1300 mm and has increased from its original thickness over time due to continued ballast replacement. Sections of the site show large amounts of deterioration, and there is a need for frequent maintenance. The deterioration is believed to be due to poor drainage, causing localized softening, fines migration into the ballast, heterogeneous dynamic sleeper support stiffness, and consequent nonuniform track settlement (see Figure 14.25). Lower bound estimates of the subgrade strength and resilient modulus values found at the site are 100 kPa and 25 MPa, respectively. The design line speed for the section concerned is 128 km/h (although there are speed restrictions in place), and the annual tonnage at the site is approximately 6 MGT/yr. Using these values, and assuming 50% of the traffic is freight, the track bed layer thicknesses recommended by the five procedures are given in Table 14.19. It can be seen from this table that the recommended thicknesses given by Li et al. and UIC 719 R are similar, but the British Rail thickness is approximately 15% greater and the Network Rail and WJRC recommendations are significantly lower. The greater thickness given by the British Rail method may be

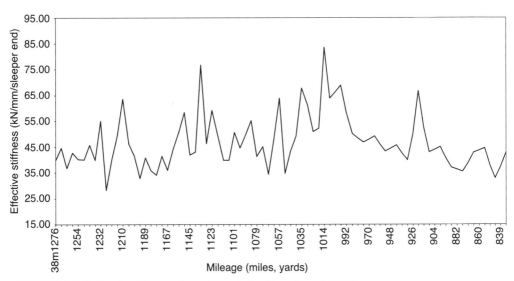

FIGURE 14.25 Track stiffness at Leominster (after Brough et al. 2003).

TABLE 14.19 Leominster Track Bed Layer Thickness

	Depth of Track Bed Layers (m)				
Li et al.	UIC 719 R	British Rail	Network Rail	Indian Railways	WJRC
0.86	0.82	0.97	0.49	1.30	0.20[a]

[a] Subgrade improvement required, as bearing capacity of soil is likely to be less than the minimum of 288 kPa.

attributed to its single-layer model of the track substructure which neglects the effect of the much higher stiffness of the track bed layer. All of the recommendations are less than, or in the case of the British Rail and Indian Railways procedures within, the 0.9- to 1.3-m track bed layer thickness found at the site. This would suggest that if the thickness of the track bed layers is the sole consideration in track bed design, then there should be no need for the excessive maintenance that has occurred. However, the fact that there has been a large amount of maintenance illustrates that other factors should be taken into account in any design process. These factors include appropriate drainage, prevention of subgrade attrition (through the use of a suitable subballast layer), and the need to ensure a uniform track stiffness (which, as Figure 14.25 illustrates, is not the case at Leominster).

14.7 Track Bed Investigation

Site investigation involves gathering sufficient information for the design of new track to identify sources of hazards and problems in existing track and to provide information such that suitable remediation techniques may be effected. In general, site investigation is a multi-stage process which normally is comprised of a desk study to identify hazards and the extent of field work required. This usually is followed by field work.

14.7.1 Desk Study

The desk study entails finding facts about the site in question and requires a review of available information (results of previous investigations) as well as gathering geological and hydrogeological information and site history data (including utilization). Together with information from a walk-over survey of the site, which should be an integral part of the study, this is used to formulate a conceptual model of the structural geology of the site, engineering and other risks posed in terms of the envisaged work, potential problems, and possible methods of effecting solutions. It normally leads to the design of the ground investigation scheme in order to acquire the required information. In addition to the ground investigation plan, this stage of the study should lead to the identification of the constraints, such as access to the site, limitations of space and time for the investigation should be conducted, health and safety risks, and requirements for compliance with regulations prior to and during the investigation. An example of the limitations of space in the railway environment is given in Figure 14.26 (Brough et al. 2003) for a typical U.K. electrified line.

In addition to space constraints on existing track, time limitations often apply. Appreciation of time constraints is crucial, as this often limits methods of investigation that may be used.

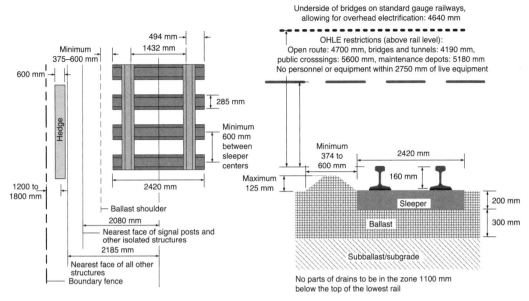

FIGURE 14.26 Typical boundary constraints for railways on electrified track in the U.K. (adapted from Brough et al. 2003). (OHLE = overhead line equipment.)

In addition to the above constraints, it should be noted that any equipment used on an existing track may require approval of the track operating companies. Thus, at the planning stage, additional time may be required to gain approval if nonapproved methods are to be trialed.

It is conventional wisdom to investigate an area larger than the site in question at this stage of the study in order to get a fuller understanding of the factors that influence site ground conditions.

14.7.2 Scope of Ground Investigation

Any study undertaken should be designed to confirm the conceptual geology of the underlying ground and determine the properties of soils so that suitable design work may be undertaken.

14.7.2.1 Depth of Investigation

In planning a ground investigation, it is important to establish beforehand the depth of the investigation required to identify soil properties so that the suitable analyses may be undertaken. For some conditions, shallow trail pits may suffice, whereas for others, it may be necessary to undertake investigation to greater depths, requiring the use of boreholes to recover specimens.

As a guide, Li and Selig (1995) show that traffic loads have a significant influence on the track bed to depths of 4.5–6.0 m. However, it is worth noting that the reduction in stress with depth should only be used as a rough guide for determining the depth of exploration. It is most important to ascertain the presence of compressible soils in the stressed zone, since these soils may exhibit significant deformation when subjected to even a small increase in pressure.

TABLE 14.20 Soil Groups and Associated Applications

Application	Problem Soil Groups	Soil Groups That Pose Slight to No Problem
Drainage	Essentially clays, silts, and organic soils *Soil groups*: GM, GC, SM, SC, ML, CL, MH, CH, OH, and PT	Essentially granular soils ranging from sand to gravel *Soil groups*: GW, GP, SW and SP
Value as filter layer	Essentially silts, clays, and organic soils *Soil groups*: GM, GC, SP, SM, SC, ML, CL, MH, CH, OH, and PT	Essentially granular soils ranging from sand to gravel *Soil groups*: GW, GP, and SW
Value as subgrade	Clay and organic soils are considered to be the worst *Soil groups*: SM, SC, ML, CL, MH, CH, OH, and PT	Essentially granular soils ranging from silt to gravel *Soil groups*: GW, GP, GM, GC, SW, and SC
Stability in compacted fill	Clay and organic soils are considered to be the worst *Soil groups*: ML, MH, and CH Organic soils (OH) and peat (PT) are not to be used	Essentially granular soils ranging from sand to gravel *Soil groups*: GW, GP, GM, GC, SW, SP, SM, SC, and CL CH may only be used on flat slopes

Adapted from AREMA (2007).

14.7.2.2 Properties of Subgrade Soils

An important task of any investigation concerns the identification of soil (material) types. This information is used to assess potential applications and/or possible problems. Soil classification tests include the determination of moisture content, both plastic and liquid limits, particle size distribution, organic content, and mass per unit volume. In terms of railways, AREMA (2007) summarizes applications for soil groups, as shown in Table 14.20. Based on such classification, appropriate measures may be taken to utilize particular soils and consider suitable remedial measures if they are unsuitable.

In addition to the classification tests, it will be necessary to undertake a suite of tests for designing the track foundation. Geotechnical requirements for seven track design methods are shown in Table 14.21. The table shows a wide range of test requirements, from just soil classification for the UIC (1994) to a comprehensive range of properties that include soil classification and resilient modulus as required by Li and Selig (1994).

Okada (2000) summarizes methods used by practitioners for estimating track design parameters (subgrade modulus) for subgrade. These are summarized in Table 14.22.

14.7.3 Ground Investigation

This stage of the investigation is designed to gather detailed information about the ground conditions, identify the causes of failure at problematic sites, and provide engineering properties of materials required to undertake the design work.

Techniques used for this phase are in general the same as those used for the investigation of sites to assess their suitability for the construction of civil engineering and building works, as described in BS 5930 (British Standards Institution 1999) and by Clayton et al. (2006) and numerous other authors. Available techniques can be divided into two categories: geophysical (noninvasive) techniques and physical ground investigation (invasive) techniques. In both cases, although many methods are available, only those relevant to railways are described below.

TABLE 14.21 Geotechnical Input Required for a Range of Design Methods

Design Method	Geotechnical Requirement	Geotechnical Input Required
AREMA (AREA 1996)	Maximum stress at subgrade	Undrained strength of subgrade
AREMA (2007)	Maximum stress at subgrade	Soil description, moisture content, liquid limit, plastic limit, particle size distribution, specific gravity of soils, compaction characteristics, swell and settlement characteristics, unconfined compressive strength, and consolidated undrained compressive strength
Raymond (1978)	Subgrade surface pressure	Soil description, moisture content, liquid limit, plastic limit, particle size distribution, compaction characteristics, and soaked CBR
British Rail (Heath et al. 1972)	Threshold strength	Soil classification, load deformation characteristics of soils subjected to repeat load cycles in a triaxial cell
Li and Selig (1998a, 1998b)	Soil type, subgrade stiffness properties, deformation properties, empirical parameters that relate to material properties	Soil classification, compressive strength and both resilient modulus and deformation together with soil-specific material properties determined from repeated load triaxial tests
UIC (1994)	Soil type	Soil classification, particle size distribution
Network Rail (2005)	Sleeper support stiffness	Undrained shear strength, dynamic sleeper support stiffness

14.7.3.1 Noninvasive Geophysical Techniques

14.7.3.1.1 Ground-Penetrating Radar

A useful summary of the use of ground-penetrating radar (GPR) for monitoring transport infrastructure is presented by Saarenketo (2006). GPR is a nondestructive technique that in essence consists of sending a pulse of radio energy into the ground and receiving a return reflection from the interfaces between underlying materials. For railways, it is used to identify boundaries between various track bed layers as well as the presence of voids, perched water and clay (arising from the subgrade soils), and/or fines arising from ballast attrition, etc. before these features manifest at the surface (Sharpe 2000). Typically, antennas ranging in frequency from 400 MHz to 1 GHz are used (Roberts et al. 2007). Lower frequencies give greater penetration, but lower resolution. Thus, a range of frequencies can be used to build a more complete picture of ground stratigraphy. An example of a typical GPR survey output that shows the presence of good ballast and contaminated ballast is given in Figure 14.27.

GPR surveys can be undertaken using perambulator-mounted devices pushed along the track at a walking pace. Although such devices are still used for local surveys, the advent of high-speed computers and large data storage and data processing capabilities has made it possible to use GPR for ballast condition monitoring using devices that can be mounted on trains or special track vehicles which can be operated at up to 200 km/h (Al-Nuaimy et al.

TABLE 14.22   Methods of Estimating Track Modulus (Okada, 2000)

Device	Type of Results	Merit	Demerit	Threshold Value	Equivalent Modulus	Country
Repeated load triaxial test	Secant modulus	Resilient modulus and shear strength can be measured	Time consuming	15 MN/m^2	15 MN/m^2	U.K.[a] U.S.
CBR test	CBR value	Widely used in highway pavement design	More related to shear stress	5%	50 MN/m^2	Holland[b]
Plate load test	Elastic modulus	Related to a CBR value	Slow to perform	70 MN/m^3 — —	16 MN/m^{2d} 20 MN/m^{2e} 45 MN/m^2	Japan[c] Germany Germany
Unconfined compression test	Compressive strength	Basic test	Measures static properties	—	[f]	U.S.

[a] Railtrack (1997).
[b] Esveld (1989).
[c] Sekine (1996)
[d] V (train speed) $\leq$ 160 km/h.
[e] V > 160 km/h.
[f] Allowable subgrade pressure is recommended as 20 psi (138 kPa).

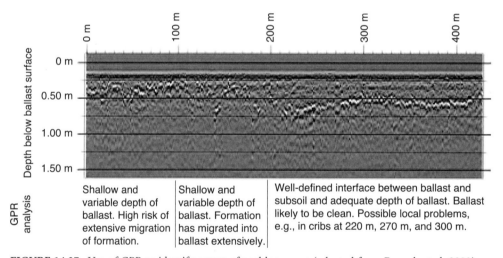

FIGURE 14.27   Use of GPR to identify extent of problem areas (adapted from Brough et al. 2003).

2004). In the U.K., Network Rail employs GPR to monitor track condition using equipment that scans the track bed at between 2.5 and 10 cm operating at 55 km/h (35 mph).

14.7.3.1.2 Continuous Surface Wave System

The continuous surface wave technique can be used to determine the small strain shear modulus of the track bed layers at strain levels of less than 0.002%. Below this level, soil stiffness is approximately linear elastic. It is possible to estimate the soil stiffness at larger stains using the technique by applying empirically founded reduction functions, as described by Heymann (2007) and Clayton and Heymann (2001).

The technique uses an electromagnetic oscillator to generate sinusoidal Rayleigh waves that propagate along the ground surface. Geophones are placed at intervals along the ground surface to collect the response of the ground to the excitation. For geotechnical applications, a frequency of vibration of between 15 and 200 Hz is used to achieve depths of penetration of up to 10 m. The lower the frequency, the higher the depth of penetration. It is worth noting that the accuracy of shear modulus depends on the knowledge of the unit weight of the relevant layers.

In railway applications, the continuous surface wave technique has been used to, among other things, assess the effectiveness of ground improvement (Sutton and Snelling 1998), and Gunn et al. (2006) successfully examined the possibility of using the technique to assess the condition of a Victorian railway embankment.

14.7.3.1.3 Soil Resistivity

Soil resistivity tests are used to help identify the presence of different materials in the track bed, particularly those that contain water. They thus can be useful in identifying potential problematic areas.

The technique measures the electrical resistance of a soil to the passage of current through a series of electrodes. The electrical conductivity of soil is dependent on a number of factors that include clay content, groundwater conductivity, degree of saturation, and soil porosity. Typical resistivity values for a range of soil and rock types are shown in Figure 14.28.

Variations in soil resistivity can be used to develop a stratigraphic map of the underlying geology. In addition, as electrical resistance is related to the moisture content of the soil, it can be used to map moisture movement in the ground. Gunn et al. (2006), for example, have demonstrated the application of this technique to identify the stratigraphy of a Victorian embankment and to monitor moisture movement throughout the year. They also showed how moisture moved through the embankment after a rainfall event.

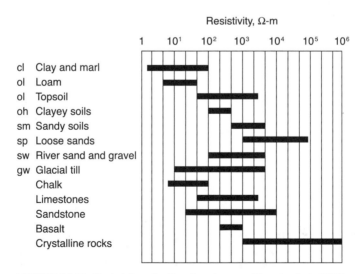

FIGURE 14.28 Resistivity of soil and rock types (Geonics Ltd. 2003).

Although resistivity surveys for stratigraphic work are time consuming, they do give an overall picture of the ground structure.

14.7.3.1.4 *Falling Weight Deflectometer*

The falling weight deflectometer (FWD) test permits the determination of track/subballast/ballast/subgrade layer stiffness without the removal of track and ballast (see Figure 14.29). It was developed for assessment of the structural condition of both road and airport pavements.

The test consists of dropping a known weight through a fixed distance onto the surface of interest. Accelerometers positioned at known distances along a line radiating from the load drop position are used to record the response of the surface to the dropped load. This information, together with the thickness of the layers, also can be used to estimate the resilient modulus of the layers (Burrow et al. 2007a).

In the U.K., the technique has been adopted for estimating sleeper support stiffness used in railway track design and for assessing the condition of in-service track (Sharpe and Collop 1998).

For railway tracks, the device is designed to apply a 125-kN load to a sleeper disconnected from the rails, via a 1.1-m-long loading beam shaped to distribute the load to both ends of the sleeper (see Figure 14.30). This loading system is considered to produce a load pulse, similar to that applied by a single axle of a train traveling at high speed. The magnitude of the applied load is measured in the center of the loading beam, and the geophones are lowered into place from the FWD apparatus onto the loaded sleeper and ballast at various distances from the center of the beam. Sensors are positioned at multiple locations, as shown in Figure 14.30, where d1 and d2 are located on the unclipped sleeper to which the load is applied.

FIGURE 14.29 The FWD.

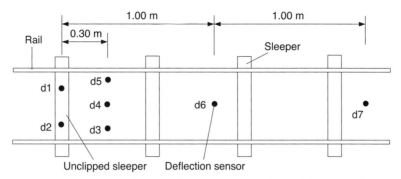

FIGURE 14.30 Layout of sensors for FWD test on railways (Sharpe 2000).

The deflections obtained using the FWD methodology on different layers of track for different types of railway lines in the U.K. are shown in Figure 14.31. From this figure, it is apparent that changes in deflections in the subgrade seem to be magnified in deflections of both the ballast and the sleepers.

Network Rail (2005) recommends that the test be conducted at intervals of between 10 and 20 m. In terms of number of tests, two individuals can conduct 20 tests per hour (after 1-h setup) for shallow depth investigations (Brough et al. 2003).

14.7.3.2 Invasive Techniques

14.7.3.2.1 Boreholes
Percussion boring techniques may be used for new-build track to obtain samples so that soils/rock strata can be identified and their properties determined for track foundation design. Particular attention should be paid to the assessment of soil types and groundwater conditions

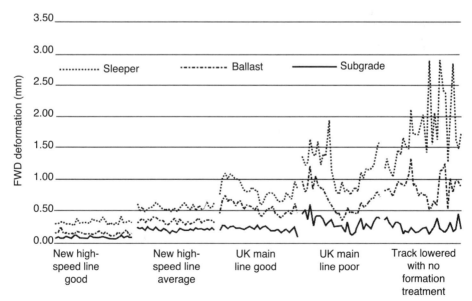

FIGURE 14.31 Deflections of track layer determined using FWD for different types of railway lines (Sharpe 2000). (Note: Data interval of 150 m is plotted for each section of track.)

depending on the susceptibility of the *in situ* soils to the problems described in Section 14.4.2. Appropriate measures then can be exercised to design a cost-effective track.

Because of cost issues, boreholes are not normally used for maintenance purposes or for shallow track foundations.

14.7.3.2.2 Trial Pit

Trail pits can be used to identify stratum, take both disturbed and undisturbed specimens testing, to recover soil samples, or to enable noninvasive tests to be carried out in the base of a pit.

For new-build track, trial pits should be constructed in accordance with BS 5930 (British Standards Institution 1999). They normally extend to between 3 and 5 m and depending on the requirements should be spaced between 10 and 30 m apart.

Trails pits in cribs normally are the most common technique for areas where there is a need to investigate shallow failures such as ballast contamination or shallow failures in the subgrade. They normally do not extend beyond about 1 m in depth and are dug manually. In the U.K., Network Rail (2005) recommends that the excavation should not penetrate the subballast sand or stone layer unless it is essential to do so. Should it be necessary to penetrate through any such layer, then the pit should be large enough to make good repairs; in the case of geotextiles, excavation should be large enough to ensure a 150-mm overlap. In terms of position, normally there are limitations for positioning them near rail joints and crossings.

In an 8-h shift, it is possible to dig 15 shallow pits (less than 0.5 m in depth) or 10 deep pits (0.5–1 m in depth), depending on ground conditions (Network Rail 2005).

14.7.3.2.3 Window Sampling

The window sampling technique, also referred to as automatic ballast sampling in the U.K., is a method that is used to obtain samples of the ballast and subballast. It consists of using a hydraulic percussive device to push a 65-mm (inside diameter) steel sampling tube with a plastic liner into the ground. Relatively undisturbed samples are obtained in approximately 1-m lengths. Samples from greater depths may be obtained by using multiple lengths of sampling tubes. A typical example of output from a window sampler is shown in Table 14.23.

Using a window sampler, two operators can obtain five 2-m-deep cores per hour (Brough et al. 2003) for maintenance-related investigation.

14.7.3.2.4 Cone Penetration Testing

In general, two types of tests are conducted: dynamic and "static" cone penetration tests. Both types measure the resistance to penetration of a cone attached to a steel rod. A variety of correlations can be applied to determine a number of other material properties.

Dynamic Penetration Test. In the dynamic penetration test, a cone is attached to the end of a steel rod and pushed into the ground by applying a fixed percussive load that falls through a fixed distance. The resistance to penetration is measured. This can be correlated to strength, elastic modulus, resilient modulus, and CBR. The two most common dynamic penetration tests are the standard penetration test and the dynamic cone penetration test.

The *standard penetration test* is perhaps one of the most commonly used tests in ground investigation and has been described by numerous authors. A detailed description of the test can be found in a variety of texts, including BS 5903 (British Standards Institution 1999) and Clayton et al. (2006). In brief, the test is conducted by dropping a fixed weight (63.5 kg) attached to drill rods through a fixed distance (762 mm). The resistance to penetration of a

TABLE 14.23 Typical Output from Window Sampler

Depth to Base of Layer[a] (m)	Layer Thickness (m)	Layer Description	Comments[b]	Window Sample (2-m Length)
0.95	0.75	Clean ballast	Good sized angular	
1.10	0.15	Slurried ballast (wet)	Ballast in wet grey clay slurry	
1.15	0.05	Slurried ballast	Ballast in soft grey clay	
1.30	0.15	Soft clay	Soft grey clay 25 kPa @ 1200 mm 55 kPa @ 1300 mm	
2.00	0.70	Soft clay	Soft brown clay 50 kPa @ 1400 mm 50 kPa @ 1500 mm 55 kPa @ 1600 mm 50 kPa @ 1700 mm 65 kPa @ 1800 mm 50 kPa @ 1900 mm	

[a] Referenced to rail level.
[b] Strength noted in the comments column was measured using hand vane.

cone or a split-spoon sampler attached to the end of the drill rods is measured in terms of N-value, which is equal to the number of blows required to penetrate 300 mm. The N-value can be used to estimate the density and angle of friction for granular soils (see Figure 14.32). The undrained shear strength of cohesive soils also can be estimated from N using relationships suggested by Stroud (1974) (see Figure 14.33). It should be noted that the soil properties are inferred and therefore should be used only as a guide. However, where it is not possible to obtain undisturbed specimens, the standard penetration test can provide useful information. The test normally is conducted using a cable percussion drilling rig. Therefore, it can be used only when there is a full procession of track.

The *dynamic cone penetration test* is a dynamic test that measures the resistance to penetration of a 20-mm-diameter 60° cone attached to a rod that can be extended to a maximum length of 1.5 m. The test is performed by dropping an 8-kg mass through 575 mm on a slide bar attached to the top of the penetration rod. A useful description of the procedure is given by Jones (2004). The measure of resistance to penetration can be related to a number of material properties (TRL 1993).

Static Cone Penetration Test. The cone penetration test consists of pushing an instrumented cone into the ground at a known rate of penetration (usually 2 cm/min). The instrumented cone is used to measure the cone tip resistance (q_c) and skin friction (f_s) of a sleeve positioned immediately behind the cone. In addition, the dynamic pore water pressure generated as the cone is pushed into the ground also can be measured.

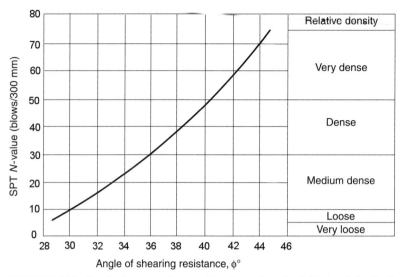

FIGURE 14.32 Standard penetration test *N*-value and angle of friction (after Peck et al. 1974).

In addition to identification of stratification, soil types can be identified using relationships such as those described by Jacobs (2004). Furthermore, using various correlations, a number of engineering properties of the soils through which the cone is pushed can be estimated, including the following:

- *Fine-grained soils (silts and clays)*—Undrained shear strength, coefficient of compressibility (m_v), undrained modulus of elasticity, and permeability
- *Granular soils (sands and gravels)*—Relative density, angle of friction, Young's modulus, constrained modulus, and shear modulus

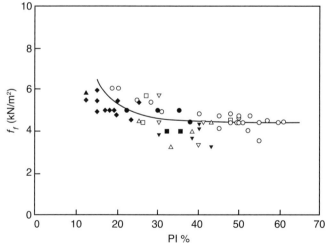

FIGURE 14.33 Variation of f_i ($= c_u/N_{60}$) with plasticity index for overconsolidated clay (after Stroud 1974).

References

Al-Nuaimy, W., Eriksen, A., and Gasgoyne, J. (2004). Train-mounted GPR for high-speed rail track bed inspection. *Tenth International Conference on Ground Penetrating Radar,* Delft, Netherlands.

AREA (1996). *Manual of Railway Engineering,* American Railway Engineering Association, Washington, D.C., chap. 1.

AREMA (2007). *Manual of Railway Engineering,* American Railway Engineering and Maintenance Right-of-Way Association, Lanham, MD, chap. 1.

ARTC (2007). *Standard Classification of Lines,* TDS 11, Australian Railway Track Corporation Engineering (Track & Civil) Standard, Adelaide, Australia.

Brandl, H. (2001a). Freeze-thawing behavior of soil and other granular material—influence of compaction. *Geotechnics for Roads, Rail Tracks and Earth Structures,* A.G. Correia and H. Brandl, Eds., A.A. Balkema, Netherlands.

Brandl, H. (2001b). Compaction of soil and other granular materials—interactions. *Geotechnics for Roads, Rail Tracks and Earth Structures,* A.G. Correia and H. Brandl, Eds., A.A. Balkema, Netherlands.

Brandl, H. (2001c). *Importance of Optimum Compaction of Soil and Granular Materials Geotechnics for Roads, Rail Tracks and Earth Structures,* A.A. Balkema Netherlands.

Brandl, H. (2001d). Geotechnics of rail track structures. *Geotechnics for Roads, Rail Tracks and Earth Structures,* A.G. Correia and H. Brandl, Eds., A.A. Balkema, Netherlands.

British Standards Institute (1990). *BS 1377: Methods of Test for Soils for Civil Engineering Purposes. Part 9. In-Situ Tests,* British Standards Institution, London.

British Standards Institution (1999). *BS 5930: Code of Practice for Site Investigations,* British Standards Institution, London.

Brough, M., Ghataora, G.S., Stirling, A.B., Madelin, K., Rogers, C.D.F., and Chapman, D.N. (2003). Investigation of railway subgrade. 1. In-situ assessment. *Proc. Transport J.,* 156(TR3):9.

Brown, S.F. (2003). Soil mechanics for pavement engineers. *Transportation Geotechnics Symposium,* Nottingham Trent University, Nottingham, U.K., Thomas Telford, London.

Burrow, M.P.N., Ghataora, G.S., and Bowness, D. (2004). Analytical track substructure design. *ISSMGE TC3 International Seminar on Geotechnics in Pavement and Railway Design and Construction,* NTUA, Athens, December 16–17.

Burrow, M.P.N., Chan, A.H.C., and Shein, A. (2007a). Falling weight deflectometer based inverse analysis of ballasted railway tracks. *Geotech. Eng.,* 160(GE3):169–177.

Burrow, M.P.N., Bowness, D., and Ghataora, G.S. (2007b). A comparison of railway track foundation design methods. *Proc. Inst. Mech. Eng. Part F: J. Rail Rapid Transit,* 221(1):1–12.

Cedergren, H.R. (1987). *Drainage of Highway and Airfield Pavement,* Robert E. Krieger, Malabar, FL.

Cedergren, H.R. (1989). *Seepage, Drainage and Flownets,* Wiley-Interscience.

Clayton, C.R.I. and Heymann, G. (2001). Stiffness of geomaterials at very small strains. *Géotechnique,* 5(3):245–256.

Clayton, C.R.I., Matthews, M.C., and Simons, N.E. (2006). *Site Investigation,* University of Surrey, U.K.

Esveld, C. (1989). *Modern Railway Track,* MRT-Productions, Zaltbommel, Netherlands.

Esveld, C. (2001). *Modern Railway Track,* 2nd edition, MRT-Productions, Zaltbommel, Netherlands.

Fox, P., Pritchard, R., and Hall, P. (2004). British Railways Locomotives and Coaching Stock 2004, Sheffield, U.K.

Freeme, C.R. and Servas, V. (1985). *Advances in Pavement Design and Rehabilitation—Accelerated Testing of Pavements,* CSIR, Pretoria, Republic of South Africa.

Geonics Ltd. (2003). TN5—Electrical conductivity of soils and rocks. Embedded in *Federal Highway Administration Report on Application of Geophysical Methods to Highway Related Problems.*

Ghataora, G.S., Burns, B., Burrow, M., and Evdorides, H. (2006). Development of an index test for assessing anti-pumping materials in railway track foundations. *First International Conference on Railway Foundations,* University of Birmingham, U.K., 355–366.

Gräbe, P.J. and Clayton, C.R.I. (2003). *Permanent Deformation of Railway Foundations Under Heavy Axle Loading,* Specialist technical session. International Heavy Haul Association, Dallas.

Gunn, D.A., Nelder, L.M., Chambers, J.E., Raines, M.R., Reeves, H.J., Boon, D., Pearson, S., Haslam, E., Carney, J., Stirling, A.B., Ghataora, G.S., Burrow, M., Tinsley, R.D., Tinsley, W.H., and Tilden-Smith, R.B. (2006). Assessment of railway embankment stiffness using continuous surface waves. *RailFound06,* G.S. Ghataora and M. Burrow, Eds., University of Birmingham, U.K.

Hausmann, M.R. (1990). *Engineering Principles of Ground Improvement,* McGraw-Hill.

Hay, W.W. (1982). *Railroad Engineering,* 2nd edition, Wiley-Interscience.

Heath, D.L., Shenton, M.J., Sparrow, R.W., and Waters, J.M. (1972). Design of conventional rail track foundations, *Proc. Inst. Civ. Eng.,* 51:251–267.

Heymann, G. (2007). Ground stiffness measurement by the continuous surface wave test. *J. S. Afr. Inst. Civ. Eng.,* 49(1):23–31.

Hornych, P., Hameury, O., and Paute, J.L. (1998). Influence de l'eau sur la comportement mécanique des graves non traitées et des sols supports de chaussées. *Symposium Int'l AIPCR sur le Drainage des Chaussées,* Grenada, Spain, PIARC/AIPCR, 249–257.

Hunt, G.A. (1993). *Optimisation of Track Formation Stiffness Report 1,* British Rail Research Track Mechanics and Systems.

Indian Railways (2003). *Guidelines for Earthwork in Railway Projects,* Guideline No. GE: G-1, Geotechnical Engineering Directorate, Ministry of Railways, Luknow, India.

Indian Railways (2004). Permanent Way Manual, *The Maintenance of Permanent Way,* Government of India, Ministry of Railways (Railways Board), chap. 2.

Jacobs, P.A. (2004). Cone penetration testing.

Jones, C. (2004). *Dynamic Cone Penetrometer Tests and Analysis,* Technical Information Report PR/INT/277/04, Crowthorne, U.K.

Li, D. and Selig, E.T. (1994). Resilient modulus of fine grained subgrade soils. *J. Geotech. Eng.,* 120(6):939–957.

Li, D. and Selig, E.T. (1995). *Evaluation of Railway Subgrade Problems,* Washington, D.C.

Li, D. and Selig, E.T. (1996). Cumulative plastic deformation for find grained soils. *J. Geotech. Eng.,* 122:(12).

Li, D. and Selig, E.T. (1998a). Method for railroad track foundation design. I. Development. *J. Geotech. Geoenviron. Eng.,* 124:(4).

Li, D. and Selig, E.T. (1998b). Method for railroad track foundation design. II. Applications. *J. Geotech. Geoenviron. Eng.,* 124(4):323–332.

Li, D., Sussman, T.R., and Selig, E.T. (1996). Procedure for Railway Track Granular Layer Thickness Determination, Pueblo, CO.

Lichtberger, B. (2005). *Track Compendium: Formation, Permanent Way, Maintenance, Economics,* Eurailpress Tetzlaff-Hestra, Hamburg.

Lord, J.A., O'Riordan, N.J., and Phear, A.G. (1999). Design and analysis of railway track formation subgrade for high speed railways. *Rail Technology for the Future,* ICE, London.

Lowson, M.V. (1998). Surface transport history in the UK: analysis and history. *Proc. Transport J.,* 129:14–19.

McElvaney, J. and Snaith, M.S. (2002). Highways, the location, design, construction and maintenance of pavements. *Analytical Design of Flexible Pavements,* C.A. O'Flaherty, Ed., Butterworth-Heinemann, Oxford, U.K., chap. 15.

National Lime Association (2001). *Guidelines for Stabilization of Soils Containing Sulfates,* Technical Memorandum, National Lime Association.

Network Rail (2000). *Track Ballast and Stoneblower Aggregate,* Network Rail Standard: NR/SP/TRK006, Network Rail, London.

Network Rail (2005). *Formation Treatment,* NR/SP/TRK/9039, Code of Practice for Track Maintenance, Network Rail, London.

Okada, K. (2000). *Assessment of Stiffness of Railway Subgrade,* MPhil thesis, University of Birmingham, U.K.

O'Riordan, N.J. and Phear, A.G. (2001). Design and construction control of ballasted track formation and subgrade for high speed lines. *Railway Engineering,* London.

Peck, R.B., Hanson, W.E., and Thornburn, T.H. (1974). *Foundation Engineering,* 2nd edition, John Wiley, New York.

Railtrack (1997). *Track Construction Standards,* Railtrack Line Specification RT/CE/S/102.

Raymond, G.P. (1978). Design for railroad ballast and subgrade support. *J. Geotech. Eng. Div. ASCE,* 10491(14).

Roberts, R., Al-Qadi, I., Tutumluer, E., and Kathage, A. (2007). Ballast Fouling Assessment Using 2GhZ Horn Antennas—GPR and Ground Track Comparison from 238 km of Track, http://www.alphageofisica.com.br/gssi/gpr_2008/RailEng%202007_br.pdf.

Saarenketo, T. (2006). *Electrical Properties of Road Materials and Subgrade Materials and Subgrade Soils and the Use of Ground Penetrating Radar in Traffic Infrastructure Surveys,* Department of Geosciences, Oulu University, Oulu, Finland.

Schofield, R. and Franklin, A. (2005). Maintaining track geometry for 300 km/h operation on CTRL. 1. Client and contractor in perfect harmony. *J. Permanent Way Inst.,* July:63–70.

Sekine, E. (1996). *Bearing Capacity of Roadbed and Its Reinforcement,* Ph.D. thesis, Nagaoka University of Technology, Nagaoka, Japan.

Selig, E.T. and Waters, J.M. (1994.) *Track Geotechnology and Substructure Management,* Thomas Telford, London.

Sharpe, P. (2000). Track bed investigation. *J. Proc. Permanent Way Inst. UK,* 118(3):238–255.

Sharpe, P. (2005). Geotechnical engineering and track bed engineering practice. Lecture to the MSc programme Railway Systems Engineering and Integration, University of Birmingham, U.K.

Sharpe, P. and Collop, A.C. (1998). Track bed investigation—a modern approach. *First International Conference on Maintenance and Renewal of Permanent Way and Structures,* Brunel University, London, Engineering Technics Press.

Shenton, M.J. (1984). Ballast deformation and track deterioration. *Track Technology,* Nottingham, U.K., Thomas Telford, London.

Stirling, A.B., Konstantelias, S., Ghataora, G.S., Brough, M., and Madelin, K.B. (2003). Improving existing railway subgrade stiffness. A case study of ground improvement techniques. *World Congress on Railway Research,* November.

Stroud, M.A. (1974). The standard penetration test in insensitive clays and soft rocks. *Proceedings of the European Symposium on Penetration Testing,* Stockholm.

Sutton, J.A. and Snelling, K. (1998). Assessment of ground improvement using the continuous surface wave method. *4th Meeting of the Environmental and Engineering Geophysical Society.*

Terzaghi, K. (1922). Der grundguch an astauwerken und seine verhütung (the failure of dams in piping and its prevention). *Die Wasserkraft,* 17:445–449.

TRL (1993). *A Guide to the Structural Design of Bitumen Surfaced, Roads in Tropical and Sub-tropical Countries,* 4th edition, Overseas Road Note 31, Transport Research Laboratories, Crowthorne, U.K.

Tung, K.W. (1989). *An Investigation of the Causes of Railroad Ballast Fouling,* University of Massachusetts, Amherst.

UIC (1989). *UIC 714 (A.A.72): Classification of Lines for the Purpose of Track Maintenance,* International Union of Railways, Paris.

UIC (1994). *UIC Code 719 R: Earthworks and Track-Bed Layers for Railway Lines,* International Union of Railways, Paris.

Ullidtz, P. (2002). Analytical tools for pavement design. Keynote address, *8th International Conference on Asphalt Pavements,* Copenhagen, Denmark.

U.S. Corps of Engineers (1984). *Engineering and Design—Pavement Criteria for Season Frost Conditions,* Report No. EM 110-3-138, Washington, D.C.

Walui, H., Matsumoto, N., and Inoue. (1997). Ladder sleeper and new track structures development. *World Congress on Railway Research,* Florence, Italy.

WJRC (2002a). *Construction and Maintenance Standards for Shinkansen Track,* West Japan Railway Company, Osaka.

WJRC (2002b). *Construction and Maintenance Standards for Commuter and Local Railway Track,* West Japan Railway Company, Osaka.

15

Special Foundations

by
R.L. Handy
Iowa State University, Ames, Iowa

David J. White
Iowa State University, Ames, Iowa

15.1 What Makes a Foundation Special?

Special foundations include foundations that have special requirements, such as for wind turbine generators, and those that do not depend on conventional materials such as wood, concrete, and steel. Pile foundations date back to Neolithic time, when tree trunks were driven into mucky lake bottom soils to support houses and walkways in what now is Switzerland. Spread foundations were used by Romans to support their roads. The Egyptian pyramids are an extreme example of a spread foundation but nothing was put on top.

Modern pile foundations still can consist of tree trunks, but are more likely to be made from steel or a combination of steel and concrete, with steel reinforcement acting to resist tensile stresses from bending or from conflicts between compression waves generated and rebounding during pile driving.

Spread foundations consisting of two layers of tree trunks with the second arranged cross-wise from the first were sometimes used during early days of the American West. A similar approach often is used to provide temporary support to track-mounted cranes. Modern spread foundations almost always are composed of Portland cement concrete or reinforced concrete.

Foundations that in this book are considered to be special foundations often employ weaker, less expensive materials such as crushed aggregate. The justification is that the weakest link in a conventional foundation system is not the concrete or steel, but the soil. Special foundations therefore also may improve the soil in order to obtain a more balanced and more efficient system. Recently, this approach has proven effective for wind turbine foundations, which are discussed later.

A special foundation may involve only a soil treatment. The oldest and simplest example is compaction, although we now recognize that compaction actually is quite complex and requires careful design, supervision, and control in order to obtain a consistent product. Compaction no longer is simply a matter of compressing soil in layers, but may involve deep compaction using a falling weight or lateral compaction from expansion of bulbs of nonpenetrating grout. More recently, high lateral pressures that can dramatically influence soil properties have been obtained through lateral expansion of aggregate piers by ramming in layers.

Soil properties also may be improved by the addition of a chemical stabilizing agent such as lime or Portland cement. Another approach is to incorporate horizontal tensile-reinforcing steel or plastic mesh in layers of soil to increase strength, while still maintaining sufficient flexibility to accommodate some settlement.

15.1.1 Proprietary Nature and Design-Build

As new foundation methods are developed, they usually are protected by patents and offered as a package that includes both design and construction. This helps to maintain a high standard and prevent misapplications and failures that would cloud the future of a method. The goal of this chapter, therefore, is a better understanding of the mechanisms involved in the various methods. Design examples may be simplified for the sake of illustration and presented as an aid to understanding and to help in evaluations of different competing methods.

15.1.2 After Design-Build

The design-build procedure may fade after a patent has expired and a method comes into the public domain. An example is auger-cast piles, where a continuous hollow auger is screwed into the ground to the full pile depth and grout is pumped to the bottom as the auger is withdrawn. This method is particularly useful in caving soils, as it does not require casing to keep a boring open. After the patent expired, royalty payments no longer were required, and employees who were expert in applications of the process formed their own businesses. Competitiveness increased and design became separated, in some cases still performed by the contracting company, but more and more with the guidance of consulting geotechnical engineers who assume the ultimate responsibility.

Special foundations include any foundations that do not fit the classical mold. Because they derive from a robust ancestry that includes piles, piers, shallow foundations, wall footings, column footings, and mats, some of the principles revealed by those relationships will be discussed first.

15.2 Classic Foundation Methods

15.2.1 Pile Foundations

When tree trunks were pounded into mucky lake soils, they most likely were driven until they either stopped or the end of the tree trunk was reached. The pile that did not stop as a result of bearing on a hard layer still could support a load, but in this case support did not come from end bearing at the tip of the pile but rather came from friction along the sides of the pile. Piles were driven upside down to take advantage of a wedging action that would tend to increase friction.

Thus were defined two distinctive types of piles: those that are supported by end bearing and those that are supported by side friction. Soil mechanics now tells us that both mechanisms can exist simultaneously in the same pile. This can be beneficial, but it also can be troublesome if soil encasing the pile settles so that skin friction acts downward and adds to the weight that must be supported by the pile. In addition, since the two mechanisms are independent, they do not develop and peak out simultaneously; skin friction generally becomes mobilized first. Thus, after application of a factor of safety, a pile that is designed with end bearing may actually be supported by skin friction.

End-bearing piles, friction piles, and larger diameter piers and caissons that act in a similar manner are collectively referred to as *deep foundations. Underpinning* is a remedial treatment that involves inserting piles underneath overstressed or failed shallow foundations.

15.2.1.1 How Deep Foundations Reduce Settlement

An unsolved mystery was why friction piles reduce settlement when all they do is transfer load deeper into what is essentially the same soil. The answer is that it is not exactly the same soil, because with increasing depth, soil usually becomes stiffer as a consequence of consolidating or densifying under its own weight. A friction pile reduces settlement by transferring load downward into a stiffer version of the same soil. A soil that has a density that is in equilibrium with its overburden pressure is said to be "normally consolidated."

15.2.1.2 Overconsolidated Soil

Field observations indicate that "normal" consolidation is not very normal because it is rare in nature. All that is necessary to convert a normally consolidated soil into an overconsolidated soil is to remove some overburden by erosion or by melting of glacial ice. Consolidation of soil under a continental glacier usually is incomplete because of excess pore water pressure that also aids sliding of the glacier.

A more subtle but nevertheless common source of overconsolidation is a cycling of a groundwater table that alternately decreases and increases buoyant support and effective stress.

A pseudo-overconsolidation is caused by shrinkage of clayey soils upon drying, in which case consolidation is orthogonal instead of one-dimensional, since shrinkage acts in all directions. In this case, the consolidating forces are internal and tensile instead of being external and compressive.

As will be shown, application of a high lateral stress can create another kind of pseudo-overconsolidation that causes significant changes in the behavior of a soil. These changes are

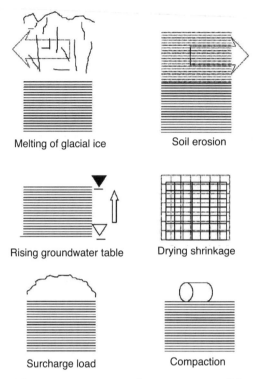

FIGURE 15.1 Some contributors to preconsolidation pressure.

consistent with and help to explain the effectiveness of Rammed Aggregate Pier® Systems for reducing foundation settlement (Handy and White 2006a, 2006b).

15.2.1.3 Soil Compressibility and the Consolidation State

Applying an additional load to a normally consolidated soil reinitiates consolidation, which proceeds according to a linear relationship between void ratio and the logarithm of pressure —the classic $e - \log p$ curve. Loading that does not reach this turning point does not reinitiate consolidation, but nevertheless can slightly compress the soil. The compression in this case is near linear elastic. It is only partly recoverable when a load is removed. For brevity in this chapter, the behavior is referred to as "elastic."

Preconsolidation pressure (see Figure 15.1) therefore is an important consideration when predicting or designing to minimize foundation loading. Temporary surcharge loading or roller compaction can be used to densify the soil and increase its preconsolidation pressure.

15.2.2 Spread Foundations

Another classical approach is to spread a load over a larger area to reduce bearing pressure. The early efforts involved laying large stones for foundations of castles and other structures in the Middle Ages and did not spread the load so much as form a stable platform upon which to build. Because construction was slow, there was sufficient time for the underlying soil to consolidate and gain strength. Soil mechanics now tells us that slow loading allowed the system work. Compression of a saturated soil under load creates excess pore water pressure that must be allowed to escape or the reduction in shear strength may cause a structure to sink into the ground or tip over. The medieval structures that remain intact are the ones that survived.

Settlement without preloading can lead to another problem: uneven settlement caused by variations in the soil and in the loading conditions. As medieval towers were constructed, tilting usually was compensated for by using thicker masonry units on the low side, which was like a hound chasing a rabbit around a circle. That is because soil under the low side would be more compressed, so the next move would be to tilt in a different direction. Several episodes of tilting have been identified from masonry layers in the famous Leaning Tower. Fortunately, the hound did not catch the rabbit.

Shallow foundations normally have an enlarged contact area that reduces bearing pressure. Spread foundations also are effective for preventing bearing capacity failures by increasing the area of potential shear surfaces. They are somewhat less effective for reducing settlement because although increasing the width of a bearing area reduces the bearing pressure, it also causes that pressure to extend deeper. This effect for long and for square foundations is illustrated in Figure 15.2, where it can be seen that a square foundation is more effective for reducing the vertical stress at a particular depth. Thus, when both types of foundations are used under one structure, as often is the case, both settlements must be minimized to reduce differential settlement.

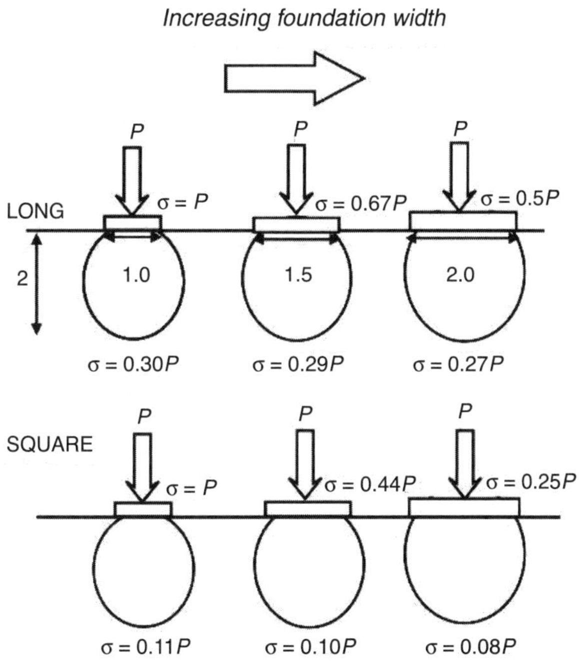

FIGURE 15.2 Illustration of how square foundations are more effective than long foundations for reducing settlement. The figure is based on the integrated Boussinesq solution assuming an elastic soil response.

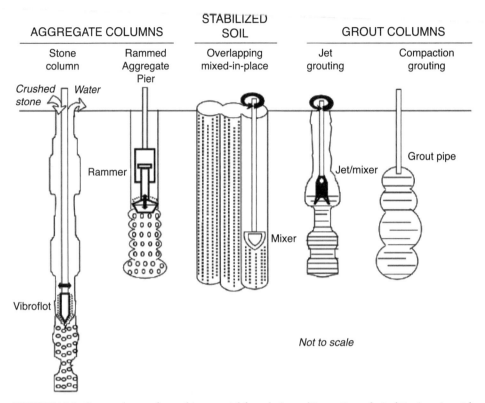

FIGURE 15.3 Some schemes for making special foundations. (From *Geotechnical Engineering, 5th ed.* by Handy and Spangler, © 2007 The McGraw-Hill Companies. Used with permission.)

15.3 Treatments and Methods Used in Special Foundations

The treatments and methods used for making the special foundations discussed in this section are depicted in Figure 15.3.

15.3.1 Compaction

Compaction literally is fundamental in road, highway, and airfield construction, and compacted soil often is used as fill or a replacement soil under other structures. Compacted soil therefore may be considered as a kind of special foundation. Properly engineered and controlled compaction increases the soil shearing strength and therefore its bearing capacity. Compaction also overconsolidates the soil and therefore reduces settlement. In addition, the denser the soil, the higher the elastic modulus when loading does not exceed the preconsolidation pressure.

The relationships among soil engineering properties, water content, and the compaction energy and delivery methods are complex and warrant laboratory evaluation because of the large range in parameter values that result. Figure 15.4 shows the relationship between shear strength and modulus for compacted glacial till as a function of compaction energy and moisture content. With standard Proctor compaction, the optimum moisture content for this soil is about 12%.

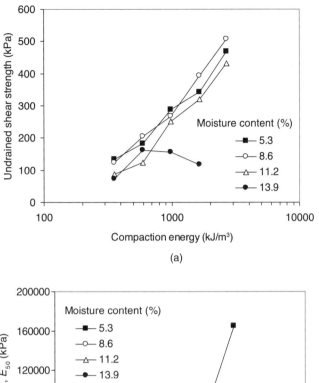

FIGURE 15.4 (a) Semilogarithmic relationship between undrained shear strength (from unconfined compressive strength tests) and compaction energy as a function of water content and (b) semilogarithmic relationship between secant modulus (from stress-strain response of unconfined compressive strength tests) and compaction energy as a function of water content (optimum about 12%) (White et al. 2005).

Generally, as the soil moisture content is reduced, the soil strength and modulus increase with increasing compaction energy. However, increasing the compactive energy with an overly wet soil can cause a sharp reduction in strength and shearing, attributed to temporary excess pore water pressure. This is called overcompaction. Shearing can permanently damage a soil through the development of shear surfaces called *slickensides* and because a residual shearing strength after remolding generally is significantly lower than the peak strength.

15.3.1.1 Surcharging

Compaction can be accomplished quasi-statically with a temporary surcharge load that is simply a mound of soil piled on a future building site. The surcharge is allowed to remain in place to give time for the underlying soil to consolidate. Drainage of excess pore water is accelerated by separating the surcharge and the soil with a layer of sand that is exposed at the edges. Drainage also is assisted by installing vertical drains. These can be *sand drains,* which are borings filled with sand, or prefabricated plastic *wick drains* that are stitched into the soil. Temporary surcharging in excess of the anticipated foundation load also speeds up consolidation.

By monitoring the settlement, a prediction curve can be obtained and used for scheduling of construction. A method recently developed to accomplish this employs a first-order rate equation and has been given the acronym FORE. A first-order rate equation states that a rate of change is proportional to the departure from a final equilibrium. This results in a linear relationship between the logarithm of the departure and time. It should be noted that this is the converse of the more common method of plotting vs. the logarithm of time. As the end value is not known, it is determined by trial and error to obtain a linear relationship, after which a regression equation is used to define settlement amounts at any particular time (Handy and Spangler 2007).

15.3.1.2 Dynamic Compaction

It was not until the 1930s that R.R. Proctor and his colleagues in the Los Angeles County engineers office established the scientific principles for soil compaction. Proctor devised the basic test that still is used and carries his name. It also has been formalized in various standard methods with numbers.

Compaction must be carefully controlled in order to achieve the desired result. If too wet, the soil is likely to shear and become overcompacted. If too dry, the soil is collapsible, meaning that it can further densify upon wetting. This can occur even though the compacted density meets specification requirements, because it is not the soil density that is the governing factor for this type of behavior—it is its content of air. Compaction is a specialized topic that is discussed elsewhere in this book.

15.3.1.3 Example Application of FORE

The data for secondary compression settlement in Figure 15.5 were obtained at the Kansai International Airport, Japan, courtesy of Professor Emeritus Koichi Akai, University of Kyoto. Approximately 33 m (110 ft) of fill was used to make the artificial island that supports the airport, so it is important to estimate how much additional fill may be needed to compensate for future settlement in low places.

Measured settlements are listed in column 2. Column 3 shows the ultimate settlement that gave the highest R^2 value between data in columns 1 and 4, and this relationship is shown in the graph. The ratio of final settlement to fill thickness at this site therefore is $0.91/33 = 0.28$. The equation in the graph can be solved for S to give a settlement-time relationship. Additional examples are described by Handy (2002).

15.3.1.4 Compressibility of Compacted Soil

The preconsolidation pressure from dynamic compaction is related to the weight and impact of the compacting element, whether it is a roller, a vibrating plate, or the tamping foot on a

| 1 | 2 | 3 | 4 |
Years	S (m)	S_u (m)	$\log(S_u - S)$
5.779037	6.71040	9.1	0.378325
6.317280	6.89256	9.1	0.343889
6.827195	7.05229	9.1	0.311268
7.365439	7.20892	9.1	0.276710

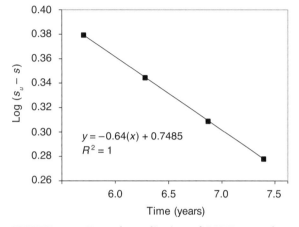

FIGURE 15.5 Example application of FORE: secondary compression data to estimate additional fill to compensate for future settlement.

roller. Because roller contact area is small in order to achieve higher pressures, there is a rapid dissipation of pressure with depth. Dynamic compaction therefore *must* proceed in relatively thin layers. Failure to do so or the use of too thick a layer results in inconsistent and poorly compacted fill that alternates between dense and loose layers, sometimes referred to as the "Oreo® cookie effect."

Because most compacted soil is used for the support of roads and highways, emphasis is on density and water content rather than strength and compressibility, and design frequently is based on complex soil classification schemes. One consequence that demonstrates a limitation of this method is the "bump at the end of a bridge." This not only is annoying, but can cause a complex dynamic loading condition that can lead to lateral abutment movements.

Consolidation tests that measure the preconsolidation pressure and compression index are a logical requirement for structural fill that is to be used for the support of buildings. Nevertheless, it often is assumed that the soil classification coupled with a moisture-density specification and testing will be adequate. For the support of foundations, a common requirement is that the unit weight equal or exceed 95% of the maximum obtained in a standard test with the moisture content within 2% of the optimum. The success of this procedure depends on limiting it to relatively light structures and particular kinds of soil.

A wide range of strength and stiffness can result even when moisture content and the final density stay within specification limits, as shown in Figure 15.4. One indicator of preconsolidation pressure induced by a compaction procedure is the pressure imposed by the compactor, but that is nebulous because of the unknown contact area and soil drainage conditions.

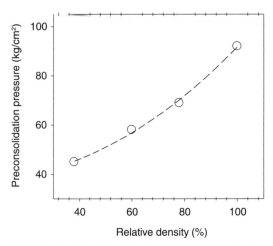

FIGURE 15.6 Relationship between relative density and preconsolidation pressure for Sacramento River sand (interpreted from Lee and Seed 1967).

Contact pressures are increased by the use of tamping foot rollers and can be doubled with vibratory rollers. Figures 15.6 and 15.7 demonstrate trends in preconsolidation pressure and elastic modulus with increasing compaction.

Without consolidation tests, the use of compacted fill for the support of foundations is largely a judgment call. The process of mixing, spreading, and compacting a soil in layers tends to remove spatial variability and reduce differential settlement if the soil and the compaction processes are consistent. More recently, implementation of rollers outfitted with accelerometers and GPS mapping capabilities is providing new insights into the spatial variability of compacted soils (e.g., White and Thompson 2008; Thompson and White 2008).

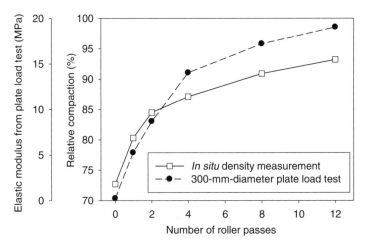

FIGURE 15.7 Relationship between relative compaction and plate load test elastic modulus for GC (A-2-6) soil (interpreted from White et al. 2007).

15.3.1.5 Deep Dynamic Compaction

In the 1930s, a method was developed in Russia to compact collapsible loess soil by repeatedly lifting and dropping a heavy weight using a crane. Essentially the same process was developed independently in France some years later and has been used on various projects around the world.

A trip mechanism is used to release the weight from the supporting cable and avoid tangles from backlash on the spinning cable drum. The weight then is reattached and raised and dropped several times at each location; then the setup is moved and the process repeated on a grid pattern that encompasses the entire area of a foundation. Weights may be as heavy as 100 tons, with a drop equal to or exceeding 100 ft (30 m). The maximum compaction depth is of the order of 40–45 ft (12–14 m).

The most common use of deep dynamic compaction is to shake down and densify potentially liquefiable sand before there is an earthquake. Such soils typically are recently deposited fill (e.g., alluvium) or sediment that has been deposited in water (e.g., in a delta). After it has collapsed and densified, a soil should resist future liquefaction under the same acceleration and overburden conditions. *In situ* tests such as cone or standard penetration tests are performed to determine suitability of the densified soil for support of a foundation. A limitation of deep dynamic compaction is low hydraulic conductivity and high groundwater table. In this case, excess pore water pressure can weaken the soil and result in burial of the drop weight.

A process similar to deep dynamic compaction but with a smaller weight is rapid impact compaction. The process provides controlled impact compaction of the earth using excavator-mounted equipment with a 5- to 9-ton weight, 7 tons being most common. The weight is dropped approximately 4 ft (1.1 m) onto a tamper 5 ft (1.5 m) in diameter that is capable at a rate of about 40–60 blows per minute. The resulting force can densify soils to depths of the order of 10–20 ft (3–7 m). The depth of compaction depends on the soil properties, groundwater conditions, and compaction energy (e.g., see Zakharenkov and Marchuk 1967; Watts and Charles 1993; Serridge and Synac 2006; Braithwaite and du Preez 1997).

15.3.2 Soil Stabilization

The benefits from compaction sometimes are augmented or preserved by the addition of a cementing agent such as Portland cement. The compacted and cured mixture is called soil-cement, which is essentially a lean concrete that has been compacted as a soil instead of being poured as a fluid. This reduces the water-cement ratio, which benefits strength and reduces the amount of cement. However, soil-cement normally is much weaker than concrete. The process is most effective with sandy soils. A spin-off from the manufacture of soil-cement is *roller-compacted concrete*. Cementation also can be achieved with asphalt.

A similar product is soil–lime–fly ash. Fly ash is the ash produced by burning powdered coal in coal-fired power plants. The ash is collected electrostatically and is a fine powder that mostly consists of tiny spheres of glass. The glass, being noncrystalline, is reactive with alkalies including lime.

Fly ash is a pozzolan, named after volcanic ash deposits near Pozzuoli, Italy, that Roman engineers mixed with lime to make concrete. The setting reaction is much slower than with Portland cement, which can be an advantage when wetting, mixing, spreading, and compacting large amounts of a fly ash–soil mixture.

"Type C" fly ash derives from burning coal that contains limestone and therefore already contains substantial amounts of lime, typically of the order of 25%. However, much of the lime occurs not as quicklime but as calcium aluminates, which in themselves are cements. Free lime in the type C ash is available for pozzolanic reactions with the glassy fraction. The manufacture of lime and Portland cement and the production of fly ash release CO_2, a principal greenhouse gas.

The use of soil-cement, soil–lime–fly ash, and soil–type C–fly ash is mainly limited to pavement foundation layers, but they also can be used to support shallow building foundations. Standard tests for highway uses emphasize the resistance to freeze-thaw and wet-dry cycles and are less relevant than strength tests such as the unconfined compressive strength test. The use of unconfined strength for design recognizes that the stabilized soils tend to develop shrinkage cracks during drying. This is particularly true for soil-cement.

15.3.2.1 Soil-Lime

Admixtures of hydrated lime are a common remedial treatment for plastic or expansive clays. The purpose is not to cement a soil and obtain a high compressive strength so much as to reduce its plasticity and expansive character. However, lime added in excess of the amount needed to modify the soil plasticity does engage reactive clay minerals in a pozzolanic reaction that very slowly cements the soil.

Expansive clays generally have a high liquid limit and contain a significant percentage of a clay mineral called smectite or montmorillonite. The typical classification in the Unified Soil Classification System is CH. Expansive clay minerals have a mica-like crystal structure, where individual sheets separate and are invaded by water. Such clays expand upon wetting and shrink upon drying. Volume changes from expansive clays can be devastating to foundations and are a major cause of foundation failures in the U.S. and around the world. Such soils must be removed and replaced or chemically treated to make them nonexpansive. The most common chemical used for this purpose is lime, which can be either mixed in or introduced in a pattern of boreholes. The required amount of lime can be determined by measuring the effects of different amounts on the plastic limit. The liquid limit is largely unaffected. The minimum lime requirement based on plasticity or pH modification is the "lime retention point."

15.3.2.2 Hydrated Lime

Hydrated lime, or $Ca(OH)_2$, is slightly soluble in water and creates a high pH that attacks a clay mineral structure. One theory is that OH^- ions pull H^+ ions out of the clay structure so it becomes more negative and therefore more attractive to the Ca^{++} ions that are provided by the lime. Regardless of the mechanism, the result is an electrostatic linking that greatly reduces the soil plasticity, primarily by increasing the plastic limit. Lime-modified soils are used extensively to support highways and foundations and have even been successfully applied to support canals built on expansive clays.

It often is supposed that lime treatment should extend to the full depth of seasonal shrinkage and swelling, but research conducted in India shows that a depth of 1 m (3 ft) is sufficient to obtain satisfactory control (Katti et al. 2002). If some uplift can be tolerated, a treatment depth of 0.3 m (1 ft) can be expected to reduce heave potential by about one-half.

Another common alternative is to control access to water by extending a concrete slab foundation outside the perimeter of a structure. However, these efforts can be sabotaged by nearby trees that take water from the soil and by a tendency for moisture to accumulate underneath an impermeable membrane that prevents evaporation. Slab-on-grade foundations therefore are reinforced to account for a loss of support around the perimeter and are limited to the support of small structures.

Expansive clays are the most problematic of problem soils and account for annual expenditures of many billions of dollars for repairs to buildings and roads.

15.3.2.3 Drilled Lime

Structures that inadvertently have been built on expansive clay sometimes can be salvaged by drilling holes underneath the foundations and filling with hydrated lime or quicklime. Quicklime poses a hazard for handling but is more effective because it takes water from the surrounding soil, expands as it hydrates, and injects into radial tension cracks created by the expansion. This procedure also is used to treat and stop landslides, but is limited to soils that contain expansive clay minerals (Handy and Williams 1967).

A quick test for viability of a drilled lime treatment is to determine if a small amount of lime increases the plastic limit above the existing moisture content, so the soil changes from plastic to solid and crumbly.

15.3.2.4 Deep Soil Mixing

A process for mixing cement with soil in an auger hole was developed in the 1950s in the U.S. by the Intrusion-Prepakt Corporation and independently developed and refined some years later in Sweden. The treated soils receive moderate compaction by reversing the auger as it is withdrawn. Soil-lime piles are used extensively to stabilize weak deltaic soils and muds. A valuable reference is Elias et al. (2001).

15.3.2.5 Jet Grouting

Jet grouting is similar to deep soil mixing, but mechanical energy is augmented by injecting water and air under high pressure. The process is applicable for a wide range of soils from gravels to clays because the high-pressure injection acts to erode the soil that then is displaced and mixed with grout. Advantages of jet grouting over other ground improvement technologies are that the grout can be designed for site-specific applications, the process is relatively fast, and it can be used around existing structures (Borden et al. 1992). Recently, technology improvements have led to what is referred to as "super" jet grouting, which can result in column diameters up to 17 ft (5 m) (Burke et al. 2000).

15.3.3 Lateral Compaction

Only recently have the benefits from lateral compaction started to be fully recognized. An early method of lateral compaction consisted of driving an array of displacement piles, when tests revealed that the pile group "reduction" factor was larger than 1.0. Another method for increasing lateral stress is compaction grouting, discussed later in this chapter. In this case, the intent is not to penetrate the soil pores but rather to push the soil aside so it compacts. However, grouting pressure should be limited to the overburden pressure or it will lift the ground surface and go into places where it will do more harm than good.

A more direct approach to lateral compaction involves ramming of aggregate layers into prebored holes or into holes created by ramming a hollow probe. In both cases, the hydraulically operated rammer is beveled so that part of the ramming energy goes outward as well as downward into the soil. Rammed Aggregate Pier® Systems currently are the most rapidly growing specialty foundation method in the world.

15.3.4 Tensile Reinforcement

Application of lateral stress increases the strength and decreases the compressibility of soil. The same effect can be achieved by incorporating horizontal tension members that act as a reinforcement. Lateral stress then is developed passively under load, because of the tendency for soil or any solid to expand laterally when subjected to a vertical load. In elastic theory, this tendency is quantified by Poisson's ratio. However, lateral bulging is greatly increased by plastic behavior in soft soils.

An early use of tensile reinforcement involved containing crushed aggregate in steel mesh boxes called gabions. The gabions are wired together, most commonly to make small retaining walls, where they combine the advantages of light weight, flexibility, and drainage.

Another type of tensile reinforcement is *Reinforced Earth*, developed in France in the 1960s by French architect-engineer Henri Vidal. Steel strips are attached to concrete facing elements in retaining walls and extend horizontally back into soil in back of the wall, where they act as tiebacks that are held by friction with the soil. Construction proceeds in layers, with each new tier attached to strips. The strips then are covered with a layer of sandy soil that then is compacted. The procedure differs from conventional tiebacks because the strips are not post-tensioned, but develop tension as the wall is constructed. A later modification involves substituting plastic grid for steel. Because stability depends in part on friction between soil and the strips, the method is intended to be employed only with sand, and misapplication to plastic clay can result in failure.

Embankments and their foundations can be similarly reinforced in two directions with plastic grids laid between soil layers as they are spread and compacted. When used to support building foundations, aggregate is substituted for ordinary soil.

The function of horizontal tensile reinforcement is analogous to the application of an external lateral pressure in a triaxial compression test, with similar results: shearing strength is dramatically increased and compressibility reduced. Figure 15.8 illustrates the transfer of tensile stresses to the reinforcement, which reduces the amount of lateral confinement needed

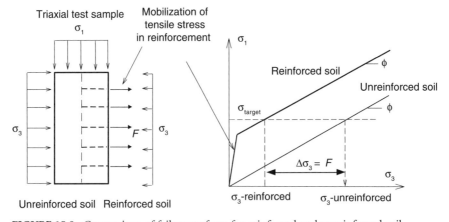

FIGURE 15.8 Comparison of failure surfaces for reinforced and unreinforced soil.

to achieve the desired shear strength. The amount of increase in stiffness depends on the mobilization of stress in the reinforcement and on the reinforcement tensile properties.

15.3.5 Compaction Grouting

Permeation grouting has been used for many years to seal off and reduce the flow of water, for example in gravel or porous rock under and around a dam. This type of grouting also can harden a soil and may be regarded as *in situ* soil stabilization, but applications are restricted by the void size and permeability of the soil.

In compaction grouting, the goal is not to invade soil pores, but rather to displace the soil and cause it to compact in the neighborhood of an expanding grout bulb. This procedure may have grown out of an earlier process called "mud jacking," where grout is pumped underneath a sagging pavement or foundation slab to bring it back to level. A similar process has been used for many years in the petroleum production industry to fracture rocks and increase the flow of oil into oil wells.

The most common compaction grout is slurry consisting of water, sand, Portland cement, and fly ash. Fly ash is used because the spherical shape of the particles aids pumpability. The effectiveness and design parameters are obtained by testing the soil after grouting has been completed.

Compaction grouting is most frequently used as a remedial treatment underneath existing structures, to reduce the susceptibility of a soil to excessive compression, liquefaction, or collapse.

15.3.6 Vibroflotation and Stone Columns

Vibroflotation, developed in the 1950s (see Barksdale and Bachus 1983), is similar in intent to compaction grouting but involves the use of vibration and water to compact sandy soil to a considerable depth. As the vibrating probe is lowered and water is added at the tip, the soil densifies and forms a cone of settlement around the probe rod. Sand then is dumped into the depression. The purpose is deep densification of loose sandy soil. After the process is completed, the soil is tested to determine its bearing capacity and to estimate settlement under load.

Stone columns are an adaptation of the vibroflot principle that involves substituting crushed rock for the fill sand to create a continuous column of compacted stone. The advantage of this procedure is that it creates a kind of aggregate pier that provides additional support for a foundation. Stone columns also are used to stop landslides but can require a considerable percent replacement of the sliding soil by stone.

15.3.7 Rammed Aggregate Pier® Systems

Rammed Aggregate Pier® Systems use high lateral stress to confine soil between the piers and change its behavior from consolidating to elastic. Design is based on elastic theory for the soil layer that is penetrated by the piers and on conventional consolidation theory for the underlying soil, usually resulting in a substantial reduction in settlement.

Rammed Aggregate Pier® Systems are compacted in layers with a hydraulic rammer. The total energies involved are of the same order of magnitude as used in deep dynamic compaction. An advantage is that the energy is distributed vertically instead of being applied at the ground surface.

Graded coarse aggregate is used to prevent invasion by the soil. In the Geopier® method of construction, measured amounts of aggregate are dumped into an open boring, and each layer is rammed to near refusal. In the Impact® Pier method, the hole is produced by ramming a hollow probe, and the probe is lifted incrementally to introduce a charge of aggregate into the hole. After each increment of lifting, the probe is driven back down with a rammer to push aside and densify the aggregate. A valving arrangement using suspended sections of chain prevents aggregate from re-entering the probe.

An important part of the process is to ram and compact the aggregate outward as well as downward in order to create a high lateral pressure. Lateral pressures of the order of 2500 lb/ft^2 (120 kPa) have been measured in soil near the pier. Lateral pressure measurements also indicate the existence of vertically oriented radial tension cracks extending outward from each pier and acting as drainage galleries. Rammed Aggregate Pier® Systems are regarded as an Intermediate Foundation® System where shallow foundations are impractical or inadequate and conventional deep foundations represent overkill.

A mechanism whereby high lateral stress can act to prevent consolidation is illustrated in Figure 15.9. If lateral stress is increased so it exceeds the *in situ* vertical stress, the directions

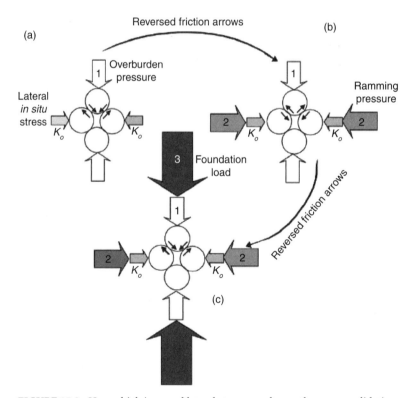

FIGURE 15.9 How a high imposed lateral stress can change the preconsolidation pressure: (a) lateral stress is low in a normally consolidated soil because of support from contact friction between grains; (b) a high lateral ramming pressure causes a reversal of friction at the contacts, which in turn (c) requires a much higher vertical pressure to again reverse the arrows and initiate consolidation; hence an increase in preconsolidation pressure so the soil behaves elastically.

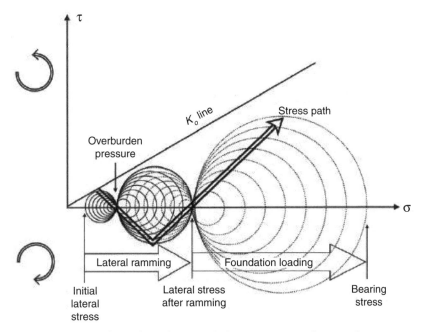

FIGURE 15.10 Mohr circles and stress path during ramming. The vertical stress must be such that the large circle will engage the K_o line before consolidation can occur. Friction reversal is indicated at the left.

of shearing stresses at grain contacts are reversed. Vertical pressure from a foundation then must be high enough that shearing stresses again are reversed before consolidation can occur. A high imposed lateral stress therefore in effect creates a preconsolidation pressure so the soil behaves elastically instead of consolidating. This has been confirmed in the laboratory and in the field and is the basis for design. The same behavior can be expected in expansive soils during an expansion cycle, temporarily increasing their stiffness.

The theory also can be illustrated by a sequence of Mohr circles, shown in Figure 15.10. The left circle depicts stresses in a normally consolidated soil. Ramming increases lateral stress but does not affect the vertical stress, so the Mohr circle radius first decreases and then increases as stresses shift to the second circle. A foundation load results in a similar shift of stresses from the second to the third circle; not shown is the influence of the additional foundation load on horizontal stress. For consolidation to initiate, the foundation pressure must be large enough that the third circle touches a consolidation stress envelope or K_o line.

15.3.7.1 New Theory or Old Soil Mechanics?

Although the theory of ramming aggregate appears to be relatively new, it may only be the implications that are new. Increasing lateral confining stress in a triaxial test is known to increase soil modulus, but testing has been hampered by test instrumentation that does not include a capability to apply lateral stresses that are in excess of vertical stresses, even though such conditions commonly exist in the field where the overconsolidation ratio is greater than 1.

15.3.7.2 Ramming Energy and Liquefaction

Temporary soil liquefaction has been suspected to act as a temporary aid to driving of piles, but has not yet been confirmed with lateral stress measurements. This hypothesis can be tested by measuring *in situ* stresses at different distances from a driven pile, since liquefaction should result in a perfect transfer of stress. If temporary liquefaction does occur, there must be an avenue for escape of excess pore water to allow the soil to solidify. Lateral compaction during pile driving also requires an escape route for water, which has been shown to occur from a rapid reduction in pore water pressure measured with a piezometer after driving stops. These measurements have led to conjecture that water must drain outward through radial tension cracks.

Recent measurements of lateral *in situ* stress in soil near Rammed Aggregate Pier® elements indicate that both of these speculations may be correct: that soil impacted by pile driving or by ramming may temporarily liquefy and that rapid drainage occurs through radial tension cracks.

The evidence favoring liquefaction is shown in Figure 15.10, where there is a perfect transfer of radial stress outward from the surface of the pier. This pattern is repeated in different test sequences. The increase in circumferential contact area requires that radial stress must be reduced unless the soil has been liquefied.

There are two pieces of evidence in support of the conjecture that radial tension cracks outside of the liquefied zone. First, radial cracking reduces tangential stresses to zero, which affects the relationship between radial stress and radial distance, which should be linear. The second piece of evidence is more subtle. As the liquefied soil drains, pore water pressure and hence total stress reduce, which relieves stress acting to support the surrounding soil. A reduction in total stress as the liquefied soil drains should relieve radial stress in the surrounding soil, but it remains constant. This may be explained if liquefied soil injects into the open tension cracks and props them open to create an arching effect.

The effect of liquefaction, therefore, is to aid the distribution of lateral stress into the soil. A more complete discussion of the liquefaction hypothesis is provided by Handy (2008).

15.3.7.3 Measuring Lateral *In Situ* Stress

Lateral stress has been called the "Holy Grail" of soil mechanics but until recently has been very difficult to accurately measure. Boring a hole reduces radial stress to zero in the vicinity of the boring and according to elastic theory will double tangential stress. If the tangential stress exceeds the unconfined compressive strength, the hole will squeeze shut. On the other hand, implanting a rigid object such as a pressure cell into soil concentrates and increases stress.

"Self-boring pressuremeters" developed simultaneously in France and in England bore a hole while simultaneously inserting a rigid shield to try and maintain the *in situ* stress. However, this is difficult in a particulate material. The K_0 *stepped blade,* developed in the U.S., introduces different levels of disturbance and extrapolates pressures to a condition of zero disturbance. The speed and accuracy of the latter have allowed investigations to proceed with special foundations. The stepped blade was developed at Iowa State University for the U.S. Department of Transportation Federal Highway Administration, with additional support from the U.S. Army Waterways Experiment Station and consulting engineering firms.

Lateral soil stress also is an important clue to the soil stress history. For example, a high lateral stress may be inherited from an earlier episode of consolidation and give an indication of the amount of preconsolidation pressure. High lateral stresses also indicate expansive clay, where stress builds up from wet-dry cycling and filling of shrinkage cracks. A low lateral stress can indicate a potentially collapsible soil that is not in equilibrium with the existing overburden pressure, or it may indicate the presence of tension cracks. A lateral stress that exactly equals the vertical stress is a clue to either existing or a prior history of liquefaction.

15.3.7.4 Relief of Lateral Stress

If lateral stress is relieved, as by trenching, will the soil return to a consolidating behavior? Field experience indicates that it does not. This is attributed to a slow recovery of strength of the remolded soil upon aging. Rammed Aggregate Pier® elements are not tested until at least two days after their installation.

A similar behavior is observed when driving pile; load tests performed after a few days or even a few hours reveal a "setup factor" that typically is around 2. Setup can freeze a pile in place if for any reason driving is interrupted.

Lateral stress can be relieved in soil under pavement edges if the road shoulders are not maintained, which can contribute to deterioration of the pavement. The effect of stress relief is amplified by destructive influences from wetting/drying and freezing/thawing.

15.4 Mechanics of Load Transfer

The different load transfer mechanisms strongly influence behaviors of different foundation systems. The distribution of stress under a shallow foundation is immediate upon application of load and is approximately in accordance with elastic theory. If the bearing stress exceeds the soil preconsolidation pressure, the soil will consolidate, adding to settlement of the foundation. This behavior is modified by incorporation of horizontal tensile reinforcement, which has an effect similar to that of a lateral confining stress.

Two types of load transfer occur with intermediate and deep foundations: side friction and end bearing. (Side friction also occurs with shallow foundations but is not considered to be significant.) The ultimate behavior is strongly influenced by compressibility of the foundation elements, whether concrete, steel, or crushed aggregate.

The upper part of an aggregate pier bulges outward either during ramming or later upon application of a foundation load. Horizontal ramming pressures often are high enough that they exceed the passive resistance of the upper part of the surrounding soil, resulting in a substantial enlargement of the pier diameter. Because the upper part is rammed, it may increase load-bearing capability, but this is not considered in the design.

Some hypothetical stress transfer mechanisms are illustrated in Figure 15.11. Dashed lines show approximate distributions of lateral stress, which in turn affects the vertical distribution of side friction.

Lateral stresses were measured in the case of rammed piers. They may be inferred by assuming $K = 1$ for poured concrete and $K > 1$ for stone columns that are vibrated into place in the presence of excess water. Ramming, conducted essentially in the dry, inflicts a lateral effective stress that is retained regardless of later submergence under a groundwater table.

The transfer of stress through side friction depends on both the contact stress and the degree of mobilization by slipping. Then, after side friction is fully mobilized, continued slipping causes it to decrease due to remolding. End bearing also reduces side friction near the bottom by pushing the soil down.

Deep foundations normally are tested to twice the design load, so a pile that develops end bearing under a test load most likely will not do so after it is placed in service. In that case, the

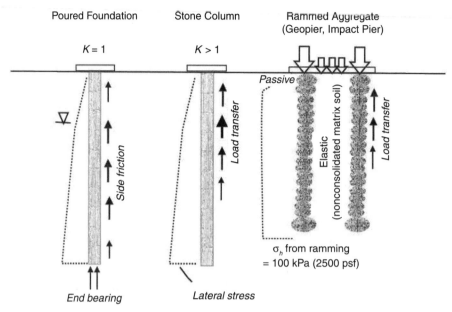

FIGURE 15.11 Comparison of load transfer and lateral stress development.

lower part of the pile has no function except as a safety factor. This is in contrast to a rammed pier, where ramming stresses induce an elastic response in neighboring soil for the entire length of a pier. The diameter of the elastic conversion varies, but has been determined to be as much as 12 ft (4 m) if stress transfer has been aided by temporary liquefaction (Handy and White 2006a, 2006b).

The compressibility of the upper part of rammed piers and the increased density and modulus of soil between the piers contribute to partial support of the foundation slabs where they are in contact with the soil.

15.5 Application of Specialty Foundations: Wind Turbines

Wind turbine foundation systems with arrays of 100 or more turbines are a relatively new development and emphasize the need for economical as well as safe foundation designs. The designs involve unique criteria because although static loads are significant, a major part of the loading is dynamic and wind related. The foundation system therefore must not only resist lateral and overturning wind loading, but also must incorporate a resistance to resonance of the soil-foundation system. Although a number of different foundation alternatives exist for the same soil profiles (see Lesny 2009), most land-based turbines are founded on a relatively simple gravity foundation, but increasingly combined with a specialty foundation system.

In many locations, wind turbine farms are located on some difficult soils, including expansive clays, soft clay soils, and collapsible loess. Expansive clays can be managed by anchoring the foundations below the active layer. Collapsible loess can be more difficult because it typically has never been fully saturated except near the base of the section where there is a perched water table. The underconsolidated condition therefore can exist to a considerable depth. Ironically, winds that deposited the loess 14,000–25,000 years ago still remain to drive the turbines.

Design to support wind turbines is based on bearing capacity and settlement using conventional methods, accounting for eccentric loading during an extreme wind event. Common allowable bearing pressures are on the order of less than 4000 lb/ft^2 (192 kPa). Because wind turbines are dynamically loaded, there are minimum requirements for rotational stiffness (see Naval Facilities Engineering Command 1983). This can introduce more uncertainty in the design analysis, especially for a specialty foundation, because of a lack of performance history and full-scale testing. One approach is to calculate rotational stiffness using a composite approach based on replacement area, but this does not include the effects discussed previously in terms of converting the soil to elastic behavior through development of high lateral stress. This is an area that will benefit from more research.

Figure 15.12 illustrates a wind turbine supported on a composite gravity foundation with a specialty foundation to reinforce the compressible soil layer. Without the specialty foundation, the turbine would need to be supported on an expensive deep foundation system or risk

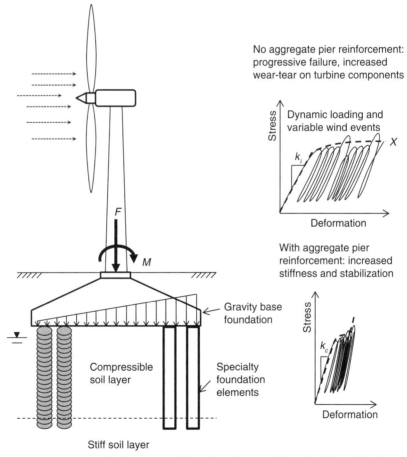

FIGURE 15.12 Wind turbine gravity base foundation reinforced with specialty foundation system.

failure from cumulative plastic deformation during cyclic loading. With the specialty foundation system, the rotation stiffness is increased and the plastic deformation is substantially reduced. New design approaches and testing are needed to further advance application of specialty foundations to support wind turbines.

15.6 Conclusions

Special foundations cover a wide range of materials and applications and have evolved from simple to more sophisticated systems and applications. A common thread between these various systems is that special foundations often use less expensive materials, such as coarse aggregate compared to concrete and steel, and can be integrated with traditional shallow foundations. The weakest link in a conventional foundation system is not the concrete or steel, but the soil.

Specialty foundation systems are ideally suited for ground conditions that are relatively soft or collapsible. High lateral stresses created during installation make the soil stiffer and more elastic. It appears that the high induced lateral stresses also can precollapse a collapsible soil.

Special foundations often start as proprietary systems that include design and the construction delivery method. Future research, particularly with respect to development of lateral stress in the soil, should aid integration into the design of more conventional shallow foundation systems. An important target area for research is gravity foundations for wind turbine generators, where only limited performance information is available relative to the control of rotational stiffness.

References

Barksdale, R.D. and Bachus, R.S. (1983). *Design and Construction of Stone Columns,* Vol. 1, Report No. FHWA/RD-83/02C, Federal Highway Administration.

Borden, R., Holtz, R.D., and Juran, I., Eds. (1992). Jet grouting. *Proceedings of the ASCE Geotechnical Engineering Specialty Conference: Grouting, Soil Improvement and Geosynthetics,* Vol. 1, Geotechnical Special Publication No. 30, ASCE, New York, 144–214.

Braithwaite, E.J. and du Preez, R.W. (1997). Rapid impact compaction in southern Africa. *Proceedings of the Conference on Geology for Engineering, Urban Planning and the Environment,* South African Institute of Engineering Geologists, November 13–14.

Burke, G.K., Cacoilo, D.M., and Chadwick, K.R. (2000). SuperJet grouting: new technology for in situ soil improvement. *Transp. Res. Rec.,* 1721:45–53.

Elias, V., Welsh, J., Warren, J., and Lukas, R. (2001). *Ground Improvement Technical Summaries,* Vol. II, Publication No. FHWA-SA-98-086R, U.S. Federal Highway Administration, Office of Infrastructure, U.S. Federal Highway Administration, Office of Bridge Technology, and Earth Engineering and Sciences, Inc.

Handy, R.L. (2002). First-order rate equations in geotechnical engineering. *J. Geotech. Geoenviron. Eng.,* 128(5):416–425.

Handy, R.L. (2008). "Liquepaction": hydraulic compaction of soil near rammed piers. *From Research to Practice in Geotechnical Engineering,* GSP 180, ASCE, New York, 251–258.

Handy, R.L. and Spangler, M.G. (2007). *Geotechnical Engineering,* 5th edition, McGraw-Hill, New York.

Handy, R.L. and White, D.J. (2006a). Stress zones near displacement piers. I. Plastic and liquefied behavior. *J. Geotech. Geoenviron. Eng.*, 132(1):54–62.

Handy, R.L. and White, D.J. (2006b). Stress zones near displacement piers. II. Radial cracking and wedging. *J. Geotech. Geoenviron. Eng.*, 132(1):63–71.

Handy, R.L. and Williams, W.W. (1967). Chemical stabilization of an active landslide (1967). *Civ. Eng.*, August:62–65.

Katti, R.K., Katti, D.R., and Katti, A.R. (2002). *Behaviour of Saturated Expansive Soil and Control Methods*, Revised and Enlarged Edition, Taylor and Francis, 1270.

Lee, K.L. and Seed, H.B. (1967). Drained strength characteristics of sands. *J. Soil Mech. Found. Div. ASCE*, 93(SM6):117–141.

Lesny, K. (2009). Offshore wind energy in Germany compared to other European areas—challenges and recent developments. *Contemporary Topics in Deep Foundations*, M. Iskander, D. Laefer, and M. Hussein, Eds., Geotechnical Special Publication No. 185, ASCE.

Naval Facilities Engineering Command (1983). *Soil Dynamics, Deep Stabilization, and Special Geotechnical Construction: Design Manual 7.3*, Alexandria, VA.

Serridge, C.J. and Synac, O. (2006). *Application of the Rapid Impact Compaction (RIC) Technique for Risk Mitigation in Problematic Soils*, IAEG2006 Paper No. 294, The Geological Society of London.

Thompson, M.T. and White, D.J. (2008). Estimating compaction of cohesive soils from machine drive power. *J. Geotech. Geoenviron. Eng.*, 134(12):1771–1777.

Watts, K.S. and Charles, J.A. (1993). Initial assessment of a new rapid impact ground compactor. *Proceedings of the Conference on Engineered Fills*, Thomas Telford, London, 399–412.

White, D.J. and Thompson, M.T. (2008). Relationship between in-situ and roller-integrated compaction measurements for cohesionless soils. *J. Geotech. Geoenviron. Eng.*, 134(12):1763–1770.

White, D.J., Jaselskis, E.J., Schaefer, V.R., and Cackler, E.T. (2005). Real-time compaction monitoring in cohesive soil from machine response. *Transp. Res. Rec.*, 1936:173–180.

White, D.J., Thompson, M.J., and Vennapusa, P.R.K. (2007). *Field Study of Compaction Monitoring Systems: Self-Propelled Non-Vibratory 825G and Vibratory Smooth Drum CS-533E Rollers*, Final Report, Iowa State University, Ames, April.

Zakharenkov, M.M. and Marchuk, A.I. (1967). Experimental radiometric investigation of compaction of settled ground by means of heavy-duty tamping equipment. *Osn. Fundam. Mekh. Gruntov*, 4(July/August):21–23.

Index